14th International Workshop on Meson Production, Properties and Interaction (MESON 2016)

EPJ Web of Conferences Volume 130 (2016)

Krakow, Poland
2 - 7 June 2016

Editors:

A. Wronska
A. Magiera
C. Guaraldo
H. Stroher

ISBN: 978-1-5108-3367-8

Printed from e-media with permission by:

Curran Associates, Inc.
57 Morehouse Lane
Red Hook, NY 12571

Some format issues inherent in the e-media version may also appear in this print version.

Printed by Curran Associates, Inc. (2016)

For additional information, please contact EDP Sciences – Web of Conferences at the address below.

EDP Sciences – Web of Conferences
17, Avenue du Hoggar
Parc d'Activité de Courtabœuf
BP 112
F-91944 Les Ulis Cedex A
France

Phone: +33 (0) 1 69 18 75 75
Fax: +33 (0) 1 69 28 84 91

contact-edps@webofconferences.org

Additional copies of this publication are available from:

Curran Associates, Inc.
57 Morehouse Lane
Red Hook, NY 12571 USA
Phone: 845-758-0400
Fax: 845-758-2633
Email: curran@proceedings.com
Web: www.proceedings.com

TABLE OF CONTENTS

MESON 2016

PROCEEDINGS OF THE

14th INTERNATIONAL WORKSHOP ON MESON PRODUCTION, PROPERTIES AND INTERACTION

2 – 7 June 2016, Kraków, Poland

organized by

Jagiellonian University, Cracow
Forschungszentrum Jülich GmbH, Jülich
INFN-LNF, Frascati
IFJ-PAN, Kraków

Edited by

A. Wrońska, A. Magiera, C. Guaraldo, H. Ströher

Preface

The 14th International Workshop on Meson Production, Properties and Interaction - MESON 2016 - took place in Cracow, Poland, from 2 to 7 June 2016. The conference with a tradition dating back to 1991 was attended by over 170 participants from 21 countries from around the world. In total, there were 37 plenary talks, 69 parallel session talks and 24 posters. Additionaly to regular conference program a special public talk on recent discovery of gravitational waves was given by Eugenio Coccia.

Objective of the MESON Conferences is to provide an overview of the present status of the physics areas in which mesons play an important role, as well as of new developments, and a preview of the forthcoming investigations. The topics covered hadronic and electromagnetic meson production in a broad energy range, meson interaction with mesons and nuclei, structure of hadrons, fundamental symmetry studies with mesons, exotic systems including the recent discovery of a dibaryon. Theoretical lectures focused mainly on the problems of low-energy QCD with use of effective field theory and lattice QCD. Widely presented were diverse approaches to the problems of dynamical chiral symmetry breaking, kaon bound states, mesic atoms and nuclei, and mesons in nuclear matter. Research related to the physics of mesons are closely linked to fundamental problems of physics. Therefore, the conference program this time was expanded to cover these items. In particular, the conference program covered recent results and future plans on investigations of electric dipole moment of hadrons and dark photon searches.

The success of the conference was made possible by the extensive work of the Advisory Committee, the Organizing Committee and the Local Organizing Committee. Special thanks are to our home institutions, organizers of the conference, Jagiellonian University, Forschungszentrum Juelich, INFN Frascati, and Institute of Nuclear Physics, Polish Academy of Science, Cracow for their generous support.

We gratefully acknowledge further financial assistance from our sponsors: Marian Smoluchowski Cracow Scientific Consortium, the Municipality of Cracow (KRK2B), CAEN, ISEG, Wiener, Nowoczesna Elektronika and European Physical Journal. Last but not least, we thank all the participants for interesting presentations and lively discussions on recent results and future projects. We appreciated very much the attendance of many young physicist who really vivified the atmosphere of the conference. The large number of participants engaged in interesting discussions, expected further discoveries in the fields presented at the conference and great ambience of Cracow direct our thoughts and motivation towards the next edition of the MESON conference, which we will organize in June 2018, at the same location.

Carlo Guaraldo, Andrzej Magiera, Hans Ströher

October 2016

ADVISORY COMMITTEE

M. Anselmino (University of Torino)
D. Bettoni (University of Ferrara, INFN)
F. Bossi (LNF-INFN Frascati)
V. Burkert (Jefferson Lab)
F. Close (University of Oxford)
A. Denig (Mainz University)
A. Gal (Hebrew University of Jerusalem)
H. Gao (Duke University, Durham)
M. Garçon (DAPNIA Saclay)
F. Hayano (University of Tokyo)
B. Krusche (University of Basel)
T. Nakano (Osaka University)
E. Oset(University of Valencia)
C.D. Roberts (ANL Argonne)
G. Rosner (University pf Glasgow, FAIR)
P. Salabura (Jagiellonian University)
C. Sfienti (University of Mainz)
E. Strokovsky (JINR Dubna)
U. Thoma (University of Bonn)
E. Widmann (Austrian Academy of Sciences)
C. Wilkin (University College London)
M. Wolke (Uppsala University)
S. Wycech (IPJ Warsaw)
T. Yamazaki (University of Tokyo, RIKEN)
B. Zou (IHEP Beijing)

ORGANIZING COMMITTEE

Carlo Guaraldo (INFN-LNF Frascati)	*Chairman*
Andrzej Magiera (UJ Cracow)	*Chairman*
Hans Ströher (FZ Jülich)	*Chairman*
Catalina Curceanu (INFN-LNF Frascati)	
Stanisław Kistryn (UJ Cracow)	
James Ritman (FZ Jülich)	
Antoni Szczurek (IFJ-PAN Cracow)	
Aleksandra Wrońska (UJ Cracow)	*Scientific Secretary*

LOCAL ORGANIZING COMMITTEE

Damian Gil (UJ Cracow)
Barbara Kozak (UJ Cracow)
Wojciech Migdał (UJ Cracow)
Agnieszka Wach (UJ Cracow)

SPONSORS

KNO Krajowy Naukowy Ośrodek Wiodący

KRAKÓW krakow.pl

KRK>2B Krakow to business

WIENER Plein & Baus Elektronik A Phoenix Mecano Company

iseg EXACTLY.

EPJ.org your physics journal

NOWOCZESNA ELEKTRONIKA

CAEN Tools for Discovery

LIST OF PARTICIPANTS

1.	Achenbach	Patrick	University of Mainz
2.	Alfs	Dominika	Jagiellonian University, Kraków
3.	Aulenbacher	Kurt	University of Mainz
4.	Bashir	Adnan	Universidad Michoacana de San Nicolás de Hidalgo, Morelia
5.	Bibrzycki	Łukasz	Pedagogical University of Cracow
6.	Biernat	Jacek	Jagiellonian University, Kraków
7.	Cardman	Lawrence	Jefferson Lab
8.	Ciborowski	Jacek	University of Warsaw
9.	Cieply	Ales	Nuclear Physics Institute CAS, Rez
10.	Cisek	Anna	University of Rzeszów
11.	Coccia	Eugenio	Gran Sasso Science Institute and University of Rome Tor Vergata
12.	Czerwiński	Eryk	Jagiellonian University, Kraków
13.	Demmich	Kay	University of Münster
14.	Denig	Achim	University of Mainz
15.	Druzhinin	Vladimir	Novosibirsk State University and Budker Institute of Nuclear Physics
16.	Duk	Viacheslav	INFN Perugia
17.	Dymov	Sergey	Forschungszentrum Jülich
18.	Elouadrhiri	Latifa	Jefferson Lab
19.	Escribano	Rafel	Universitat Autònoma de Barcelona
20.	Espadanal	João	LIP Lisbon
21.	Fedotovich	Gennady	Budker Institute of Nuclear Physics, Novosibirsk
22.	Fernandez-Soler	Pedro	Centro Mixto Universidad de Valencia, CSIC
23.	Friedman	Eliahu	Hebrew University, Jerusalem
24.	Gajos	Aleksander	Jagiellonian University, Kraków
25.	Gal	Avraham	Hebrew University, Jerusalem
26.	Ghosh	Sudeep	Indian Institute of Technology Indore
27.	Giacosa	Francesco	UJK Kielce
28.	Gianotti	Paola	INFN - LNF
29.	Gil	Damian	Jagiellonian University, Kraków
30.	Gladilin	Leonid	Lomonosov Moscow State University
31.	Glazier	Derek	University of Glasgow
32.	Goerlich	Lidia	Institute of Nuclear Physics PAS, Kraków
33.	Golak	Jacek	Jagiellonian University, Kraków
34.	Goldenbaum	Frank	Forschungszentrum Jülich
35.	Gonzàlez-Solís	Sergi	Institut de Física d'Altes Energies, Barcelona
36.	Goswami	Ankita	Indian Institute of Technology Indore
37.	Goudzovski	Evgueni	University of Birmingham
38.	Grube	Boris	TU Munich
39.	Guaraldo	Carlo	INFN - LNF
40.	Gumberidze	Małgorzata	TU Darmstadt
41.	Guo	Feng-Kun	Institute of Theoretical Physics CAS, Beijing
42.	Guryn	Włodek	Brookhaven National Laboratory
43.	Haberzettl	Helmut	The George Washington University
44.	Hartmann	Jan	HISKP, University of Bonn
45.	Hrtankova	Jaroslava	Nuclear Physics Institute CAS, Rez
46.	Husek	Tomáš	Charles University in Prague
47.	Itahashi	Kenta	RIKEN
48.	Iwasaki	Masahiko	RIKEN
49.	Jackura	Andrew	Indiana University
50.	Janek	Marian	Zilina University

51.	Jarczyk	Lucjan	Jagiellonian University, Kraków
52.	Kachaev	Igor	IHEP Protvino
53.	Kadavy	Tomas	Charles University in Prague
54.	Kamiński	Robert	Institute of Nuclear Physics PAS, Kraków
55.	Kamys	Bogusław	Jagiellonian University, Kraków
56.	Karliner	Marek	Tel Aviv University
57.	Khreptak	Oleksandr	Jagiellonian University, Kraków
58.	Kistryn	Stanisław	Jagiellonian University, Kraków
59.	Kornakov	Georgy	TU Darmstadt
60.	Kozak	Barbara	Jagiellonian University, Kraków
61.	Kozela	Adam	Institute of Nuclear Physics PAS, Kraków
62.	Krusche	Bernd	University of Basel
63.	Krzemień	Wojciech	National Centre for Nuclear Research
64.	Kubicz	Ewelina	Jagiellonian University, Kraków
65.	Kubis	Bastian	HISKP, University of Bonn
66.	Kulessa	Paweł	Institute of Nuclear Physics PAS, Kraków
67.	Kulessa	Reinhard	Jagiellonian University, Kraków
68.	Kunkel	Michael	Forschungszentrum Jülich
69.	Kycia	Radosław	Cracow University of Technology
70.	Kłos	Barbara	University of Silesia, Katowice
71.	Kłusek-Gawenda	Mariola	Institute of Nuclear Physics PAS, Kraków
72.	Lang	Michael	HISKP, University of Bonn
73.	Lavagno	Andrea	Politecnico di Torino
74.	Lebiedowicz	Piotr	Institute of Nuclear Physics PAS, Kraków
75.	Lenisa	Paolo	University of Ferrara and INFN
76.	Lersch	Daniel	Forschungszentrum Jülich
77.	Levi Sandri	Paolo	INFN - LNF
78.	Leśniak	Leonard	Jagiellonian University, Kraków
79.	Liang	Wei-Hong	Guangxi Normal University
80.	Lieb	Joseph	George Mason University
81.	Lukin	Peter	Budker Institute of Nuclear Physics and Novosibirsk State University
82.	Łabaza	Adam	Institute of Nuclear Physics PAS, Kraków
83.	Łuszczak	Marta	University of Rzeszów
84.	Maciuła	Rafał	Institute of Nuclear Physics PAS, Kraków
85.	Magiera	Andrzej	Jagiellonian University, Kraków
86.	Malec	Leszek	Jagiellonian University, Kraków
87.	Mares	Jiri	Nuclear Physics Institute CAS, Rez
88.	Mascagna	Valerio	University of Brescia - INFN Pavia
89.	Mathieu	Vincent	Indiana University
90.	Meyer	Curtis	Carnegie Mellon University
91.	Migdał	Wojciech	Jagiellonian University, Kraków
92.	Min	Tianjue	IHEP Beijing
93.	Mohammed	Muhsin	Jagiellonian University, Kraków
94.	Moir	Graham	University of Cambridge
95.	Moskal	Paweł	Jagiellonian University, Kraków
96.	Moussallam	Bachir	Universite Paris-sud
97.	Mueller	Dominik	University of Manchester
98.	Münzer	Robert	Goethe University of Frankfurt
99.	Nanova	Mariana	University of Giessen
100.	Nazari	Vahabeddin	Institute of Nuclear Physics PAS, Kraków
101.	Nedolya	Anatoliy	Zaporizhzhya National University
102.	Nowakowski	Krzysztof	Jagiellonian University, Kraków
103.	Nowak	Wolf-Dieter	University of Mainz and DESY
104.	Oset	Eulogio	University of Valencia

105.	Ozvenchuk	Vitalii	Institute of Nuclear Physics PAS, Kraków
106.	Palczewski	Tomasz	University of California and LBNL
107.	Parol	Wiktor	Institute of Nuclear Physics PAS, Kraków
108.	Passemar	Emilie	Indiana University and Jefferson Lab
109.	Pattky	Robert	iseg Spezialelektronik GmbH
110.	Pauk	Vladiszlav	Jefferson Lab
111.	Piasecki	Krzysztof	University of Warsaw
112.	Pilloni	Alessandro	Jefferson Lab
113.	Popczak	Arkadiusz	Jagiellonian University, Kraków
114.	Preghenella	Roberto	CERN and INFN Bologna
115.	Price	Darren	University of Manchester
116.	Przygoda	Witold	Jagiellonian University, Kraków
117.	Qin	Sixue	Argonne National Laboratory
118.	Redmer	Christoph Florian	University of Mainz
119.	Ritman	James	Forschungszentrum Jülich
120.	Ronchese	Paolo	University and INFN Padova
121.	Rozp edzik	Dagmara	Jagiellonian University, Kraków
122.	Rudy	Zbigniew	Jagiellonian University, Kraków
123.	Rundel	Oleksandr	Jagiellonian University, Kraków
124.	Rusiecka	Katarzyna	Jagiellonian University, Kraków
125.	Ryabchikov	Dmitry	TU Munich and IHEP Protvino
126.	Ryan	Sinead	Trinity College Dublin
127.	Rybicki	Andrzej	Institute of Nuclear Physics PAS, Kraków
128.	Salabura	Piotr	Jagiellonian University, Kraków
129.	Schaefer	Wolfgang	Institute of Nuclear Physics PAS, Kraków
130.	Schröer	Daniel	University of Münster
131.	Schumacher	Reinhard	Carnegie Mellon University
132.	Scordo	Alessandro	INFN - LNF
133.	Scozzi	Federico	TU Darmstadt and IPN Orsay
134.	Segovia	Jorge	TU Munich
135.	Shekhovtsova	Olga	Institute of Nuclear Physics PAS, Kraków
136.	Shen	Chengping	Beihang University
137.	Silarski	Michał	INFN - LNF
138.	Skoupil	Dalibor	Nuclear Physics Institute CAS, Rez
139.	Skurzok	Magdalena	Jagiellonian University, Kraków
140.	Skwarnicki	Tomasz	Syracuse University
141.	Skwira-Chalot	Izabela	University of Warsaw
142.	Smyrski	Jerzy	Jagiellonian University, Kraków
143.	Strakovsky	Igor	The George Washington University
144.	Strandberg	Bruno	University of Glasgow
145.	Strokovsky	Eugene	JINR Dubna
146.	Ströher	Hans	Forschungszentrum Jülich
147.	Surovtsev	Yury	BLTP of Joint Institute for Nuclear Research
148.	Szczepaniak	Adam	Indiana University and Jefferson Lab
149.	Szczurek	Antoni	Institute of Nuclear Physics PAS, Kraków and Rzeszów University
150.	Tanaka	Yoshiki	GSI Darmstadt
151.	Tatsuno	Hideyuki	Lund University
152.	Thomas	Andreas	University of Mainz
153.	Tiator	Lothar	University of Mainz
154.	Trassinelli	Martino	Institute of NanoScience of Paris CNRS
155.	Unal	Yasemin	Canakkale Onsekiz Mart University
156.	Wach	Agnieszka	Jagiellonian University, Kraków
157.	Walford	Natalie	University of Basel
158.	Wang	Zhiyong	IHEP Beijing
159.	Warczak	Andrzej	Jagiellonian University, Kraków
160.	Wharton	Andy	Lancaster University

161.	Wieczorek	Anna	Jagiellonian University, Kraków
162.	Wilkin	Colin	University College London
163.	Witała	Henryk	Jagiellonian University
164.	Wrońska	Aleksandra	Jagiellonian University, Kraków
165.	Wycech	Sławomir	National Centre for Nuclear Research
166.	Xie	Ju-Jun	Institute of Modern Physics CAS
167.	Yazkov	Valeriy	Lomonosov Moscow State University
168.	Zejma	Jacek	Jagiellonian University, Kraków
169.	Zhemchugov	Evgenii	ITEP Moscow
170.	Zieliński	Marcin	Jagiellonian University, Kraków

edp sciences

EPJ Web of Conferences

Statement of Peer review

In submitting conference proceedings to EPJ Web of Conferences, I certify to the Publisher that I adhere to the **Policy on Publishing Integrity** of the journal in order to safeguard good scientific practice in publishing.

1. All articles have been subjected to peer review administered by the proceedings editors.
2. Reviews have been conducted by expert referees, who have been requested to provide unbiased and constructive comments aimed, whenever possible, at improving the work.
3. Proceedings editors have taken all reasonable steps to ensure the quality of the materials they publish and their decision to accept or reject a paper for publication has been based only on the merits of the work and the relevance to the journal.

Title, date and place of the conference

MESON2016 – 14th International Workshop on Meson Production, Properties and Interaction
Kraków, Poland, 2–7 June 2016

Proceedings editor(s):

Aleksandra Wrońska
Andrzej Magiera
Carlo Guaraldo
Hans Ströher

Date and editor's signature 7.11.2016

17, avenue du Hoggar - PA de Courtabœuf – BP 112 - 91944 Les Ulis Cedex A (France)
Tél. : 33 (0)1 69 18 75 75 - Fax : 33(0)1 69 07 45 17 – www.edpsciences.org

EPJ Web of Conferences **130**, 01001 (2016) DOI: 10.1051/epjconf/201613001001
MESON 2016

Exotics at BESIII

Zhiyong Wang
for the BESIII Collaboration

[1] *Institute of High Energy of Physics, Chinese Academy of Sciences*

Abstract. Using the 1.31×10^{10} J/ψ events and the data above open charm collected at BESIII detector, we studied and observed a lot of exotics, including $X(3872), X(3823), Y(2175), Y(4140), Z_c(3900), Z_c(4020)$, and so on. Some of them are confirmed, some of them are observed for the first time, and some of them are searched, but with no obvious signal.

1 Introduction

The exotics are quite different from the common hadrons though they are not forbidden by QCD theory. The study on these exotics is very helpful to understand the nature of hadrons. A lots of exotics, some of them are called X, Y, and Z states, are observed in different experiments. However, our understanding to these exotics is quite limited. For example, the Y(4260) is discovered in ISR process $e^+e^- \to \gamma\pi^+\pi^+ J/\psi$, and despite its subsequent observation, its nature has remained a mystery. Considering that its mass above open charm and its strong coupling to $\pi^+\pi^+ J/\psi$, it is regarded as a non-conventional state of charmonium.

In this paper, we report the study on X, Y, and Z states at BESIII detector.

2 X states

2.1 $X(3872)$

The $X(3872)$ was first observed 10 years ago by Belle [1] in $B^+ \to K^\pm\pi^+\pi^+ J/\psi$ decays; it subsequently confirmed by several other experiments [2–4]. Its J^{PC} is determined to be 1^{++} by other decay modes and angular information. In order to understand the puzzle of $Y(4260)$ and $Y(4360)$, an investigation of their other decay processes, such as their radiative decay to the low lying charmonium or charmoniumlike states is important. The process of $Y(4260)/Y(4360) \to \gamma X(3872)$ is unique due to the exotic feature of $X(3872)$ and $Y(4260)$ or $Y(4360)$ resonances. We study the process of $e^+e^- \to \gamma X(3872), X(3872) \to \pi^+\pi^+ J/\psi, J/\psi \to l^+l^-$ ($l^+l^- = e^+e^-$ or $\mu^+\mu^-$) using the data collected at $\sqrt{s}$=4.009 GeV and 4.42 GeV. Figure 1 (top-left) shows the fit to $M_{\pi^+\pi^+ J/\psi}$ with sum of all energies. A clear X(3872) signal is seen with a statistical significance of 6.3σ. The product of the Born-order cross section times the branching fraction of $X(3872) \to \pi^+\pi^+ J/\psi$ is calculated at 4 energies. Figure 1 (bottom-left) shows the fit to the energy-dependent Born cross section with a Y(4260) resonance, a linear continuum, or a E1-transition phase space. The Y(4260) resonance describes better than the other two options.

2.2 $X(3823)$

The lightest charmonium state above $D\bar{D}$ is $\psi(3770)$, which is currently defined as 1^3D_1 state [5]. Until now, there have been no definitive observations its two D-wave spin-triplet partner states. i.e., 1^3D_2 and 1^3D_3 states. Phenomenological models predict that the 1^3D_2 charmnium state has large decay widths to $\gamma\chi_{c1}$ and $\gamma\chi_{c2}$. Recently, the Belle Collaboration reported evidence for a narrow resonance $X(3823) \to \gamma\chi_{c1}$ in B meson decays with 3.8σ significance and mass $3823.1 \pm 1.8 \pm 0.7$ MeV/c^2, and suggested that this is a good candidate for the 1^3D_2 charmonium state [6]. We performed search for the production of the 1^3D_2 state via the process $e^+e^- \to \pi^+\pi^-X$, using 4.67 fb^{-1} data at $\sqrt{s}$ =4.19-4.60 GeV. The 1^3D_2 candidates are reconstructed in their $\gamma\chi_{c1,2}$ decay modes. Figure 1 (top-left) shows the fit to $M_{recoil}(\pi^+\pi^-)$ of $\gamma\chi_{c1,2}$, respectively. The statistical significance of the $X(3823)$ signal in $\gamma\chi_{c1}$ mode is estimated to be 6.2σ. For the $\gamma\chi_{c2}$ mode, we don't observe an $X(3823)$ signal. The $X(3823)$ is a candidate for the 1^3D_2 charmonium state with $J^{PC} = 2^{--}$. The product of the Born-order cross section and the branching ratio of $X(3823) \to \gamma\chi_{c1,2}$ is calculated. We fit the energy-dependent cross sections of $e^+e^- \to \pi^+\pi^-X(3823)$ with the $Y(4360)$ shape or the $\psi(4415)$ shape with their resonance parameters fixed to the PDG values. Figure 1 (bottom-right) shows the fit results. Due to the low statistics, we accept both the $Y(4360)$ and $\psi(4415)$ hypotheses at 90% C.L.

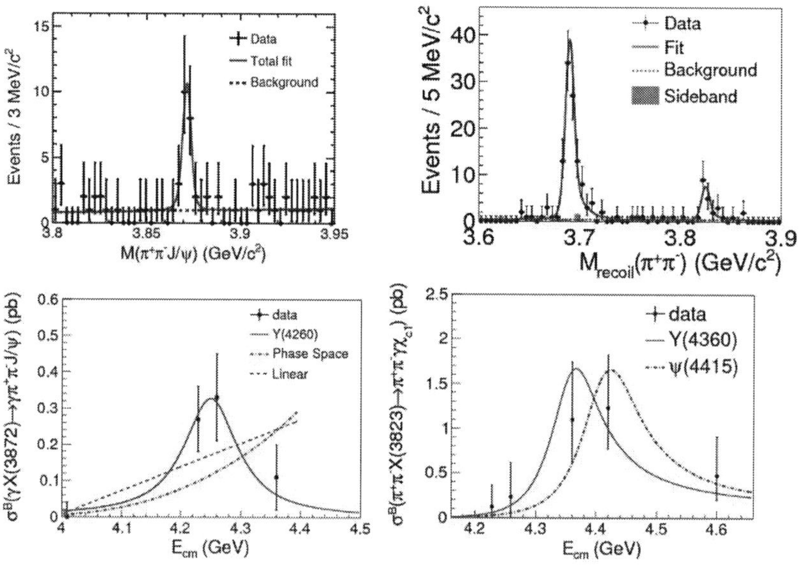

Figure 1. (top-left) $X(3872)$ signal. (top-right) $X(3823)$ signal. (bottom-left) Born cross-section of $e^+e^- \to \gamma X(3872)$. (bottom-right) Born cross-section of $e^+e^- \to \gamma X(3823)$

3 Y state

3.1 $Y(2175)$

The $Y(2175)$ was first observed by the *BABAR* experment [7] in the $e^+e^- \to \gamma_{ISR}\phi f_0(980)$ process. It was later confirmed by the BESII experiment in $J/\psi \to \eta\phi f_0(980)$ decays [8] and other experiments. Since the $Y(2175)$ has $J^{PC} = 1^{--}$, one speculates whether it may be an s-quark counterpart to the

$Y(4260)$. A number of different interpretations have been proposed for the $Y(2175)$. The confirmation and study of the $Y(2175)$ in $J/\psi \to \eta\phi\pi^+\pi^-$ with a large data sample is necessary for clarifying its nature. Figure 2 (left) shows the fit results to $\phi f_0(980)$ invariant mass spectrum. The solid curve is the overall fit projection and a clear $Y(2175)$ signal is seen [9]. Its mass and width are determined to be $M = 2200 \pm 6$ MeV/c^2 and $\Gamma = 104 \pm 15$ MeV, respectively.

3.2 $Y(4140)$

The CDF experiment observed the $Y(4140)$ in the decay of $B^+ \to \phi J/\psi K^+$ with a statistical significance greater than 5σ [10]. However, this new state was not confirmed by the Belle and LHCb collaborations, but confirmed by CMS and D0 Collaborations. The $Y(4140)$ is the first charmoniumlike state decaying into two vector mesons consisting of $c\bar{c}$ and $\Sigma^0\bar{\Sigma}^0$ pairs. Since the $\phi J/\psi$ system has positive C-parity, and can be searched for through radiative transitions of $Y(4260)$ or other 1^{--} charmonium or charmoniumlike states. Here, we report a search for the radiative transition $Y(4260) \to \gamma Y(4140) \to \gamma\phi J/\psi$ with the data collected at $\sqrt{s} = 4.23, 4.26$, and 4.36 GeV. As a result, no obvious signal is found in $\phi J/\psi$ mass spectrum (see Fig. 2 (right)) and the upper limit is given [11].

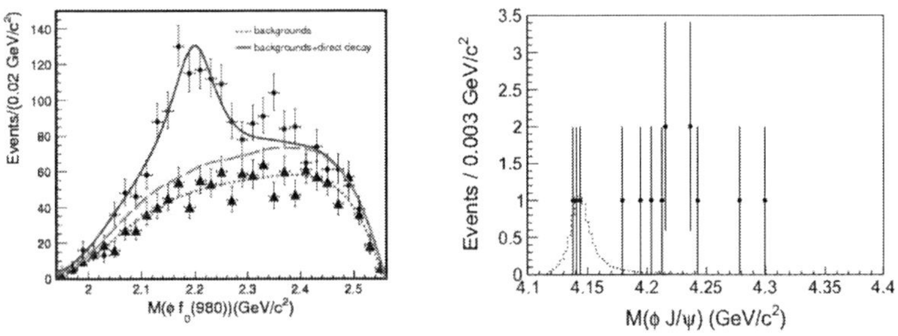

Figure 2. (left) The invariant mass of $\phi f_0(980)$ in $J/\psi \to \pi^+\pi^-\phi\eta$. (right) The invariant mass of $\phi J/\psi$ in $e^+e^- \to \gamma K^+K^-J/\psi$.

3.3 $e^+e^- \to \omega\chi_{cJ}$

The authors of Ref. [12] predict a sizeable coupling between the $Y(4260)$ and the $\omega\chi_{c0}$ channel by the considering the threshold effect that plays a role in reducing the decay rates into open-charm channels. By adopting the spin rearrangement scheme in the heary quark limit and the experimental information, Ref. [13] predicts the ratio of the decays $Y(4260) \to \omega\chi_{cJ}(J = 0, 1, 2)$ to be 4:3:5.

Enhancements can be seen in the line shapes of $\omega\chi_{c0}$ mode. It is fitted with a phase-space modified BW function, and the fit results for the structure parameters are $\Gamma_{ee}\mathcal{B}(\omega\chi_{c0}) = (2.8 \pm 0.5 \pm 0.4)$ eV, $M = (4226 \pm 8 \pm 6)$ MeV/c^2, and $\Gamma_t = (39 \pm 12 \pm 2)$ MeV, where M, Γ_{ee}, Γ_t are mass, total width, e^+e^- partial for the potential Y state. In the $e^+e^- \to \omega\chi_{c2}$ cross section, an enhancement is seen around 4.146 GeV, so we use a coherent sum of the $\psi(4415)$ BW function and a phase-space term to fit the cross section. There are two solutions, constructive and destructive [14]. The cross section of $e^+e^- \to \omega\chi_{c1}$ seems to be raising near 4.6 GeV. The different line shapes observed for $\omega\chi_{cJ}$ might indicate that the production mechanism is different.

4 Z states

4.1 $Z_c(3900)$ and $Z_c(3885)$

The nature of $Y(4260)$ has remained a mystery since its discovery. Unlike other charmonium states with the same quantum numbers and in the same mass region, the $Y(4260)$ state does not have a natural place within the quark model of charmonium. Furthermore, it shows strong coupling to the $\pi^+\pi^+ J/\psi$ final state, but relatively small coupling to open charm decay modes. A similar situation has recently become apparent in the bottomonium system above the $b\bar{b}$ threshold. The observed substructure in these two decay modes indicates the possible existence of charged bottomniumlike state [15], which must have a t least four constituent quarks. By analogy, this suggests there may exist interesting substructure in the $Y(4260) \to \pi^+\pi^+ J/\psi$ process in the charmonium region. Figure 3 (left) shows the distribution of $M_{max}(\pi^\pm J/\psi)$, the larger one of the two mass combinations $M(\pi^+ J/\psi)$ and $M(\pi^- J/\psi)$ in each event. A clear structure near 3.9 GeV is seen An unbinned maximum likelihood fit yields a mass of (3899.0 ± 3.6) MeV/c^2, and a width of (46 ± 10) MeV [16]. Later a search of its neutral isospin partner, $Z_c(3900)^0$, was also performed in $e^+e^- \to \pi^0\pi^0 J/\psi$ process in BESIII [17]. The $Z_c(3900)^0$ mass and width values with statistical and systematic errors are consistent with the charged $Z_c(3900)^\pm$. Meanwhile, we also calculated the Born cross section of $e^+e^- \to \pi^+\pi^+ J/\psi$ at different energies. Compared with the measured Born cross sections of $e^+e^- \to \pi^+\pi^+ J/\psi$ by Belle, we found that the ratio between charged and neutral is consistent with the isospin symmetry expectation for resonances

A peculiar feature of $Y(4260)$ is the absence of any apparent corresponding structure in the cross sections for $e^+e^- \to D^{(*)}\bar{D}^{(*)}(\pi)$ in the same energy region. This implies a partial-width lower-limit of $\Gamma(Y(4260) \to \pi^+\pi^+ J/\psi) > 1$ MeV that is 1 order of magnitude larger than is typical for conventional charmonium meson transitions, and indicates that the $Y(4260)$ is probably not a conventional quarkonium state. The mass of observed $Z_c(3900)$ in $e^+e^- \to \pi^+\pi^+ J/\psi$ is 24 MeV/c^2 above the $D\bar{D}^*$ mass threshold, which is suggestive of a virtual $D\bar{D}^*$ moleculelike structure, i.e., a charmed-sector analogy of the $Z_b(10650)$. We observed a peak in the $(D\bar{D}^*)^-$ invariant mass distribution in $e^+e^- \to \pi^+(D\bar{D}^*)^-$ annihilation events at $\sqrt{s}$=4.26 GeV [18]. Figure 3 (right) shows the distribution of D^0D^{*-} invariant masses. There is a distinct peak near the $m_D + m_{\bar{D}^*}$ mass threshold. The solid curves show the fit results and the dashed curves show the non-resonant background. The $Z_c(3885)$ signal significance for each fit is greater than 18σ. Later, its neutral isospin partner is also observed in the processes $e^+e^- \to D^+D^{*-}\pi^0$ +c.c. and $e^+e^- \to D^0\bar{D}^{*0}\pi^0$+ c.c. at $\sqrt{s}$ =4.226 and 4.257 GeV [19]. The fitted mass and width to the enhancement in neutral mode are very consistent with those of the charged $Z_c(3885)$. Meanwhile, we also determined the ratio, $\mathcal{R} = \mathcal{B}_{D^+D^{*-}}/\mathcal{B}_{D^0\bar{D}^{*0}}$ to be 0.96 ± 0.18, which is very close to the expectation value from isospin conservation.

Assuming the $Z_c(3885)$ structure is due to the $Z_c(3900)$, the ratio of partial decay widths is determined to be $(\Gamma(Z_c(3885) \to D\bar{D}^*)/\Gamma(Z_c(3900) \to \pi J/\psi)) = 6.2 \pm 1.1 \pm 2.7$. This ratio is much smaller than typical values for decays of conventional charmonium states above the open charm threshold. This suggests the influence of very different dynamics in the $Y(4260) - Z_c(3900)$ system.

4.2 $Z_c(4025)$ and $Z_c(4020)$

The mass of observed $Z_b(10610)$ and $Z_b(10650)$ at the Belle experiment [15] are close to the $B\bar{B}^*$ and $B^*\bar{B}^*$ thresholds, respectively, which supports a molecular interpretation of Z_b's as $B\bar{B}^*$ and $B^*\bar{B}^*$ bound states. One intriguing suggestion is to look for corresponding particles in the charmonium sector. As anticipated, a charged charmoniumlike structure, $Z_c(3900)$, has been observed in the $\pi^\pm J/\psi$ mass spectrum by three experiments. Therefore, a search of Z_c candidates via their direct decays

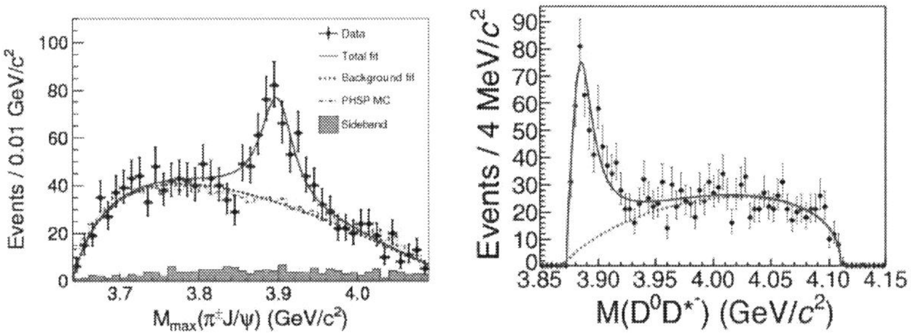

Figure 3. (left) The invariant mass of $\pi^{\pm}J/\psi$ in $e^+e^- \to \pi^+\pi^+J/\psi$. (right) The invariant mass of $D^0\bar{D}^{*-}$ in $e^+e^- \to \pi^+D^0\bar{D}^{*-}$.

into $D^*\bar{D}^*$ pairs is strongly motivated. We study the process $e^+e^- \to (D^*\bar{D}^*)^{\pm}\pi^{mp}$ at a center-of-mass energy $\sqrt{s}$=4.26 GeV, where $(D^*\bar{D}^*)$ refers to the sum of the $D^{*+}\bar{D}^{*0}$ and its charge conjugate $\bar{D}^{*+}D^{*0}$ states. Here, we use a partial reconstruction technique to identify the $D^{*+}\bar{D}^{*0}\pi^-$ final states This technique requires that only the pi^- from the primary decay (denoted as the bachelor π^-), the D^+ decaying from $D^{*+} \to D^+\pi^0$, and at least one soft π^0 from D^{*+} or $\bar{D}^{*0}$ decay are reconstructed. By reconstructing the D^+ particle the charges of its other particle D^{*+} and the bachelor π^- can be unambiguously identified. Therefore, possible combinatoric backgrounds are suppressed with respect to the signals. Figure 4 (left) shows the recoil mass of bachelor π^- with the events in $D^{*+}\bar{D}^{*0}$ mass region. An enhancement near $D^{*+}\bar{D}^{*0}$ threshold is seen [20]. We assume that this enhancement is due to a particle, Labeled as $Z_c(4025)^+$, and parameterized its line shape by the product of an S-wave BW shape and a phase space factor. An unbinned maximum likelihood fit to the π^- recoil mass spectrum yields the mass $m = (4026.3 \pm 2.6)$ MeV/c^2 and $\Gamma = (24.8 \pm 5.6)$ MeV. Later its neutral isospin partner, $Z_c(4025)^0$ is also observed in $e^+e^- \to (D^*\bar{D}*)\pi^0$. A clear enhancement near $D^*\bar{D}^*$ mass threshold is seen. A simultaneous unbinned maximum likelihood fit yields the pole position $m_{pole} = (4025.5^{+2.0}_{-4.7})$ MeV/c^2 and width at pole $\Gamma_{pole} = (23.0 \pm 6.0)$ MeV [21], which are consistent with those of the charged $Z_c(4025)$.

After the report of the observation of charged $Z_c(4025)$, someone points that it may couple to $\pi^{\pm}h_c$. Thus it can be searched for in $e^+e^- \to \pi^+\pi^-h_c$. We present such a search at 13 energies from 3.900 to 4.420 GeV. Here, the h_c is reconstructed via its electric-dipole transition $h_c \to \gamma\eta_c$ with η_c decays to 16 exclusive hadronic final states. Figure 4 (right) shows the fit to the invariant mass of $M_{\pi^{\pm}h_c}$ (two entries per event) distribution. The fit yields a mass of (4020 ± 0.8) MeV/c^2 and a width of (7.9 ± 2.7) MeV. We also tried to add a $Z_c(3900)$ with the mass and width fixed to the measured value. It is found that the statistical significance of $Z_c(3900)$ is 2.9σ. Later, with similar technique, we observed the neutral charmoniumlike state $Z_c(4020)^0$ in $e^+e^- \to \pi^0\pi^0h_c$ process [23]. An unbinned maximum likelihood fit with the fixed width to its charged partner yields a $Z_c(4020^0$ mass of 4023.9 ± 2.2 MeV/c^2, which agrees with that of its charged partner. We also measured the Born cross section of $e^+e^- \to \pi\pi h_c$ and $e^+e^- \to \pi Z_c$ for charged and neutral modes. No obvious isospin violation are found for both processes.

References

[1] S.K. Choi *et al.* (Belle Collaboration), Phys. Rev. Lett. **91**, 262001 (2003).
[2] D. Acosta *et al.* (CDF Colloaboration), Phys. Rev. Lett. **93**, 072001 (2004).
[3] V. M. Abazor *et al.* (D0 Collaboration), Phys. Rev. Lett. **93**, 162002 (2004).

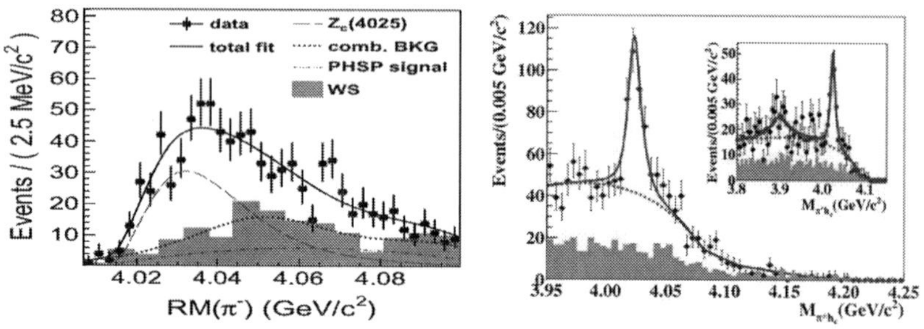

Figure 4. (left) The recoil mass of π^- in $e^+e^- \to (D^*\bar{D}^*)^+\pi^-$. (right) The recoil mass of $\pi^\pm$ in $e^+e^- \to \pi^+\pi^-h_c$.

[4] B. Aubert *et al.* (*BARBAR* Collaboration) Phys. Rev. D **71**, 071103 (2005).
[5] E. Eichen, K. Gottfried, T. Kinoshita, K. D. Lane, and T. M. Yan, Phys. Rev. D **17**, 3092 (1978); 21, 203 (1980).
[6] V. Bhardwaj *et al.* (Belle Collaboration), Phys. Rev. Lett. **111**, 032001 (2013).
[7] B. Aubert *et al.* (*BABAR* Collaboration), Phys. Rev. D **74**, 091103 (R) (2006).
[8] M. Ablikim *et al.* (BES Collaboration), Phys. Rev. Lett. **100**, 102003 (2008).
[9] M. Ablikim *et al.* (BESIII Collaboration), Phys. Rev. D **91**, 052017 (2015).
[10] T. Aaltonen *et al.* (CDF Collolaboration), Phys. Rev. Lett. **102**, 242002 (2009).
[11] M. Ablikim *et al.* (BES Collaboration), Phys. Rev. D **91**, 032002 (2015).
[12] L. Y. Dai, M. Shi, G. Y. Tang, and H. Q. Zheng, Phys. Rev. D **92** 014020 (2015).
[13] L. Ma, X. H. Liu, X. Liu, and S. L. Zhu, Phys. Rev. D **91** 034032 (2015).
[14] M. Ablikim *et al.* (BES Collaboration), Phys. Rev. D **93**, 011102 (R) (2016).
[15] A. Bondar *et al.* (Belle Collaboration), Phys. Rev. Lett. **108**, 122001 (2012).
[16] M. Ablikim *et al.* (BES Collaboration), Phys. Rev. Lett. **110**, 252001 (2013).
[17] M. Ablikim *et al.* (BES Collaboration), Phys. Rev. Lett. **115**, 112003 (2015).
[18] M. Ablikim *et al.* (BES Collaboration), Phys. Rev. Lett. **112**, 222001 (2014).
[19] M. Ablikim *et al.* (BES Collaboration), Phys. Rev. Lett. **115**, 222002 (2015).
[20] M. Ablikim *et al.* (BES Collaboration), Phys. Rev. Lett. **112**, 132001 (2014).
[21] M. Ablikim *et al.* (BES Collaboration), Phys. Rev. Lett. **115**, 182002 (2015).
[22] M. Ablikim *et al.* (BES Collaboration), Phys. Rev. Lett. **111**, 242001 (2013).
[23] M. Ablikim *et al.* (BES Collaboration), Phys. Rev. Lett. **113**, 212002 (2014).

Hadron spectroscopy from lattice quantum chromodynamics

Sinéad M. Ryan[1,⋆]

[1] *School of Mathematics, Trinity College, Dublin 2, Ireland*

Abstract. Lattice calculations of hadron spectroscopy are discussed. A brief introduction to numerical simulations of QCD, with a focus on spectroscopy is given. Results for spectroscopy of low-lying, excited and exotic states including hybrids and the X, Y, Z states in charmonium are described. New developments that have allowed for determinations of scattering parameters, including in a coupled-channel analysis are briefly described providing a roadmap for future lattice calculations of excited and exotic states.

1 Introduction

The study of hadronic energies and strong decay properties is a flourishing field of study. In recent years, hadron spectroscopy has been invigorated by the experimental discovery of unpredicted states in the charmonium and bottomonium systems, some of which are manifestly exotic and cannot be described by quark models. In addition, the unexpected properties of some open-charm mesons, which are narrower and lighter than predicted in quark models has reinvigorated the theoretical study of this sector. We are fortunate that hadron spectroscopy is begin explored by many experiments that can probe these new states with complementary focus and that planned experiments at GlueX and PANDA will bring additional experimental insight and discovery.

In this talk I will discuss lattice hadron spectroscopy with a focus on exotic and hybrid states, in particular in the meson sector where much of the progress and many recent results can be found. In quark models of QCD, the allowed states (the natural spin-parity series) have $P = (-1)^J, S = 1, CP = +1$ whereas states with $P = (-1)^J$ but $CP = -1$ are forbidden. These so-called *exotic* states have corresponding $J^{PC} = 0^{+-}, 0^{--}, 1^{-+}, 2^{+-}, \ldots$ and while they do not appear in quark models they are allowed by QCD and can be studied on the lattice. Hybrid mesons and baryons are characterised by excited gluonic content, and can have exotic or conventional quantum numbers. When charge conjugation is not a good quantum number (e.g. open charm) there are no exotic states but hybrids are permitted, and can be determined in a lattice calculation, as we will see below.

2 Lattice QCD

The formulation of QCD on a four-dimensional hypercubic lattice with a lattice spacing, a and a finite extent, L, provides a nonperturbative regulator for the theory and facilitates numerical simulation of QCD observables. This *ab initio* approach starts from the QCD lagrangian $\mathcal{L} = \bar{\psi}\left(i\gamma_\mu D_\mu - m\right) -$

⋆e-mail: ryan@maths.tcd.ie

$\frac{1}{4}F^a_{\mu\nu}F^{\mu\nu}_a$. Quark fields ψ are placed on sites of the lattice and carry colour, flavour and Dirac indices, as in the continuum. The lattice gluon fields are described by $SU(3)$ matrices - the links of the hypercube - and derivatives in the continuum theory are replaced with finite differences. Observables, O are then determined from the (euclidean) QCD path integral

$$\langle O \rangle = \frac{1}{Z}\int \mathcal{D}U_\mu \mathcal{D}\bar{\psi}\mathcal{D}\psi O e^{iS_{\rm QCD}} \stackrel{t\to i\tau}{\longrightarrow} \frac{1}{Z}\int \mathcal{D}U_\mu \mathcal{D}\bar{\psi}\mathcal{D}\psi O e^{-S_{\rm QCD}}, \quad (1)$$

where the rotation to Euclidean time ensures a non-negative weight that can be used for importance sampling. The quark fields are analytically integrated which introduces the fermion determinant, M in the gauge generation. Practically, calculations proceed by generating an ensemble of gauge field configurations, according to a probability weight given by the gluon action and for a fermion determinant, M. Observables, O are evaluated on each gauge field and the ensemble average gives the quantity's value in that simulation with an associated statistical error determined from the standard deviation.

2.1 Computing correlation functions

For meson and baryon spectroscopy we consider two-point correlation functions. As an example, consider a local meson operator $\phi(x) = \bar{\psi}_a(x)\Gamma\psi_b(x)$, where Γ is an element of the Dirac algebra with possible displacements and flavour indices a and b. A simple two-point function is $C(t) = \langle \phi(x)\phi^\dagger(0)\rangle$. Exploiting euclidean time evolution and inserting a complete set of states yields $C(t) = \sum_k |\langle 0|\phi|k\rangle|^2 e^{-E_k t}$. and the ground state energy, E_0 is determined at large euclidean time t by an exponential fit to the correlator. To access higher-lying energies, a matrix of correlators is constructed from single meson, $\bar{\psi}\Gamma\psi$ and multi-hadron e.g. $\phi = (\bar{\psi}\Gamma_1\psi)(\bar{\psi}\Gamma_2\psi)$, operators and energies are determined by solving a generalised eigenvalue problem $C_{ij}(t)v_j^{(n)} = \lambda^{(n)}(t)C_{ij}(t_0)v_j^{(n)}$, yielding eigenvalues $\lambda^{(n)}(t) \sim e^{-E_n t}\left[1 + O(e^{-\Delta E t})\right]$ and eigenvectors $v_j^{(n)}$ related to the operator overlaps, $Z_j^{(n)} = \langle 0|\phi|n\rangle$ [1].

The practicalities of lattice calculations, mean that simulations are frequently performed at heavier-than-physical pion masses and at a single lattice spacing. This is particularly relevant for recent pioneering results on scattering and coupled-channel analyses and a full systematic error budget can be expected as methods mature.

3 Results

There has been remarkable progress in lattice spectroscopy in recent years with effort focused both on the low-lying stable states and on excited and exotic hadrons, including resonances. The results discussed here are a snapshot of recent activity - more comprehensive reviews and discussion can be found in Ref. [2]. Much of the focus in lattice studies of exotics and resonances has to date been in the meson sector and the results shown here will reflect that.

3.1 Stable hadrons below thresholds

Considering first the spectra of stable states, lying safely below decay thresholds. Many lattice calculations now address all systematic uncertainties by simulating at physical values of the quark masses, with dynamical quarks ($N_f = 2, 2+1, 2+1+1$) and including continuum extrapolations and in some cases QED effects. Fig. 1 shows a compilation of results of low-lying states (upper plot, taken from Ref. [3]) and a recent result including QED effects (lower plot, taken from Ref. [4]). In the calculations shown there is a robust error determination and impressive agreement between different lattice calculations. The calculation of baryon electromagnetic splittings includes predictions subsequently verified by experiment.

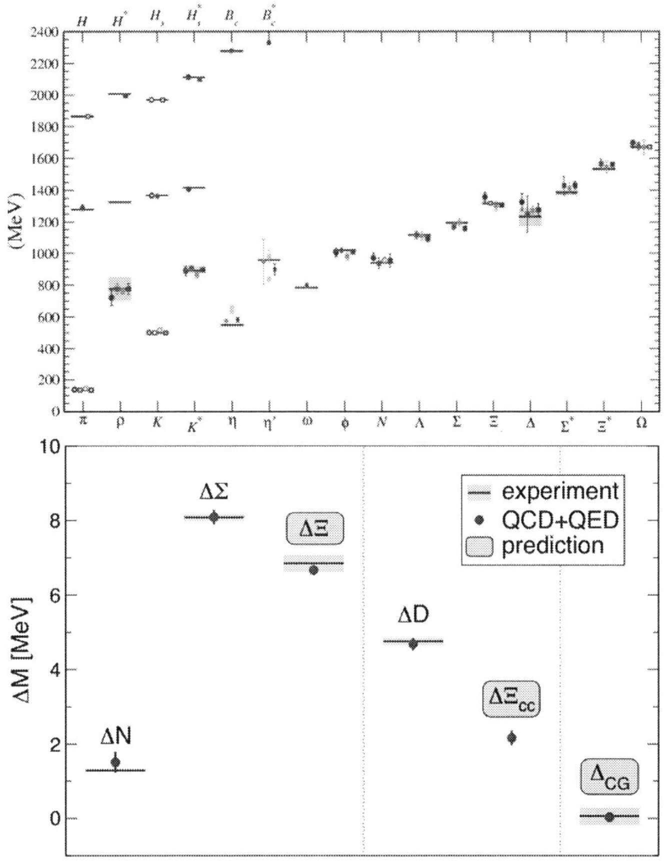

Figure 1. The upper plot shows precision lattice results for a range of low-lying states with all systematic uncertainties under control. The lower plot shows a calculation of electromagnetic splittings in the light baryon sector, with experimental results.

3.2 Excited and exotic hadrons: as stable states

The first step in understanding the spectrum of states above decay thresholds - which includes exotic and hybrid states - is to demonstrate that lattice methods for extracting energies and spin identification are working well, yielding precise and robust results for $J \geq 2$ as well as for ground states. The Hadron Spectrum Collaboration has pioneered a suite of methods to do this, enabled by distillation for quark propagation [5]. This approach has been successfully applied to light, heavy-light and heavy hadron spectroscopy [6–14], illustrated in Fig. 2 for the light isoscalar and charmonium spectrum, respectively [7, 10]. Note that in each case a number of mesons with manifestly exotic quantum numbers are determined - highlighted in the rightmost panel in each plot.

3.2.1 Hybrids

The hybrid mesons discussed in the introduction are a focus for experiments including GlueX and PANDA. Such states, with both exotic and conventional quantum numbers can be determined on the lattice and the most extensive study has been carried out by the Hadron Spectrum Collaboration for

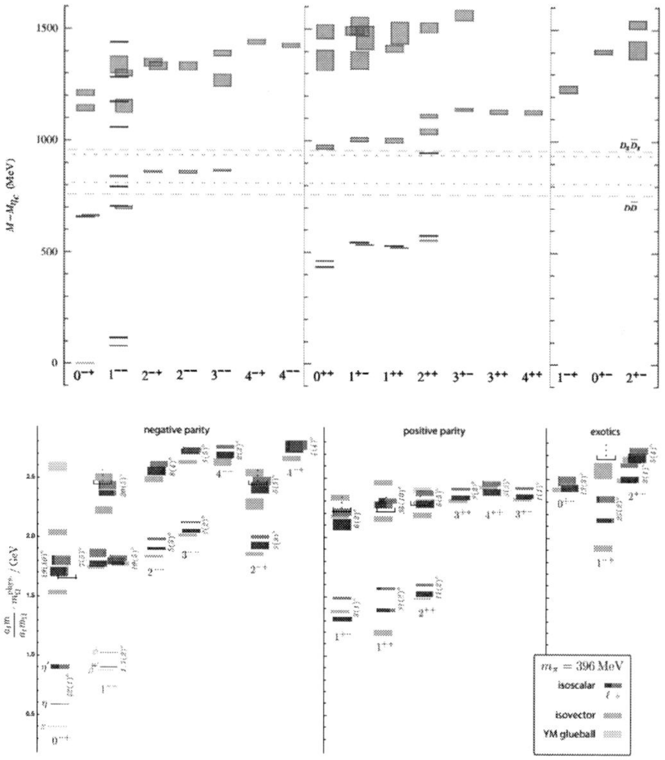

Figure 2. The charmonium (upper plot) and light meson isoscalar (lower plot) spectra determined by the Hadron Spectrum Collaboration [7, 10]. Colored boxes are the lattice results including statistical uncertainty and experimental results, where available, are black lines.

meson and baryon hybrids as part of spectroscopy calculations described above. In a lattice calculation, a state is identified as a candidate hybrid when it has a relatively large overlap onto an operator that is proportional to the field strengh tensor (the commutator of two covariant derivatives on the lattice). Studies with both heavy and light quarks have found the lightest exotic meson in the associated spectrum is a candidate hybrid and is nearly degenerate with three states in the negative partity sector also identified as (non-exotic) hybrids. As a representative result, Fig. 3 shows the light baryon (left) and the charmonium (right) spectrum for the subset of J^{PC} in which candidate hybrids were identified in Refs. [7, 15]. A similar pattern of states and energy scale above the lightest conventional state in the spectrum is observed in light, heavy-light and heavy mesons and baryons and is discussed in more detail in Ref. [16]. The calculations shown here are carried out at a relatively heavy pion mass, $M_\pi \sim 400$ MeV and at fixed lattice spacing. Nevertheless, the common features found for a range of quark mass and for both mesons and baryons are striking.

3.3 Scattering and resonance states

Most hadrons are resonances, located close to or above strong decay thresholds and in a lattice calculation the relevant scattering parameters must be extracted indirectly. The most widely used approach

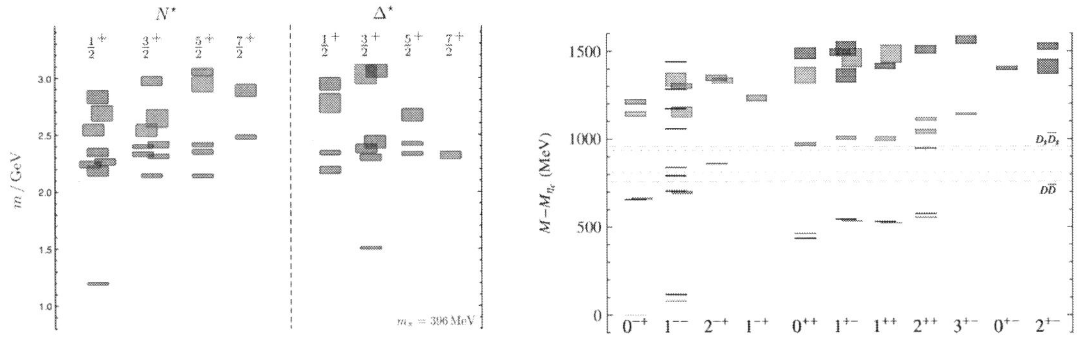

Figure 3. The left pane shows states identified as hybrids, colored blue, in the light baryon spectrum. The right pane shows charmonium hybrids, colored red and blue. The red states are identified as a hybrid supermultiplet, as described in the text, and the blue as hybrids that fall into an excited supermultiplet structure.

relates energies determined in a finite volume (the lattice simulation) to the infinite-volume scattering matrix amplitudes. The formalism, by Lüscher for $\pi\pi$ scattering in the elastic regime [17] has been generalised for moving frames, non-identical particles, multiple two-particle channels and particles with spin - see e.g. Ref. [18] and references therein. An alternative approach, by HAL QCD relies on potentials, determined in a lattice calculation, to extract the required scattering parameters - see Ref. [19] and references therein. While the Lüscher approach is used by a number of groups and has been extended and extensively tested including a recent coupled-channel calculation in the open-charm sector [20], the potential method remains at any earlier stage and is not, to date, widely used.

3.3.1 A first look - X(3872)

Early lattice investigations of the $X(3872)$ are described in Refs. [21–23] and summarised in Figure 4. These studies considered $D\bar{D}^*$ scattering using the Lüscher approach and find a pole just below threshold, in agreement with the experimental picture. A combination of single-meson and multi-meson operators were included and further study of the operator dependence [22] found that the threshold state (interpreted as the $X(3872)$) is not found if the single meson operators, corresponding to the $\chi_{c1}(1P)$, are omitted, implying that $c\bar{c}$ components are more important in the basis than the diquark-antidiquark components. It is important to note that these are exploratory calculations which should be improved before any definitive statements can be made.

3.3.2 A first look - $Z_c^+(3900)$

The charged state Z_c^+ is manifestly exotic and does not fit in quark models. The precise nature of the state is as yet undetermined and it is hoped that lattice QCD can help to resolve this puzzle. However, a number of challenges must be faced: the Z_c^+ lies above several thresholds and so decays to several two-meson final states. In principle this requires a coupled-channel analysis and in a lattice calculation the number of relevant coupled-channels is large for high energies. The state has been studied using the Lüscher approach by a number of groups [23–25] who find no Z_c^+ candidate, below 4.2GeV in their analyses. It should be noted that these do not include coupled channel analyses and the dependence on the operator bases used has not been explored. More recently, the HAL QCD collaboration have

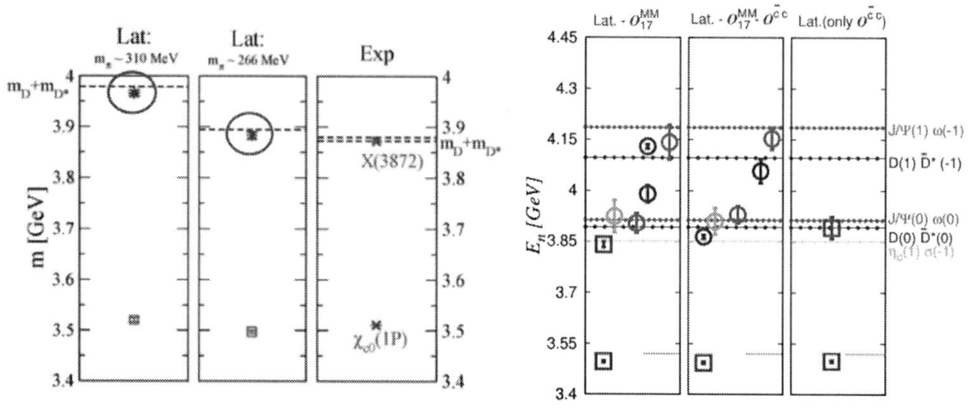

Figure 4. From Ref. [22]. The left pane shows the lattice determination of a state close to DD^* threshold and the right pane illustrates the dependence on the operator basis: the state is missing when $c\bar{c}$ operators are removed.

studied this state in a coupled-channel analysis in their potential approach [26] and suggest the Z_c^+ appears as a threshold cusp, albeit for relatively heavy pions in their simulation. It is still too early for a definitive statement from lattice and more work is certainly motivated by these results.

4 Summary and outlook

I have highlighted the recent progress in lattice hadron spectroscopy. Precision calculations of the low-lying mesons and baryons, with all systematic uncertainties under control are now available. However, the most dramatic progress has been in calculations of excited and exotic states and the treatment of resonances and scattering. As methods are developed and refined sophisticated determinations of resonance properties can be expected - and are already available in some cases, as discussed here. While the numerical precision in current calculations is already very impressive controlling the "traditional" systematic uncertainties remains a challenge - as an example, scattering calculations with physical pion masses presents a challenge as the lighter quark masses open more decay channels which should be included. In charmonium, a proliferation of thresholds and coupled-channels complicates the lattice calculations and the inclusion of disconnected contributions remains a technical difficulty in that system. Nevertheless, theoretical developments continue e.g. in the treatment of three-coupled channel systems and work is ongoing on a numerical implementation. I expect that lattice calculations will be able to contribute meaningful and precise information about the nature of exotic states in QCD.

Acknowledgements

I would like to thank the organisers for the invitation to speak at this conference. I am grateful to my colleagues in the Hadron Spectrum Collaboration and acknowledge support of Science Foundation Ireland [RFP-PHY-3201].

References

[1] B. Blossier, M. Della Morte, G. von Hippel, T. Mendes, R. Sommer, JHEP **04**, 094 (2009)
[2] D. Wilson, C. Liu (2016), `Invited plenary talks at Lattice 2016`
[3] A.S. Kronfeld, in *100 Years of Subatomic Physics*, edited by E.M. Henley, S.D. Ellis (2013)

[4] S. Borsanyi et al., Science **347**, 1452 (2015)
[5] M. Peardon et al. (Hadron Spectrum), Phys. Rev. **D80**, 054506 (2009)
[6] H.W. Lin et al. (Hadron Spectrum), Phys. Rev. **D79**, 034502 (2009)
[7] L. Liu et al., JHEP **07**, 126 (2012)
[8] G. Moir, M. Peardon, S.M. Ryan, C.E. Thomas, L. Liu, JHEP **05**, 021 (2013)
[9] J.J. Dudek, R.G. Edwards, M.J. Peardon, D.G. Richards, C.E. Thomas, Phys. Rev. Lett. **103**, 262001 (2009)
[10] J.J. Dudek et al., Phys. Rev. **D83**, 111502 (2011)
[11] J.J. Dudek, R.G. Edwards, P. Guo, C.E. Thomas (Hadron Spectrum), Phys. Rev. **D88**, 094505 (2013)
[12] R.G. Edwards, N. Mathur, D.G. Richards, S.J. Wallace (Hadron Spectrum), Phys. Rev. **D87**, 054506 (2013)
[13] M. Padmanath, R.G. Edwards, N. Mathur, M. Peardon, Phys. Rev. **D91**, 094502 (2015)
[14] M. Padmanath, R.G. Edwards, N. Mathur, M. Peardon, Phys. Rev. **D90**, 074504 (2014)
[15] J.J. Dudek, R.G. Edwards, Phys. Rev. **D85**, 054016 (2012)
[16] J.J. Dudek, Phys. Rev. **D84**, 074023 (2011)
[17] M. Luscher, Nucl. Phys. **B364**, 237 (1991)
[18] R.A. Briceno, Phys. Rev. **D89**, 074507 (2014)
[19] S. Aoki, B. Charron, T. Doi, T. Hatsuda, T. Inoue, N. Ishii, Phys. Rev. **D87**, 034512 (2013)
[20] G. Moir, M. Peardon, S.M. Ryan, C.E. Thomas, D.J. Wilson, JHEP **10**, 011 (2016)
[21] S. Prelovsek, L. Leskovec, Phys. Rev. Lett. **111**, 192001 (2013)
[22] M. Padmanath, C.B. Lang, S. Prelovsek, Phys. Rev. **D92**, 034501 (2015)
[23] S.h. Lee, C. DeTar, H. Na, D. Mohler (Fermilab Lattice, MILC) (2014)
[24] S. Prelovsek, C.B. Lang, L. Leskovec, D. Mohler, Phys. Rev. **D91**, 014504 (2015)
[25] Y. Chen et al., Phys. Rev. **D89**, 094506 (2014)
[26] Y. Ikeda, S. Aoki, T. Doi, S. Gongyo, T. Hatsuda, T. Inoue, T. Iritani, N. Ishii, K. Murano, K. Sasaki (2016), 1602.03465

Pentaquarks and doubly heavy exotic mesons

Marek Karliner[1,*]

[1] *Raymond and Beverly Sackler School of Physics and Astronomy*
Tel Aviv University, Tel Aviv, Israel

Abstract. I discuss the experimental evidence for and theoretical interpretation of the new mesons and baryons with two heavy quarks. These include doubly-heavy baryons, exotic hadronic quarkonia and most recently a manifestly exotic pentaquark-like doubly heavy baryon with a minimal quark content $uud\bar{c}$ discovered by LHCb, whose mass, decay mode and width are in agreement with a prediction based on a physical picture of a deuteron-like $\Sigma_c\bar{D}^*$ "hadronic molecule".

1 First observation of manifestly exotic hadrons

In late 2007 the Belle Collaboration reported [1] anomalously large rate partial widths of $\Upsilon(5S) \to \Upsilon(2S)$ and $\Upsilon(5S) \to \Upsilon(1S)$, two orders of magnitude larger than the analogous decays of $\Upsilon(3S)$. Soon afterward Harry Lipkin and I proposed [2] that a four-quark exotic resonance $[b\bar{b}u\bar{d}]$ might mediate these decays through the cascade $\Upsilon(mS) \to [b\bar{b}u\bar{d}]\pi^- \to \Upsilon(nS)\pi^+\pi^-$. We suggested looking for the $[b\bar{b}u\bar{d}]$ resonance in these decays as peaks in the invariant mass of $\Upsilon(1S)\pi$ or $\Upsilon(2S)\pi$ systems. Subsequently Belle collaboration confirmed this prediction, with the observation [3, 4] of two charged bottomonium-like resonances Z_b as narrow structures in $\pi^\pm\Upsilon(nS)$ $(n = 1, 2, 3)$ and $\pi^\pm h_b(mP)$ $(m = 1, 2)$ mass spectra that are produced in association with a single charged pion in $\Upsilon(5S)$ decays. The measured masses of the two structures averaged over the five final states are $M_1 = 10608.4 \pm 2.0$ MeV, $M_2 = 10653.2 \pm 1.5$ MeV, both with a width of about 15 MeV. Interestingly enough, the two resonances are only a few MeV away from the respective $B^*\bar{B}$ and $B^*\bar{B}^*$ thresholds.

This strongly suggests a parallel with $X(3872)$, whose mass is almost exactly at the $D^*\bar{D}$ threshold. It also raises the possibility that such states might have a complementary description as deuteron-like "molecule" of two heavy mesons quasi-bound by pion exchange [5–7].

The attraction due to π exchange is 3 times weaker in the $I{=}1$ channel than in the $I{=}0$ channel. Consequently, in the charm system the $I{=}1$ state is expected to be well above the $D^*\bar{D}$ threshold and the $I{=}0$ $X(3872)$ is at the threshold.[1] In the bottom system the attraction due to π exchange is essentially the same, but the kinetic energy is much smaller by a factor of $\sim m(B)/m(D) \approx 2.8$. Therefore the net binding in the bottom system is stronger than in the charm system.

This expectation has been confirmed by the discovery by BESIII [8] of charged exotic charmonium $Z_c(3900)$ at $3899.0 \pm 3.6 \pm 4.9$ MeV with $\Gamma = 46 \pm 10 \pm 20$ MeV, 27 MeV above $\bar{D}D^*$ threshold.

*e-mail: marek@proton.tau.ac.il

[1] For simplicity we treat $X(3872)$ as an isoscalar, since it has no charged partners, and we ignore here the issue of isospin breaking in its decays. A more refined treatment results in the same conclusions.

The narrow widths of these exotic resonances implies that they are deuteron-like loosely bound states and/or near threshold resonances in the two heavy-meson system. Such "molecular" states, $\bar{D}D^*$, etc., were introduced in Ref. [5]. They were later extensively discussed [6, 7, 9, 10] in analogy with the deuteron which binds via exchange of pions and other light mesons, and were referred to as *"deusons"*.

Since the quarks are heavy, we can treat their kinetic energy as a perturbation depending linearly on a parameter inversely proportional to μ_{red}, the reduced mass of the two meson system, which scales like the mass of the heavy quark with the hamiltonian [11], $H = a \cdot p^2 + V$, where $a = 1/\mu_{red} \sim 1/m_Q$. We can use the existing data in order to make a very rough estimate of the isovector binding potential in the $m_Q \to \infty$ limit. We have two data points: $Z_c(3900)$ 27 MeV above $\bar{D}D^*$ threshold and $Z_b(10610)$ approximately 3 MeV above $\bar{B}B^*$ threshold. Linear extrapolation then yields approximately −20 MeV as the $\bar{B}B^*$ binding energy in the isoscalar channel.

The upshot is that the $Z_c(3900)$ isovector resonance confirms and refines the estimates in [9, 10] for the mass of the putative $\bar{B}B^*$ isoscalar bound state. This immediately lead to several predictions [11]:
(a) two $I = 0$ narrow resonances X_b in the bottomonium system, about 20 MeV below the corresponding $\bar{B}B^*$ and $\bar{B}^*B^*$ thresholds;
(b) an $I = 1$ resonance above $\bar{D}^*D^*$ threshold;
(c) an $I = 0$ resonance near $\bar{D}^*D^*$ threshold.

Soon after these predictions [11], the BESIII collaboration observed what looks just like (b) above, namely a new charmonium-like charged narrow resonance $Z_c(4025/4020)$, slightly above the $\bar{D}^*D^*$ threshold [12, 13]. Fig. 1 provides a concise summary of the experimental information about the masses of doubly-heavy exotics observed so far, together with our predictions for additional states, as discussed above.

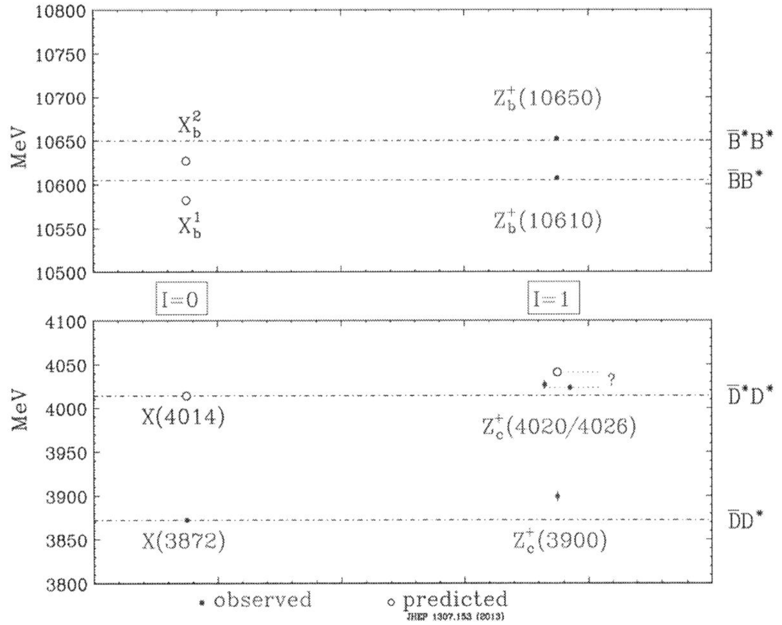

Figure 1. Masses of the doubly-heavy exotic quarkonia vs. two-meson thresholds. The states observed so far are shown in red, the predicted states are shown in blue. $I = 0$ resonances are shown on the left and isovectors are shown on the right. Note the proximity of all the states to the corresponding two-meson thresholds.

The X_b states can most likely be observed through the decays $X_b \to \Upsilon\omega$ or $X_b \to \chi_b\pi\pi$. The decay to $\Upsilon\pi\pi$ is forbidden by G-parity conservation [14]. The seemingly analogous decay $X(3872) \to J/\pi\pi$ is only possible because isospin is strongly broken between D^+ and D^0, and because $X(3872)$ is at the $\bar{D}D^*$ threshold. In contradistinction, in the bottomonium system isospin is almost perfectly conserved. *Thus the null result in CMS search [15] for* $X_b \to \Upsilon(1S)\pi^+\pi^-$ *does not tell us if* X_b *exists.* In fact, X_b might have been already observed by ATLAS, D0 and LHCb [16–19], camouflaging as run-of-the-mill excited bottomonium $\chi_b(3P)$ [20].

When discussing molecular states (of which $X(3872)$ is at least a partial example), one notices that narrow resonances have been discovered very close to $\bar{D}D^*$, $\bar{D}^*D^*$ and $\bar{B}B^*$, $\bar{B}^*B^*$ thresholds, but no analogous resonances have been observed close to $\bar{D}D$ or $\bar{B}B$ threshold; cf. Table 1. The absence of narrow resonances at $\bar{D}D$ and $\bar{B}B$ thresholds provides a strong hint that forces such as pion exchange play a role in the formation of these resonances, because one-pion exchange cannot bind two pseudoscalars. Moreover, the heavy-light mesons D, D^*, B and B^* contain only one light quark, so their coupling to pions is significantly smaller than $g_{\pi NN}$. Therefore in molecular systems containing two heavy-light mesons two-pion exchange is suppressed in comparison with one-pion exchange.

Strong support for the molecular interpretation of the states listed in Table 1 comes from the fact that they are quite narrow, despite many hundreds of MeV available for their decay into quarkonium and pion(s). The width is a product of the available (very large) phase space and the square of the absolute value of the matrix element between the initial and the final state. Thus the experiments tell us that the overlap of the resonance wave function with quarkonium must be very small. In a tightly-bound pentaquark the the c and $\bar{c}$ quarks are both within one confinement volume, so the overlap with the J/ψ wave function is then generically large. On the other hand, in the molecular picture such a small overlap of the wave functions is automatic. This is because in a hadronic molecule the two heavy quarks spend most of their time far from each other. This point is schematically illustrated in Fig. 2.

Table 1. Five narrow exotic states close to meson-meson thresholds.

State	Mass^a (MeV)	Width^a (MeV)	$\bar{Q}Q$ decay mode	Phase space^b (MeV)	Nearby threshold	ΔE (MeV)
$X(3872)$	3872	< 1.2	$J/\psi\,\pi^+\pi^-$	495	$\bar{D}D^*$	< 1
$Z_b(10610)$	10608	21	$\Upsilon\pi$	1008	$\bar{B}B^*$	2 ± 2
$Z_b(10650)$	10651	10	$\Upsilon\pi$	1051	$\bar{B}^*B^*$	2 ± 2
$Z_c(3900)$	3900	$24 - 46$	$J/\psi\,\pi$	663	$\bar{D}D^*$	24
$Z_c(4020)$	4020	$8 - 25$	$J/\psi\,\pi$	783	$\bar{D}^*D^*$	6
×					$\bar{D}D$	
×					$\bar{B}B$	

^{a}Masses and widths approximate

^{b}Quarkonium decay modes listed have maximum phase space

One then can infer from Table 1 the *necessary* conditions for existence of a near-threshold resonance of this type:

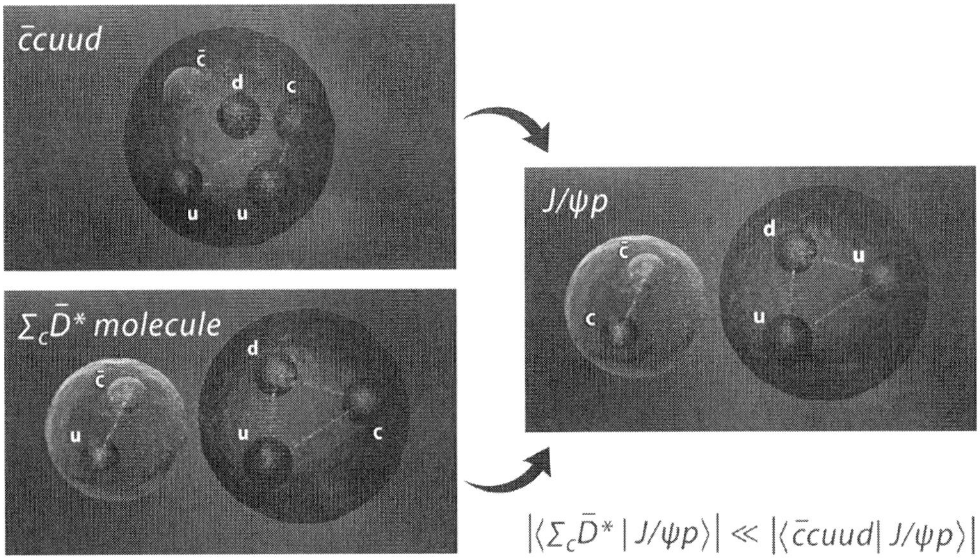

Figure 2. Schematic comparison of decay of a tightly-bound pentaquark vs. hadronic molecule into $J/\psi\ p$. The overlap of the hadronic molecule wave function with the $J/\psi\ p$ final state is generically much smaller than for a tightly-bound pentaquark.

(a) The state contains two heavy hadrons. They have to be heavy, as the repulsive kinetic energy is inversely proportional to the reduced mass.

(b) The two hadrons carry isospin, so that they can couple to pions. Channels like $\Sigma_c\bar{\Lambda}_c$, in which one of the particles has zero isospin, can exchange a pion to become the equal-mass channel $\Lambda_c\bar{\Sigma}_c$.

(c) The spin and parity of the two hadrons have to be such that they can bind through single pion exchange.

(d) The hadrons making up the molecule have to be sufficiently narrow, as the molecule's width cannot be smaller than the sum of its constituents' widths.

A crucial step is then the realization that *the pion-exchange binding mechanism can in principle apply to any two heavy hadrons which satisfy conditions (a)–(d) above, be they mesons or baryons.*

A quantitative understanding of these effects then can be applied to other bound or resonant states of two heavy hadrons. For example, the discovery of the Z_b and Z_c states and their probable interpretation as $B^*\bar{B}/B^*\bar{B}^*$ and $D^*\bar{D}/D^*\bar{D}^*$ bound by pion exchange led us to propose that a weakly bound $\Sigma_c\bar{D}^*$ deuteron-like state might exist [21]. The narrow pentaquark resonance discovered by LHCb [22] has the right properties to be that state. Many more analogous loosely bound states of two heavy hadrons are predicted [21]. They are listed in Table 2. Note in particular the bottom quark analogue of the resonance discovered by LHCb [22], expected somewhat below the $\Sigma_b B^*$ threshold at 11140 MeV. Note also that in the open flavor sector we expect a D^*B^* resonance, but no DB^*. This is because a bound state of a pseudoscalar and a vector of different flavors cannot bind by one pion exchange. The pentaquark states observed by LHCb in Λ_b decay are excellent candidates for photoproduction off a proton target [23–25]. The size of the resonant cross sections is illustrated by Fig. 3 which shows the cross section for resonant photoproduction $\gamma p \to J/\psi\, p \to P_c(4450) \to J/\psi\, p$, as a function of the incident photon laboratory energy E_γ. The relevant experiments can be carried out at JLab. One of them has already been approved [26]. In principle, one can also search this way for the bottom analogue of $P_c(4450)$ [25].

Table 2. Thresholds for $Q\bar{Q}'$ doubly-heavy molecular states

Channel	Minimum isospin	Minimal quark content[a,b]	Threshold (MeV)[c]	Example of decay mode
$D\bar{D}^*$	0	$c\bar{c}q\bar{q}$	3875.8	$J/\psi\,\pi\pi$
$D^*\bar{D}^*$	0	$c\bar{c}q\bar{q}$	4017.2	$J/\psi\,\pi\pi$
D^*B^*	0	$c\bar{b}q\bar{q}$	7333.8	$B_c^+\pi\pi$
$\bar{B}B^*$	0	$b\bar{b}q\bar{q}$	10604.6	$\Upsilon(nS)\pi\pi$
$\bar{B}^*B^*$	0	$b\bar{b}q\bar{q}$	10650.4	$\Upsilon(nS)\pi\pi$
$\Sigma_c\bar{D}^*$	1/2	$c\bar{c}qqq'$	4462.4	$J/\psi\, p$
$\Sigma_c B^*$	1/2	$c\bar{b}qqq'$	7779.5	$B_c^+ p$
$\Sigma_b\bar{D}^*$	1/2	$b\bar{c}qqq'$	7823.0	$B_c^- p$
$\Sigma_b B^*$	1/2	$b\bar{b}qqq'$	11139.6	$\Upsilon(nS)p$
$\Sigma_c\bar{\Lambda}_c$	1	$c\bar{c}qq'\bar{u}\bar{d}$	4740.3	$J/\psi\,\pi$
$\Sigma_c\bar{\Sigma}_c$	0	$c\bar{c}qq'\bar{q}\bar{q}'$	4907.6	$J/\psi\,\pi\pi$
$\Sigma_c\bar{\Lambda}_b$	1	$c\bar{b}qq'\bar{u}\bar{d}$	8073.3[d]	$B_c^+\pi$
$\Sigma_b\bar{\Lambda}_c$	1	$b\bar{c}qq'\bar{u}\bar{d}$	8100.9[d]	$B_c^-\pi$
$\Sigma_b\bar{\Lambda}_b$	1	$b\bar{b}qq'\bar{u}\bar{d}$	11433.9	$\Upsilon(nS)\pi$
$\Sigma_b\bar{\Sigma}_b$	0	$b\bar{b}qq'\bar{q}\bar{q}'$	11628.8	$\Upsilon(nS)\pi\pi$

[a]Ignoring annihilation of quarks. [b]Plus other charge states when $I \neq 0$.
[c]Based on isospin-averaged masses. [d]Thresholds differ by 27.6 MeV

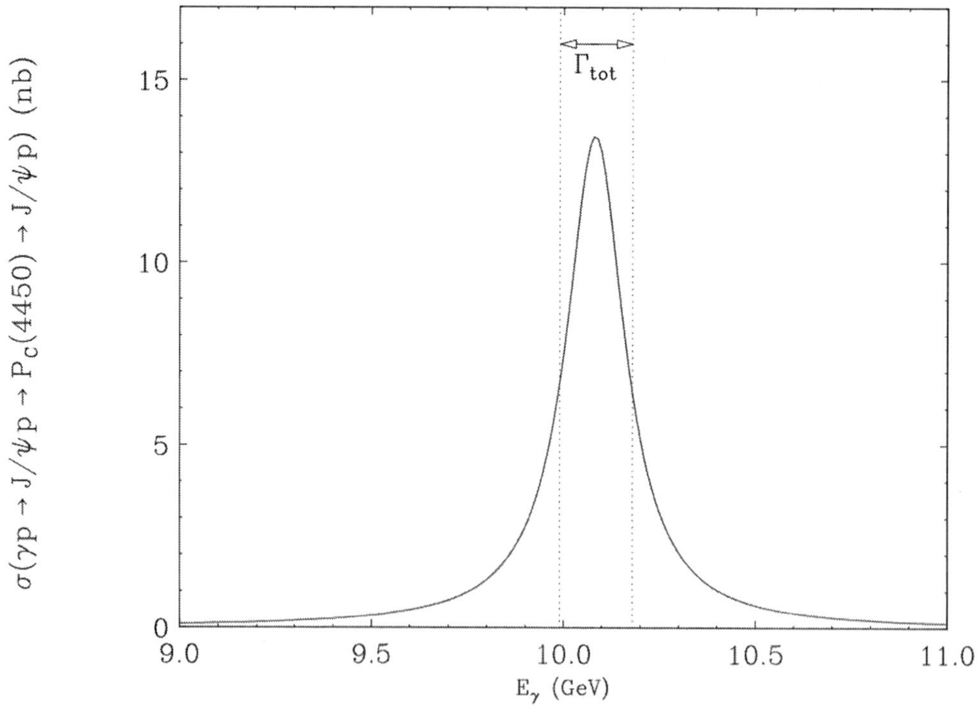

Figure 3. $\sigma(\gamma p \to J/\psi\, p \to P_c(4450) \to J/\psi\, p)$ resonant photoproduction cross section vs. the incident photon energy E_γ, assuming $B_{\rm out} = 0.1$ (see text). The width of the $P_c(4450)$ resonance is indicated by the vertical dotted lines.

One can extend the molecular model to doubly heavy hadrons which do not contain light quarks, with binding provided by η exchange [27], leading .e.g. to $D_s\bar{D}_s^*$ molecules, which might manifest themselves as $J/\psi\phi$ resonances, such as those recently observed by LHCb [28].

A $(\Sigma_b^+\Sigma_b^-)$ *beauteron* dibaryon?

In addition to the states listed in Table 2, there is an interesting possibility of genuine di-baryons, e.g. a *strongly bound* $\Sigma_b^+\Sigma_b^-$, $\Sigma_b^\pm\Sigma_c^\mp$ and $\Sigma_c^+\Sigma_c^-$ *deuteron-like states* [9]. The Σ_b is about 500 MeV heavier than B^*. The $\Sigma_b\Sigma_b$ kinetic energy is therefore significantly smaller than that of $B\bar{B}^*$ or $B^*\bar{B}^*$. Moreover, Σ_b with $I{=}1$ couples more strongly to pions than B and B^* with $I = \frac{1}{2}$. The opposite electric charges of Σ_b^+ and Σ_b^- provide additional 2-3 MeV of binding energy. Analogous considerations apply to $\Sigma_b^\pm\Sigma_c^\mp$ and $\Sigma_c^+\Sigma_c^-$ states. The heavy dibaryon bound state might be sufficiently long-lived to be observed experimentally. A possible decay mode of the beauteron is $(\Sigma_b^+\Sigma_b^-) \to \Lambda_b\Lambda_b\pi^+\pi^-$, which might be observable in LHCb. It should also be seen in lattice QCD.

References

[1] K. F. Chen *et al.* [Belle Coll.], Phys. Rev. Lett. **100**, 112001 (2008)
[2] M. Karliner and H. J. Lipkin, arXiv:0802.0649 [hep-ph]
[3] I. Adachi *et al.*, Belle Coll., arXiv:1105.4583 [hep-ex]
[4] A. Bondar *et al.* [Belle Coll.], Phys. Rev. Lett. **108**, 122001 (2012)
[5] M.B. Voloshin and L.B. Okun, JETP Lett. **23**, 333 (1076) [Pisma Zh.Eksp.Teor.Fiz. **23**, 369 (1976)]
[6] N. A. Tornqvist, Phys. Rev. Lett. **67**, 556 (1991); Z. Phys. C **61**, 525 (1994)
[7] N. A. Tornqvist, Phys. Lett. B **590**, 209 (2004)
[8] M. Ablikim *et al.* [BESIII Coll.], arXiv:1303.5949 [hep-ex]
[9] M. Karliner, H. J. Lipkin, N. A. Tornqvist, arXiv:1109.3472 [hep-ph]
[10] M. Karliner, H. J. Lipkin, N. A. Tornqvist, Nucl. Phys. Proc. Suppl. **225-227**, 102 (2012)
[11] M. Karliner and S. Nussinov, JHEP **1307**, 153 (2013)
[12] M. Ablikim *et al.* [BESIII Coll.], arXiv:1308.2760 [hep-ex]
[13] M. Ablikim *et al.* [BESIII Coll.], arXiv:1309.1896 [hep-ex]
[14] I thank Roman Mizuk for discussion on this point
[15] S. Chatrchyan *et al.* [CMS Coll.], arXiv:1309.0250 [hep-ex]
[16] G. Aad *et al.* (ATLAS Coll.), Phys. Rev. Lett. **108**, 152001 (2012)
[17] V. M. Abazov *et al.* [D0 Coll.], Phys. Rev. D **86**, 031103 (2012)
[18] R. Aaij *et al.* (LHCb Coll.), arXiv:1407.7734
[19] R. Aaij *et al.* (LHCb Coll.), arXiv:1409.1408
[20] M. Karliner and J. L. Rosner, Phys. Rev. D **91**, 014014 (2015)
[21] M. Karliner and J. L. Rosner, Phys. Rev. Lett. **115**, 122001 (2015)
[22] LHCb Coll., R. Aaij *et al.*, Phys. Rev. Lett. **115**, 072001 (2015)
[23] Q. Wang, X. H. Liu and Q. Zhao, Phys. Rev. D **92**, 034022 (2015)
[24] V. Kubarovsky and M.B. Voloshin, Phys. Rev. D **92**, 031502 (2015)
[25] M. Karliner and J. L. Rosner, Phys. Lett. B **752**, 329 (2016)
[26] Z.-E. Meziani *et al.*, arXiv:1609.00676 [hep-ex]
[27] M. Karliner and J. L. Rosner, Nucl. Phys. A **954**, 365 (2016)
[28] R. Aaij *et al.* [LHCb Coll.], arXiv:1606.07895 [hep-ex] and arXiv:1606.07898 [hep-ex]

Overview of the status/results in the exotics sector

Chengping Shen[1,]* and Changsheng Li[1] for the Belle Collaboration

[1] *School of Physics and Nuclear Energy Engineering, Beihang University, Beijing, People's Republic of China*

Abstract. Many exotic hadronic states beyond the conventional quark model (called charmoniumlike/bottomoniumlike states or XYZ particles) were found in the B-factories or energy scans of the cross sections of e^+e^- annhilation into conventional quarkonia and light hadrons. Their nature properties were proposed including glueballs, hybrids, multiquark states, hadron molecules, etc. Dramatic progress was made in the study of the them after the running of the B-factories. In this report, I present the most recent results on the XYZ results from Belle experiment.

1 Introduction

During the past thirteen years many exotic states decaying into conventional quarkonia and light hadrons, the so-called "XYZ" particles, have been reported in many different reactions [1]. They correspond to high excitations and have masses above the $D\bar{D}$ or $B\bar{B}$ thresholds. Most cannot be described well by quarkonium potential models [1–3]. Their unusual properties have stimulated considerable theoretical interest and various interpretations have been proposed, including tetraquarks, molecules, hybrids, or hadrocharmonia [1, 3, 4]. To distinguish among these explanations, more experimental information is needed. Here, I present the most recent results on the study of the XYZ states from the Belle experiment.

2 The X states

The observation of the $X(3872)$ [5] in 2003 kicked off the door of searching for exotic states in experiment. Considerable efforts by both experimentalists and theorists have been invested to clarify its nature. As a result, we know precisely its mass (3871.69±0.17) MeV$/c^2$, have a stringent limit on its width (less than 1.2 MeV at 90% confidence level (C.L.)) and have a definitive J^{PC} assignment of 1^{++}. Due to the close mass to $D^{*0}\bar{D}^0$ threshold, it was a considerable speculation from the beginning that the $X(3872)$ might be a molecule-like bound state of D^{*0} and $\bar{D}^0$ meson. The generally accepted interpretation at present for the still-fascinating $X(3872)$ is a mixture of a charmonium state $\chi_{c1}(2P)$ and an S-wave $D^{*0}\bar{D}^0$ molecule. More experimental information on the production and decays of the $X(3872)$ will shed additional light on its nature.

New interesting information on the $X(3872)$ is from BESIII by measuring the process $e^+e^- \to \gamma X(3872) \to \gamma\pi^+\pi^- J/\psi$ at e^+e^- center-of-mass energies from $\sqrt{s} = 4.009$ GeV to 4.420 GeV [6]. The observed statistical significance of $X(3872)$ in the $M(\pi^+\pi^- J/\psi)$ distribution reaches 6.3σ. Figure 1(a)

*e-mail: shencp@buaa.edu.cn

shows the energy-dependent cross section together with the fitted results with a $Y(4260)$ resonance, a linear continuum, or a $E1$-transition phase space term. The $Y(4260)$ resonance describes the data better than the other two options, which strongly supports the existence of the radiative transition process $Y(4260) \to \gamma X(3872)$. This is the first time that we observe a relation between the X and Y states.

It is natural to search for a similar state X_b with $J^{PC} = 1^{++}$ in the bottomonium system. The search for X_b supplies important information about the discrimination of a compact multiquark configuration and a loosely bound hadronic molecule configuration for the $X(3872)$. The CMS Collaboration ever reported a null search for such a state in the $\pi^+\pi^-\Upsilon(1S)$ final state [7]. However, unlike the $X(3872)$, whose decays exhibit large isospin violation, the X_b would decay preferably into $\pi^+\pi^-\pi^0\Upsilon(1S)$ rather than $\pi^+\pi^-\Upsilon(1S)$ if it exists. So Belle did a search for an X_b signal decaying to $\omega\Upsilon(1S)$ in $e^+e^- \to \gamma X_b \to \gamma\pi^+\pi^-\pi^0\Upsilon(1S)$ at a center-of-mass energy of 10.867 GeV based on a 118 fb^{-1} data sample [8].

Figure 1(b) shows the final $\omega\Upsilon(1S)$ invariant mass distribution with the requirement of $M(\pi^+\pi^-\pi^0)$ within the ω signal region; the X_b is searched for from 10.55 to 10.65 GeV/c^2. The dots with error bars are from data, the solid histogram is from the normalized contribution of $e^+e^- \to \omega\chi_{bJ}$ (J = 0, 1, 2) and the blank histogram is from the normalized ω mass sideband, defined as 0.54 GeV/c^2 < $M(\pi^+\pi^-\pi^0)$ < 0.72 GeV/c^2. No obvious X_b signal is observed. The upper limit on the yield of the X_b signal events is 4.0 at 90% C.L. with a mass at 10.6 GeV/c^2. The product branching fraction is determined to be $\mathcal{B}(\Upsilon(5S) \to \gamma X_b)\mathcal{B}(X_b \to \omega\Upsilon(1S)) < 2.9 \times 10^{-5}$ at 90% C.L.

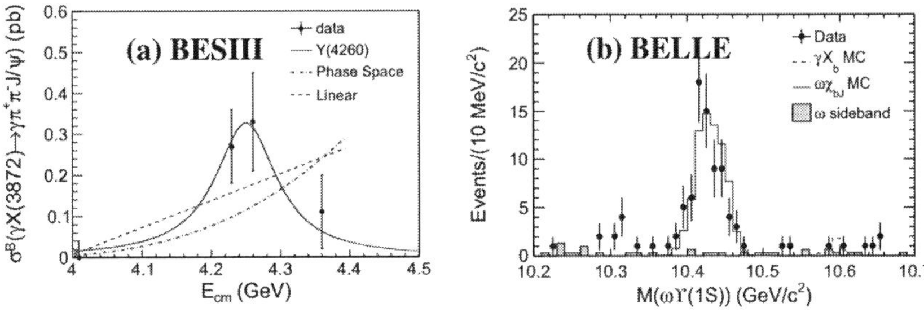

Figure 1. Results on the $X(3872)$ and its similar state in the bottomonium system X_b from BESIII and Belle. (a) The $\sigma^B[e^+e^- \to \gamma X(3872)] \times \mathcal{B}[X(3872) \to \pi^+\pi^- J/\psi]$ distribution and fit results with different line shapes from BESIII; (b) The $\omega\Upsilon(1S)$ invariant mass distribution in $e^+e^- \to \gamma\omega\Upsilon(1S)$ at $\sqrt{s}$ = 10.867 GeV from Belle measurement.

3 The Y states

States with $J^{PC} = 1^{--}$ can be studied with initial state radiation (ISR) in Belle's and BaBar's large $\Upsilon(4S)$ data samples or via direct production in e^+e^- collisions at BESIII.

The $Y(4260)$ state was first observed by BaBar in $e^+e^- \to \pi^+\pi^- J/\psi$ [9] and this was confirmed by the CLEO [10] and Belle experiments [11] with the same technique. Moreover, Belle reported a broad structure near 4.0 GeV called $Y(4008)$ [12]. Later, the BaBar Collaboration reported updated results with a 454 fb^{-1} data sample and a modified approach for the background description [13]; the $Y(4260)$ state was observed with improved significance, but the $Y(4008)$ structure was not confirmed. Instead, they attributed the structure below the $Y(4260)$ to exponentially falling non-resonant $\pi^+\pi^- J/\psi$ production. In 2013, Belle updated $e^+e^- \to \pi^+\pi^- J/\psi$ process again with a 967 fb^{-1} data sample. Not only the $Y(4260)$ state was observed clearly, but also the Y(4008) was confirmed [14].

A better understanding of the structures observed in $e^+e^- \to \pi^+\pi^- J/\psi$ would benefit from improved measurements in the future.

In an analysis of the $e^+e^- \to \pi^+\pi^-\psi(2S)$ process, BaBar found a structure near 4.32 GeV/c^2 [15], while Belle observed two resonant structures at 4.36 and 4.66 GeV/c^2 [16]. The latest $e^+e^- \to \pi^+\pi^-\psi(2S)$ analysis from the BaBar experiment with its full data sample confirmed the existence of the $Y(4660)$ state [17]. To characterize more precisely the properties of the $Y(4360)$ and $Y(4660)$, Belle updated the $e^+e^- \to \pi^+\pi^-\psi(2S)$ process with the full Belle data 980 fb^{-1} [18]. Figure 2 shows the invariant mass distribution of $\pi^+\pi^-\psi(2S)$ from the updated measurement. Besides the $Y(4360)$ and $Y(4660)$, there are a few events in the vicinity of the $Y(4260)$ mass, so an alternative fit with a coherent sum of $Y(4260)$, $Y(4360)$, and $Y(4660)$ amplitudes was performed as shown in Fig. 2. But the signal significance of the $Y(4260)$ is only 2.4σ. With more statistic at BelleII, it will be clear if the $Y(4260)$ can decay to $\pi^+\pi^-\psi(2S)$.

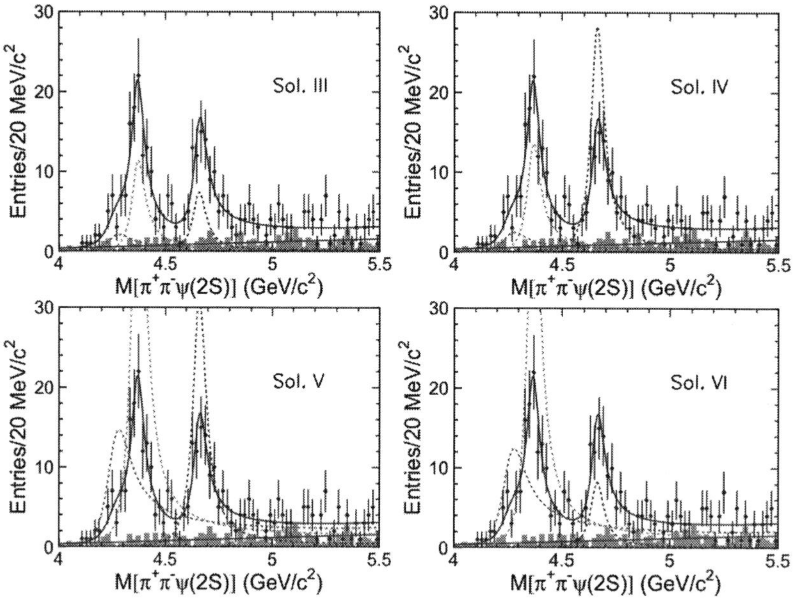

Figure 2. The four solutions from the fit to the $\pi^+\pi^-\psi(2S)$ invariant mass spectra with the $Y(4260)$ included. The curves show the best fit and the dashed curves show the contributions from the three Breit-Wigner components.

BESIII accumulated large data samples between 4.0 and 4.6 GeV for the study of the charmoniumlike states. With these data samples, BESIII measured some interesting processes, for examples, $e^+e^- \to \pi^+\pi^- h_c$ [19], $\omega\chi_{c0}$ [20], $\pi^+\pi^-\psi_2$ [21]. In the future BelleII needs to check $e^+e^- \to \pi^+\pi^- h_c$, $\pi^+\pi^- h_c(2P)$, $K^+K^- J/\psi$, $K^+K^-\psi(2S)$, $\pi^+\pi^-\psi(2S)$, $\pi^+\pi^-\psi_2$, $\omega\chi_{cJ}$, $\phi\chi_{cJ}$, etc., to search for more new vector charmoniumike states or new decay modes of the discovered Y states.

4 The Z states

After the charged charmoniumlike state $Z_c(3900)$ was observed by BESIII and Belle experiments [14, 22] BESIII and Belle observed a series of charged Z_c states including $Z_c(4020)$ [19], $Z_c(4200)$ [23], and $Z_c(3885)$ [24]. These states seem to indicate that a new class of hadrons has been observed. As there are at least four quarks within these Z_c states, they have been interpreted either as tetraquark states, molecular states, or other configurations.

In the Belle updated $e^+e^- \to \pi^+\pi^-\psi(2S)$ analysis [18] there is an excess evident at around 4.05 GeV/c^2 in the $\pi^\pm\psi(2S)$ invariant mass distributions in the $Y(4360)$ subsample defined as $4.0 < M_{\pi^+\pi^-\psi(2S)} < 4.5$ GeV/c^2. An unbinned maximum-likelihood fit is performed on the distribution of $M_{\rm max}(\pi^\pm\psi(2S))$, the maximum of $M(\pi^+\psi(2S))$ and $M(\pi^-\psi(2S))$. The excess is parameterized with a Breit-Wigner function and the non-resonant non-interfering background with a second-order polynomial function. The fit yields a mass of (4054 ± 3(stat.) ± 1(syst.)) MeV/c^2, a width of (45 ± 11(stat.) ± 6(syst.)) MeV and a 3.5σ significance. This evident structure is called $Z_c(4050)$, which is a possible excited state of the $Z_c(3900)$.

Two charged bottomoniumlike resonances, $Z_b(10610)$ and $Z_b(10650)$, have been observed by the Belle Collaboration in $e^+e^- \to \Upsilon(nS)\pi^+\pi^-$, $n = 1,2,3$ and $e^+e^- \to h_b(m\mathrm{P})\pi^+\pi^-$, $m = 1,2$ [25, 26]. Several models suggested that the $Z_b(10610)$ and $Z_b(10650)$ states might be loosely bound $B\bar{B}^*$ and $B^*\bar{B}^*$ systems, respectively. If so, it is natural to expect the Z_b states to decay to final states with $B^{(*)}$ mesons at substantial rates. Using a data sample with an integrated luminosity of 121.4 fb^{-1} collected near the peak of the $\Upsilon(5S)$ resonance, Belle reconstructed three-body $B^{(*)}\bar{B}^{(*)}\pi$ combinations with a charged primary pion [27], where the set of $B^+\bar{B}^0\pi^-$ and $B^-B^0\pi^+$ final states is referred to as $BB\pi$; the set of $B^+\bar{B}^{*0}\pi^-$, $B^-B^{*0}\pi^+$, $B^0B^{*-}\pi^+$ and $\bar{B}^0B^{*+}\pi^-$ final states is referred to as $BB^*\pi$; and the set of $B^{*+}\bar{B}^{*0}\pi^-$ and $B^{*-}B^{*0}\pi^+$ final states is denoted as $B^*B^*\pi$. B mesons are reconstructed in 18 decay channels totally. There are 12263 ± 168 fully reconstructed B mesons. In the missing mass distribution of $B\pi$, peaks corresponding to the $BB^*\pi$ and $B^*B^*\pi$ signals are evident. The $M_{\rm miss}(\pi)$ distributions are shown in Fig. 3, where three models are used to fit: only the $Z_b(10610)$ [$Z_b(10650)$] amplitude [Model-0], an additional possible non-resonant component [Model-1], and a combination of two Z_b amplitudes [Model-2]. The fit results are shown in Fig. 3. The fit to the $BB^*\pi$ data with Model-0 gives 10605 ± 6 MeV/c^2 and 25 ± 7 MeV for the $Z_b(10610)$ mass and width, respectively, and the fit to the $B^*B^*\pi$ data gives 10648 ± 13 MeV/c^2 and 23 ± 8 MeV for the $Z_b(10650)$ mass and width, respectively.

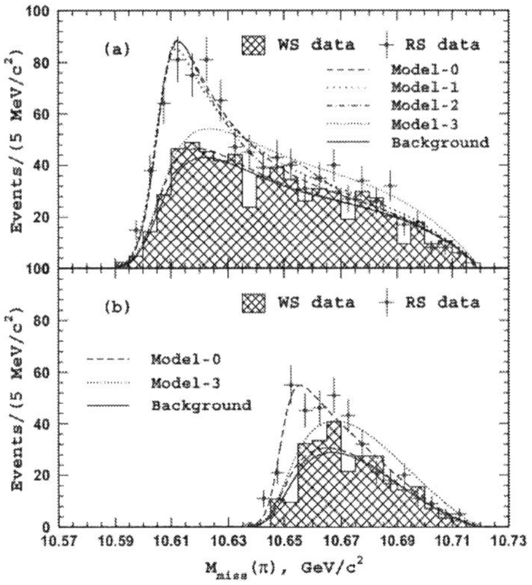

Figure 3. The $M_{\rm miss}(\pi)$ distribution for the (a) $BB^*\pi$ and (b) $B^*B^*\pi$ candidate events.

The calculated ratio of the branching fractions are $\mathcal{B}(Z_b(10610) \to B\bar{B}^* + c.c.)/\mathcal{B}(Z_b(10610) \to \text{bottomonium}) = 4.76 \pm 0.64 \pm 0.75$ and $\mathcal{B}(Z_b(10650) \to B^*\bar{B}^*)/\mathcal{B}(Z_b(10650) \to \text{bottomonium}) = 2.40 \pm 0.44 \pm 0.50$. Assuming that the $Z_b(10610)$ and $Z_b(10650)$ are saturated by the already observed $\Upsilon(n\mathrm{S})\pi$, $h_b(m\mathrm{P})\pi$, and $B^*B^{(*)}$ channels, the decay rates are summarized in Table 1.

Table 1. Branching fractions for the $Z_b^+(10610)$ and $Z_b^+(10650)$ decays.

Channel	Fraction, %	
	$Z_b(10610)$	$Z_b(10650)$
$\Upsilon(1S)\pi^+$	$0.60 \pm 0.17 \pm 0.07$	$0.17 \pm 0.06 \pm 0.02$
$\Upsilon(2S)\pi^+$	$4.05 \pm 0.81 \pm 0.58$	$1.38 \pm 0.45 \pm 0.21$
$\Upsilon(3S)\pi^+$	$2.40 \pm 0.58 \pm 0.36$	$1.62 \pm 0.50 \pm 0.24$
$h_b(1P)\pi^+$	$4.26 \pm 1.28 \pm 1.10$	$9.23 \pm 2.88 \pm 2.28$
$h_b(2P)\pi^+$	$6.08 \pm 2.15 \pm 1.63$	$17.0 \pm 3.74 \pm 4.1$
$B^+\bar{B}^{*0} + \bar{B}^0 B^{*+}$	$82.6 \pm 2.9 \pm 2.3$	–
$B^{*+}\bar{B}^{*0}$	–	$70.6 \pm 4.9 \pm 4.4$

Very recently Belle measured the energy dependence of the $e^+e^- \to h_b(nP)\pi^+\pi^-$ ($n = 1, 2$) cross sections from thresholds up to 11.02 GeV using 121.4 fb^{-1} of on-resonance $\Upsilon(5S)$ data, as well as 1 fb^{-1} of data taken at each of 19 different energies between 10.77 and 11.02 GeV [28]. Clear $\Upsilon(10860)$ and $\Upsilon(11020)$ peaks with little or no continuum contribution are observed. The distributions of $M_{\rm miss}(\pi)$ for the yields of $\pi^+\pi^- h_b(1P)$ and $\pi^+\pi^- h_b(2P)$ are not distributed uniformly in phase space; they populate the $Z_b(10610)$ and $Z_b(10650)$ mass region. The model with the $Z_b(10610)$ and $Z_b(10650)$ signals describes the data well: the confidence levels of the fits are 65% and 77% for the $h_b(1P)$ and $h_b(2P)$, respectively. The phase space hypothesis is excluded at the 3.6σ and 4.5σ levels in the $h_b(1P)\pi^+\pi^-$ and $h_b(2P)\pi^+\pi^-$ channels, respectively. The single $Z_b(10610)$ hypothesis is excluded at the 3.3σ level in the $h_b(1P)\pi^+\pi^-$ channel, while the single $Z_b(10650)$ hypothesis cannot be excluded at a significant level.

5 Other topics

Using a 5.74 fb^{-1} data sample collected at the peak of the $\Upsilon(1S)$ resonance, containing 102×10^6 $\Upsilon(1S)$ decays, Belle searched for some of the XYZ states in $\Upsilon(1S)$ inclusive decays for the first time [29]. In these searches, fourteen decay modes are considered: $X(3872)$ and $Y(4260)$ to $\pi^+\pi^- J/\psi$; $Y(4260)$, $Y(4360)$ and $Y(4660)$ to $\pi^+\pi^-\psi(2S)$; $Y(4260)$ to $K^+K^- J/\psi$; $Y(4140)$ and $X(4350)$ to $\phi J/\psi$; $Z_c(3900)^\pm$, $Z_c(4200)^\pm$ and $Z_c(4430)^\pm$ to $\pi^\pm J/\psi$; $Z_c(4050)^\pm$ and $Z_c(4430)^\pm$ to $\pi^\pm\psi(2S)$; and a predicted $Z_{cs}^\pm$ state with mass (3.97 ± 0.08) GeV$/c^2$ and width (24.9 ± 12.6) MeV [30] to $K^\pm J/\psi$. All the invariant mass distributions from a $J/\psi(\psi(2S))$ and up to two charged light hadrons ($K^\pm/\pi^\pm$) are checked. No evident signal is found for any of them and 90% C.L. upper limits are set on the product branching fractions and listed in Table III in Ref. [29]. The tagged J/ψ and $\psi(2S)$ signals are observed clearly and the corresponding branching fractions are measured to be $\mathcal{B}(\Upsilon(1S) \to J/\psi + \text{anything}) = (5.25 \pm 0.13(\text{stat.}) \pm 0.25(\text{syst.})) \times 10^{-4}$ and $\mathcal{B}(\Upsilon(1S) \to \psi(2S) + \text{anything}) = (1.23 \pm 0.17(\text{stat.}) \pm 0.11(\text{syst.})) \times 10^{-4}$ with substantially improved precision.

The LHCb collaboration reported the observation of two pentaquark states, denoted as $P_c(4380)^+$ and $P_c(4450)^+$, in the $J/\psi p$ system in $\Lambda_b^0 \to J/\psi K^- p$ [31]. The first strong experimental evidence for a pentaquark state $\Theta(1540)^+$ was reported in the reaction $\gamma n \to nK^+K^-$ in the LEPS experiment [32]. However, it was not confirmed in larger-statistics data samples in the same experiment and was most probably not a genuine state. Very recently Belle did the first measurement $\gamma\gamma \to p\bar{p}K^+K^-$ and searched for novel exotic baryons, denoted as $\Theta(1540)^0 \to pK^-$ and $\Theta(1540)^{++} \to pK^+$ which are similar to $\Theta(1540)^+$, in intermediate processes [33]. The results are based on an analysis of a 980 fb^{-1} data sample taken at or near the $\Upsilon(nS)$ ($n = 1, ..., 5$) resonances with the Belle detector. Clear $\gamma\gamma \to p\bar{p}K^+K^-$ signals are observed and the corresponding cross section is measured. No clear structure is seen in the corss section distribution and the largest value is around 40 pb, which is lower than the theoretical rough estimate of 100 pb. While the $\Lambda(1520)^0$ signals in pK^- invariant mass spectrum are also observed, no evidence for any exotic baryon is seen in the pK^- or pK^+ invariant mass spectrum. For all of the above-mentioned processes, the cross sections are measured for the first time.

Supported in part by the National Natural Science Foundation of China under Contract No. 11575017 and National Key Basic Research Program of China under Contact No. 2015CB856701.

References

[1] N. Brambilla *et al.*, Eur. Phys. J. C **71**, 1534 (2011)
[2] S. Godfrey and S. L. Olsen, Annu. Rev. Nucl. Part. Sci. **58**, 51 (2008)
[3] N. Brambilla *et al.*, Eur. Phys. J. C **74**, 2981 (2014)
[4] C. Z. Yuan, Int. J. Mod. Phys. A **29**, 1430046 (2014)
[5] S. K. Choi *et al.* (Belle Collaboration), Phys. Rev. Lett. **91**, 262001 (2003)
[6] M. Ablikim *et al.* (BESIII Collaboration), Phys. Rev. Lett. **112**, 092001 (2014)
[7] S. Chatrchyan *et al.* (CMS Collaboration), Phys. Lett. B **727**, 57 (2013)
[8] X. H. He *et al.* (Belle Collaboration), Phys. Rev. Lett. **113**, 142001 (2014)
[9] B. Aubert *et al.* (BaBar Collaboration), Phys. Rev. Lett. **95**, 142001 (2005)
[10] Q. He *et al.* (CLEO Collaboration), Phys. Rev. D **74**, 091104(R) (2006)
[11] K. Abe *et al.* (Belle Collaboration), hep-ex/0612006
[12] C. Z. Yuan *et al.* (Belle Collaboration), Phys. Rev. Lett. **99**, 182004 (2007)
[13] J. P. Lees *et al.* (BaBar Collaboration), Phys. Rev. D **86**, 051102(R) (2012)
[14] Z. Q. Liu *et al.* (Belle Collaboration), Phys. Rev. Lett. **110**, 252002 (2013)
[15] B. Aubert *et al.* (BaBar Collaboration), Phys. Rev. Lett. **98**, 212001 (2007)
[16] X. L. Wang *et al.* (Belle Collaboration), Phys. Rev. Lett. **99**, 142002 (2007)
[17] B. Aubert *et al.* (BaBar Collaboration), Phys. Rev. D **89**, 111103 (2014)
[18] X. L. Wang *et al.* (Belle Collaboration), Phys.Rev. D **91**, 112007 (2015)
[19] M. Ablikim *et al.* (BESIII Collaboration), Phys. Rev. Lett. **111**, 242001 (2013)
[20] M. Ablikim *et al.* (BESIII Collaboration), Phys. Rev. Lett. **114**, 092003 (2015)
[21] M. Ablikim *et al.* (BESIII Collaboration), Phys. Rev. Lett. **115**, 011803 (2015)
[22] M. Ablikim *et al.* (BESIII Collaboration), Phys. Rev. Lett. **110**, 252001 (2013)
[23] K. Chilikin *et al.* (Belle Collaboration), Phys. Rev. D **90**, 112009 (2014)
[24] M. Ablikim *et al.* (BESIII Collaboration), Phys. Rev. Lett. **112**, 022001 (2014)
[25] A. Bondar *et al.* (Belle Collaboration), Phys. Rev. Lett. **108**, 122001 (2012)
[26] A. Garmash *et al.* (Belle Collaboration), Phys. Rev. D **91**, 072003 (2015)
[27] A. Garmash *et al.* (Belle Collaboration), Phys. Rev. Lett. **116**, 212001 (2016)
[28] R. Mizuk *et al.* (Belle Collaboration), Phys. Rev. Lett. **117**, 142001 (2016)
[29] C. P. Shen *et al.* (Belle Collaboration), Phys. Rev. D **93**, 112013 (2016)
[30] S. H. Lee, M. Nielsen and U. Wiedner, J. Korean Phys. Soc. **55**, 424 (2009); J. M. Dias, X. Liu and M. Nielsen, Phys. Rev. D **88**, 096014 (2013)
[31] R. Aaij *et al.* (LHCb Collaboration), Phys. Rev. Lett. **115**, 072001 (2015)
[32] T. Nakano *et al.* (LEPS Collaboration), Phys. Rev. Lett. **91**, 012002 (2003)
[33] C. P. Shen *et al.* (Belle Collaboration), Phys. Rev. D **93**, 112017 (2016)

EPJ Web of Conferences **130**, 01005 (2016) DOI: 10.1051/epjconf/201613001005
MESON 2016

Review of dark photon searches

Achim Denig[1,⋆]

[1] *Institute for Nuclear Physcis and PRISMA Cluster of Excellence, Johannes Gutenberg University Mainz, Germany*

Abstract. Dark Photons are hypothetical extra-U(1) gauge bosons, which are motivated by a number of astrophysical anomalies as well as the presently seen deviation between the Standard Model prediction and the direct measurement of the anomalous magnetic moment of the muon, $(g-2)_\mu$. The Dark Photon does not serve as the Dark Matter particle itself, but acts as a messenger particle of a hypothetical Dark Sector with residual interaction to the Standard Model. We review recent Dark Photon searches, which were carried out in a global effort at various hadron and particle physics facilities. We also comment on the perspectives for future invisble searches, which directly probe the existence of Light Dark Matter particles.

1 Motivation

Dark Photons are hypothetical particles of an extra U(1) gauge group. Such extra U(1) gauge groups are predicted by almost any extension of the Standard Model. Indeed, the related extra U(1) gauge bosons are searched for from the lowest (e.g. searches for axion-like particles) up to the highest energies at the LHC. Recently, the mass range for a vector particle in the MeV to GeV scale has been in the focus of interest due to the observation by Arkani Hamed *et al.* [1] that particles of such a mass scale might explain a surprisingly large number of astrophysical anomalies. In addition, it was realized that a Dark Photon of such a mass might explain the presently seen deviation between the direct measurement and the Standard Model prediction of the anomalous magnetic moment of the muon, $(g-2)_\mu$ [2].

Fig. 1 shows two Feynman diagrams with Dark Photon interactions. The upper diagram shows the kinetic mixing between the Standard Model photon and the hypothetical Dark Photon γ' (aka A', Z', or U boson) via a loop of charged leptons, L. Such a kinetic mixing term was propsed already in the 80's by Holdom [3]. The lower diagram might give rise to the positron excess presently seen in the spectrum of cosmic rays, which is indeed one of the most striking astrophysical anomalies. Dark Matter (DM) particles annihilate via a Dark Photon into an e^+e^- pair. Within the kinetic mixing model two free parameters remain: the mass of the Dark Photon, $m_{\gamma'}$, and the mixing parameter ϵ, which parametrizes the strength of the coupling of the Dark Photon to ordinary Standard Model matter: $\epsilon^2 = \alpha'/\alpha$, where α is the QED fine structure constant.

The possible relation of the Dark Photon to the Dark Matter problem as well as the fact that it might explain the $(g-2)_\mu$ puzzle, triggered an enormous theoretical and experimental interest in the particle

⋆e-mail: denig@kph.uni-mainz.de

and hadron physics community. In the following we will discuss electron scattering experiments as well as results from hadron and e^+e^- accelerators. Very recently, an anomaly was observed in a nuclear decay of Berryllium [4]. This could be a first hint for a 17 MeV Dark Photon signal, however the confirmation of the anomaly by an independent group is pending.

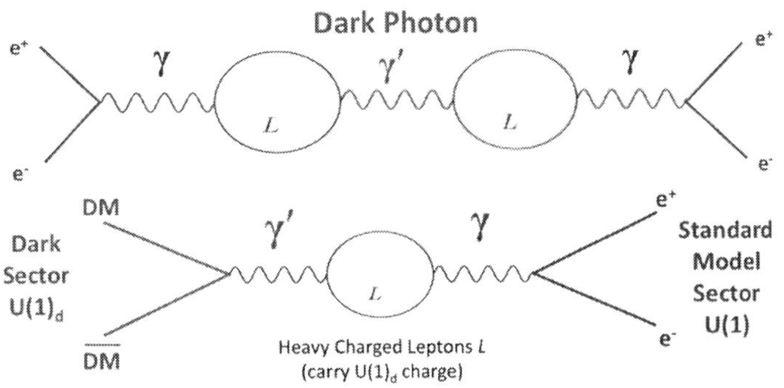

Figure 1. Feynman graphs depicting interactions via a hypothetical Dark Photon γ'. Up: Kinetic mixing model; Down: Interaction between the Standard Model sector and the Dark Sector via a Dark Photon. The loop denotes a pair of charged leptons, which couple both to the Standard Model photon as well as the Dark Photon.

2 Searches using fixed-target experiments

As Bjoerken and collaborators [5] have pointed out, low-energy electron accelerators in combination with high-resolution detectors are very well suited for Dark Photon searches. By scattering the electron beam on a nuclear target, the Dark Photon may be emitted in the initial or final state, see Fig. 2. Its coupling to an e^+e^- pair allows for an identification by looking for a bump in the e^+e^- invariant mass. The huge background is almost entirely given by QED contributions, such as for instance Bethe-Heitler processes. In order to achieve competitive results in these kind of experiments, the following conditions need to be fulfiled:

- High intensity electron beams (in the order of hundreds of microamps) are needed.
- An excellent mass resolution is needed as obtained for instance with high resolution magnetic spectrometers.
- The kinematics of the signal process requires a detection of electrons and positrons in the final state at small angles.

Successful pilot experiments have been carried out in 2011 at MAMI in Mainz [6] (A1 experiment) and JLAB [7] (APEX experiment) with electron beam energies of 0.9 GeV and 2.3 GeV, respectively. These runs could improve upon existing Dark Photon limits from BaBar (2009 results) in the mass range around 200 MeV. More recently, a very wide parameter range between approximately 40 MeV and 200 MeV was tested at MAMI with the high resolution spectrometer (HRS) setup A1 [8]. No significant signal was found and the ϵ parameter range down to 10^{-3} was excluded, constraining a large part of the parameter range motivated by $(g-2)_\mu$, see Fig. 4.
For the near and mid-term future several dedicated experiments are in preparation at JLAB. The APEX experiment [9], which is using an existing HRS setup at JLAB, will extend the mass range covered by A1/MAMI towards higher masses and lower ϵ values. The HPS experiment [10] will exploit a

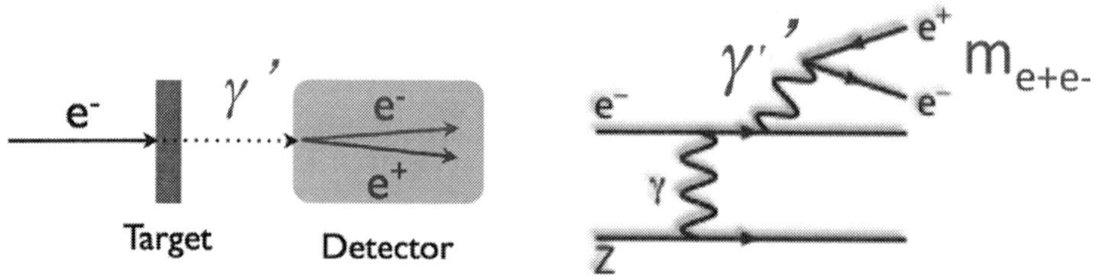

Figure 2. Poduction of a Dark Photon in a fixed-target environment. Left: Sketch of the experiment. Right: Feynman diagram for the production of the Dark Photon. One important background process is given by exchanging the Dark Photon with a QED photon.

displaced vertex technique, which allows to test even lower values of ϵ. First data has been collected with the HPS setup in 2016. Finally, the Dark Light [11] experiment at the FEL accelerator at JLAB aims for testing the low mass region below the results already covered by A1. The new spectrometer setup MAGIX at the MESA accelerator [12] in Mainz will also be able to cover this mass range. Both MAGIX and Dark Light are planned to operate at modern Energy-Recovering-Linac accelerators which provide the highest intensities. High resolution measurements can be expected due to the fact that windowless gas targets will be used, which minimize effects of multiple scattering.

3 Data mining campaigns using existing data sets

The recent interest in MeV to GeV scale Dark Photons triggered also a number of searches at electron-positron as well as hadron accelerators using already existing data sets, which were collected for completely different purposes (data mining campaigns). Essentially, any process in which within the Standard Model a (virtual) photon couples to a pair of electrons, muons, or hadrons, can be used as a probe to search for Dark Photons. The following investigations came out to be most successful in producing competitive exclusion limits:

- **Radiative decays of vector resonances:** The BABAR collaboration had first explored this technique by looking for bumps in the invariant mass spectrum of muon pairs in radiative decays of the $\Upsilon(3S)$ resonance [13]. Later, the KLOE collaboration investigated decays of the ϕ resonance into an η meson and a pair of e^+e^- [14].
- **Dalitz decays of the π^0 and η mesons:** This allows to explore the low mass range, see also Fig. 3 (left). The method was applied by the NA48 collaboration at the SPS/CERN [15], by WASA at COSY [16], HADES at GSI [17], as well as PHENIX at RHIC [18].
- **Initial State Radiation (ISR):** A wide range of Dark Photon masses can be probed by looking for Dark Photons, which are produced via the ISR process, see Fig. 3 (right). KLOE has searched for the Dark Photon below 1 GeV in the $\mu^+\mu^-$ [19], the e^+e^- [20], and more recently also in the $\pi^+\pi^-$ final states [21]. A particulrily competitive parameter range was probed by BABAR below 10 GeV by searching for peaks in the radiative $\mu^+\mu^-$ and e^+e^- spectrum [22].

A compilation of all exclusion plots obtained in the last years can be found in Fig. 4. Constraints from old axion searches (experiments E141 and E774) as well as a constraint from the precision observables (g-2) of the electron, are also shown. The red band depicts the parameter range, in which a Dark Photon could explain the presently seen deviation of 3.6 standard deviations between the

DOI: 10.1051/epjconf/201613001005

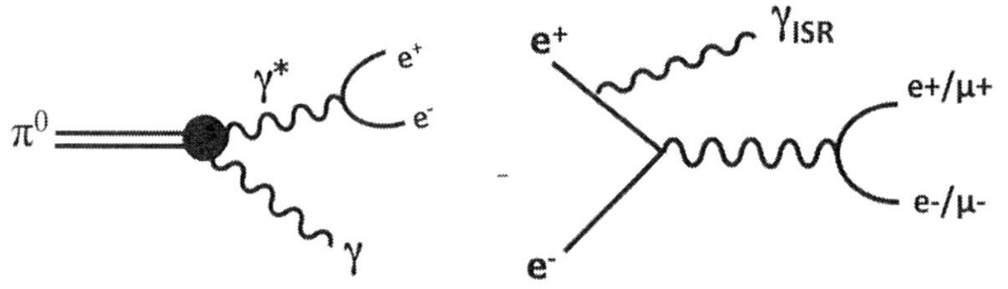

Figure 3. A search for the Dark Photon can be carried out in any process, in which a virtual photon couples to a lepton or quark pair. By exchanging the QED photon with a Dark Photon, one can look for a bump in the invariant mass spectrum of the lepton/quark pair. Left: π^0 Dalitz decay; Right: ISR production of the Dark Photon.

Standard Model prediction and the direct measurement of the $(g-2)$ of the muon. In the meantime, the $(g-2)_\mu$ band is completely covered by experiments.

For the future, the BELLE II project will collect a statistics, which will exceed the BABAR data set by almost two orders of magnitude, which will further constrain the Dark Photon parameter range towards lower ϵ values. At low Dark Photon masses, it can be expected that Dalitz decay studies at LHCb or elsewhere as well as the already mentioned low-energy fixed-target experiments will yield competitive results. The HPS project is designed to resolve displaced vertices, which will test islands of very low ϵ at masses below 500 MeV. A recent summary of the already published searches as well as the future experiments can be found in Ref. [23].

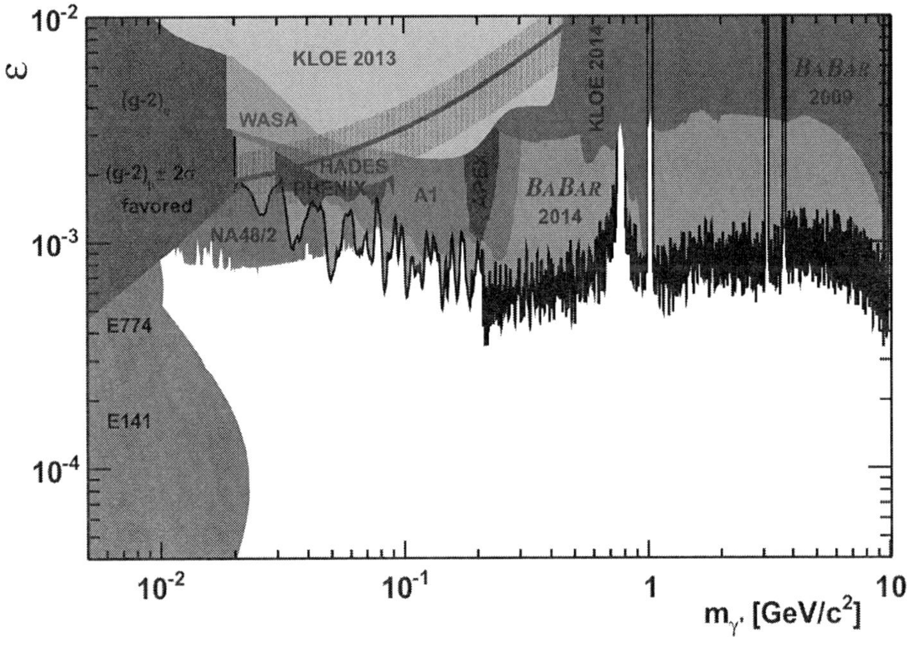

Figure 4. Compilation of existing Dark Photon constraints in the parameter range Dark Photon coupling, ϵ, versus the mass of the Dark Photon, $m_{\gamma'}$.

4 Invisible dark photon decays

The kinetic mixing model discussed so far, predicts a Dark Photon, which kinematically mixes with the ordinary Standard Model photon. As mentioned in the beginning, such a Dark Photon not necessarily needs to be related to the Dark Matter problem, but astrophysical observations motivate the existence of a GeV scale Dark Photon. In the kinematic mixing model, the Dark Photon has a mass, which is small compared to the Dark Matter candidate. A feature of the kinematic mixing model is that the Dark Photon decays to Standard Model particles (leptons and quarks), however suppressed by a factor of ϵ in comparison to the ordinary photon coupling.

We would like to discuss also a second model, in which the Dark Photon is actually heavier than the Dark Matter particle. In this case the Dark Photon can decay to the Dark Matter candidate directly and hence does not primarily decay to Standard Model particles. Fig. 5 shows the production and decay of such a Dark Photon in a fixed target scattering experiment. Due to the fact that the Dark Photon decays invisibly in the experiment, we call this model the Invisible Dark Photon model. It should be stressed that theoretically Light (i.e. MeV scale) Dark Matter is a vital option for the Dark Matter problem. Present direct Dark Matter searches are however not sensitive to such light particles.

A Dark Photon in the Invisible Dark Photon model is experimentally much less constraint than a Dark Photon in the kinematic mixing model. This calls for new experiments and in fact quite a number of them are actually prepared at various laboratories worldwide (Cornell, JLAB, Mainz, Frascati, CERN, Novosibirsk) [23]. A common feature of these projects is the search for large missing mass and/or missing momentum in scattering events. Again, several experimental environments can be exploited, ranging from electron-positron annihilation to fixed target experiments.

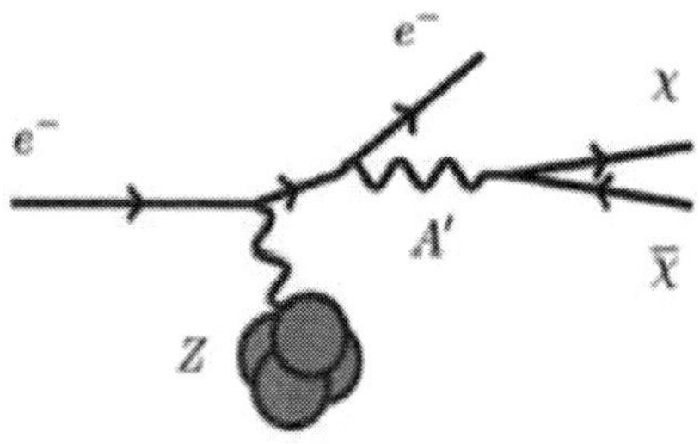

Figure 5. Invisible Dark Photon decay into a pair of Dark Matter candidates, $\chi\bar{\chi}$. Differently from previous plots the Dark Photon is shere denoted as A'.

An exciting idea is to search for the Dark Matter candidates directly in a fixed target experiments as depicted in Fig. 5. This might be achieved very effectively by carrying out these searches behind the beam dumps of high intensity electron accelerators, so-called Beam Dump Experiments (BDX) Ref. [24]. High-intensity accelerators, as the ones operated at JLAB or Mainz, have the potential to produce per year in the order of 10^{21} electrons on target (EOT). For such kind of searches the experimental techniques to detect the LDM candidate could be similar to the techniques used in classical Dark Matter experiments. It yet remains to be seen whether the beam-related background (e.g. neutrons) is low enough to successfully carry out the experiments.

5 Summary and outlook

Searches for Dark Photons, which are hypothetical vector particles of an extra U(1) gauge sector, are motivated by the fact that they might explain a number of astrophysical anomalies, such as for instance the excess of positrons in cosmic rays. The Dark Photon is not a candidate for the Dark Matter particle itself, but serves as a gauge particle of a hypothetical Dark Sector. In the last years, we have witnessed a huge experimental activity worldwide to search for such a Dark Photon. Presently, the only evidence stems from a nuclear transition in Beryllium, in which an evidence for a 17 MeV particle is claimed. it will be most exciting to see whether this evidence is confirmed by yet another group.

Recent Dark Photon searches were also motivated by the fact that a MeV to GeV scale Dark Photon could serve as an explanation for the deviation presently seen between the Standard Model prediction and the direct measurement of $(g-2)_\mu$. In the meantime, such a possibility is ruled out in the kinetic mixing model, in which the Dark Photon couples primarily to lepton pairs (or depending on its mass also to quark pairs).

A new generation of experiments is in preparation to explore Dark Photon searches, in which the Dark Photon decays primarily to Dark Matter itself. Such kind of experiments open the avenue to search for Light Dark Matter, which so far is not constraint by classical Dark Matter experiments. In this Invisible Dark Photon model an explanation for the $(g-2)_\mu$ discrepancy is still given. A recent summary report describes in detail the future directions in the field of Dark Photon searches and contains a full list of references of Dark Photon experiments and of theoretical papers related to the subject. We invite the interested reader to have a look at this summary report Ref. [23].

References

[1] N. Arkani-Hamed, D. P. Finkbeiner, T. R. Slatyer and N. Weiner, Phys. Rev. D **79** 015014 (2009)
[2] M. Pospelov, Phys. Rev. D **80** 095002 (2009)
[3] B. Holdom, Phys. Lett. B **166** 196 (1986)
[4] A. J. Krasznahorkay *et al.*, Phys. Rev. Lett. **116** 042501 (2016)
[5] J. D. Bjorken, R. Essig, P. Schuster and N. Toro, Phys. Rev. D **80** 075018 (2009)
[6] H. Merkel *et al.* [A1 Collaboration], Phys. Rev. Lett. **106** 251802 (2011)
[7] S. Abrahamyan *et al.* [APEX Collaboration], Phys. Rev. Lett. **107** 191804 (2011)
[8] H. Merkel *et al.* [A1 Collaboration], Phys. Rev. Lett. **106** 251802 (2011)
[9] R. Essig, P. Schuster, N. Toro and B. Wojtsekhowski, JHEP **1102** 009 (2011)
[10] http://nuclear.unh.edu/HPS/HPS_proposal.pdf.
[11] http://dmtpc.mit.edu/DarkLight/DarkLightProposal_PAC39.pdf.
[12] K. Aulenbacher, M. Dehn, H. J. Kreidel, R. Heine and R. Eichhorn, ICFA Beam Dyn. Newslett. **58** 145 (2012)
[13] B. Aubert *et al.* [BaBar Collaboration], Phys. Rev. Lett. **103** 081803 (2009)
[14] D. Babusci *et al.* [KLOE-2 Collaboration], Phys. Lett. B **720** 111 (2013)
[15] J. R. Batley *et al.* [NA48/2 Collaboration], Phys. Lett. B **746** 178 (2015)
[16] P. Adlarson *et al.* [WASA-at-COSY Collaboration], Phys. Lett. B **726** 187 (2013)
[17] G. Agakishiev *et al.* [HADES Collaboration], Phys. Lett. B **731** 265 (2014)
[18] A. Adare *et al.* [PHENIX Collaboration], Phys. Rev. C **91** 031901 (2015)
[19] D. Babusci *et al.* [KLOE-2 Collaboration], Phys. Lett. B **736** 459 (2014)
[20] A. Anastasi *et al.*, Phys. Lett. B **750** 633 (2015)
[21] A. Anastasi *et al.* [KLOE-2 Collaboration], Phys. Lett. B **757** 356 (2016)
[22] J. P. Lees *et al.* [BaBar Collaboration], Phys. Rev. Lett. **113** 201801 (2014)
[23] J. Alexander *et al.*, arXiv:1608.08632 [hep-ph]
[24] E. Izaguirre, G. Krnjaic, P. Schuster and N. Toro, Phys. Rev. D **90** 014052 (2014)

Pion–nucleon scattering: from chiral perturbation theory to Roy–Steiner equations

Bastian Kubis[1,⋆], Martin Hoferichter[2], Jacobo Ruiz de Elvira[1], and Ulf-G. Meißner[1,3]

[1] *HISKP (Th) and BCTP, Universität Bonn, 53115 Bonn, Germany*
[2] *INT, University of Washington, Seattle, WA 98195-1550, USA*
[3] *IKP, IAS, and Jülich Center for Hadron Physics, FZ Jülich, 52425 Jülich, Germany*

Abstract. Ever since Weinberg's seminal predictions of the pion–nucleon scattering amplitudes at threshold, this process has been of central interest for the study of chiral dynamics involving nucleons. The scattering lengths or the pion–nucleon σ-term are fundamental quantities characterizing the explicit breaking of chiral symmetry by means of the light quark masses. On the other hand, pion–nucleon dynamics also strongly affects the long-range part of nucleon–nucleon potentials, and hence has a far-reaching impact on nuclear physics. We discuss the fruitful combination of dispersion-theoretical methods, in the form of Roy–Steiner equations, with chiral dynamics to determine pion–nucleon scattering amplitudes at low energies with high precision.⋆⋆

1 Introduction

Pion–nucleon (πN) scattering is one of the simplest processes to study chiral dynamics involving nucleons. The leading order (LO) in the chiral expansion, i.e., in the expansion in pion masses and momenta, results in the well-known low-energy theorems (LETs) for the S-wave scattering lengths, the amplitudes evaluated at threshold [3, 4]:

$$a_{0+}^{-} = \frac{M_\pi m_N}{8\pi(m_N + M_\pi)F_\pi^2} + O(M_\pi^3), \qquad a_{0+}^{+} = O(M_\pi^2). \tag{1}$$

The isospin-odd scattering length is hence predicted solely in terms of the pion (M_π) and nucleon (m_N) masses as well as the pion decay constant F_π, while the isospin-even one is suppressed. The strength of the Born term amplitudes (that do not contribute at threshold to leading order) is given in terms of the pion–nucleon coupling constant g, which is related to the axial coupling g_A by the Goldberger–Treiman relation, $g = g_A m_N / F_\pi$, up to higher orders.

Already at next-to-leading order (NLO; $O(p^2)$ in the chiral counting), the πN scattering amplitude depends on a list of low-energy constants (LECs), conventionally denoted by c_{1-4}, which are less readily determined from phenomenology. These NLO contributions tend to be large: three of the couplings (c_{2-4}) incorporate the leading low-energy effects of the $\Delta(1232)$, the lowest-lying resonant excitation of the nucleon. As the mass gap is small, $m_\Delta - m_N \sim 2M_\pi$, and as the Δ couples strongly to

⋆e-mail: kubis@hiskp.uni-bonn.de

⋆⋆These proceedings borrow heavily from a previous conference contribution [1], as well as from our original article [2].

the πN system, the numerical values of the NLO LECs are somewhat larger than expected from naive dimensional analysis. To pin down these LECs accurately and consistently is an important task also in view of many nuclear physics applications: πN amplitudes constitute an important contribution to the two-pion exchange in nucleon–nucleon scattering potentials, and determine the leading long-range three-nucleon force.

A further strong incentive to study pion–nucleon scattering derives from its relation to the pion–nucleon σ-term $\sigma_{\pi N}$, defined via the scalar form factor of the nucleon

$$\sigma(t) = \frac{1}{2m_N}\langle N(p')|\hat{m}(\bar{u}u + \bar{d}d)|N(p)\rangle, \qquad t = (p'-p)^2, \qquad \hat{m} = \frac{m_u + m_d}{2}, \qquad \sigma_{\pi N} \equiv \sigma(0). \quad (2)$$

Through the Feynman–Hellmann theorem [5, 6], $\sigma_{\pi N}$ determines the light-quark content of the nucleon mass,

$$\sigma_{\pi N} = \hat{m}\frac{\partial m_N}{\partial \hat{m}} = -4c_1 M_\pi^2 + O(M_\pi^3), \quad (3)$$

where we have already indicated the leading term in the chiral expansion of this quantity: it is given in terms of the LEC c_1, which in turn should be fixed from πN scattering. This is the leading approximation to the Cheng–Dashen LET [7, 8], which we will discuss in more detail below. The σ-term has garnered strong interest beyond the hadron physics community due to its relation to the scalar couplings of the nucleon, which are prerequisite for a consistent interpretation of direct-detection dark matter searches [9–11].

2 Roy–Steiner equations for πN scattering

The predictive power of chiral symmetry can be vastly increased by combining chiral perturbation theory (ChPT) with dispersive techniques. In particular, for $\pi\pi$ scattering, the use of Roy equations [12] has led to a determination of the low-energy $\pi\pi$ scattering amplitude with unprecedented accuracy [13, 14]. In the case of πN scattering, a full system of partial-wave dispersion relations (PWDRs) has to include dispersion relations for two distinct physical processes, $\pi N \to \pi N$ (s-channel) and $\pi\pi \to \bar{N}N$ (t-channel), and the use of $s \leftrightarrow t$ crossing symmetry will intertwine s- and t-channel equations. Roy–Steiner (RS) equations are a set of coupled PWDRs that, in contrast to $\pi\pi$ Roy equations, are derived from hyperbolic dispersion relations [15], which automatically relate the different channels in the πN system. Solving the RS equations for πN, in particular once combined with the pionic-atom constraints on the scattering lengths [16, 17], can provide a remarkably precise representation of the πN amplitude at low energies. The construction of a complete system of RS equations for πN scattering has been presented in detail in [18]. For the s-channel partial waves, they read [15]

$$\begin{aligned} f_{l+}^I(W) &= N_{l+}^I(W) + \frac{1}{\pi}\int_{4M_\pi^2}^{\infty} dt' \sum_J \Big\{G_{lJ}(W,t')\,\mathrm{Im}\, f_+^J(t') + H_{lJ}(W,t')\,\mathrm{Im}\, f_-^J(t')\Big\} \\ &+ \frac{1}{\pi}\int_{W_+}^{\infty} dW' \sum_{l'=0}^{\infty}\Big\{K_{ll'}^I(W,W')\,\mathrm{Im}\, f_{l'+}^I(W') + K_{ll'}^I(W,-W')\,\mathrm{Im}\, f_{(l'+1)-}^I(W')\Big\}, \end{aligned} \quad (4)$$

where due to G-parity only even/odd angular momenta J contribute for isospin $I = +/-$, respectively. The kernels $K_{ll'}^I(W,W)$, $G_{lJ}(W,t)$, and $H_{lJ}(W,t)$ are known analytically, and $N_{l+}^I(W)$ denotes the partial-wave projections of the pole terms.

The strategy for the solution of the RS equations is as follows. In the s-channel, the six S- and P-waves $f_{l\pm}^I$, with $I = \pm$ for the isospin index and orbital angular momentum l, are considered

 DOI: 10.1051/epjconf/201613001006

dynamically below a matching point s_m, whereas the imaginary parts of higher partial waves for all s, the imaginary parts of the S- and P-waves above s_m, and, potentially, inelasticities below s_m are required as input. We choose the matching point at its optimal value $s_m = (1.38\,\text{GeV})^2$ [18]. In contrast, there are only three S- and P-waves in the t-channel, $f_\pm^J$, with total angular momentum J and the subscript referring to parallel/antiparallel antinucleon–nucleon helicities.

Given that data in the t-channel reaction $\pi\pi \to \bar{N}N$ become available only above the two-nucleon threshold, the solution of the t-channel equations is subject to the additional complication of the large pseudophysical region $4M_\pi^2 \leq t \leq 4m_N^2$ in this reaction. The amplitudes required for the t-channel integrals need to be reconstructed from unitarity. While for every partial wave $\pi\pi$ intermediate states generate by far the dominant contribution, a coupled-channel $\pi\pi/\bar{K}K$ treatment is required for the S-wave [19], which accounts for the occurrence of the $f_0(980)$ resonance.

Once the t-channel problem is solved, the resulting t-channel partial waves are used as input for the s-channel problem, which then reduces to the form of conventional $\pi\pi$ Roy equations. Eventually, a full solution of the system can be obtained by iterating this procedure until all partial waves and parameters are determined self-consistently. The s-channel phase shifts below the matching point are represented in suitable parametrizations whose free parameters, together with the subtraction constants, are determined by minimizing the difference between the left-hand side (LHS) and right-hand side (RHS) of (4). We found that the solution is stabilized substantially when the S-wave scattering lengths, known very precisely from pionic atoms [2, 16, 17], are imposed as constraints on the system,

$$a_{0+}^{1/2} = (169.8 \pm 2.0) \times 10^{-3} M_\pi^{-1}, \qquad a_{0+}^{3/2} = (-86.3 \pm 1.8) \times 10^{-3} M_\pi^{-1}. \tag{5}$$

The minimization then provides us with a new set of subthreshold parameters and S- and P-wave phase shifts.

We have performed a full error analysis, where the uncertainty estimates include a number of effects [2]: first, we vary the input for the matching condition as well as for the energy region above the matching point and higher partial waves, both regarding different partial-wave analyses [20–23] and truncations of the partial-wave expansion. Furthermore, we vary the πN coupling constant within $g^2/(4\pi) = 13.7(2)$ [16, 17] and investigate the sensitivity to the parametrization of the low-energy phase shifts used in the solution. Second, we observe that there are important correlations in the phase space of parameters, leading to flat fit minima, but with significantly different values for the subthreshold parameters. To account for this effect, we generated a statistical ensemble of solutions exploring these shallow minima while imposing sum rules for the higher subthreshold parameters, and took the observed distribution as an additional source of uncertainty. Third, we propagate the errors in the scattering lengths, which crucially enter as constraints in the minimization, to the results for the subthreshold parameters.

The corresponding results for the s-channel partial-wave phase shifts are plotted in Fig. 1; the solutions for the t-channel are shown in [2]. The complete list of resulting subthreshold parameters is also given and compared to the Karlsruhe–Helsinki (KH80) values [20, 21] in [2]; here we just quote the two most important ones in connection to the σ-term,

$$\begin{aligned} d_{00}^+ &= (-1.361 \pm 0.032) M_\pi^{-1} &\qquad& (\text{KH80:} \quad (-1.46 \pm 0.10) M_\pi^{-1}), \\ d_{01}^+ &= (1.155 \pm 0.016) M_\pi^{-3} &\qquad& (\text{KH80:} \quad (1.14 \pm 0.02) M_\pi^{-3}). \end{aligned} \tag{6}$$

We also keep track of the correlations between subthreshold parameters, obtaining a 13×13 covariance matrix that encodes uncertainties and correlations of the 13 subthreshold parameters, which is relevant for the matching to ChPT.

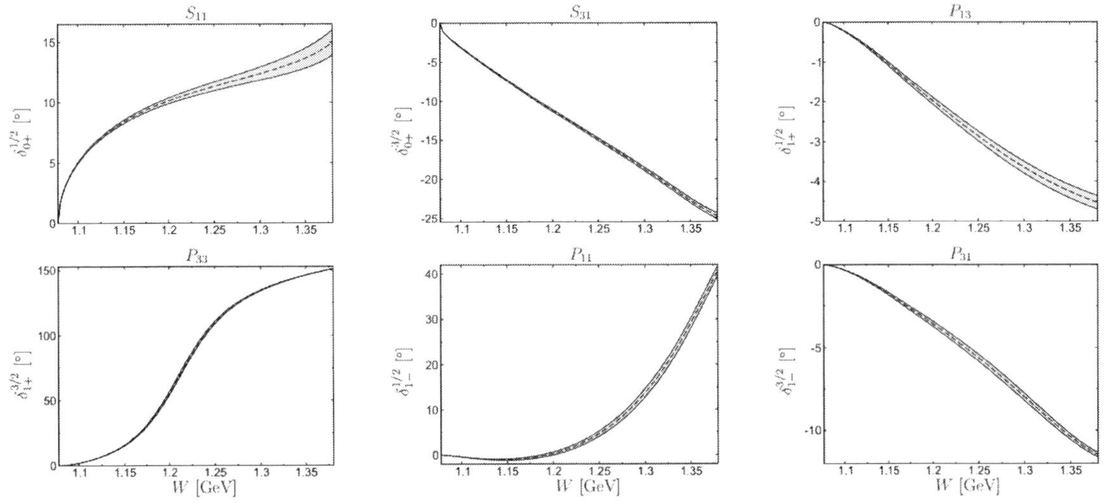

Figure 1. Error bands for the πN phase shifts; dashed lines indicate the central curves. Figures taken from [2].

3 Consequences for the πN σ-term

The Cheng–Dashen LET [7, 8] relates the Born-term-subtracted isoscalar amplitude evaluated at the Cheng–Dashen point ($\nu = 0, t = 2M_\pi^2$) to the scalar form factor of the nucleon, evaluated at momentum transfer $t = 2M_\pi^2$,

$$\bar{D}^+(0, 2M_\pi^2) = \sigma(2M_\pi^2) + \Delta_R, \tag{7}$$

where Δ_R represents higher-order corrections in the chiral expansion. These corrections are expected to be very small: we use the estimate $|\Delta_R| \lesssim 2\,\text{MeV}$ [24], derived from resonance saturation for the $O(p^4)$ LECs. In practice, the relation (7) is often rewritten as

$$\sigma_{\pi N} = \sigma(0) = \Sigma_d + \Delta_D - \Delta_\sigma - \Delta_R, \tag{8}$$

with correction terms $\Delta_\sigma = \sigma(2M_\pi^2) - \sigma_{\pi N}$, $\Delta_D = \bar{D}^+(0, 2M_\pi^2) - \Sigma_d$, $\Sigma_d = F_\pi^2(d_{00}^+ + 2M_\pi^2 d_{01}^+)$. Δ_σ measures the curvature in the scalar form factor, while Δ_D parametrizes contributions to the πN amplitude beyond the first two terms in the subthreshold expansion. As shown in [25], although these corrections are large individually due to strong rescattering in the isospin-0 $\pi\pi$ S-wave, they cancel to a large extent in the difference. For the numerical analysis we will use $\Delta_D - \Delta_\sigma = (-1.8 \pm 0.2)\,\text{MeV}$ [19]. Based on (7) the RS results for the subthreshold parameters translate immediately to a corresponding value of $\sigma_{\pi N}$. To illustrate the dependence of the σ-term on the scattering lengths used as input to the solution, we expand Σ_d linearly around the central values and find

$$\Sigma_d = (57.9 \pm 0.9)\,\text{MeV} + \sum_{I_s} c_{I_s} \Delta a_{0+}^{I_s}, \qquad c_{1/2} = 0.24\,\text{MeV}, \qquad c_{3/2} = 0.89\,\text{MeV}, \tag{9}$$

where $\Delta a_{0+}^{I_s}$ gives the deviation from the scattering lengths extracted from hadronic atoms in units of $10^{-3} M_\pi^{-1}$. (9) leads to $\Sigma_d = (46 \pm 4)\,\text{MeV}$ if the KH80 scattering lengths are used, in excellent agreement with the original KH80 value $\Sigma_d = (50 \pm 7)\,\text{MeV}$. In contrast, our central solution corresponds to $\Sigma_d = (57.9 \pm 1.9)\,\text{MeV}$, and thus to a significant increase compared to the early estimates. Including also isospin-breaking corrections [26–28], the final result [29]

$$\sigma_{\pi N} = (59.1 \pm 3.5)\,\text{MeV}, \tag{10}$$

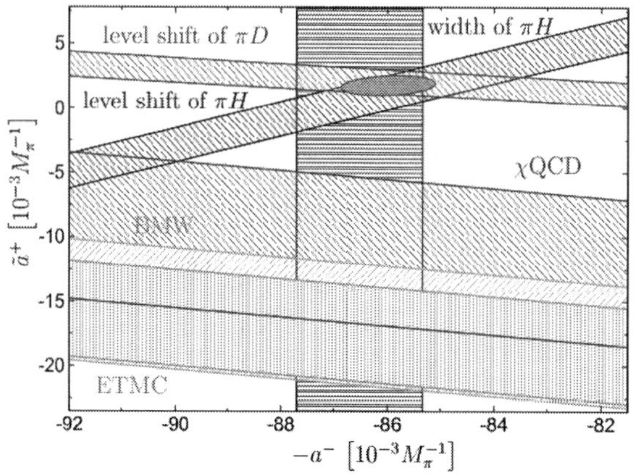

Figure 2. Constraints on the πN scattering lengths from pionic atoms and lattice σ-terms. Figure taken from [38].

does amount to a significant increase compared to the "canonical value" of $\sigma_{\pi N} \sim 45\,\text{MeV}$. Already 4.2 MeV are due to new corrections to the LET (and thereof 3.0 MeV from isospin breaking); the remaining increase of nearly 10 MeV is dictated by experiment: the new scattering lengths from pionic atoms determine the position of the σ-term on the curve approximately described by (9).

Our result for the σ-term (10) seems to be somewhat at odds with a series of recent lattice calculations of the same quantity, performed near or at physical pion masses, which yield values

$$\begin{aligned} &\sigma_{\pi N} = 38(3)(3)\,\text{MeV} \quad (\text{BMW [30]}), &&\sigma_{\pi N} = 44.4(3.2)(4.5)\,\text{MeV} \quad (\chi\text{QCD [31]}),\\ &\sigma_{\pi N} = 37.2(2.6)\left({}^{+1.0}_{-0.6}\right)\text{MeV} \quad (\text{ETMC [32]}), &&\sigma_{\pi N} = 35.0(6.1)\,\text{MeV} \quad (\text{RQCD [33]}). \end{aligned} \tag{11}$$

Such smaller values are more consistent with analyses of flavor $SU(3)$ breaking in the baryon spectrum and the OZI rule for scalar strangeness matrix elements of the nucleon (see the discussion in [34] and references therein). However, these lattice results are in significant tension with the pionic-atoms spectroscopy measurements [35–37], interpreted in [16, 17]: the relation (9) can be inverted in a way that every value of $\sigma_{\pi N}$ corresponds to a linear constraint in the plane of πN S-wave scattering lengths [38]. The constraints corresponding to the lattice results of [30–32] are shown in Fig. 2, compared to the bands extracted from pionic atoms, illustrating the disagreement quite clearly. A lattice calculation of the πN scattering lengths may be a good way to illuminate the cause of this discrepancy.

4 Matching to chiral perturbation theory

The matching to ChPT is one of the most fundamental applications of the RS solution, since it offers a unique opportunity for a systematic determination of πN LECs [39]. One expects the chiral expansion to work best in the subthreshold region, far away from singularities, where the amplitude can be described solely by a polynomial in the Mandelstam variables. The matching is thus most conveniently performed by equating the chiral expansion for the subthreshold parameters to the RS results.

The πN amplitude at N^{3}LO, $O(p^4)$, involves four NLO LECs, c_i, four (combinations of) N^{2}LO, and five N^{3}LO LECs [40], corresponding to the 13 subthreshold parameters that receive contributions from LECs in a fourth-order calculation. Inverting the expressions for the subthreshold parameters, we obtain all 13 LECs including the corresponding correlation coefficients [39]; Table 1 only shows the NLO LECs explicitly. Comparing the different extractions up to N^{3}LO, the convergence pattern

Table 1. Results for the NLO πN LECs at different orders in the chiral expansion [2, 39].

	NLO	N^2LO	N^3LO
c_1 [GeV^{-1}]	-0.74 ± 0.02	-1.07 ± 0.02	-1.11 ± 0.03
c_2 [GeV^{-1}]	1.81 ± 0.03	3.20 ± 0.03	3.13 ± 0.03
c_3 [GeV^{-1}]	-3.61 ± 0.05	-5.32 ± 0.05	-5.61 ± 0.06
c_4 [GeV^{-1}]	2.17 ± 0.03	3.56 ± 0.03	4.26 ± 0.04

for the c_i looks reasonably stable. In contrast, the convergence of some of the higher-order LECs proves to be more problematic, which can partly be explained by loop contributions enhanced by large $\Delta(1232)$ effects. The errors for the LECs at a given chiral order are negligible compared to the uncertainties to be attached to the chiral expansion itself.

We would like to thank the organizers of MESON 2016 in Kraków for the invitation to a wonderful conference. Financial support by the DFG (SFB/TR 16, "Subnuclear Structure of Matter") is gratefully acknowledged.

References

[1] B. Kubis, J. Ruiz de Elvira, M. Hoferichter and U.-G. Meißner, PoS **CD 15**, 021 (2015)
[2] M. Hoferichter, J. Ruiz de Elvira, B. Kubis and U.-G. Meißner, Phys. Rept. **625**, 1 (2016)
[3] S. Weinberg, Phys. Rev. Lett. **17**, 616 (1966)
[4] Y. Tomozawa, Nuovo Cim. A **46**, 707 (1966)
[5] H. Hellmann, *Einführung in die Quantenchemie* (Franz Deuticke, Leipzig, 1937)
[6] R. P. Feynman, Phys. Rev. **56**, 340 (1939)
[7] T. P. Cheng and R. F. Dashen, Phys. Rev. Lett. **26**, 594 (1971)
[8] L. S. Brown, W. J. Pardee and R. D. Peccei, Phys. Rev. D **4**, 2801 (1971)
[9] A. Bottino, F. Donato, N. Fornengo and S. Scopel, Astropart. Phys. **13**, 215 (2000)
[10] J. R. Ellis, K. A. Olive and C. Savage, Phys. Rev. D **77**, 065026 (2008)
[11] A. Crivellin, M. Hoferichter and M. Procura, Phys. Rev. D **89**, 054021 (2014)
[12] S. M. Roy, Phys. Lett. B **36**, 353 (1971)
[13] B. Ananthanarayan, G. Colangelo, J. Gasser and H. Leutwyler, Phys. Rept. **353**, 207 (2001)
[14] R. García-Martín, R. Kamiński, J. R. Peláez, J. Ruiz de Elvira and F. J. Ynduráin, Phys. Rev. D **83**, 074004 (2011)
[15] G. E. Hite and F. Steiner, Nuovo Cim. A **18**, 237 (1973) [CERN-TH-1590 for app. D and E]
[16] V. Baru *et al.*, Phys. Lett. B **694**, 473 (2011)
[17] V. Baru *et al.*, Nucl. Phys. A **872**, 69 (2011)
[18] C. Ditsche, M. Hoferichter, B. Kubis and U.-G. Meißner, JHEP **1206**, 043 (2012)
[19] M. Hoferichter, C. Ditsche, B. Kubis and U.-G. Meißner, JHEP **1206**, 063 (2012)
[20] R. Koch and E. Pietarinen, Nucl. Phys. A **336**, 331 (1980)
[21] G. Höhler, *Pion–Nukleon-Streuung: Methoden und Ergebnisse*, in Landolt-Börnstein, **9b2**, ed. H. Schopper (Springer Verlag, Berlin, 1983)
[22] R. A. Arndt, W. J. Briscoe, I. I. Strakovsky and R. L. Workman, Phys. Rev. C **74**, 045205 (2006)

[23] R. L. Workman, R. A. Arndt, W. J. Briscoe, M. W. Paris and I. I. Strakovsky, Phys. Rev. C **86**, 035202 (2012)
[24] V. Bernard, N. Kaiser and U.-G. Meißner, Phys. Lett. B **389**, 144 (1996)
[25] J. Gasser, H. Leutwyler and M. E. Sainio, Phys. Lett. B **253**, 260 (1991)
[26] J. Gasser, M. A. Ivanov, E. Lipartia, M. Mojžiš and A. Rusetsky, Eur. Phys. J. C **26**, 13 (2002)
[27] M. Hoferichter, B. Kubis and U.-G. Meißner, Phys. Lett. B **678**, 65 (2009)
[28] M. Hoferichter, B. Kubis and U.-G. Meißner, Nucl. Phys. A **833**, 18 (2010)
[29] M. Hoferichter, J. Ruiz de Elvira, B. Kubis and U.-G. Meißner, Phys. Rev. Lett. **115**, 092301 (2015)
[30] S. Dürr *et al.* [BMW Collaboration], Phys. Rev. Lett. **116**, 172001 (2016)
[31] Y. B. Yang, A. Alexandru, T. Draper, J. Liang and K. F. Liu [χQCD Collaboration], Phys. Rev. D **94**, 054503 (2016)
[32] A. Abdel-Rehim *et al.* [EMT Collaboration], Phys. Rev. Lett. **116**, 252001 (2016)
[33] G. S. Bali *et al.* [RQCD Collaboration], Phys. Rev. D **93**, 094504 (2016)
[34] H. Leutwyler, PoS **CD 15**, 022 (2015)
[35] T. Strauch *et al.*, Eur. Phys. J. A **47**, 88 (2011)
[36] M. Hennebach *et al.*, Eur. Phys. J. A **50**, 190 (2014)
[37] D. Gotta *et al.*, Lect. Notes Phys. **745**, 165 (2008)
[38] M. Hoferichter, J. Ruiz de Elvira, B. Kubis and U.-G. Meißner, Phys. Lett. B **760**, 74 (2016)
[39] M. Hoferichter, J. Ruiz de Elvira, B. Kubis and U.-G. Meißner, Phys. Rev. Lett. **115**, 192301 (2015)
[40] N. Fettes, U.-G. Meißner, M. Mojžiš and S. Steininger, Annals Phys. **283**, 273 (2000) [Annals Phys. **288**, 249 (2001)]

The legacy of the hadron physics programme at COSY

Colin Wilkin[1,*]

[1] *UCL, Gower Street, London WC1E 6BT, U.K.*

Abstract. A brief review is presented of some elements of the twenty years of hadronic physics research at the COoler SYnchrotron (COSY).

1 Introduction

At the end of 2014 the priorities of the Institut für Kernphysik Jülich switched from the investigation of hadronic processes to the study of a variety of symmetry-breaking reactions. A review of the activities at the COSY accelerator and storage ring is currently being prepared [1] and this will culminate by highlighting ten specific programmes, viz.

1. Proton-proton elastic scattering
2. The WASA dibaryon
3. Neutron-proton elastic scattering
4. Large acceptance hyperon production
5. The hyperon cusp
6. η-mesic nuclei
7. Non-strange meson production in nucleon-nucleon collisions
8. Kaon pair production
9. Determination of the mass of the η meson
10. Amplitude analysis of $NN \to \{pp\}_S\pi$ at 353 MeV

In the time available, I cannot discuss all of them — the rest will be found, along with many many references, in the review article [1].

2 Proton-proton elastic scattering

There have been stupendous advances at COSY in the study of pp elastic scattering using the EDDA, ANKE, and KOALA detectors. These all involved measurements with very thin targets inside the COSY ring, where double-scattering experiments were impractical. Hence only initial spin degrees of freedom could be studied. EDDA identified elastic scattering by demanding that the laboratory polar angles of fully coplanar events satisfy

$$\cot\theta^1_{\text{lab}}\cot\theta^2_{\text{lab}} = 1 + T_p/2m_p, \tag{1}$$

*A European Physical Journal A special lecture. e-mail: c.wilkin@ucl.ac.uk where T_p is the beam laboratory kinetic energy and m_p the proton mass.

Measurements of the differential cross section, proton analysing power, and spin-correlations were carried out during acceleration (and deceleration) over a continuum of energies from 230 to 2590 MeV. But, since both protons were detected, data were only available for $\theta_{cm} \geq 35°$. Excitation functions for these observables at fixed angle are shown in Fig. 1.

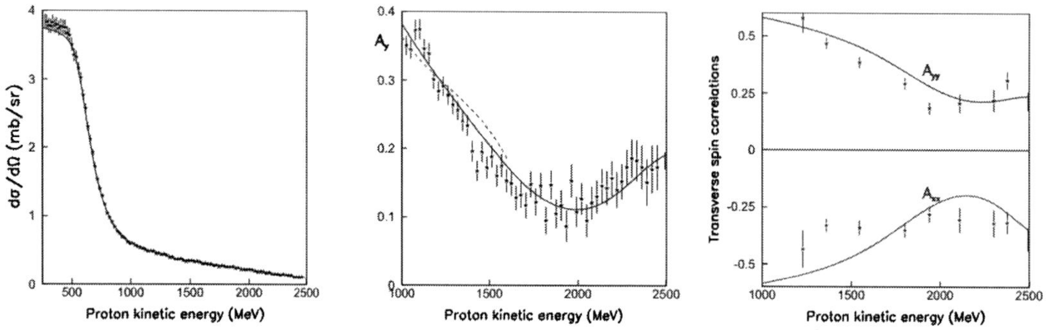

Figure 1. Left: EDDA measurements of the differential cross section for pp scattering at $\theta_{cm} = 89°$ [2]. Middle: Analysing power in pp elastic scattering measured with EDDA at $\theta_{cm} = 56°$ [4]. The dashed (solid) curve represents SAID solutions [3] before (after) the EDDA measurements. Right: Transverse spin correlations at $\theta_{cm} = 57.5°$ [5] compared to the SAID SP07 solution [3].

ANKE investigated pp elastic scattering at smaller angles than EDDA by measuring one final proton and its energy/momentum. The results shown in Fig. 2 were obtained by detecting the fast proton in a magnetic spectrometer or the slow recoil in a pair of silicon tracking telescopes [6, 7]. Even smaller momentum transfers could be investigated using the KOALA detector [8], though the data shown in Fig. 2 are not yet reliably normalised. Nevertheless, it is clear that COSY has completely revolutionised the understanding of pp scattering up to 3 GeV.

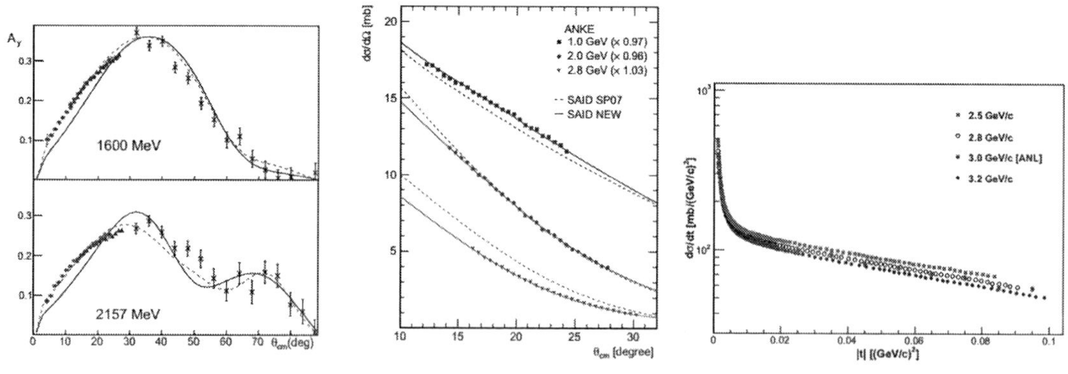

Figure 2. Left: ANKE measurements of the analysing power in pp elastic scattering at two energies (circles/triangles) [6] compared to EDDA (crosses) data [4] and the SAID SP07 fit (solid line) [3] and a revised one (dashed). Middle: ANKE pp differential cross sections [7], scaled by factors within the overall uncertainties, compared to the SAID SP07 solution (dashed) [3] and modified fit (solid line). Right: Preliminary values of the pp differential cross section measured with the KOALA detector [8] at very small momentum transfers t.

 DOI: 10.1051/epjconf/201613001007

3 The WASA dibaryon

The search for dibaryons has a long and generally frustrating history. The inspiration came from six-quark bag models that predict several states. But the only confirmed dibaryon was the deuteron, where the relevant degrees of freedom are (probably) nucleons and pions. The WASA collaboration at CELSIUS and COSY measured the total cross section for quasi-free $np \to d\pi^0\pi^0$ by using a deuteron beam or a deuterium target. The c.m. energy W is spread by the Fermi momentum but, by reconstructing the whole event, W could be evaluated with some precision. The very impressive peak of Fig. 3 at $W = 2.38$ GeV with $\Gamma = 70$ MeV was obtained in three experiments [9–11]. It was suggested that this was a dibaryon, $d^*(2380)$. As shown in the middle panel, the $d^*(2380)$ is seen also in $pn \to d\pi^+\pi^-$ [10]. The right panel shows that the inclusion of the $d^*(2380)$ improves considerably the SAID description of the analysing power in $\vec{n}p$ quasi-elastic scattering [12].

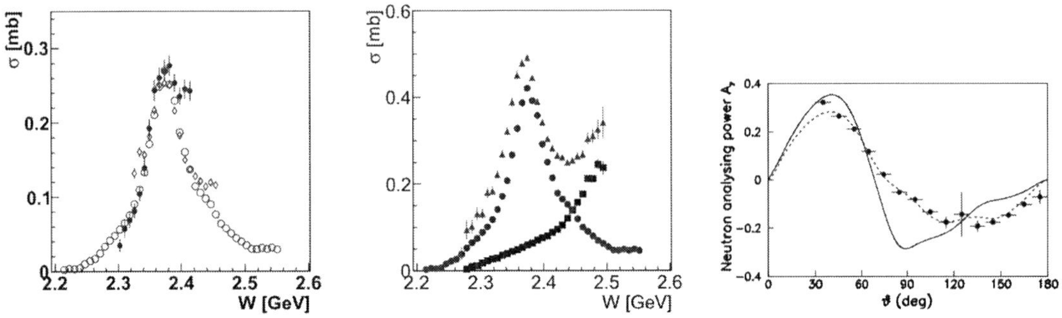

Figure 3. Left: Total cross section for the quasi-free $pn \to d\pi^0\pi^0$ reaction as a function of the total energy $W = \sqrt{s}$ [9–11]. Middle: Total cross section for $pn \to d\pi^+\pi^-$ (red triangles) and its decomposition into isoscalar (blue circles) and isovector parts (black squares) [10]. Right: Angular distribution of the np quasi-elastic analysing power in the $W = 2.377$ GeV region [12]. The (black) solid line is the suitably averaged SP07 prediction [3], whereas a SAID solution that includes the dibaryon is shown by the (blue) dashed line.

Though there is still some lingering doubt whether the $d^*(2380)$ is really a dibaryon resonance, it is a good working hypothesis that must be tested further. If it exists, is it a 6-quark state or is it a bound state of $\Delta(1232)\Delta(1232)$? If $\Delta\Delta$, why is the width so narrow? Even if it turns out not to be a dibaryon, it is still a very important observation in our field and it should be tested in an independent experiment with an independent analysis.

4 Hyperon production

By detecting the K^+ and p from $pp \to K^+pX$ one can see peaks from Λ and Σ^0 production — but there are significant physics backgrounds due to the misidentification of the direct proton. Thus the proton could have arisen from the decay of a Λ, a Σ^0, or even a Σ^+. These contributions are less troublesome near threshold where COSY-11 and ANKE have reasonable acceptance. At high excess energy Q (the total c.m. kinetic energy in the final state), the larger acceptance of COSY-TOF is required and this division of effort between low and high Q can be seen in Fig. 4 (left).

The curve for Σ^0 production corresponds to pure three-body phase space whereas that for the Λ shows evidence for a strong Λp final state interaction (FSI). This is seen more clearly in the right

 DOI: 10.1051/epjconf/201613001007

panel, which shows the ratio of Λ to Σ^0 production near threshold. The FSI hypothesis suggests that this should vary as $C\Big/\left(1+\sqrt{1+Q/B}\right)^2$ and the data are well fit with $B \approx 5.2$ MeV.

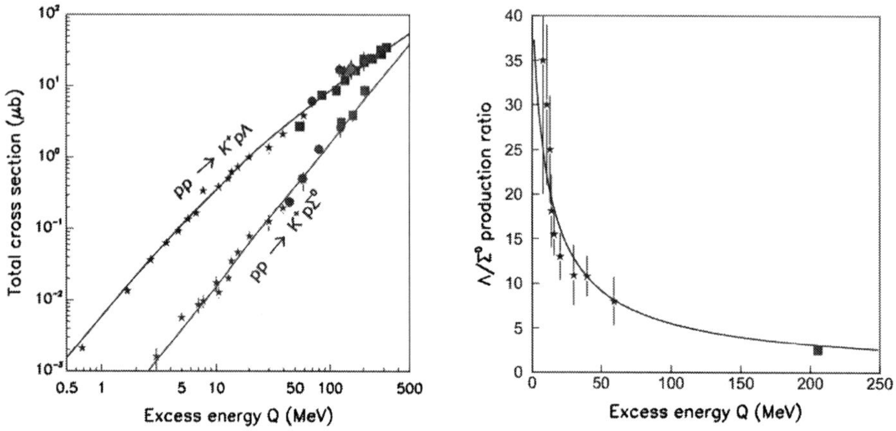

Figure 4. Left: Total cross sections for the $pp \to K^+p\Lambda$ and $pp \to K^+p\Sigma^0$. The stars at low Q were obtained at COSY-11 whereas the squares at high Q were taken with the higher acceptance COSY-TOF detector. The curves indicate a significant FSI in the Λp system. This is made even clearer in the ratio of the production cross sections shown in the right panel.

The high geometric acceptance of the COSY-TOF detector allowed the collaboration to investigate a wealth of differential distributions for hyperon production, including polarisation measurements but the one that really fascinates me is the cusp effect shown in Fig. 5.

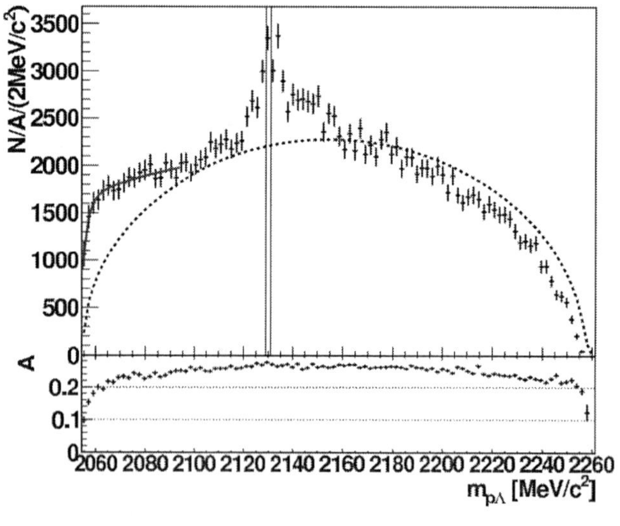

Figure 5. The distribution in $p\Lambda$ invariant mass from the $pp \to K^+p\Lambda$ reaction at 2.95 GeV/c [13]. The vertical lines indicate the positions of the two ΣN thresholds. The solid (red) line at low invariant masses represents the data used in the COSY-TOF scattering length determination.

In unambiguous $pp \to K^+p\Lambda$ events the COSY-TOF collaboration found a very pronounced peak in the vicinity of $m_{p\Lambda} \approx m_p + m_{\Sigma^0}$ or $m_n + m_{\Sigma^+}$. Since the acceptance A is smooth in this region, this must be a physics effect, linked to the coupling between the Λp and ΣN channels. One is therefore seeing the effects of $pp \to K^+N\Sigma$ followed by $N\Sigma \to p\Lambda$. This contribution interferes with direct $pp \to K^+p\Lambda$ production and, since it is an interference effect, there is no reason for it to look like a simple resonance peak.

Such a cusp is very common in reactions involving strange particles but they are very hard to model. If it is only the K^+ that is detected it is not possible to distinguish true Σ production from Λ production where there is a cusp at the ΣN thresholds. For example, the HIRES collaboration measured $pp \to K^+X$ at $\theta_K = 0°$ and found a big jump at $M_X \approx M_\Sigma + M_N$ [14]. How much of this was due to Σ production – impossible to tell in an inclusive measurement.

5 η-mesic nuclei

It was suggested many years ago [15] that the strong attraction between the isoscalar η meson and nucleons might be sufficient to let the η bind to some nuclei. These would be at best quasi-bound states because they could decay through pion or nucleon emission. The clearest signal for such a possible state is to be found in measurements of the $dp \to {}^3\mathrm{He}\,\eta$ cross section carried out near threshold at COSY-11 [16] and ANKE [17], where the data suggest that there is a pole in the $\eta^3\mathrm{He}$ elastic scattering amplitude in the complex Q plane for $|Q| < 0.5$ MeV.

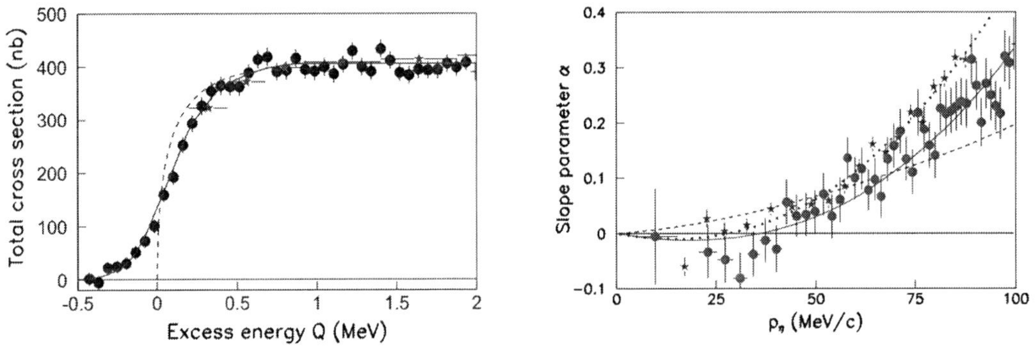

Figure 6. Left: Near-threshold $dp \to {}^3\mathrm{He}\,\eta$ total cross sections measured at ANKE (black circles) [17] and COSY-11 (red stars) [16]. The solid curve is the FSI fit to the ANKE data with a 171 keV smearing in Q. The dashed curve shows what the data would look like if all sources of smearing could be eliminated. Right: Slope parameter as a function of the η c.m. momentum. The ANKE (red circles) and COSY–11 (blue stars) data are compared to fits (solid red curve and blue dots) where the phase variation of the s–wave amplitude is taken into account [18]. If the phase variation is neglected, the best fit (black dashed curve) fails to describe the data.

As can be seen from Fig. 6, the cross section jumps to its plateau value for $Q \lesssim 0.5$ MeV and the jump is even more abrupt if the effects of the smearing of the beam momentum are taken into account. These data suggest that there is a pole at an η c.m. momentum of $p_1 \approx (-5 \pm i19)$ MeV/c. The $\pm$ sign here reflects the fact that above-threshold data can never distinguish between a "bound" and a "virtual" state.

The rapid change of phase due to the pole in the s-wave amplitude affects the s–p interference and hence the relative slope of the differential cross section at 90°. It is shown in the right panel of Fig. 6 that the momentum dependence of the slope is better described if the pole is included [18]. This is by far the best signal in the literature for nearby η-nucleus poles but it doesn't tell us whether it lies on the bound or virtual sheet.

The same reaction was used as the basis of a measurement of the mass of the η meson [19]. For this purpose the momentum of the circulating deuteron beam in COSY was measured with a relative precision of about 3×10^{-5} by inducing an artificial depolarising resonance, whose position could be evaluated very accurately. The value obtained, $m_\eta = (547.873 \pm 0.005_{\text{stat}} \pm 0.023_{\text{syst}})$ MeV$/c^2$, is marginally better than any of the other modern measurements.

6 Kaon pair production

The measurements of $pp \to ppK^+K^-$ by the COSY-11 collaboration [20] suggested that the K^- is preferentially attracted to one (or both) of the final protons. This was repeated with higher statistics at ANKE [21], where the initial drive came from the study of ϕ production in $pp \to pp(\phi \to K^+K^-)$. To see the effect more clearly, define the ratios of cross sections with respect to the $K^\pm p$ and $K^\pm pp$ invariant masses:

$$R_{Kp} = \frac{d\sigma/dM_{K^-p}}{d\sigma/dM_{K^+p}}, \qquad R_{Kpp} = \frac{d\sigma/dM_{K^-pp}}{d\sigma/dM_{K^+pp}}, \tag{2}$$

whose experimental values at $Q = 24$ MeV are shown in Fig. 7 [21].

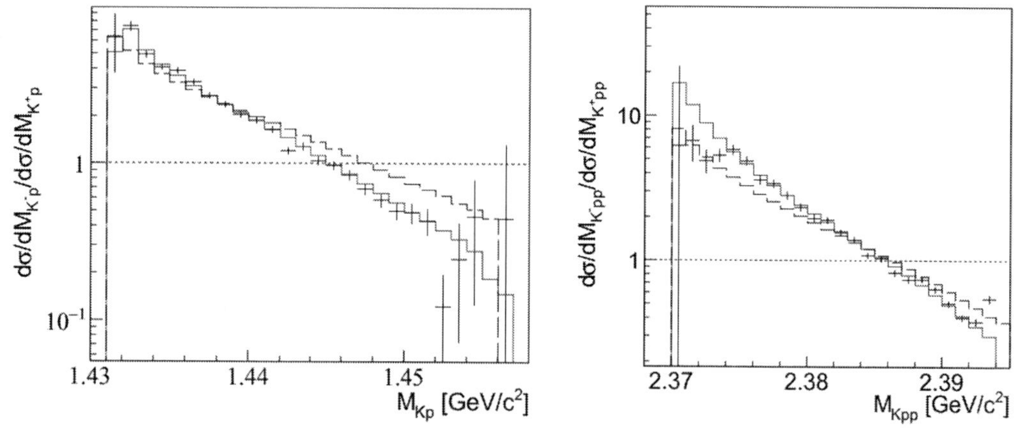

Figure 7. The ratios R_{Kp} and R_{Kpp} of Eq. (2) at $Q = 24$ MeV [21]. The red solid and broken black histograms represent estimations that take into account K^-p, pp and K^+K^- final state interactions with $a_{K^-p} = 2.45i$ fm and $a_{K^-p} = 1.5i$ fm, respectively.

There can be no rigorous evaluation of an enhancement factor when three or more particles interact in the final state. The estimates made in Fig. 7 assume that the overall enhancement is the product of simple FSI factors in the different kinematic variables. With a reasonable value of the effective K^-p scattering length, such a simple model does explain the enormous variations of the ratios as functions of the Kp and Kpp masses. Any relationship between this and a possible bound K^-pp system is a matter for speculation. There is, however, almost certainly a connection to the $pp \to K^+p\Lambda(1405)$ reaction, which has also been studied at COSY [22].

7 Conclusions

The results presented here represent only a very small part of the successes in hadronic physics achieved at COSY over the last twenty years. Their compression into a few pages will certainly lead to distortions in the text and the references. It is to be hoped that the full review article of Ref. [1] will have fewer flaws.

Support from both the conference organisers and the European Physical Journal are gratefully acknowledged.

References

[1] C. Wilkin, Invited review for Eur. Phys. J. *in preparation*
[2] D. Albers *et al.*, Eur. Phys. J. A **22**, 125 (2004)
[3] R.A. Arndt, W.J. Briscoe, I.I. Strakovsky, R.L. Workman, Phys. Rev. C **76**, 025209 (2007)
[4] M. Altmeier *et al.*, Eur. Phys. J. A **23**, 351 (2005)
[5] F. Bauer *et al.*, Phys. Rev. C **71**, 054002 (2005)
[6] Z. Bagdasarian *et al.*, Phys. Lett. B **739**, 152 (2014)
[7] D. Mchedlishvili *et al.*, Phys. Lett. **755**, 92 (2016)
[8] Q. Hu *et al.*, Eur. Phys. J. A **50**, 156 (2014)
[9] P. Adlarson *et al.*, Phys. Rev. Lett. **106**, 242302 (2011)
[10] P. Adlarson *et al.*, Phys. Lett. B **721**, 229 (2013)
[11] P. Adlarson *et al.*, Eur. Phys. J. A **52**, 147 (2016)
[12] P. Adlarson *et al.*, Phys. Rev. Lett. **112**, 202301 (2014)
[13] M. Röder *et al.*, Eur. Phys. J. A **49**, 157 (2013)
[14] A. Budzanowski *et al.*, Phys. Lett. B **692**, 10 (2010)
[15] Q. Haider and L.C. Liu, Phys. Lett. B **172**, 257 (1986)
[16] J. Smyrski *et al.*, Phys. Lett. B **649**, 258 (2007)
[17] T. Mersmann *et al.*, Phys. Rev. Lett. **98**, 242301 (2007)
[18] C. Wilkin *et al.*, Phys. Lett. B **654**, 92 (2007)
[19] P. Goslawski *et al.*, Phys. Rev. D **85**, 112011 (2012)
[20] P. Winter *et al.*, Phys. Lett. B **635**, 23 (2006)
[21] Q.J. Ye *et al.*, Phys. Rev. C **87**, 065203 (2013)
[22] I. Zychor *et al.*, Phys. Lett. B **660**, 167 (2008)

Recent result from the A2 collaboration at MAMI

Andreas Thomas[1,*]
for the A2 Collaboration

[1] *Institut für Kernphysik, Universität Mainz*

Abstract. The A2 Collaboration at the Mainz Microtron MAMI is measuring photon absorption cross section using circularly and linearly polarized photons up to energies of 1.6 GeV. The photons are produced in the 'Bremsstrahlungs' process, the energy is determined by a dedicated tagging system. The Crystal Ball-TAPS detector system with its high capability to cope with multi photon final states is used to acquire data with a variety of nonpolarized and spin polarized targets. Physical goals are the investigation of the nucleons excitation spectrum via single and double meson photoproduction and in addition a detailed determination of meson decays in precision experiments. We have started a program to measure double polarised Compton scattering to determine the nucleons scalar and spin polarisibilities. In this proceedings recent results from A2 collaboration are presented.

1 Experimental Apparatus

1.1 The A2 Photon Beam

The MAMI accelerator with its source of polarised electrons, based on the photoeffect on a strained GaAs crystal, routinely delivers polarised beam with a maximum energy of 1604 MeV and a degree of polarisation of approximately 85%. The last accelerator stage, MAMI C, is realized as a Harmonic Double Sided Microtron (HDSM). Details of the machine can be found in reference [1].
The A2-Glasgow-Mainz tagging facility [2] stands out due to its high photon intensity. The beam is derived from the production of Bremsstrahlung photons during the passage of the MAMI electron beam through a thin radiator. The resulting photons can be circularly polarised, with the application of a polarised electron beam, or linearly polarised, in the case of a crystalline radiator. The tagger focal plane is segmented into 352 scintillation detectors. Each counter can operate reliably to a rate of $\sim$1 MHz, giving a photon flux of $2.5 \cdot 10^5$ photons per MeV. Photons can be tagged in the momentum range from 4.7 to 93.0% of E_0.

1.2 The A2 Detector Setup

The central detector system consists of the Crystal Ball calorimeter combined with a barrel of scintillation counters for particle identification and two coaxial multiwire proportional counters for charged

*e-mail: thomand@uni-mainz.de

particle tracking. This central system provides position, energy and timing information for both charged and neutral particles in the region between 21° and 159° in the polar angle (θ) and over almost the full azimuthal (ϕ) range. At forward angles, less than 21°, reaction products are detected in the TAPS forward wall. The full, almost hermetic, detector system is shown schematically in Figure 1.

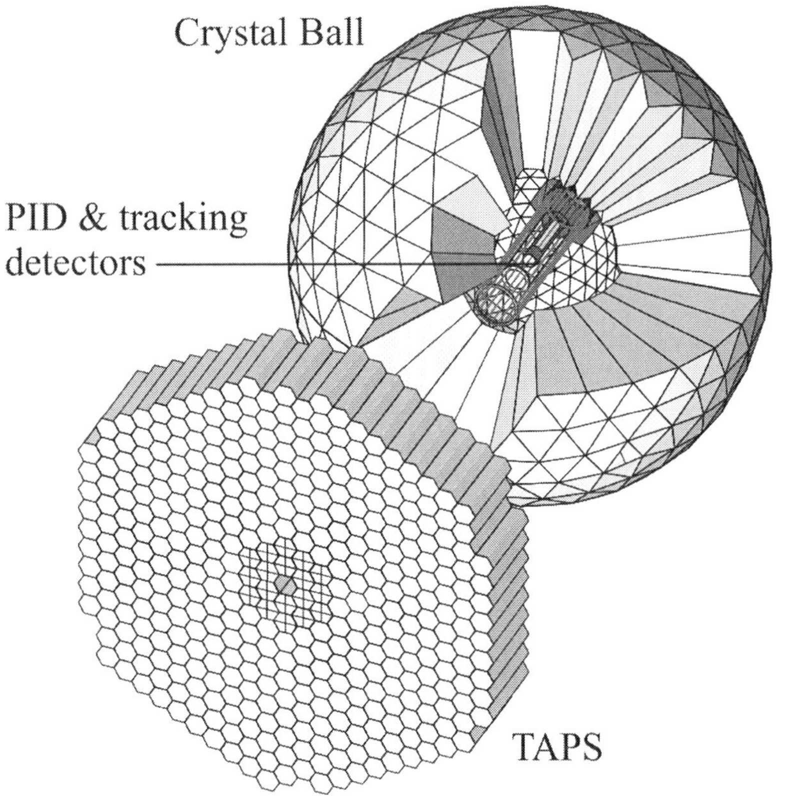

Figure 1. The A2 detector setup: The Crystal Ball calorimeter, with cut-away section showing the inner detectors, and the TAPS forward wall.

The Crystal Ball detector (CB) is a highly segmented 672-element NaI(Tl), self triggering photon spectrometer constructed at SLAC in the 1970's. Each element is a truncated triangular pyramid, 41 cm (15.7 radiation lengths) long. The Crystal Ball has an energy resolution of $\Delta E/E = 0.020 \cdot E[GeV]^{0.36}$ and angular resolutions of $\sigma_\theta = 2\ldots3°$ and $\sigma_\phi = \sigma_\theta/\sin\theta$ for electromagnetic showers. The TAPS forward wall is composed of 384 BaF_2 elements, each 25 cm in length (12 radiation lengths) and hexagonal in cross section, with a diameter of 59 mm. The front of every TAPS element is covered by a 5 mm thick plastic veto scintillator.

The tracks of charged particles emitted within the angular and momentum acceptance of the CB detector are reconstructed using two coaxial cylindrical multiwire proportional chambers (MWPCs) with cathode strip readout.

1.3 The A2 Targets

A variety of unpolarised and polarised targets can be implemented into the A2 setup. A standard liquid hydrogen or deuterium target is used for unpolarised and beam polarised experiments , another specially adopted liquid ^{3}Helium and ^{4}Helium target is available. In addition, solid nuclear targets are used in several experiments. For double polarisation measurements on polarised protons and deuterons a large horizontal ^{3}He/^{4}He dilution refrigerator was built in cooperation with the Joint Institute for Nuclear Research (JINR) Dubna. It has minimum limitations for the particle detection and fits into the central core of the inner Particle Identification Detector (PID). This was achieved by using the frozen spin technique with the new concept of placing a thin superconducting holding coil inside the polarisation refrigerator, longitudinal and transverse polarisations are possible. Highest nucleon polarisation in solid-state target materials is obtained by a microwave pumping process, known as 'Dynamic Nucleon Polarisation' (DNP). Several polarised proton and deuteron materials are available such as alcohols and deuterated alcohols (e.g. butanol C_4H_9OH), NH_3, ND_3 or ^{6}LiD. The most important criteria in the choice of material suitable for particle physics experiments are the degree of polarisation P and the ratio k of free polarisable nucleons to the total number of nucleons. Further requirements on polarised target materials are a short polarisation build-up time and a simple, reproducible target preparation. Another important consideration is, if the chemically bound background nuclei are spin-polarised or not, especially for our exsperiments at moderate enrgies. Taking all properties together, butanol and deuterated butanol are the best material for this experiment. For protons and deuterons we could achieve maximum polarisations in the order of $P = 80\%$ and an average polarisation of $P = 70\%$ in the frozen spin mode at 25mKelvin and spin relaxation times of more than 1000 hours. Figure 2 shows the Dubna-Mainz dilution cryostat with internal longitudinal superconducting coil.

Figure 2. The Dubna-Mainz dilution cryostat with internal longitudinal superconducting coil.

2 Recent results

2.1 Unpolarised cross sections on protons and deuterons

Differential cross sections for the $\gamma p \rightarrow p\pi^0$ reaction have been measured up to the center-of-mass energy W = 1.9 GeV [3]. The new results, obtained with a fine energy and angular binning, increase the existing quantity of n π^0 photoproduction data by 47%. Owing to the unprecedented statistical accuracy and the full angular coverage, the results are sensitive to high partial-wave amplitudes. Figure 3 shows the high quality of the new data set.

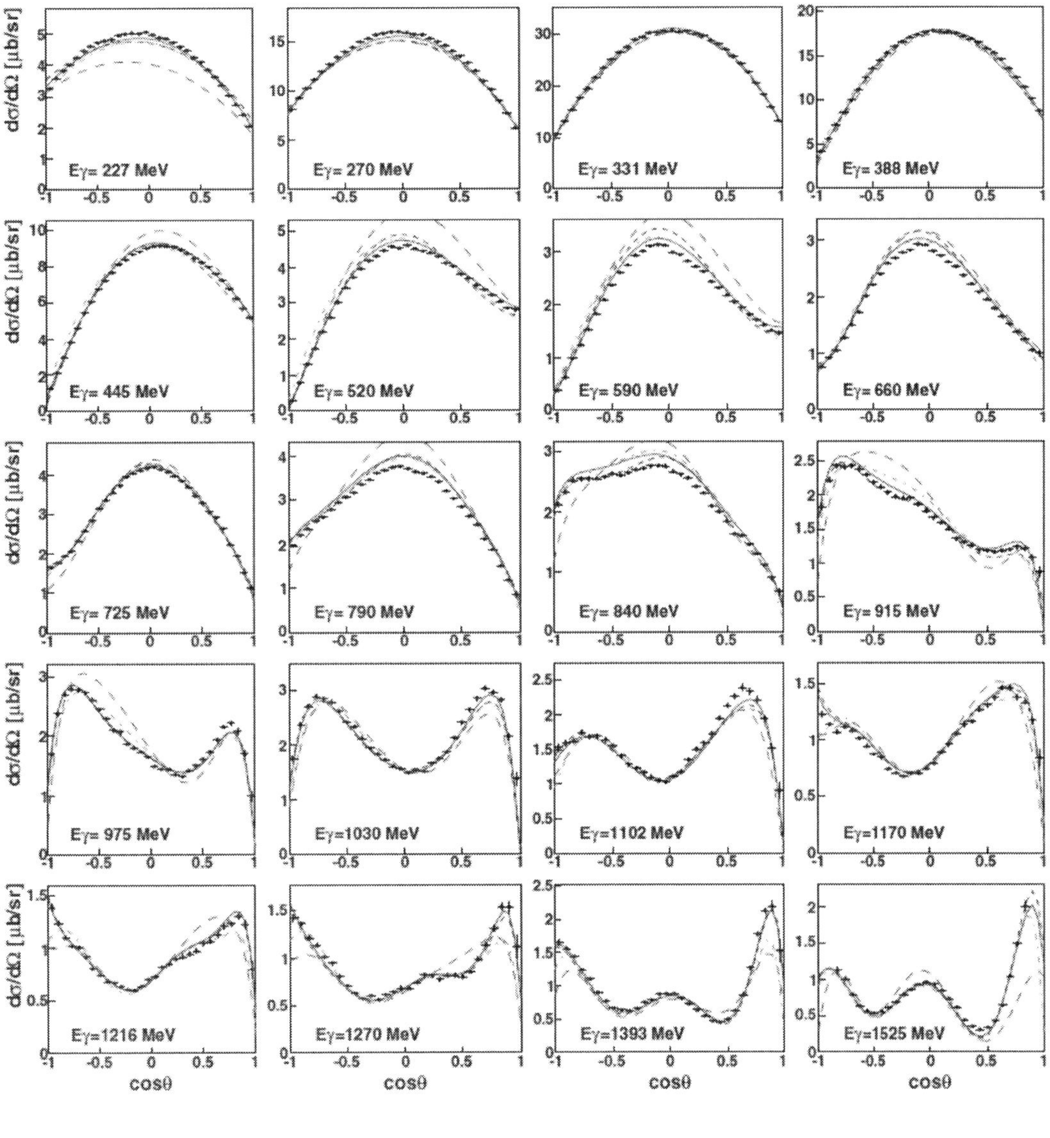

Figure 3. Selected results of the differential cross sections from the our work ([3] black points) with most recent partial wave analyses.

DOI: 10.1051/epjconf/201613001008

Another experiment was performed with an unpolarised liquid deuteron target for incident photon energies between 0.45 and 1.4 GeV [4]. Precise angular distributions have been measured for the first time for the photoproduction of π^0 mesons off neutrons bound in the deuteron. The effects from nuclear Fermi motion have been eliminated by a complete kinematic reconstruction of the final state.

2.2 Double polarized meson production

Since many years measurements to accomplish a complete set of polarisation observables in meson photoproduction were carried out. With the new quality of polarised beams and targets we could determine experimentally many of these observables in the recent experiments. The corresponding polarisation parameters are defined e.g. in ref [5]. Fig. 4 illustrates the geometrical orientation of the target and the corresponding photon beam polarisation.

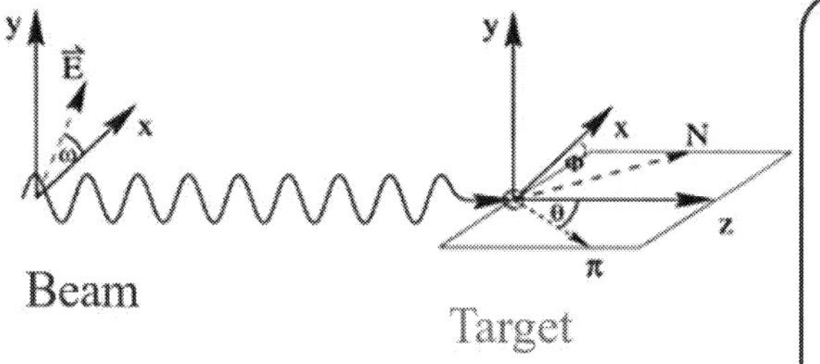

Target \ Beam	γ_{unpol}	P_γ^{lin} $\left(0, \frac{\pi}{2}\right)$	P_γ^{lin} $\left(+\frac{\pi}{4}, -\frac{\pi}{4}\right)$	P_γ^{circ}
P_{unpol}	$\left(\frac{d\sigma}{d\Omega}\right)$	$\Sigma(\theta)$	-	-
P_x	-	-	$H(\theta)$	$F(\theta)$
P_y	$T(\theta)$	$P(\theta)$	-	-
P_z	-	-	$G(\theta)$	$E(\theta)$

Figure 4. Polarisation observables in single meson production.

2.2.1 *Transverse polarisation observables T and F*

The reaction $\gamma p \rightarrow p\pi^0$ was studied at laboratory photon energies from 425 to 1445 MeV with a transversely polarized target and a longitudinally polarized beam [6]. The beam-target asymmetry F was measured for the first time and new high precision data for the target asymmetry T were obtained. In addition, the first data on target and beam-target asymmetries for the $\gamma p \rightarrow p\eta\pi^0$ reaction at photon energies from 1050 up to 1450 MeV were published [7].

2.2.2 *Threshold π^0 photoproduction*

Polarisation-dependent differential cross sections associated with the target asymmetry T have been measured in a dedicated experiment for the reaction $\gamma p \rightarrow p\pi^0$ with transverse target polarisation from threshold to photon energies of 190MeV [8]. These results have been used in combination with our previous measurements of the unpolarised cross section and the beam asymmetry for a model-independent extraction of S-and P-wave multipoles in the π^0-threshold region, which includes for the first time a direct determination of the imaginary part of the E_{0+}multipole.

2.2.3 Longitudinal polarisation observables E and G

A set of data for the longitudinal polarisation observable E and G were taken in the last year at the A2 facility and is under analysis. We could show in that experiments that a simultaneous measurement of both observables is possible by impinging with a longitudinaly polarised electron beam on a diamond radiatior. First preliminary results have been shown on this conference.

2.3 Double polarized Compton scattering

Beside the meson production program we have started with the experimental determination of double polarised Compton scattering. Since the cross section is two orders of magnitude smaller compared to pion-production and in addtion the final state requires an open trigger the experiment is very challenging. Our first results are published in [9].

The spin polarizabilities of the nucleon describe how the spin of the nucleon responds to an incident polarized photon. The most model-independent way to extract the nucleon spin polarizabilities is through polarized Compton scattering. Double-polarized Compton scattering asymmetries on the proton were measured in the $\Delta(1232)$ region using circularly polarized incident photons and a transversely polarized proton target. Fits to asymmetry data were performed using a dispersion model calculation and a baryon chiral perturbation theory calculation, and a separation of all four proton spin polarizabilities in the multipole basis was achieved. Our program to measure the Compton observables will be continued in the next years with other beam and target polarisations, which will lead finally to a more precise determination of the nucleons scalar and spin polarisibilities.

Acknowledgement

The authors wish to acknowledge the excellent support of the accelerator group and operators of MAMI. This work was supported by the Deutsche Forschungsgemeinschaft (SFB1044), the European Community-Research Infrastructure Activity under the FP6 "Structuring the European Research Area" programme (Hadron Physics, Contract No. RII3- CT-2004-506078), Schweizerischer Nationalfonds (Contract Nos. 200020-156983, 132799, 121781, 117601, 113511), the UK Science and Technology Facilities Council (STFC 57071/1, 50727/1), the US Department of Energy (Offices of Science and Nuclear Physics, Award Nos. DE-FG02-99- ER41110, DE-FG02-88ER40415, DE-FG02-01-ER41194), and National Science Foundation (Grant Nos. PHY-1039130, IIA-1358175), INFN (Italy), and NSERC (Canada). We thank the undergraduate students of Mount Allison University and The George Washington University for their assistance.

References

[1] A. Jankowiak, Eur. Phys. J. A **28**, s01, 149 (2006)
[2] J.C. McGeorge *et al.*, Eur. Phys. J. A **37**, 129 (2008)
[3] P.Adlarson *et al.*, Phys. Rev. C **92**, 024617 (2015)
[4] M.Dieterle *et al.*, Phys. Rev. Lett. **112**, 142001 (2014)
[5] I.S.Barker *et al.*, Nucl. Phys. B **95**, 347 (1975)
[6] J.R.Annand *et al.*, Phys. Rev. C **93**, 055209 (2016)
[7] J.R.Annand *et al.*, Phys. Rev. C **91**, 055208 (2015)
[8] S.Schumann *et al.*, Phys. Lett. B **750**, 252 (2015)
[9] P.P.Martel *et al.*, Phys. Rev. Lett. **91**, 112501 (2015)

EPJ Web of Conferences **130**, 01009 (2016) DOI: 10.1051/epjconf/201613001009
MESON 2016

Modelling glueballs

Francesco Giacosa[1,2,*]

[1] *Institute of Physics, Jan Kochanowski University, ul. Swietokrzyska 15, Kielce, Poland*
[2] *Institute for Theoretical Physics, J.W. Goethe University, Max-von-Laue-Str. 1, Frankfurt am Main, Germany*

Abstract. Glueballs are predicted in various theoretical approaches of QCD (most notably lattice QCD), but their experimental verification is still missing. In the low-energy sector some promising candidates for the scalar glueball exist, and some (less clear) candidates for the tensor and pseudoscalar glueballs were also proposed. Yet, for heavier gluonic states there is much work to be done both from the experimental and theoretical points of view. In these proceedings, we briefly review the current status of research of glueballs and discuss future developments.

1 Introduction

The fundamental particles of Quantum Chromodynamics (QCD) are quarks and gluons. Both are colored: quarks in the fundamental representation of the color group $SU_c(3)$ (red, green, blue), gluons in the adjoint representation (color-anticolor, minus the white configuration). The fundamental principle on which QCD is built is the invariance under local color transformations.

Color is not directly seen: confinement implies that the physical states emerging from QCD are 'white'. For instance all conventional quark-antiquark ($\bar{q}q$) states have the white wave function $\sqrt{1/3}(\bar{R}R + \bar{B}B + \bar{G}G)$. Such states constitute the majority of the mesonic resonances listed in the PDG [1], see also the review papers [2] and the predictions of the quark model [3].

From the early days of QCD [4–6] it was clear that bound states made solely of gluons, called glueballs, might exist. In fact, gluons interact strongly with themselves. The existence of glueballs was also predicted by various methods, most notably by lattice QCD, e.g. Refs. [7–9], both in the quenched and unquenched approximations. Yet, while their existence seems compelling from the theoretical point of view, up to now no resonance was found which can be *unambiguously* recognized as a predominantly glueball state. While in the low-energy sector (below 2.6 GeV) some candidates exist, in the high-energy sector no candidate is known. Experimental searches at low energies in the very soon upcoming experiments GlueX [10] and CLAS12 [11] at Jefferson Lab and at high energy at the ongoing BESIII [12, 13] and at the future PANDA [14] experiments are expected to improve our understanding.

The theoretical and experimental work on glueballs has been huge: up to now (status on 26/9/2015), there are 1404 papers which contain the world glueball in the title (for reviews, see Refs. [15–17]). In these proceedings we present some recent developments on this fascinating and still unsolved piece of QCD.

*e-mail: fgiacosa@ujk.edu.pl

Table 1. Central values of glueball masses from lattice (from [8])

J^{PC}	Value [GeV]
0^{++}	1.70
2^{++}	2.39
0^{-+}	2.55
1^{-+}	2.96
2^{-+}	3.04
3^{+-}	3.60
3^{++}	3.66
1^{--}	3.81
2^{--}	4.0
3^{--}	4.19
2^{+-}	4.22
0^{+-}	4.77

2 Existence and masses of glueballs

Photons do not interact with each other at tree-level. (A quartic photon interaction emerges through a fermionic loop, whose amplitude is suppressed by α^2.) Gluons are completely different: they shine, already at the leading order, in their own light. This fact, together with confinement, naturally leads to the idea that bound states made of solely gluons should exist.

The early studies of glueballs were performed within bag models, e.g. Refs. [4–6]. In particular, in Ref. [6] various microscopic currents were introduced and a glueball spectrum was shown. The lightest states were the scalar and tensor glueballs (at about 1 GeV), followed by pseudoscalar and pseudotensor ones.

The development of lattice QCD allowed to perform quantitative and model independent studies of the QCD spectrum. Already in 1999 a complete spectrum of glueballs (in the quenched approximations, i.e. without quarks) was presented [7]. The lightest state is a scalar glueball with a mass of about 1.7 GeV, followed by the tensor and the pseudoscalar states. This result has been confirmed by numerous and more recent lattice calculations, see Ref. [8], which is currently cited in the PDG in the review of the quark model [1]. The results are reported in Table 1.

Calculations within unquenched lattice QCD (i.e., with quark fluctuations) basically confirmed the same trend [9], in turn meaning that the mixing and the decays of glueballs should not be too large. This is indeed a very good information for model builders and experimental searchers.

While lattice QCD is the best theoretical proof of the existence of glueballs and the most reliable calculation of masses, other approaches were also used by theoreticians: QCD sum rules [18], Hamiltonian QCD [19], flux-tube model [20], anti De-Sitter approaches [21], and Bethe-Salpeter equations [22]. All of them find glueballs and the scalar state is the lightest.

In conclusion, there is nowadays a great confidence about the existence of glueballs and about the qualitative form of the spectrum. Nevertheless, the identification of glueballs has still to come.

3 Decays of glueballs

In the following we describe some general decay properties of glueballs and then we study separately some specific examples.

Large-N_c

According to the famous large-N_c limit [23] (simplifications occurs when the number of colors N_c is artificially increased to large values), glueballs' masses scale with N_c^0, just as $\bar{q}q$ masses. The decay of glueballs into mesons scales as N_c^{-2}, which is even more suppressed than regular $\bar{q}q$ states (that scale as N_c^{-1}). It is then expected that glueballs are narrow. This theoretical consideration is particularly important for the future PANDA project [14], which will search for glueballs between 2.2-5 GeV. Namely, only if glueballs are sufficiently narrow, they can be discovered experimentally.

Flavour and chiral blindness

Glueballs are flavour-invariant, hence they should decay in a flavour-blind way. For instance, for a glueball decaying into two pseudoscalar states and neglecting phase space, one obtains the ratios $\pi\pi : KK : \eta\eta : \eta'\eta' : \eta\eta' = 3 : 4 : 1 : 1 : 0$. In addition, the decays of a glueball are also chirally invariant, since it couples with the same strength to all chiral partners (such as $\rho\rho$ and $a_1(1230)a_1(1230)$).

Scalar glueball

The ground-state scalar glueball is undoubtedly the most studied gluonium. In the literature, many different scenarios concerning the identification of the scalar glueball in the realm of scalar states listed in the PDG were proposed. In most cases, the result was that the largest gluonic amount is either in $f_0(1500)$ or in $f_0(1710)$, e.g. Refs. [24–33]. A very short summary of the historical development is the following: in the pioneering work of Amsler and Close [24], later on confirmed by Close and Kirk [25], the largest gluonic amount sits in $f_0(1500)$. This conclusion was reached analyzing the decays of the three resonances $f_0(1370)$, $f_0(1500)$, and $f_0(1710)$ into two pseudoscalar states using a $^3P_0$ approach. On the other hand, Lee and Weingarten [26] used a lattice QCD approach to study the mass of the scalar glueball and its couplings to pions and kaons: the outcome was that $f_0(1710)$ is mostly gluonic. In Ref. [27], Giacosa et al. used an hadronic model inspired by ChPT in which also the other members of the scalar nonet were included, $K_0^*(1430)$ and $a_0(1450)$. The fit to all decays showed the existence of two solutions, one in which $f_0(1500)$ is predominantly a glueball, and on in which $f_0(1710)$ is such. Shortly after, Cheng et al [28] also found a phenomenological solution in which $f_0(1710)$ is predominantly a glueball. Various other studies were performed which involved constituents quarks and gluons, e.g. Ref. [29], or which involved the decay of the j/ψ meson, e.g. Ref. [30].

The scalar glueball is also linked to the anomalous breaking of dilatation symmetry (at the composite level, a dilaton/glueball field is introduced [34, 35]). In Ref. [36] a peculiar fact was shown. Using the dilaton potential from Refs. [34, 35], the decay of the glueball into pions turned out to be about 4 GeV, hence definitely too large to be detected. The numerical value is obtained by assuming that the dilaton saturates the gluon condensate [37]. If this were true, large-N_c would badly fail in the scalar sector and one could never find such a broad glueball.

As discussed in Ref. [31], the determination of the parameters of the dilaton potential through the gluon's condensate is not necessarily true. More in detail, in Ref. [31] the glueball was studied within the so-called extended Linear Sigma Model (eLSM). This is an hadronic model based on chiral symmetry and dilatation invariance together with their explicit and spontaneous breaking. The eLSM, first developed for two flavours [38], has shown to be capable to describe masses and decays of mesons up to 1.7 GeV, as the three-flavour study of Ref. [39] shows. The glueball as a dilaton is naturally included in this model. Quite remarkably, there is only a solution within the eLSM: $f_0(1710)$ is mostly gluonic. This result is in agreement with the original claim of Ref. [26], but also with the recent lattice study of Ref. [32], in which the decay $j/\psi \to \gamma G$ is numerically evaluated. Moreover, the very same conclusion has been reached in Ref. [33] by using an approach based on the AdS/QCD correspondence. Future information from the GlueX experiment is expected [40].

In conclusion, while a final assignment cannot yet be done, there is mounting evidence from different directions that $f_0(1710)$ is mostly gluonic.

Tensor glueball

According to lattice, the tensor glueball has a mass of about 2.2 GeV (it is second lightest). In Ref. [41] it was pointed out that the resonance $f_J(2220)$ does not lie on the Regge trajectories. Moreover, the state is very narrow, the $\pi\pi/KK$ ratio is in agreement with flavour blindness [42], and no $\gamma\gamma$ decay was seen. A necessary improvement would be the experimental assessment of this resonance. In particular, it is not yet clear if J is 2 or 4. Nevertheless, this is a promising starting point for future studies (for instance, employing the eLSM).

Pseudoscalar glueball

The pseudoscalar glueball has been also investigated in a variety of scenarios, see Ref. [43] for a review. One has investigated the gluonic content of the resonance η', e.g. Refs. [44] and [45]. In various other works, e.g. Ref. [46], the pseudoscalar glueball was assigned to the resonance $\eta(1405)$, while $\eta(1295)$ and $\eta(1475)$ are $\bar{q}q$ states. Such a scenario is controversial for two reasons: (i) At present, it is not clear if $\eta(1405)$ and $\eta(1475)$ are two independent states. (ii) The mass of the pseudoscalar glueball as predicted by lattice QCD is about 2.6 GeV, i.e. 1 GeV heavier.

In Ref. [47] the eLSM has been used to study the decays of an hypothetical pseudoscalar glueball (linked to the chiral anomaly [48]) with a mass of about 2.6 GeV, in agreement with lattice. The outcome was that the decay channels into $KK\pi$ and $\eta\pi\pi$ are dominant, while $\pi\pi\pi$ should vanish. A possible experimental candidate is the state $X(2370)$ measured by BES [13], yet future measurements on its decay rates are needed.

Other glueballs

The other glueballs listed in Table 1 need further studies. Very recently, two steps have been performed: (i) in Ref. [49] the decays of a pseudotensor glueball has been studied in a flavour-invariant hadronic model: sizable decay into $K_2^*(1430)K$ and $a_2(1320)\pi$ and a vanishing decay into $\rho\pi$ are predicted. (ii) The decays of a vector glueball in a fully chirally invariant approach (using the eLSM) have been investigated in Ref. [50]: a sizable decay into $\omega\pi\pi$ (both direct and indirect through $b_1\pi$) and into $\pi KK^*(892)$ are expected to be the main signal of a vector gluonium. Such simple predictions may help future experimental searches.

4 Conclusions

Glueballs are expected to exist but were not yet found in experiments. While GlueX and CLAS12 can help our understanding in the light sector, BESIII and, in the future, PANDA can search for glueballs in the heavy sector. Definitely, more work is needed: predictions about the decay channels of glueballs might be particularly helpful in the process of identifications of possible candidates. The aim is to close the gap between a basic theoretical expectation of QCD and the present experimental status.

Support from the Polish National Science Centre NCN through the OPUS project nr. 2015/17/B/ST2/01625 is acknowledged.

References

[1] K. A. Olive et al. (Particle Data Group), Chin. Phys. **C38**, 090001 (2014).

[2] C. Amsler and N. A. Tornqvist, Phys. Rept. **389**, 61 (2004); E. Klempt and A. Zaitsev, Phys. Rept. **454** 1 (2007); F. E. Close and N. A. Tornqvist, J. Phys. G **28**, R249 (2002).
[3] S. Godfrey and N. Isgur, Phys. Rev. D **32** 189 (1985).
[4] A. Chodos, R. L. Jaffe, K. Johnson, C. B. Thorn and V. F. Weisskopf, Phys. Rev. D **9** 3471 (1974).
[5] R. L. Jaffe and K. Johnson, Phys. Lett. B **60** 201 (1976).
[6] R. L. Jaffe, K. Johnson and Z. Ryzak, Annals Phys. **168** 344 (1986).
[7] C. J. Morningstar and M. J. Peardon, Phys. Rev. D **60,** 034509 (1999).
[8] Y. Chen *et al.*, Phys. Rev. D **73,** 014516 (2006).
[9] E. Gregory*et al.*, JHEP **1210,** 170 (2012).
[10] H. Al Ghoul *et al.* [GlueX Collaboration], AIP Conf. Proc. **1735** 020001 (2016). B. Zihlmann [GlueX Collaboration], AIP Conf. Proc. **1257** 116 (2010) .
[11] A. Rizzo [CLAS Collaboration], PoS CD **15** 060 (2016).
[12] G. Mezzadri, PoS EPS **-HEP2015** 423 (2015). S. Marcello [BESIII Collaboration], JPS Conf. Proc. **10** 010009 (2016).
[13] M. Ablikim *et al.* [BESIII Collaboration], Phys. Rev. Lett. **106** 072002 (2011) =.
[14] M. F. M. Lutz *et al.* [PANDA Collaboration], arXiv:0903.3905 [hep-ex]].
[15] W. Ochs, J. Phys. G **40** 043001 (2013).
[16] V. Mathieu, N. Kochelev and V. Vento, Int. J. Mod. Phys. E **18** 1 (2009).
[17] V. Crede and C. A. Meyer, Prog. Part. Nucl. Phys. **63** 74 (2009).
[18] H. G. Dosch and S. Narison, Nucl. Phys. Proc. Suppl. **121** 114 (2003); H. Forkel, Phys. Rev. D **71** 054008 (2005).
[19] A. P. Szczepaniak and E. S. Swanson, Phys. Lett. B **577** 61 (2003).
[20] N. Isgur, R. Kokoski and J. Paton, Phys. Rev. Lett. **54** 869 (1985).
[21] R. C. Brower, S. D. Mathur and C. I. Tan, Nucl. Phys. B **587** 249 (2000).
[22] H. Sanchis-Alepuz, C. S. Fischer, C. Kellermann and L. von Smekal, Phys. Rev. D **92** 034001 (2015).
[23] G. t Hooft, Nucl. Phys. B **72** 461 (1974); E. Witten, Nucl. Phys. B **160** 57 (1979).
[24] C. Amsler and F. E. Close, Phys. Rev. D **53** 295 (1996).
[25] F. E. Close and A. Kirk, Eur. Phys. J. C **21**, 531 (2001).
[26] W. J. Lee and D. Weingarten, Phys. Rev. D **61**, 014015 (2000).
[27] F. Giacosa*et al.*, Phys. Rev. D **72**, 094006 (2005); F. Giacosa *et al.* , Phys. Lett. B **622**, 277 (2005).
[28] H. Y. Cheng, C. K. Chua and K. F. Liu, Phys. Rev. D **74**, 094005 (2006).
[29] M. Strohmeier-Presicek *et al.* , Phys. Rev. D **60** 054010 (1999); F. Giacosa, T. Gutsche and A. Faessler, Phys. Rev. C **71** 025202 (2005).
[30] P. Chatzis *et al.*, Phys. Rev. D **84,** 034027 (2011); E. Close and Q. Zhao, Phys. Rev. D **71**, 094022 (2005).
[31] S. Janowski, F. Giacosa and D. H. Rischke, Phys. Rev. D **90** no. 11 114005(2014).
[32] L. -C. Gui *et al.*, Phys. Rev. Lett. **110** 021601 (2013).
[33] F. Brünner, D. Parganlija and A. Rebhan, Phys. Rev. D **91** no. 10 106002 (2015); F. Brünner and A. Rebhan, Phys. Rev. Lett. **115** no.13 131601 (2015); F. Brünner and A. Rebhan, Phys. Rev. D **92** no.12 121902 (2015).
[34] A. A. Migdal and M. A. Shifman, Phys. Lett. B **114** 445 (1982).
[35] A. Salomone, J. Schechter and T. Tudron, Phys. Rev. D **23** 1143 (1981).

[36] J. R. Ellis and J. Lanik, Phys. Lett. B **150** 289 (1985).

[37] V. A. Novikov, L. B. Okun, M. A. Shifman, A. I. Vainshtein, M. B. Voloshin and V. I. Zakharov, Phys. Rept. **41** 1 (1978); E. I. Lashin, Int. J. Mod. Phys. A **21** 3699 (2006).

[38] D. Parganlija, F. Giacosa and D. H. Rischke, Phys. Rev. D **82** 054024 (2010); S. Gallas, F. Giacosa and D. H. Rischke, Phys. Rev. D **82** 014004 (2010). S. Janowski *et al.*, Phys. Rev. **D84 054007** (2011).

[39] D. Parganlija *et al.*, Phys. Rev. D **87** no. 1 014011 (2013).

[40] T. Gutsche *et al.*, Phys. Rev. D **94** no.3 034010 (2016) .

[41] V. V. Anisovich, JETP Lett. **80** 715 (2004) [Pisma Zh. Eksp. Teor. Fiz. **80** (2004) 845]; V. V. Anisovich, M. A. Matveev, J. Nyiri and A. V. Sarantsev, Int. J. Mod. Phys. A **20** 6327 (2005) [Int. J. Mod. Phys. A **20** (2005) 0502842].

[42] F. Giacosa, T. Gutsche, V. E. Lyubovitskij and A. Faessler, Phys. Rev. D **72** 114021 (2005).

[43] A. Masoni, C. Cicalo and G. L. Usai, J. Phys. G **32** R293 (2006).

[44] G. Amelino-Camelia *et al.*, Eur. Phys. J. C **68** 619 (2010).

[45] R. Escribano and J. Nadal, JHEP **0705** (2007) 006.

[46] T. Gutsche, V. E. Lyubovitskij and M. C. Tichy, Phys. Rev. D **80** 014014 (2009).

[47] W. I. Eshraim *et al.*, Phys. Rev. D **87** no. 5 054036 (2013); W. I. Eshraim *et al.* , Acta Phys. Polon. Supp. **5** 1101 (2012).

[48] C. Rosenzweig, A. Salomone and J. Schechter, Phys. Rev. D **24** 2545 (1981).

[49] A. Koenigstein and F. Giacosa, arXiv:1608.08777 [hep-ph].

[50] F. Giacosa, J. Sammet and S. Janowski, arXiv:1607.03640 [hep-ph].

DOI: 10.1051/epjconf/201613001010

Search for electric dipole moments in storage rings

Paolo Lenisa[1,⋆]

[1] *University of Ferrara and INFN, Ferrara, Italy*

Abstract. The JEDI collaboration aims at making use of storage ring to provide the most precise measurement of the electric dipole moments of hadrons. The method makes exploits a longitudinal polarized beam. The existence an electric dipole moment would generate a torque slowly twisting the particle spin out of plan of the storage ring into the vertical direction. The observation of non zero electric dipole moment would represent a clear sign of new physics beyond the Standard Model. Feasiblity tests are presently undergoing at the COSY storage ring Forschungszentrum Jülich (Germany), to develop the novel techniques to be implemented in a future dedicated storage ring.

1 Introduction

The fact that we and the world around us are made of matter and only a minimal amount of antimatter is observed, constitutes one of the fundamental puzzles in modern physics, motivating a variety of theoretical speculations and experimental investigations. The combined standard models of cosmology and particle physics suggest that at the end of the inflation epoch immediately following the Big Bang, the number of particles and antiparticles were almost in precise balance. Yet the laws of physics contrived to act differently on matter and antimatter to generate the apparently large imbalance that we observe today. One of the necessary mechanisms required for this to happen – namely CP violation – is very small in the Standard Model of particle physics and therefore only able to account for a tiny fraction of the actual imbalance. New sources of CP violation are needed, and one such potential signature would be the appearance of electric dipole moments (EDMs) in fundamental particles.

An EDM originates from a permanent charge separation inside the particle. In its centre-of-mass frame, the ground state of a subatomic particle has no direction at its disposal except its spin, which is an axial vector, while the charge separation (EDM) corresponds to a polar vector (see Fig. 1). Therefore, if such a particle with nonzero mass and spin possesses an EDM, it must violate both parity (P) and time-reversal (T) invariance. If the combined CPT symmetry is to be valid, T violation also implies breaking of the combined CP symmetry. The Standard Model predicts the existence of EDMs, but their sizes (in the range of 10^{-31} to 10^{-33} e·cm for nucleons) fall many orders of magnitude below the sensitivity of current measurements and still far below the expected levels of projected experiments. An EDM observation at a much higher value would therefore be a clear and convincing sign of new physics beyond the current Standard Model (BSM).

BSM theories such as supersymmetry (SUSY), technicolour, multi-Higgs models and left–right symmetric models generally predict nucleon EDMs in the range of 10^{-24} to 10^{-28} e·cm (part of the

⋆e-mail: lenisa@fe.infn.it

DOI: 10.1051/epjconf/201613001010

upper region of this range is already excluded by experiment). Although tiny, EDMs of this size would be large enough to be observed by a new generation of highly sensitive accelerator based experiments with charged particles such as the proton and deuteron. In this respect, EDMs offer a complementary approach to searches for BSM physics at collider experiments, probing scales far beyond the reach of present high-energy machines such as the LHC. For example, in certain SUSY scenarios the present observed EDM limits provide information about physics at the TeV or even PeV scales, depending on the mass scale of the supersymmetric mechanisms and the strength of the CP-violating SUSY phase parameters (Fig.2).

Researchers have been searching for EDMs in neutral particles, especially neutrons, for more than 50 years, by trapping and cooling particles in small volumes and using strong electric fields. Despite an enormous improvement in sensitivity, however, these experiments have only produced upper bounds. The current upper limit of approximately 10^{-26} e·cm for the EDM of the neutron is an amazingly accurate result: if we had inflated the neutron so that it had the radius of the Earth, the EDM would correspond to a separation between positive and negative charges of about 1 mm. An upper limit of less than 10^{-29} e·cm has also been reported for a special isotope of mercury, but the Coulomb screening by number to the permanent EDMs of the neutrons and protons in its nucleus. For the electron, meanwhile, the reported EDM limits on more complicated polar molecules can be used to deduce a bound of about 10^{-28} e·cm – which is even further away from the Standard Model prediction (10^{-38} e·cm) than is the case for the neutron.

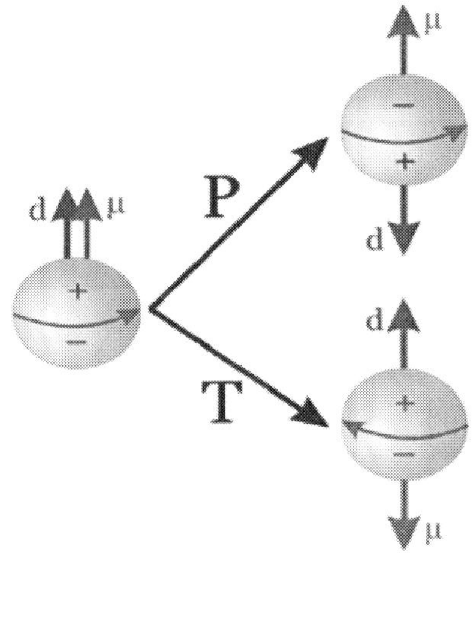

Figure 1. Naively, an electric dipole moment (d) and a magnetic dipole moment (μ) transfer differently under P and T. In a fundamental particle, both quantities are proportional to the spin vector (s). Therefore, the interaction term (ds·E) is odd under P and T, whereas (μs·B) is even under these transformations.

2 Perspectives for storage rings

Although these experiments provide useful constraints on BSM theories, a new class of experiments based on storage rings is needed to measure the electric dipole moment of charged particles (such as the proton, deuteron or helium-3). These highly sensitive acceleratorbased experiments will allow the EDM of charged particles to be inferred from their very slow spin precession in the presence of large electric fields, and promise to reach a sensitivity of 10^{-29} e·cm. This is due mainly to the larger number of particles available in a stored beam compared with the ultra-cold neutrons usually found in trap experiments, and also the potentially longer observation time possible because such experiments are not limited by the particle decay time. Storage-ring experiments would span the range of EDM sizes where new CP violation is expected to lie. Furthermore, the ability to measure EDMs of more

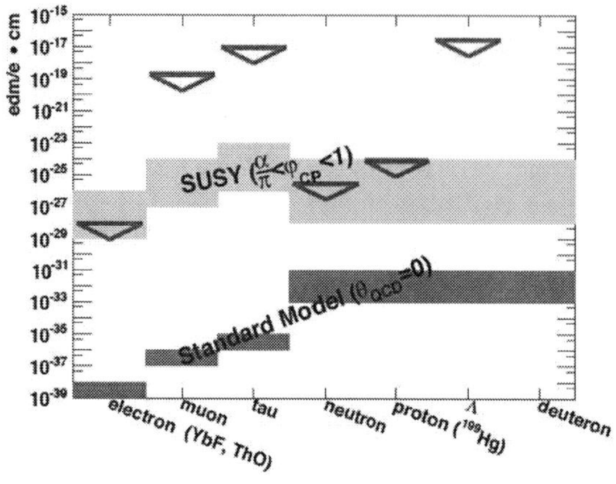

Figure 2. Experimental upper limits for the EDMs of different particles (red arrows) plotted together with the prediction from SUSY (blue bands) and the Standard Model (grey). No experimental limit exists yet for the deuteron.

than one type of particle will help to constrain the origin of the CP-violating source because not all particles are equally sensitive to the various CP-violating mechanisms.

At the Cooler Synchrotron "COSY" located at the Forschungszentrum Jülich (FZJ), Germany (Fig.3), the JEDI (Jülich Electric Dipole moment Investigations) Collaboration [1] is working on a series of feasibility studies for such a measurement using an existing conventional hadron storage ring. COSY, which is able to store both polarised proton and deuteron beams with a momentum up to 3.7 GeV/c, is an ideal machine for the development and commissioning of the necessary technology. This R&D work has recently replaced COSY's previous hadron-physics programme of particle production and rare decays, although some other service and user activities continue.

Figure 3. The COSY storage ring at the Forschungszentrum Jülich in Germany.

[1] http://collaborations.fz-juelich.de/ikp/jedi/index.shtml

A first upper limit for an EDM directly measured in a storage ring was obtained for muons at the (g–2) experiment at Brookhaven National Laboratory (BNL) in the US, but the measurement was not optimised for sensitivity to the EDM. Subsequently, scientists at BNL began to explore what would be needed to fully exploit the potential of a storage-ring experiment. While much initial discussion for an EDM experiment also took place at Brookhaven, commitments to the Relativistic Heavy Ion Collider (RHIC) operation and planning for a potential Electron–Ion Collider have prevented further development of such a project there. Therefore the focus shifted to FZJ and the COSY storage ring in Germany, where the JEDI collaboration was formed in 2011 to address the EDM opportunity.

The measuring principle is straightforward: a radial electric field is applied to an ensemble of particles circulating in a storage ring with their polarisation vector (or spin) initially aligned with their momentum direction. Maintaining the polarisation in this direction requires a storage ring in which the bending elements are a carefully matched set of both vertical magnetic fields and radial electric fields. The field strengths must be chosen such that the precession rate of the polarisation matches the circulation rate of the beam (called the "frozen spin"). For particles such as the proton with a positive gyromagnetic anomaly, this can be achieved by using only electric fields and choosing just the right "magic" momentum value (around 0.7 GeV/c). For deuterons, which have a negative gyromagnetic anomaly, a combination of electric and magnetic fields is required, but in this case the frozen spin condition can be achieved for a wide range of momentum and electric/magnetic field combinations. Such combined fields may also be used for the proton and would allow the experiment to operate at momenta other than the magic value.

The existence of an EDM would generate a torque that slowly rotates the spin out of the plane of the storage ring and into the vertical plane (see panel opposite). This slow change in the vertical polarisation is measured by sampling the beam with elastic scattering off a carbon target and looking for a slowly increasing left–right asymmetry in the scattered particle flux. For an EDM of 10^{-29} e·cm and an electric field of 10 MV/m, this would happen at an angular velocity of $3{\cdot}10^{-9}$rad$\cdot s^{-1}$ (about 1/100th of a degree per day of continuous operations). This requires the measurement to be sensitive at a level never reached before in a storage ring. To obtain a statistically significant result, the polarisation in the ring plane must last for approximately 1000 s during a single fill of the ring, while the scattering asymmetry from the carbon target must reach levels above 10^{-6} to be measurable within a year of running.

3 Achievements at COSY

Following the commissioning of a measurement system that stores the clock time of each recorded event in the beam polarimeter with respect to the start of the accelerator cycle, the JEDI collaboration has passed a series of important milestones in recent years. Working with the deuteron beam at COSY, these time stamps make it possible to unfold for the first time the rapid rotation of the polarisation in the ring plan (which has a frequency of around 120 kHz) that arises from the gyromagnetic anomaly. In a one-second time interval, the number of polarisation revolutions may be counted and the final direction of the polarisation known to better than 0.1 rad (Fig. 4)[1]. The magnitude of the polarisation may decline slowly due to decoherence effects in storage ring, as can be seen in subsequent polarisation measurements within a single fill.

Maintaining the polarisation requires the cancellation of effects that may cause the particles in the beam to differ from one another. Bunching and electron cooling serves to remove much of this spurious motion, but particle path lengths around the ring may differ if particles in the beam have transverse oscillations with different amplitudes. Recently, we demonstrated that the effect of these differences on polarisation decoherence can be removed by applying correcting sextupole fields to the ring. As a result, we can now achieve polarisation lifetimes in the horizontal plane of more than 1000 s – as required for the EDM experiment (Fig. 5) [2]. In the past year, the JEDI group has also shown that by determining errors in the polarisation direction and feeding this back to make small changes in the ring's radio-frequency, the direction of the polarisation may be maintained at the level of 0.1 rad during any chosen time period. This is a further requirement for managing the polarisation in the ring for the EDM measurement.

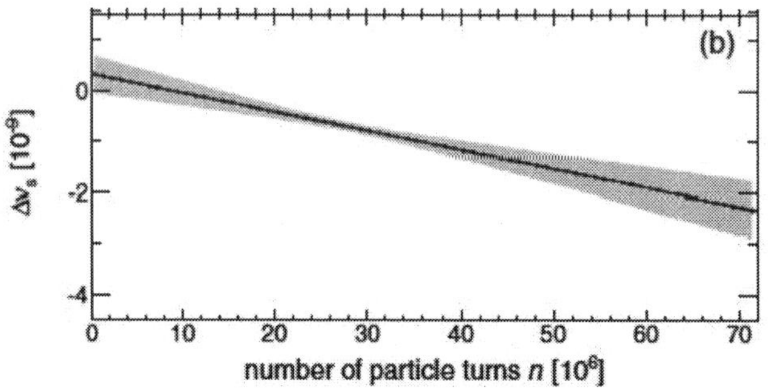

Figure 4. Deviation of the spin tune ν_s, which is defined as the number of spin precessions per turn, as a function of the number of turns in the ring. At t = 38 s (about 28 × 10^6 turns), the interpolated spin tune amounts to 16097540628.3 ± 9.7 × 10^{-11} , which represents the most precise measurement of this quantity ever performed. The previous best measurement, performed for the muon at the (g–2) experiment, had a precision of 3 × 10^{-8} per year. The higher precision achieved here is mainly attributed to the much longer measurement time of 100 s compared with 600 μ in the (g–2) experiment.

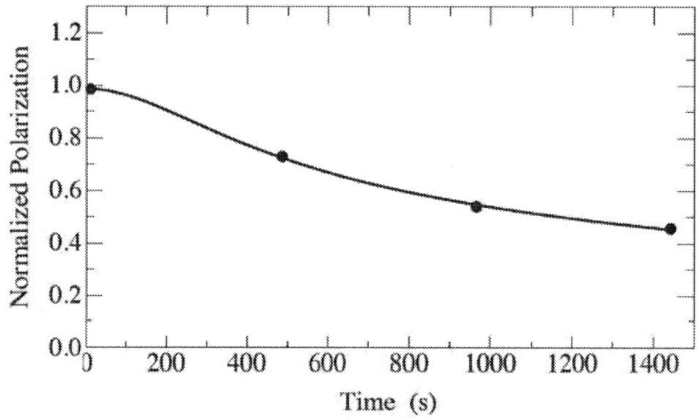

Figure 5. One of the longest polarisation lifetimes recorded for the COSY ring. Measurements made at four separate times (to conserve beam) are matched to a depolarisation curve that assumes a Gaussian distribution of transverse oscillation amplitudes. The half-life of the polarisation is 1173±172 s, which is three orders of magnitude longer than previous results using electron beams.

COSY is also an important test facility for many EDM-related technologies, among them new beam-position monitoring, control and feedback systems. High electric fields and combined electric/ magnetic deflectors may also find applications in other fields, such as accelerator science. Many checks for systematic errors will be undertaken, and a technical design report for a future dedicated storage ring will be prepared. The most significant challenges will come from small imperfections in the placement and orientation of ring elements, which may cause stray field components that generate the accumulation of an EDM-like signal. The experiment is most sensitive to radial magnetic fields and vertical electric fields. Similar effects may arise through the non-commutativity of spurious rotations within the ring system, and efforts are under way to model these effects via spin tracking supported with beam testing. Eventually, many such effects may be reduced or eliminated by comparing the signal accumulation rates seen with beams travelling in opposite directions in the storage ring. During the next decade, this will allow researchers to approach the design goals of the EDM search using a storage ring, adding a new opportunity to unveil physics beyond the Standard Model.

4 ERC-Advanced grant

In early 2016, the European Research Council awarded an advanced research grant to the Juelich group to support further developmental efforts. The five-year grant, starting in October, will support a consortium that also includes RWTH Aachen University in Germany and the University of Ferrara in Italy. The goal of the project is to conduct the first measurement of the deuteron EDM. Since the COSY polarisation cannot be maintained parallel to its velocity (because no combined electric and magnetic bending elements exist), a novel device called a radiofrequency Wien filter will be installed in the ring to slowly accumulate the EDM signal (the filter influences the spin motion without acting on the particle's orbit) [3]. The idea is to exploit the electric fields created in the particle rest system by the magnetic fields of the storage-ring dipoles, which would allow the first ever measurement of the deuteron EDM.

References

[1] D. Eversmann *et al.*, Phys. Rev. Lett. **115**, 094801 (2015)
[2] G. Guidoboni *et al.*, Phys. Rev. Lett. **116**, 054801 (2016)
[3] J. Slim *et al.*, Nucl. Instr. and Meth. A **828**, 116 (2016)

EPJ Web of Conferences **130**, 01011 (2016) DOI: 10.1051/epjconf/201613001011
MESON 2016

Double polarisation experiments in meson photoproduction

Jan Hartmann[1,*]
for the CBELSA/TAPS Collaboration

[1] *Helmholtz-Institut für Strahlen- und Kernphysik, Universität Bonn, Germany*

Abstract. One of the remaining challenges within the standard model is to gain a good understanding of QCD in the non-perturbative regime. A key step towards this aim is baryon spectroscopy, investigating the spectrum and the properties of baryon resonances. To gain access to resonances with small πN partial width, photoproduction experiments provide essential information. Partial wave analyses need to be performed to extract the contributing resonances. Here, a complete experiment is required to unambiguously determine the contributing amplitudes. This involves the measurement of carefully chosen single and double polarisation observables. In a joint endeavour by MAMI, ELSA, and Jefferson Laboratory, a new generation of experiments with polarised beams, polarised proton and neutron targets, and 4π particle detectors have been performed in recent years. Many results of unprecedented quality were recently published by all three experiments, and included by the various partial wave analysis groups in their analyses, leading to substantial improvements, e.g. a more precise determination of resonance parameters. An overview of recent results is given, with an emphasis on results from the CBELSA/TAPS experiment, and their impact on our understanding of the nucleon excitation spectrum is discussed.

1 Introduction

The spectrum of excited nucleon states reflects the dynamics of QCD in the non-perturbative regime. It has been studied for many years using pion beams. However, the spectrum of known nucleon resonances is in conflict with predictions from quark models [1, 2]. Most obvious is the missing resonance problem: the fact that more states are predicted by models at higher masses than have been observed experimentally. But also the ordering of excited states with positive and negative parity is partly in disagreement with experiments, the most prominent example being the $N(1440)\,1/2^+$ which is predicted by most quark models to be heavier than the $N(1535)\,1/2^-$. QCD calculations on the lattice [3], though using unphysically large quark masses, yield a similar pattern as the non-relativistic quark model. Measuring the properties of the known resonances more precisely and searching for new resonances is essential to understanding the discrepancies between theory and experiment.

Photoproduction experiments allow access to resonances with small πN couplings and therefore have great potential to observe the missing resonances. The contributing resonances are extracted from the measured data in a partial wave analysis (PWA). To determine the contributing amplitudes in an unambiguous way, a complete experiment [4] is needed, which requires the measurement of

[*] e-mail: hartmann@hiskp.uni-bonn.de

DOI: 10.1051/epjconf/201613001011

polarisation observables. In this paper, a selection of recent measurements of single and double polarisation observables accessible with linearly or circulary polarised photon beam and a longitudinally or transversely polarised proton target is shown, and their impact on the PWA is discussed.

2 Experimental setup

The data presented here were obtained with the CBELSA/TAPS experiment at ELSA [5]. A linearly polarised photon beam was produced from the incident 3.2 GeV electron beam via coherent bremsstrahlung off a carefully aligned diamond crystal [6]. For a coherent edge at $E_\gamma = 950$ MeV a maximum polarisation of 65 % was reached. A circularly polarised photon beam was produced from a longitudinally polarised 2.4 GeV electron beam provided by ELSA. The polarisation was measured with a Møller polarimeter, a maximum electron polarisation of 64 % was reached. The electrons, after hitting the radiator target, passed through a magnet onto a tagging hodoscope which defined the energy of the bremsstrahlung photons. The photon beam impinged on a frozen spin butanol target [7] providing transversely polarised protons with an average target polarisation degree of 74 %.

The detector system, which is shown in Figure 1, consisted of two electromagnetic calorimeters, the Crystal Barrel [8] and the MiniTAPS detector [9], together covering the polar angle range from 1° to 156° and the full azimuthal angle. For charged particle identification, a three-layer scintillating fibre detector [10] surrounding the target, and plastic scintillators in forward direction could be used. The detector setup provided a high detection efficiency for photons and is therefore ideally suited to measure single and double polarisation observables in reactions with neutral mesons decaying into photons in the final state.

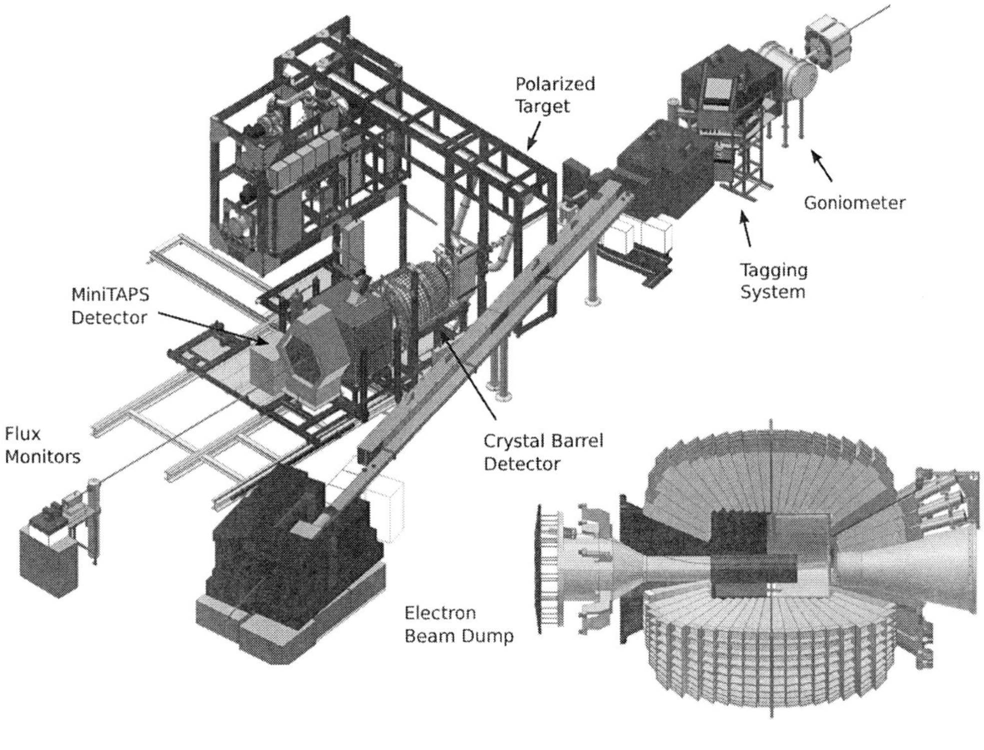

Figure 1. The experimental setup of the CBELSA/TAPS experiment.

 DOI: 10.1051/epjconf/201613001011

3 Results

The cross-section for photoproduction of single pseudoscaler mesons with a polarised beam off a polarised target can be written in the form [11]

$$\frac{\mathrm{d}\sigma}{\mathrm{d}\Omega} = \left(\frac{\mathrm{d}\sigma}{\mathrm{d}\Omega}\right)_0 \cdot \Big[1 - \Sigma\delta_\ell \cos(2\phi) + T\Lambda_y - P\Lambda_y\delta_\ell \cos(2\phi) - E\Lambda_z\delta_\odot + F\Lambda_x\delta_\odot + G\Lambda_z\delta_\ell \sin(2\phi) + H\Lambda_x\delta_\ell \sin(2\phi)\Big], \quad (1)$$

where $\left(\frac{\mathrm{d}\sigma}{\mathrm{d}\Omega}\right)_0$ is the unpolarised cross section, δ_ℓ ($\delta_\odot$) denotes the linear (circular) polarisation degree of the photon beam, and $\Lambda_{x,y,z}$ is the polarisation degree of the target protons in the centre-of-mass (CM) coordinate frame. In a polarised butanol (C_4H_9OH) target, not only the polarised protons, but also unpolarised nucleons in the C and O nuclei contribute to the count rate. This effectively reduces the target polarisation and can be taken into account by the dilution factor d: $\Lambda \to d\,\Lambda$. To determine d, the background contribution from bound nucleons needs to be determined, e.g. using a background measurement with a carbon target. For details, see e.g. [12, 13].

3.1 π^0 photoproduction

With circularly polarised photons and longitudinally polarised protons, the helicity asymmetry E is accessible. It can be determined from the asymmetry of events with parallel ($N_{3/2}$) and anti-parallell ($N_{1/2}$) spins of the initial-state particles as

$$E = \frac{N_{1/2} - N_{3/2}}{N_{1/2} + N_{3/2}} \cdot \frac{1}{d\,\Lambda_z\,\delta_\odot}. \quad (2)$$

With linearly polarised photons and logitudinally polarised protons, the double polarisation observable G can be deduced from the correlation between the photon polarisation plane and the scattering plane. The number of events N as a function of the azimuthal angle ϕ between the two planes is given by

$$N(\phi)/N_0 = 1 - \delta_\ell\,\Sigma_{\text{eff}} \cos(2\phi) + d\,\Lambda_z\,\delta_\ell\,G \sin(2\phi), \quad (3)$$

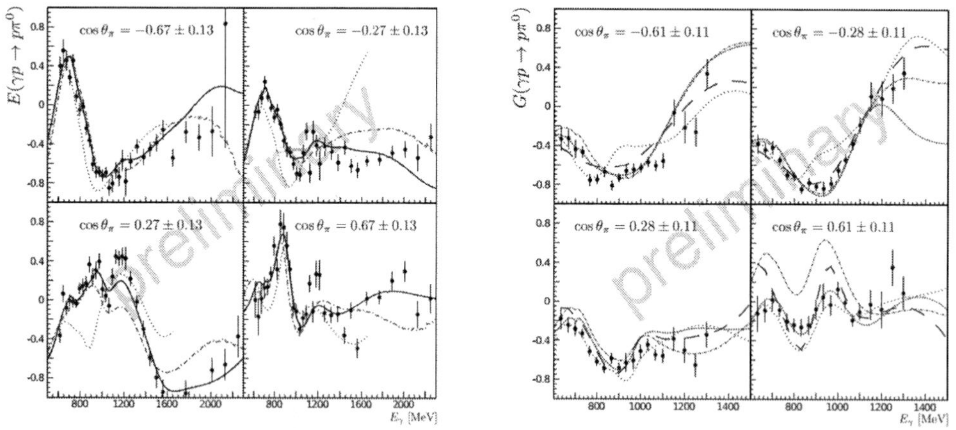

Figure 2. Results for the observables E (left) [12] and G (right) [14, 15] for four different angular bins, compared to different PWA predictions: MAID (green, dotted) [16], SAID CM12 (red, dashed-dotted) [17], BnGa 2011-02 (black, solid) [18], and JüBo 13-01 (blue, dashed) [19].

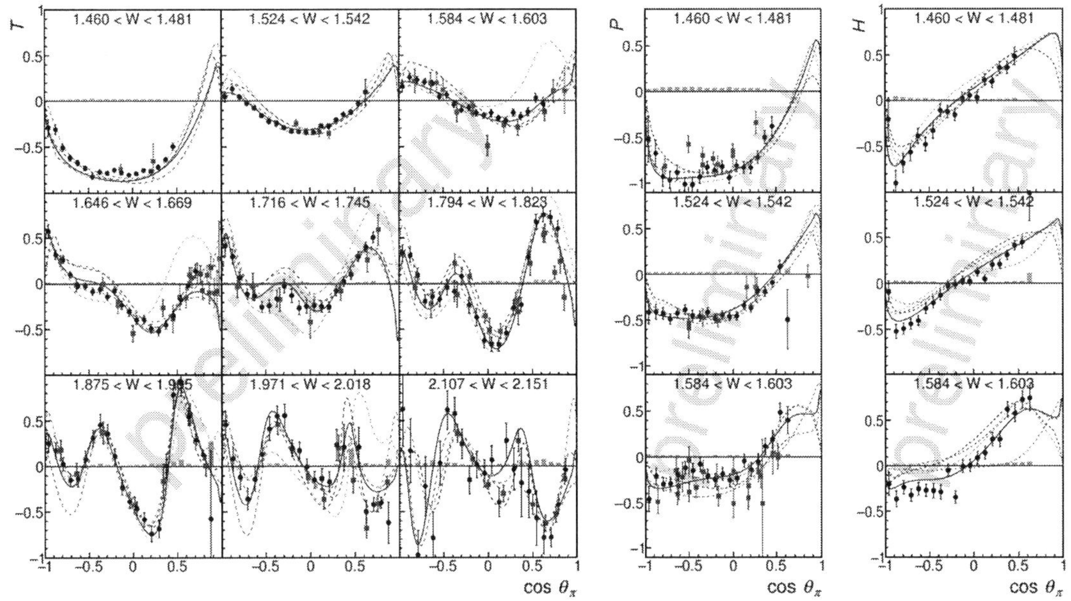

Figure 3. Results for the observables T, P, and H [13, 20] as a function of the scattering angle and the γp invariant mass W (in GeV, only every third bin is shown here). References to earlier data (red points) are given in [18], Refs. [49-71] therein. The solid black line represents the BnGa2014 fit [13]. The data are compared to predictions (dashed curves) from BnGa2011-02 (red) [18], MAID (green) [16], SAID CM12 (blue) [17], and JüBo 2015 (magenta) [21].

where N_0 is given by averaging $N(\phi)$ over ϕ. Σ_{eff} mixes the beam asymmetry from free and bound nucleons. Results for both E and G are shown in Fig. 2 in comparison to different PWA predictions.

With linearly polarised photons and transversely polarised protons, the observables T, P, and H can be measured simultaneously. In this case, the azimuthal distribution of events is given by

$$\begin{aligned} N(\phi)/N_0 = 1 &- \delta_\ell \Sigma_{\text{eff}} \cos(2\phi) + d\,\Lambda_T\, T \sin(\phi - \varphi) \\ &- d\,\Lambda_T\, \delta_\ell P \cos(2\phi) \sin(\phi - \varphi) + d\,\Lambda_T\, \delta_\ell H \sin(2\phi) \cos(\phi - \varphi), \end{aligned} \tag{4}$$

where φ is the azimuthal angle between the target polarisation vector and the photon polarisation plane. Results are shown in Fig. 3.

Our data up to $E_\gamma = 930$ MeV were used as a basis for an energy-independent PWA [20], allowing for the determination of the $N(1520)\,3/2^-$ helicity amplitudes with minimal model dependence. All the data were included in the BnGa multi-channel PWA, together with further data on other channels.[1] Starting from the previous solutions BnGa2011-01 and BnGa2011-02 [18] all parameters were re-optimised. The newly determined multipoles are compatible with the previous ones at the 2σ level over the full mass range. The errors are significantly reduced, on average by a factor of 2.25 [13]. The impact of the new data on the BnGa, SAID, and JüBo analyses was recently investigated in a joint effort of the analysis groups [22]. To quantify the discrepancy between different analyses, the variance of the 16 (complex) multipole amplitudes $\mathcal{M}$ up to $L = 4$ can be used, defined as

$$\text{var}(1,2) = \frac{1}{2} \sum_{i=1}^{16} (\mathcal{M}_1(i) - \mathcal{M}_2(i))(\mathcal{M}_1^*(i) - \mathcal{M}_2^*(i)). \tag{5}$$

[1] For a complete list, see [13], Ref. [25] therein.

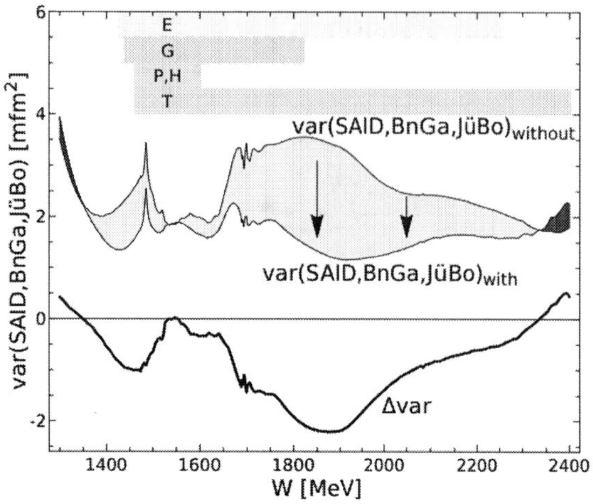

Figure 4. The variance of all three PWAs summed over all $\gamma p \to \pi^0 p$ multipoles up to $L = 4$ [22]. The range covered by the new double-polarisation observables is indicated by shaded areas. Over the largest part of the energy range the new data have enforced an improvement of the overall consistency. The improvement is displayed as a light green area and, separately, as difference of the variance. Ranges with an overall deterioration are marked in red.

Once the new data are included, the discrepancy between all three analyses is significantly reduced, as can be seen in Fig. 4.

3.2 η photoproduction

Preliminary results for the polarisation observables T, P, and H are shown in Figure 5. Large deviations from the data are observed for the predictions from MAID [16], SAID [25], BnGa2011 [18], Gießen [26], and the JüBo model [21], emphasising how important these new data are to constrain the amplitudes for η photoproduction.

The analysis by the BnGa group of our new data on T, P, and H, together with not yet published data on E and G and further data from Mainz (T, F) [24] and Jefferson Laboratory [27] (E) is presently ongoing. The results will be published in the near future [28].

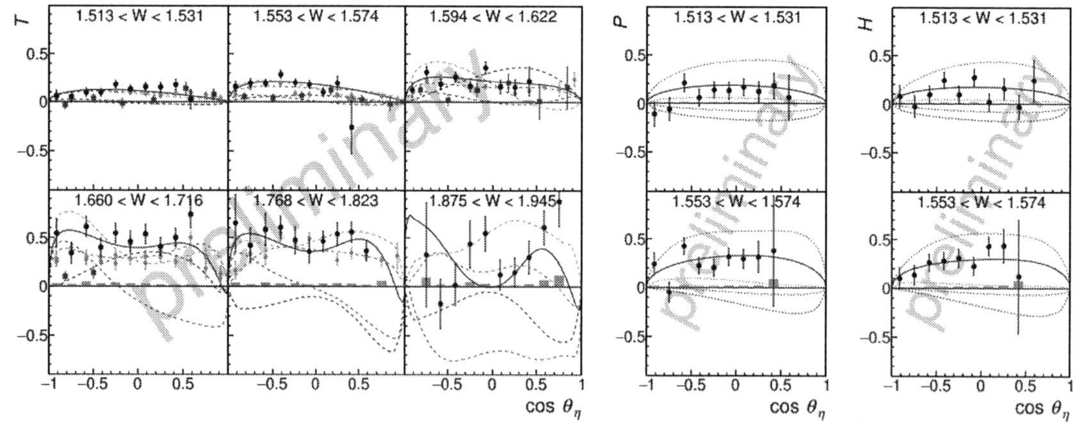

Figure 5. Preliminary results for the observables T, P, and H as a function of the scattering angle and the γp invariant mass W (in GeV, only every second bin is shown here). Earlier ELSA data (red) [23] and recent MAMI results (green) [24] are shown for comparison. The solid black line represents a preliminary BnGa fit. The data are compared to predictions (dashed curves) from BnGa2011-02 (red) [18], MAID (green) [16], SAID GE09 (blue) [25], and JüBo 2015 (magenta) [21].

4 Summary and outlook

In π^0 photoproduction, the unprecedented precision of the data significantly reduces the errors of the PWAs, leading to a more precise determination of resonance parameters. In η photoproduction, where several observables are now measured for the first time, the new data are crucial to constrain the photoproduction amplitudes. Further reaction channels are also being investigated. In particular multi-meson final states like $p\pi^0\pi^0$ or $p\pi^0\eta$ are sensitive to cascade decays of higher-mass resonances via intermediate N^* and Δ^* states [29–31].

We thank the technical staff of ELSA, the polarised target group, and the participating institutions for their invaluable contributions to the success of the experiment. We acknowledge support from the *Deutsche Forschungsgemeinschaft* (SFB/TR16) and *Schweizerischer Nationalfonds*.

References

[1] N. Isgur, G. Karl, Phys. Lett. B **72**, 109 (1977)
[2] U. Loring, B.C. Metsch, H.R. Petry, Eur. Phys. J. A **10**, 395 (2001)
[3] R.G. Edwards, J.J. Dudek, D.G. Richards, S.J. Wallace, Phys. Rev. D **84**, 074508 (2011)
[4] W.T. Chiang, F. Tabakin, Phys. Rev. C **55**, 2054 (1997)
[5] W. Hillert, Eur. Phys. J. A **28S1**, 139 (2006)
[6] D. Elsner *et al.*, Eur. Phys. J. A **39**, 373 (2009)
[7] C. Bradtke, H. Dutz *et al.*, Nucl. Instrum. Meth. A **436**, 430 (1999)
[8] E. Aker *et al.*, Nucl. Instrum. Meth. A **321**, 69 (1992)
[9] R. Novotny, IEEE Trans. Nucl. Sci. **38**, 379 (1991)
[10] G. Suft *et al.*, Nucl. Instrum. Meth. A **538**, 416 (2005)
[11] I.S. Barker, A. Donnachie, J.K. Storrow, Nucl. Phys. B **95**, 347 (1975)
[12] M. Gottschall *et al.*, Phys. Rev. Lett. **112**, 012003 (2014)
[13] J. Hartmann *et al.*, Phys. Lett. B **748**, 212 (2015)
[14] A. Thiel *et al.*, Phys. Rev. Lett. **109**, 102001 (2012)
[15] A. Thiel *et al.* (2016), `arXiv:1604.02922`
[16] D. Drechsel, S.S. Kamalov, L. Tiator, Eur. Phys. J. A **34**, 69 (2007)
[17] R.L. Workman, M.W. Paris, W.J. Briscoe, I.I. Strakovsky, Phys. Rev. C **86**, 015202 (2012)
[18] A.V. Anisovich *et al.*, Eur. Phys. J. A **48**, 15 (2012)
[19] D. Rönchen *et al.*, Eur. Phys. J. A **50**, 101 (2014)
[20] J. Hartmann *et al.*, Phys. Rev. Lett. **113**, 062001 (2014)
[21] D. Rönchen *et al.*, Eur. Phys. J. A **51**, 70 (2015)
[22] A. Anisovich *et al.*, Eur. Phys. J. A **52**, 284 (2016)
[23] A. Bock *et al.*, Phys. Rev. Lett. **81**, 534 (1998)
[24] C.S. Akondi *et al.*, Phys. Rev. Lett. **113**, 102001 (2014)
[25] E.F. McNicoll *et al.*, Phys. Rev. C **82**, 035208 (2010)
[26] V. Shklyar, H. Lenske, U. Mosel, Phys. Rev. C **87**, 015201 (2013)
[27] I. Senderovich *et al.*, Phys. Lett. B **755**, 64 (2016)
[28] J. Müller, J. Hartmann, M. Grüner *et al.*, publication in preparation (2016)
[29] V. Sokhoyan *et al.*, Phys. Lett. B **746**, 127 (2015)
[30] V. Sokhoyan *et al.*, Eur. Phys. J. A **51**, 95 (2015)
[31] P. Mahlberg, T. Seifen, JPS Conf. Proc. **10**, 032012 (2016)

 DOI: 10.1051/epjconf/201613001012

The $\overline{\text{P}}$ANDA physics program

P. Gianotti[1,★]
for the $\overline{\text{P}}$ANDA Collaboration

[1] *Laboratori Nazionali di Frascati of INFN, via E. Fermi 40, 00044 Frascati, Italy*

Abstract. The understanding of the QCD in the non-perturbative regime, is one of the key issues to have a complete picture of strong interactions. Recent findings of new and unexpected resonances, with unresolved properties, show that the hadron spectrum is not yet completely understood. This is also underlined by the ongoing discussion on multi-quark states, and on other exotic states with gluonic degrees of freedom. The $\overline{\text{P}}$ANDA experiment, one of the biggest enterprises at the FAIR facility, aims at exploring this field thanks to the gluon rich environment offered by the annihilation of antiprotons. A general overview of the $\overline{\text{P}}$ANDA physics program is given in this paper.

1 Introduction

To have a complete picture of strong interactions, a better understanding of Quantum Chromo Dynamics (QCD) in the non-perturbative regime is necessary. To understand how quarks and gluonic degrees of freedom lead to hadronic states, and how hadrons get mass and all the properties we observe, still deserve a deeper study. These are the most burning questions of fundamental research, strongly linked to the issue about the existence of other forms of matter different from mesons and baryons (*i.e.* glueballs, hybrids, etc..). These unusual states have been theoretically predicted by QCD since its formulation, but experimentally their existence has been established only recently. Some of the so-called X, Y, Z states, that have been found by currently ongoing experiments, cannot be "conventional" antiquark-quark or three quark combinations. In order to understand their precise nature a campaign of dedicated measurements is desirable.

Antiproton-proton annihilation proceeds via two- or three-gluon processes, and this gluon-rich environment is perfect to investigate the nature of all kind of hadronic states. $\overline{\text{P}}$ANDA is a next generation hadron physics experiment whose name stands for antiProton ($\bar{p}$) ANnihilations in DArmstadt, and it will be located at the High Energy Storage Ring (HESR) of the Facility for Antiproton and Ion Research (FAIR) in Darmstadt (Germany) [1]. A state of the art, almost 4π fixed target detector (see Fig. 1), will be combined with a high intensity and high momentum resolution antiproton beam, in the momentum range from 1.5 to 15 GeV/c. $\overline{\text{P}}$ANDA will address at best many open and burning questions in the energy domain where QCD sees the transition from the perturbative to the non-perturbative regime.

The field of hadron physics is often subdivided into the following branches:

- hadron spectroscopy;

★e-mail: gianotti@lnf.infn.it

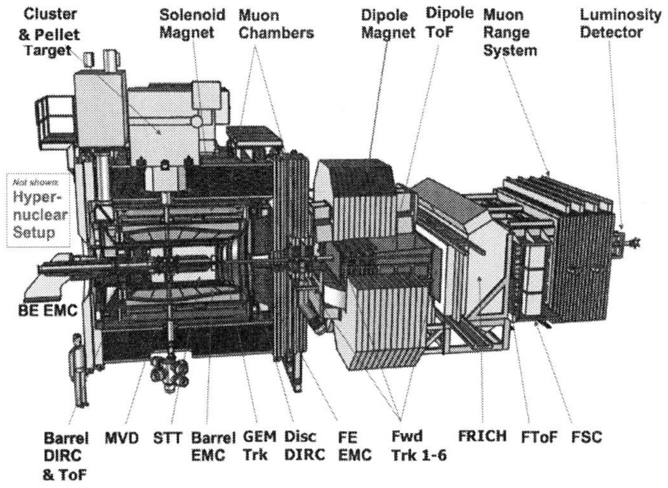

Figure 1. Schematic layout of the $\overline{\text{P}}$ANDA detector; a dedicated setup (not shown in the picture) will be added in the central region for hypernuclear physics measurements. More details about detector sub-components can be found in Ref. [2]

- hadron structure;
- interaction of hadrons.

The $\overline{\text{P}}$ANDA detector [2] has been especially designed, together with the HESR accelerator, to be a perfect instrument to shed new light on the complete list of items above mentioned. This multi-task approach is new in the study of strong interactions and promises to deliver world class results.

In the following sections, more details on some topics of the $\overline{\text{P}}$ANDA physics program are given. A complete overview of the scientific program of the $\overline{\text{P}}$ANDA experiment, with detailed simulation results obtained with a realistic layout of the detector, can be found in Ref. [3].

2 Understanding QCD by means of antiprotons

$\overline{\text{P}}$ANDA aims to perform a vast program of hadron spectroscopy of strange and charmed states to firmly identify exotics. Experimentally, the search for exotics in the light quark sector is a challenging task due to their mixing with nearby ordinary states having the same quantum numbers. In this respect, the charmonium energy region is favorable because the density of states is lower and also their widths are narrower. This is probably the reason why the first unambiguous exotic candidates have been detected in this energy region. To draw definitive conclusions, it is important to map out the entire spectrum of states and to study many different decay channels and different production processes. $\overline{\text{P}}$ANDA will complement the efforts at facilities like JLab, BESIII, COMPASS and J-PARC by utilizing the environment of the antiproton-proton annihilation.

In the spectroscopy sector, antiproton-proton annihilations enable two modes to investigate final states (see Fig. 2):

- **Formation:** a single resonance is directly produced in the annihilation process, in this case the J^{PC} quantum numbers accessible to a fermion-antifermion pair can be obtained;

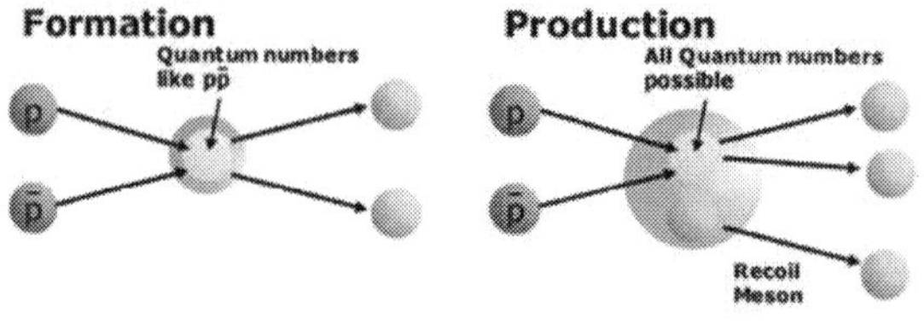

Figure 2. The cartoon is showing the two different production mechanisms by which a certain final state (yellow ball) can be studied via $\bar{p}p$ annihilation: **Formation** and **Production** (see the text for more details).

- **Production:** at least one additional particle is produced together with the one under study. Here restrictions on J^{PC} do not apply.

The comparison of the two production mechanisms helps to classify resonances and to identify those with an exotic nature. Furthermore, the high momentum resolution of the antiprotons achievable in the HESR ($\Delta p/p$ up to 2×10^{-5}), will allow mass resolutions better than $\Delta m/m < 10^{-4}$ that cannot be obtained in the production via the decay chain of heavier particles, where the detector resolution becomes a limiting factor. This feature will allow to measure precisely the widths and the line shapes of all poorly known conventional and exotic states in the centre of mass (c.m.) range from 2.2 to 5.5 GeV/c^2.

To give an example of the possibilities of the $\overline{\text{P}}$ANDA experiment, let's consider the $X(3872)$ resonance. This is the first of the new resonances in the series of unresolved states in the charmonium energy domain discovered by the Belle collaboration in 2003 [4] and then confirmed by many other experiments. Assuming a cross section of 100 nb for the channel $\bar{p}p \rightarrow X(3872)$, $\overline{\text{P}}$ANDA could produce about 860.000 events/day when the HESR will reach the top luminosity, but already at the initial luminosity of $\mathcal{L} = 10^{31}\text{cm}^{-2}\text{s}^{-1}$, foreseen for the FAIR starting phase (MSV0-3), the number of produced $X(3872)$ per day will still be about 43.000. No other running or planned experiment could have such a rate.

Another peculiar aspect of the hadronic studies performed via $\bar{p}p$ annihilations, resides in the unique possibility to measure precisely the resonance parameters, namely the mass M, the width Γ, and the line shape. These can be extracted by measuring the formation rate for that resonance as a function of the c.m. energy [3]. As an illustration of this technique, Fig. 3 shows a scan of the χ_{c1} resonance carried out at the Fermilab antiproton accumulator by the E835 experiment [5] using the process $\bar{p}p \rightarrow \chi_{c1} \rightarrow J/\psi\gamma$. The value obtained for the resonance width $\Gamma = 0.881 \pm 0.052 \pm 0.026$ MeV is still the best ever reached.

The beam energy distribution available at the HESR will be a factor from two to ten times better than that available at Fermilab, allowing to measure widths down to tens of keV. Again, in the case of the $X(3872)$ where there is only an upper limit for the width ($\Gamma < 1.2$ MeV/c^2 [6]), $\overline{\text{P}}$ANDA will be the only experiment able to perform a precise determination of this important parameter. The vicinity of the $X(3872)$ to the value of the sum of the masses $M_{D^0} + M_{D^{*0}}$ as well as the relative decay magnitudes to charmonium and open charm channels has inspired several models interpreting the state either as a loosely bound $D^0 - D^{*0}$ molecule or a virtual scattering state effectively created by threshold dynamics. The line shape of the state differs dramatically depending on the hypothesis, therefore a precise determination of the resonance line shape, will allow to disentangle between the two possibilities.

Complementary information on QCD dynamics can also be obtained from the measurement of the excitation spectrum of strange and charmed baryons. Baryons with strangeness extend the study

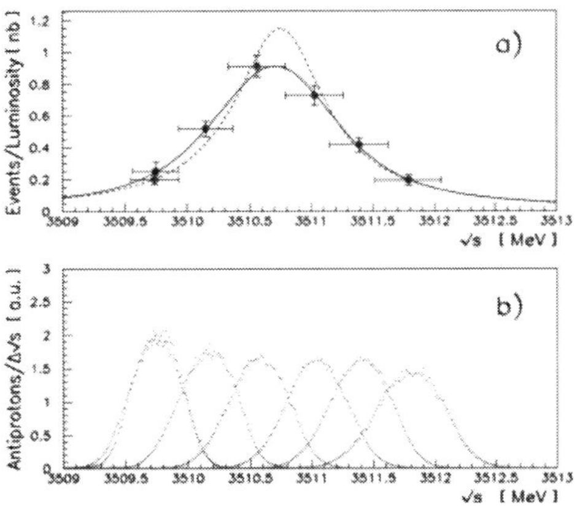

Figure 3. Resonance scan at the χ_{c1} carried out at Fermilab (a) and the beam energy distribution in each data point (from Ref. [5]).

of the nucleon structure to the three-flavor sector and via their self-analyzing weak decays one can access to spin properties of baryons. $\overline{\text{P}}$ANDA has unique opportunities in this field since the cross section for baryon-antibaryon final states, both ground state and excited, is large. Therefore, already in the starting phase of the experiment it would be possible to study the excitation spectrum of Ξ and Ω baryons.

The high production rate of hyperon pairs in antiproton annihilations will also allow to carry on a novel program at the border with nuclear physics: that of doubly-strange nuclear systems. The goal is to expand the nuclear chart in the strangeness dimension, thereby providing data that would help to study two- and many-body baryon-baryon forces that have astrophysical implications *i.e.* in the modelling of neutron stars. $\overline{\text{P}}$ANDA focuses on systems with strangeness $S = -2$: Ξ–atoms and $\Lambda - \Lambda$ hypernuclei, for which high resolution gamma spectroscopy will be performed for the first time. Details and plans for this research field can be found in Ref. [7].

3 Hadron's structure functions

Nucleon structure investigations have been performed since long time using electron beams. Electron scattering allows to access nucleon structure matrix elements in the region of negative momentum transfer of the intermediate virtual photon (space-like region) for many processes like elastic scattering or deep inelastic exclusive or inclusive scattering. The availability of a high intense antiproton beam annihilating on a proton target in combination with the $\overline{\text{P}}$ANDA detector offers the unique opportunity to extend these investigations to the region of positive momentum transfer of the photon (time-like region). Due to its analyticity, space-like and time-like observables are intimately connected by the application of dispersion relations. Perturbative QCD makes predictions for the large q^2 behavior of the connection between space-like and time-like regions that would be worth testing. Furthermore, recent measurements performed at JLab [8] using the polarization transfer and target asymmetry method, have shown that the ratio of the electric (G_E) and the magnetic (G_M) components of the proton form factors $R = |G_E|/|G_M|$ deviates from unity. This is inconsistent with the results derived from the use of the Rosenbluth separation technique. This surprising result has reopened the question on the determination of G_E and G_M, which in the time-like domain are complex functions. It is important to check whether the behaviour of G_E and G_M in the two domains are coherent.

The extension of the form factor measurements to the time-like region and the separate measurement of G_M and G_E can be improved almost of one order of magnitude compared to existing world data by using the antiproton beam at HESR. Figure 4 shows the statistical precision on the form factor ratio obtained using a detailed Monte Carlo simulation compared to the world database. $\overline{\text{P}}$ANDA will improve the precision on this ratio at lower q^2 in comparison to other experiments, and will provide additional measurements in a higher energy region [9]. Furthermore, the $\overline{\text{P}}$ANDA experiment offers

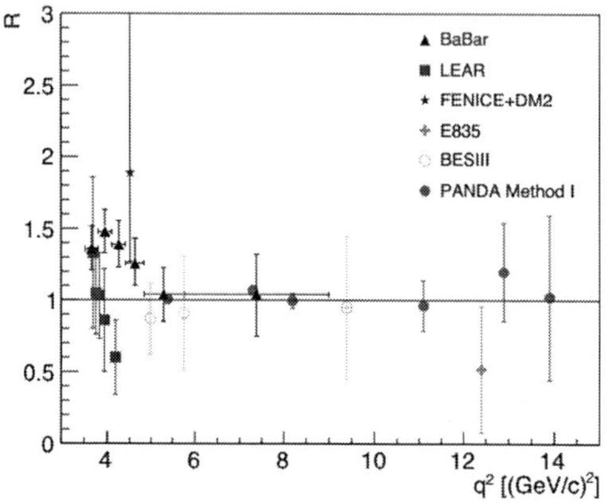

Figure 4. World data set of measurements of the ratio $R = |G_E|/|G_M|$ performed in the time-like region. Monte Carlo simulations of expected $\overline{\text{P}}$ANDA results are also shown (Details can be found in Ref. [9]).

a unique chance to determine the moduli of the complex form factors in the time-like domain, by measuring the angular distribution of the process $\bar{p}p \rightarrow e^+e^-$ in a q^2 range from about 5 (GeV/c)^2 up to 14 (GeV/c)^2. The $\overline{\text{P}}$ANDA detector will provide unprecedented luminosity and rich particle identification capabilities, which are necessary in order to discriminate against the very large hadronic background which is of the order of 10^6 times higher in cross section. Any possible two-photon exchange contribution in the time-like domain, which is regarded to be one of the radiative correction processes responsible for the discrepancy between the Rosenbluth and the polarization transfer methods in the space-like regime, can be detected in the same measurement. This effect will introduce a forward-backward asymmetry in the angular distribution which otherwise is symmetric if only one-photon exchange holds. Moreover, $\overline{\text{P}}$ANDA could access for the first time proton form factors in the muonic channel $\bar{p}p \rightarrow \mu^+\mu^-$, and make measurements in the unphysical region below the threshold through the reaction $\bar{p}p \rightarrow e^+e^-\pi^0$ at low beam energies, where the cross section is high.

Not only form factors, but also a wide set of electromagnetic observables which have been only studied in space-like momentum-transfer range can be addressed at $\overline{\text{P}}$ANDA: Transition Distribution Amplitudes via lepton pair production with an associated meson; Generalized Distribution Amplitudes with hard exclusive processes like Wide Angle Compton Scattering; Transverse Momentum Parton Distribution Functions via the measurement of final state asymmetries in Drell-Yan Production are in the plans of the experiment. For all the cases mentioned above, a $10^5 - 10^6$ times larger hadronic background is expected. Therefore feasibility studies of all signal channels are ongoing. The results are extremely promising thanks to the large solid angle and the PID-capabilities of the $\overline{\text{P}}$ANDA detector that helps to suppress this hadronic background.

4 Data acquisition strategy

$\bar{\text{P}}$ANDA will be a fixed target experiment to measure $\bar{p}p$ and $\bar{p}A$ collisions up to rate of $2 \cdot 10^7$ events/s. To achieve this goal, it will need a data acquisition concept matching the complexity of the experiment and the diversity of physics objectives. Every sub-detector unit will be a self-triggering entity. Signals will be detected autonomously by each sub-systems and pre-processed in order to transmit only the physically relevant information. The final event selection will occur in computing nodes which first isolate events, and then filter physical signatures of interest. This new concept will provide a high degree of flexibility in the development of trigger algorithms, and it will allow trigger conditions which are outside the capabilities of the standard approaches.

5 Summary

The $\bar{\text{P}}$ANDA research program at FAIR will explore a wide range of important questions for the understanding of strong interactions.

The $\bar{\text{P}}$ANDA collaboration aims to connect the perturbative and the non-perturbative QCD regions by means of high precision measurements that cannot be obtained elsewhere. The international collaboration ($\sim$ 400 scientists from all over the world) combines researchers from the nuclear and the particle physics communities that come together with the intention to have the most open-minded approach to the unsettled questions of subatomic physics. The examples given in the above sections represent only some highlights of the physics program that will be exploited. In fact, thanks to the completeness and versatility of the $\bar{\text{P}}$ANDA detector under construction, a wider spectrum of measurements would be possible. To accomplish at best this task, even the data acquisition system will be flexible and as much as possible unbiased, to allow for many measurements to be done in parallel. All these elements make $\bar{\text{P}}$ANDA to be more than a single experiment but rather as a universal facility for QCD physics.

References

[1] http://www.fair-center.de.

[2] https://panda.gsi.de/article/panda-detector-overview;
$\bar{\text{P}}$ANDA Electromagnetic Calorimeter, 2008, FAIR TDR 4_01, arXiv:0810.1216 [physics.ins-det];
$\bar{\text{P}}$ANDA Magnets, 2009, FAIR TDR 4_02, arXiv 0907.0169 [physics.ins-det];
$\bar{\text{P}}$ANDA Targets, 2012, FAIR TDR 4_05;
$\bar{\text{P}}$ANDA Micro Vertex Detector, 2012, FAIR TDR 4_03, arXiv:1207.6581v2 [physics.ins-det];
$\bar{\text{P}}$ANDA Straw Tube Tracker, 2012, FAIR TDR 4_04, Eur. Phys. J. **A49**, 25 (2013);
$\bar{\text{P}}$ANDA Muon System, 2013, FAIR TDR 4_06;
$\bar{\text{P}}$ANDA Forward Spectrometer Calorimeter, 2015, FAIR TDR 4_08.

[3] W. Erni *et al.* ,"Physics Performance Report for $\bar{\text{P}}$ANDA, Strong Interaction Studies with Antiprotons", arXiv:093.3905 [hep-ex].

[4] S.-K. Choi *et al.*, Phys. Rev. Lett. **91**, 26 (2003).

[5] M. Andreotti *et al.* , Nucl. Phys. **B717**, 34 (2005).

[6] S.-K. Choi *et al.*, Phys. Rev. **D84**, 052004 (2011).

[7] B. Singh *et al.*, Nucl. Phys. **A954**, 323 (2016).

[8] I. A. Qattan *et al.*, Phys. Rev. Lett. **94**, 142301 (2005).

[9] B. Singh *et al.*, arXiv:1606.01118 [hep-ex], in press on EPJ A.

EPJ Web of Conferences **130**, 01013 (2016)
MESON 2016
DOI: 10.1051/epjconf/201613001013

Hadronic inputs to the $(g-2)_\mu$ puzzle

C. F. Redmer[1,⋆]

[1] *Institut für Kernphysik, Johannes Gutenberg-Universität Mainz, J.-J.-Becher-Weg 45, 55128 Mainz, Germany*

Abstract. There is a long standing discrepancy of 3 – 4 standard deviations between the direct measurement and the Standard Model prediction of the anomalous magnetic moment of the muon $(g-2)_\mu$. While new direct measurements have been proposed to clarify the situation, theory calculations are completely limited by the accuracy of the hadronic contributions to $(g-2)_\mu$. In order to increase their precision, experimental information can be used as input. In this presentation we will discuss recent and future measurements of relevant hadronic cross sections and transition form factors.

1 Introduction

The anomalous magnetic moment of the muon, or rather the muon anomaly $a_\mu = (g-2)_\mu/2$, is one of the most precisely measured observables of the Standard Model. It has been measured directly in storage ring experiments since the 1960s with continuously increasing accuracy. The latest measurement was performed in Brookhaven and yielded $a_\mu^{exp} = (11659208.9 \pm 6.3) \times 10^{-10}$ [1]. In the experiment, a beam of longitudinally polarized muons was circulated in a storage ring with a precisely known, constant magnetic field. The polarization of the muons is achieved through the parity violating decay $\pi^+ \rightarrow \mu^+\nu_\mu$ of previously produced pions. The muons circulate in the magnetic field $\vec{B}$ of the storage ring with the cyclotron frequency $\vec{\omega}_c$ and their spin precesses with the frequency $\vec{\omega}_s$, resulting in the anomalous precession frequency

$$\vec{\omega}_a = \vec{\omega}_s - \vec{\omega}_c = -\frac{q}{m_\mu}\left[a_\mu\vec{B} - \left(a_\mu\frac{1}{\gamma^2-1}\right)\frac{\vec{\beta}\times\vec{E}}{c}\right].$$

The dependence on $\vec{\omega}_a$ on the electrical field $\vec{E}$ can be canceled by the choice of the value of $\gamma = 29.3$, the so called "magic momentum" of the muons of 3.094 GeV/c. The muons eventually decay through their weak decay $\mu^+ \rightarrow e^+\bar{\nu}_\mu\nu_e$, where the direction of the emitted positron is correlated with the spin direction of the muon. Thus, the e^+ rate is modulated with $\vec{\omega}_a = -a_\mu\frac{q\vec{B}}{m_\mu}$ and is used to determine a_μ.

Two new measurements have been proposed at Fermilab, Chicago, USA [2] and at J-PARC, Tokai, Japan [3], which aim at a fourfold improvement of the current experimental precision. While the Fermilab experiment is a continuation of the BNL experiments with major improvements to the apparatus, the J-PARC approach does not rely on the "magic momentum", but avoids the need of focusing electric fields by using an ultra-cold muon beam. Thus, the two experiments are fully complementary.

⋆e-mail: redmer@uni-mainz.de

The motivation for new direct measurements comes from a difference of three to four standard deviations between the direct measurement and the Standard Model prediction of a_μ, which might be considered as a hint for New Physics. The theory prediction of the anomalous magnetic moment of the muon consists of three parts. The largest part comes from the QED contribution a_μ^{QED}. It has been determined with high accuracy in perturbation theory up to the 10th order [4]. Another contribution comes from the weak interaction a_μ^{weak}. It is small and its uncertainty is well under control by means of perturbation theory [5]. The third contribution stems from the strong interaction a_μ^{had}. At the relevant energies, perturbative approaches cannot be applied to determine the strong contribution. Thus, though its absolute value is comparably small, the uncertainty of a_μ^{had} completely dominates the total uncertainty of the Standard Model prediction of a_μ [6, 7]. However, experimental information can be used as input to the calculations in order to improve the prediction.

Currently, the largest contribution to the absolute value and to the uncertainty of a_μ^{had} comes from the hadronic vacuum polarization. It can be related to hadronic cross sections measured in e^+e^- annihilations, which can be used to systematically improve the prediction. The second largest contribution to a_μ^{had} comes from the hadronic Light-by-Light scattering. In contrast to the hadronic vacuum polarization, it cannot be directly related to experimental observables. So far, hadronic models have been used to evaluate the contribution, using experimental data on transition form factors of light mesons for validation. The absolute value of the hadronic Light-by-Light contribution is comparably small. Its uncertainty is model dependent and can be as large as the uncertainty due to the hadronic vacuum polarization.

2 The hadronic Vacuum Polarization

The hadronic vacuum polarization can be directly related to the experimentally measurable hadronic cross sections $\sigma_{e^+e^-\to had}$ at e^+e^- colliders. The contribution to the anomalous magnetic moment of the muon can be calculated in leading order with the dispersion integral

$$a_\mu^{hVP,LO} = \int ds K(s)\sigma_{e^+e^-\to had}(s),$$

where $K(s)$ is the Kernel function, which is proportional to $1/s$ [8]. The energy dependence of the hadronic cross sections and of the Kernel function make cross sections measured at low center of mass energies most important for $a_\mu^{hVP,LO}$ [7]. In fact, the evaluation of the dispersion integral shows that $a_\mu^{hVP,LO}$ is determined to more than 75% by the hadronic cross sections measured below 1 GeV/c^2, which are dominated by the ρ-resonance and its decay into two pions. For the determination of the uncertainty $\delta a_\mu^{hVP,LO}$ the cross sections of higher pion multiplicities, like $e^+e^- \to \pi^+\pi^-\pi^0$ or $e^+e^- \to \pi^+\pi^-\pi^0\pi^0$, and the production of kaon pairs play an almost equally important role. Thus, the measurement of these hadronic cross sections with high accuracy allows to systematically improve the theory prediction of a_μ^{had} with experimental input.

Two methods have been established to measure hadronic cross sections at e^+e^- machines. Traditionally, the energy dependence of hadronic cross sections is studied in energy scan experiments, where the accelerators have to be adjusted for the measurement of every individual data point. Alternatively, events with Initial State Radiation (ISR) can be exploited. The irradiated photon reduces the effective center-of-mass energy, which is available for the reaction and, thus, allows to study processes at energies different than the nominal energy of the machine. This method has proven to be very effective at high luminosity machines, which run at fixed energies, like ϕ, τ-charm, and B factories.

As the cross section $\sigma(e^+e^- \to \pi^+\pi^-)$ dominates $a_\mu^{hVP,LO}$, precise knowledge is mandatory. In the past, it has been measured by several experiments over a wide energy range. Reference [7] provides a compilation of the relevant measurements. The most precise measurements with systematic uncertainties of less than 1% have been provided from the BaBar collaboration [9, 10], the KLOE collaboration [11–13], and the CMD, CMD-2, and SND collaborations [14–19]. Since the accuracy of the Novosibirsk experiments is currently limited by statistics, the world average is dominated by the BaBar and KLOE results. However, even though both experiments claim a total uncertainty of less than 1%, their results differ by more than 5% at and above the ρ-peak region. The difference reflects directly in the uncertainty of $a_\mu^{hVP,LO}$.

Recently, the BESIII collaboration measured $\sigma(e^+e^- \to \pi^+\pi^-)$ in the mass region between 600 and 900 MeV/c^2[20]. As the mass region contains the dominating ρ-peak, it makes up for 70% of the total $\pi^+\pi^-$ contribution and determines $a_\mu^{hVP,LO}$ to 50%. The BESIII measurement is based on 2.9 fb^{-1} of data taken at the $\psi(3770)$ resonance and exploits the ISR technique to determine the cross section at lower energies. The dominating background from radiative di-muon production is successfully suppressed by applying an artificial neural network for particle identification. The careful evaluation of all systematic uncertainties results in a total uncertainty of the cross section measurement of 0.9%. Fig. 1 shows the comparison of $a_\mu^{hVP,LO}$ evaluated in the mass range of this measurement with the corresponding results from BaBar and KLOE. Good agreement with the KLOE measurements is observed, while the BaBar result is shifted to higher values of $a_\mu^{hVP,LO}$. The BESIII result confirms the deviation between the direct measurement of a_μ and its Standard Model prediction on the level of three – four standard deviations.

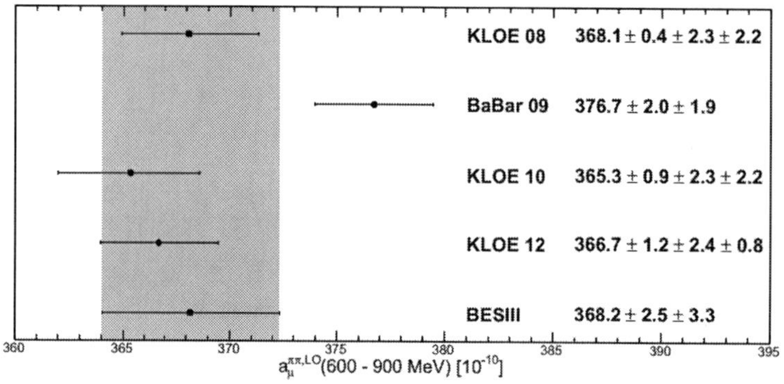

Figure 1. Comparison of $a_\mu^{\pi\pi,LO}(600 - 900\text{MeV})$ from the measurements of KLOE, BaBar, and BESIII. The yellow band shows the error of the BESIII result. Error bars include statistical and systematical errors. Figure is taken from Ref. [20].

3 The hadronic Light-by-Light scattering

Improving the accuracy of the hadronic Light-by-Light scattering contribution to the muon anomaly a_μ^{hLbL} is a more complex endeavour. It cannot be related directly to a measurable quantity. The interactions of real and virtual photons with virtual mesons, which are part of a_μ^{hLbL} can be described at lowest energies by means of chiral perturbation theory and at highest energies by means of perturbative QCD. The relevant energy scale for the muon anomaly, however, is the intermediate energy regime, where perturbative approaches fail.

The classic approaches to evaluate a_μ^{hLbL} are based on hadronic models. The two calculations mostly referred to are the so-called "Glasgow Consensus" $a_\mu^{hLbL} = (10.5 \pm 2.6) \times 10^{-10}$ [21] and the more conservative estimate of Jegerlehner and Nyffeler $a_\mu^{hLbL} = (11.6 \pm 4.0) \times 10^{-10}$ [6]. The difference in the uncertainties δa_μ^{hLbL} is model dependent. Experimental input is used for the validation of the models. The relevant observables are transition form factors (TFF), which describe the coupling of photons and hadrons.

The information for which mesons the TFF should be measured in which energy range can be deduced from theoretical evaluations of the hadronic Light-by-Light term. A counting scheme proposed by de Rafael [22] shows that pion loops and pseudoscalar meson exchange are the dominating contributions to a_μ^{hLbL}. Thus, data on the TFF of single π^0 and of pion pairs are needed. Knecht and Nyffeler provide a two-dimensional integral representation of the π^0 exchange contribution to the hadronic Light-by-Light term [23]. In the representation, the term factorizes into so-called universal weighting functions $w_i(Q_1, Q_2)$, which contain the model independent parts of a_μ^{hLbL,π^0}, and functions $f_i(Q_1, Q_2)$, which contain the form factor information and, thus, the model dependence of different approaches. From the momentum dependence of the $w_i(Q_1, Q_2)$, it has been shown for VMD-like models that for the determination of a_μ^{hLbL} it is most relevant to determine the TFF of π^0 for space-like momentum transfers between $0.05 \le Q^2[\mathrm{GeV}^2] \le 1.5$.

Experimentally, space-like momentum transfers are accessible in two-photon reactions at e^+e^- colliders. In contrast to the annihilation reaction, the exchange of two photons in the collision allows for the direct production of pseudoscalar, scalar, and tensor states. The cross section for the reaction is directly proportional to the TFF of the produced mesons. The measurement of two-photon reactions is challenging since the cross section is peaked towards small scattering angles of the lepton. To study TFFs at arbitrary values of momentum transfers would require special tagging detectors, which are installed close to the beam pipes. In conventional detector setups at e^+e^- machines these detectors are not available. A trade-off can be found in single-tag measurements, where only one of the two scattered leptons and the produced meson are reconstructed. By energy and momentum conservation the unmeasured lepton is required to be scattered along the beam direction and, thus, to have very small momentum transfer. Instead of a TFF depending on two arbitrary momentum transfers $F(Q_1^2, Q_2^2)$, the dependence on only one momentum transfer $F(Q^2)$ is studied. The information is still valuable for a_μ^{hLbL}, as the TFF should factorize for small energies as $F(Q_1^2, Q_2^2) \sim F(Q_1^2, 0) \times F(0, Q_2^2)$.

Currently, the world data base on the space-like TFF of the π^0 is dominated by the recent measurements of the B-factories, BaBar and Belle [24, 25]. The results cover a wide range of momentum transfers of $4 \le Q^2[\mathrm{GeV}^2] \le 40$, and show a discrepancy, which is often referred to as "Belle-BaBar Puzzle". For the evaluation of the hadronic Light-by-Light scattering these data are, however, less important as the provided information is well outside of the relevant region of momentum transfer. At lower momentum transfers only the results from CELLO [26] and CLEO [27] are available. They provide information of the TFF down to $Q^2 = 0.5$ GeV^2. However, they suffer from low statistics.

New measurements of space-like TFF have been announced by the BESIII and KLOE-2 collaborations. At BESIII, 2.9 fb^{-1} of data taken at $\sqrt{s} = 3.773$ GeV/c^2 are used to measure the TFF of π^0 at momentum transfers of $0.3 \le Q^2[\mathrm{GeV}^2] \le 3.1$ [28, 29]. Estimates based on Monte Carlo studies suggest an unprecedented statistical accuracy for $Q^2 > 1.5$ GeV^2 and competitive results at higher momentum transfers. The KLOE-2 experiment has been equipped with special tagging detectors, which allow to measure the space-like TFF for $0.01 \le Q^2[\mathrm{GeV}^2] \le 0.1$ [30]. The ongoing data taking aims at collecting 5 fb^{-1} at the peak of the ϕ resonance, which allows for a statistical accuracy of the TFF measurement of 6%.

Recently, the Belle collaboration reported the first single-tag measurement of neutral pion pairs [31]. The $\pi^0\pi^0$ system was studied for momentum transfers between $3 \le Q^2[\mathrm{GeV}^2] \le 30$, for invariant masses of the pion pairs between $0.5 \le m_{\pi^0\pi^0}[\mathrm{GeV/c}^2] \le 2.1$, and with complete coverage of the helicity angle. In a partial wave analysis the $|D_0|^2$, $|D_2|^2$, and $|S|^2$ wave amplitudes have been determined, and the TFF of the $f_0(980)$ resonance and the three possible helicity states of the $f_2(1270)$ resonance were measured and compared to different predictions [32, 33]. Similar to the measurement of the π^0 TFF by the Belle collaboration, this measurement does not provide information in a Q^2 region relevant for a_μ^{hLbL}. The measurement of the $\pi^+\pi^-$ and $\pi^0\pi^0$ pair production in $\gamma\gamma$ collisions announced by the BESIII collaboration [29] will provide information at $0.2 \leq Q^2[\text{GeV}^2] \leq 2$, for values of $m_{\pi^0\pi^0}$ threshold to 2 GeV/c^2, and with full coverage of the helicity angle.

Despite of the existing and upcoming measurements there remains the issue of model dependence in the determination of a_μ^{hLbL}. In the recent years data driven approaches have been proposed in order to reduce the model dependence. Escribano, Masjuan and colleagues proposed to parameterize the TFFs with Padé approximants and fit the free parameters to data [34, 35]. The method provides an estimate of the systematic uncertainty and allows for space-like and time-like data to be used in order to obtain an analytic description of the TFF, which can be used in the calculations of a_μ^{hLbL}.

The development of another approach has recently been started by theory groups in Bern [36–39] and Mainz [40, 41]. The aim is to provide a data driven estimate of a_μ^{hLbL} based on dispersion relations. The approaches of the two groups are fully complementary in the way they relate the dominating contributions of a_μ^{hLbL} to measurable quantities. The goal is to achieve a reliable error estimate in the order of 10% to 20%. The input from experiments needed in the approaches are the space-like TFF of π^0 for arbitrary virtualities $F_{\pi\gamma^*\gamma^*}(Q_1^2, Q_2^2)$ and the partial waves of the two-pion system in $\gamma^*\gamma^* \to \pi\pi$. Since the currently available data cannot provide the required information, the Bern group proposed a scheme to reconstruct the TFF and the partial waves dispersively from available time-like information of hadronic and Dalitz decays of vector mesons and hadronic cross sections in e^+e^- reactions [37]. Among the various available experimental results, which can be evaluated in this reconstruction, also the recent A2 result on $\omega \to \pi^0 e^+e^-$ [42] and the KLOE-2 result on $\phi \to \pi^0 e^+e^-$ [43] become relevant. The latter does not only provide a significant improvement on the precision of the branching ratio, but also provides the first measurement of the time-like TFF.

4 Summary

The current limitation of the Standard Model prediction of a_μ comes from the uncertainties of the hadronic contributions. The dispersive approach to systematically improve the contribution of the hadronic vacuum polarization a_μ^{hVP} requires precision data on hadronic cross sections. The recent measurement of $\sigma(e^+e^- \to \pi^+\pi^-)$ by BESIII has yet to be included into the world average. Further high precision measurements of the relevant hadronic cross sections will be provided by the BaBar, KLOE, BESIII, and CMD-3 collaborations.

In order to achieve a reliable error estimate for the hadronic Light-by-light contribution, dispersive approaches have been proposed, which relate the dominating contributions to a_μ^{hLbL} to experimental results on the TFF of π^0 and $\pi\pi$ partial waves. New measurements by KLOE, BESIII, and Belle will provide the data. In the meantime a procedure has been proposed to dispersively reconstruct the information from other, already available experimental inputs. In view of the announced precision of the new direct measurements of a_μ similar efforts have to be made for the contributions of the η and η' meson exchange in a_μ^{hLbL}, which are of the same size [23].

It is considered as a realistic outcome of the ongoing efforts to reduce the uncertainty of the Standard Model prediction of a_μ by a factor two in the near future [44].

References

[1] G.W. Bennett *et al.* (Muon g-2), Phys. Rev. D **73**, 072003 (2006)
[2] G. Venanzoni (Muon g-2), Nucl. Phys. Proc. Suppl. **225-227**, 277 (2012)
[3] T. Mibe (J-PARC g-2), Nucl. Phys. Proc. Suppl. **218**, 242 (2011)

[4] T. Aoyama, M. Hayakawa, T. Kinoshita, M. Nio, Phys. Rev. Lett. **109**, 111808 (2012)
[5] C. Gnendiger, D. Stöckinger, H. Stöckinger-Kim, Phys. Rev. D **88**, 053005 (2013)
[6] F. Jegerlehner, A. Nyffeler, Phys. Rept. **477**, 1 (2009)
[7] M. Davier, A. Hoecker, B. Malaescu, Z. Zhang, Eur. Phys. J. C **71**, 1515 (2011), [Erratum: Eur. Phys. J.C72,1874(2012)]
[8] S.J. Brodsky, E. De Rafael, Phys. Rev. **168**, 1620 (1968)
[9] B. Aubert *et al.* (BaBar), Phys. Rev. Lett. **103**, 231801 (2009)
[10] J. Lees *et al.* (BaBar), Phys. Rev. D **86**, 032013 (2012)
[11] F. Ambrosino *et al.* (KLOE), Phys. Lett. B **670**, 285 (2009)
[12] F. Ambrosino *et al.* (KLOE), Phys. Lett. B **700**, 102 (2011)
[13] D. Babusci *et al.* (KLOE), Phys. Lett. B **720**, 336 (2013)
[14] L.M. Barkov *et al.*, Nucl. Phys. B **256**, 365 (1985)
[15] R.R. Akhmetshin *et al.* (CMD-2), Phys. Lett. B **578**, 285 (2004)
[16] V.M. Aulchenko *et al.* (CMD-2), JETP Lett. **82**, 743 (2005)
[17] R.R. Akhmetshin *et al.*, JETP Lett. **84**, 413 (2006)
[18] R.R. Akhmetshin *et al.* (CMD-2), Phys. Lett. B **648**, 28 (2007)
[19] M.N. Achasov *et al.*, J. Exp. Theor. Phys. **103**, 380 (2006)
[20] M. Ablikim *et al.* (BESIII), Phys. Lett. B **753**, 629 (2016)
[21] J. Prades, E. de Rafael, A. Vainshtein, Adv. Ser. Direct. High Energy Phys. **20**, 303 (2009)
[22] E. de Rafael, Phys. Lett. B **322**, 239 (1994)
[23] M. Knecht, A. Nyffeler, Phys. Rev. D **65**, 073034 (2002)
[24] B. Aubert *et al.* (BaBar), Phys. Rev. D **80**, 052002 (2009)
[25] S. Uehara *et al.* (Belle), Phys. Rev. D **86**, 092007 (2012)
[26] H. Behrend *et al.* (CELLO), Z. Phys. C **49**, 401 (1991)
[27] J. Gronberg *et al.* (CLEO), Phys. Rev. D **57**, 33 (1998)
[28] C.F. Redmer (BESIII), EPJ Web Conf. **72**, 00021 (2014)
[29] C.F. Redmer (BESIII), Proceedings, 10th International Workshop on e+e- collisions from Phi to Psi (PHIPSI15): Hefei, China, September 23-26 (2015)
[30] D. Babusci, H. Czyz, Gonnella *et al.*, Eur. Phys. J. C **72**, 1917 (2012)
[31] M. Masuda *et al.* (Belle), Phys. Rev. D **93**, 032003 (2016)
[32] G.A. Schuler, F.A. Berends, R. van Gulik, Nucl. Phys. B **523**, 423 (1998)
[33] V. Pascalutsa, V. Pauk, M. Vanderhaeghen, Phys. Rev. D **85**, 116001 (2012)
[34] P. Masjuan, Phys. Rev. D **86**, 094021 (2012)
[35] R. Escribano, P. Masjuan, P. Sanchez-Puertas, Eur. Phys. J. C **75**, 414 (2015)
[36] G. Colangelo, M. Hoferichter, M. Procura, P. Stoffer, JHEP **1409**, 091 (2014)
[37] G. Colangelo, M. Hoferichter, B. Kubis *et al.*, Phys. Lett. B **738**, 6 (2014)
[38] M. Hoferichter, B. Kubis, S. Leupold *et al.*, Eur. Phys. J. C **74**, 3180 (2014)
[39] G. Colangelo, M. Hoferichter, M. Procura, P. Stoffer, JHEP **09**, 074 (2015)
[40] V. Pauk, M. Vanderhaeghen, Phys. Rev. D **90**, 113012 (2014)
[41] V. Pauk, M. Vanderhaeghen (2014), 1403.7503
[42] P. Adlarson *et al.* (2016), 1609.04503
[43] A. Anastasi *et al.* (KLOE-2), Phys. Lett. B **757**, 362 (2016)
[44] T. Blum, A. Denig, I. Logashenko, E. de Rafael, B. Lee Roberts, T. Teubner, G. Venanzoni (2013), 1311.2198

Overview of the CMD-3 recent results at e^+e^- collider VEPP-2000

G. V. Fedotovich[1,2,⋆], R. R. Akhmetshin[1], A. V. Anisenkov[1,2], V. M. Aulchenko[1,2], V. S. Banzarov[1], N. S. Bashtovoy[1], D. E. Berkaev[1,2], A. V. Bragin[1], S. I. Eidelman[1,2], D. A. Epifanov[1,5], L. B. Epshteyn[1,3], A. L. Erofeev[1,2], S. E. Gayazov[1,2], A. A. Grebenuk[1,2], D. N. Grigoriev[1,2,3], E. M. Gromov[1], F. V. Ignatov[1], V. L. Ivanov[1,2], S. V. Karpov[1], V. F. Kazanin[1,2], B. I. Khazin[1,2], A. N. Kirpotin[1], I. A. Koop[1,2], O. A. Kovalenko[1,2], A. N. Kozyrev[1], E. A. Kozyrev[1,2], P. P. Krokovny[1,2], A. E. Kuzmenko[1,3], A. S. Kuzmin[1], I. B. Logashenko[1,2], P. A. Lukin[1,2], A. P. Lysenko[1], K. Yu. Mikhailov[1,2], V. S. Okhapkin[1], Yu. N. Pestov[1], E. A. Perevedentsev[1,2], A. S. Popov[1,2], G. P. Razuvaev[1,2], Yu. A. Rogovsky[1], A. L. Romanov[1], A. A. Ruban[1], N. M. Ryskulov[1], A. E. Ryzhenenkov[1,2], V. E. Shebalin[1,2], D. N. Shemyakin[1,2], B. A. Shwartz[1,2], D. B. Shwartz[1,2], A. L. Sibidanov[4], P. Yu. Shatunov[1], Yu. M. Shatunov[1], E. P. Solodov[1,2], V. M. Titov[1], A. A. Talyshev[1,2], A. I. Vorobiov[1], Yu. V. Yudin[1], Yu. M. Zharinov[1]

[1]*Budker Institute of Nuclear Physics, SB RAS, Novosibirsk, 630090, Russia*
[2]*Novosibirsk State University, Novosibirsk, 630090, Russia*
[3]*Novosibirsk State Technical University, Novosibirsk, 630092, Russia*
[4]*University of Sydney, School of Physics, Falkiner High Energy Physics, NSW 2006, Sydney, Australia*
[5]*University of Tokyo, Department of Physics, 7-3-1 Hongo Bunkyo-ku Tokyo, 113-0033, Japan*

Abstract. Since December 2010, the CMD-3 detector has collected data at the electron-positron collider VEPP-2000. The sample of the accumulated data corresponds to about 60 pb^{-1} of integrated luminosity in the c.m. energy from 0.32 up to 2 GeV. Preliminary results of the analysis of various processes e^+e^- annihilation to hadrons are presented. It is shown the processes with multihadron events have several intermediate states which must be taken into account to correctly describe the angular and invariant mass distributions as well as cross section dependence versus energy.

1 Introduction

The electron-positron collider VEPP-2000 [1] operates at the Budker Institute of Nuclear Physics since December 2010. The collider is designed to provide luminosity up to 10^{32}cm^{-2}s^{-1} at the maximum center-of-mass energy $\sqrt{s}$ = 2 GeV. The current integrated luminosity accumulated by the CMD-3 is about 60 pb^{-1}. One of the main physical task is to measure the hadronic cross sections in the wide energy range, and searches for the new vector and scalar mesons and as well as for the exotic hadrons.

The precision data of the hadronic cross sections are required for various applications, in particular, to evaluate the anomalous magnetic moment (AMM) of muon, $a_\mu = (g-2)_\mu/2$. The VEPP-2000

⋆e-mail: fedotovich@inp.nsk.su

 DOI: 10.1051/epjconf/201613001014

energy range gives the major hadronic contribution to AMM ($a_{\mu}^{had} \sim 92\%$) both to the hadronic vacuum polarization itself and to its uncertainty [2].

The precision measurement of luminosity is a key ingredient of many experiments which study the hadronic cross sections at e^+e^- colliders. So far it is very important to have several QED processes such as $e^+e^- \to e^+e^-, \mu^+\mu^-, \gamma\gamma$ to have the cross check as it was done by CLEO [3].

The general purpose detector CMD-3 has been described in detail elsewhere [4]. Its tracking system consists of a cylindrical drift chamber (DC) and double-layer multiwire proportional Z-chamber, both also used for a trigger, and both inside a thin (0.2 X_0) superconducting solenoid with a field of 1.3 T. The liquid xenon (LXe) barrel calorimeter with 5.4 X_0 thickness has fine electrode structure, providing 1-2 mm spatial resolution, and shares the cryostat vacuum volume with the superconducting solenoid. The barrel CsI crystal calorimeter [5] with thickness of 8.1 X_0 is placed outside the LXe calorimeter, and the end-cap BGO calorimeter with a thickness of 13.4 X_0 is placed inside the solenoid. The luminosity is measured using events of Bhabha scattering and $\gamma\gamma$ at large angles with about 1% accuracy.

The beam energy has been monitored by measuring the current in the dipole magnets of the main ring, and for a few energy points by using the Back-Scattering-Laser-Light system [6]. Using measured average momentum from Bhabha events, and average momentum of proton-antiproton pairs from the $e^+e^- \to p\bar{p}$ process, we determine E_cm for each energy point with about 1 MeV accuracy.

To study a detector responce and to obtain a registration efficiency, we have developed a Monte Carlo (MC) simulation of our detector based on GEANT4 package, and all simulated events pass all our reconstruction and selection procedures. The MC simulation includes photon radiation by initial electron or positron, calculated according to Ref. [7].

2 Luminosity measurement

The energy range from 1 to 2 GeV was scanned upward and downward with a step of 50 MeV. At each energy point the integrated luminosity of about 500 nb^{-1} was accumulated. In the case of scanning downward, the energy points have been shifted by 25 MeV with respect to the previous case. The experiments were performed from January to June 2011 and, similarly, in 2012.

Two types the «CHARGED» and the «NEUTRAL» first level triggers were used while data have been taking. A combinations of the signals from DC cells and ZC sectors, which roughly reproduce "track", start a special processor «TRACKFINDER» (TF). The combinations of signals from calorimeters with different energy thresholds actuated the «CLUSTERFINDER» (CF) processor. A positive decision of either processor allowed the recording of current event onto a hard disk with capacity about 2 TB. In the course of data accumulation, the mean trigger frequency ranged in ~200 ÷ 400 Hz. The sample of collinear Bhabha events e^+e^- were selected for luminosity determination.

The process $e^+e^- \to \gamma\gamma$ was also used, since it has essential advantages [8] with respect to the first one. It is free of radiation of the final state particles and its Coulomb interaction, the corresponding Feynman graphs do not contain photon propagators affected by the vacuum polarization effects. These advantages are the main motivation to exploit this process as an independent tool for luminosity.

The collected integrated luminosity above the ϕ mass is about 34.5 pb^{-1}, 8.3 and 8.4 pb^{-1} at the ω and ϕ resonances, respectively, and 9.4pb^{-1} from a scan below the ϕ. The peak luminosity $\sim 2 \cdot 10^{31}$ $cm^{-2}s^{-1}$ is limited while by a positrons injection rate and it will gain by a factor of ten after upgrade of the injection facility.

The relative difference of the luminosities determined with two processes versus energy is presented in Fig. 1, where only statistical errors are shown (SCAN 2011). The horizontal line is a fit for this ratio and in average it is about 0.2±0.3%. The main sources which contribute to systematic

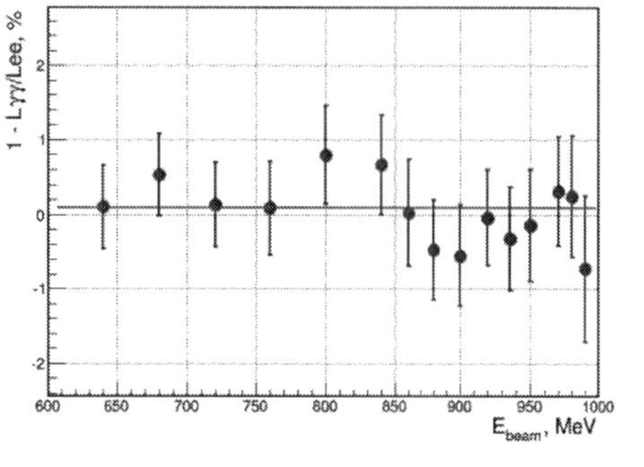

Figure 1. The ratio of the relative difference of the luminosities vs beam energy (scan 2011). Full circles - scan up, triangles - scan down. Line - fit: 0.2 ± 0.3%

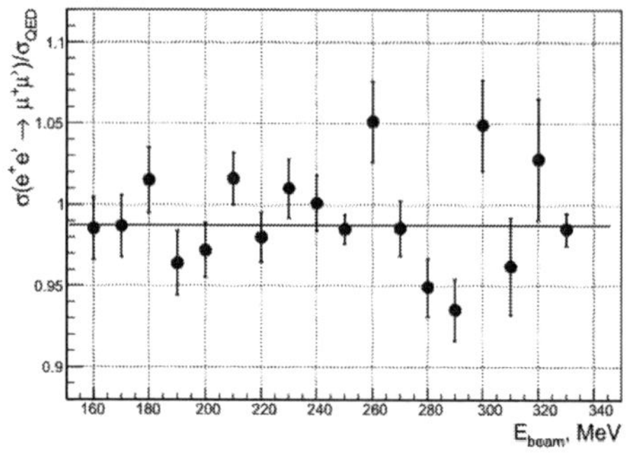

Figure 2. Result of the measurement of muon pair production in comparison with the QED prediction. Horizontal line - fit: 0.995 ± 0.005

error are: contribution due to the different angular resolutions for Bhabha events and $\gamma\gamma$ is estimated as ~0.8%; correction, which takes into account inclination of the beam axis with respect to detector is about ~0.4%; DC z-coordinate calibration contributes about 0.3%; radiative corrections are calculated according to [9] with the accuracy about 0.2%. Presently we estimate the current luminosity systematic accuracy as ~1% for energies higher than 1 GeV.

3 Processes with multihadrons in final states

One of the main goal of the CMD-3 experiment is to reduce a systematic uncertainty of the cross section of two pion production to 0.3-0.4%. The $\pi^+\pi^-$ events are separated either using the particles momentum or their energy deposition in EM-calorimeter. Two ways of event separation will provide cross-check and is expected allow to keep the systematic error under control. The first energy scan below 1 GeV was performed in 2013. The collected statistics is a few times higher than we had in the previous CMD-2 measurements and it is at the level of ISR statistics accumulated by the BaBar and KLOE. The process $e^+e^- \to \mu^+\mu^-$ is very importance since it provides overall systematic test of the event separation accuracy. Preliminary results for the cross section $\sigma(e^+e^- \to \mu^+\mu^-)$ are shown in Fig. 2 with respect to the QED prediction. Horizontal line is a fit for the double ratio $\sigma^{exp}_{\mu\mu}/\sigma^{QED}_{\mu\mu}/\sigma^{exp}_{ee}/\sigma^{QED}_{ee}$ which was found to be 0.995±0.005. At the moment this result demonstrates our potential power of the event separation procedure. Study of different systematic uncertaintices is ongoing.

3.1 Study of the processes $e^+e^- \to K_S K_L$ and $e^+e^- \to K^+K^-$.

It is known the CMD-2 and BaBar results for cross sections at the ϕ-peak region disagree at the level ~4% for charged channel, so a new measurement are required. The $e^+e^- \to K^0_S K^0_L$ and $e^+e^- \to K^+K^-$ cross sections were measured in the c.m. energy range 1.004-1.060 GeV at 25 energy points. The neutral mode detection is based on the search of two central tracks with common vertex in DC from the $K^0_S \to \pi^+\pi^-$ decay. The number of events is defined by the fit of two pions invariant mass distribution [10].

The registration of the charged mode is based on the search of two central collinear tracks of kaons with defined momentum in DC. Each track should has ionization losses significantly larger than mip due to relatively small velocity of kaons under study. After these requirements the level of remaining background is less than 0.5%. The detection efficiency of each kaon was determined with data and as well as with MC and deliver a deviation less 1.5%.

The obtained cross sections for the neutral (published) and charged mode (preliminary) are presented in Fig. 3 and Fig. 4, correspondingly. The measured cross section is approximated according to VDM model as a sum of ϕ, ω, ρ-like amplitudes and their excitations. The interference of non-resonant amplitudes with the amplitude of ϕ meson scales in ten times and shown too at the bottom of graphs. The neutral and charged channels were approximated simultaneously, as a result the following values of the ϕ meson parameters have been obtained: $m_\phi = 1019.464 \pm 0.060$ MeV/c^2, $\Gamma_\phi = 4.240 \pm 0.017$ MeV, $\frac{B_{\phi\to K^+K^-}}{B_{\phi\to K^0_S K^0_L}} = 1.573 \pm 0.06$. The obtained parameters have accuracy comparable or better than it was obtained in previous experiments.

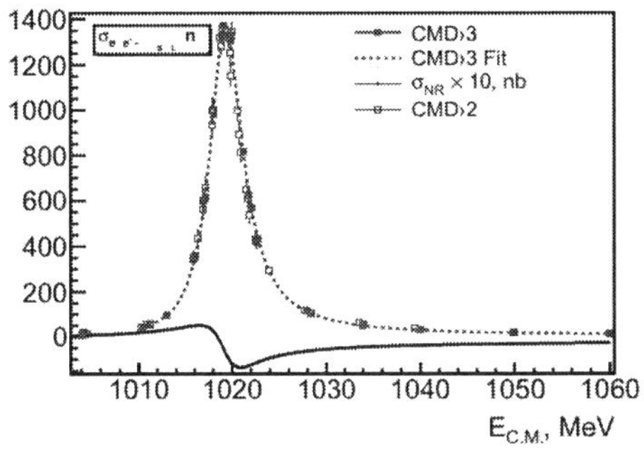

Figure 3. The cross section of the process $e^+e^- \to K_LK_S$ around ϕ-meson energy region. CMD-2, CMD-3 and BaBar data are presented.

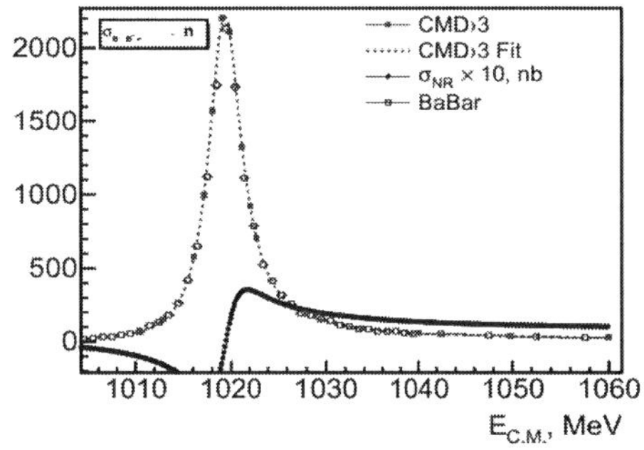

Figure 4. The cross section of the process $e^+e^- \to K^+K^-$ around ϕ-meson energy region. CMD-2, CMD-3 and BaBar data are presented.

3.2 Study of the process $e^+e^- \to K^+K^-\pi^+\pi^-$

The cross section measurement of the process $e^+e^- \to K^+K^-\pi^+\pi^-$ is based on the integrated luminosity of 22 pb^{-1} in the c.m. energy range from 1.5 to 2.0 GeV and early was measured by the BaBar via ISR. Nevertheless the direct measurements are very important, since some contributes to a_μ are based on isospin relations of various $K\bar{K} + n\pi$ final states. Any uncertainty of this approach will be crucial for the a_μ accuracy in the future calculations.

The signal events should have three or four tracks in the DC coming from interaction region and obey to the energy-momentum conservation. Two tracks corresponding to kaons should have specific ionization losses dE/dx in DC. Fig. 5 shows the difference between measured total energy and c.m. energy $\Delta E_4 = E_{tot} - E_{c.m.}$ vs the total momentum for all events with four tracks. The signal events are located near in origin of coordinates. The cluster of events with a zero total momentum but shifted up along the vertical axis, corresponds to $\pi^+\pi^-\pi^+\pi^-$ events.

The similar procedure was used to select signal events with the three-track in DC. As a result, ~13300 four-track events and ~16000 three-track events were selected. To calculate a detection efficiency, the $K^+K^-\pi^+\pi^-$ events were simulated with a primary generator using the GEANT4 package and then reconstructed with the same software as experimental data.

The cross section as a function of energy shown in Fig. 6, and well agrees with the previous BaBar measurement [11] presented by open circles. Systematic error was studied in detail and currently is

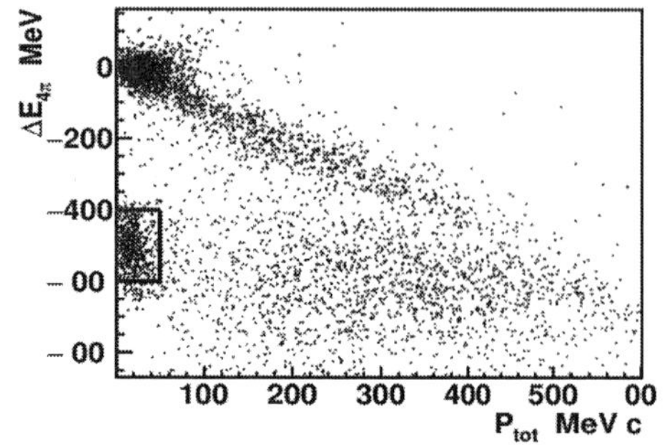

Figure 5. The difference between the total energy and c.m. energy (ΔE_4) versus the total momentum for the four-track events. The upper cluster of dots represents $\pi^+\pi^-\pi^+\pi^-$ while the lower one - $K^+K^-\pi^+\pi^-$ events.

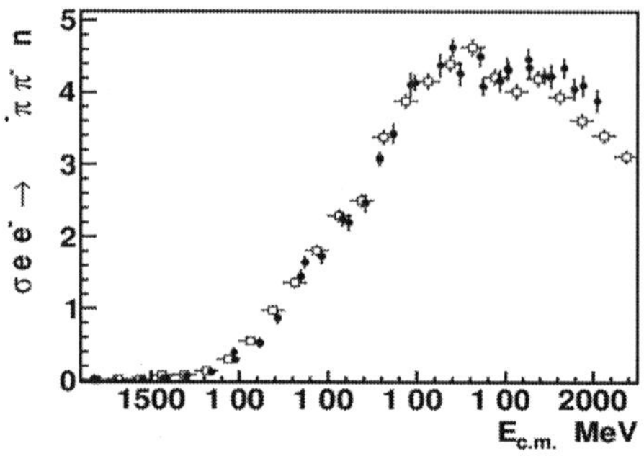

Figure 6. Dots - the $e^+e^- \to K^+K^-\pi^+\pi^-$ cross section measured with the CMD-3. The BaBar results are shown by open circles.

estimated as 6% and mainly due to model dependence of the detection efficiency. More detail analyses can be found in publication [12].

3.3 Six pion production

Production of the six charged pions in e^+e^- annihilation was studied at DM2 [13] and BaBar [14]. The DM2 experiment observed a "dip" in the cross section of the process $e^+e^- \to 3(\pi^+\pi^-)$ near 1.9 GeV, confirmed later by the BaBar. The origin of the "dip" remains unclear, but the most popular explanation is based on a hypothesis of a presence of the under-threshold ($p\bar{p}$) resonance discussed in many theoretical papers [15].

The analysis is based on 22 pb^{-1} of the integrated luminosity collected in the c.m. energy range from 1.5 to 2 GeV. Candidates for the process under study are required to have five or six tracks in DC. For six- or five-track candidates the total energy and total momentum are calculated, assuming all tracks to be pions. To estimate the background MC simulation of the major processes $2(\pi^+\pi^-\pi^0)$ and $2(\pi^+\pi^-)\pi^0$ was performed and was found to be smaller than 1%. The analysis in detail for this channel can be found in [16].

To measure the cross section of the process $e^+e^- \to 2(\pi^+\pi^-\pi^0)$ the sample of events with the four charged and two neutral pions were selected. To select neutral pions the spectrum of invariant mass of all two photon combinations was studied inside energy gap from $60 < m_{\gamma\gamma} < 200$ MeV/c^2 and combination with the nearest to the pion mass is chosen. The number of events under study at each energy point was determined by fit of the distributions shown in Fig. 7 with a sum of three Gaussian functions for signal events and quadratic polynomial for background. The cross section calculates according to the number of determine events and takes into account RC and detection efficiency. The preliminary results for the cross section are presented in Fig. 8. The analysis of the data is in progress.

3.4 Cross sections measurement of the $e^+e^- \to \eta\pi^+\pi^-$ and $e^+e^- \to \omega\pi^+\pi^-$ processes

The process $e^+e^- \to \eta\pi^+\pi^-$ was studied in two decay modes of η: 2γ and $\pi^+\pi^-\pi^0$. The signal events should have two tracks and at least two photons in EM-calorimeter. The shape of two photons invariant mass distribution was fix from MC and used to determine the number of the $\eta\pi^+\pi^-$ events at

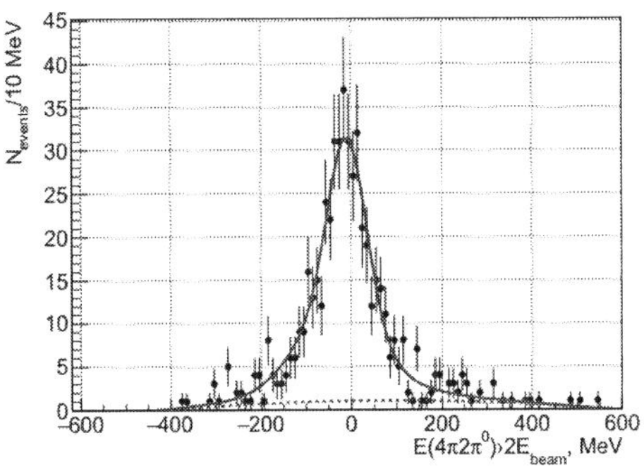

Figure 7. The number of events vs the total energy of the system $2(\pi^+\pi^-\pi^0)$. Signsl- fit with a sum of three Gaussian functions, background - fit with quadratic polynomial.

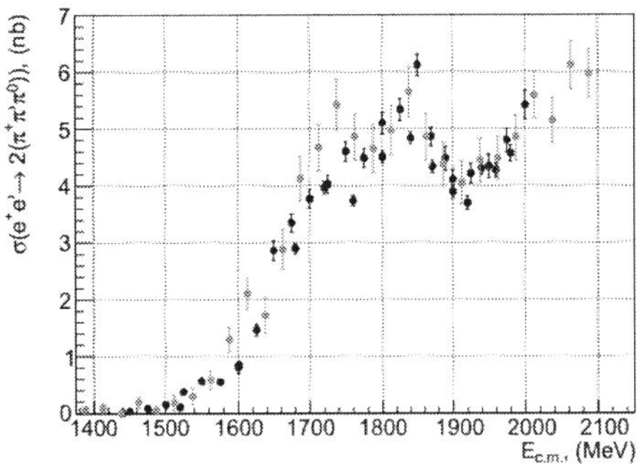

Figure 8. . Cross section of the process $e^+e^- \to 2(\pi^+\pi^-\pi^0)$ vs energy. Black points - CMD3, green - BaBar.

each energy point. The result of such fit at 1500 MeV is shown in Fig. 9. The preliminary results for the Born cross section are presented in Fig. 10. The systematic uncertainty for this process is about 5.2% and mainly due to uncertainty of detection efficiency, which depends on angular distribution of the final particles, i.e. depends on the intermediate states (theoretical models) through this process goes.

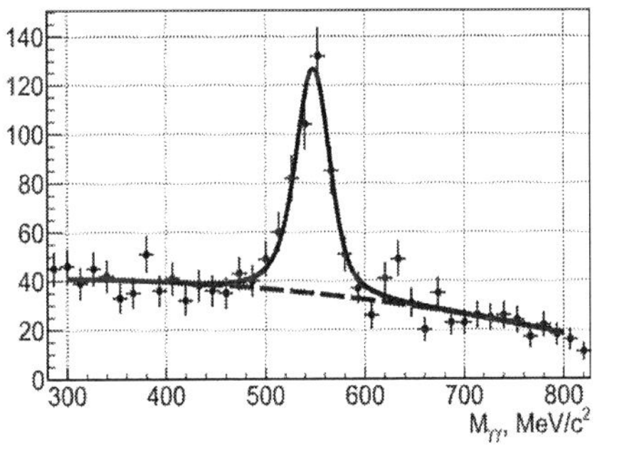

Figure 9. Two photons invariant mass at the energy of 1500 MeV.

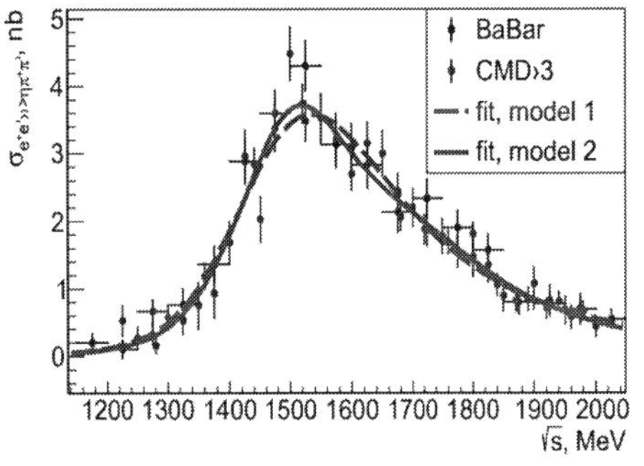

Figure 10. The $e^+e^- \to \eta\pi^+\pi^-$ Born cross section measured in the $\eta \to \gamma\gamma$ channel. The results are presented together with BaBar data.

These two processes were studied too when η and ω decay to the three pions: $\pi^+\pi^-\pi^0$. To determined the signal events the shape of the $\pi^+\pi^-\pi^0$ invariant mass distribution has been studied using Monte Carlo simulation and was used to count the number of the signal events. The preliminary results for the Born cross sections are consistent with other experiments. The current systematic uncertainty for these channels we estimated as 15% and mainly driven by the difference between angular distributions of charged particles in simulation and experiment. The analysis of systematic errors is ongoing.

3.5 Study of the processes $e^+e^-\to K^+K^-\eta$ and $e^+e^-\to K^+K^-\omega$

The analysis of the process $e^+e^-\to K^+K^-\eta$ was based on 19 pb^{-1} of an integrated luminosity collected by the CMD-3 detector in 2011–2012 in the $E_{\text{c.m.}}$ range from 1.59 to 2.01 GeV. On the base of these statistics we observed the contribution of $\phi(1020)\eta$ intermediate state only. Candidates for the events of the signal process were required to have two, three or four tracks in the DC, coming out of the beams intersection point. The kaon/pion separation was performed with the use of $f_{K/\pi}(p, dE/dx)$ functions [12], representing the probability density for charged kaon/pion with the momentum p to produce the energy losses dE/dx in the DC. We considered η-meson as a recoil particle, which allowed us to avoid the loss of statistics due to the selection of the specific η decay mode. But such an inclusive approach lead to the complication of the signal/background separation. Therefore the major background processes were studied and were found to be $e^+e^-\to K^+K^-\omega(782)$, $e^+e^-\to K^+K^-\pi^+\pi^-$, $e^+e^-\to\phi f_0(500)$, $K^{*\pm}(892)K^{\mp}\pi^0\to K^+K^-\pi^0\pi^0$, $e^+e^-\to 2\pi^+2\pi^-2\pi^0$. We perform the signal/background separation and 1296±43 signal events were extracted. The resulting $e^+e^-\to\phi(1020)\eta$ cross section is shown in the Fig. 11 along with the BaBar results. The overall systematic uncertainty of the cross section measurement was estimated to be 6%. Via the cross section approximation the $\phi(1680)$-meson parameters have been determined.

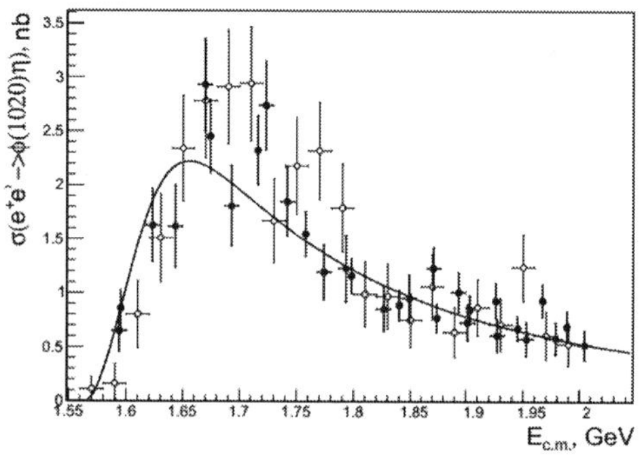

Figure 11. The BaBar (open circles) and CMD-3 (filled circles) results for the measurement of $e^+e^-\to\phi(1020)\eta$ cross section in the range from 1.59 to 2.01 GeV. The approximation of CMD-3 data together with the BaBar data in the range from 2.3 to 3.46 GeV is also shown.

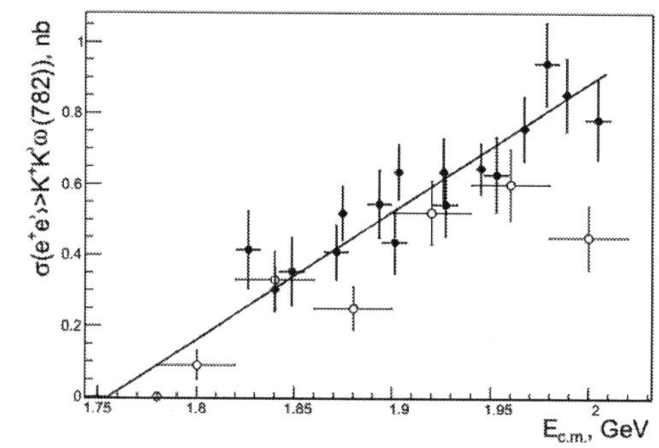

Figure 12. The BaBar (open circles) and CMD-3 (filled circles) results for the measurement of $e^+e^-\to K^+K^-\omega(782)$ cross section in the range from 1.8 to 2.01 GeV. The approximation of CMD-3 data is also shown.

The analysis of the process $e^+e^-\to K^+K^-\omega(782)$ was based on 12 pb^{-1} of the integrated luminosity collected by the CMD-3 detector in 2011–2012 in the $E_{\text{c.m.}}$ range from 1.8 to 2.01 GeV. Candidates for the events of the signal process were required to have three or four tracks in the DC, flying out of the area of the beams intersection. The kaon/pion separation was performed in the same way, as in the $e^+e^-\to K^+K^-\eta$ process analysis. We studied the process in the $\omega(782)\to\pi^+\pi^-\pi^0$ decay mode, considering π^0 as a recoil particle. The major background processes were found to be $e^+e^-\to K^+K^-\eta$ and $e^+e^-\to K^+K^-\pi^+\pi^-$, but their contribution was almost completely suppressed by the cuts on the invariant and missing masses of kaons and pions. After background suppression we performed a direct subtraction of the estimated number of background events and in total in the experiment we obtained 886±30 signal events. The resulting $e^+e^-\to K^+K^-\omega(782)$ cross section is shown in the Fig. 12 along with the BaBar results. The overall systematic uncertainty of the cross section measurement was estimated to be 6%.

4 Summary and conclusion

CMD-3 detector will operate with a goal to get $\sim 1fb^{-1}$ in 5-10 years and provides the new precise results on the hadron production. The current integrated luminosity was measured using two well known QED processes $e^+e^- \to e^+e^-, \gamma\gamma$ and systematic accuracy is estimated as 1%. Two type of the first level triggers «CHARGED» and «NEUTRAL» deliver the independent information that allowed to determine the detection efficiencies and to estimate their uncertainties. Data analysis is in progress, the already collected data sample provides the same or better statistical precision for the hadronic cross sections than in previous experiments were achieved.

Acknowledgements

The authors are grateful to A.I. Milstein for help with a theoretical interpretation and development of the models. We thank the VEPP-2000 team for excellent machine operation.

This work is supported in part by the Russian Education and Science Ministry, by FEDERAL TARGET PROGRAM Scientific and scientific-pedagogical personnel of innovative Russia in 2009-2013, by agreement 14.B37.21.07777, by the Russian Fund for the Basic Research grants.

Part of this work related to the photon reconstruction algorithm in the electromagnetic calorimeter is supported by the Russian Science Foundation (project No. 14-50-00080).

RFBR 10-02-00695-a, RFBR 10-02-00253-a, RFBR 11-02-00328-a, RFBR 11-02-00112-a, RFBR 12-02-31501-mol-a, RFBR 12-02-31499-mol-a, RFBR 12-02-31498-mol-a, RFBR 12-02-01032-a, RFBR 13-02-00215-a.

References

[1] I. Koop *et al.*, Nucl.Phys.B, Proc.Suppl. **181**, 371 (2008)
[2] M. Davier *et al.*, Eur. Phys. J. C **31**, 503 (2003)
K. Hagiwara *et al.*, J. Phys. G **38**, 085003 (2011)
[3] G. Grawford *et al.*, Nucl. Instr. Meth. A **345**, 429 (1994)
[4] B. I. Khazin *et al.*, Nucl.Phys.B, Proc. Suppl. **181-182**, 376 (2008)
[5] V. M. Aulchenko *et al*, 2015 JINST 10 P10006 (2015)
[6] E. V. Abakumova *et al.*, Phys. Rev. Lett. **110**, 140402 (2013)
[7] E. A. Kuraev and V. S. Fadin, Yad. Fiz., vol.3 **41**, 733 (1984)
[8] S. I. Eidelman *et al.*, Eur. Phys. J. C **71**, 1597 (2011)
[9] A. B. Arbuzov *et al.* Eur. Phys. J. C **46**, 689 (2006)
S. Actis *et al.*, Eur. Phys. J. C **66**, 585 (2010)
[10] E. A. Kozyrev *et al.*, Phys. Lett. B **760**, 314 (2016)
[11] B. Aubert et al. (BaBar Collaboration), Phys. Rev. D **76**, 012008 (2007)
J. P . Lees *et al.* (BaBar Collaboration), Phys. Rev. D **86**, 012008 (2012)
[12] D. N. Shemyakin *et al.*, Phys.Lett. B **756**, 153-160 (2016)
[13] R. Baldini *et al.*, reported at the "Fenice" Workshop, Frascati (1988);
A. B. Clegg and A. Donnachie, Z. Phys. C **45**, 677 (1990);
M. R. Whalley, J. Phys. G **29**, A1 (2003)
[14] B. Aubert *et al.*, (BaBar Collaboration), Phys. Rev. D **73**, 052003 (2006)
[15] A. Sibirtsev and J. Haidenbauer, Phys. Rev. D **71**, 054010 (2005)
A. Antonelli et al. (FENICE Collaboration), Phys. Lett. B **365**, 427 (1989)
[16] R. R. Akhmetshin *et al.*, Phys.Lett. B **723**,82 (2013)

The investigation of $K^+\pi^-$, π^+K^- and $\pi^+\pi^-$ atoms

Valeriy Yazkov[1,★]
for the DIRAC Collaboration

[1] *Skobeltsyn Institute of Nuclear Physics, Lomonosov Moscow State University, Moscow, Russia*

Abstract. Theory, using Low Energy QCD, predicts with high precision the pion-pion and pion-kaon scattering lengths. There is accurate relation between $K^+\pi^-$ and π^+K^- atoms lifetime and pion-kaon S-wave scattering lengths with isospin 1/2 and 3/2. Experiment DIRAC at CERN PS detects 345 ± 61 pairs from $K^+\pi^-$ and π^+K^- atoms breakup. It allows to achieve the first observation of exotic atoms consisted of pion and kaon. Measured values of πK atom lifetime and corresponding pion-kaon scattering length difference are presented. It is shown, that experimental accuracy for pion-kaon scattering length difference could be significantly improved with an experiment at SPS energy.

1 Introduction

Chiral Perturbation Theory (ChPT) describes QCD processes at low energies. ChPT in 1-loop approximation predicts S-wave pion-kaon scattering lengths with isospin 1/2 ($a_0^{1/2}$) and 3/2 ($a_0^{3/2}$) to be [1, 2]:

$$a_0^{1/2} = 0.19 \pm 0.2,\ a_0^{3/2} = -0.05 \pm 0.02,\ a_0^{1/2} - a_0^{3/2} = 0.23 \pm 0.01\,. \tag{1}$$

ChPT with $L^{(2)}$, $L^{(4)}$, $L^{(6)}$ in 2-loop approximation predicts scattering length difference to be [3]:

$$a_0^{1/2} - a_0^{3/2} = 0.267\,. \tag{2}$$

Scattering length difference also has been predicted, using Roy-Steiner equations [4]:

$$a_0^{1/2} - a_0^{3/2} = 0.269 \pm 0.015\,. \tag{3}$$

In the framework of lattice QCD, predictions for πK scattering length and their combination a_0^- have been obtained in [5] (Eq. 4), [6] (Eq. 5), [7] (Eq. 6):

$$a_0^{1/2} = 0.1725^{+0.0026}_{-0.0157},\ a_0^{3/2} = -0.0574^{+0.0029}_{-0.0060}, \tag{4}$$

$$a_0^{1/2} = 0.183 \pm 0.039,\ a_0^{3/2} = -0.0602 \pm 0.0040, \tag{5}$$

$$a_0^- = \frac{1}{3}(a_0^{1/2} - a_0^{3/2}) = 0.0811 \pm 0.0143\,. \tag{6}$$

★e-mail: valeri.yazkov@cern.ch

The measurement of the S-wave πK scattering lengths would test our understanding of the chiral $SU(3)_L \times SU(3)_R$ symmetry breaking of QCD (u, d and s quarks), while the measurement of $\pi\pi$ scattering lengths checks only the $SU(2)_L \times SU(2)_R$ symmetry breaking (u, d quarks). This is the principal difference between $\pi\pi$ and πK scattering.

Experimental data on the πK low-energy phases are absent. The only experimental pion-kaon scattering length measurement has been done with estimation of πK atom lifetime [8]:

$$|a_0^-|M_\pi = 0.107^{+0.093}_{-0.035}\,. \tag{7}$$

2 Method of πK atom observation and investigation

πK-atom ($A_{\pi K}$) is a hydrogen-like atom consisting of K^+ (K^-) and π^- (π^+) mesons. The πK-atom lifetime (ground state 1S), $\tau = \frac{1}{\Gamma}$ is dominated by the annihilation process into $\pi^0 K^0$. There is a relation between the width of $A_{\pi K}$ decay and S-wave πK scattering lengths for isospin 1/2 and 3/2 [9]:

$$\Gamma_{1S,\pi^0 K^0} = 8\alpha^3\mu^2 p^*(a_0^-)^2(1 + \delta_K). \tag{8}$$

Here α is the fine structure constant, μ is the reduced mass of the $\pi^\pm K^\mp$ system, p^* is the outgoing π^0 momentum in the πK atom system, and δ_K accounts for corrections, due to isospin breaking, at order α and quark mass difference $(m_u - m_d)$.

Prediction of scattering length difference in Eq. (3) provides an estimation of lifetime of $A_{\pi K}$ in ground state to be: $\tau = (3.5 \pm 0.4) \times 10^{-15}$.

A method of investigation for $\pi^+\pi^-$, πK and other atoms, consisted of two oppositely charged mesons, has been proposed in [10]. Pairs of K^+ (K^-) and π^- (π^+) mesons are producing in proton-target interactions. Pairs, which are generated from fragmentation and strong decays ("short-lived" sources), are affected by Coulomb interaction in the final state. Some of them form Coulomb bound states — atoms, other are generated as free pairs ("Coulomb pairs"). Number of produced atoms (N_A) is proportional to a number of "Coulomb pairs" (N_C) with low relative momentum Q in a pair C.M. system: $N_A = K \cdot N_C$. The coefficient K is calculated with an accuracy better than 1% [11].

If at least one meson is generated from long-lived sources (electromagnetically or weakly decaying mesons or baryons: $\eta, \eta', K_s^0, \ldots$), then such pairs ("non-Coulomb pairs") are not affected by interaction in the final states.

After production, $A_{\pi K}$ travel through the target and could to annihilate into $\pi^0 K^0$, or to be ionised due to interaction with the target matter, producing specific "atomic pairs". These pairs have small relative momentum ($Q < 3$ MeV/c) and a number of such pairs n_A could be measured experimentally. Ratio of "atomic pair" number to a number of atom produced is a breakup probability: $P_{\rm br}(\tau) = n_A/N_A = n_A/(K \cdot N_C)$ [12, 13]. In Fig 1 dependence of $A_{\pi K}$ breakup probability is shown for two nickel target are used in experiment DIRAC for pair laboratory momentum range 5.1 ÷ 8.5 GeV/c. Value is averaged, using experimentally measured spectrum of atoms.

3 DIRAC setup

DIRAC setup was created to detect $\pi^+\pi^-$ with small relative momenta [14]. In 2004-2006 it has been modified in order to detect both $\pi^+\pi^-$ and πK pairs. New detectors for particle identification have been added: Cherenkov counters with heavy gas and aerogel for identification of K-mesons among background of pions and protons, correspondingly. Taking into account kinematic of πK "atomic pairs", new detectors cover only internal parts of each arm (see Fig. 2).

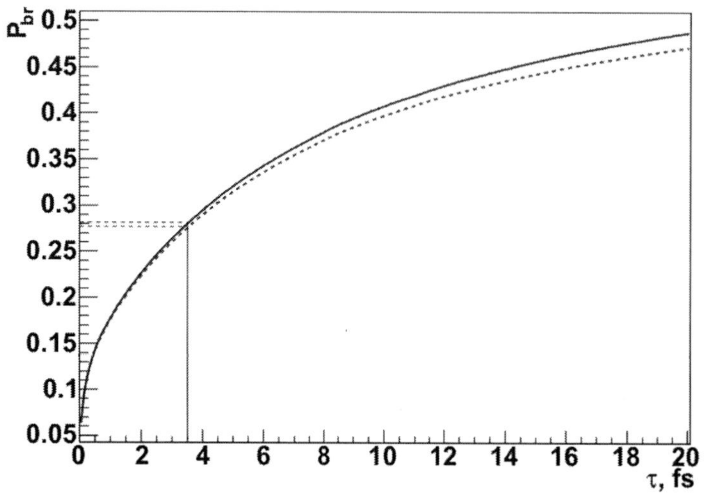

Figure 1. Dependence of the breakup probability P_{br} on $A_{\pi K}$ lifetime for 108μm (solid blue line) and 98μm (dashed red line) nickel targets, and an example how lifetime could be obtained from experimentally measured breakup probability

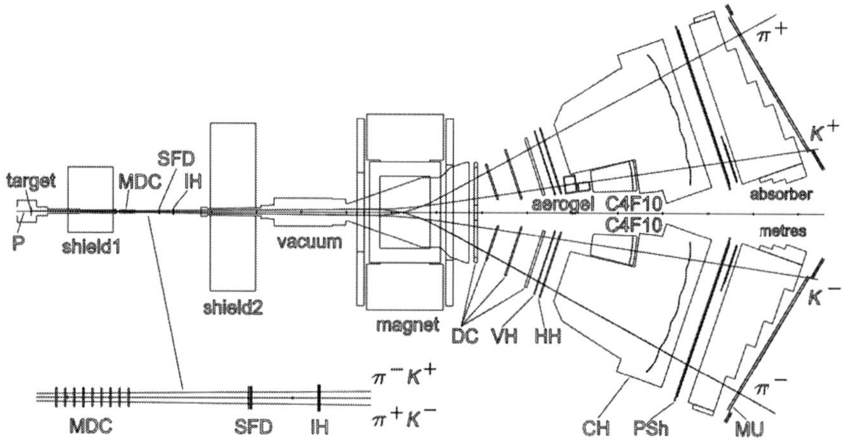

Figure 2. Upgraded DIRAC setup: MDC are microdrift gas chambers, SFD is a scintillating fiber detector and IH is a scintillation ionisation hodoscope. Downstream the spectrometer magnet there are drift chambers (DC), vertical (VH) and horizontal (HH) scintillation hodoscopes. Cherenkov detectors contain nitrogen (CH), heavy gas C4F10 and aerogel radiators. Shower detectors (PSh) and scintillation muon detectors (MU).

4 Observation of π^+K^- and $K^+\pi^-$ atoms

Analysis procedure selects events which have signals of detectors expected for π^+K^- and $K^+\pi^-$ pairs. Criteria on Q-projections are $|Q_L| < 15$ MeV/c, $Q_T < 4$ MeV/c. Distributions of experimental data over relative momentum Q and its projections Q have been fitted by a sum of simulated distributions of "atomic", "Coulomb" and "non-Coulomb" pairs. Contributions of simulated distributions are free parameters of fit. In order to reproduce distribution of experimental pairs over relative momentum Q and its projections, simulation procedure takes into account resolution and efficiency of the setup detectors, multiplicity of background particles and noise signals, multiple scattering in Platinum (run 2007) and Nickel (2008-2010) targets, detector planes and partitions. Results are presented in Fig. 3.

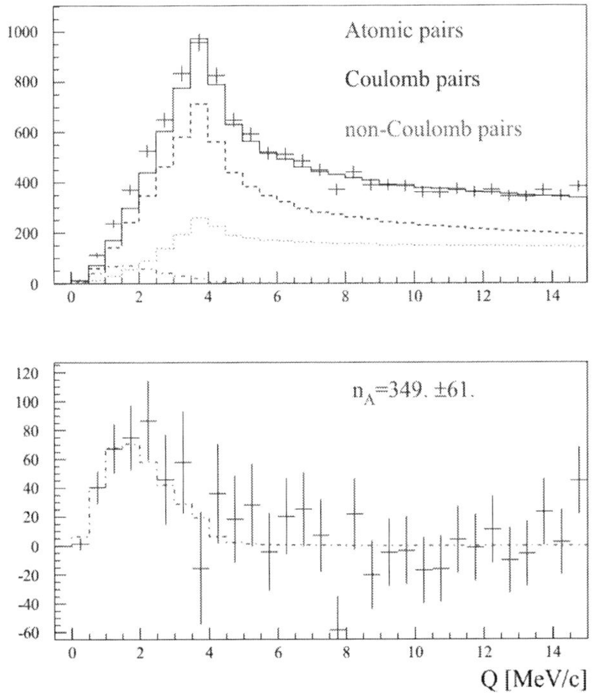

Figure 3. Distribution of π^+K^- and $K^+\pi^-$ pairs over Q (upper pictures), shown by points with error bars, is fitted by a sum of simulated distributions of "atomic" (red dotted-dashed), "Coulomb" (blue dashed) and "non-Coulomb" (magenta dotted) distributions ($\chi^2/n = 41/37$, n = number of degrees of freedom). A sum of background distributions ("Coulomb" and "non-Coulomb") is shown by a solid black line. Differences of experimental and background distributions are shown in lower picture together with simulated distributions of "atomic pairs".

Table 1. Atomic pair numbers n_A by analysing the 1-dimensional Q and $|Q_L|$ distributions and the 2-dimensional ($|Q_L|, Q_T$) distribution. Only statistical errors are given.

Analysis	π^-K^+	π^+K^-	π^-K^+ and π^+K^-?
Q	243 ± 52 (4.7σ)	106 ± 32 (3.3σ)	349 ± 61 (5.7σ)
$\|Q_L\|$	164 ± 79 (2.1σ)	67 ± 47 (1.4σ)	230 ± 92 (2.5σ)
$\|Q_L\|, Q_T$	237 ± 50 (4.7σ)	78 ± 32 (2.5σ)	314 ± 59 (5.3σ)

Numbers of π^+K^- and $K^+\pi^-$ "atomic pairs" obtained with analysis of one-dimensional distributions over Q, $|Q_L|$ and two-dimensional ($|Q_L|, Q_T$) distribution are presented in Table 1 (Ni and Pt target together). The best statistical accuracy are achieved by analysis of Q and ($|Q_L|, Q_T$) distributions. Signal to error ratio is more than 5. The 1-dimensional $|Q_L|$ analysis for all πK data yields $n_A = 230 \pm 92$, which does not contradict the values, obtained in the other two statistically more precise analyses.

Compared to the previous investigation [15], the Pt data was analysed including the upstream detectors. The consequence is a decrease of the statistics, but on the other hand an increase of the Q_T resolution. This better resolution improves the quality of data. Concerning the Ni target, the increase of n_A, compared to [8], is caused by optimizing the time-of-flight criteria, which decreases atomic pair losses for the same fraction of background in the final distributions.

The evaluation of the atomic pair number n_A is affected by several sources of systematic errors [16, 17]. These uncertainties lead to differences in the shapes of experimental and MC distributions for "atomic", "Coulomb" and to a much lesser extent for "non-Coulomb" pairs. The shape differences induce a bias in the value of the fit parameter n_A, corresponding to a systematic error of the atomic pair number. Sources of systematic error and estimation of error values are listed in Table 2.

Table 2. Estimations of systematic errors, which are induced by different sources, for analysis of data distribution over relative momentum Q, its longitudinal projection $|Q_L|$ and two dimensional distribution over $(|Q_L|, Q_T)$.

Sources of systematic errors	σ_Q^{syst}	$\sigma_{Q_L}^{syst}$	$\sigma_{\|Q_L\|,Q_T}^{syst}$
Uncertainty in Λ width correction	0.8	3.0	2.0
Uncertainty of multiple scattering in Ni (Pt) target	4.4	0.7	2.7
Accuracy of SFD simulation	0.2	0.0	0.1
Correction of Coulomb correlation function due to finite size production region	0.0	0.2	0.1
Uncertainty in πK pair laboratory momentum spectrum	3.3	5.4	7.8
Uncertainty in laboratory momentum spectrum of background pairs	6.6	1.6	5.4
Total	8.6	6.4	10.1

Table 3. Estimation of time needed for measurement a_0^- with statistical accuracy $\delta_{a_0^-}$ for present DIRAC setup and beam condition, and for versions Mod1 and Mod2, modified for proton beam energy E_p = 450 GeV and intensity I_B. It is assumed, that at 450 GeV beam the setup would obtain 3000 spills (4.5s) per day.

Setup	E_p GeV	I_b p/s	Beam time s	Run time months	$\delta_{a_0^-}$ %
Present	24	$2.7 \cdot 10^{11}$	$1.2 \cdot 10^6$	14.5	43.
Present	24	$2.7 \cdot 10^{11}$	$6.0 \cdot 10^7$	715.6	5.
Mod1	450	$1.0 \cdot 10^{11}$	$5.8 \cdot 10^6$	14.3	5.
Mod2	450	$1.0 \cdot 10^{12}$	$7.4 \cdot 10^5$	1.9	5.

Taking into account both statistical and systematic errors, the one-dimensional $\pi^\mp K^\pm$ analysis in Q yields $n_A = 349 \pm 61(stat) \pm 9(syst) = 349 \pm 62(tot)$ atomic pairs (5.6 σ) for both combinations of charge and two-dimensional analysis in $(|Q_L|, Q_T)$ yields $n_A = 314 \pm 59(stat) \pm 10(syst) = 314 \pm 60(tot)$ atomic pairs (5.2 σ). This is the first statistically significant observation of the strange dimesonic πK atom.

Experimentally measured numbers of "atomic pairs" n_A and produced atoms N_A allow (see section 2) to obtain new estimation of $A_{\pi K}$ lifetime in ground state, based on two-dimensional analysis in $(|Q_L|, Q_T)$:

$$\tau = (3.8^{+3.3}_{-2.0}|_{stat}\ {}^{+1.0}_{-0.6}|_{syst})\ \text{fs} = (3.8^{+3.5}_{-2.1}|_{tot})\ \text{fs}, \tag{9}$$

which corresponds to isospin-odd πK scattering length estimation to be:

$$|a_0^-| M_\pi = 0.087^{+0.043}_{-0.024}|_{tot}. \tag{10}$$

This result has been used for estimation of time which is needed for measurement a_0^- with accuracy 5% [18]. Table 3 presents expected beam time and run time are needed for achievement of this accuracy for present (Nickel target only) and modified DIRAC setup. It is seen that experiment at SPS energy allows to measure S-wave pion-kaon scattering length difference with sufficient accuracy for checking theoretical predictions made by ChPT and LQCD.

5 Investigation of $\pi^+\pi^-$ atoms

DIRAC experiment collected data with Be-Pt target in 2012. These data allow to make observation of $A_{2\pi}$ in states with non-zero orbital momentum (long-lived atoms): $n_A^L = 436 \pm 61$ [19]. Lifetime of long-lived atoms ($\tau \geq 1.17 \cdot 10^{-11}$ s) is sensitive to energy splitting between ns and np atomic states.

It provides possibility to plan experiments for measurement of "Lamb shift like" effect in $\pi^+\pi^-$ system and new combination of S-wave pion-pion scattering lengths: $2 \cdot a_0^0 + a_0^2$.

6 Summary

In the DIRAC experiment at CERN, the dimesonic Coulomb bound states involving strangeness, the π^-K^+ and π^+K^- atoms, were observed for the first time with reliable statistics. The one-dimensional $\pi^\mp K^\pm$ analysis in Q yields $349 \pm 62(tot)$ atomic pairs ($5.6\,\sigma$) for both combinations of charge. Analogously, a two-dimensional analysis in $(|Q_L|, Q_T)$ was performed with the result of $314 \pm 60(tot)$ atomic pairs ($5.2\,\sigma$).

DIRAC-like experiment at SPS energy provides possibility to check prediction of ChPT and LQCD for pion-kaon scattering lengths with accuracy at the level 5%.

Observation of "atomic pairs" from $\pi^+\pi^-$ atoms in long-lived states ($n_A^L = 436 \pm 61(tot)$) provides possibility to plan experiments for measurement of "Lamb shift like" effect in $\pi^+\pi^-$ system and new combination of S-wave pion-pion scattering lengths: $2 \cdot a_0^0 + a_0^2$.

We are grateful to CERN for support and the PS team for the excellent performance of the accelerator. This work was funded by CERN, INFN (Italy), INCITE and MICINN (Spain), IFIN-HH (Romania), the Ministry of Education and Science and RFBR grant 01-02-17756-a (Russia), the Grant-in-Aid from JSPS and Sentanken-grant from Kyoto Sangyo University (Japan).

References

[1] V. Bernard *et al.*, Phys. Rev. D **43**, 2757 (1991)
[2] A. Roessl, Nucl. Phys. B **555**, 507 (1999)
[3] J. Bijnens *et al.*, JHEP **0405**, 036 (2004)
[4] P. Buttiker *et al.*, Eur. Phys. J. C **33**, 409 (2004)
[5] S.R. Beane *et al.*, Phys. Rev. D **77**, 094507 (2008)
[6] C.B. Lang *et al.*, Phys. Rev. D **86**, 054508 (2012)
[7] K. Sasaki *et al.*, Phys. Rev. D **89**, 054502 (2014)
[8] B. Adeva *et al.*, Phys. Lett. B **735**, 288 (2014)
[9] J. Schweizer, Phys. Lett. B **587**, 33 (2004)
[10] L. Nemenov, Sov. J. Nucl. Phys. **41**, 629 (1985)
[11] L. Afanasyev and O. Voskresenskaya, Phys. Lett. B **453**, 302 (1999)
[12] L. Afanasyev and A. Tarasov, Phys. At. Nucl. **59**, 2130 (1996)
[13] M. Zhabitsky, Phys. At. Nucl. **71**, 1040 (2008)
[14] B. Adeva *et al.*, Nucl. Instrum. Methods A **515**, 467 (2003)
[15] B. Adeva *et al.*, Phys. Lett. B **674**, 11 (2009)
[16] A. Benelli and V. Yazkov, DN-2016-01, cds.cern.ch/record/2137645
[17] V. Yazkov and M. Zhabitsky, DN-2013-06, cds.cern.ch/record/1628544
[18] V. Yazkov, DN-2016-05 (DN = DIRAC Note), http://cds.cern.ch/record/2207227
[19] B. Adeva *et al.*, Phys. Lett. B **751**, 12 (2015)

EPJ Web of Conferences **130**, 01016 (2016) DOI: 10.1051/epjconf/201613001016
MESON 2016

Investigation of the low-energy kaons hadronic interactions in light nuclei by AMADEUS

A. Scordo[1,⋆], M. Cargnelli[2], C. Curceanu[1], L. Fabbietti[3,4], J. Marton[2], K. Piscicchia[1,5], D. Sirghi[1], I. Tucakovic[1], O. Vazquez Doce[3,4], S. Wycech[6], J. Zmeskal[2], G. Mandaglio[7,8], M. Martini[1,9], and P. Moskal[10]

[1] *INFN, Laboratori Nazionali di Frascati, 00044 Frascati, Italy*
[2] *Stefan-Meyer-Institut für Subatomare Physik, 1090 Wien, Austria*
[3] *Excellence Cluster 'Origin and Structure of the Universe', 85748 Garching, Germany*
[4] *Physik Department E12, Technische Universität München, 85748 Garching, Germany*
[5] *Museo Storico della Fisica e Centro Studi e Ricerche Enrico Fermi, Italy*
[6] *National Centre for Nuclear Research, 00681 Warsaw, Poland*
[7] *Dipartimento M.I.F.T. dell'Università di Messina, 98166 Messina, Italy*
[8] *INFN Gruppo collegato di Messina, 98166 Messina, Italy*
[9] *Dipartimento di Scienze e Tecnologie applicate, Università 'Guglielmo Marconi', 00193 Roma, Italy*
[10] *Institute of Physics, Jagiellonian University, 30-059 Cracow, Poland*

Abstract. The AMADEUS experiment deals with the investigation of the low-energy kaon-nuclei hadronic interaction at the DAΦNE collider at LNF-INFN. This study is fundamental to solve longstanding questions concerning interactions of strange quarks in the non-perturbative QCD. AMADEUS step 0 consisted in the reanalysis of the 2004/2005 KLOE data, exploiting K^- absorptions in H, ^{4}He, ^{9}Be and ^{12}C, leading to the first invariant mass spectroscopy study with very low momentum (100 MeV/c) in-flight K^- captures. In this paper, we present an overview of the analysis strategy, with particular emphasis on the results obtained in the analyses of the events with correlated Σ^0 and p.

1 Introduction

The AMADEUS experiment [1] studies the low-energy interactions of the negatively charged kaons with light nuclei. Such type of physics, important for the understanding of the non-perturbative QCD in the strangeness sector, has an important impact on hadron and nuclear physics and astrophysics. In this context, useful information can be obtained from the strength of the K^- binding in nuclei. The study of the K^--nucleus interaction at low energies is of interest for quantifying the meson–baryon potential with strange content [2], and for its impact on models describing the structure of neutron stars (NS) [3], especially after the recent observations of heavy NS [5]. One of the most useful methodes to acquire information on the strength of the KN interaction is represented by the investigation of the absorption of the K^- from one or more nucleons. Since one or more nucleons could be kept together by the strong attractive interaction between antikaons and nucleons, the formation of the so-called kaonic

⋆e-mail: alessandro.scordo@lnf.infn.it

bound states, as ppK^- or $ppnK^-$, might occur. The observation of such states and the measurement of their binding energies and widths would provide a measurement of the K-nucleon interaction in vacuum, as an important reference for the investigation of the in-medium properties of kaons. For the di-baryonic kaonic bound state ppK^- theoretical predictions deliver a wide range of binding energies and widths [6] and experimental results are contradictory [7–13]. For the search of such states in K^- absorption experiments, the competing multi-nucleonic absorption plays a fundamental role, but the contribution of the different single- and multi-nucleon absorption components has not been evidenced yet. Absorption processes also play an important role in the understanding of kaonic atoms, where a substantial multi-nucleon component is put forward by some theoretical models [14]. Measurements of these processes have been carried out and reported in [15, 16]. From the analysis of the KLOE data from 2004-2005, information on both the strength of the K^- binding in nuclei and the in-medium modification of the Σ^* and Λ^* resonances properties can be extracted by analysing, respectively, the $\Lambda/\Sigma - p, d, t$ correlated pairs and the resonances decay channels $\Lambda/\Sigma - \pi$. In this paper, we present an overview of the analysis procedure and of different ongoing analyses. Particular emphasis is put on the results of the analysis of the $\Sigma^0 p$ final state produced in absorption processes of K^- on two or more nucleons, occurring in the entrance wall of the KLOE drift chamber (DC), and on the search for a signature of the $ppK^- \to \Sigma^0 p$ kaonic bound state.

2 AMADEUS step 0

The AMADEUS experiment [1] studies the low-energy interactions of the negatively charged kaons with light nuclei; it is conceived to integrate the high acceptance and momentum resolution of the KLOE detector with the low momentum K^- beam of the DAΦNE collider in a future dedicated setup. DAΦNE [17] (Double Anular Φ-factory for Nice Experiments) is a double ring e^+e^- collider, designed to work at the center of mass energy of the Φ particle; the Φ meson decay produces charged kaons with low momentum ($\simeq$ 127 MeV/c) which is ideal either to stop them, or to explore the products of their low-energy nuclear absorptions. The KLOE detector [18] is centered around the interaction region of DAΦNE and is characterised by a $\simeq 4\pi$ geometry and an acceptance of $\simeq 98\%$; it consists of a large cylindrical Drift Chamber (DC) and a fine sampling lead-scintillating fibers calorimeter, all immersed in the axially simmetric magnetic field with intensity of 0.52 T, provided by a superconducting solenoid. The DC entrance wall composition is 750 μm of carbon fibre and 150 μm of aluminum foil. Dedicated GEANT Monte Carlo simulations of the KLOE apparatus show that out of the total number of kaons interacting in the DC entrance wall, about 81% are absorbed in the carbon fibre component and remaining 19% in the aluminum foil. The KLOE DC is filled with a mixture of helium and isobutane (90% in volume ^{4}He and 10% in volume C_4H_{10}). As a step 0 of AMADEUS, we analysed the 2004-2005 KLOE data, for which an important contribution of in-flight K^- nuclear captures, in different nuclear targets from the KLOE materials, was evidenced and characterised, enabling to perform invariant mass spectroscopy of in-flight K^- nuclear captures [19].

3 The particle identification procedure

The investigation of the negatively charged kaons interactions in nuclear matter is performed through the reconstruction of hyperon-pion and hyperon-nucleon/nucleus correlated pairs productions, following the K^- absorptions in H, ^{4}He, ^{9}Be and ^{12}C. The investigation of the K^- multi-nucleon absorptions and the properties of possible antikaon multi-nucleon bound states proceeds through the analyses of the $\Lambda/\Sigma - p, d, t$, correlations; this last channel is, in particular, extremely promising for the search and characterisation in different nuclear targets of the extremely rare four nucleon absorption process. The search for the $\Lambda(1405)$ is performed through its decay into $\Sigma^0\pi^0$ (purely isospin I=0) and $\Sigma^+\pi^-$. Given the excellent momentum resolution for charged particles, also the analysis of the $\Lambda\pi^-$ (isospin I=1) production, both from direct formation process and from internal conversion of a primary produced Σ hyperon ($\Sigma N \rightarrow \Lambda N'$) is presently ongoing. The aim is to measure, for the first time, the module of the non-resonant transition amplitude (compared with the resonant Σ^{*-}) below threshold. The presence of a hyperon always represents the signature of a K^- hadronic interaction inside the KLOE setup materials. All the analyses introduced before start with the identification of a $\Lambda(1116)$, through the reconstruction of the $\Lambda \rightarrow p + \pi^-$ (BR = 63.9 ±0.5%) decay vertex. The reconstructed $M_{p\pi^-}$ invariant mass shows a mean value of 1115.753±0.002 MeV/c^2, with a resolution of σ=0.5 MeV/c^2, in agreement with the PDG value. The particle identification is done using energy information coming from the Drift Chamber wires and, only for the protons, also from the Calorimeter, while the momenta are reconstructed from the tracks collected in the DC; the final selection is performed via the two-dimensional plot of energy and momentum. A common vertex between the Λ candidate and an additional proton or deuteron or triton track is then searched for. As an example, the obtained resolution on the radial coordinate ($\rho_{\Lambda p}$) for the Λp vertex is 12 mm, and this topological variable is used to select the K^- absorption processes inside the DC wall. The Σ candidates are identified through their decay into $\Lambda\gamma$ or pπ pairs. The photon selection is carried out via its identification in the EMC. Photon candidates are selected by applying a cut on the difference between the EMC time measurement and the expected time of arrival of the photon within $-1.2 < t < 1.8$ ns. Using the time and energy information from the electromagnetic calorimeter, π^0 particles can be reconstructed through their $\gamma\gamma$ decay. In order to select K^- interactions in the various targets, cuts on the radial position (ρ_Λ) of the Λ decay vertex are optimised in order to separate the two samples of K^- absorption events occurring in the DC wall and the DC gas: $\rho_\Lambda = 25 \pm 1.2$ cm and $\rho_\Lambda > 33$ cm, respectively. The ρ_Λ limits were set based on MC simulations and a study of the Λ decay path. In particular, the $\rho_\Lambda = 25 \pm 1.2$ cm cut guarantees, for the first sample, a contamination of K^- interactions in gas as low as ($5.5^{+1.3}_{-1.8}$%). More details on the particle identification procedure are given in [20]

4 K^- absorption on two nucleons in the $\Sigma^0 p$ final state

The first result we report on is the search for K^- absorption on two nucleons in the $\Sigma^0 p$ final state. The analysed data corresponds to a total integrated luminosity of 1.74 fb^{-1} collected in 2004–2005 with the KLOE detector. The number of stopped K^- in the DC entrance wall is calculated by combining the experimental K^+ tagging efficiency, the luminosity information and a Monte Carlo simulation to determine the rate of K^- stopped in the DC wall. The decay nearly at rest of the Φ meson allows to tag K^- events by the identification of a K^+ track in the opposite hemisphere of the DC. The total number of stopped K^- obtained in this way is equal to $(3.25 \pm 0.06) \times 10^8$. This value is used to normalise the measured yields of the different absorption processes. The following kinematic distributions are used simultaneously in a global fit to extract the contributions of the different absorption processes: the $\Sigma^0 p$ invariant mass, the relative angle of the Σ^0 and proton in the laboratory system $\cos(\theta_{\Sigma^0 p})$, the Σ^0 and the proton momenta. The processes that are taken into account in the fit of the experimental data are: (1) $K^- A \rightarrow \Sigma^0 - (\pi) p_{spec}(A')$, (2) $K^- pp \rightarrow \Sigma^0 p$ (2NA), (3) $K^- ppn \rightarrow \Sigma^0 pn$ (3NA), (4) $K^- ppnn \rightarrow \Sigma^0 pnn$ (4NA). This list includes the K^- absorption on two nucleons (2NA) with (2NA-FSI) and without (2NA-QF) final state interaction for the $\Sigma^0 p$ state and processes involving more than two nucleons in the initial state (3NA, 4NA). For all the details on how the distributions of the different contributions are modelled and sampled please refer to [20]. Panels (a)–(d) in Fig. 1 show the experimental distributions for the $\Sigma^0 p$ invariant mass, the $\cos(\theta_{\Sigma^0 p})$, and the Σ^0 and proton momenta, together with the fit results. The black points represent the experimental data after the subtraction

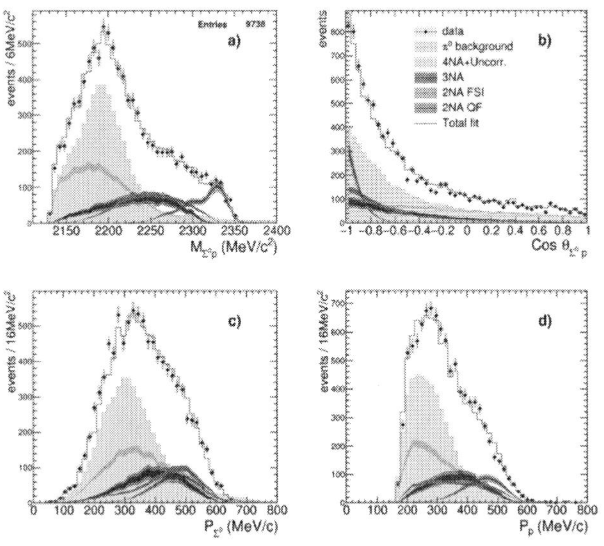

Figure 1. (Colour online.) Experimental distributions of the $\Sigma^0 p$ invariant mass, $\cos(\theta_{\Sigma^0 p})$, Σ^0 momentum and proton momentum together with the results of the global fit. The experimental data after the subtraction of the machine background are shown by the black circles, the systematic errors are represented by the boxes and the coloured histograms correspond to the fitted signal distributions where the light-coloured bands show the fit errors and the darker bands represent the symmetrised systematic errors. The gray line shows the total fit distributions (see text for details).

of the machine background with the systematic errors shown by boxes. The gray-filled histogram represents the $\pi^0 p$ background. The cyan distributions show the sum of the 4NA simulation together with the uncorrelated production of the $\Sigma^0 p$ final state. The blue distribution represents the 3NA and the magenta histogram the 2NA-FSI. The red distribution shows the 2NA-QF. The gray line is the resulting total fit. For each fitted distribution the light error band corresponds to the statistical error resulting from the fit, while the darker band visualises the symmetrised systematic error (refer to [20] for more details). The fit results lead to the first measurements of the genuine 2NA-QF for the final state $\Sigma^0 p$ in reactions of stopped K^- on targets of ^{12}C and ^{27}Al. This contribution is found to be about 9% of the total absorption cross-section including 2NA, 3NA and 4NA processes with also the contribution of the uncorrelated background leading to a $\Sigma^0 p$ final state taken into account. The last step of the analysis consists in the search for the ppK^- bound state produced in K^- interactions with nuclear targets, decaying into a $\Sigma^0 p$ pair. The ppK^- are simulated similarly to the 2NA-QF process but sampling the mass of the ppK^- state with a Breit-Wigner distribution, rather than the Fermi momenta of the two nucleons in the initial state. The event kinematic is implemented by imposing the momentum conservation of the ppK^- -residual nucleus system. Different values for the binding energy and width varying within 15-75 MeV/c^2 and 30-70 MeV/c^2 in steps of 15 and 20 MeV/c^2 , respectively, are tested. This range is selected according to theoretical predictions [6] and taking into account the experimental resolution. The global fit is repeated adding the ppK^- state to the processes 1-4. The best fit (χ^2 / ndf = 0 . 807) is obtained for a ppK^- candidate with a binding energy of 45 MeV/c^2 and a width of 30 MeV/c^2 , respectively. The resulting yield normalised to the number of stopped K^- is $ppK^-/K^-_{stop} = (0.044 \pm 0.009 stat^{+0.004}_{-0.005} syst) \times 10^{-2}$. Considering also that the improvement of the χ^2 is only marginal, an F-Test is carried out to compare the two models: with

DOI: 10.1051/epjconf/201613001016

and without ppK^- bound state. This test consists in evaluating the statistical significance of the model with the ppK^-, accounting for the additional fit parameter, by comparing the residuals and numbers of degrees of freedom of two models. The results of this F-Test shows that even the best fit corresponds to a p-value equal to 0.25 and hence to a significance of 1σ (see [20] for more details). This implies that although the fit favours the presence of an additional component that can be parametrised with a Breit-Wigner distribution with a certain mass and width, its significance is not sufficient to claim the observation of the bound state.

5 KN interaction studies in other channels

5.1 4NA process in Λt events

The experimental search for a Λ and a triton in the final state of K^- induced interactions on helium, representing the golden channel for the study of the 4NA process, was published by two experiments [21, 22] but the contribution from 4NA K^- capture was not clearly identified. Analysing the KLOE 2004-2005 data, the highest statistics ever of correlated Λt pairs from K^- nuclear absorption in ^{4}He (150 events) can be obtained; in addition, as shown in Fig. 2, for the first time the 4NA process events can be kinematically evidenced. Following energy and momentum conservations, the events related to

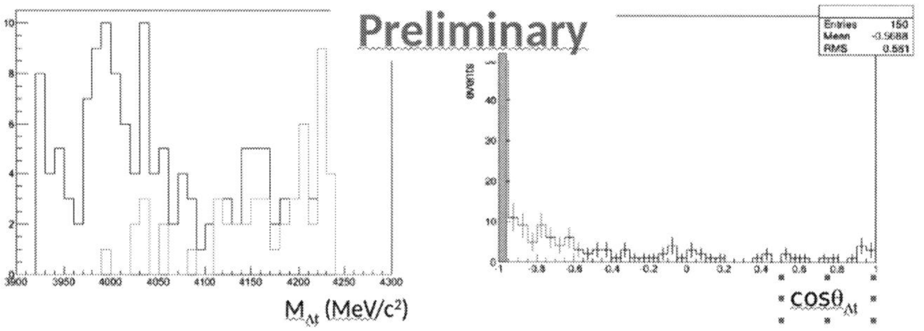

Figure 2. (Colour online.) Λt invariant mass and $\cos\theta_{\Lambda t}$; in green, events for $cos\theta_{\Lambda t} < -0.95$ are shown.

the simultaneous absorption of a K^- on the 4 nucleons of the ^{4}He are characterised by an almost back-to-back configuration. The simultaneous fit of Λt invariant mass, momentum and angular correlation, presently undergoing, will enabled to isolate for the first time the contribution of the $K^- +^4 He \rightarrow \Lambda t$ 4NA process.

5.2 The key role of the $\Lambda(1405)$ and its detection through the $\Sigma^+\pi^-$ decay

The full understanding of the $\Lambda(1405)$ structure is one of the hottest topics in the low energy KN interactions domain. In order to understand its nature, new measurements involving non-zero momentum K^- absorptions are mandatory [24]. Analysing the events occurring in the DC entrance wall it is possible to identify a subsample of K^- interacting on the hydrogen and the carbon contained in it. These events provide, indeed, the proof of the presence in the 2002-2005 KLOE data of both in-flight and at-rest interactions. To confirm this hypothesis, in 2012 a dedicated run with a pure carbon target was performed with the KLOE detector at DAΦNE; in this configuration, all the kaons are stopped in the carbon target, so only at-rest events can be found. Fig. 3 shows the $\Sigma^+\pi^-$ invariant mass spectra for the 2004-2005 (left) compared to the 2012 (right) data. In 2004-2005 data, in-flight (green), at-rest (violet) and total (blue) events are shown together with interactions with hydrogen atoms present in the DC entrance wall. In the right panel, 2012 data (blue) are compared with the 2004-2005 (black) data to show the difference due to in-flight events [24]. These spectra enable unique possibility to

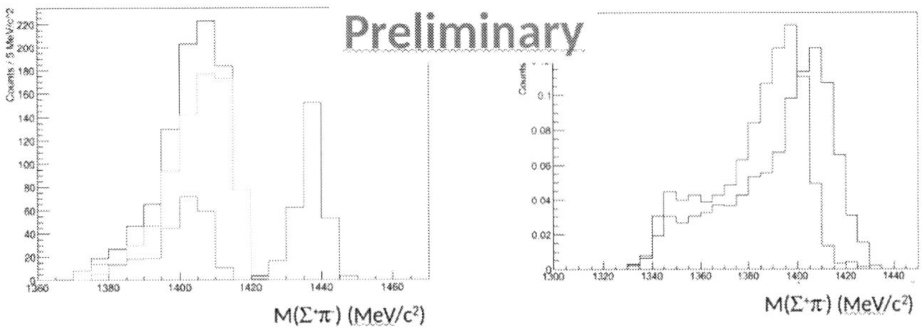

Figure 3. (Colour online.) Left panel: 2004-2005 data in-flight (green), at-rest (violet) and total (blue) events are shown together with events occurring on hydrogen atoms also present in the DC entrance wall. Right panel, 2012 data (blue) are compared with the 2004-2005 (black) data to show the difference due to in-flight events.

explore higher invariant mass region, not accessible untill now in experiments using kaons.

In addition to the analysis shown above, other interesting processes can be studied. Multi-nucleon absorption can be also studied in the $\Sigma^0 t$ channel. Analysis on $\Lambda/\Sigma p, d$, and Σt channels are also ongoing. The $\Lambda(1405)$ resonance can be studied in its pure I=0 $\Sigma^0\pi^0$ decay channel, whilst the ratio between the resonant and non-resonant contribution to the $\Sigma\pi$ channel can be studied starting from the pure I=1 $\Lambda\pi^-$ decay of the $\Sigma^*(1385)$ [25].

6 Conclusions and outlook

The broad experimental program of AMADEUS, dealing with the non-perturbative QCD in the strangeness sector, is supported by the quest for high precision and statistics measurements, able to set more stringent constraints on the existing theoretical models. We demonstrated the capabilities of performing high-quality research on nuclear interactions of strange particles, taking advantage of the KLOE detector and the DAΦNE factory. Our investigations, presently spread on a wide spectrum of physical processes like $\Lambda/\Sigma - p, d, t$ correlations and $\Lambda^*(1405)/\Sigma^*(1385)$ properties investigations, represent the most ambitious and systematic effort in the field of low energy KN interactions. In particular, after general considerations on the analysis procedures and a brief overview on the $K^- +^4 He \rightarrow \Lambda t$ and $K^- p \rightarrow \Sigma^+\pi^-$ analyses, in this report we have presented the results of the K^- absorption processes leading to the $\Sigma^0 p$ final state measured with the KLOE detector. The absorption on two nucleons without final state interaction (2NA-QF) is isolated and its yield normalised to the number of absorbed K^- is presented. The possibility to accomodate in the final spectra a ppK^- bound state decaying into a $\Sigma^0 p$ pair has been also carried out and the best value of the total reduced χ^2 is achieved for the hypothesis of a ppK^- with a binding energy of 45 MeV/c^2 and a width of 30 MeV/c^2. A F-test is conducted to compare the simulation models with and without the ppK^- signal and to extract the significance of the result. A significance of only 1σ is obtained for the ppK^- yield result. For the future we plan to realise a dedicated AMADEUS setup, where to explore more in details the various processes with a broad range of dedicated targets.

Acknowledgement

We acknowledge the KLOE Collaboration for their support and for having provided us the data and the tools to perform the analysis presented in this paper.

References

[1] The AMADEUS collaboration, LNF preprint, LNF/9607/24(IR).
AMADEUS Letter of Intent:
http://www.lnf.infn.it/esperimenti/siddharta/LOI_AMADEUS_March2006.pdf
[2] D. Cabrera *et al.*, Phys. Rev. C **90**, 0552017 (2014)
[3] A. E. Nelson *et al.*, Phys. Lett. B **192**, 193 (1987)
[4] M. Bazzi *et al. (SIDDHARTA Coll.)*, Phys. Lett. B **704**, 113 (2011)
[5] P. Demorest *et al.*, Nature **467**, 1081 (2010)
[6] T. Yamazaki, *et al.*, Phys. Rev. C **76** 045201 (2007)
A. Doté, *et al.*, Phys. Rev. C **79** 014003 (2009)
S. Wycech, *et al.*, Phys. Rev. C **79** 014001 (2009)
N. Barnea, *et al.*, Phys. Lett. B **712** 132 (2012)
N.V. Shevchenko, *et al.*, Phys. Rev. Lett. **98** 082301 (2007)
Y. Ikeda, *et al.*, Phys. Rev. C **79** 035201 (2009)
E. Oset, *et al.*, Nucl. Phys. A **881** 127 (2012)
[7] G. Agakishiev, *et al.*, HADES Coll., Phys. Lett. B **742** 242 (2015)
[8] M. Agnello, *et al. (FINUDA Coll.)*, Phys. Rev. Lett. **94** 212303 (2005)
[9] T. Yamazaki, *et al.*, Phys. Rev. Lett. **104** 132502 (2010)
[10] Y. Ichikawa, *et al.*, Prog. Theor. Exp. Phys. **021D01** (2015)
[11] A. O. Tokiyasu *et al.*, Phys.Lett. B **728**, 616 (2014)
[12] L. Fabbietti *et al.*, Nucl. Phys. A **914**, 60 (2013)
[13] T. Hashimoto *et al.*, Prog. Theor. Exp. Phys. 061D01 (2015)
[14] A. Gal, Nucl. Phys. A **914** 270 (2013)
[15] P. A. Katz *et al.*, Phys. Rev. D **1** 1267 (1970)
[16] M. Agnello, *et al.*, Phys. Rev. C **92** 045204 (2015)
[17] INFN (1990) 371 p. and Frascati INFN - LNF-90-031(R) (90,rec.Jun.) 371 p
[18] F. Bossi *et al. (KLOE coll.)*, Nuovo Cim. **31** 531-623 (2008)
[19] K. Piscicchia *et al.* PoS Bormio2013 **034** (2013)
[20] O. Vazquez Doce *et al.* Phys. Lett. B **758** 134-139 (2016)
[21] R. Roosen, J. H. Wickens, Il Nuovo Cim. A Series 11, **66** 101 (1981)
[22] FINUDA collaboration, Phys. Lett. B, **669** 229 (2008)
[23] J. Beringer, *et al.*, (Particle Data Group), Phys. Rev. D **86**, 010001 (2012)
[24] A. Scordo, *et al.*, PoS Bormio2014 **039** (2014)
[25] K. Piscicchia, S. Wycech, Nucl. Phys. A **954**, 75-93 (2016)

Precision spectroscopy of pionic atoms and chiral symmetry in nuclei

Kenta Itahashi[1,⋆], DeukSoon Ahn[1], Georg P.A. Berg[2], Masanori Dozono[1], Daijiro Etoh[3], Hiroyuki Fujioka[4], Naoki Fukuda[1], Nobuhisa Fukunishi[1], Hans Geissel[5], Emma Haettner[5], Tadashi Hashimoto[1], Ryugo S. Hayano[6], Satoru Hirenzaki[7], Hiroshi Horii[6], Natsumi Ikeno[8], Naoto Inabe[1], Masahiko Iwasaki[1], Daisuke Kameda[1], Shouichiro Kawase[9], Keiichi Kisamori[1,9], Yu Kiyokawa[9], Toshiyuki Kubo[1], Kensuke Kusaka[1], Masafumi Matsushita[9], Shin'ichiro Michimasa[9], Go Mishima[6], Hiroyuki Miya[9], Daichi Murai[1], Hideko Nagahiro[7], Takahiro Nishi[1], Shinsuke Ota[9], Naruhiko Sakamoto[1], Kimiko Sekiguchi[3], Hiroshi Suzuki[1], Ken Suzuki[10], Motonobu Takaki[9], Hiroyuki Takeda[1], Yoshiki K. Tanaka[5,6], Tomohiro Uesaka[1], Yasumori Wada[3], Yuni N. Watanabe[6], Helmut Weick[5], Hiroki Yamakami[4], Yoshiyuki Yanagisawa[1], and Koichi Yoshida[1]

[1]*Nishina Center for Accelerator-Based Science, RIKEN, Saitama 351-0198, Japan*
[2]*JINA and Department of Physics, University of Notre Dame, Indiana 46556, USA*
[3]*Department of Physics, Tohoku University, Sendai 980-8578, Japan*
[4]*Department of Physics, Kyoto University, Kyoto 606-8502, Japan*
[5]*GSI Helmholtzzentrum für Schwerionenforschung GmbH, D-64291 Darmstadt, Germany*
[6]*Department of Physics, University of Tokyo, Tokyo 113-0033, Japan*
[7]*Department of Physics, Nara Women's University, Nara 630-8506, Japan*
[8]*Department of Regional Environment, Tottori University, Tottori 680-8551, Japan*
[9]*Center for Nuclear Study, University of Tokyo, Tokyo 113-0033, Japan*
[10]*Stefan Meyer Institute, Austrian Academy of Sciences, A-1090 Vienna, Austria*

Abstract. We conduct an experimental project to make spectroscopy of deeply bound pionic atoms systematically over wide range of nuclei. We aim at studying the strong interaction in the low energy region, which has close connection to spontaneous chiral symmetry breaking and its partial restoration in nuclear matter. First experimental results show improved spectral resolution and much better statistical sensitivity than previous experiments. Present status of the experiment is reported.

1 Introduction

Strong interaction in low-energy region exhibits various interesting aspects. One of the particularly interesting aspects is spontaneous breakdown of chiral symmetry, which is characterised by the order parameter of the symmetry, chiral condensate. Pionic atoms, bound systems of a negatively charged pion to a nucleus, can be used for investigating the chiral symmetry in the nucleus [1–3], where the high density results in partial restoration of the chiral symmetry. Theoretically, interaction strength

⋆e-mail: itahashi@riken.jp

between pion and nucleus has a close connection with the partial restoration [4, 5], and precision determination of the isovector interaction sets quantitative constraints to the chiral condensate $\langle \bar{q}q \rangle$ in the nuclear medium.

Orbitals of pions in "deeply bound pionic atoms" are largely overlapping with the nuclei, and the pionic levels and widths reflect the s-wave pion-nucleus interaction [2, 3], which is phenomenologically formulated in the Ericson-Ericson optical potential [6]. The s-wave part of the optical potential is described in the following form;

$$\begin{aligned} -2\mu U_{s\text{-wave}}(r) &= 4\pi\epsilon_1\{b_0\rho(r) + b_1\Delta\rho(r)\} + 4\pi\epsilon_2 B_0\rho^2(r), \\ \rho(r) &= \rho^{(\mathrm{n})}(r) + \rho^{(\mathrm{p})}(r), \\ \Delta\rho(r) &= \rho^{(\mathrm{n})}(r) - \rho^{(\mathrm{p})}(r), \end{aligned}$$

where $\epsilon_1 = 1 + \mu/M = 1.147$, $\epsilon_2 = 1 + \mu/2M = 1.073$ with M being the nucleon mass and μ being the reduced mass of the pion. b_0, b_1 and B_0 are the s-wave potential parameters. $\rho^{(n)}(r)$ and $\rho^{(p)}(r)$ are radial neutron and proton density distributions of a nucleus, respectively.

Among the parameters to parametrize the potential, isovector parameter b_1 is known to have close-connection with the chiral condensate at a finite density through the Gell-Mann–Oaks–Renner relation [7] and Tomozawa-Weinberg relation [8, 9]. Systematic measurement of pionic tin isotopes was performed in GSI, and the energy levels and widths of $1s$ states in the pionic tin 115, 119, and 123 isotopes were measured [1]. The measured energy levels and widths were analyzed to deduce the b_1 parameter in medium leading to quantitative evaluation of $\langle \bar{q}q \rangle$ in nuclear medium to be reduced by 30% compared with that in vacuum although the deduced value was associated with relatively large errors. In order to deduce the chiral condensate with better precision, high precision measurement of pionic atoms was awaited.

2 Experiment

Aiming at unprecedented spectral resolution of 300 keV (FWHM), we conducted an experimental spectroscopy of pionic tin isotopes at RI Beam Factory, RIKEN, Japan [10], using $(d, {}^3\mathrm{He})$ nuclear reaction for the production of the pionic atoms as the method of the reaction spectroscopy was established in the former experiments [11–14]. The Q-value of the reaction was precisely measured to deduce the mass of the reaction product, pionic atoms, in the missing mass spectroscopy. The produced pionic atoms are observed as peak structures in the Q-value spectrum or in the excitation spectrum of the reaction product below the free pion emission threshold. In the $(d, {}^3\mathrm{He})$ nuclear reaction, a neutron is picked up from the target nucleus hence the produced pionic bound states are coupling with a neutron hole state in the core nucleus.

We employed 500 MeV deuteron beam accelerated by the Superconducting Ring Cyclotron (SRC). A remarkable characteristics of the RI Beam Factory used as "Pionic Atom Factory (piAF)" is its high current primary beam with an intensity of $\geq 10^{12}$/s, which helps measurement with a good statistical sensitivity. As for the targets, we chose ${}^{122}\mathrm{Sn}$ and ${}^{117}\mathrm{Sn}$ and placed them at the nominal target position. There are theoretical calculations for the production cross section of pionic atoms in the $(d, {}^3\mathrm{He})$ reactions [15–17]. An isotope with an odd neutron number was chosen for the first time. We expected good separation of spectral strengths with valence neutron pick-up from those with core-neutron pick-up [18]. The emitted ${}^3\mathrm{He}$ in the $(d, {}^3\mathrm{He})$ reaction were momentum analyzed by BigRIPS projectile fragment separator used as a forward spectrometer with the momentum resolving power of $\delta p/p \sim 3000$.

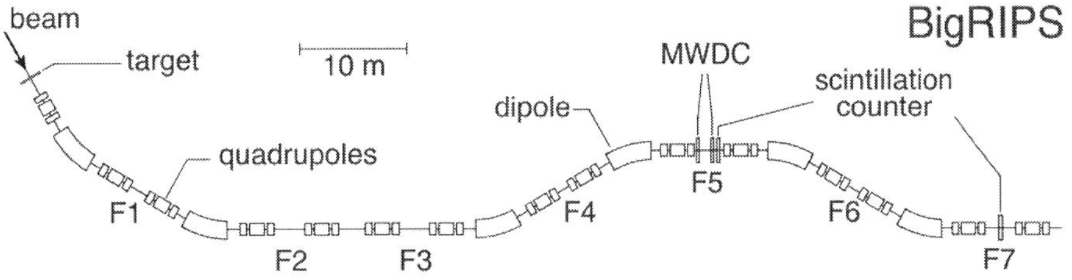

Figure 1. Experimental setup at BigRIPS. The 500 MeV deuteron beam was extracted from SRC impinging on the target. We installed two sets of MWDCs at F5 dispersive focal plane and sets of scintillation counters at F5 and F7.

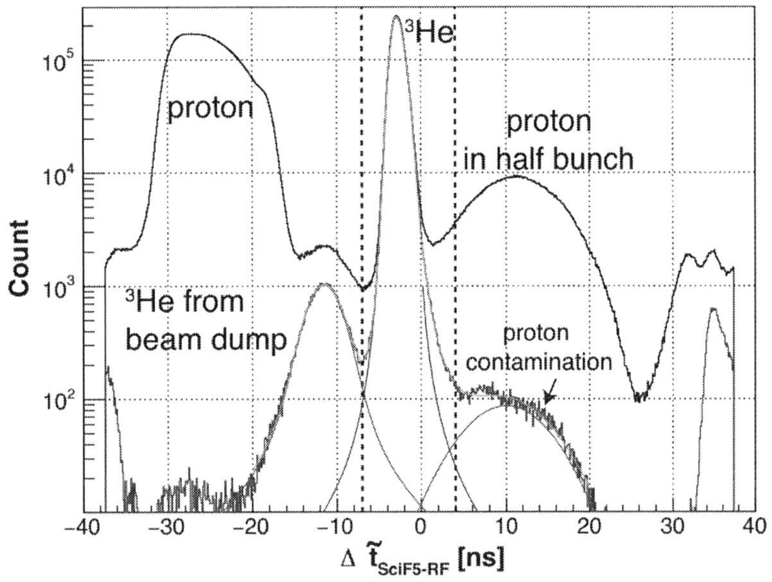

Figure 2. Measured relative timing of particles at F5 to RF phase. The central peak corresponds to ^{3}He. A cut condition has been applied to select ^{3}He. Proton contamination level is estimated to be $\leq 10^{-3}$ after combining with other cut conditions.

Figure 1 schematically depicts the experimental setup. We placed two sets of multi-wire drift chambers (MWDCs) at the dispersive focal plane (F5) and measured the ^{3}He tracks for the momentum analysis. We also placed a segmented scintillation counter at F5 and a scintillation counter at an achromatic focal plane F7 to identify the particles by the energy loss in the scintillation counters, by the time of flight between F5 and F7, and by relative timing to the radio frequency (RF) of SRC.

Figure 2 shows a measured relative timing spectrum of particles at F5 to the RF phase. Since the particles impinging on the target are bunched and synchronized to the RF phase, the relative timing reflects the time-of-flight from the target to F5. The central peak corresponds to ^{3}He. We applied a cut condition indicated by the vertical dashed lines to select ^{3}He. Proton contamination level is controlled to be $\leq 10^{-3}$ after combining with other cut conditions.

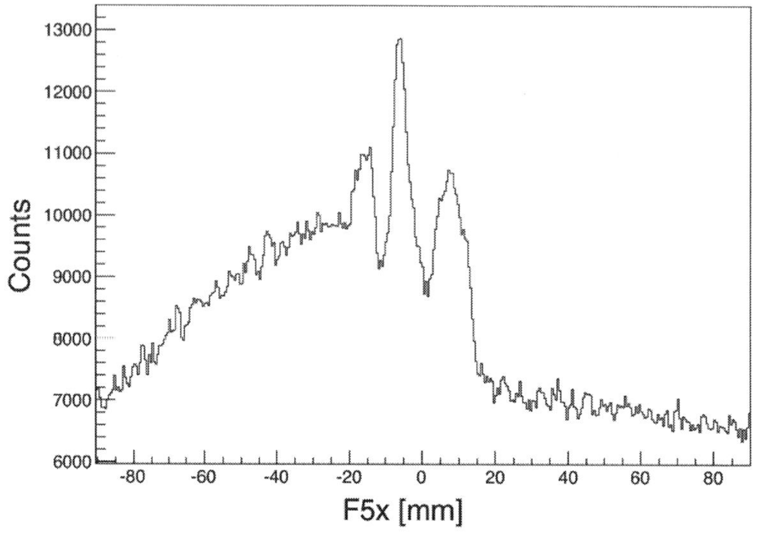

Figure 3. Measured acceptance uncorrected position spectrum at F5. Right side corresponds to lower excitation energy of the reaction products. Nearly linear background is observed in the region 20 - 80 mm, which is due to nuclear excitation without pion production. At left side −80 - −30mm, we observe quasifree pion production continuum. Peak structures near the center corresponds to production of pionic atoms. The right-most peak is due to $1s$ pionic atom formation coupling with a neutron hole state of $3s_{1/2}$. The central peak is mainly due to pionic $2p$ states.

Figure 3 shows a typical acceptance-uncorrected position spectrum measured at F5. The right side of the spectrum corresponds to higher ^{3}He momentum. Spectral strengths are standing on a nearly linear background of nuclear excitation unassociated with pion production. The peaks near the center of the spectrum are originating in the formation of pionic atoms, and the right-most peak is associated with the pionic $1s$ state coupling mainly with $3s_{1/2}$ neutron hole state. The achieved experimental resolution is estimated to be as good as ~300 keV (FWHM). Further analysis to improve the experimental resolution and to deduce the binding energies and widths is ongoing.

3 Conclusion

The present piAF project makes full use of the advantages of the facility and aims at performing systematic high precision spectroscopy of pionic atoms. Recent experiments achieved unprecedented resolution and precision of the measurement by optimization of the experimental conditions including partial application of the dispersion-matching ion-optics. A simultaneous measurement of $1s$ and $2p$ pionic levels will also contribute to reduction of the systematic uncertainties in the relative energy scale. Considering possible perturbative effects of pionic orbitals to nuclei, systematic study of pionic atoms over wide range of nuclei will also benefit in comprehensive understanding of the strong interaction and the nuclear structure.

Acknowledgments

This work is supported by MEXT Grants-in-Aid for Scientific Research on Innovative Areas (Nos. JP22105517, JP24105712 and JP15H00844).

References

[1] K. Suzuki *et al.*, Phys. Rev. Lett. **92**, 072302 (2004)
[2] R. S. Hayano and T. Hatsuda, Rev. Mod. Phys. **82**, 2949 (2010)
[3] T. Yamazaki *et al.*, Phys. Rept. **514**, 1 (2012)
[4] E. E. Kolomeitsev, N. Kaiser, and W. Weise, Phys. Rev. Lett. **90**, 092501 (2003)
[5] D. Jido, T. Hatsuda and T. Kunihiro, Phys. Lett. B **670**, 109 (2008)
[6] M. Ericson and T. E. O. Ericson, Ann. Phys. **36**, 323 (1966)
[7] M. Gell-Mann, R. J. Oakes, and B. Renner, Phys. Rev. **175**, 2195 (1968)
[8] Y. Tomozawa, Nuovo Cimento A **46**, 707 (1966)
[9] S. Weinberg, Phys. Rev. Lett. **17**, 616 (1966)
[10] Y. Yano, Nucl. Instr. Meth. Phys. Res. B **261**, 1009 (2007)
[11] T. Yamazaki *et al.*, Z. Phys. A **355**, 219 (1996)
[12] H. Gilg *et al.*, Phys. Rev. C **62**, 025201 (2000)
[13] K. Itahashi *et al.*, Phys. Rev. C **62**, 025202 (2000)
[14] H. Geissel *et al.*, Phys. Rev. Lett. **88**, 122301 (2002)
[15] N. Ikeno *et al.*, Prog. Theor. Phys. **126**, 483 (2011)
[16] N. Ikeno, H. Nagahiro and S. Hirenzaki, Euro. Phys. J. A **47**, 161 (2011)
[17] N. Ikeno *et al.*, Prog. Theor. Exp. Phys. **2015**, 033D01 (2015)
[18] N. Ikeno *et al.*, Prog. Theor. Exp. Phys. **2013**, 063D01 (2013)

EPJ Web of Conferences **130**, 01018 (2016) DOI: 10.1051/epjconf/201613001018
MESON 2016

Future projects of light kaonic atom X-ray spectroscopy

H. Tatsuno[1,⋆], M. Bazzi[2], G. Beer[3], G. Bellotti[4,5], C. Berucci[2,6], A.M. Bragadireanu[2,7], D. Bosnar[8], M. Cargnelli[6], C. Curceanu[2], A.D. Butt[4,5], A. d'Uffizi[2], C. Fiorini[4,5], F. Ghio[9], C. Guaraldo[2], R.S. Hayano[10], M. Iliescu[2], T. Ishiwatari[6], M. Iwasaki[11], P. Levi Sandri[2], J. Marton[6], S. Okada[12], D. Pietreanu[2,7], K. Piscicchia[2,13], A. Romero Vidal[14], E. Sbardella[2], A. Scordo[2], H. Shi[2], D.L. Sirghi[2,7], F. Sirghi[2,7], O. Vazquez Doce[15], E. Widmann[6], and J. Zmeskal[6]
for the SIDDHARTA Collaboration

[1]*Department of Chemical Physics, Lund University, Lund, Sweden*
[2]*INFN, Laboratori Nazionali di Frascati, Frascati (Roma), Italy*
[3]*Department of Physics and Astronomy, University of Victoria, Victoria BC, Canada*
[4]*Politecnico di Milano, Dipartimento di Elettronica e Informazione, Milano, Italy*
[5]*INFN Sezione di Milano, Milano, Italy*
[6]*Stefan-Meyer-Institut für subatomare Physik, Wien, Austria*
[7]*IFIN-HH, Institutul National pentru Fizica si Inginerie Nucleara Horia Hulubbei, Magurele, Romania*
[8]*Department of Physics, Faculty of Science, University of Zagreb, Zagreb, Croatia*
[9]*INFN Sezione di Roma I and Istituto Superiore di Sanità, Roma, Italy*
[10]*Department of Physics, School of Science, The University of Tokyo, Tokyo, Japan*
[11]*RIKEN Nishina Center, RIKEN, Wako, Japan*
[12]*AMO physics laboratory, RIKEN, Wako, Japan*
[13]*Museo Storico della Fisca e Centro Studi e Ricerche "Enrico Fermi", Roma, Italy*
[14]*Universidade de Santiago de Compostela, Santiago de Compostela, Spain*
[15]*Excellence Cluster Universe, Technische Universität München, Garching, Germany*

Abstract. X-ray spectroscopy of light kaonic atoms is a unique tool to provide precise information on the fundamental $\bar{K}N$ interaction at the low-energy limit and the in-medium nuclear interaction of K^-. The future experiments of kaonic deuterium strong-interaction shift and width (SIDDHARTA-2 and J-PARC E57) can extract the isospin dependent K^-N interaction at threshold. The high-resolution X-ray spectroscopy of kaonic helium with microcalorimeters (J-PARC E62) has the possibility to solve the long-standing potential-strength problem of the attractive K^--nucleus interaction. Here, the recent experimental results and the future projects of X-ray spectroscopy of light kaonic atoms are presented.

1 Introduction

The kaon is one of the pseudo-scalar mesons, the mass (495 MeV) lies between that of pion (140 MeV) and eta (510 MeV) mesons, and the anti-kaon $\bar{K}$ ($\bar{K}^0$ and K^-) includes a strange quark ($\bar{d}s$ and $\bar{u}s$). Therefore the anti-kaon is a unique and suitable particle to study the role of explicit chiral

⋆e-mail: hideyuki.tatsuno@gmail.com,

symmetry breaking due to the relatively heavier strange quark and the dynamics of low-energy QCD in the strangeness sector. The in-medium nuclear interaction of the K^- is also important to understand the structure of exotic nuclear states such as the $\bar{K}N$ sub-threshold resonance $\Lambda(1405)$, the nuclear bound state of K^-, as well as the high density of matter in the universe (neutron stars).

The precise study of the $\bar{K}N$ strong interaction has started from X-ray spectroscopy of kaonic atoms. The kaonic atom is a Coulomb-bound system of a K^-, electrons, and a nucleus. The transition X-ray energy is typically several keV for light kaonic atoms due to the K^-/electron mass ratio of ~ 1000. The $\bar{K}N$ interaction is attractive at short range, thus the strong interaction effects appear on the atomic states as the perturbation; an energy shift from the QCD calculation and a life-time broadening due to nuclear absorption of K^-. The $1s$-state strong-interaction shifts and widths of kaonic hydrogen (K^--p) and kaonic deuterium (K^--d) can be used to directly extract the isospin dependent K^-N interaction at threshold by the Deser-Trueman type formula [1]. They are necessary for precise calculations of chiral SU(3) meson-baryon effective-field theory in low-energy QCD. Furthermore, the K^-N threshold information provides the basic constraint on the structure of the $\Lambda(1405)$ [2, 3]. The X-ray spectroscopy of other kaonic atoms (helium to uranium) enables the study of the strong interaction of the K^- in dense nuclei. The bound state of a K^- in a nucleus has been predicted using the strongly attractive interaction [4], and recently the K^- multi-nucleon cluster system like K^-pp is one of the hot topics in this field.

Currently, the strong-interaction shifts and widths of kaonic atoms from hydrogen to uranium have been measured. The K^-p scattering length has been determined, and also the K^--nucleus interaction is qualitatively understood. A systematic analysis with a phenomenological density-dependent optical potential can accurately describe the trend of shift and width of kaonic atoms and predict the deep potential depth (Re(V_0) ~ -180 MeV) [5, 6]. In addition to that, a chiral SU(3) calculation which is based on the experimental data of the $\bar{K}N$ interaction can also describe the experimental data of shifts and widths and predict the relatively shallow potential depth (Re(V_0) ~ -50 MeV) [7]. This quantitative disagreement, so-called 'deep-or-shallow problem' is fundamentally correlated to the characteristic and structure of the K^- multi-nucleon cluster states. Hence, a more precise and decisive experimental measurement has been awaited.

We present here the key experiments and the future projects of light kaonic atom X-ray spectroscopy:

1. past and recent results of kaonic hydrogen $2p \rightarrow 1s$ X-ray measurement by KEK-PS E228 [8], DEAR [9], and SIDDHARTA [10],

2. first measurement of kaonic deuterium $2p \rightarrow 1s$ X-rays planned by the SIDDHARTA-2 [11] and the J-PARC E57 [12] groups,

3. high-resolution x-ray spectroscopy of kaonic helium-3 and helium-4 with superconducting transition-edge-sensors (TESs) by the HEATES and J-PARC E62 groups [13], which can solve the deep-or-shallow problem on the K^--nucleus potential depth.

2 Kaonic hydrogen

The primary study of the $\bar{K}N$ interaction based on the K^-p scattering experiments in the 1960s-1980s. The threshold information was extrapolated with the cross section data with large uncertainties. At that time, the precise measurement technique of kaonic hydrogen X-rays was not yet established; therefore even the sign of the $\bar{K}N$ interaction (attractive or repulsive) was not determined. Later, the kaonic hydrogen X-ray measurement with new background suppression technique and with new X-ray detectors was performed by three experimental groups. The details of experiments are shown in Table 1.

The reliable $\bar{K}N$ threshold information was first provided in the 1990s. The KEK-PS E228 group measured the strong-interaction shift and width of K^--p $1s$ state at KEK-PS in Tsukuba, Japan [8]. The $2p \rightarrow 1s$ X-ray energy is about 6.2 keV, and the strong interaction shift is ~ -300 eV (repulsive) and the width is $\sim$ 400 eV. The K^- beam extracted through the K3 secondary beam line was slowed down with carbon degraders and stopped on the hydrogen gaseous target. They measured the X-rays with Si(Li) detectors installed inside of the target chamber and also tracked the coincident secondary charged particles generated from the reaction of K^- and nucleus with a cylindrical drift chamber system. The event selection of specific reaction modes and the timing selection of stopped K^- successfully reduced the synchronous background, and they clearly observed the clear peaks of K^--p X-rays. Though the background was small, the statistical uncertainties of the shift and width were large (see Table 1), because the energy resolution of the Si(Li) detectors was 360 eV full-width-at-half-maximum (FWHM) at 6 keV (this is relatively worse than recent silicon detectors of 150 eV FWHM). However, it was enough to determine the sign of the $\bar{K}N$ interaction (attractive) and laid the basis of theory. The attractive $\bar{K}N$ interaction led to the prediction of the deeply-bound K^- nuclear states [4]. The established event selection technique at the high intensity hadron beam line has been transitioned to other kaonic atom measurements at KEK and J-PARC in Tokai, Japan [12, 14].

In 2005, the DEAR group measured the strong-interaction shift and width of kaonic hydrogen at the DAΦNE positron-electron collider in Frascati, Italy [9]. DAΦNE produced almost at rest $\phi(1020)$ mesons with 510 MeV e^+ and e^- beams. The low kinetic energy K^-s from $\phi \rightarrow K^+K^-$ decay were easily stopped in the gaseous target. They measured the K^--p X-rays with the CCD detectors with 180 eV FWHM energy resolution at 6 keV. Although the accumulated number of events was high, the background was also large due to no timing information of the CCDs. The background level of a lepton collider is usually lower than that of a hadron beam line. However, the signal-to-background ratio depends on the yield of the interesting event, and in the DEAR case, the K^--p X-ray yield is $\sim$ 1 % per stopped kaon in the 24 ρ_{STP} density hydrogen target (ρ_{STP} is the density at the standard temperature and pressure). They subtracted the large background and characteristic X-ray peaks of ion and manganese which were overlaid with the K^--p X-rays, and then determined the shift and width. After the DEAR result was published, the KEK-PS E228 and DEAR results suffered from theoretical ambiguities even with considering their experimental uncertainties [15].

The most precise and new results were provided by the SIDDHARTA group in 2011 at DAΦNE. They used newly developed silicon drift detectors (SDDs) with 150 eV FWHM energy resolution at 6 keV. They detected the decay charged kaons and created the hardware 'kaon' trigger. The trigger system and the sub-micro second timing resolution of SDDs successfully worked to reduce the background, and the significantly clear K^--p X-ray peaks were observed. One of the most remarkable things was the global analysis with K^--p and K^--d X-ray spectra. The measured X-ray spectrum with the deuterium target was useful to confirm the background shape and X-ray peaks of other kaonic atoms (K^--C, K^--N, K^--O, and K^--Al), because the K^--d X-ray yield was under the background level. The precise K^-p scattering length of the SIDDHARTA results provided new basic constrains for the $\bar{K}N$ interaction of chiral approach calculations (e.g., [3]).

3 Kaonic deuterium

The SIDDHARTA results of kaonic hydrogen give the strong constraint for the K^-p scattering length at the threshold. On the other hand, the calculation of K^-n scattering length (pure isospin=1 term) is

Table 1. Experimental details of kaonic hydrogen X-ray spectroscopy [8–10, 16], where ρ_{STP} is the density at the standard temperature and pressure, ΔE_{FWHM} is the full-width-at-half-maximum energy resolution, and Δt_{FWHM} is the time resolution. The first error of shift and width is statistical and second one is systematic.

	KEK-PS E228 (1997)	DEAR (2005)	SIDDHARTA (2011)
Place	KEK-PS K3 (Japan)	DAΦNE (Italy)	DAΦNE (Italy)
H_2 gas density	10 ρ_{STP}	24 ρ_{STP}	15 ρ_{STP}
X-ray detector	Si(Li) 120 cm^2	CCD 116 cm^2	SDD 114 cm^2
ΔE_{FWHM} at 6 keV	360 eV (beam off)	180 eV	150 eV
Δt_{FWHM}	290 ns	–	800 ns
1s shift (eV)	$-323 \pm 63 \pm 11$	$-193 \pm 37 \pm 6$	$-283 \pm 36 \pm 6$
1s width (eV)	$407 \pm 208 \pm 100$	$249 \pm 111 \pm 30$	$541 \pm 89 \pm 22$
$2p \to 1s$ X-ray yield	1.5±0.5 % / stopped K^-	–	$1.2^{+0.4}_{-0.3}$ % / stopped K^-

totally varied for theories [3]. To determine the isospin=1 term of the scattering length, the essential measurement of kaonic deuterium 1s-state strong-interaction shift and width has been awaited for a long time. There is no evidence of the K^--d X-rays so far due to the low X-ray yield (~ 10 % of kaonic hydrogen) and the broad width (650–1000 eV predicted by theory) [17, 18]. In order to measure such a low-yield and broad X-ray peaks, a high-efficient and low-background measurement is required. Aimed at the first observation of the K^--d X-rays, two experiments are planned at DAΦNE and J-PARC. Table 2 shows the expected parameters of the future experiments.

The SIDDHARTA-2 group plans to measure the K^--d X-rays from a gaseous deuterium target at DAΦNE [11]. They improve the event selection efficiency and the background suppression. The efficiency of event selection is improved by a K^+-tag counter to identify the K^- from $\phi \to K^+K^-$ decay. The synchronous background is suppressed by VETO counters to catch the coincident secondary charged particles generated from the K^--nucleus reactions. They are also developing new SDDs (total 200 cm^2) with 130 eV FWHM energy resolution at 6 keV and 400 ns timing resolution. Finally, these updates will achieve a 10-times higher signal-to-background ratio than the SIDDHARTA measurement.

The J-PARC E57 experiment will measure the K^--d X-rays from a gaseous deuterium target at J-PARC [12]. They use the high-intensity K^- beam at the K1.8 BR beam line. The K^--d X-rays are detected with new SDDs (total 246 cm^2) with conjugation of a cylindrical drift chamber system. The background suppression by the event selections of stopped-K^- and the K^-d reaction modes will result in a comparable signal-to-background ratio of SIDDHARTA-2.

4 Kaonic helium

Helium is a nucleus and its nuclear density can be treated as a medium for the K^-. The kaonic helium (K^--^{4}He and K^--^{3}He) is one of the good probes to investigate the medium effect of the K^- and the exotic K^--bound states in nuclei. While the direct information of the K^--nucleus interaction could be extracted from the strong-interaction 1s-state shift and width of kaonic helium, the $2p \to 1s$ X-ray has not been observed due to the low X-ray yield. This indicates that the K^- is most likely absorbed before the transition to the 1s state. Hence, the observable strong-interaction shift and width of kaonic helium are on the 2p state.

The kaonic helium 2p-state strong-interaction shift is predicted to be ~ 0 eV, and the width to be a few eV by both the phenomenological 'deep' optical potential model [5, 6] and the chiral unitary 'shallow' potential model [7]. Recently, a possibility to solve the deep-or-shallow problem has been

 DOI: 10.1051/epjconf/201613001018

Table 2. Expected experimental details of kaonic deuterium X-ray spectroscopy [11, 12].

	SIDDHARTA-2 (DAΦNE)	J-PARC E57 (K1.8 BR)
D_2 gas density	3 % of liquid D_2	5 % of liquid D_2
X-ray yield	0.1 % / stopped K^-	0.1 % / stopped K^-
X-ray detector	SDD 200 cm^2	SDD 246 cm^2
ΔE_{FWHM} at 6 keV	130 eV	130 eV
Δt_{FWHM}	400 ns	400 ns
Signal/BG	1/3	1/3
Precision 1s shift	±30 eV	±60 eV
Precision 1s width	±70 eV	±130 eV
Assumed 1s width	750 eV	800 eV
Feature	high geometrical efficiency, new trigger and VETO counters	high beam intensity, event selection with cylindrical drift chamber system

Table 3. Experimental results and details of kaonic helium X-ray spectroscopy [14, 19–22]. The first error of shift and width is statistical and second one is systematic.

	KEK-PS E570 (2007)	SIDDHARTA (2011)	J-PARC E62
Place	KEK-PS K5	DAΦNE	J-PARC K1.8 BR
Target	liquid ^{4}He	^{4}He gas / ^{3}He gas	liquid ^{4}He / liquid ^{3}He
X-ray detector	SDD 8 cm^2	SDD 144 cm^2	240-pixel TES 23 mm^2
ΔE_{FWHM} at 6 keV	190 eV	150 eV	5 eV
Δt_{FWHM}	380 ns	800 ns	1 μs
2p shift	$+2 \pm 2 \pm 2$ eV (^{4}He)	$+5 \pm 3 \pm 4$ eV (^{4}He) $-2 \pm 2 \pm 4$ eV (^{3}He)	goal ±0.2 eV (^{4}He–^{3}He)
2p width	≤ 18 eV (90 % CL)	$14 \pm 8 \pm 5$ eV (^{4}He) $6 \pm 6 \pm 7$ eV (^{3}He)	sensitive to determine

discussed with new theoretical calculations [23]. The *isotope shift* of K^--^{4}He and K^--^{3}He 2p states (the difference of two shifts) is calculated to be 0.6 eV for the deep potential [24] and 0.0 eV for the shallow one [25]. Therefore, the two models can be distinguished experimentally with the precision of 0.2 eV. On the other hand, the current best precision of the 2p-state shift is ±2 eV (stat.) ±2 eV (syst.) [14]. At least one order of magnitude higher precision is necessary to distinguish the predictions. In order to achieve such a high precision, new measurement technique and low noise X-ray detectors are needed.

The HEATES group has attempted to apply the superconducting TES microcalorimeters to high-resolution hadronic atom X-ray spectroscopy. The TES is an ideal detector to measure a thin X-ray line like kaonic helium $3d \to 2p$ X-ray and to determine the X-ray energy precisely. According to the results of feasibility test experiment [26], the 240-pixel TES array (23 mm^2 effective area) has achieved 5 eV FWHM energy resolution at 6 keV with 1 μs time resolution, and they successfully measured the pionic carbon $4f \to 3d$ X-ray energy with precision of ≤ 0.2 eV [27].

The J-PARC E62 group plans the measurement of both the K^--^{4}He and K^--^{3}He X-rays with the TESs at the J-PARC K1.8 BR beam line [13]. The successive measurement can minimize the systematic uncertainty of *isotope shift*, and it is free from the theoretical uncertainty associated with the error of charged kaon mass. The recent results and the future experiment of kaonic helium X-ray spectroscopy are summarized in Table 3.

5 Summary

The experimental results and the future projects of light kaonic atom spectroscopy are presented with the details of experimental methods.

The precise measurement of kaonic hydrogen X-rays by the SIDDHARTA group provided a unique possibility to determine the K^-p scattering lengths. This is one of the most important observables to investigate the chiral SU(3) dynamics in low-energy QCD. For further study of the isospin dependence of the K^-N interaction, the SIDDHARTA-2 and the J-PARC E57 groups plan the kaonic deuterium X-ray measurement with new SDDs. These measurements will provide constraints on the structure of the $\Lambda(1405)$ and the K^- multi-nucleon clusters.

The long-standing problem of the optical potential depth of the K^--nucleus interaction could be solved by the high-resolution kaonic helium X-ray measurement with superconducting microcalorimeters. The J-PARC E62 group plans the measurement of both the K^--^{4}He and K^--^{3}He X-rays with a 240-pixel TES array. The difference of $2p$-state strong-interaction shifts of K^--^{4}He and K^--^{3}He will be determined with the precision of 0.2 eV. The precise result can give a decisive settlement of the deep-or-shallow problem. This will be an important complement of the experimental direct search for the K^--bound states in nuclei.

Acknowledgments We thank C. Capoccia, G. Corradi, B. Dulach, and D. Tagnani from LNF-INFN; and H. Schneider, L. Stohwasser, and D. Stuükler from Stefan-Meyer-Institut, for their fundamental contribution in designing and building the SIDDHARTA setup. We thank as well the DAΦNE staff for the excellent working conditions and permanent support. Part of this work was supported by the European Community-Research Infrastructure Integrating Activity "Study of Strongly Interacting Matter" (Hadron-Physics2, Grant Agreement No. 227431, and Hadron-Physics3 (HP3) Contract No. 283286) under the Seventh Framework Programme of EU; Hadron-Physics I3 FP6 European Community program, Contract No. RII3-CT-2004-506078; Austrian Science Fund (FWF) (P24756-N20); Austrian Federal Ministry of Science and Research BMBWK 650962/0001 VI/2/2009; Romanian National Authority for Scientific Research, Contract No. 2-CeX 06-11-11/2006; the Grant-in-Aid for Specially Promoted Research (20002003), MEXT, Japan; and the Croatian Science Foundation, under project HRZZ 1680.

References

[1] S. Deser, M.L. Goldberger, K. Baumann, W. Thirring, Phys. Rev. **96**, 774 (1954); U.-G. Meißner, U. Raha, A. Rusetsky, Eur. Phys. J. C **35**, 349 (2004)
[2] A. Gal, Nucl. Phys. A **914**, 270 (2013)
[3] A. Cieplý, M. Mai, U.-G. Meißner, J. Smejkal, Nucl. Phys. A **954**, 17 (2016)
[4] Y. Akaishi, T. Yamazaki, Phys. Rev. C **65**, 044005 (2002)
[5] C.J. Batty, E. Friedman, A. Gal, Phys. Rep. **287**, 385 (1997)
[6] E. Friedman, Int. J. Mod. Phys. A **26**, 468 (2011)
[7] S. Hirenzaki, Y. Okumura, H. Toki, E. Oset, A. Ramos, Phys. Rev. C **61**, 055205 (2000)
[8] M. Iwasaki *et al.*, Phys. Rev. Lett. **78**, 3097 (1997); T. M. Ito *et al.*, Phys. Rev. C **58**, 2366 (1998)
[9] G. Beer *et al.*, Phys. Rev. Lett. **94**, 212302 (2005)
[10] SIDDHARTA Collaboration, M. Bazzi, *et al.*, Phys. Lett. B **704**, 113 (2011)
[11] SIDDHARTA-2 Collaboration, Proposal of Laboratori Nazionali di Frascati of INFN, The upgrade of the SIDDHARTA apparatus for an enriched scientific case, 2010

[12] J. Zmeskal *et al.*, Proposal for J-PARC 30-GeV Proton Synchrotron, Measurement of the strong interaction induced shift and width of the 1st state of kaonic deuterium at J-PARC (J-PARC E57), 2014
[13] R.S. Hayano *et al.*, Proposal for J-PARC 30-GeV Proton Synchrotron, Precision spectroscopy of kaonic atom X-rays with TES (J-PARC E62), 2015
[14] S. Okada, *et al.*, Phys. Lett. B **653**, 387 (2007)
[15] B. Borasoy, R. Nißler, W. Weise, Phys. Rev. Lett. **94**, 213401 (2005); B. Borasoy, U.-G. Meißner, R. Nißler, Phys. Rev. C **74**, 055201 (2006)
[16] SIDDHARTA Collaboration, M. Bazzi, *et al.*, Nucl. Phys. A **954**, 7 (2016)
[17] SIDDHARTA Collaboration, M. Bazzi, *et al.*, Nucl. Phys. A **907**, 69 (2013)
[18] M. Döring, U.-G. Meißner, Phys. Lett. B **704**, 663 (2011); T. Mizutani, C. Fayard, B. Saghai, K. Tsushima, Phys. Rev. C **87**, 035201 (2013); N.V. Shevchenko, Nucl. Phys. A **890–891**, 50 (2012)
[19] SIDDHARTA Collaboration, M. Bazzi, *et al.*, Phys. Lett. B **681**, 310 (2009)
[20] SIDDHARTA Collaboration, M. Bazzi, *et al.*, Phys. Lett. B **697**, 199 (2011)
[21] SIDDHARTA Collaboration, M. Bazzi, *et al.*, Phys. Lett. B **714**, 40 (2012)
[22] H. Tatsuno, *Precision X-ray spectroscopy of kaonic helium atom*, Ph.D. Thesis, The University of Tokyo, 2011
[23] J. Yamagata-Sekihara, S. Hirenzaki, E. Hiyama *et al.* (private communication)
[24] J. Mareš, E. Friedman, A. Gal, Nucl. Phys. A **770**, 84 (2006)
[25] A. Ramos, E. Oset, Nucl. Phys. A **671**, 481 (2000)
[26] H. Tatsuno *et al.*, J. Low. Temp. Phys. **184**, 930 (2016)
[27] S. Okada *et al.*, Prog. Theor. Exp. Phys., accepted; S. Okada *et al.*, arXiv:1608.05436

EPJ Web of Conferences **130**, 01019 (2016) DOI: 10.1051/epjconf/201613001019
MESON 2016

Kaon experiments at CERN: recent results and prospects

Evgueni Goudzovski[1,⋆]

[1] *School of Physics and Astronomy, University of Birmingham, B15 2TT, United Kingdom*

Abstract. The NA48/2 and NA62-R_K experiments at the CERN SPS collected large samples of charged kaon decays in flight in 2003–07. The data analysis is still on-going (with over 20 publications produced so far); the recent results from these experiments are presented. A new upper limit on the rate of a lepton number violating decay $K^\pm \to \pi^\mp\mu^\pm\mu^\pm$ is reported: $\mathcal{B} < 8.6\times10^{-11}$ at 90% CL. Searches for heavy sterile neutrino N_4 and neutral scalar resonances (χ) in $K^\pm \to \pi\mu\mu$ decays are reported. Upper limits on the products $\mathcal{B}(K^\pm \to \mu^\pm N_4)\mathcal{B}(N_4 \to \pi^\mp\mu^\pm)$ and $\mathcal{B}(K^\pm \to \pi^\pm\chi)\mathcal{B}(\chi \to \mu^+\mu^-)$ are set in the range 10^{-10} to 10^{-9} for resonance lifetimes up to 100 ps. A preliminary measurement of the electromagnetic transition form factor slope of the π^0 from 1.05×10^6 fully reconstructed $\pi^0 \to \gamma e^+e^-$ decays is presented: the obtained result $a = (3.70 \pm 0.53_{\rm stat} \pm 0.36_{\rm syst}) \times 10^{-2}$ represents the first observation of a non-zero slope in the time-like region of momentum transfer.

⋆For the NA48/2 and NA62 collaborations: F. Ambrosino, A. Antonelli, G. Anzivino, R. Arcidiacono, W. Baldini, S. Balev, J.R. Batley, M. Behler, S. Bifani, C. Biino, A. Bizzeti, B. Bloch-Devaux, G. Bocquet, V. Bolotov, F. Bucci, N. Cabibbo, M. Calvetti, N. Cartiglia, A. Ceccucci, P. Cenci, C. Cerri, C. Cheshkov, J.B. Chèze, M. Clemencic, G. Collazuol, F. Costantini, A. Cotta Ramusino, D. Coward, D. Cundy, A. Dabrowski, G. D'Agostini, P. Dalpiaz, C. Damiani, H. Danielsson, M. De Beer, G. Dellacasa, J. Derré, H. Dibon, D. Di Filippo, L. DiLella, N. Doble, V. Duk, J. Engelfried, K. Eppard, V. Falaleev, R. Fantechi, M. Fidecaro, L. Fiorini, M. Fiorini, T. Fonseca Martin, P.L. Frabetti, A. Fucci, S. Gallorini, L. Gatignon, E. Gersabeck, A. Gianoli, S. Giudici, A. Gonidec, E. Goudzovski, S. Goy Lopez, E. Gushchin, B. Hallgren, M. Hita-Hochgesand, M. Holder, P. Hristov, E. Iacopini, E. Imbergamo, M. Jeitler, G. Kalmus, V. Kekelidze, K. Kleinknecht, V. Kozhuharov, W. Kubischta, V. Kurshetsov, G. Lamanna, C. Lazzeroni, M. Lenti, E. Leonardi, L. Litov, D. Madigozhin, A. Maier, I. Mannelli, F. Marchetto, G. Marel, M. Markytan, P. Marouelli, M. Martini, L. Masetti, P. Massarotti, E. Mazzucato, A. Michetti, I. Mikulec, M. Misheva, N. Molokanova, E. Monnier, U. Moosbrugger, C. Morales Morales, M. Moulson, S. Movchan, D.J. Munday, M. Napolitano, A. Nappi, G. Neuhofer, A. Norton, T. Numao, V. Obraztsov, V. Palladino, M. Patel, M. Pepe, A. Peters, F. Petrucci, M.C. Petrucci, B. Peyaud, R. Piandani, M. Piccini, G. Pierazzini, I. Polenkevich, I. Popov, Yu. Potrebenikov, M. Raggi, B. Renk, F. Retière, P. Riedler, A. Romano, P. Rubin, G. Ruggiero, A. Salamon, G. Saracino, M. Savrié, M. Scarpa, V. Semenov, A. Sergi, M. Serra, M. Shieh, S. Shkarovskiy, M.W. Slater, M. Sozzi, T. Spadaro, S. Stoynev, E. Swallow, M. Szleper, M. Valdata-Nappi, P. Valente, B. Vallage, M. Velasco, M. Veltri, S. Venditti, M. Wache, H. Wahl, A. Walker, R. Wanke, L. Widhalm, A. Winhart, R. Winston, M.D. Wood, S.A. Wotton, O. Yushchenko, A. Zinchenko, M. Ziolkowski. Email: eg@hep.ph.bham.ac.uk; supported by ERC Starting Grant 336581.

Introduction

The NA48/2 experiment at the CERN SPS collected a large sample of charged kaon decays in 2003–04 (corresponding to about 2×10^{11} $K^\pm$ decays in the vacuum decay volume). The experiment used simultaneous K^+ and K^- beams and was optimized for the search for direct CP violating charge asymmetries in the $K^\pm \to 3\pi^\pm$ decays [1]. Its successor, the NA62-R_K experiment, collected a 10 times smaller $K^\pm$ decay sample with low intensity beams and minimum bias trigger conditions in 2007 using the same detector [2]. The large data samples accumulated by the two experiments have allowed precision studies of rare $K^\pm$ decays and searches for new physics. Recent results from these experiments are reported.

1 Beam, detector and data sample

The NA48/2 and NA62-R_K experiments used simultaneous K^+ and K^- beams produced by 400 GeV/c primary CERN SPS protons impinging on a beryllium target. Charged particles in a narrow momentum band were selected by an achromatic system of four dipole magnets which split the two beams in the vertical plane and recombined them on a common axis. The beams then passed through collimators and a series of quadrupole magnets, and entered a 114 m long cylindrical vacuum tank with a diameter of 1.92 to 2.4 m containing the fiducial decay region.

The vacuum tank was followed by a magnetic spectrometer housed in a vessel filled with helium at nearly atmospheric pressure, separated from the vacuum by a thin (0.3% X_0) Kevlar® window. An aluminium beam pipe of 158 mm outer diameter traversing the centre of the spectrometer (and all the following detectors) allowed the undecayed beam particles to continue their path in vacuum. The spectrometer consisted of four drift chambers (DCH) with an octagonal transverse width of 2.9 m: DCH1, DCH2 located upstream and DCH3, DCH4 downstream of a dipole magnet that provided a horizontal transverse momentum kick of 120 MeV/c for charged particles (for NA48/2). Each DCH was composed of eight planes of sense wires. The DCH space point resolution was 90 μm in both horizontal and vertical directions, and the momentum resolution was $\sigma_p/p = (1.02 \oplus 0.044 \cdot p)\%$, with p expressed in GeV/c (for NA48/2). The spectrometer was followed by a plastic scintillator hodoscope (HOD) with a transverse size of about 2.4 m, consisting of a plane of vertical and a plane of horizontal strip-shaped counters arranged in four quadrants (each logically divided into four regions). The HOD provided time measurements of charged particles with 150 ps resolution. It was followed by a liquid krypton electromagnetic calorimeter (LKr), an almost homogeneous ionization chamber with an active volume of 7 m^3 of liquid krypton, 27 X_0 deep, segmented transversally into 13248 projective $\sim 2{\times}2$ cm^2 cells. The LKr energy resolution was $\sigma_E/E = (3.2/\sqrt{E} \oplus 9/E \oplus 0.42)\%$, the spatial resolution for an isolated electromagnetic shower was $(4.2/\sqrt{E} \oplus 0.6)$ mm in both horizontal and vertical directions, and the time resolution was 2.5 ns/$\sqrt{E}$, with E expressed in GeV. The calorimeter was followed by a muon system consisting of three scintillator planes, with an iron wall installed in front of each plane. A detailed description of the beamline, detector and trigger is given in Refs. [1–3].

2 Searches for lepton number violation & resonances in $K^\pm \to \pi\mu\mu$ decays

Lepton number violation (LNV) has not been observed. Heavy Majorana neutrinos, which are possible mediators of LNV processes, arise in particular in the νMSM model [4] which introduces three heavy sterile neutrinos. The lightest of them has a mass of O(keV) and is a dark matter candidate, while the other two have masses ranging from 100 MeV to a few GeV. The model explains baryon asymmetry through CP-violating sterile neutrino oscillations and mixing with active neutrinos, and

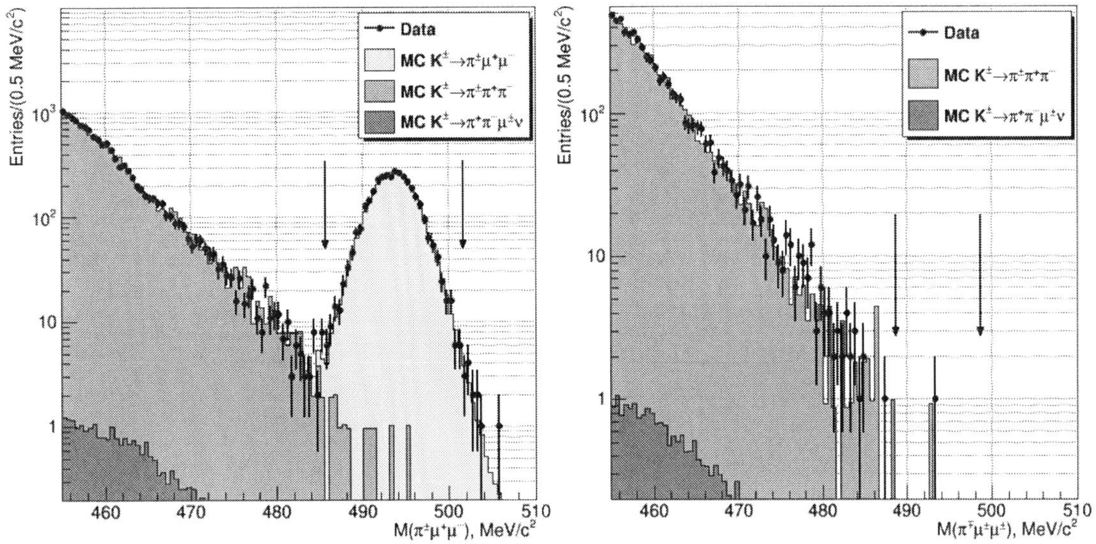

Figure 1. Reconstructed $\pi\mu\mu$ invariant masses for LFC and LFV decays $K^\pm \to \pi^\pm\mu^+\mu^-$ and $K^\pm \to \pi^\mp\mu^\pm\mu^\pm$. Data and simulated events are both shown. The signal mass regions are indicated with vertical arrows.

the low neutrino masses through see-saw mechanism. A simple extension of this model with a scalar field (the inflaton, χ) also explains the homogeneity and isotropy of the universe on large scales and the existence of structures on smaller scales [5].

A particular case of the LNV decay $K^\pm \to \pi^\mp\mu^\pm\mu^\pm$ has been investigated with the NA48/2 data sample. This decay can be mediated by the sterile neutrinos produced via $K^\pm \to \mu^\pm N$ and subsequently decaying via $N \to \pi^\mp\mu^\pm$. For an off-shell neutrino N with a mass m_4, the accessible mass range is $m_\pi + m_\mu < m_4 < m_K - m_\mu$, and the branching fraction for this channel would be

$$\mathcal{B}(K^\pm \to \pi^\mp\mu^\pm\mu^\pm) = \mathcal{B}(K^\pm \to \mu^\pm N) \times \mathcal{B}(N \to \pi^\mp\mu^\pm) \sim \left|U_{\mu 4}\right|^4,$$

where $U_{\mu 4}$ is the corresponding mixing parameter. The most stringent published limit on this process is $\mathcal{B}(K^\pm \to \pi^\mp\mu^\pm\mu^\pm) < 1.1 \times 10^{-9}$ at 90% CL [6]. The neutrino decaying as $N \to \pi^\pm\mu^\mp$ can also be observed in the lepton number conserving (LNC) channel $K^\pm \to \pi^\pm\mu^+\mu^-$. The inflaton (or another scalar resonance) in the mass range $2m_\mu < m_\chi < m_K - m_\pi$ can be produced in a similar way in the decay $K^\pm \to \pi^\pm\chi$ and subsequently decay into a muon pair: $\chi \to \mu^+\mu^-$.

The rates of $K^\pm \to \pi^\mp\mu^\pm\mu^\pm$ and $K^\pm \to \pi^\mp\mu^+\mu^-$ decays (collectively referred to as $K_{\pi\mu\mu}$) are measured with respect to the abundant $K^\pm \to \pi^\pm\pi^+\pi^-$ ($K_{3\pi}$) decays. The $K_{3\pi}$ decay also represents the main background via pion decays in flight and mis-identification. The selection of both $K_{\pi\mu\mu}$ and $K_{3\pi}$ decays is based on a three-track vertex topology, with the vertex reconstructed inside the fiducial decay volume. The total reconstructed momentum of the three tracks is required be compatible with the nominal beam momentum, with no transverse momentum with respect to the nominal beam direction. Particle identification is performed using the ratio E/p of the energy deposit measured in the LKr calorimeter to the momentum reconstructed in the spectrometer, as well as the response of the downstream muon detector. The mass spectra of the selected LNC and LFV candidates and the background spectra evaluated with simulations are shown in Fig. 1.

In the LNV channel, a blind analysis is performed, with the selection optimized and the background simulation validated in the control mass region defined as $M_{\pi\mu\mu} < 480$ MeV$/c^2$, where $M_{\pi\mu\mu}$

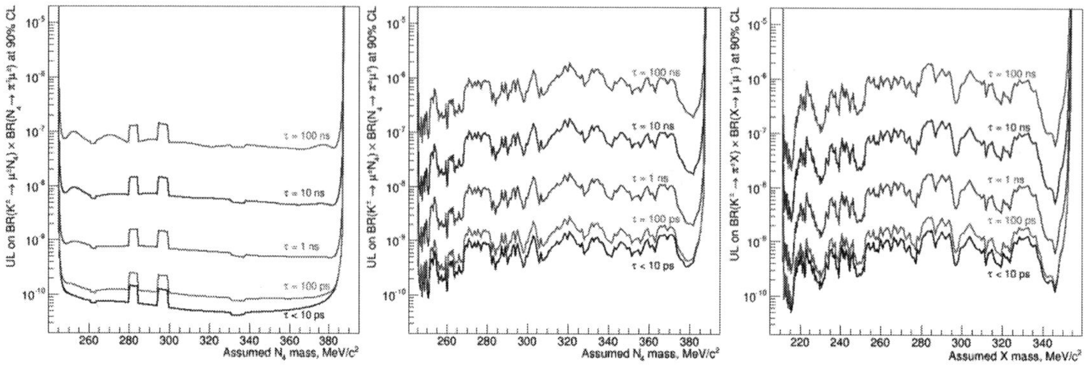

Figure 2. Obtained upper limits at 90% CL on the products of branching fractions depending on the assumed resonance masses and lifetimes. Left: $\mathcal{B}(K^{\pm} \to \mu^{\pm}N_4)\times\mathcal{B}(N_4 \to \pi^{\mp}\mu^{\pm})$; centre: $\mathcal{B}(K^{\pm} \to \mu^{\pm}N_4)\times\mathcal{B}(N_4 \to \pi^{\pm}\mu^{\mp})$; right: $\mathcal{B}(K^{\pm} \to \pi^{\pm}N_4) \times \mathcal{B}(N_4 \to \mu^{\pm}\mu^{\mp})$.

is the reconstructed invariant mass of the pion and the two muons. The signal region is defined as $|M_{\pi\mu\mu} - m_K| < 5$ MeV$/c^2$. For the LNC, the signal region is defined as $|M_{\pi\mu\mu} - m_K| < 8$ MeV$/c^2$, and 3489 candidates of the well established flavour changing neutral current decay are observed with sub-percent background contamination. One $K^{\pm} \to \pi^{\mp}\mu^{\pm}\mu^{\pm}$ candidate event is observed in the signal region while the expected number of background events is $N_{\text{exp}} = 1.163 \pm 0.867_{\text{stat}} \pm 0.021_{\text{ext}} \pm 0.116_{\text{syst}}$. The upper limit on the branching fraction of the LFV decay is then established to be $\mathcal{B}(K^{\pm} \to \pi^{\mp}\mu^{\pm}\mu^{\pm}) < 8.6 \times 10^{-11}$ at 90% CL. This limit represents an improvement by more than an order of magnitude with respect to the previous search [6].

A scan for a resonance of mass M_{res} in the reconstructed $\pi^{\pm}\mu^{\mp}$ invariant mass $M_{\pi\mu}$ of both LNV and LNC channels, and in the $\mu^{+}\mu^{-}$ invariant mass $M_{\mu\mu}$ of the LNC channel is performed. The step and width of the search windows are determined by the mass resolution $\sigma(M_{\text{res}})$ at the tested value. The mass step between two mass hypotheses is $0.5\sigma(M_{\text{res}})$ and the window centered on M_{res} has a width of $2\sigma(M_{\text{res}})$. The upper limit on the product of branching fractions at 90% CL is established for each resonance mass and lifetime hypothesis. As a result of the three-track vertex selection requirement, the acceptances (evaluated with simulations) are inversely proportional to the assumed resonance lifetime for sufficiently large lifetimes.

A total of 284 mass hypotheses are tested on the LNV $M_{\pi\mu}$ spectrum to search for $K^{\pm} \to \mu^{\pm}N_4$; $N_4 \to \pi^{\mp}\mu^{\pm}$ decays. In this case two choices are possible to build $M_{\pi\mu}$, selecting one of the muons or the other. The possibility closest to M_{res} is chosen. No signals are observed, and upper limits of $O(10^{-10})$ are set on the product of the branching ratios for heavy Majorana neutrinos with lifetimes $\tau_{N_4} < 100$ ps. The LNC $M_{\pi\mu}$ spectrum is tested with 280 hypotheses, and no $K^{\pm} \to \mu^{\pm}N_4$; $N_4 \to \pi^{\pm}\mu^{\mp}$ signals are observed. The obtained upper limits are $O(10^{-9})$ for sterile neutrinos with lifetimes $\tau_{N_4} < 100$ ps. Finally, a scan is performed on the $M_{\mu\mu}$ spectrum of the LNC channel to search for $K^{\pm} \to \pi^{\pm}\chi$; $\chi \to \mu^{\pm}\mu^{\mp}$ with 267 hypotheses, and no signals are observed. The upper limits are $O(10^{-9})$ for resonance lifetimes $\tau_{\chi} < 100$ ps. The obtained ULs depending on the assumed resonance masses and lifetimes are shown in Fig. 2.

3 π^0 electromagnetic transition form factor slope

As a π^0 is produced in four of the six main $K^{\pm}$ decays, the NA62-R_K experiment exposed to about 2×10^{10} kaon decays in flight in the vacuum fiducial decay region with minimum bias trigger conditions is an ideal environment to study the neutral pion physics. The Dalitz decay $\pi^0_D \to \gamma e^+ e^-$ proceeds through the $\pi^0\gamma\gamma$ vertex with an off-shell photon. The commonly used kinematic variables defined in terms of the particle four-momenta are:

$$x = \left(\frac{M_{ee}}{m_{\pi^0}}\right)^2 = \frac{(p_{e^+} + p_{e^-})^2}{m^2_{\pi^0}}, \qquad y = \frac{2\, p_{\pi^0} \cdot (p_{e^+} - p_{e^-})}{m^2_{\pi^0}(1-x)},$$

where p_{π^0}, p_{e^+}, p_{e^-} are respectively the π^0 and $e^\pm$ four-momenta, m_{π^0} is the mass of the π^0, and M_{ee} is the e^+e^- invariant mass. The physical region is given by

$$r^2 = \left(\frac{2m_e}{m_{\pi^0}}\right)^2 \le x \le 1, \quad |y| \le \sqrt{1 - \frac{r^2}{x}}\,.$$

The π^0_D differential decay width normalised to the $\pi^0_{2\gamma} \to \gamma\gamma$ decay width reads

$$\frac{1}{\Gamma(\pi^0_{2\gamma})}\frac{d^2\Gamma(\pi^0_D)}{dx dy} = \frac{\alpha}{4\pi}\frac{(1-x)^3}{x}\left(1 + y^2 + \frac{r^2}{x}\right)(1 + \delta(x,y))\, |\mathcal{F}(x)|^2,$$

where $\mathcal{F}(x)$ is the semi-off-shell electromagnetic transition form factor (TFF) of the π^0 and $\delta(x, y)$ encodes the radiative corrections.

The TFF is usually expanded as $\mathcal{F}(x) = 1 + ax$ where a is the form factor slope parameter. This approximation is justified by the smallness of this parameter. In the vector meson dominance (VMD) model, it is dominated by the ρ and ω mesons, resulting in a slope of $a \approx m^2_{\pi^0}\left(m_\rho^{-2} + m_\omega^{-2}\right)/2 \approx 0.03$.

A crucial aspect of measuring the π^0 TFF with the π^0_D decays is control of the radiative corrections in the differential rate: their effect on the differential decay rate exceed that of the TFF. The first study of the corrections on the differential rate was performed in [7] in the soft-photon approximation, and extended in [8]. A recent improved computation [9] largely triggered by and used for this analysis includes additional second-order contributions and accounts for the emission of radiative photons, i.e. the internal bremsstrahlung contribution.

A sample of 1.05×10^6 $K^\pm \to \pi^\pm\pi^0_D$ decays with negligible background has been selected in the NA62-R_K data using a spectrometer three-track vertex selection and reconstructing the photon in the LKr calorimeter. A χ^2 fit to the reconstructed x distribution of the data π^0_D candidates with equipopulous binning has been performed to extract the TFF slope value (Fig. 3). The main systematic uncertainties arise from the simulation of the beam spectrum and from the calibration of the spectrometer global momentum scale. The preliminary result is

$$a = (3.70 \pm 0.53_{\text{stat}} \pm 0.36_{\text{syst}}) \times 10^{-2},$$

with $\chi^2/\text{ndf} = 52.5/49$ corresponding to a p-value of 0.34. This measurement represents an observation of a positive π^0 electromagnetic TFF slope (with more than 5σ significance) in the time-like region of momentum transfer.

It is interesting to note that the π^0_D spectrum has been used to search for "dark sector" Standard Model extension. In particular, a search for the dark photon (DP) production in the $\pi^0 \to \gamma A'$ decay followed by the prompt $A' \to e^+e^-$ decay has been performed with the π^0_D data sample collected by the NA48/2 experiment [10]. No DP signal has been observed, providing improved upper limits on the mixing parameter ε^2 in the mass range 9–70 MeV/c^2. In combination with other experimental searches, this result ruled out the DP as an explanation for the muon $(g-2)$ measurement under the assumption that the DP couples to quarks and decays predominantly to SM fermions.

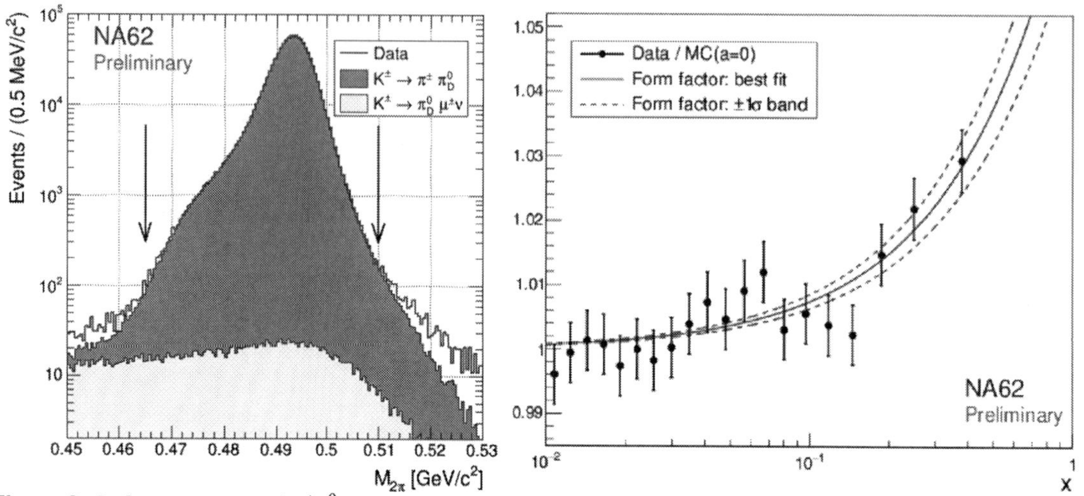

Figure 3. Left: reconstructed $\pi^{\pm}\pi^0_D$ mass spectrum of the data and simulated samples. The residual difference is mainly due to the non-inclusion of the $K^{\pm} \to \pi^0_D e^{\pm}\nu$ component of the signal into the simulation. Right: Illustration of π^0 the TFF fit result showing the data/MC ratio, with the MC sample weighted to obtain $a = 0$. The events are divided into 20 equipopulous bins, and the markers are located at the bin barycentres of each bin. The solid line represents the TFF function with a slope value equal to the fit central value. The dashed lines indicate the 1σ band.

Conclusions

Searches for the LNV $K^{\pm} \to \pi^{\mp}\mu^{\pm}\mu^{\pm}$ decay and resonances in $K^{\pm} \to \pi\mu\mu$ decays using the 2003–04 data sample collected by the NA48/2 experiment at CERN are presented. A new upper limit at 90% CL on the branching ratio of the LNV channel is established: $\mathcal{B} < 8.6 \times 10^{-11}$, improving the previous best limit by more than an order of magnitude. Upper limits are set on the products of branching ratios $\mathcal{B}(K^{\pm} \to \mu^{\pm}N_4)\mathcal{B}(N_4 \to \pi^{\mp}\mu^{\pm})$, $\mathcal{B}(K^{\pm} \to \mu^{\pm}N_4)\mathcal{B}(N_4 \to \pi^{\pm}\mu^{\mp})$ and $\mathcal{B}(K^{\pm} \to \pi^{\pm}\chi)\mathcal{B}(\chi \to \mu^+\mu^-)$ as functions of the assumed resonance masses and lifetimes. These limits are in the 10^{-9} to 10^{-10} range for resonance lifetimes below 100 ps.

A preliminary result from a measurement of the electromagnetic transition form factor slope of the π^0 with 1.05×10^6 $\pi^0_D \to e^+e^-\gamma$ reconstructed decays collected at the NA62 experiment at CERN in 2007 is also reported. The obtained result $a = (3.70 \pm 0.64) \times 10^{-2}$ is the first observation of positive TFF slope in the time-like transfer momentum region.

References

[1] J. R. Batley *et al.*, Eur. Phys. J. **C52**, 875 (2007)
[2] C. Lazzeroni *et al.*, Phys. Lett. **B719**, 326 (2013)
[3] V. Fanti *et al.*, Nucl. Instrum. Methods **A574**, 433 (2007)
[4] T. Asaka and M. Shaposhnikov, Phys. Lett. **B620**, 17 (2005)
[5] T. Asaka and M. Shaposhnikov, Phys. Lett. **B639**, 414 (2006)
[6] J. R. Batley *et al.*, Phys. Lett. **B697**, 107 (2011)
[7] B. E. Lautrup and J. Smith, Phys. Rev. **D3**, 1122 (1971)
[8] K. O. Mikaelian and J. Smith, Phys. Rev. **D5**, 1763 (1972)
[9] T. Husek *et al.*, Phys. Rev. **D92**, 054027 (2015)
[10] J. R. Batley *et al.*, Phys. Lett. **B746**, 178 (2015)

EPJ Web of Conferences **130**, 01020 (2016)
MESON 2016
DOI: 10.1051/epjconf/201613001020

The KLOE-2 experiment at DAΦNE

M. Silarski[1,⋆]
for the KLOE-2 Collaboration

[1] *LNF-INFN, Via E. Fermi 40, 00044 Frascati (RM), Italy*

Abstract. The KLOE-2 experiment at the INFN Laboratori Nazionali di Frascati (LNF) is currently taking data at the upgraded e^+e^- DAFNE collider. Present Run II follows a development phase to assess the feasibility of a long term acquisition program, Run I, which successfully ended in July 2015 with 1 fb^{-1} integrated luminosity collected in less than eight months. KLOE-2 represents the continuation of the KLOE experiment with a new physics program. The KLOE detector has undergone several upgrades including state-of-the-art cylindrical GEM Inner Tracker, electron-positron taggers for the $\gamma\gamma$-physics studies and new calorimeters around the interaction point. In this article we briefly present the overview of the KLOE-2 experiment including the present status and achievements together with the physics plans.

1 Introduction

The KLOE-2 experiment operates at the upgraded DAΦNE e^+e^- collider in Frascati aiming to complete and extend the KLOE physics program with more than 5 fb^{-1} of integrated luminosity collected in the next 2-3 years [1]. For the first time the "'Crab-Waist"' interaction scheme developed in Frascati has been applied in the presence of a high-field detector solenoid, where the transverse dimensions of the beams and their crossing angle are tuned to maximize the machine luminosity [2]. KLOE-2 achieved a record performance in terms of 2 x 10^{32} cm^{-2}s^{-1} peak luminosity and 12 pb^{-1} maximum daily integrated luminosity with this innovative scheme of beam collisions, which will be employed in the upgrade of the B-factory currently under construction at the KEK Laboratory in Japan, and is also considered a valid option in several future projects.

The main components of the KLOE-2 detector are the Drift Chamber, one of the biggest ever built, and a lead-scintillating fiber Electromagnetic Calorimeter, one of the best ones for energy and timing performance at low energies. The KLOE drift chamber provides three–dimensional tracking with perpendicular decay vertex resolution of the order of 1 mm and transverse momentum accuracy better than 0.4% [3]. The electromagnetic calorimeter consists of a 24-module barrel surrounding the drift chamber and two endcap detectors which together cover in around 98% of the solid angle. It provides determination of energy and time with accuracies of $\sigma_E/E = 5.7\%/\sqrt{E[\mathrm{GeV}]}$ and $\sigma(t) = 57\mathrm{ps}/\sqrt{E[\mathrm{GeV}]} \oplus 100$ ps, respectively [4]. To improve vertex reconstruction capabilities near the interaction region KLOE-2 was equipped with an inner tracker and it is the first high-energy experiment using the GEM technology with a cylindrical geometry, a novel idea that was developed

⋆e-mail: michal.silarski@lnf.infn.it

at LNF exploiting the kapton properties to build a transparent and compact tracking system [5]. To study $\gamma\gamma$-physics the detector has been upgraded with two pairs of electron-positron taggers. The Low Energy Tagger (LET) is a small calorimeter placed inside KLOE near the interaction piont, consisting of LYSO crystals read out by silicon photomultipliers. This subdetector will serve to measure electrons and positrons from $\gamma\gamma$ interaction with energy up to around 400 MeV [6]. The second tagger is called High Energy Tagger (HET) which provides measurement of the displacement of the scattered leptons with respect to the main orbit. Therefore, it was inserted inside the machine lattice 11 m away from the interaction point. This position sensitive detector consists of 30 small BC418 scintillators 3x3x5 mm^3, which provide a spatial resolution of 2 mm (corresponding to momentum resolution of ~1 MeV/c) [7]. In parallel to the ongoing data taking campaign we have completed several analyses using the old KLOE data sample. Some of these results will be briefly described in the next section.

2 Recent KLOE-2 results

Apart from running the KLOE-2 experiment and activities devoted to commissioning of the new subdetectors there are still several ongoing, or recently finished, physics analyses based on the KLOE data, including search for dark forces at 1 GeV, tests of the CPT and Lorentz invariance in the neutral kaon system and new precision measurements in hadronic physics at low energy.

2.1 CPT tests in the neutral kaon system

An unique feature of the Φ-factory is the production of neutral kaon pairs in a pure quantum state, thus at KLOE we can study quantum interference effects and tag pure monochromatic K_S and K_L beams. Due to the quantum entanglement the kaons double differential rate of decay into two final states f_1 and f_2 takes the following form:

$$I(f_1, f_2, \Delta t) \sim |\eta_1|^2 e^{-\Gamma_L \Delta t} + |\eta_2|^2 e^{-\Gamma_S \Delta t} - 2|\eta_1|^2|\eta_2|^2 e^{-\frac{(\Gamma_L+\Gamma_S)}{2}\Delta t}\cos(\Delta m \Delta t + \phi_2 - \phi_1) \qquad (1)$$

where Δt denotes the proper decay times difference and η_1 and η_2 are the two decay amplitude ratios: $\eta_i = |\eta_i| e^{i\phi_i} = \frac{\langle f_i|T|K_L\rangle}{\langle f_i|T|K_S\rangle}$. Γ_L and Γ_S denote the widths of K_L and K_S meson, respectively. If both kaons decay to the same final state the interference pattern in Eq. 1 is very sensitive to any deviation from unity of the ratio η_1/η_2 in the interference region of $\Delta t \approx 0$. Such deviations may be present due to the CPT violation which in the framework of SME manifests to lowest order only in the parameter δ_K exhibiting dependence on the 4-momentum of the kaon [9]:

$$\delta_K \approx i \sin\phi_{SW} e^{i\phi_{SW}} \gamma_K (\Delta a_0 - \vec{\beta_K} \cdot \Delta\vec{a})/\Delta m \,, \qquad (2)$$

where ϕ_{SW} is the superweak phase, γ_K and $\vec{\beta_K}$ are the boost factor and velocity of kaon and Δm denotes the difference between K_L and K_S mass. The four Δa_μ parameters represent the differences of couplings to valence quark and antiquark of the meson. Since the Earth is rotating the CPT and Lorentz violating parameters are usually considered in the fixed stars reference frame which makes δ_K dependent also on the sidereal time t_s. At KLOE a search for the Lorentz and CPT symmetries violation have been performed by measurements of the interference pattern for both kaons decaying to $\pi^+\pi^-$ final states using a sample of 1.7 fb^{-1} of integrated luminosity [8]. The data sample was divided into two subsets: one containing events where kaon emitted in the forward hemisphere in the KLOE reference frame was moving along with the ϕ momentum and the second one with events where kaon emitted in the same hemisphere was moving in the opposite direction. Moreover, the data were divided into four bins of sidereal time. For every subset of the data distribution of the decay

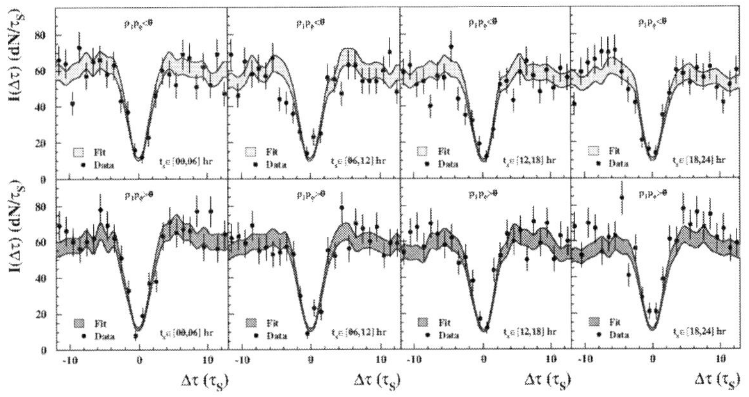

Figure 1. Distributions of the double differential decay rates into $\pi^+\pi^-, \pi^+\pi^-$ final states measured with the KLOE detector. The top panel contains distributions for four bins of sidereal time when kaon emitted in the forward hemisphere ($\cos\theta > 0$) in the KLOE reference frame was moving along with the ϕ momentum (i.e. $\vec{p}_{1(2)} \cdot \vec{p}_\phi > 0$). The lower panel shows distributions for a subset of data where kaon emitted in the forward hemisphere was moving in the opposite direction (i.e. $\vec{p}_{1(2)} \cdot \vec{p}_\phi < 0$). Figure taken from [8].

intensity in function of Δt, expressed in units of the K_S lifetime, was determined. The resulting eight distributions are presented in Fig. 1. The distributions were fitted simultaneously with the interference pattern allowing possible modulation effects induced by the CPT violation. The results of the fit are indicated in Fig. 1 together with uncertainties by colored bands. They do not reveal any CPT or Lorentz symmetries violation and constitute the first estimation of Δa_μ parameters in the kaon sector with sensitivity of about 10^{-18} GeV [8].

2.2 Search for dark forces

In the framework of the Dark Matter searches it was postulated, that there may exist a secluded gauge sector mediated by a new vector gauge boson U, called also the dark photon [10]. The dark photon could couple with ordinary photon via mixing term and subsequently decay into Standard Model particles giving opportunity to be observed in high-energy experiments. Direct searches of these dark matter mediators are performed with several experiments including KLOE-2. We have explored three different processes: Dalitz decays of the ϕ meson: $\phi \to \eta U$, $U \to e^+e^-$ with $\eta \to 3\pi$ [11, 12], $e^+e^- \to U\gamma$ with U decaying to lepton or pion pairs [13–15] and the dark Higgsstrahlung process $e^+e^- \to Uh_0$, $U \to \mu^+\mu^-$ with h_0 invisible [16]. Search for the dark photon in the ϕ Dalitz dacays were performed using an integrated luminosity of 1.7 fb^{-1} and 1.5 fb^{-1} for $\eta \to 3\pi^0$ and $\eta \to \pi^+\pi^-\pi^0$, respectively. The irreducible background from the $\phi \to \eta\gamma^* \to \eta e^+e^-$ decay was directly extracted from the data by a fit to the di-lepton invariant mass distribution parameterized according to the Vector Meson Dominance model. No resonant signal was observed in the e^+e^- invariant mass distributions for both η decays. As a result we have estimated upper limit on the kinetic mixing parameter as a function of the U boson mass which is shown in Fig. 2 as the KLOE$_{(1)}$ region.

We have searched for the $e^+e^- \to U\gamma$ analysing three possible U decays into e^+e^-, $\mu^+\mu^-$ and $\pi^+\pi^-$. Again the signature of the dark photon is the narrow structure of the lepton-antilepton or $\pi^+\pi^-$ invariant mass distribution. In case of the $\mu^+\mu^-\gamma$ and $\pi^+\pi^-\gamma$ we were looking for both charged particles emitted with large polar angle and an undetected photon at a small angle which resulted in

DOI: 10.1051/epjconf/201613001020

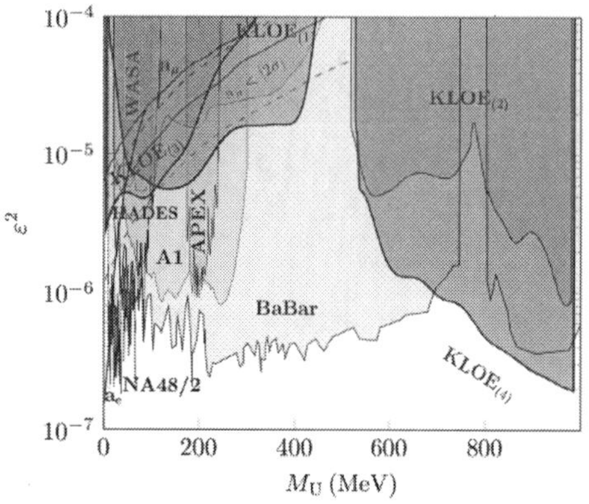

Figure 2. 90% CL exclusion plot for the U boson coupling ε^2 as a function of the U-boson mass. The limits obtained by KLOE-2 are denoted as: KLOE$_{(1)}$ (ϕ Dalitz dacay studies [11, 12]), KLOE$_{(2)}$ ($e^+e^- \to e^+e^-\gamma$ [13]), KLOE$_{(3)}$ ($e^+e^- \to \mu^+\mu^-\gamma$ [14]) and KLOE$_{(4)}$ ($e^+e^- \to \pi^+\pi^-\gamma$ [15]). The limits from the A1 [17], APEX [18], the WASA [19], HADES [20], BaBar [21] and NA48/2 [22] experiments are also shown. The gray dashed lines shows U boson parameters that could explain the observed muon anomalous magnetic moment [23]. Figure adapted from [16].

a strong suppression of the resonant $\phi \to \pi^+\pi^-\pi^0$ and FSR background. The residual background components were simulated and fitted to the invariant mass spectra. The measured distributions are in perfect agreement with the simulations done with the PHOKHARA generator [24] and does not show any peak structure. The exclusion limits estimated based on these studies are shown in Fig. 2 as KLOE$_{(2)}$, KLOE$_{(3)}$ and KLOE$_{(4)}$ regions. A natural consequence of massive dark photon would be the existence of associated Higgs-like particle h', so-called dark Higgs, coupling to U with coupling constant α_D. Thus, the U boson could manifest itself in the e^+e^- collisions through the Higgsstrahlung process: $e^+e^- \to Uh'$ where U decays to the $\mu^+\mu^-$ pair. If the mass of h' exeeds mass of the U boson the transition $h' \to U$ is allowed and one can search for the Standard Model decay products of both U bosons produced (e.g. two di-lepton or di-pion pairs) [25, 26]. The KLOE-2 analysis was done assuming $m_{h'} < m_U$, therefore we were looking for the missing mass in the final state of $\mu^+\mu^-$ [16]. The expected signal would show up as a sharp enhancement in the two dimensional distribution of $\mu^+\mu^-$ missing mass versus the $\mu^+\mu^-$ invariant mass. In order to better understand backgrounds, two data samples were analysed: sample gathered at the ϕ mass peak (1.65 fb^{-1}) and the off-peak sample at $\sqrt{s}$ = 1000 MeV (0.2 fb^{-1}), where backgrounds are strongly reduced. The 90% confidence level limits of $\alpha_D \cdot \varepsilon^2$ obtained by KLOE-2 for combined on- and off-peak data samples are at the level of $10^{-9} - 10^{-8}$ depending on the masses of U and h' [16].

2.3 New results in hadronic physics

Recently KLOE-2 published new, independent result on the $\eta \to \pi^+\pi^-\pi^0$ Dalitz plot measurement with the highest statistics in the world ($\sim 4.48 \cdot 10^6$ events) [27]. This isospin-breaking decay is of particular interest because it provides information about the light-quark mass difference. Moreover, analysis of the $\eta \to \pi^+\pi^-\pi^0$ Dalitz plot gives possibility to test the charge conjugation symmetry. At KLOE the η meson is produced in the radiative ϕ decay together with a mono-energetic photon with energy E $\sim$ 363 MeV. The candidate events are searched for in a preselected sample with at least three prompt photons and two tracks with opposite curvature reconstructed. The main background originate from the $e^+e^- \to \omega\pi^0$ reaction and the Bhabha scattering. The resulting $\eta \to \pi^+\pi^-\pi^0$ Dalitz plot was bin-by-bin background subtracted and fitted with the decay squared amplitude $|A|^2$ parametrized with a polynomial expansion:

$$|A(X,Y)|^2 \simeq N(1 + aY + bY^2 + cX + dX^2 + eXY + fY^3 + gX^2Y + \ldots), \quad (3)$$

where X and Y are the normalized Dalitz plot variables expressed by the kinetic energies of all the particles in the final state [27]. We have obtained the most precise estimations of the Dalitz parameters,

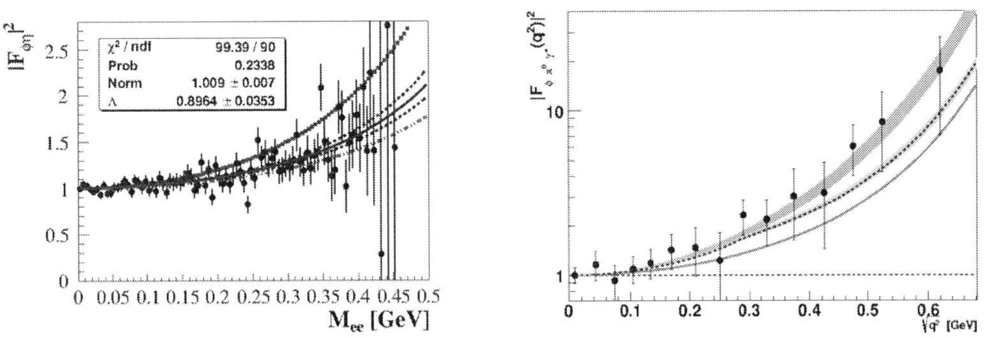

Figure 3. Left: $F_{\phi\eta}$ distribution as a function of the e^+e- invariant mass. The blue curve shows the results of fit of the VMD model parametrization to the data. VMD theoretical expectations are superimposed in pink dashed curve while the $F_{\phi\eta}$ dependence obtained from the Terschluesen/Leupold model is reported as red points. Right: Measured $F_{\phi\pi^0}$ in function of modulus of the 4-momentum transfer $\sqrt{q^2}$ (black points) and the theoretical predictions based on: the dispersive analysis of Ref. [32] (orange and cyan bands) and Ref. [33] (blue dashed line), the chiral theory approach [34] (green band), and the one-pole VMD model [33] (solid red line).

for the first time including also the g parameter. The final values of the charge asymmetries are all consistent with zero and were obtained with the best sensitivity in the world [27].

Precise determination of the transition form factors (TFF) of mesons is crucial in many fields of particle physics, such as determination of the light-by-light contribution to the anomalous magnetic moment of the muon or understanding the low-energy structure of hadrons. KLOE-2 has obtained new results on TFFs and branching fractions for the $\phi \to \eta e^+e^- \to 3\pi^0 e^+e^-$ and $\phi \to \pi^0 e^+e^-$ decays [28, 29]. In the case of $\phi \to \eta e^+e^-$ decay the Form Factor $F_{\phi\eta}$ was extracted from the e^+e^-invariant mass spectrum by dividing the measured invariant mass spectrum by the spectrum of reconstructed MC signal events generated with a constant $F_{\phi\eta}$, after all the analysis cuts. The resulting $F_{\phi\eta}$ distribution has been fitted as a function of the invariant mass with the Vector Meson Dominance model (VMD) parametrization [30]. The result of the fit is shown in the left panel of Fig. 3 together with theoretical predictions from the VMD (pink dashed curve) and Terschluesen/Leupold models [31]. The value of the TFF slope obtained with the fit amounts to $b_{\phi\eta} = 1.28 \pm 0.10^{+0.09}_{-0.08}$ GeV^{-2} [28] and is in agreement with the VMD predictions [30]. An analogus analysis of the $\phi \to \pi^0 e^+e^-$ decay led to the determination of the modulus square of the $F_{\phi\pi^0}$ Transition Form Factor in a function of the 4-momentum modulus $\sqrt{q^2}$ below 700 MeV [29] shown in the right panel of Fig. 3 together with theoretical expectations. As one can see the chiral theory approach [34] is favoured by the data. The TFF slope parameter extracted from this dependence amounts to $b_{\phi\pi^0} = 2.02 \pm 0.11$ GeV^{-2}.

3 Summary

The KLOE-2 experiment constitute the continuation of KLOE with a new physics program mainly focused on the study of K_s, η and η' decays as well as on kaon interferometry, test of discrete symmetries and search for physics beyond the Standard Model [1]. The new data taking campaign aiming to collect more than 5 fb^{-1} integrated luminosity in the next 2-3 years, will allow to perform CPT symmetry and quantum coherence tests using entangled neutral kaons with an unprecedented precision, high precision studies of $\gamma\gamma$-physics processes like $e^+e^- \rightarrow e^+e^-\pi^0(\gamma\gamma \rightarrow \pi^0)$, and the search for signals of a hidden dark-matter sector, among the fields to be addressed.

This work was supported in part by the EU Integrated Infrastructure Initiative Hadron Physics Project under contract number RII3-CT- 2004-506078; by the European Commission under the 7th Framework Programme through the 'Research Infrastructures' action of the 'Capacities' Programme, Call: FP7-INFRASTRUCTURES-2008-1, Grant Agreement No. 227431; by the Polish National Science Centre through the Grants No. 2011/03/N/ST2/02652, 2013/08/M/ST2/00323, 2013/11/B/ST2/04245, 2014/14/E/ST2/00262, 2014/12/S/ST2/00459.

References

[1] G. Amelino-Camelia *et al.,* Eur. Phys. J. C **68**, 619 (2010)
[2] M. Zobov *et al.,* Phys. Rev. Lett. **104**, 174801 (2010)
[3] M. Adinolfi *et al.,* Nucl. Instrum. Meth. A **461**, 25 (2001)
[4] F. Ambrosino *et al.,* Nucl. Instrum. Meth. A **598**, 239 (2009)
[5] A. Di Cicco and G. Morello, Acta Phys. Polon. B **46**, 73 (2015) 73
[6] D. Babusci *et al.*, Acta Phys. Polon. B **46**, 81 (2015)
[7] F. Archilli *et al.*, Nucl. Instrum. Meth. A **617**, 266 (2010)
[8] D. Babusci *et al.*, Phys. Lett. B **730**, 89 (2014)
[9] D. Calladay, V.A. Kostelecky, Phys. Rev. D **55**, 6760 (1997)
[10] B. Holdom, Phys. Lett. B **166** 196 (1986)
[11] F. Archilli *et al.*, Phys. Lett. B **706**, 251 (2012)
[12] D. Babusci *et al.*, Phys. Lett. B **720**, 111 (2013)
[13] D. Babusci *et al.*, Phys. Lett. B **736**, 459 (2014)
[14] A. Anastasi *et al.*, Phys. Lett. B **750**, 633 (2015)
[15] A. Anastasi *et al.*, Phys. Lett. B **757**, 356 (2016)
[16] A. Anastasi *et al.*, Phys. Lett. B **747**, 365 (2015)
[17] H. Merkel *et al.*, Phys. Rev. Lett. **112**, 221802 (2014)
[18] S. Abrahamyan *et al.*, Phys. Rev. Lett. **107**, 191804 (2011)
[19] P. Adlarson *et al.*, Phys. Lett. B **726**, 187 (2013)
[20] G. Agakishiev *et al.*, Phys. Lett. B **731**, 265 (2014)
[21] J. P. Lees *et al.*, Phys. Rev. Lett. **113**, 201801 (2014)
[22] J. R. Batley *et al.*, Phys. Lett. B **746**, 178 (2015)
[23] M. Pospelov, Phys. Rev. D **80**, 095002 (2009)
[24] H. Czyż, A. Grzelińska, J. H. Kühn, G. Rodrigo, Eur. Phys. J. C **39**, 411 (2005)
[25] I. Jaegle, Nucl. Phys. Proc. Suppl. **234**, 33 (2013)
[26] J. P. Lees *et al.*, Phys. Rev. Lett. **108**, 211801 (2012)
[27] A. Anastasi *et al.*, JHEP **1605**, 019 (2016)
[28] D. Babusci *et al.*, Phys. Lett. B **742**, 1 (2015)
[29] A. Anastasi *et al.*, Phys. Lett. B **757**, 362 (2016)
[30] A. Faessler, C. Fuchs, M. I. Krivoruchenko, Phys. Rev. C **61**, 035206 (2000)
[31] C. Terschluesen, S. Leupold, Phys. Lett. B **691**, 191 (2010)
[32] S. P. Schneider, B. Kubis, F. Niecknig, Phys. Rev. D **86**, 054013 (2012)
[33] I. Danilkin, *et al.*, Phys. Rev. D **91**, 094029 (2015)
[34] S. Ivashyn, Probl. At. Sci. Tech. 2012 (1), 179 (2012)

EPJ Web of Conferences **130**, 01021 (2016) DOI: 10.1051/epjconf/201613001021
MESON 2016

Production and decay of baryonic resonances in pion induced reactions

Witold Przygoda[1,*]
for the HADES Collaboration

[1] *Smoluchowski Institute of Physics, Jagiellonian University, 30-048 Krakow, Poland*

Abstract. Pion induced reactions give unique opportunities for an unambiguous description of baryonic resonances and their coupling channels. A systematic energy scan and high precision data, in conjunction with a partial wave analysis, allow for the study of the excitation function of the various contributions. A review of available world data unravels strong need for modern facilities delivering measurements with a pion beam. Recently, HADES collaboration collected data in pion-induced reactions on light (^{12}C) and heavy (^{74}W) nuclei at a beam momentum of 1.7 GeV/c dedicated to strangeness production. It was followed by a systematic scan at four different pion beam momenta (0.656, 0.69, 0.748 and 0.8 GeV/c) in $\pi^- - p$ reaction in order to tackle the role of $N(1520)$ resonance in conjunction with the intermediate ρ production. First results on exclusive channels with one pion ($\pi^- p$) and two pions ($n\pi^+\pi^-$, $p\pi^-\pi^0$) in the final state are discussed.

1 Probing baryon resonances

One of the key issues in the research of nature over the years was the investigation of baryon resonances, i.e. short-lived excited states of nucleons. From experimental point of view a resonance manifests as a peak located around a certain energy in differential cross section. Its composite nature is probed in scattering experiments and is characterized not only by the complex pole position of the scattering amplitude but also by the couplings to the various channels and hence decay branching ratios. Particularly the important region of 1-2 GeV is occupied by light-flavoured baryons. Such baryons are built out of three quarks (u, d, s) with possible spin either 1/2 or 3/2. Accounting for orbital motion, a supermutiplet SU(6)×O(3) defines all combinations resulting in 434 possible resonances. It highly exceeds the number of 112 identified resonances (Particle Data Group [1]) many of which do not have well determined parameters. The resonance properties were tackled by many models, including quark models (for a review see [2] Sec. 6.1), Dyson-Schwinger approach [3] and last but not least, lattice QCD calculations [4]. Those models are still confronting the fundamental question of "missing resonances". A possible explanation (as suggested in [5]) lies in the fact that missing states can couple weakly to the $N\pi$ channel, which is a predominant source of knowledge about baryon properties. Additional difficulty stems from the fact that resonance spectral functions are usually quite broad and various resonances overlap. This challenge triggered great experimental effort resulting in collection of high precision and quality data in meson photo- and electroproduction [6]. In contrary

*e-mail: witold.przygoda@uj.edu.pl

the world data from pion-induced reactions are often outdated and of poor quality. This discrepancy calls for new facilities (pion beam factories) to provide the missing data, expressed by many scientists in "Physics opportunities with meson beams" [7]. Considering the large number of resonant and non-resonant mechanisms involved, the combined analysis of various exit channels (coupled channels models, see i.e. [8]) with the partial wave analysis approach is mandatory.

2 World pion-beam data

Presently, all information on the N^* and Δ resonances listed in the PDG [1] originate from the partial wave analysis of πN total, elastic and charge exchange-scattering data performed by three major groups. Two of them (Karlsruhe-Helsinki [9] and Carnegie-Mellon-Berkeley [10]) were using only data from experiments conducted before 1980. The available statistics was 10 kevents for each of the $\pi^{\pm}p \rightarrow \pi^{\pm}p$ reaction, 1.5 kevents for $\pi^{-}p \rightarrow \pi^{0}n$, with only 17% of polarized data. The George Washington University (SAID) is the only group providing the partial wave analysis for the πN elastic scattering data. They use also the updated database (13 kevents of each of $\pi^{\pm}p \rightarrow \pi^{\pm}p$, 3 kevents of $\pi^{-}p \rightarrow \pi^{0}n$ and 250 events of $\pi^{-}p \rightarrow \eta n$; 25% of polarized data) [11] and continuously include the new available data sets. For example, $\pi^{\pm}p$ elastic scattering by EPECUR [12], added to the SAID database, demonstrates how big improvements in statistics are possible with the modern experiments and, in fact, demanded for the low energy data in order to construct the unbiased partial wave amplitudes. Another very important channel is $\pi\pi N$, since many well established resonances couple to it. The most extensive study in the resonance energy region 1.32-1.93 GeV was made by Manley *et al.* [13] (with an update [14]) within the isobar approximation. The analysis relied on 241 214 bubble chamber events, they were divided into 22 energy bins and the simultaneous fit to 2-pion channels (no $\pi^{-}p \rightarrow \pi^{0}\pi^{0}n$ channel) was performed. The single energy solution was extracted for each energy bin and every isobar channel, delivering knowledge on N^* coupling to ρN, $\Delta\pi$ and σN channels. This complete work had great impact on resonance parametrizations used in various transport models (GiBUU [15, 16] and UrQMD [17]). Since then new experimental data became available in the $\pi^{-}p \rightarrow \pi^{0}\pi^{0}n$ channel from the Crystal Ball collaboration (349 611 events at the energy range W = 1213-1527 GeV [18]) and in the $\pi^{-}p \rightarrow \pi^{+}\pi^{-}n$ channel from the CHAOS collaboration (20 kevents, low energies, W = 1257-1302 GeV, [19]) and from ITEP (40 kevents, high energy W = 2060 GeV [20]). The database for $\pi^{+}\pi^{-}$ in the energy range W = 1.3-2.0 GeV remains scarce and often lack differential distributions. More generally, there is a strong need for detailed new measurements in all charged channels.

3 HADES spectrometer and pion beam facility

The High-Acceptance Di-Electron Spectrometer (HADES) [21] is installed at GSI Darmstadt on SIS18. It provides high acceptance, good charged particle identification ($p/K/\pi/e$) and good mass resolution of $2-3\%$ for $e^{+}e^{-}$ pairs in the light vector meson mass range. It allows to study both hadron and rare dilepton production in $N+N$, $p+A$, $A+A$ in the beam energy range of a few AGeV. Recently, also pion-induced collisions on proton and nuclei delivered high statistics data sets with $\pi^{+}\pi^{-}$ and first world measurements of $e^{+}e^{-}$. The idea of a pion beam facility at SIS18 dates back to the beginning of the HADES project (ealy 1990s). The first commissioning tests took place in 1999 [22, 23]. A primary ^{14}N beam is provided by the SIS18 synchrotron with an intensity of $0.8-1.0\times10^{11}$ ions/spill on beryllium (^{9}Be) target, where the secondary pion beam is generated. Pions are transported within the beam line, equipped with 9 quadrupole and 2 dipole magnets, to the HADES target located 33 meters downstream from the production point (Fig. 1). The maximum pion

DOI: 10.1051/epjconf/201613001021

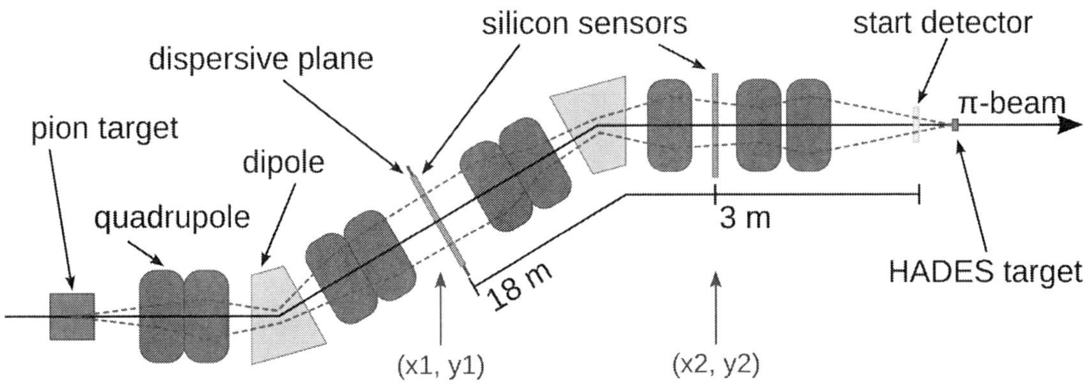

Figure 1. Pion beam line from the production point (pion target) to the HADES target position. Quadrupole and dipole magnets guide the pions (the dashed line) through the in-beam tracking detectors (indicated by the arrows) [26].

intensity of 10^6 pions/spill was achieved at a momentum $p = 1.0$ GeV/c and decreased by a factor 2 for the momenta $p = 0.7$ GeV/c or 1.3 GeV/c. The dedicated tracking system CERBEROS [26] allows for a reconstruction of the momentum of each pion beam particle with a resolution of 0.3% (for details, see [24]).

4 Pion beam results

The pion beam facility has been operated in 2014 for data taking during an effective period of two weeks. The first run was dedicated to strangeness production in pion-induced reactions on light (^{12}C) and heavy (^{74}W) nuclei at a beam momentum of 1.7 GeV/c. The goal of the second run was to measure the excitation function of two-pion production in the $\pi^- - p$ reaction around the pole of the $N(1520)D_{13}$ resonance. Data at four different pion beam momenta (656, 690, 748 and 800 MeV/c) were collected with the largest statistics in the case of the momentum of 690 MeV/c, aimed for dilepton production.

Below, first results of the analysis of three channels (elastic scattering $\pi^- p \to \pi^- p$ and two inelastic channels $\pi^- p \to n\pi^+\pi^-$, $\pi^- p \to p\pi^-\pi^0$) are presented. In all measurements two targets were used: polyethylene $(C_2H_4)_n$ and carbon (C) respectively, with the aim to subtract events from scattering on carbon and identify pure contribution from scattering on protons. The subtraction procedure involves first establishing a ratio between events collected with the polyethylene target and with the carbon target based on the $\pi^- p$ missing mass squared.

4.1 $\pi^- p$ elastic scattering

$\pi^- p$ elastic scattering was measured in the same experimental run, serving as the normalization for all other channels. First, events were pre-selected with the help of a cut on coplanar pion and proton reconstructed tracks (±5°) and $\tan\theta_{\pi^-} \cdot \tan\theta_p > 1$ selection. Events originating from scattering on polyethylene target (composed of both carbon and proton atoms in the ratio 1 : 2) and carbon target are depicted in Fig. 2 (left) showing $\pi^- p$ missing mass squared. The proportion of carbon events (red curve) was scaled such that the left tail, far from the dominating elastic scattering peak on protons, should match the tail from polyethylene events (black curve). This scaling factor is in agreement with the number of all events recorded in the START detector [27] during the measurement with

 DOI: 10.1051/epjconf/201613001021

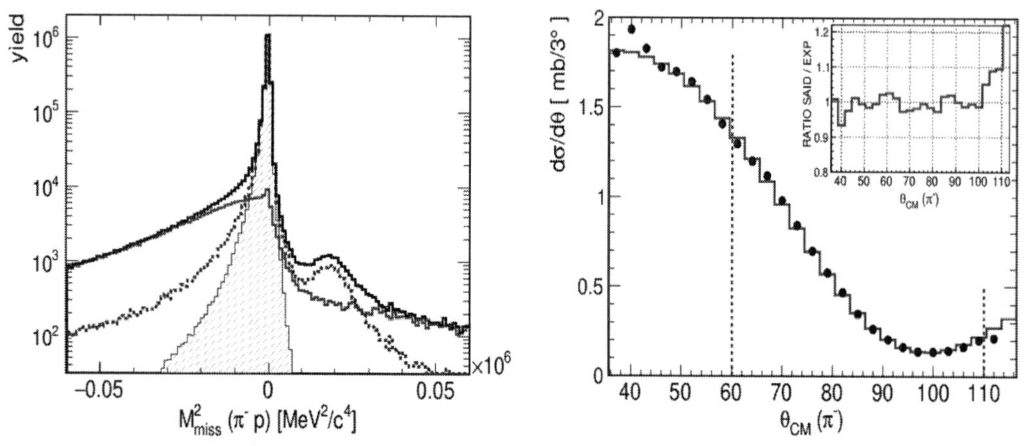

Figure 2. (Color online) Example of $\pi^- p \to \pi^- p$ elastic scattering at nominal beam momentum 800 MeV/c (791.1 MeV/c measured with the pion tracker). Left: uncorrected data, $\pi^- p$ missing mass squared with events from $(C_2H_4)_n$ (polyethylene) target (black histogram), C (carbon) target rescaled (red histogram) and their difference (blue dotted curve) - signal obtained by matching scattering on carbon in polyethylene target with scattering on carbon target on the event by event basis. Finally, elastic scattering selection (dark red dashed area). Right: angular projection of π^- in c.m.s. from $\pi^- p$ elastic scattering, corrected for the reconstruction inefficiencies and acceptance (black dots) compared with the distribution from the SAID database [29] within the fiducial volume $60° < \theta^{CM}_{\pi^-} < 110°$ (limited by the vertical blue dashed lines). Inset: ratio of SAID data to reconstructed elastic events (red histogram).

the polyethylene target and the carbon target, respectively (after correction for the data acquisition dead time and taking into account different carbon densities of both targets). Scaling factor for the polyethylene and carbon target events, deduced in the elastic scattering channel, holds to all inelastic channels under investigation.

The next step, performed for all investigated channels separately, was matching events of the same kinematic characteristics from polyethylene target and carbon target. Events were grouped into bins of similar missing mass values, then for all combinations among tracks measured with the polyethylene and the carbon target those with the lowest χ^2 were paired together. The calculation of the χ^2 correlation included a) the momentum and b) polar angle of the measured particles (π^+, π^-, proton) and c) invariant mass ($\pi^+\pi^-$ and $p\pi^-$, respectively). This method delivers pure pion-proton reaction candidates. The event by event tagging and rejection of carbon events allows also for the further cut refinement, i.e. elastic scattering selection what is not possible when having also events with the pion scattering on carbon. A few approaches were investigated (i.e. two-dimensional graphical cuts on angle and momentum of p and π^- or condition on the same momentum of particles in the c.m.s.) in order to estimate the systematic error of the elastic scattering identification.

To describe elastic scattering, events were generated in the framework of the PLUTO event generator [25] with the angular parameterization taken from [28]. Then, the full GEANT simulation of the detector response, followed by the same analysis strategy as used for the experimental data, were performed. Simulated elastic scattering events describe the measured data within the HADES acceptance very well and are used for the one-dimensional acceptance correction. The contribution from quasi-elastic scattering from C (red histogram) has been subtracted based on the missing mass dis-

 DOI: 10.1051/epjconf/201613001021

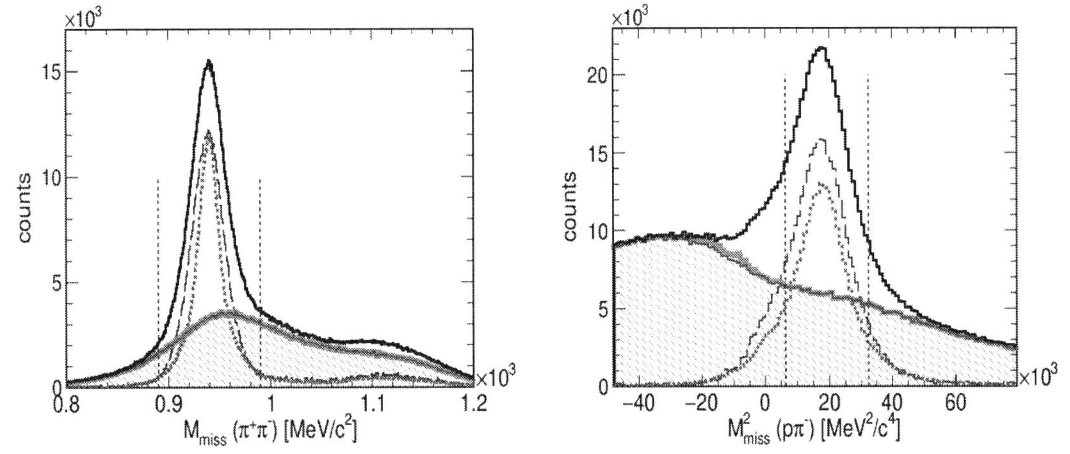

Figure 3. (Color online) Example of missing masses for pion beam scattering at a beam momentum 800 MeV/c. Both figures: uncorrected data, black curve - events from polyethylene target, red curve and red dashed area - events from carbon target, green thick curve (behind red) - events from polyethylene target matching events from carbon target, blue dashed curve - signal from scattering on protons from polyethylene target, magenta dotted curve - signal calculated based on information on momentum measured by the pion tracker. The vertical blue dashed lines limit a window (± 50 MeV/c^2 around the neutron missing mass) for selection of events for the partial wave analysis. Left: $\pi^+\pi^-$ missing mass. Right: $p\pi^-$ missing mass squared.

tributions measured with polyethylene and carbon separately (as described above) and the difference agrees with the line shape expected for π^-p events obtained from the full Monte-Carlo simulations. The simulations took into account the momentum reconstructed with the help of the pion tracker, with the most probable values of 654.1 MeV/c, 683.5 MeV/c, 738.9 MeV/c and 791.1 MeV/c. The corrected data are normalized to the distribution from the SAID database [29] of the known differential π^-p elastic scattering. It amounts to 3.01 mb, 3.10 mb, 3.08 mb and 2.59 mb for the subsequent pion beam momenta in the polar angle range between 60° – 110° in c.m.s. The corrected data for the pion beam momentum 791.1 MeV/c (nominal value 800 MeV/c) are presented in Fig. 2. The deviation between SAID and experimental values is in the range 2-3% (root-mean-square error).

4.2 $n\pi^+\pi^-$ and $p\pi^-\pi^0$ channels

Identification of two charged pions ($\pi^+\pi^-$) or proton and pion (pπ^-) in the final state allows for the reconstruction of the kinematically complete exclusive channels via cuts in the missing mass distributions around the value of the not detected neutron (Fig. 3 left) or neutral pion (Fig. 3 right). Events from the reactions on proton are deduced after subtraction of the reactions on carbon, as described earlier in this paper. The obtained resolution of the $\pi^+\pi^-$ missing mass peak amounts to 1.5% (1σ) for a fixed beam momentum or even less, 0.8-1.0%, when using the beam momentum reconstructed in the pion tracker for each event. In both cases the peak position is at the expected value of the neutron mass with an accuracy of ± 1 MeV/c^2. The resolution of the $p\pi^-$ missing mass squared is above 40% (1σ) and above 30% for the pion tracker momentum used in the reconstruction. The peak position is slightly lower (3-4 MeV/c^2) than 135 MeV/c^2 (π^0 mass) and close (1-2 MeV/c^2 below) in the case of the pion tracker momentum.

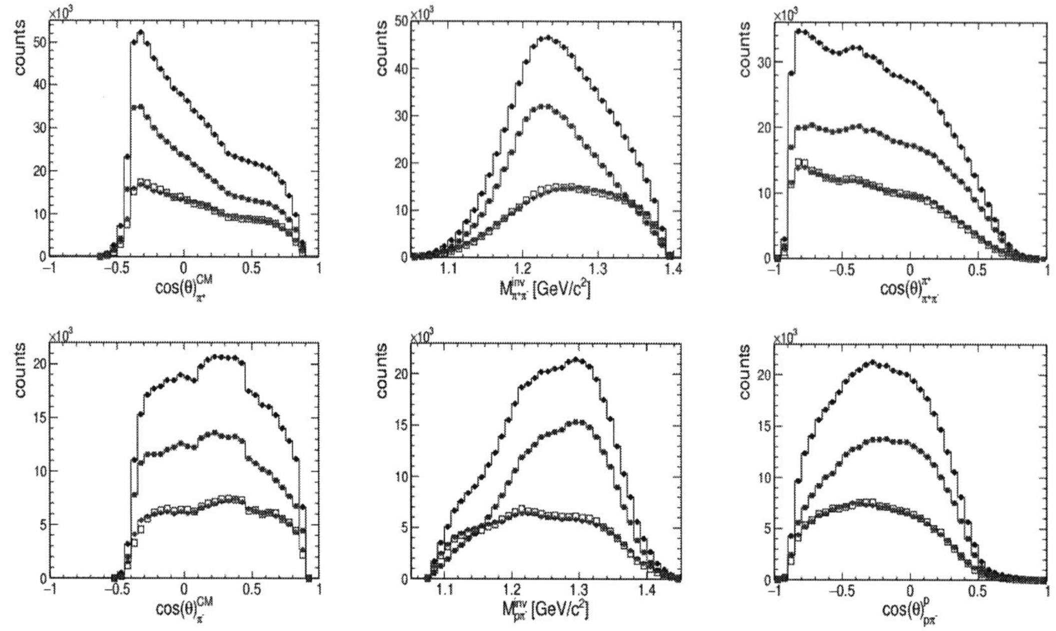

Figure 4. (Color online) Selected projections of $\pi^- p$ inelastic scattering at 800 MeV/c beam momentum in two pion channels. Events from $(C_2H_4)_n$ (polyethylene) target (black histogram), C (carbon) target (red histogram), events from scattering on protons from polyethylene target (blue histogram), events from polyethylene target matching events from carbon target (black open squares). Upper row: $n\pi^+\pi^- n$ channel with events from missing mass window (see Fig. 3 left), from left to right, π^+ angular distribution in c.m.s., $\pi^+\pi^-$ invariant mass, π^- angular distribution in Godfried-Jackson (GJ) reference frame of $\pi^+\pi^-$. Lower row: $p\pi^-\pi^0$ channel with events from missing mass window (see Fig. 3 right), from left to right, π^- angular distribution in c.m.s., $p\pi^-$ invariant mass, p angular distribution in Godfried-Jackson (GJ) reference frame of $p\pi^-$.

Fig. 4 presents a verification that the selected carbon events from the polyethylene target match events from carbon target not only in the missing mass spectrum (Fig. 3) but also in various projections, i.e. angular distributions and invariant mass. This is the proof that the subtraction of carbon-like events from the polyethylene sample allows for the identification of scattering on protons only. The number of reconstructed events in the $n\pi^+\pi^-$ exclusive channel amounts to more than 400 000 (beam momentum 656 MeV/c), 7 900 000 (690 MeV/c), 815 000 (748 MeV/c) and 526 000 (800 MeV/c). The number of reconstructed events in the $p\pi^-\pi^0$ exclusive channel amounts to more than 122 000 (beam momentum 656 MeV/c), 3 100 000 (690 MeV/c), 490 000 (748 MeV/c) and 368 000 (800 MeV/c). Those numbers dramatically improve the world data base of two-pion production in pion-induced experiments in the energy range around the $N(1520)$ resonance. These data will allow for investigation of the branching ratios of this resonance, with a special interest for the ρN decay.

4.3 Analysis strategy

The two-pion data samples, measured at the four above mentioned pion beam momenta, have been included into the multichannel partial wave analysis developed by the Bonn-Gatchina group [30, 31]. This analysis includes pion induced reactions (including Crystal Ball data [18] with neutral two-pion channel) and a large number of pseudoscalar meson photoproduction data taken with polarized beam and target. The analysis also includes the information about photoproduction of vector mesons. This work is in progress.

In parallel to hadronic channels, the identification of dilepton pairs in the ne^+e^- exclusive channel within the neutron missing mass window and e^+e^- invariant masses above the π^0 mass, in conjunction with the results from the $\pi^+\pi^-$ analysis, will shed more light on the ρ meson production mechanism (see Federico Scozzi contribution [32] to this conference).

5 Future plans

The pion beam program will be continued in 2018 once the SIS18 will have been upgraded to serve as an injector for the FAIR facility. In the year 2016 the HADES spectrometer will be extended by a large area electromagnetic calorimeter (ECAL) [33–36], replacing the PreSHOWER detector [37]. The ECAL will enable measurements of real photons emitted from nuclear matter and neutral meson production via their photonic decays (e.g. $\pi^0/\eta \to \gamma\gamma$). It will also improve the electron-hadron separation of the spectrometer for the purpose of future experiments of HADES at FAIR in elementary and heavy ion collisions in the energy range of 2-10 AGeV. The reconstruction of neutral mesons is essential also for the complete partial wave analysis of $\pi^- N$ reactions.

The HADES Collaboration gratefully acknowledges the support by the grants PTDC/FIS/113339/2009 LIP Coimbra, NCN 2013/10/M/ST2/00042 SIP JUC Cracow, Helmholtz Alliance HA216/EMMI GSI Darmstadt, VH-NG-823, Helmholtz Alliance HA216/EMMI TU Darmstadt, 283286, 05P12CRGHE HZDR Dresden, Helmholtz Alliance HA216/EMMI, HIC for FAIR (LOEWE), GSI F&E Goethe-University, Frankfurt VH-NG-330, BMBF 06MT7180 TU München, Garching BMBF:05P12RGGHM JLU Giessen, Giessen UCY/3411-23100, University Cyprus, CNRS /IN2P3, IPN Orsay, Orsay MSMT LG 12007, AS CR M100481202, GACR 13-06759S NPI AS CR, Rez EU Contract No. HP3-283286.

References

[1] K.A. Olive *et al.* (Particle Data Group), Chin. Phys. C **38**, 090001 (2014)
[2] V. Crede and W. Roberts, Rep. Prog. Phys. **76**, 076301 (2013)
[3] C.D. Roberts and A.G. Williams, Prog. Part. Nucl. Phys. **33**, 477 (1994)
[4] R. Edwards *et al.*, Phys. Rev. D **84**, 074508 (2011)
[5] R. Koniuk and N. Isgur, Phys. Rev. Lett. **44**, 2720 (1980)
[6] I.G. Aznauryan and V.D. Burkert, Prog. Part. Nucl. Phys. **67**, 1 (2012)
[7] W.J. Briscoe *et al.*, Eur. Phys. J. A **51**, 129 (2015)
[8] V. Shklyar, H. Lenske and U. Mosel, Phys. Rev. C **93**, 045206 (2016)
[9] G. Höhler, Pion-Nucleon Scattering, Landold-Börnstein (1983) vol I/9b2
[10] R.E. Cutkosky *et al.*, Phys. Rev. D **20**, 2839 (1979)
[11] R.A. Arndt *et al.*, Phys. Rev. C **74**, 045205 (2006)
[12] I.G. Alekseev *et al.* (EPECUR), Phys. Rev. C **91**, 025205 (2015)

DOI: 10.1051/epjconf/201613001021

[13] D.M. Manley, R.A. Arndt, Y. Goradia, V.L. Teplitz, Phys. Rev. D **30**, 904 (1984)
[14] D.M. Manley and E.M. Saleski, Phys. Rev. D **45**, 4002 (1992)
[15] O. Buss *et al.*, Phys. Rept. **512**, 1 (2012)
[16] J. Weil, H. van Hees and U. Mosel, Eur. Phys. J. A **48**, 111 (2012)
[17] S. A. Bass *et al.*, Prog. Part. Nucl. Phys. **41**, 255 (1998)
[18] S. Prakhov *et al.* (Crystal Ball), Phys. Rev. C **69**, 045202 (2004)
[19] M. Kermani *et al.* (CHAOS), Phys. Rev. C **58**, 3431 (1998)
[20] I.G. Alekseev *et al.*, Phys. Atom. Nucl. **61**, 174 (1998)
[21] G. Agakishiev *et al.* (HADES), Eur. Phys. J. A **41**, 243 (2009)
[22] R.S. Simon, Prog. Part. Nucl. Phys. **42**, 247 (1999)
[23] J. Díaz *et al.*, Nucl. Instrum. Meth. A **478**, 511 (2002)
[24] W. Przygoda, JPS Conf. Proc. **10**, 010013 (2016)
[25] I. Fröhlich *et al.*, PoS ACAT2007 **076** (2007)
[26] R. Lalik *et al.*, 2013 IEEE Nuclear Science Symposium and Medical Imaging Conference (2013), DOI: 10.1109/NSSMIC.2013.6829440
[27] J. Pietraszko, L. Fabbietti, W. Koenig, M. Weber (HADES), Nucl. Instr. Meth. A **618**, 121 (2010)
[28] A.D. Brody *et al.*, Phys. Rev. D **3**, 2619 (1971)
[29] SAID database: http://gwdac.phys.gwu.edu/
[30] A. Anisovich, E. Klempt, A.V. Sarantsev and U. Thoma, Eur. Phys. J. A **24**, 111 (2005)
[31] A.V. Sarantsev, JPS Conf. Proc. **10**, 010005 (2016)
[32] F. Scozzi, "Studying ρN couplings with HADES in pion-induced reactions", MESON 2016
[33] W. Czyzycki *et al.*, arXiv:1109.5550 [nucl-ex]
[34] O. Svoboda *et al.* (HADES), JINST **9**, C05002 (2014)
[35] P. Rodríguez-Ramos *et al.* (HADES), EPJ Web Conf. **81**, 06009 (2014)
[36] O. Svoboda *et al.*, J. Phys. Conf. Ser. **599**, 012026 (2015)
[37] A. Balanda *et al.* (HADES), Nucl. Instrum. Meth. A **531**, 445 (2004)

Measurement of the charged pion mass using a low-density target of light atoms

M. Trassinelli[1,*], *D.F. Anagnostopoulos*[2], *G. Borchert*[3,**], *A. Dax*[4], *J.-P. Egger*[5], *D. Gotta*[3***], *M. Hennebach*[3,****], *P. Indelicato*[6], *Y.-W. Liu*[4,†], *B. Manil*[6,‡], *N. Nelms*[7,§], *L.M. Simons*[4], and *A. Wells*[7]

[1] *Institut des NanoSciences de Paris, CNRS, Sorbonne Universités, UPMC Univ Paris 06, F-75005, Paris, France*
[2] *Dept. of Materials Science and Engineering, University of Ioannina, GR-45110 Ioannina, Greece*
[3] *Institut für Kernphysik, Forschungszentrum Jülich GmbH, D-52425 Jülich, Germany*
[4] *Laboratory for Particle Physics, Paul Scherrer Institut, CH 5232-Villigen PSI, Switzerland*
[5] *Institut de Physique de l' Université de Neuchâtel, CH-2000 Neuchâtel, Switzerland*
[6] *Laboratoire Kastler Brossel, UPMC-Sorbonne Universités, CNRS, École Normale Supérieure-PSL Research University, Collège de France, F-75005 Paris, France*
[7] *Dept. of Physics and Astronomy, University of Leicester, Leicester LE17RH, England*

Abstract. We present a new evaluation of the negatively charged pion mass based on the simultaneous spectroscopy of pionic nitrogen and muonic oxygen transitions using a gaseous target composed by a N_2/O_2 mixture at 1.4 bar. We present the experimental set-up and the methods for deriving the pion mass value from the spatial separation from the $5g-4f$ πN transition line and the $5g-4f$ μO transition line used as reference. Moreover, we discuss the importance to use dilute targets in order to minimize the influence of additional spectral lines from the presence of remaining electrons during the radiative emission. The occurrence of possible satellite lines is investigated via hypothesis testing methods using the Bayes factor.

1 Introduction

The first estimation of the charged pion mass came with its discovery from cosmic rays traces in photographic plates [1, 2]. Counting the photographic emulsion grains in the particle trajectory, Powell and his group estimated the pion-to-muon mass ratio to be about 1.5. With the pion production from accelerators [3], the charged pion mass has been measured from the deflection trajectory in a magnetic field [4–6] but also from the energy of the gamma-ray produced by the reaction $\pi^- + p \rightarrow n + \gamma$ [7] (see Fig. 1). The first mass measurement employing pionic atoms was performed in 1954 by using $4f \rightarrow 3d$ X-ray emission of several light pionic atoms with the critical absorption edge technique reaching an accuracy of 0.5% [8]. With the increase of intensity of pion beams, the use of crystal spectrometers has become possible. In 1967, the spectroscopic measurement of pionic calcium and titanium reached the relative accuracy of 10^{-4} [9]. As shown in Fig. 1, in the following years the pion mass has been re-measured several times using X-ray spectroscopy of pionic atoms with continuously improved accuracy [10–16]. In parallel, additional results for the charged pion mass were obtained from the measurement of the muon momentum in the decay $\pi^+ \rightarrow \mu^+ \nu_\mu$ [17–19].

With increasing experimental accuracy, in the '80s discrepancies between pionic atoms and pion decay results showed up. The disagreement was due to the ambiguous assumption on the remaining electrons in pionic atoms when produced in solid-state targets. In the πMg experiment of Jeckelmann and collaborators [13, 14], where the $4f-3d$ transition (25.9 keV) was measured with a transmission X-ray DuMond spectrometer, different assumptions on the K electron population lead to a difference in the pion mass of 16 ppm between the two possible interpretations, called A and B [15]. In particular, solution A is in complete disagreement with evaluations obtained from the pion decay measurement at rest [17–19].

To solve this dilemma, in the '90s our collaboration designed and realized a new experiment with a gaseous nitrogen target for having X-rays emitted from a purely hydrogen-like pion-nucleus system. In this experiment, the energy of the 4 keV X rays of the $5g \rightarrow 4f$ line was measured with a reflection spectrometer set up in Johann geometry. The spectrometer was previously developed for light antiprotonic atom spectroscopy [20]. This experiment's limitation was the calibration line, the Cu Kα fluo-

* e-mail: martino.trassinelli@insp.jussieu.fr
** present address: TU Munich, D-85747 Garching, Germany
*** e-mail: d.gotta@fz-juelich.de
**** present address: DAHER NUCLEAR TECHNOLOGIES GmbH, D-63457 Hanau, Germany
† present address: Phys. Depart., National Tsing Hua Univ., Hsinchu 300, Taiwan
‡ present address: Lab. de Physique des Lasers, Université Paris 13, Sorbonne Paris Cité, CNRS, France
§ present address: ESA-ESTEC, PO Box 299, 2200 AG, Noordwijk, The Netherlands

DOI: 10.1051/epjconf/201613001022

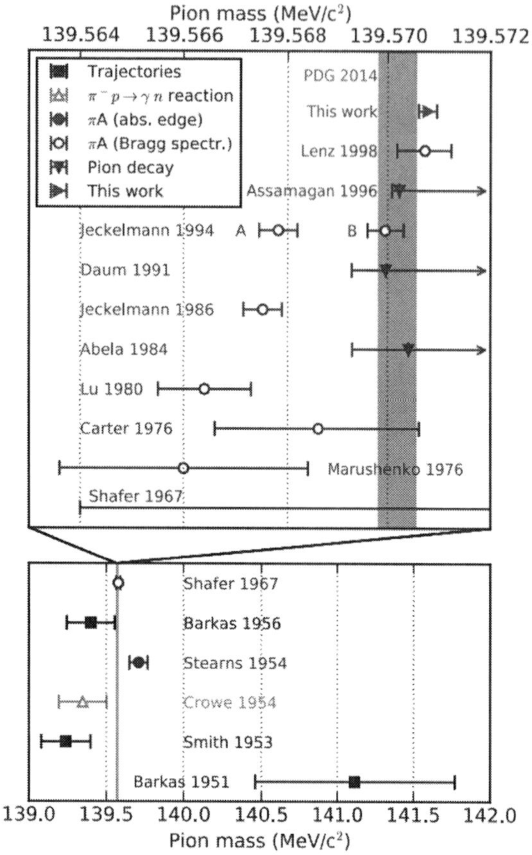

Figure 1. Results of for the mass of the charged pion from various experimental methods since its discovery. The world average value "PDG 2014" [21], indicated by the magenta region, is calculated from "Lenz 1998" [16] and solution B of "Jeckelmann 1994" [14]. Other results are taken from Refs. 4–14, 17, 18.

rescence radiation, whose large natural width (4 eV) prevents an energy determination at ultimate precision.

The present pion-mass value given by the Particle Data Group (PDG) [21] has an accuracy of 2.5 ppm and is the result of the average of solution B of the measurement of pionic magnesium [15] and the one obtained from pionic nitrogen spectroscopy [16].

The experiment described here resumes the strategy of gas targets, but exploits (i) the high precision of 0.033 ppm for the mass of the positively charged muon being $m_{\mu^+} = (105.6583715 \pm 0.0000035)\,\mathrm{MeV/c^2}$ [21] and (ii) the unique feature that in πN and μO transition energies almost coincide (Fig. 2). When using a gas mixture, the simultaneous measurement of πN and μO lines becomes possible with the muonic transition serving as an on-line calibration.

The experiment and some aspects of the data analysis have been described in Ref. 22. Here, we present additional aspects. In particular, we discuss in detail the possibility of remaining electrons during the radiative dominated part of the cascade (Sec. 2) applying specific analy-

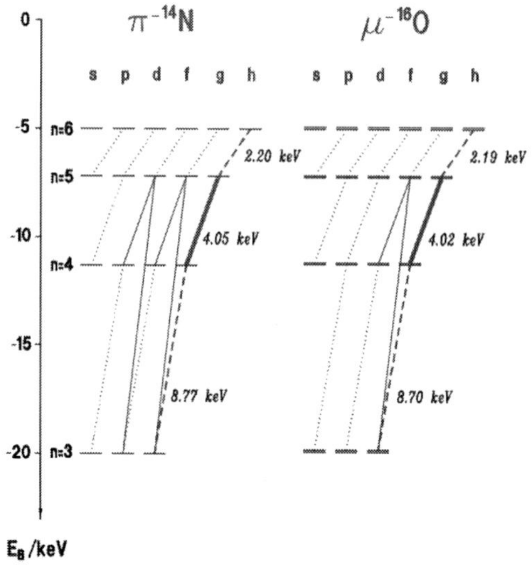

Figure 2. Comparison of the intermediate parts of the cascade for pionic nitrogen and muonic oxygen.

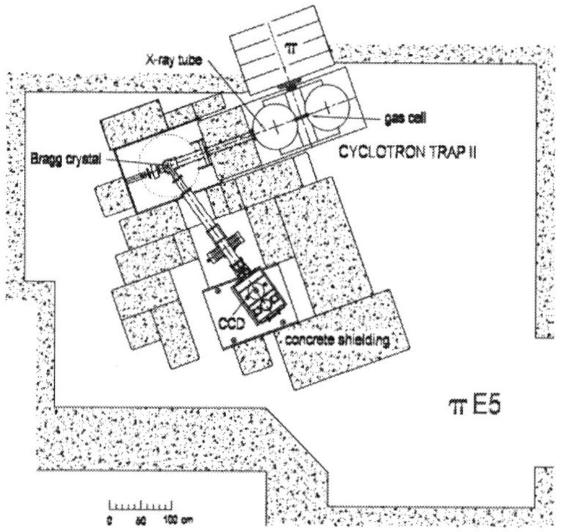

Figure 3. Experimental set-up with at the πE5 beam line of the Paul Scherrer Institut (Villigen, Switzerland). The pions enter the cyclotron trap from top perpendicular to its magnetic field.

ses and statistical tests to our data (Sec. 5). We present a brief general description of the experiment (Sec. 3) and the detailed formulas used for extracting the pion mass from the line positions (Sec. 4).

2 Pionic atoms production and atomic cascade

The experiment was performed by using the intense pion beam at the πE5 beam line of the Paul Scherrer Institut. A general overview of the set-up is presented in Fig. 3. The pions, with an initial momentum of 112 MeV/c, are captured and slowed down in the so-called cyclotron trap II. The cyclotron trap consists of a superconducting split-coil magnet with a field perpendicular to the pion trajectories. In a set of plastic degraders, pions are slowed down in order to be stopped in the gas contained in a target cell at center of the trap. A first version of this trap had been initially developed to efficiently decelerate and trap antiprotons for antiprotonic atoms production [23]. Lenz *et. al.* [16] used this trap to decelerate and stop pions.

The cyclotron trap II used in the present experiment was specifically designed with a larger gap between the magnet coils. This allows a more efficient trapping of muons produced in the decay of slow pions, which have to be captured in the bottle field of the magnet before being stopped in the target [24]. 1 − 3% of the initial pions are stopped in the gas inside the target cell. The stop rate of the decay muons is about 10% of the one of the pions.

For the simultaneous measurement of $5g - 4f$ transitions from pionic nitrogen and muonic oxygen, comparable count rates are required. This was achieved with a N_2/O_2 mixture of 10%/90%. The gas was kept at a pressure of 1.4 bar and room temperature.

Pions and muons are captured where the overlap of the wave-functions of the outermost electrons and of the pion or muon is largest. Electronic quantum numbers n_e correspond to highly exited states with an initial quantum number $n_i \sim n_e \times 16$ (14) for pions (muons), from where a quantum de-excitation cascade starts [25].

For exotic atoms with atomic number $Z > 2$, the formation and the first steps of the de-excitation proceed via Auger emission and self-ionization of the target atom or molecule. Acting at the femtosecond time scale, this process quickly leads to a high degree of ionization.

Auger emission determines the upper part of the atomic de-excitation cascade where the radiative emission is dominating for lower levels. The Auger transition probability Γ_A is proportional to $1/\sqrt{2\Delta E + 37.8\,\text{eV}}$, where ΔE is the transition energy, favouring $\Delta n = 1$ transitions with the selection rule $\Delta\ell = 0, \pm 1$ [26, 27]. The radiative transition probability Γ_X is proportional to ΔE^3 with selection rules $\Delta\ell = \pm 1$ and, hence, maximal Δn de-excitation steps are preferred. Such a dependence on ΔE efficiently populates $|n, \ell = n-1\rangle$ circular states allowing subsequently only slow radiative $(n, \ell = n-1) \to (n-1, \ell = n-2)$ transitions after depletion of the electron shells. Therefore, the atomic cascade duration is essentially determined by the first radiative transitions.

If the target consists of molecules, the initial ionisation causes a repulsive Coulomb force between the atomic cores leading to an acceleration (Coulomb explosion) where the captured pion or muon is attached to one of the atoms. After Coulomb explosion, Auger emission continues until the complete depletion of the electron shells if capture from neighbouring target atoms is avoided.

If a low-density target ($\lesssim 1 - 2$ bar) is used, electron recapture from external atoms is unlikely because the probability for having a collision with another atom of the target is low even in the presence of Coulomb explosion. This is proven by the appearance of X-ray lines at $n \geq 5$, which otherwise would be converted into Auger transitions [26, 28–31]. In the case of solid targets, however, electron refilling is unavoidable.

The circular transitions $(n, \ell = n-1) \to (n-1, \ell = n-2)$ are by far the most intense X-ray lines and, therefore, to be used for the measurement. For these transitions, strong interaction effects in the case of pionic atoms are minimal but Coulomb explosion leads to a significant Doppler broadening of the X-ray lines [32]. More details on the atomic cascade in exotic atoms can be found in Refs. [25, 33–38].

In our experiment, electron recapture during the cascade is not expected when using a N_2/O_2 mixture with a pressure of 1.4 bar at room temperature. This hypothesis is explicitly tested in the analysis presented here (Sec. 5).

3 X-ray spectrometer and data acquisition

The exotic atom transition energies are precisely measured using Bragg diffraction spectroscopy. The crystal spectrometer is set up in Johann geometry [39] and has been specifically designed for high-accuracy X-ray spectroscopy of light exotic atoms [40, 41] and was initially used for antiprotonic hydrogen and deuterium spectroscopy at LEAR [20]. The Johann configuration allows the simultaneous measurement of two different energies. The limit of the acceptable energy difference is given by the extension of the X-ray source (target) and/or of the detector on the focus position.

The Bragg crystal was made from a silicon crystal disk cut along the 220 plane of 290 μm thickness with a diameter of 100 mm. The disk is attached to a high-quality polished glass lens defining a spherical segment. Its curvature was determined to be (2981.3 ± 0.3) mm. Spherical bending leads to a partial vertical focusing [42] which increases the count rate.

For the simultaneous detection of the pionic and muonic atoms lines, the spectrometer has been equipped with a large position-sensitive detector, which is composed of an array of charge-coupled devices (CCDs) corresponding to a total sensitive area of about 48×72 mm^2 [43, 44].

The Kapton window of the target cell towards the crystal spectrometer had a diameter of 54 mm. For the geometry as given here, the overall efficiency of the crystal set-up is $\approx 5 \cdot 10^{-8}$. About 85% of the reflected intensity is covered by the detector area.

After 5 weeks of data acquisition, about 9000 events were collected of the $(5g - 4f)$ transitions for each element. The final spectrum is obtained from the projection of the two-dimensional hit pattern on the CCD onto the axis of dispersion. The result of the projection is presented in Fig. 4.

The accuracy of the pion mass measurement essentially depends on the precision of the spectral lines position determination. This requires both the knowledge of the response function of the spectrometer $F_{\text{spectr.}}$ and the control of the effects that the atomic cascade can induce to the line shape via Doppler broadening. $F_{\text{spectr.}}$ depends on the characteristic crystal rocking curve but also to aberrations due to the spherical bending of the crystal and the positioning of X-ray source and detector. The imaging properties of the spectrometer were calculated by means of a

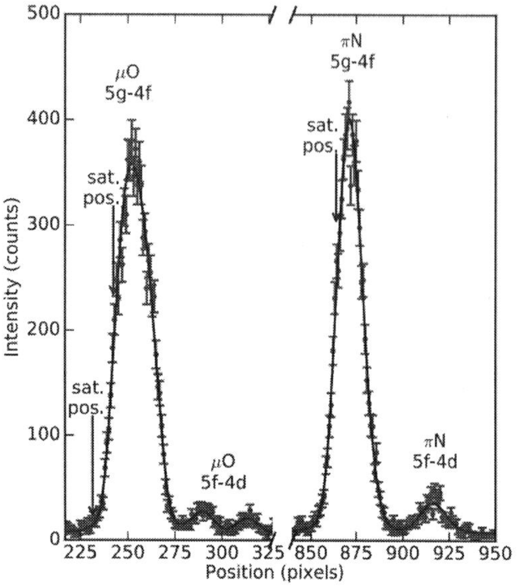

Figure 4. Simultaneously measured spectra of $5g$–$4f$ transitions in muonic oxygen (energy calibration) and pionic nitrogen. The parallel transitions $5f-4d$ and possible additional transitions from the presence of one remaining electron in the K shell are indicated ("sat. pos."). For μO, the larger width of the $5g-4f$ transition is due to the non-resolved fine structure splitting, which is resolved for the $5f-4d$ transition.

Monte-Carlo simulation which has been validated by a series of experiments with highly charged ions where natural line widths are negligibly small and high X-ray intensities are available [41, 45–48].

Concerning the atomic cascade processes, the dominant effect on the line width for a molecular target is Coulomb explosion. The Doppler broadening and a possible additional broadening due to imperfections of the diffraction crystal are determined from the analysis of a dedicated measurement optimized for pion stops, where in total 60000 events were accumulated in the πN $5g-4f$ transition. From this analysis, we found kinetic energies up to 146^{+6}_{-7} eV, which corresponds to molecule fragments having about 3.3 elementary charges in agreement with previous results [32]. More details of this procedure can be found in Refs. [49, 50].

4 Pion mass evaluation

The pion mass is deduced from the measurement of the $5g-4f$ πN transition energy using the $5g-4f$ μO transition as reference. The benefits of the choice of this reference are multiple: (i) the reference energy can be calculated very precisely due to the high accuracy of the muon mass value [21] and (ii) the two lines have almost coinciding energies, *i. e.* they can be simultaneously measured in the same order of reflection, which reduces drastically the effect of many systematic corrections.

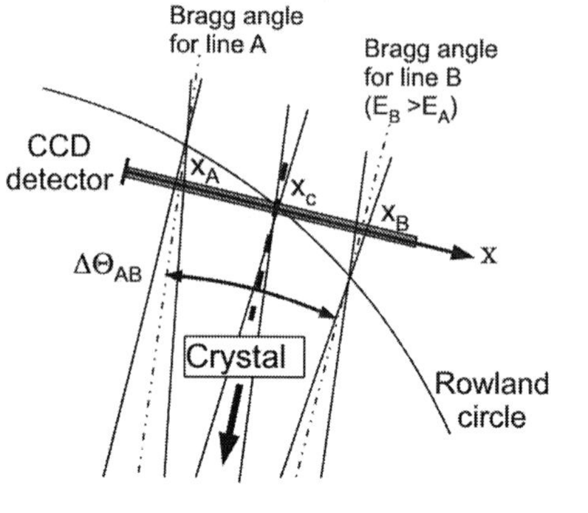

Figure 5. Scheme of the X-ray detector position with respect to the Rowland circle (the focus location of the spectrometer) and two diffracted lines A and B with different energy $E_B > E_A$. The (small) displacements of the two lines from their individual focal conditions is taken into account in the calculation of the spectrometer response by means of a Monte-Carlo ray tracing calculation.

We present here in detail the formulas for deducing the pion mass from the position difference between the spectral line of a pionic atom transition with unknown energy E_B and a reference spectral line with energy E_A (in the same diffraction order). The basic formula is the Bragg law that relates the X-ray energies to diffraction angles Θ:

$$E_A = n\frac{hc}{2d\sin\Theta_A}, \tag{1}$$

where h and c are the Planck and speed of light constant, d is the lattice spacing between crystal planes of the considered reflection direction with order n (an integer number) and Θ_A is the Bragg angle.

The energy E_B is related to E_A by

$$E_B = E_A\frac{1}{\cos\Delta\Theta_{AB} - \cot\Theta_A\sin\Delta\Theta_{AB}}, \tag{2}$$

where $\Delta\Theta_{AB} = \Theta_A - \Theta_B$ is the difference between the two Bragg angles, where a small corrections is applied because of the slightly different index of refraction. $\Delta\Theta_{AB}$ is deduced from the spectral line positions x_A and x_B on the detector plane (see Fig. 5), which is positioned at a distance D from the crystal. We have

$$\Delta\Theta_{AB} = \arctan\left(\frac{x_B - x_c}{D}\right) - \arctan\left(\frac{x_A - x_c}{D}\right), \tag{3}$$

where x_c is the position corresponding to the center of the detector.

From Eq. (2) we see that $\cot\Theta_A$ is the only quantity where the values of hc and d play a role according to Eq. (1).

The pion mass m is deduced from the relation of the transition energy E_B

$$E_B = \mu_{\pi\mathrm{N}}\, c^2\frac{(Z\alpha)^2}{2}\left(\frac{1}{n_f^2} - \frac{1}{n_i^2}\right) + O\left[(Z\alpha)^4\right] \tag{4}$$

to the reduced mass of the pion-nuclear system

$$\mu_{\pi N} = \frac{m_\pi}{1 + \frac{m_\pi}{M}}, \tag{5}$$

with M being the nuclear mass of nitrogen, n_i and n_f the quantum number of the initial and final state of the transition and α the fine structure constant. The relationship between E_B and $\mu_{\pi N}$ contains corrections, not shown explicitly here, that include relativistic effects (additional terms due to the Klein-Gordon equation in our cases), quantum electrodynamics effects (vacuum polarization, self-energy, ...), recoil corrections, etc. and are summarized here by the term $O\left[(Z\alpha)^4\right]$.

For the limited range of pion-mass values, we can assume a linear dependency between E_B and $\mu_{\pi N}$. The proportionality factor is calculated by the MCDF program [51–53] considering the charged pion mass value from the Particle Data Group m_π^{PDG} [21] and the corresponding computed transition energy E_B^{PDG}. In this approximation, the value of m_π is given by

$$m = \frac{\mu_{\pi N}^{PDG} \frac{E_B}{E_B^{PDG}}}{1 - \frac{\mu_{\pi N}^{PDG}}{M} \frac{E_B}{E_B^{PDG}}}. \tag{6}$$

From the line positions and the evaluation of the systematic effects and uncertainties, a pion mass value of $(139.57077 \pm 0.00018)\,\mathrm{MeV/c^2}$ (± 1.3 ppm) is determined. Additional details of the evaluation and of the estimation of the systematic uncertainties are given in Ref. 22. In the next section we specifically discuss the effect of possibly remaining electrons.

5 Determination of remaining electrons

The possible presence of remaining electron(s) in the exotic atom may induce an important systematic energy shift of the X-ray energies and, consequently, of the pion mass. One (or two) remaining electron(s) in the K shell in pionic nitrogen can generate satellite lines having energies 0.45 eV (0.81 eV) lower than the main transition $5g - 4f$ [51, 52]. Such weak satellite lines cannot be resolved from the main transitions, in particular, as they are expected to be of very low intensity (see Fig. 4).

As discussed in Sec. 2, for light pionic atoms production in a gaseous target of moderate density, the possibility that electrons are present is small when X-ray emission starts. However, it cannot be excluded beforehand that small fractions of pions or muons may arrive at the $5g$ level by $\Delta n \gg 1$ transitions from low angular momentum states immediately after capture.

To estimate the probability for the occurrence of satellite lines we use the evaluation of the Bayes factor [54–57] relative to two hypothesis: the presence or not of satellite lines. The first hypothesis is associated to the model M_0 of the spectra without additional satellite lines. In the model

Table 1. Bayes factor (in logarithmic scale) and the average P_0 of the model M_0 from the analysis of the different spectra. We also present the possible range of P_0 tanking into account the uncertainty of $\ln B_{01}$.

Spectrum	$\ln B_{01}$	P_0	P_0^{min}	P_0^{max}
high-stat. πN	6.6 ± 1.8	99.98%	99.86%	100%
low-stat. μO	-0.3 ± 0.4	42.52%	32.70%	52.98%

M_1 associated to the second hypothesis, we consider the presence of additional satellite lines with fixed positions with respect to the main transition but keeping the intensity as a free parameter (two lines in the case of muonic oxygen, one per each fine structure main component).

For each model we sample the likelihood function for different values of parameters to evaluate the Bayesian evidence [55, 58, 59], which corresponds to the integral over the parameter space of the likelihood function times the prior probability distributions of the parameters. The Bayes factor B_{01}, the ratio of the Bayesian evidence of the two models, is then calculated as well as the relative probability of the two models. The evidence is calculated using an homemade code based on the nested sampling algorithm developed by John Skilling in 2004 [55, 60] (see also Refs. 49, 61–63 for more details on the calculation method).

An important feature of the evidence calculation is, that contrary to maximum likelihood and minimum χ^2 methods which provide for each parameter only the most probable value and the standard deviation, a probability distribution is established for each model parameter (as well as joint probability distributions).

For the calculation of B_{01}, two data sets were used: the high-statistics spectra of πN and the low-statistics spectra of μO (shown in Fig. 4). The results are summarized by Figs. 6, 7 and Table 1. Due to the low statistics, the results from muonic oxygen cannot be used as test against one of the two models considering the associated uncertainty, because with $\ln B_{01}$ being too close to zero the probability for the two models is similar. For the high-statistics pionic nitrogen spectra, the Bayes factor is significantly different to the unity and the relative probability of the two models can be reliably calculated. The value $\ln B_{01} = 6.6$ indicates a decisive support for the M_0 hypothesis for any Bayes factor scale considered ("decisive" for Jeffreys scale [54], "very strong" for the Kass scale[56] or "strong" for the Gordon-Trotta scale, equivalent to a p-value of about 10^{-5} for M_1 [57]). Model M_0 and M_1 relative probabilities are 99.98% and 0.02%, respectively. Though being small, the effect of such a non-zero probability for M_1 on the pion mass can be evaluated.

When the model M_1 is considered, for both sets a satellite amplitude of about 1% of the main line (see Fig. 6) is found. As expected and clearly visible in Fig. 7, for πN the satellite amplitude is strongly correlated to the main line position. We found a shift of the main line of $(\delta x)_1 = 0.08$ pixels with respect to the case where satel-

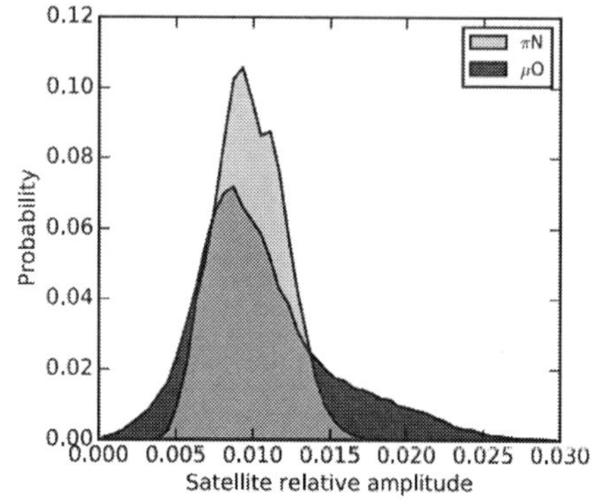

Figure 6. Probability distribution of the amplitudes relative to the main line intensity of the possible satellite line due to the presence of one remaining electron in the K shell.

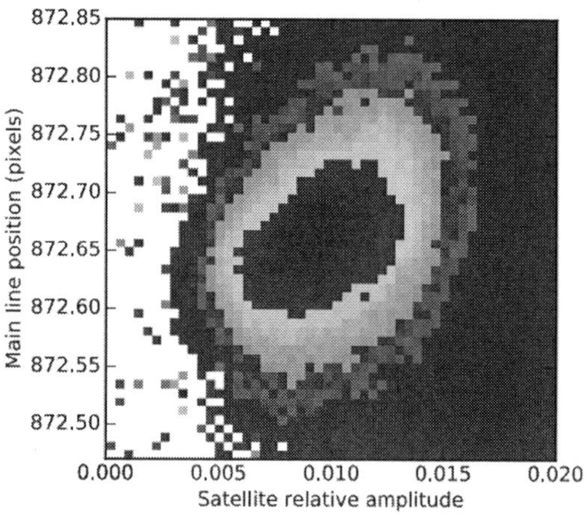

Figure 7. Joint probability distribution of the relative satellite amplitude and the position of the main line $5g \rightarrow 4f$ in pionic nitrogen. The accumulation on the diagonal shows the correlation between satellite intensity and main line position. Assuming no satellite line, the position of the main line for this set of data is 872.58 ± 0.04 pixels.

lite lines are not taken into account. This is equivalent to about 1 ppm of the pion mass. This shift cannot be considered in total but has to be weighted by the probability of the two different models using the values of Table 1 and where $(\delta x)_0 = 0$. The expected shift of the main line is then

$$\delta x = 0 \times P_0 + (\delta x)_1 \times P_1 = 1.8 \times 10^{-5} \text{ pixels} \quad (7)$$

corresponding to a systematic uncertainty of $^{+0}_{-0.0002}$ ppm. When the Bayes factor accuracy is taken into account (Table 1) and the worst case is considered, the probability of M_0 drops to $P_0^{min} = 99.86\%$ and the systematic uncertainty increases to $^{+0}_{-0.0014}$ ppm which is still completely negligible with respect to the statistical error and other systematic effects.

6 Conclusions

In conclusion, we present a new measurement of the negatively charged pion mass based on Bragg spectroscopy of pionic nitrogen and muonic oxygen using a gaseous target. The use of the low-density target with light atoms prevents recapture of electrons from neighbouring molecules during the atomic cascade after the formation of the exotic atom. Consequently, the X rays emitted in the last steps of the cascade stem from a purely hydrogen-like system without systematic effects from unresolved satellite lines due to remaining electrons. From our high-statistics pionic nitrogen measurement, the probability of presence of satellite peaks has been found to be about 0.02% confirming the hypothesis on the absence of electrons. Such a contribution introduces a negligible systematic uncertainty on the pion mass of less than one part per billion.

We are grateful to N. Dolfus, H. Labus, B. Leoni and K.-P. Wieder for solving numerous technical problems. We thank the PSI staff for providing excellent beam conditions and appreciate the support by the Carl Zeiss AG, Oberkochen, Germany, which fabricated the Bragg crystals. We thank Prof. Dr. E. Förster and his collaborators at the University of Jena, and A. Freund and his group at ESRF, for the help in characterising the crystal material as well as A. Blechmann for a careful study of the CCD performance. We are indebted to PSI for supporting the stay during the run periods (D. F. A.). This work is part of the PhD thesis of B. M. (Université Pierre et Marie Curie, 2001), N. N. (University of Leicester, 2002) and M. T. (Université Pierre et Marie Curie, 2005).

References

[1] C.M.G. Lattes, G.P.S. Occhialini, C.F. Powell, Nature **160**, 453 (1947)
[2] C.M.G. Lattes, G.P.S. Occhialini, C.F. Powell, Nature **160**, 486 (1947)
[3] E. Gardner, C.M.G. Lattes, Science **107**, 270 (1948)
[4] W.H. Barkas, F.M. Smith, E. Gardner, Phys. Rev. **82**, 102 (1951)
[5] F.M. Smith, W. Birnbaum, W.H. Barkas, Phys. Rev. **91**, 765 (1953)
[6] W.H. Barkas, W. Birnbaum, F.M. Smith, Phys. Rev. **101**, 778 (1956)
[7] K.M. Crowe, R.H. Phillips, Phys. Rev. **96**, 470 (1954)
[8] M. Stearns, M.B. Stearns, S. DeBenedetti, L. Leipuner, Phys. Rev. **95**, 1353 (1954)
[9] R.E. Shafer, Phys. Rev. **163**, 1451 (1967)
[10] V.N. Marushenko, A.F. Mezentsev, A.A. Petrunin, S.G. Skornyakov, A.I. Smirnov, JETP Lett. **23**, 72 (1976)
[11] A.L. Carter, M.S. Dixit, M.K. Sundaresan, J.S. Wadden, F.J. Watson, C.K. Hargrove, E.P. Hincks, R.J.

McKee, H. Mes, H.L. Anderson et al., Phys. Rev. Lett. **37**, 1380 (1976)

[12] D.C. Lu, D. Delker, G. Dugan, C.S. Wu, A.J. Caffrey, Y.T. Cheng, Y.K. Lee, Phys. Rev. Lett. **45**, 1066 (1980)

[13] B. Jeckelmann, T. Nakada, W. Beer, G. de Chambrier, O. Elsenhans, K.L. Giovanetti, P.F.A. Goudsmit, H.J. Leisi, A. Rüetschi, O. Piller et al., Phys. Rev. Lett. **56**, 1444 (1986)

[14] B. Jeckelmann, W. Beer, G. De Chambrier, O. Elsenhans, K.L. Giovanetti, P.F.A. Goudsmit, H.J. Leisi, T. Nakada, O. Piller, A. Rüetschi et al., Nucl. Phys. A **457**, 709 (1986)

[15] B. Jeckelmann, P.F.A. Goudsmit, H.J. Leisi, Phys. Lett. B **335**, 326 (1994)

[16] S. Lenz, G. Borchert, H. Gorke, D. Gotta, T. Siems, D.F. Anagnostopoulos, M. Augsburger, D. Chatellard, J.P. Egger, D. Belmiloud et al., Phys. Lett. B **416**, 50 (1998)

[17] R. Abela, M. Daum, G.H. Eaton, R. Frosch, B. Jost, P.R. Kettle, E. Steiner, Phys. Lett. B **146**, 431 (1984)

[18] M. Daum, R. Frosch, D. Heter, M. Janousch, P.R. Kettle, Phys. Lett. B **265**, 425 (1991)

[19] K. Assamagan, C. Brnnimann, H.F. M. Daum, R. Frosch, P. Gheno, R. Horisberger, M. Janousch, P.R. Kettle, T. Spirig, C. Wigger, Phys. Rev. D **53**, 6065 (1996)

[20] D. Gotta, D.F. Anagnostopoulos, M. Augsburger, G. Borchert, C. Castelli, D. Chatellard, J.P. Egger, P. El-Khoury, H. Gorke, P. Hauser et al., Nucl. Phys. A **660**, 283 (1999)

[21] K.A. Olive, G. Particle Data, Chinese Phys. B **38**, 090001 (2014)

[22] M. Trassinelli, D.F. Anagnostopoulos, G. Borchert, A. Dax, J.P. Egger, D. Gotta, M. Hennebach, P. Indelicato, Y.W. Liu, B. Manil et al., Phys. Lett. B **759**, 583 (2016)

[23] L.M. Simons, Phys. Scripta **T22**, 90 (1988)

[24] L.M. Simons, Hyperfine Interact. **81**, 253 (1993)

[25] J.S. Cohen, Rep. Prog. Phys. **67**, 1769 (2004)

[26] G.R. Burbidge, A.H. de Borde, Phys. Rev. **89**, 189 (1953)

[27] M. Leon, H.A. Bethe, Phys. Rev. **127**, 636 (1962)

[28] P. Vogel, Phys. Rev. A **22**, 1600 (1980)

[29] R. Bacher, D. Gotta, L.M. Simons, J. Missimer, N.C. Mukhopadhyay, Phys. Rev. Lett. **54**, 2087 (1985)

[30] R. Bacher, P. Blüm, D. Gotta, K. Heitlinger, M. Schneider, J. Missimer, L.M. Simons, Phys. Rev. A **39**, 1610 (1989)

[31] K. Kirch, D. Abbott, B. Bach, P. Hauser, P. Indelicato, F. Kottmann, J. Missimer, P. Patte, R.T. Siegel, L.M. Simons et al., Phys. Rev. A **59**, 3375 (1999)

[32] T. Siems, D.F. Anagnostopoulos, G. Borchert, D. Gotta, P. Hauser, K. Kirch, L. Simons, P. El-Khoury, P. Indelicato, M. Augsburger et al., Phys. Rev. Lett. **84**, 4573 (2000)

[33] T.S. Jensen, V.E. Markushin, Eur. Phys. J. D **19**, 165 (2002)

[34] T.S. Jensen, V.E. Markushin, Eur. Phys. J. D **21**, 261 (2002)

[35] S. Kilic, J.P. Karr, L. Hilico, Phys. Rev. A **70**, 042506 (2004)

[36] V.N. Pomerantsev, V.P. Popov, Phys. Rev. A **73**, 040501 (2006)

[37] T. Jensen, V. Popov, V. Pomerantsev, arXiv preprint arXiv:0712.3010 (2007)

[38] V. Popov, V. Pomerantsev, arXiv preprint arXiv:0712.3111 (2007)

[39] H.H. Johann, Z. f. Physik **69**, 185 (1931)

[40] D. Gotta, Progress in Particle and Nuclear Physics **52**, 133 (2004)

[41] D.E. Gotta, L.M. Simons, Spectrochim. Acta, Part B **120**, 9 (2016)

[42] J. Eggs, K. Ulmer, Zf. angew. Phys. **20**, 118 (1965)

[43] N. Nelms, D.F. Anagnostopoulos, O. Ayranov, G. Borchert, J.P. Egger, D. Gotta, M. Hennebach, P. Indelicato, B. Leoni, Y.W. Liu et al., Nucl. Instrum. Methods A **484**, 419 (2002)

[44] P. Indelicato, E.O. Le Bigot, M. Trassinelli, D. Gotta, M. Hennebach, N. Nelms, C. David, L.M. Simons, Rev. Sci. Instrum. **77**, 043107 (10) (2006)

[45] S. Biri, L. Simons, D. Hitz, Rev. Sci. Instrum. **71**, 1116 (2000)

[46] D.F. Anagnostopoulos, S. Biri, V. Boisbourdain, M. Demeter, G. Borchert, J.P. Egger, H. Fuhrmann, D. Gotta, A. Gruber, M. Hennebach et al., Nucl. Instrum. Methods B **205**, 9 (2003)

[47] D.F. Anagnostopoulos, S. Biri, H. Fuhrmann, D. Gotta, A. Gruber, P. Indelicato, B. Leoni, L.M. Simons, L. Stingelin, A. Wasser et al., Nucl. Instrum. Methods A **545**, 217 (2005), `physics/0408081`

[48] D.S. Covita, Ph.D. thesis, University of Coimbra (2008)

[49] M. Theisen, Diplomarbeit, Fakultät für Mathematik, Informatik uns Naturwisssenschaften der RWTH Aachen (2013)

[50] D. Gotta, F.D. Amaro, D.F. Anagnostopoulos, P. Bühler, H. Gorke, D.S. Covita, H. Fuhrmann, A. Gruber, M. Hennebach, A. Hirtl et al., Hyperfine Interact. **234**, 105 (2015)

[51] J. Desclaux, J. Dolbeault, M. Esteban, P. Indelicato, E. Séré, in *Computational Chemistry*, edited by C. Le Bris, M. De Franceschi (Elsevier, Amsterdam, 2003), Vol. X of *Handbook of Numerical Analysis*, p. 1032

[52] J.P. Santos, F. Parente, S. Boucard, P. Indelicato, J.P. Desclaux, Phys. Rev. A **71**, 032501 (8) (2005)

[53] M. Trassinelli, P. Indelicato, Phys. Rev. A **76**, 012510 (2007)

[54] H. Jeffreys, *Theory of Probability*, 3rd edn. (Oxford University Press, Oxford, U.K., 1961)

[55] D.S. Sivia, J. Skilling, *Data analysis: a Bayesian tutorial*, 2nd edn. (Oxford University Press, 2006)

[56] R.E. Kass, A.E. Raftery, J. Acoust. Soc. America **90**, 773 (1995)

[57] C. Gordon, R. Trotta, Mon. Not. R. Astron. Soc. **382**, 1859 (2007)

[58] E. Jaynes, G. Bretthorst, *Probability Theory: The Logic of Science* (Cambridge University Press, 2003), ISBN 9781139435161

[59] R. Trotta, Contemp. Phys. **49**, 71 (2008)

[60] J. Skilling, Bayesian Anal. **1**, 833 (2006)

[61] P. Mukherjee, D. Parkinson, A.R. Liddle, Astrophys. J. Lett. **638**, L51 (2006)

[62] F. Feroz, M.P. Hobson, M. Bridges, Mon. Not. R. Astron. Soc. **398**, 1601 (2009)

[63] J. Veitch, A. Vecchio, Phys. Rev. D **81**, 062003 (2010)

Kaonic nuclear state search at J-PARC

Masahiko Iwasaki[1,*]

[1] *RIKEN, Wako-shi, Saitama 351-0198, Japan*

Abstract. At J-PARC, there are two experimental programs, E15 and E27, to search for the simplest kaonic bound nuclear state in $\bar{K}NN$ system. These two experiments were proposed based preceding experiments reporting candidates of such state, although the experimental results are not quite consistent to each other. These two experimental programs utilizing different reaction channels, ^{3}He(K^-, n) reaction by K^- momentum at 1 GeV/c for E15, and $d(\pi^+, K^+)$ reaction by π^+ momentum at 1.7 GeV/c for E27, to resolve this puzzling situation. In this paper, we overview these two experimental results as flat as possible.

1 Introduction

The $\bar{K}N$ interaction is known to be strongly attractive from low-energy scattering data [1] and X-ray spectroscopy of kaonic atoms [2]. On the other hand, there exist a well known pole, $\Lambda(1405)$, slightly below the $\bar{K}N$ mass threshold. This $\Lambda(1405)$ resonance is known that it is very difficult to be treated as a simple excited state of Λ hyperon, and it is widely accepted that the $\Lambda(1405)$ is K^-p bound state / penta-quark like structure or couple to these states. If that is true, the natural extension is that the kaon will form a nuclear bound state, helped by the existence of the $\Lambda(1405)$. Accordingly, such states are predicted and a high density matter formation, exceeding the normal nuclear density, is expected in such states [3]. Therefore, observation of a kaonic nuclear bound state would provide definitive information on tthe nature of $\Lambda(1405)$, and nuclear physics at very high density.

Both theoretical and experimental studies have been made in the last decade for the simplest kaonic nuclear state $\bar{K}NN$. However, the results do not seem to be consistent each other. Theoretically, all calculations predict the existence of a bound state, but the predicted $\bar{K}NN$ pole positions are scattered depending on $\bar{K}N$ interaction models. For the energy-independent model, the binding energy is reaching up to 50 ~ 100 MeV, while in energy-dependent case, it becomes weaker to be 10 ~ 30 MeV. The widths are also widely scattered over 30 ~ 110 MeV. Experimentally, there are few positive reports on peak structure observations as for the candidate at ~100 MeV below $\bar{K}NN$ threshold. The first report from FINUDA group showing a peak structure in the back-to-back Λp invariant mass spectra via the stopped kaon reaction on ^{6}Li, ^{7}Li, and ^{12}C targets [4], having binding energy (B.E.) ~ 115 MeV, and a width (Γ) ~ 70 MeV. The DISTO group conducted a pp collision experiment. observed $\bar{K}NN$ candidate at B.E. ~ 100 MeV, having Γ ~ 120 MeV/c^2 [5]. Conversely, no significant structure was observed in a SPring-8/LEPS γ induced inclusive experiment [6] or in a proton-proton interaction by the HADES/GSI collaboration [7].

*e-mail: masa@riken.jp

In these experimental spectra, the most significant structure was found in DISTO. They examined $<K^-pp>$ formation using $pp \to K^+ <K^-pp>$ reaction at $T_p = 2.85$ GeV. The momentum transfers q_K to $<K^-pp>$ system, viewed from pp rest-frame (CM-frame), are as low as ~300 and ~430 MeV/c at B.E. ~0 and ~100 MeV, respectively. The low q_K will enhance the K^- sticking probability to pp system, helped by the $\Lambda(1405)$ pole just below the K^-p mass threshold. If the decay branch, $<K^-pp> \to \Lambda p$, is relatively large, then the signal observation through $K^+\Lambda p$ final state, by the Λp invariant mass (or K^+ missing mass).

There are, however, many concerns in DISTO result. Their pp collision energy is just above the production threshold of N^*, which decay strongly to $K^+\Lambda$ (such as $N^*(1710)$). In the reaction near the formation threshold, such N^*s are nearly at-rest in the CM-frame, so it is rather natural to have kinematical structure in the invariant mass spectrum of Λp (or the missing mass of K^+), in the chain reaction of $pp \to N^{*+}p \to K^+\Lambda p$, even without K^-pp resonance of $pp \to K^+ <K^-pp> \to K^+\Lambda p$. The Λp invariant mass peak stand just on the $\Sigma\pi p$ mass threshold having no threshold effect, which indicates that the state do not couple to the $\Sigma\pi p$ channel. This is rather strange, because $<K^-pp>$ should strongly couples with $\Lambda(1405)$-p, and the major decay channel of $\Lambda(1405)$ is $\Sigma\pi$. The spectrum is also insensitive to the K^-pp mass threshold, while it is more natural to have non-resonant process above the threshold (quasi-free K^-pp formation), because the reaction chain of $pp \to K^+K^-pp$ (or $K^+Y^{(*)}p) \to K^+\Lambda p$ could happen at relatively low momenta of those particles in CM-frame. And finally, in contrast to the DISTO result, no peak structure found in same pp collision at different kinetic energy $T_p = 3.5$ GeV by HADES. The q_K is ~550 and ~650 MeV/c viewed from CM-frame at B.E. ~0 and ~100 MeV, respectively. The transfer of ~ 650 MeV/c is rather large to form a bound state efficiently, but it would be still sizable if the B.E. of $<K^-pp>$ is as large as ~ 100 MeV in comparison with that of ~ 430 MeV/c at DISTO.

2 Kaonic nuclear state search at J-PARC

To clarify the situation, two experimental groups, E15 and E27, conducted experiments at J-PARC searching for $\bar{K}NN$ bound state. J-PARC E15 and E27 utilizing different reaction channels. The E15 is utilizing ^{3}He$(K^-, n)X$ reaction at K^- momentum of 1 GeV/c, while the E27 is utilizing $d(\pi^+, K^+)X$ reaction at π^+ momentum of 1.69 GeV/c. E27 is conducted in 2012 earlier than E15, because high intensity π^+ beam of 1.69 GeV/c can be easily obtained at K1.8 beam line at J-PARC. E15 was assigned as *Day-One-Experiment* at J-PARC, but the first run (E15$^{1st.}$) was executed only in 2013, since it requires high intensity kaon beam.

2.1 J-PARC E27

The E27 was conducted to examine the striking result of DISTO, that the $<K^-pp>$ could be bound as astonishingly strong as ~ 100 MeV, having large cross section, and having a relatively large decay branch to Λp [8]. The $\bar{s}s$-pair is produced at the reaction as it is like in pp collision, but the kinematics is much different. In the $\pi^+d \to K^+K^-pp$ (or K^+X) reaction, the two proton in $X = <K^-pp>$ come from the deuteron target, so the q_K to the $<K^-pp>$ system should be calculated in the deuteron rest-frame (Lab.-frame). The difference of the kinematics makes energy dependence to the q_K opposite, namely 730 MeV/c for B.E. = 0, and 570 MeV/c for B.E. = 100 MeV. Thus, E27 is kinematically more like HADES rather than DISTO.

In E27, forward K^+ from $d(\pi^+, K^+)X$ reaction is identified and momentum analyzed by SKS, a large acceptance (~ 100 msr) superconducting spectrometer covering from 0.8 to 1.3 GeV/c in K^+ momentum, and from 2 to 16 degree in angle in the Lab.-frame. To detect decay protons from X for $p_p > 250$ MeV/c, E27 equipped 6 sets of range counter arrays (RCA) as shown in figure 1.

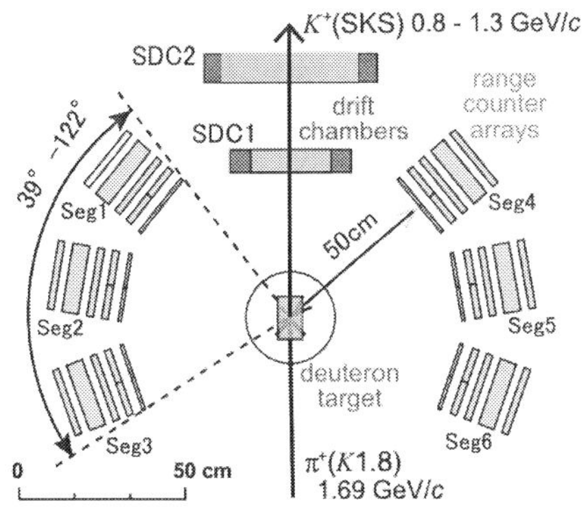

Figure 1. The proton RCA is made of five layers of plastic scintillation counters of about 12 g/cm^2 in total (1 + 2 + 2 + 5 + 2 cm in thickness). Three sets of the range counters are placed in the horizontal plane both left and right side, viewed from incident π^+ beam.

In figure 2, the K^+ missing mass spectra are shown. Fig. 2 (left) is the spectrum for inclusive $d(\pi^+, K^+)X$ reaction. As it is naturally expected, $d(\pi^+, K^+)X$ reaction is sensitive to the $K^+Y^{(*)}p$ formation. In fact, quasi-free hyperon formations are clearly seen as $K^+\Lambda p$ and $K^+\Sigma N$ final states below ~2.2 GeV/c^2. It also shows that the yields associated with additional pions, channel open at ~2.2 and ~2.27 GeV/c^2, are relatively week, when one require the forward K^+ formation as it is the case in E27. Fig. 2 (right) is that for one proton-tagged event in RCA. In the spectrum, $K^+\Lambda p$ channel is suppressed as shown in the figure, while $K^+\Sigma N$ channel still remains especially near the ΣN threshold. The rest of $K^+\Sigma N$ channel is also tagged more efficiently than $K^+\Lambda N$, helped by the energy transfer to the spectator nucleon, in the successive $\Sigma N \rightarrow \Lambda N$ conversion.

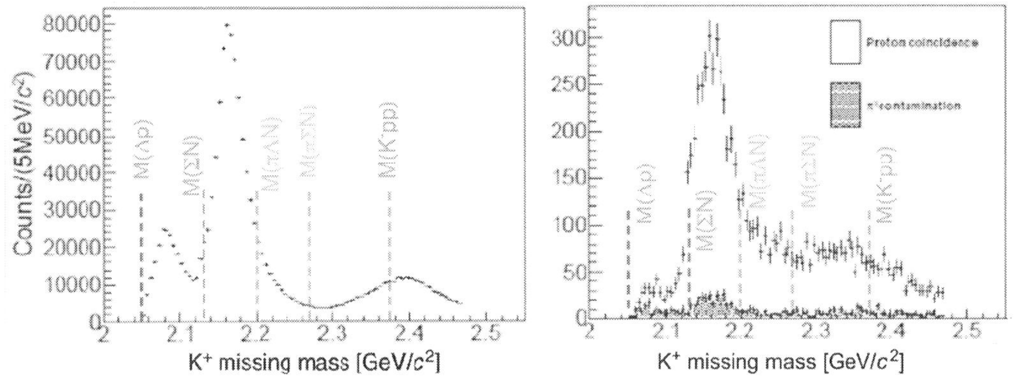

Figure 2. Left) A K^+ missing mass $M(X)$ spectrum for the inclusive $d(\pi^+, K^+)X$ reaction. Threshold energy for several related channels, where each channel open energetically, are plotted in dashed lines. Right) A subset of the same inclusive spectrum, in which at least one proton hits in one of RCA.

The other difference is the spectral strength in the energy region from $M(\pi\Lambda p)$ to $M(K^-pp)$. To understand the difference, missing mass analysis was applied for $d(\pi^+, K^+pp)Y$ to the two proton identified events on RCA. If $K^+\Lambda p$ is the final state of π^+d reaction, then the missing mass analysis of the $d(\pi^+, K^+pp)Y$ kinematics should yield that $Y = \pi$. If the final state is $K^+\Sigma^0 p$, then the analysis gives that $Y = \pi\gamma$. Thus, one can specify the final state of the reaction using kinematics, although the statistics of the pp-coincidence events is drastically reduced due to the limited RCA solid angle. One should also be careful to the kinematical deformation due to the RCA coverage both in angle and energy threshold.

 DOI: 10.1051/epjconf/201613001023

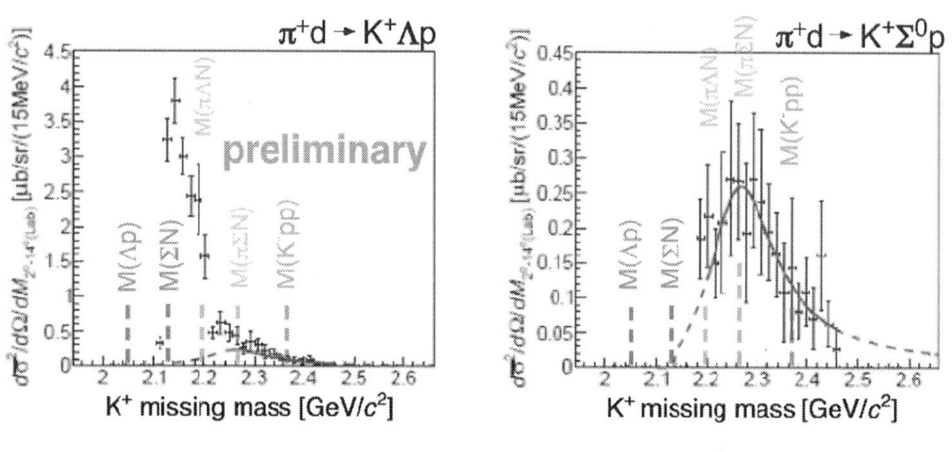

Figure 3. The K^+ missing mass spectrum of $d(\pi^+, K^+)X$ reaction for $K^+\Lambda p$ final state (left), and that for $K^+\Sigma^0 p$ final state (right),

Fig. 3 shows the K^+ missing mass spectra for $K^+\Lambda p$ final state (left) and that for $K^+\Sigma^0 p$ final state (right). In Fig. 3 (left), there is a structure from $M(\Sigma N)$ up to $M(\pi\Lambda N)$, which was interpreted that the structure is formed due to the ΣN-cusp and $\Sigma N \to \Lambda N$ conversion. In Fig. 3 (right), there is a wide event distribution above $M(\pi\Lambda N)$, which extended smoothly even beyond the $M(K^-pp)$ threshold. This broad structure in both spectra were fitted by a Breit-Wigner formula in the energy region indicated by the solid curve. The result of the peak position and the width are roughly consistent with DISTO, which gives extremely deep binding energy of about 100 MeV. On the other hand, all the concerns to the DISTO result would remain as they were. There would be another concern, why $\pi^+ d$ reaction is sensitive to ΣN-cusp and $\Sigma N \to \Lambda N$ conversion, while it is not in pp collision.

2.2 J-PARC E15

As for the other attempt to for $< K^- pp >$ much more definitively, E15 experiment on the K^- + ^{3}He reaction is underway at J-PARC in a stepwise manner. The experimental approach of E15 is much different from other experimental researches. Firstly, the strangeness is introduced as a K^- beam at 1 GeV/c, where the $\overline{K}N$ elastic cross sections are quite large (*e.g.* $n(K^-, n)K^- \sim 5$ mb @ 1 GeV/c). Therefore, elastic ^{3}He$(K^-, n)X$ reaction channel, a neutron knockout reaction, is dominant. In this reaction, the q_K to $< K^- pp >$ system, viewed from spectator protons' rest-frame (Lab.-frame) in K^- ^{3}He $\to n + K^- pp$, is as small as 230 MeV/c for B.E. = 0, and it is still 330 MeV/c even for B.E. = 100 MeV. The q_K is smaller by about 100 MeV/c than those in DISTO. Thus, one can expect even more efficient formation of the $< K^- pp >$ system in E15 than DISTO. The experimental setup is consist from three components; forward neutron counter array (NC), forward proton spectrometer using beam sweeping magnet, and cylindrical detector system (CDS) surrounding the liquid ^{3}He target. In the E15 kinematics, $< K^- pp >$ system is formed almost at-rest (as it is described, 230 MeV/c backward at B.E. = 0), so the decay products can be efficiently detected by the CDS [9]. Another remarkable point in this reaction is that it is naturally insensitive to the production and the decay of $N^* \to K^+\Lambda$, because of the Q-value.

The first paper was published on the semi-inclusive neutron spectrum ^{3}He$(K^-, n)X$ in the forward direction [10]. The missing mass spectrum of the neutron is shown in Fig. 4. As it is expected, kaon

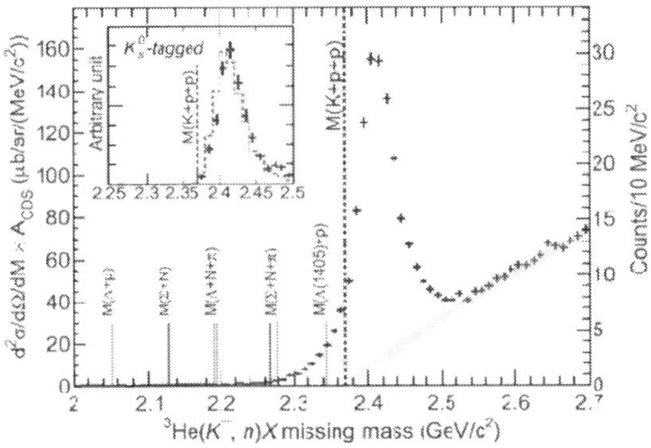

Figure 4. The semi-inclusive neutron spectrum of the ^{3}He(K^-, n)X reaction in the forward direction. To identify the kaon reaction point, one charged particle track is requested in CDS. In the semi-inclusive spectrum, a long tail is observed below $M(Kpp)$ threshold. To show pure kaon quasi-elastic scattering process, K^- "p" $\rightarrow \overline{K}^0 n$, out from the inclusive reactions, K_s^0 tagged spectrum is shown as inset. The quasi-elastic channel start right on the $M(Kpp)$ threshold, having a width caused by the Fermi-motion of the incident proton in the nucleus.

quasi-elastic channel is clearly seen at around 2.4 GeV/c^2. In spite of the lower q_K to $<K^-pp>$ system, no significant peak structure is observed around $M(\pi\Sigma p)$ energy region, as it is the case in DISTO or E27. Instead, there is a long tail below $M(Kpp)$ threshold, extended to $M(\pi\Sigma p)$ beyond $M(\Lambda(1405)p)$ threshold. This tail cannot be explained by a simple extrapolation from the high excitation continuum (large missing mass region), as shown in the figure. This indicate that there is a strong attractive interaction between $\overline{K}N$, since such a long tail ($\sim$ 100 MeV) is very difficult to be formed simply by the imaginary part of the $\overline{K}N$ interaction. In contrast to E27 spectrum given in Fig. 2 (left), no strong yield is observed near the $M(\Lambda p)$ nor $M(\Sigma N)$, which indicating that the kaon two-nucleon-absorption process (2NA) is rather weak at p_K = 1 GeV/c.

The second paper from E15 is focused on the Λpn final state [11]. Since $<K^-pp>$ system is produced almost at-rest, the decay process to Λp can be studied only by CDS information without using NC for forward neutron. By Λp 4-momenta, one can reconstruct reaction if the event can be kinematically identified as ^{3}He($K^-, \Lambda p$)n by the missing mass. Fig. 5 (left) shows the Λp invariant mass

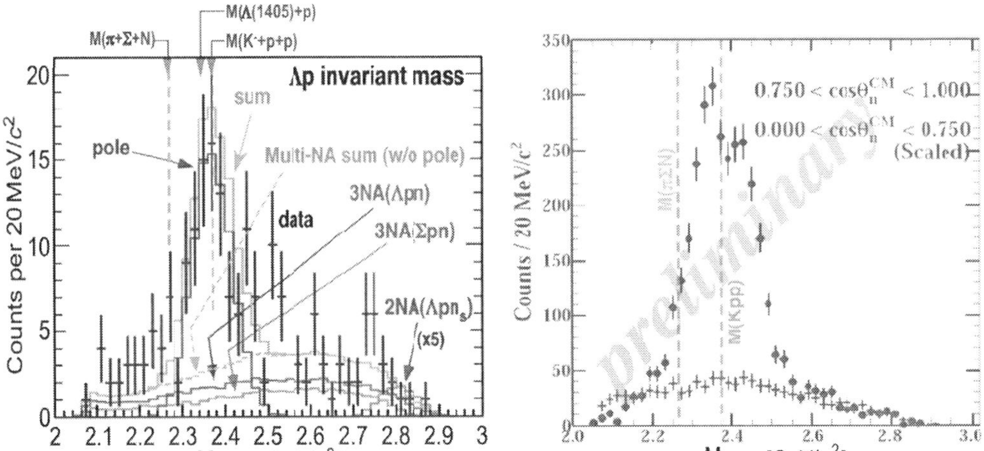

Figure 5. Left) The Λp invariant mass spectrum for ^{3}He($K^-, \Lambda p$)n events of the E15$^{1st.}$ data without selecting neutron emission angle. The spectrum was fitted with single pole together with the multi-nucleon absorption processes represented by #n-body phase space. Right) Same spectra with 30 times more data using E15$^{2nd.}$ data. The peak structure is observed in forward neutron emission events ($\cos\theta_n^{CM}$ > 0.75), while very smooth spectrum is observed in large-angle neutron emission events ($0 < \cos\theta_n^{CM} < 0.75$ region scaled by $\sim$ 1/2).

spectrum for ^{3}He(K^-, Λp)n events of the E15$^{1st.}$ data. Two spectra, given in Fig. 4 and Fig. 5 (left), are observing essentially same quasi-elastic kaon scattering reaction, except for the two differences. One is the final state selection, another is the emission angle of the neutron. Fig. 4 is semi-inclusive by limiting neutron emission angle to be zero by NC, while Fig. 5 is selective to Λpn final state without limiting emission angle, however, the spectral difference is quite clear. Although the statistics in Fig. 5 (left) is quite limited, it is clear that the structure near the $M(Kpp)$ threshold observed in Fig. 5 (left) cannot be explained by the simple quasi-elastic kaon scattering like in Fig. 4. Thus, a simplest pole existence is assumed in [11], together with the Gaussian form factor of $\left|\exp\left(-q_K^2/2Q_K^2\right)\right|^2$, where q_K is the q_K to $<K^-pp>$ system and Q_K is the fitting parameter. The fit result was $\sim$ 15 MeV for B.E, width of $\sim$ 100 MeV, and $Q_K \sim$ 400 MeV/c.

To understand the observed peak structure in more detail, E15$^{2nd.}$ run was conducted in 2015, succeeding in accumulating $\sim$ 30 time more data on Λpn final state, as shown in Fig. 5 (right). In the high statistics spectra, internal structure near the $M(Kpp)$ become very clear, both below and above the $M(Kpp)$. In the E15$^{1st.}$ run, the contribution of quasi-elastic kaon reaction above the $M(Kpp)$ is assumed to be small because of the final state. However, in comparison with Fig. 4, it is clear that the non-resonant nuclear absorption can happen even above the $M(Kpp)$ threshold in the high statistics data of E15$^{2nd.}$. The peak structure sandwiched in between $M(\pi\Sigma N)$ and $M(Kpp)$ (kaon bound region) is well understood if there exist a kaonic nuclear bound state. The sudden drop off of yield at $M(\pi\Sigma N)$ is consistent that the $< K^-pp >$ couples strongly to $\Lambda(1405)$-p, and decay to $\pi\Sigma N$ channel. In this region, free $\Lambda(1405)$ could not contribute by definition. The knockout nucleon exceeds $M(Kpp)$ threshold energy, only when $\Lambda(1405)$ forms a bound state with the other spectator proton nearby, and decay to a Λp pair. Fig. 5 (right) also showing that two structures near the $M(Kpp)$ threshold are formed only associated with forward neutron emission $\cos\theta_n^{CM} > 0.75$, where q_K is small ($Q_K \sim$ 400 MeV/c). This is consistent that present peak structures are observed only in E15 experiment, in which lowest q_K of $\sim$ 230 MeV/c is achieved.

E15 group is analyzing the data in detail, to check if the above interpretation has any further concern. In that analysis, the pole position, width and the q_K dependence, must be revisited again, because the simple single-pole assumption in [11] is not valid any more.

Acknowledgement

The author is grateful to J-PARC E15 group, E27 group and all the J-PARC/KEK staff, who contributed those experimental programs.

References

[1] A. D. Martin, Nucl. Phys. B **179**, 33 (1981)
[2] M. Bazzi *et al.*, Phys. Lett. B **704**, 113 (2011)
[3] Y. Akaishi and T. Yamazaki, Phys. Rev. C **65**, 044005 (2002)
[4] M. Agnello *et al.*, Phys. Rev. Lett. **94**, 212303 (2005)
[5] T. Yamazaki *et al.*, Phys. Rev. Lett. **104**, 132502 (2010)
[6] A. O. Tokiyasu *et al.*, Phys.Lett. B **728**, 616 (2014)
[7] L. Fabbietti *et al.*, Nucl. Phys. A **914**, 60 (2013)
[8] Y. Ichikawa *et al.*, Prog. Theor. Exp. Phys. 021D01 (2015)
[9] K. Agari *et al.*, Prog. Theor. Exp. Phys. 02B011 (2012)
[10] T. Hashimoto *et al.*, Prog. Theor. Exp. Phys. 061D01 (2015)
[11] Y. Sada *et al.*, Prog. Theor. Exp. Phys. 051D01 (2016)

Meson spectroscopy at COMPASS

Boris Grube[1,*]
for the COMPASS Collaboration

[1] *Institute for Hadronic Structure and Fundamental Symmetries, Physik-Department, Technische Universität München*

Abstract. The goal of the COMPASS experiment at CERN is to study the structure and dynamics of hadrons. The two-stage spectrometer used by the experiment has large acceptance and covers a wide kinematic range for charged as well as neutral particles and can therefore measure a wide range of reactions. The spectroscopy of light mesons is performed with negative (mostly π^-) and positive (p, π^+) hadron beams with a momentum of 190 GeV/c. The light-meson spectrum is measured in different final states produced in diffractive dissociation reactions with squared four-momentum transfer t to the target between 0.1 and 1.0 (GeV/c)^2. The flagship channel is the $\pi^-\pi^-\pi^+$ final state, for which COMPASS has recorded the currently world's largest data sample. These data not only allow to measure the properties of known resonances with high precision, but also to observe new states. Among these is a new axial-vector signal, the $a_1(1420)$, with unusual properties. Novel analysis techniques have been developed to extract also the amplitude of the $\pi^-\pi^+$ subsystem as a function of 3π mass from the data. The findings are confirmed by the analysis of the $\pi^-\pi^0\pi^0$ final state.

1 Introduction

The COMPASS experiment [1] has recorded large data sets of the diffractive dissociation reaction $\pi^- + p \to (3\pi)^- + p_{\text{recoil}}$ using a 190 GeV/c pion beam on a liquid-hydrogen target. In this process, the beam hadron is excited to some intermediate three-pion state X^- via t-channel Reggeon exchange with the target. At 190 GeV/c beam momentum, Pomeron exchange is dominant. Diffractive reactions are known to exhibit a rich spectrum of intermediate states X^- and are a good place to search for states beyond the naive constituent-quark model. In the past, several candidates for so-called spin-exotic mesons, which have J^{PC} quantum numbers that are forbidden in the non-relativistic quark model, have been reported in pion-induced diffraction [2, 3].

The scattering process is characterized by two kinematic variables: the squared total center-of-mass energy s, which is fixed by the beam energy, and the squared four-momentum transfer to the target $t = (p_{\text{beam}} - p_X)^2 < 0$. It is customary to use the reduced four-momentum transfer squared $t' \equiv |t| - |t|_{\text{min}}$ instead of t, where $|t|_{\text{min}}$ is the minimum value of $|t|$ for a given invariant mass of X^-. The analysis is performed in the range $0.1 < t' < 1.0$ (GeV/c)^2.

In addition to the three final-state pions from the X^- decays, also the recoiling proton is measured. This helps to suppress backgrounds and ensures an exclusive measurement by applying energy and

*e-mail: bgrube@tum.de

 DOI: 10.1051/epjconf/201613001024

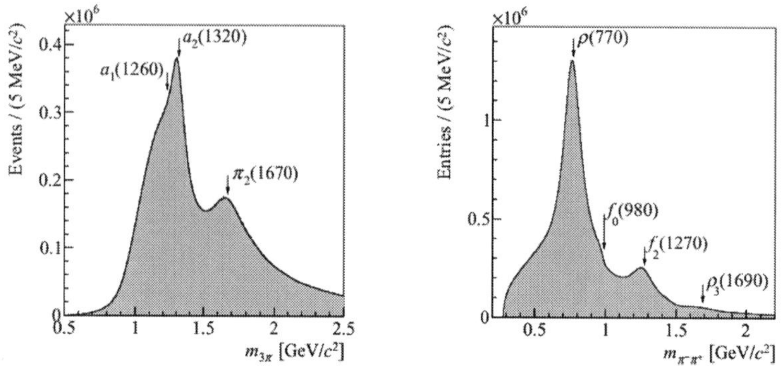

Figure 1. Left: $\pi^-\pi^-\pi^+$ invariant mass spectrum in the analyzed range; Right: invariant mass distribution of the $\pi^-\pi^+$ subsystem (two entries per event). Figures from Ref. [4].

momentum conservation in the event selection. After all selection cuts, the 3π data samples consist of 46×10^6 $\pi^-\pi^-\pi^+$ and 3.5×10^6 $\pi^-\pi^0\pi^0$ exclusive events in the analyzed kinematic region of three-pion mass, $0.5 < m_{3\pi} < 2.5\,\text{GeV}/c^2$. Figure 1 shows the $\pi^-\pi^-\pi^+$ invariant mass spectrum together with that of the $\pi^-\pi^+$ subsystem. The known pattern of resonances $a_1(1260)$, $a_2(1320)$, and $\pi_2(1670)$ is seen in the 3π system along with $\rho(770)$, $f_0(980)$, $f_2(1270)$, and $\rho_3(1690)$ in the $\pi^-\pi^+$ subsystem.

2 Partial-wave decomposition

In order to disentangle the different contributing intermediate states X^-, a partial-wave analysis (PWA) is performed. The PWA of the $(3\pi)^-$ final states is based on the isobar model, which assumes that the X^- decays first into an intermediate resonance, which is called the isobar, and a "bachelor" pion (π^- for the $\pi^-\pi^-\pi^+$ final state; π^- or π^0 for $\pi^-\pi^0\pi^0$). In a second step, the isobar decays into two pions. In accordance with the $\pi^-\pi^+$ invariant mass spectrum shown in Fig. 1 right and with analyses by previous experiments, we include $[\pi\pi]_S$, $\rho(770)$, $f_0(980)$, $f_2(1270)$, $f_0(1500)$, and $\rho_3(1690)$ as isobars into the fit model. Here, $[\pi\pi]_S$ represents the broad component of the $\pi\pi$ S-wave. Based on the six isobars, we have constructed a set of partial waves that consists of 88 waves in total, including one non-interfering isotropic wave representing three uncorrelated pions. This constitues the largest wave set ever used in an analysis of the 3π final state. The partial-wave decomposition is performed in narrow bins of the 3π invariant mass. Since the data show a complicated correlation of the $m_{3\pi}$ and t' spectra, each $m_{3\pi}$ bin is further subdivided into non-equidistant bins in the four-momentum transfer t'. For the $\pi^-\pi^-\pi^+$ channel 11 bins are used, for the $\pi^-\pi^0\pi^0$ final state 8 bins. With this additional binning in t', the dependence of the individual partial-wave amplitudes on the four-momentum transfer can be studied in detail. The details of the analysis model are described in Ref. [4].

The partial-wave amplitudes are extracted from the data as a function of $m_{3\pi}$ and t' by fitting the five-dimensional kinematic distributions of the outgoing three pions. The amplitudes do contain information not only about the partial-wave intensities, but also about the relative phases of the partial waves. The latter are crucial for resonance extraction. The three-pion partial waves are defined by the quantum numbers of the X^- (spin J, parity P, C-parity, absolute value M of the spin projection), the naturality $\varepsilon = \pm 1$ of the exchange particle, the isobar, and the orbital angular momentum L between the isobar and the bachelor pion. These quantities are summarized in the partial-wave notation J^{PC} M^{ε} [isobar] π L. Since at the used beam energies Pomeron exchange is dominant, 80 of the 88 partial waves in the model have $\varepsilon = +1$.

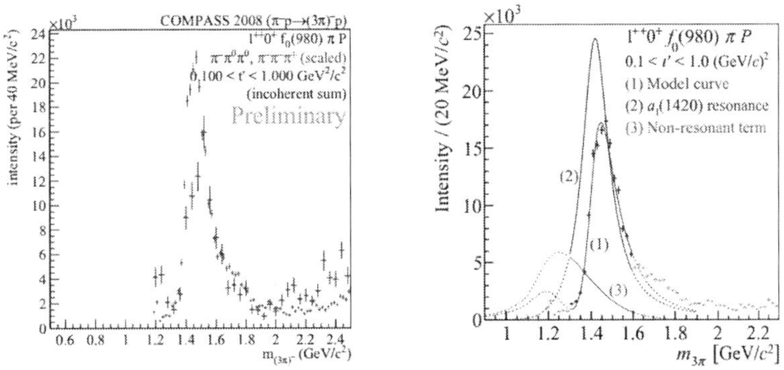

Figure 2. Left: Intensity of the $1^{++}\,0^+\,f_0(980)\,\pi\,P$ wave summed over all t' bins for the $\pi^-\pi^0\pi^0$ (blue) and the $\pi^-\pi^-\pi^+$ (red, scaled to the intensity integral of the $\pi^-\pi^0\pi^0$ channel) final states. Right: Result of a resonance-model fit to the $\pi^-\pi^-\pi^+$ data [5]. The data points correspond to the red points in the left figure.

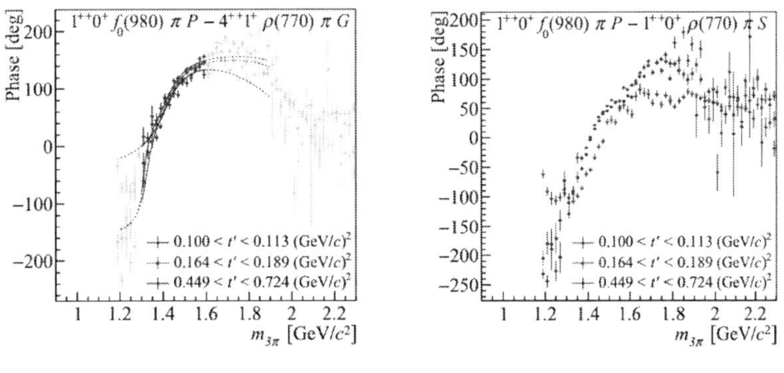

Figure 3. Examples for relative phases of the $1^{++}\,0^+\,f_0(980)\,\pi\,P$ wave with respect to the $4^{++}\,1^+\,\rho(770)\,\pi\,G$ (left) and the $1^{++}\,0^+\,\rho(770)\,\pi\,S$ wave (right). The phases are shown for three different t' regions indicated by the color. Figures from Ref. [5].

2.1 The $a_1(1420)$

A surprising find in the COMPASS data is a pronounced narrow peak at about $1.4\,\text{GeV}/c^2$ in the $1^{++}\,0^+\,f_0(980)\,\pi\,P$ wave (see Fig. 2). The peak is observed with similar shape in the $\pi^-\pi^-\pi^+$ and $\pi^-\pi^0\pi^0$ channels and is robust against variations of the PWA model. In addition to the peak in the partial-wave intensity, rapid phase variations with respect to most waves are observed in the $1.4\,\text{GeV}/c^2$ region (see Fig. 3). The phase motion as well as the peak shape change only little with t'.

In order to test the compatibility of the signal with a Breit-Wigner resonance, a resonance-model fit was performed using a novel method, where the intensities and relative phases of three waves [$1^{++}\,0^+\,f_0(980)\,\pi\,P$, $2^{++}\,1^+\rho(770)\,\pi\,D$, and $4^{++}\,1^+\rho(770)\,\pi\,G$] were fit simultaneously in all 11 t' bins [5]. Forcing the resonance parameters to be the same across all t' bins leads to an improved separation of resonant and non-resonant contribution as compared to previous analyses that did not incorporate the t' information. The Breit-Wigner model describes the peak in the $1^{++}\,0^+\,f_0(980)\,\pi\,P$ wave well and yields a mass of $m_0 = (1414\,^{+15}_{-13})\,\text{MeV}/c^2$ and a width of $\Gamma_0 = (153\,^{+8}_{-23})\,\text{MeV}/c^2$ for the $a_1(1420)$. Due to the high statistical precision of the data, the uncertainties are dominated by systematic effects.

The $a_1(1420)$ signal is remarkable in many ways. It appears in a mass region that is well studied since decades. However, previous experiments were unable to see the peak, because it contributes only 0.25 % to the total intensity. The $a_1(1420)$ is very close in mass to the 1^{++} ground state, the $a_1(1260)$. But it has a much smaller width than the $a_1(1260)$. The $a_1(1420)$ peak is seen only in the $f_0(980)\,\pi$ decay mode of the 1^{++} waves and lies suspiciously close to the $K\bar{K}^*(892)$ threshold.

The nature of the $a_1(1420)$ is still unclear and several interpretations were proposed. It could be the isospin partner to the $f_1(1420)$. It was also described as a two-quark-tetraquark mixed state [6] and a tetraquark with mixed flavor symmetry [7]. Other models do not require an additional resonance: the authors of Refs. [8, 9] propose resonant re-scattering corrections in the Deck process as an explanation, whereas Ref. [10] suggests a branching point in the triangular rescattering diagram for $a_1(1260) \to K\bar{K}^*(892) \to K\bar{K}\pi \to f_0(980)\,\pi$. The results of the latter calculation were confirmed by the authors of Ref. [11]. Triangle singularities were also proposed as an explanation for the narrow $\eta(1405)$ [12, 13], for some of the near-threshold XYZ heavy-quark states (see e.g. Ref. [14]), and for the pentaquark candidate $P_c(4450)$ recently found by LHCb [15, 16]. More detailed studies are needed in order to distinguish between the different models for the $a_1(1420)$.

2.2 Extraction of $\pi\pi$ S-wave isobar amplitudes from data

The PWA of the 3π system is based on the isobar model, where fixed amplitudes are used for the description of the $\pi^-\pi^+$ intermediate states. However, we cannot exclude that the fit results are biased by the employed isobar parametrizations. This is true in particular for the isoscalar $J^{PC} = 0^{++}$ isobars. In the PWA model, a broad $\pi\pi$ S-wave component is used, the parametrization of which is extracted from $\pi\pi$ S-wave elastic-scattering data [17]. In addition, the $f_0(980)$, described by a Flatté form [18], and the $f_0(1500)$, parametrized by a relativistic Breit-Wigner amplitude, are included as isobars. In order to study possible bias due to these parametrizations and to ensure that the observed $a_1(1420)$ signal is truly related to the narrow $f_0(980)$, a novel analysis method inspired by Ref. [19] was developed [4]. In this so-called *freed-isobar* analysis, the three fixed parametrizations for the 0^{++} isobar amplitudes are replaced by a set of piecewise constant complex-valued functions that fully cover the allowed two-pion mass range. This way the whole 0^{++} isobar amplitude is extracted as a function of the 3π mass. In contrast to the conventional isobar approach, which uses the same isobar parametrization in different partial waves, the freed-isobar method permits different isobar amplitudes for different intermediate states X^-. A more detailed description of the analysis method can be found in Ref. [4].

The freed-isobar method leads to a reduced model bias and gives additional information about the $\pi^-\pi^+$ subsystem at the cost of a considerable increase in the number of free parameters in the PWA fit. Thus, even for large data sets, the freed-isobar approach can only be applied to a subset of partial waves. We performed a freed-isobar PWA, where the fixed parametrizations of the broad $\pi\pi$ S-wave component, of the $f_0(980)$, and of the $f_0(1500)$ were replaced by piece-wise constant isobar amplitudes for the 3π partial waves $0^{-+}\,0^+\,[\pi\pi]_{0^{++}}\,\pi\,S$, $1^{++}\,0^+\,[\pi\pi]_{0^{++}}\,\pi\,P$, and $2^{-+}\,0^+\,[\pi\pi]_{0^{++}}\,\pi\,D$. Figure 4 left shows the two-dimensional intensity distribution of the $1^{++}\,0^+\,[\pi\pi]_{0^{++}}\,\pi\,P$ wave as a function of $m_{\pi^-\pi^+}$ and $m_{3\pi}$. The distribution exhibits a broad maximum around $m_{3\pi} = 1.2\,\mathrm{GeV}/c^2$ and between 0.6 and $0.8\,\mathrm{GeV}/c^2$ in $m_{\pi^-\pi^+}$, which shows a pronounced t' dependence and therefore is probably mainly of non-resonant origin. A smaller peak is observed in the $f_0(980)$ region at $m_{3\pi} \approx 1.4\,\mathrm{GeV}/c^2$. This peak is more obvious in Fig. 4 center, which shows the intensity distribution summed over the two-pion mass interval around the $f_0(980)$ as indicated by the pair of horizontal dashed lines in Fig. 4 left. The peak is similar in position and shape to the $a_1(1420)$ peak in the $1^{++}\,0^+\,f_0(980)\,\pi\,P$ wave (cf. Fig. 2). The resonant nature of the $f_0(980)$ becomes apparent in Fig. 4 right, which shows the $m_{\pi^-\pi^+}$ dependence of the extracted amplitude at the $a_1(1420)$ peak in form of an Argand diagram. The phase is measured with respect to the $1^{++}\,0^+\,\rho(770)\,\pi\,S$ wave. The $f_0(980)$ contribution shows up as a semicircle-like structure

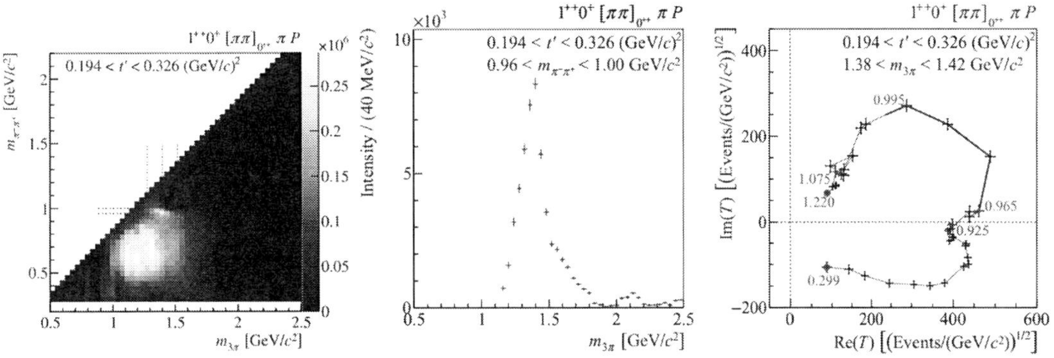

Figure 4. Left: Intensity of the $\pi\pi$ S-wave component of the $J^{PC}\,M^{\varepsilon} = 1^{++}\,0^{+}$ partial wave resulting from the freed-isobar fit as a function of $m_{\pi^-\pi^+}$ and $m_{3\pi}$. Center: intensity as a function of $m_{3\pi}$ summed over the $m_{\pi^-\pi^+}$ interval around the $f_0(980)$ indicated by the pair of horizontal dashed lines in the top figure. Right: Argand diagram representing the $m_{\pi^-\pi^+}$ dependence of the partial-wave amplitude for the 3π mass bin at the $a_1(1420)$ measured with respect to the $1^{++}\,0^{+}\,\rho(770)\,\pi\,S$ wave. Figures from Ref. [4].

(highlighted by the blue line) with a shifted origin. This demonstrates that the observed $a_1(1420)$ signal in the $f_0(980)\,\pi$ decay mode is not an artifact of the 0^{++} isobar parametrizations used in the conventional PWA method. More results of the freed-isobar PWA are discussed in Ref. [4].

2.3 The $J^{PC} = 1^{-+}$ spin-exotic wave

The 88-wave model also contains waves with exotic J^{PC} quantum numbers, that are forbidden in the non-relativistic quark model. The most interesting of these waves is the $1^{-+}\,1^{+}\,\rho(770)\,\pi\,P$ wave, which contributes less than 1 % to the total intensity. Previous analyses claimed a resonance, the $\pi_1(1600)$, at about $1.6\,\mathrm{GeV}/c^2$ in this channel [20, 21]. Figure 5 left shows the intensity sum over all t' bins of this partial wave for the two final states ($\pi^-\pi^-\pi^+$ in red, $\pi^-\pi^0\pi^0$ in blue). The two distributions are scaled to have the same integral. Both decay channels are in fair agreement and exhibit a broad enhancement extending from about 1.0 to $1.8\,\mathrm{GeV}/c^2$ in $m_{3\pi}$. In the 1.0 to $1.2\,\mathrm{GeV}/c^2$ mass range, the intensity depends strongly on the details of the fit model. Peak-like structures in this region are probably due to cross talk induced by imperfections of the applied PWA model.

A remarkable change of the shape of the intensity spectrum of the $1^{-+}\,1^{+}\,\rho(770)\,\pi\,P$ wave with t' is observed (see dark blue points in Fig. 5 center and right). At values of t' below about $0.3\,(\mathrm{GeV}/c)^2$, we observe no indication of a resonance peak around $m_{3\pi} = 1.6\,\mathrm{GeV}/c^2$, where we would expect the $\pi_1(1600)$. However, for the t' bins in the interval $0.449 < t' < 1.000\,(\mathrm{GeV}/c)^2$, the observed intensities exhibit a very different shape as compared to the low-t' region, with a peak structure emerging at about $1.6\,\mathrm{GeV}/c^2$ and the intensity at lower masses disappearing rapidly with increasing t'. This is in contrast to clean resonance signals like the $a_2(1320)$ in the $2^{++}\,1^{+}\,\rho(770)\,\pi\,D$ wave, which, as expected, do not change their shape with t'. The observed t' behavior of the 1^{-+} intensity is therefore a strong indication that non-resonant contributions play a dominant role.

It is believed that the non-resonant contribution in the 1^{-+} wave originates predominantly from the Deck effect, in which the incoming beam pion dissociates into the isobar and an off-shell pion that scatters off the target proton to become on-shell [22]. As a first step towards a better understanding of the non-resonant contribution, Monte-Carlo data were generated that are distributed according to a model of the Deck effect. The model employed here is very similar to that used in Ref. [23]. The

 DOI: 10.1051/epjconf/201613001024

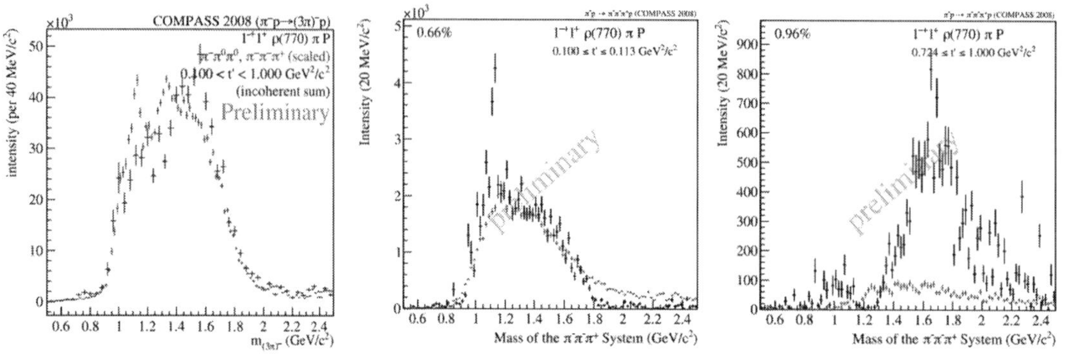

Figure 5. Intensity of the $1^{-+}\,1^+\rho(770)\,\pi\,P$ wave. Left: summed over all t' bins for the $\pi^-\pi^-\pi^+$ (red) and the $\pi^-\pi^0\pi^0$ (blue) final state. Center and right panels show the intensity for the $\pi^-\pi^-\pi^+$ final state (dark blue) in different regions of t' (center: low t'; right: high t'). The partial-wave projections of Monte-Carlo data generated according to a model of the Deck effect are overlaid in green.

partial-wave projection of these Monte Carlo data is shown as green points in Fig. 5 center and right. In order to compare the intensities of real data and the Deck-model pseudo data, the Monte Carlo data are scaled to the t'-summed intensity of the 1^{-+} wave as observed in real data. At values of t' below about 0.3 $(\mathrm{GeV}/c)^2$, the intensity distributions of real data and Deck Monte Carlo exhibit strong similarities suggesting that the observed intensity might be saturated by the Deck effect. Starting from $t' \approx 0.4\,(\mathrm{GeV}/c)^2$, the spectral shapes for Deck pseudo data and real data deviate from each other with the differences increasing towards larger values of t'. This leaves room for a potential resonance signal. It should be noted, however, that the Deck pseudo data contain no resonant contributions. Therefore, potential interference effects between the resonant and non-resonant amplitudes cannot be assessed in this simple approach.

Acknowledgements

This work was supported by the BMBF, the Maier-Leibnitz-Laboratorium (MLL), the DFG Cluster of Excellence Exc153 "Origin and Structure of the Universe", and the computing facilities of the Computational Center for Particle and Astrophysics (C2PAP).

References

[1] P. Abbon et al. (COMPASS Collaboration), Nucl. Instrum. Methods Phys. Res., Sect. A **779**, 69 (2015)
[2] C. Meyer et al., Phys. Rev. C **82**, 025208 (2010)
[3] E. Klempt et al., Phys. Rept. **454**, 1 (2007)
[4] C. Adolph et al. (COMPASS Collaboration), arXiv:1509.00992, *submitted to* Phys. Rev. D (2015)
[5] C. Adolph et al. (COMPASS Collaboration), Phys. Rev. Lett. **115**, 082001 (2015)
[6] Z.G. Wang, arXiv:1401.1134 (2014)
[7] H.X. Chen et al., Phys. Rev. D **91**, 094022 (2015)
[8] J.L. Basdevant, E.L. Berger, Phys. Rev. Lett. **114**, 192001 (2015)
[9] J.L. Basdevant, E.L. Berger, arXiv:1501.04643 (2015)

[10] M. Mikhasenko et al., Phys. Rev. D **91**, 094015 (2015)
[11] F. Aceti, L.R. Dai, E. Oset, arXiv:1606.06893 (2016)
[12] J.J. Wu, X.H. Liu, Q. Zhao, B.S. Zou, Phys. Rev. Lett. **108**, 081803 (2012)
[13] X.G. Wu, J.J. Wu, Q. Zhao, B.S. Zou, Phys. Rev. D **87**, 014023 (2013)
[14] A.P. Szczepaniak, Phys. Lett. B **747**, 410 (2015)
[15] R. Aaij et al. (LHCb Collaboration), Phys. Rev. Lett. **115**, 072001 (2015)
[16] F.K. Guo, U.G. Meißner, W. Wang, Z. Yang, Phys. Rev. D **92**, 071502 (2015)
[17] K.L. Au, D. Morgan, M.R. Pennington, Phys. Rev. D **35**, 1633 (1987)
[18] M. Ablikim et al. (BES Collaboration), Phys. Lett. B **607**, 243 (2005)
[19] E.M. Aitala et al. (E791 Collaboration), Phys. Rev. D **73**, 032004 (2006)
[20] S.U. Chung et al. (E852 Collaboration), Phys. Rev. D **60**, 092001 (1999)
[21] M. Alekseev et al. (COMPASS Collaboration), Phys. Rev. Lett. **104**, 241803 (2010)
[22] R.T. Deck, Phys. Rev. Lett. **13**, 169 (1964)
[23] C. Daum et al. (ACCMOR Collaboration), Nucl. Phys. B **182**, 269 (1981)

Light meson spectroscopy at BESIII

Tianjue Min[1,*]
for the BESIII Collaboration

[1] *Institute of High Energy Physics, Beijing 100049, People's Republic of China*

Abstract. The BESIII collaboration has collected a sample of 1.311 billion J/ψ events since 2009. In this talk, we will introduce four recent analyses on light meson spectroscopy at BESIII, including two studies on the $X(1835)$ and two partial wave analyses that are related to searching for and studying glueball candidates.

1 Introduction

Quantum Chromodynamics is one of the fundamental theories in modern high energy physics. Light meson spectroscopy plays a crucial role in examining and understanding the QCD theory in non-pertubative energy region. Decays of the J/ψ meson, being the lowest lying 1^{--} $c\bar{c}$ states, provide an ideal laboratory for light meson spectroscopy.

The BESIII detector collected 225 million J/ψ events in 2009 and 1.086 billion J/ψ events in 2012, which gives us a very good opportunity to study light mesons through J/ψ decays. In this presentation, we will introduce some recent results from BESIII that are related to light meson spectroscopy.

2 Recent results from BESIII

2.1 Observation of $X(1835)$ in $J/\psi \to \gamma K^0_S K^0_S \eta$

The state $X(1835)$ was first observed by the BESII experiment as a peak in the $\eta'\pi^+\pi^-$ invariant mass distribution in $J/\psi \to \gamma\eta'\pi^+\pi^-$ decays [1]. This observation was later confirmed by BESIII studies of the same process [2] with a mass and width measured to be $M = 1836.5 \pm 3^{+5.6}_{-2.1}$ MeV$/c^2$ and $\Gamma = 190 \pm 9^{+38}_{-36}$ MeV$/c^2$. An anomalously strong enhancement at the proton-antiproton ($p\bar{p}$) mass threshold, dubbed $X(p\bar{p})$, was first observed by BESII in $J/\psi \to \gamma p\bar{p}$ decays [3]; this observation was confirmed by BESIII [4] and CLEO [5]. This enhancement structure was subsequently determined to have spin-parity $J^P = 0^-$ by BESIII [6]. Among the various theoretical interpretations on the nature of the $X(1835)$ and $X(p\bar{p})$, a particularly intriguing one suggests that the two structures originate from a $p\bar{p}$ bound state [7–10].

To understand the nature of the $X(1835)$, it is crucial to measure its J^{PC} and to search for new decay modes. Recently, BESIII analyzed the $J/\psi \to \gamma K^0_S K^0_S \eta$ process with 1.311 billion J/ψ data sample collected since 2009 [11]. The scatter plot of the invariant mass of $K^0_S K^0_S$ versus that of $K^0_S K^0_S \eta$ is shown in Fig. 1(a). A clear accumulation of events is seen around the intersection of the

*e-mail: mintj@ihep.ac.cn

$f_0(980)$ on $M_{K_S^0K_S^0}$ and the structure around 1.85 GeV$/c^2$ on $M_{K_S^0K_S^0\eta}$. This indicates that the structure around 1.85 GeV$/c^2$ is strongly correlated with the $f_0(980)$. A partial wave analysis (PWA) of events satisfying $M_{K_S^0K_S^0\eta} < 2.8$ GeV$/c^2$ and $M_{K_S^0K_S^0} < 1.1$ GeV$/c^2$ is performed to determine the parameters of the structure around 1.85 GeV$/c^2$. The $K_S^0K_S^0\eta$ and $K_S^0K_S^0$ mass spectra are shown in Fig. 1(b) and (c). The PWA fit requires a contribution from $X(1835) \to K_S^0K_S^0\eta$ with a statistical significance greater than 12.9σ, where the $X(1835) \to K_S^0K_S^0\eta$ is dominated by $f_0(980)$ production. The spin parity of the $X(1835)$ is determined to be 0^{-+}, which is significantly better than the 1^{++} or 2^{-+} hypotheses. The mass and width of the $X(1835)$ are measured to be $1844 \pm 9(\text{stat})^{+16}_{-25}(\text{syst})$ MeV$/c^2$ and $192^{+20}_{-17}(\text{stat})^{+62}_{-43}(\text{syst})$ MeV$/c^2$, respectively. The corresponding product branching fraction is measured to be $(3.31^{+0.33}_{-0.30}(\text{stat})^{+1.96}_{-1.29}(\text{syst})) \times 10^{-5}$. The mass and width of the $X(1835)$ are consistent with the values obtained from the decay $J/\psi \to \gamma\eta'\pi^+\pi^-$ by BESIII [1]. Another 0^{-+} state, the $X(1560)$, also is observed in data with a statistical significance larger than 8.9σ and is seen to interfere with the $X(1835)$. The mass and width of the $X(1560)$ are consistent with those of the $\eta(1405)$ and $\eta(1475)$ as given in [12] within 2.0σ and 1.4σ, respectively. Present statistics do not allow us to conclusively determine if the $X(1560)$ is the same state as the $\eta(1405)/\eta(1475)$ or a new meson.

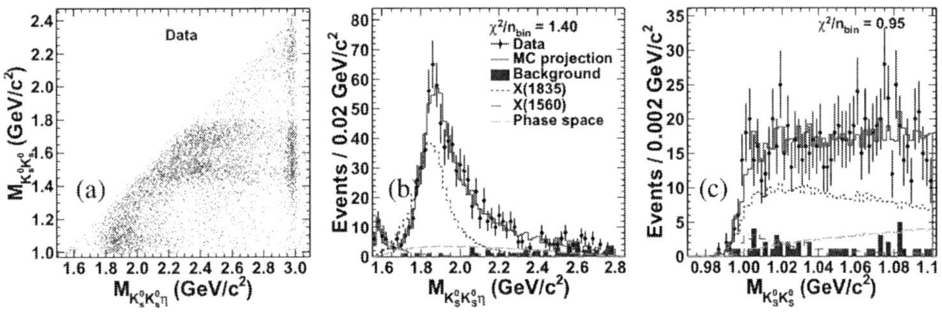

Figure 1. (a) Scatter plot of $M_{K_S^0K_S^0}$ versus that of $M_{K_S^0K_S^0\eta}$; (b) and (c) are the comparisons of invariant mass distributions of $K_S^0K_S^0\eta$ and $K_S^0K_S^0$ between data and PWA fit projections.

2.2 Anomalous line shape of $\eta'\pi^+\pi^-$ near the $p\bar{p}$ mass threshold in $J/\psi \to \gamma\eta'\pi^+\pi^-$

From Sect. 2.1 we know that the $X(1835)$ has the same J^{PC} number as the $X(p\bar{p})$ does. If the $X(1835)$ is really a $p\bar{p}$ bound state, it should have a strong coupling to 0^- $p\bar{p}$ systems, in which case the line shape of the $X(1835)$ at the $p\bar{p}$ mass threshold would be affected by the opening of the $X(1835) \to p\bar{p}$ decay mode. A study of the $\eta'\pi^+\pi^-$ line shape of $X(1835)$ with high statistical precision therefore provides valuable information that helps clarify the nature of the $X(1835)$ and $X(p\bar{p})$.

With 1.086 billion J/ψ data collected in 2012, BESIII re-analysed the $J/\psi \to \gamma\eta'\pi^+\pi^-$ process [13] in which the $X(1835)$ was observed [1, 2]. The η' is reconstructed in its two major decay modes: $\eta' \to \gamma\pi^+\pi^-$ and $\eta' \to \eta\pi^+\pi^-, \eta \to \gamma\gamma$. As shown in Fig. 2, we observed a significant abrupt change in slope of the $X(1835) \to \eta'\pi^+\pi^-$ line shape at the $p\bar{p}$ mass threshold. Study of background shows this abrupt change in line shape is not caused by background processes.

A simultaneous fit to the $\eta'\pi^+\pi^-$ mass distribution between 1.3 and 2.25 GeV$/c^2$ for both event samples is performed. We find that a simple Breit-Wigner function cannot describe the $X(1835)$ line shape near the $p\bar{p}$ mass threshold (Fig. 3(a)). Typically, there are two circumstances where an abrupt distortion of a resonance's line shape shows up: a threshold effect caused by the opening of

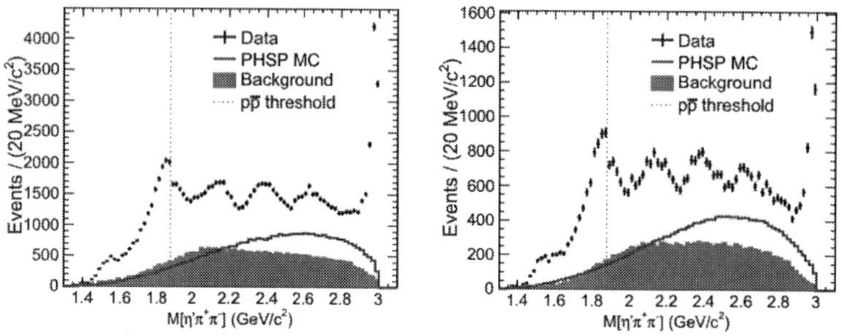

Figure 2. The $\eta'\pi^+\pi^-$ invariant mass spectra after the application of all selection criteria. The plot on the left side shows the spectrum for events with the $\eta' \to \gamma\pi^+\pi^-$ channel, and that on the right shows the spectrum for the $\eta' \to \eta\pi^+\pi^-, \eta \to \gamma\gamma$ channel.

an additional decay mode, or interference between two resonances. We try to fit the data for both of these possibilities.

In the first model, we assume the state around 1.85 GeV$/c^2$ couples to the $p\bar{p}$ and use a Flatté formula [14] to describe the $X(1835)$ line shape:

$$T = \frac{\sqrt{\rho_{\rm out}}}{\mathcal{M}^2 - s - i\sum_k g_k^2 \rho_k} \approx \frac{\sqrt{\rho_{\rm out}}}{\mathcal{M}^2 - s - i g_0^2 \left(\rho_0 + \frac{g^2_{p\bar{p}}}{g_0^2}\rho_{p\bar{p}}\right)} \tag{1}$$

Here, T is the decay amplitude, $\sqrt{\rho_{\rm out}}$ is the phase space for $J/\psi \to \gamma\eta'\pi^+\pi^-$, $\mathcal{M}$ is a parameter with the dimension of mass, s is the square of the $\eta'\pi^+\pi^-$ system's mass, ρ_k is the phase space for decay mode k, and g_k^2 is the corresponding coupling strength. The term $\sum_k g_k^2\rho_k$ describes how the decay width varies with s, and can be replaced by $g_0^2\left(\rho_0 + \frac{g^2_{p\bar{p}}}{g_0^2}\rho_{p\bar{p}}\right)$ approximately, where g_0^2 is the sum of g^2 of all decay modes other than the $X(1835) \to p\bar{p}$, ρ_0 is the maximum two-body decay phase space volume [12], and $g^2_{p\bar{p}}/g_0^2$ is the ratio between the coupling strength to the $p\bar{p}$ channel and the sum of all other channels. The fit results for this model are shown in Fig. 3(b). The Flatté model fit yields $\mathcal{M} = 1638.0 \pm 121.9(\text{stat})^{+127.8}_{-254.3}(\text{syst})$ MeV$/c^2$, $g_0^2 = 93.7 \pm 35.4(\text{stat})^{+47.6}_{-43.9}(\text{syst})$ (GeV$/c^2)^2$, $g^2_{p\bar{p}}/g_0^2 = 2.31 \pm 0.37(\text{stat})^{+0.83}_{-0.60}(\text{syst})$, a product branching fraction of $\mathcal{B}(J/\psi \to \gamma X)\cdot\mathcal{B}(X \to \eta'\pi^+\pi^-) = (3.93 \pm 0.38(\text{stat})^{+0.31}_{-0.84}(\text{syst})) \times 10^{-4}$. The value of $g^2_{p\bar{p}}/g_0^2$ implies that the couplings between the state around 1.85 GeV$/c^2$ and the $p\bar{p}$ final states is very large, the significance of $g^2_{p\bar{p}}/g_0^2$ being non-zero is larger than 7σ. Following the definitions given in [15], the pole nearest to the $p\bar{p}$ mass threshold is found to be $M_{\rm pole} = 1909.5 \pm 15.9(\text{stat})^{+9.4}_{-27.5}(\text{syst})$ MeV$/c^2$ and $\Gamma_{\rm pole} = 273.5 \pm 21.4(\text{stat})^{+6.1}_{-64.0}(\text{syst})$ MeV$/c^2$. In this fit, an additional Breit-Wigner resonance (denoted as "$X(1920)$" in Fig. 3(b)) is needed with 5.7σ statistical significance.

In the second model, we assume the existence of a narrow resonance near the $p\bar{p}$ threshold and that the interference between this resonance and the $X(1835)$ produces the line shape distortion. Here we denote this narrow resonance as "$X(1870)$." For this case we represent the line shape in the vicinity on 1835 MeV$/c^2$ by the square of T, which is the coherent sum of two Breit-Wigner amplitudes. The fit results for this model are shown in Fig. 3(c). The $X(1835)$ mass is $1825.3 \pm 2.4(\text{stat}){+17.3}_{-2.4}(\text{syst})$ MeV$/c^2$ and width is $245.2 \pm 13.1(\text{stat})^{+4.6}_{-9.6}(\text{syst})$ MeV$/c^2$; The $X(1870)$ mass is $1870.2 \pm 2.2(\text{stat})^{+2.3}_{-0.7}(\text{syst})$ MeV$/c^2$ and width is $13.0 \pm 6.1(\text{stat})^{+2.1}_{-3.8}(\text{syst})$ MeV$/c^2$, with a statistical significance that is larger than 7σ. In the parameterization of this model, there exist two solutions

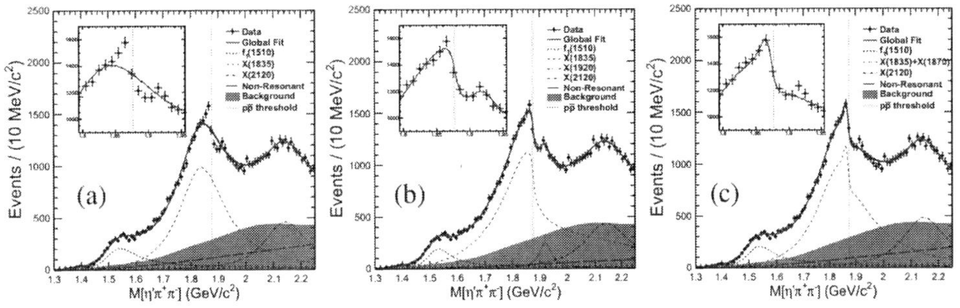

Figure 3. Fit results of using (a) simple Breit-Wigner formula; (b) the Flatté formula; (c) coherent sum of two Breit-Wigner amplitudes.

which only differ in branching fractions of the two Breit-Wigner functions with each other [16]. The product branching fractions with constructive interference are $\mathcal{B}[J/\psi \to \gamma X(1835)]\cdot\mathcal{B}[X(1835) \to \eta'\pi^+\pi^-] = (3.01 \pm 0.17(\text{stat})^{+0.26}_{-0.28}(\text{syst})) \times 10^{-4}$ and $\mathcal{B}[J/\psi \to \gamma X(1870)]\cdot\mathcal{B}[X(1870) \to \eta'\pi^+\pi^-] = (2.03 \pm 0.12(\text{stat})^{+0.18}_{-0.35}(\text{syst})) \times 10^{-7}$, while the solution with destructive interference gives $\mathcal{B}[J/\psi \to \gamma X(1835)]\cdot\mathcal{B}[X(1835) \to \eta'\pi^+\pi^-] = (3.72 \pm 0.21(\text{stat})^{+0.43}_{-0.70}(\text{syst})) \times 10^{-4}$, and $\mathcal{B}[J/\psi \to \gamma X(1870)]\cdot\mathcal{B}[X(1870) \to \eta'\pi^+\pi^-] = (1.57 \pm 0.09(\text{stat})^{+0.49}_{-0.86}(\text{syst})) \times 10^{-5}$. In this model, the $X(1920)$ is not included in the fit because its significance is just 3.9σ.

With current data, both models fit the data well with fit qualities, more sophisticated models such as a mixture of above two models cannot be ruled out. But both fits do suggest the existence of either a $p\bar{p}$ molecule-like or bound state.

2.3 Model independent partial wave analysis of $J/\psi \to \gamma\pi^0\pi^0$

The J/ψ radiative decays to two pesudoscalars are very important channels for identifying scalar and tensor glueballs. Recently, BESIII studied the $J/\psi \to \gamma\pi^0\pi^0$ process with model independent partial wave analysis [17]. In the PWA result, two types of ambiguities are present. Trivial ambiguities arise due to the possibility of the overall amplitude in each bin to be rotated by π or to be reflected over the real axis in the complex plane. These may be partially addressed by applying a phase convention to the results of the fits. Non-trivial ambiguities arise from the freedom of amplitudes with the same quantum numbers to have different phases. The non-trivial ambiguities represent a greater challenge to the analysis and cannot be eliminated without introducing model dependencies. The intensity for each amplitude as a function of $M_{\pi^0\pi^0}$ is shown in Fig. 4. For 2^{++} contributions, the phase differences with respect to the reference amplitude ($2^{++}E1$), is constrained to be real. Above the $K\bar{K}$ threshold, two distinct sets of solutions are apparent in most bins as expected. These results may be combined with those of similar reactions for a more comprehensive study of the light scalar meson spectrum. The branching fraction of radiative J/ψ decays to $\pi^0\pi^0$ is measured to be $(1.15 \pm 0.05) \times 10^{-3}$, where the error is systematic only and the statistical error is negligible. This is the first measurement of this branching fraction.

2.4 Partial wave analysis of $J/\psi \to \gamma\phi\phi$

The mass of the lowest lying pseudoscalar glueball is predicted to be around 2.3 – 2.6 GeV/c^2 by Lattice QCD [18–20]. Aside from the $\eta(2225)$, very little is known in the pseudoscalar sector above

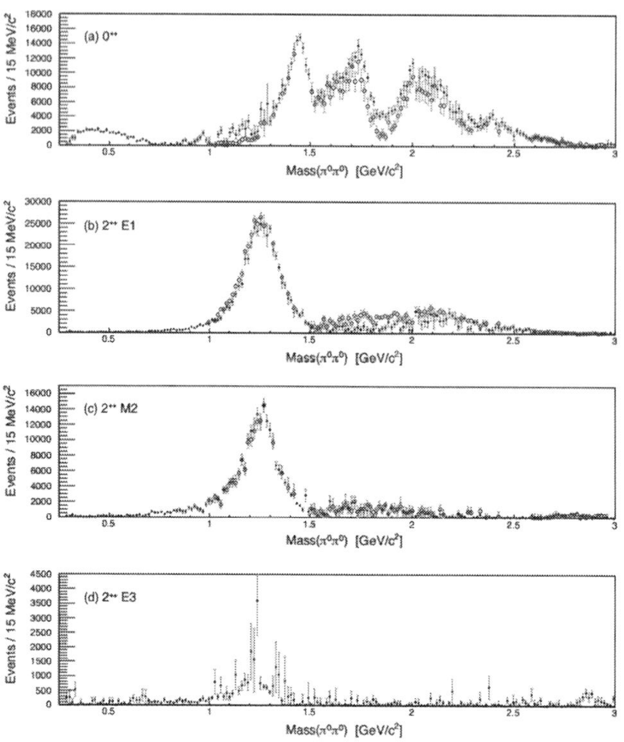

Figure 4. The intensities for the (a) 0^{++}, (b) 2^{++} E1, (c) 2^{++} M2 and (d) 2^{++} E3 amplitudes. The solid black markers show the phase differences calculated from one set of solutions, while the open red markers represent the ambiguous partner solutions.

2 GeV/c^2. The J/ψ radiative decays to two vector particles processes provide opportunities to study pseudoscalar particles. Recently, BESIII studied the $J/\psi \to \gamma\phi\phi$ process with a PWA [21]. The results are shown in Fig. 5. 0^{-+} states are dominant components in the PWA solution. The existence of the $\eta(2225)$ is confirmed and two additional pseudoscalar states, $\eta(2100)$ with a mass of 230^{+64}_{-35}(stat)$^{+77}_{-26}$(syst) MeV/c^2 and a width of 250^{+36}_{-30}(stat)$^{+187}_{-164}$(syst) MeV/c^2, and $X(2500)$ with a mass of 2470^{+15}_{-19}(stat)$^{+63}_{-23}$(syst) MeV/c^2 and a width of 230^{+64}_{-35}(stat)$^{+53}_{-33}$(syst) MeV/c^2, are obtained. The new experimental results are helpful for mapping out pseudoscalar excitations and searching for a 0^{-+} glueball. The three tensors $f_2(2100)$, $f_2(2300)$, and $f_2(2340)$ which were observed in πN scattering are also observed in $J/\psi \to \gamma\phi\phi$.

3 Summary

With the largest J/ψ data sample in the world, the BESIII collaboration has significant contributions in light meson spectroscopy. For the first time, the $X(1835)$ is determined to be a 0^{-+} state; an anomalous line shape of $X(1835) \to \eta'\pi^+\pi^-$ at the $p\bar{p}$ mass threshold is observed, which suggests the existence of a $p\bar{p}$ molecule like or bound state; sophisticated model independent PWA of $J/\psi \to \gamma\pi^0\pi^0$ improves our understanding of the rich structures in the $\pi\pi$ system and provides valuable comparison with the $\eta\eta$ [22] and KK (both model dependent and independent PWA are ongoing at BESIII) systems; the $\phi\phi$ system has been investigated to search for pseudoscalar and tensor glueballs. With the high statistics data accumulated at the BESIII detector, more interesting results are expected in the future.

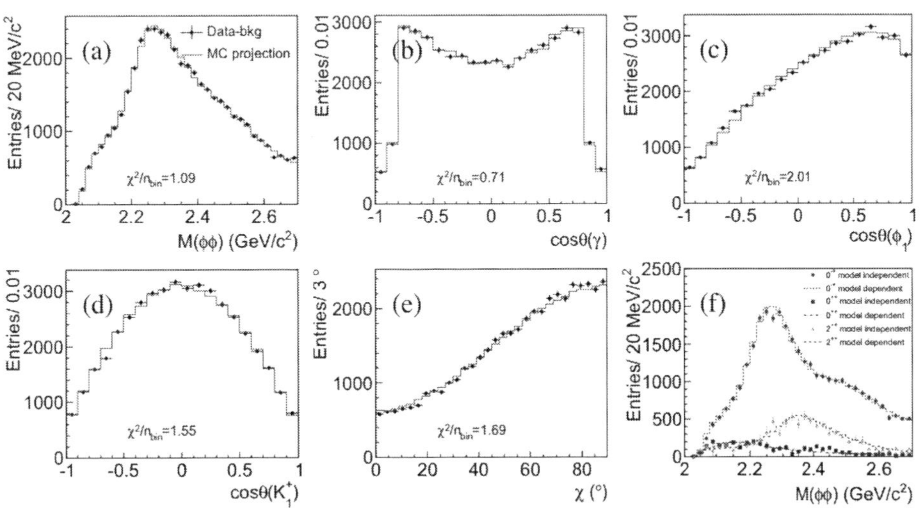

Figure 5. Superposition of data and the PWA fit projections for: (a) invariant mass distributions of $\phi\phi$; (b) $\cos\theta$ of γ in the J/ψ rest frame; (c) $\cos\theta$ of ϕ_1 in the X rest frame; (d) $\cos\theta$ of K^+ in the ϕ_1 rest frame; (e) the azimuthal angle between the normals to the two decay planes of ϕ in the X rest frame; (f) Intensities of individual J^{PC} components.

References

[1] M. Ablikim *et al.* (BES Collaboration), Phys. Rev. Lett. **95**, 262001 (2005)
[2] M. Ablikim *et al.* (BESIII Collaboration), Phys. Rev. Lett. **106**, 072002 (2011)
[3] J. Z. Bai *et al.* (BES Collaboration), Phys. Rev. Lett. **91**, 022001 (2003)
[4] M. Ablikim *et al.* (BESIII Collaboration), Chin. Phys. C **34**, 421 (2010)
[5] J. P. Alexander *et al.* (CLEO Collaboration), Phys. Rev. D **82**, 092002 (2010)
[6] M. Ablikim *et al.* (BESIII Collaboration), Phys. Rev. Lett. **108**, 112003 (2012)
[7] S. L. Zhu and C. S. Gao, Commun. Theor. Phys. **46**, 291 (2006)
[8] J. P. Dedonder, B. Loiseau, B. El-Bennich, and S. Wycech, Phys. Rev. C **80**, 045207 (2009)
[9] C. Liu, Eur. Phys. J. C **53**, 413 (2008)
[10] Z. G. Wang and S. L. Wan, J. Phys. G **34**, 505 (2007)
[11] M. Ablikim *et al.* (BESIII Collaboration), Phys. Rev. Lett. **115**, 091803 (2015)
[12] K. A. Olive *et al.* (Particle Data Group), Chin. Phys. C **38**, 090001 (2014)
[13] M. Ablikim *et al.* (BESIII Collaboration), Phys. Rev. Lett. **117**, 042002 (2016)
[14] S. M. Flatté, Phys. Lett. B **63**, 224 (1976)
[15] B. S. Zou and D. V. Bugg, Phys. Rev. D **48**, R3948 (1993)
[16] K. Zhu, X. H. Mo, C. Z. Yuan, and P. Wang, Int. J. Mod. Phys. A **26**, 4511 (2011)
[17] M. Ablikim *et al.* (BESIII Collaboration), Phys. Rev. D **92**, 052003 (2015)
[18] G. S. Bali *et al.* (UKQCD Collaboration), Phys. Lett. B **309**, 378 (1993)
[19] C. J. Morningstar and M. Peardon, Phys. Rev. D **60**, 034509 (1999)
[20] Y. Chen *et al.*, Phys. Rev. D **73**, 014516 (2006)
[21] M. Ablikim *et al.* (BESIII Collaboration), Phys. Rev. D **93**, 112011 (2016)
[22] M. Ablikim *et al.* (BESIII Collaboration), Phys. Rev. D **87**, 092009 (2013)

EPJ Web of Conferences **130**, 01026 (2016) DOI: 10.1051/epjconf/201613001026
MESON 2016

Charmed meson production at LHCb

Dominik Müller[1,*]
for the LHCb Collaboration

[1] *School of Physics and Astronomy, The University of Manchester, Manchester, United Kingdom*

Abstract. Measurements of charm meson production are important tests for QCD predictions and LHCb is uniquely suited to perform these measurements in the forward region. This paper summarises recent charm meson production measurements performed by LHCb of J/ψ and open charm mesons and the associated production of Υ and open charm mesons. The J/ψ and open charm meson measurements are performed with data recorded in Run 2 of the Large Hadron Collider. With proton-proton collisions at $\sqrt{s} = 13\,\text{TeV}$, these open a new regime in which QCD predictions for charm meson production may be precisely tested. Furthermore, ratios of cross-sections at different centre-of-mass energies benefit of cancellation of both experimental and theoretical uncertainties, providing a new sensitive test of the QCD calculations. Measurements of Υ and open charm meson associated production are performed using $\sqrt{s} = 7\,\text{TeV}$ and $\sqrt{s} = 8\,\text{TeV}$ data and constitute the first observation of this production channel.

1 Introduction

Measurements of charm production cross-sections in proton-proton collisions are important tests of perturbative quantum chromodynamics [1–3], and LHCb can contribute unique measurements in the forward region. The LHCb detector is a single-arm forward spectrometer covering a pseudorapidity range of $2 < \eta < 5$ and is described in detail in Ref. [4, 5]. Recently, LHCb measured the production of J/ψ [6] and open charm mesons [7] at $\sqrt{s} = 13\,\text{TeV}$ and the associated production of Υ and open charm mesons at $\sqrt{s} = 7\,\text{TeV}$ and $\sqrt{s} = 8\,\text{TeV}$ [8]. This paper summarises these measurements.

Differential measurements of J/ψ and open charm meson production test the prediction of quantum chromodynamics to very high values of rapidity, y, and allow to probe a region where the momentum fraction, x, of the initial state partons can reach values below 10^{-4}. In this region the uncertainties on the gluon parton density functions are large, exceeding 30% [1, 9], and LHCb measurements can be used to constrain them. This is especially true for measurements of cross-section ratios between different center-of-mass energies which benefit from large cancellations in both theoretical and experimental uncertainties. Measurements of D^0, D^+, D_s^+, and $D^*(2010)^+$ (henceforth denoted as D^{*+}), generally referred to as D in the following, have previously been performed by the LHCb experiment at $\sqrt{s} = 7\,\text{TeV}$ [10] and of J/ψ at 2.76 TeV [11], 7 TeV [12] and 8 TeV [13].

Measurements of the associated production of two particles in the same pp interaction allow to study the contributions of different production mechanisms. Associated production can occur via

*e-mail: dominik.muller@cern.ch

single parton scattering (SPS) and double parton scattering (DPS). In the former, only one parton of each colliding proton interacts to produce both particles while in the latter, both particles are produced in independent parton interactions and the cross-section for the associated production approximately factorises. LHCb has previously measured the associated production of J/ψ and open charm mesons [14, 15] and the measured cross-sections significantly exceed the SPS expectations but agree with estimates for DPS.

2 Analysis strategy

The presented analyses reconstruct the final states $J/\psi \to \mu^-\mu^+$, $\Upsilon \to \mu^-\mu^+$, $D^0 \to K^-\pi^+$, $D^+ \to K^-\pi^+\pi^+$, $D_s^+ \to K^-K^+\pi^+$ and $D^{*+} \to D^0\pi^+$ with $D^0 \to K^-\pi^+$. A detailed description of the construction and selection of candidates as well the determination of the reconstruction and selection efficiencies is given in Refs [6–8]. In case of the associated production measurements, candidates for D and Υ mesons are treated separately with the exception that both must be consistent with originating from a common vertex. Differential cross-sections are then obtained from the relation

$$\frac{\mathrm{d}\sigma_i(M)}{\mathrm{d}a} = \frac{1}{\Delta a_i} \cdot \frac{N_i(M \to f + \mathrm{c.c.})}{\varepsilon_{i,\mathrm{tot}}(M \to f)\mathcal{B}(M \to f)\mathcal{L}_{\mathrm{int}}}, \tag{1}$$

where M refers to any of the studied mesons or meson pairs, Δa_i are the widths of a bin i of the dependent variable a, $N_i(M \to f + \mathrm{c.c.})$ is the measured signal yield for decays of M to the final state f in bin i plus the charge-conjugated decay, $\varepsilon_{i,\mathrm{tot}}(M \to f)$ is the total efficiency for observing the signal decay in bin i, and $\mathcal{B}(M \to f)$ is the branching ratio of the decay to the specific final state, taken from Ref. [16]. The total integrated luminosity, $\mathcal{L}_{\mathrm{int}}$, differs for the different sets of results presented here. The open charm meson measurement uses 5 pb^{-1} and the J/ψ analysis 3 pb^{-1} of 13 TeV data while the measurement of the associated production of D and Υ mesons uses 1 fb^{-1} of 7 TeV and 2 fb^{-1} of 8 TeV data.

3 Results for J/ψ production at 13 TeV

Results for J/ψ production in different J/ψ (p_T, y) bins are presented both for prompt production at the primary vertex (PV) of the pp collision as well as secondary J/ψ from the decay of heavier, long-lived particles (in the following referred to as from-b). The two contributions are separated from each other and from backgrounds in each (p_T, y) bin with a likelihood fit in the invariant dimuon mass $m(\mu^+\mu^-)$ and the measured decay time along the beam axis $t_z = (z_{J/\psi} - z_{\mathrm{PV}})M_{J/\psi}/p_z$. As J/ψ decays instantaneously, the measured decay time for prompt J/ψ is compatible with 0 while the measured pseudo decay time for J/ψ from-b follows an exponential distribution consistent with the decay of long-lived b-hadrons.

The main measurements are the double differential cross-sections $\mathrm{d}^2\sigma/\mathrm{d}p_\mathrm{T}\mathrm{d}y$ for prompt and from-b J/ψ which are given in Fig. 1. From these results, several additional results are derived, most notably single differential cross-sections integrated over y and their ratios between 13 TeV and 8 TeV [13]. Both of these are shown in Fig. 2 for prompt J/ψ and compared to theory predictions obtained from non-relativistic QCD calculations (NRQCD) [17]. Overall, the measured and predicted values are in agreement albeit the measured cross-sections are at the edge of the theory predictions, and the measured ratios are found to be consistently above the predictions. Furthermore, integrated cross-sections are computed for prompt and from-b J/ψ in the LHCb acceptance:

$$\begin{aligned} \sigma_{\mathrm{prompt}} &= 15.30 \pm 0.03\,(\mathrm{stat}) \pm 0.86\,(\mathrm{sys})\,\mu\mathrm{b} \text{ and} \\ \sigma_{\mathrm{from}\text{-}b} &= 2.34 \pm 0.01\,(\mathrm{stat}) \pm 0.13\,(\mathrm{sys})\,\mu\mathrm{b}. \end{aligned} \tag{2}$$

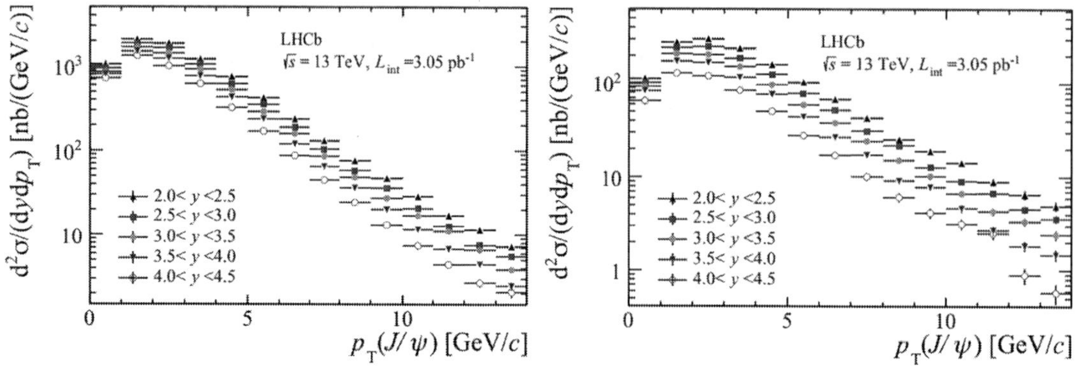

Figure 1: Measured double differential production cross-sections in p_T and y for prompt J/ψ (left) and J/ψ from-b (right).

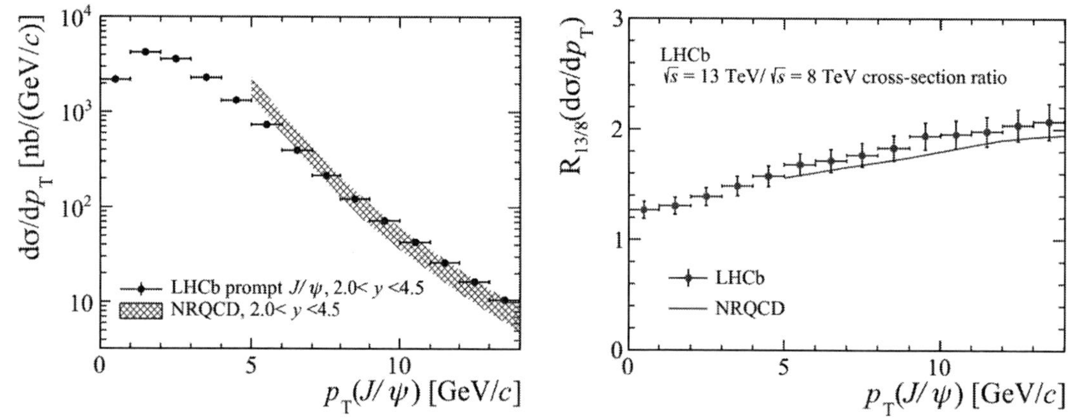

Figure 2: Production cross-sections in bins of p_T and integrated over $2 < y < 4.5$ for prompt J/ψ (left) and their ratios between 13 and 8 TeV (right). Also shown are NRQCD calculations [17].

4 Results for open charm production at 13 TeV

Results for open charm production at 13 TeV are obtained for charm mesons produced in the interaction vertex only. Secondary charm mesons are treated as background. To determine the number of promptly produced charm mesons, a two step fit is performed. The combinatorial background contamination is estimated and constrained from a fit to the invariant mass of the D candidate. Prompt and secondary contributions are then differentiated in a second fit to $\ln \chi^2_{\text{IP}}$ where χ^2_{IP} is defined as the difference in χ^2 of the reconstructed PV with and without the D candidate. As for J/ψ, the fit is performed in bins of (p_T, y) and the main results are again the differential cross-sections $d^2\sigma/dp_T dy$ measured in (p_T, y) bins. In the following, only results for $D^0 \to K^-\pi^+$ are shown and the full set of measurements are given in Ref. [7]. Fig. 3 shows the measured D^0 cross-section in (p_T, y) bins together with three sets of theory predictions [1–3]. Fig. 4 shows the cross-section ratios between 13 and 7 TeV for D^0 as well as the integrated cross-sections in the LHCb acceptance. Together, a similar pattern to the one seen in the J/ψ measurement emerges: the results are in agreement with the pre-

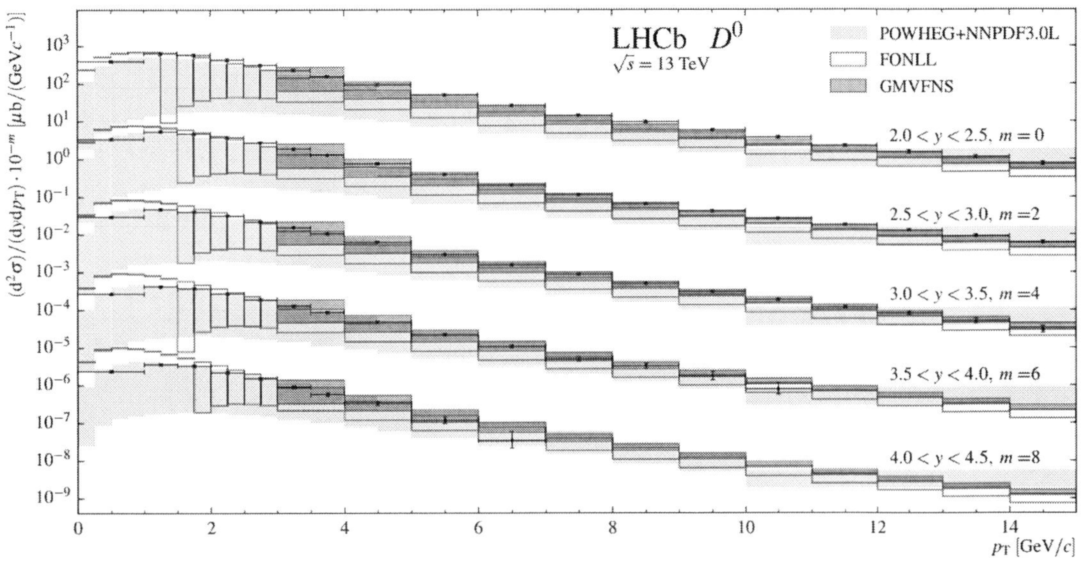

Figure 3: Differential D^0 cross-sections measured in (p_T, y) bins and compared with POWHEG+NNPDF3.0L [1], FONLL [2] and GMVFNS [3] predictions.

dictions though generally at the edge if the theoretical uncertainties. Using measured fragmentation fractions $f(c \to D)$ [18], estimates for the total $c\bar{c}$ cross-section are computed per meson species and averaged to obtain

$$\sigma(pp \to c\bar{c}X)_{p_T < 8\ \text{GeV}/c,\, 2.0 < y < 4.5} = 2840 \pm 3\,(\text{stat}) \pm 170\,(\text{syst}) \pm 150\,(\text{frag})\,\mu\text{b}, \tag{3}$$

which is compared to theory predictions in Fig. 4.

5 Results for associated open charm and Υ production at 7 and 8 TeV

The total number of signal events is extracted from a two-dimensional likelihood fit to the invariant masses of the Υ and D candidates. The requirement that the Υ and D candidate must be consistent with originating from a common vertex significantly suppresses any contributions from secondary charm mesons and hence the yields obtained from the fit are assumed to be prompt. The results are limited by the sample size but constitute the first observation of $\Upsilon(1S)D^0$, $\Upsilon(2S)D^0$, $\Upsilon(1S)D^+$, $\Upsilon(2S)D^+$ and $\Upsilon(1S)D_s^+$ associated production with more than five standard deviations significance. For the $\Upsilon(1S)$ modes, the integrated production cross-sections are calculated:

$$\begin{aligned} \mathcal{B}(\Upsilon \to \mu^+\mu^-) \cdot \sigma_{7\,\text{TeV}}^{\Upsilon(1S)D^0} &= 155 \pm 21\,(\text{stat}) \pm 7\,(\text{sys})\ \text{pb and} \\ \mathcal{B}(\Upsilon \to \mu^+\mu^-) \cdot \sigma_{8\,\text{TeV}}^{\Upsilon(1S)D^0} &= 250 \pm 28\,(\text{stat}) \pm 11\,(\text{sys})\ \text{pb}. \end{aligned} \tag{4}$$

To estimate the relative contributions of SPS and DPS, the $_s\mathcal{P}lot$ technique is used to obtain background subtracted differential cross-section measurements. Fig. 5 shows the normalised cross-section as a function of $\Delta\phi$, the azimuthal angle between the D^0 and $\Upsilon(1S)$. The measured cross-section is in good agreement with the expected distribution for the DPS case, indicating that the SPS contribution,

DOI: 10.1051/epjconf/201613001026

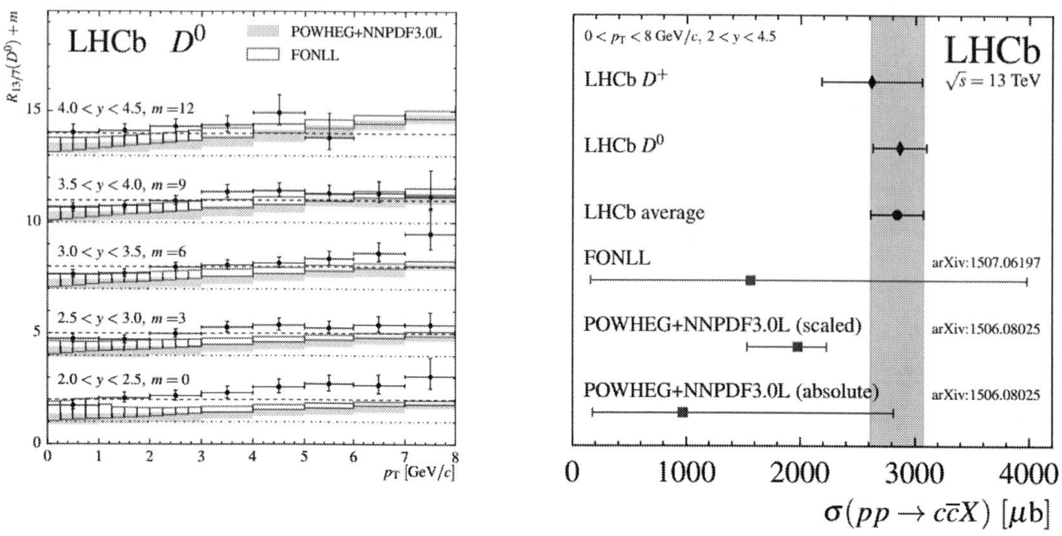

Figure 4: Ratio between D^0 cross-sections measurements at 13 and 7 TeV in (p_T, y) bins and integrated D^0 and D^+ cross-sections, both compared with POWHEG+NNPDF3.0L [1] and FONLL [2] predictions.

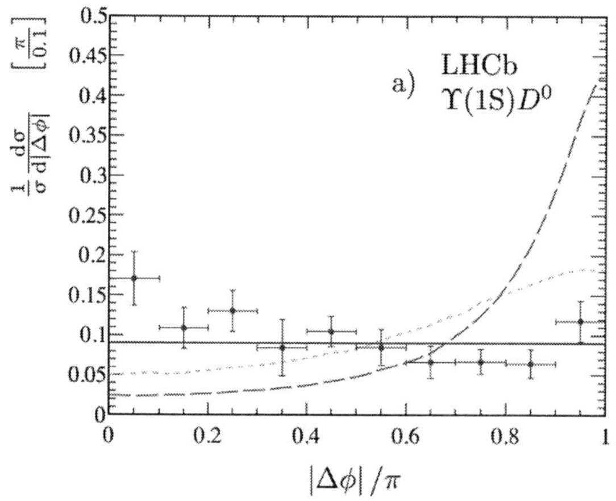

Figure 5: Normalised differential $\Upsilon(1S)D^0$ cross-section with respect to $\Delta\phi$, the difference in the azimuthal angles of $\Upsilon(1S)$ and D^0. The solid blue line shows a uniform distribution expected for uncorrelated production via DPS and the dashed magenta and orange lines show SPS calculations based on collinear approximation and k_T factorisation, respectively [8].

where an increase is expected for $\Delta\phi \approx \pi$, is small. In the DPS case, the cross-section factorises into the individual cross-sections:

$$\sigma(pp \to \Upsilon c\bar{c} + X) \approx \frac{\sigma(pp \to \Upsilon + X) \cdot \sigma(pp \to c\bar{c} + X)}{\sigma_{\text{eff}}}, \tag{5}$$

where σ_{eff} is an effective cross-section which provides the proper normalisation and is expected to be universal. To extract a value for σ_{eff}, the presented results for $\sigma(pp \to \Upsilon D + X)$ are divided by the fragmentation fractions to obtain $\sigma(pp \to \Upsilon c\bar{c} + X)$. Values for $\sigma(pp \to \Upsilon + X)$ and $\sigma(pp \to c\bar{c} + X)$ are taken from previous measurements [10, 19]. Averaging the individual results for ΥD^0 and ΥD^+ at both 7 and 8 TeV, the effective cross-section

$$\sigma_{\text{eff}} = 18.0 \pm 1.3\,(\text{stat}) \pm 1.2\,(\text{sys})\ \text{mb} \tag{6}$$

is computed and the result is in good agreement with previous measurements.

6 Summary

This paper summarises the recent production measurements of charm mesons at LHCb. Measured open charm and J/ψ production cross-sections at 13 TeV indicate an underestimation by the predictions for heavy-quark production, albeit the measurements are still within the theoretical uncertainties. The measurement of the associated production of open charm and Υ at 7 TeV and 8 TeV indicates that this process is dominated by the double parton scattering. Future measurements will benefit from the increased dataset available in Run 2 of the Large Hadron Collider and can potentially determine the fractions of SPS and DPS contributions.

References

[1] R. Gauld, J. Rojo, L. Rottoli, J. Talbert, JHEP **11**, 009 (2015)
[2] M. Cacciari, M.L. Mangano, P. Nason, Eur. Phys. J. **C75**, 610 (2015)
[3] B. Kniehl, G. Kramer, I. Schienbein, H. Spiesberger, Eur. Phys. J. **C72**, 2082 (2012)
[4] A.A. Alves Jr. et al. (LHCb collaboration), JINST **3**, S08005 (2008)
[5] R. Aaij et al. (LHCb collaboration), Int. J. Mod. Phys. **A30**, 1530022 (2015)
[6] R. Aaij et al. (LHCb collaboration), JHEP **10**, 172 (2015)
[7] R. Aaij et al. (LHCb collaboration), JHEP **03**, 159 (2016), Erratum-ibid **09**, 013 (2016)
[8] R. Aaij et al. (LHCb collaboration), JHEP **07**, 052 (2016)
[9] O. Zenaiev et al. (PROSA Collaboration), Eur. Phys. J. **C75**, 396 (2015)
[10] R. Aaij et al. (LHCb collaboration), Nucl. Phys. **B871**, 1 (2013)
[11] R. Aaij et al. (LHCb collaboration), JHEP **02**, 041 (2013)
[12] R. Aaij et al. (LHCb collaboration), Eur. Phys. J. **C71**, 1645 (2011)
[13] R. Aaij et al. (LHCb collaboration), JHEP **06**, 064 (2013)
[14] R. Aaij et al. (LHCb collaboration), Phys. Lett. **B707**, 52 (2012)
[15] R. Aaij et al. (LHCb collaboration), JHEP **06**, 141 (2012)
[16] K.A. Olive et al. (Particle Data Group), Chin. Phys. **C38**, 090001 (2014), and 2015 update
[17] H.S. Shao, H. Han, Y.Q. Ma, C. Meng, Y.J. Zhang, K.T. Chao, JHEP **2015**, 1 (2015)
[18] C. Amsler et al. (Particle Data Group), Phys. Lett. **B667**, 1 (2008)
[19] R. Aaij et al. (LHCb collaboration), JHEP **11**, 103 (2015)

EPJ Web of Conferences **130**, 01027 (2016) DOI: 10.1051/epjconf/201613001027
MESON 2016

Hadronic molecules with hidden charm and bottom

Feng-Kun Guo[1,*]

[1] *CAS Key Laboratory of Theoretical Physics, Institute of Theoretical Physics, Chinese Academy of Sciences, Beijing 100190, China*

Abstract. Many of the new structures observed since 2003 in experiments in the heavy quarkonium mass region, such as the $X(3872)$ and $Z_c(3900)$, are rather close to certain thresholds, and thus can be good candidates of hadronic molecules, which are loose bound systems of hadrons. We will discuss the consequences of heavy quark symmetry for hadronic molecules with heavy quarks. We will also emphasize that the hadronic molecular component of a given structure can be directly probed in long-distance processes, while the short-distance processes are not sensitive to it.

1 Introduction

Hadronic molecules are analogues of the light nuclei. They can be treated to a good approximation as composed systems made of two or more hadrons which are bound together via the strong interactions. Thus, their size, $\sim 1/\sqrt{2\mu E_B}$ with μ and E_B the reduced mass of the constituent hadrons and the binding energy, respectively, should be large so that the constituent hadrons can be recognized separately. They are thus located close to the corresponding thresholds of the strongly coupled hadrons. Hadronic molecules are interesting for several reasons:

(1) They are one type of realizations of color-neutral objects and the investigation of how hadronic molecular spectrum is organized would lead to a better understanding of the strong interactions at low energies.

(2) One can get important information for low-energy hadron-hadron interaction.

(3) Since they are close to thresholds, one has a good separation of energy scales $\sqrt{2\mu E_B} \ll \Lambda_{\text{hadron}}$, where Λ_{hadron} is typically a hadronic scale of the order of a few hundreds of MeV, so that they can be studied using techniques of effective field theory (EFT). One can make model-independent statements about the probability of a near-threshold structure to be a bound state of the corresponding hadrons if they are in an S-wave.

(4) There have been a lot of new structures observed since 2003 in the heavy quarkonium mass region which do not fit in the conventional quarkonium spectrum, see Fig. 1. They are generally dubbed as XYZ states. Some of them are close to open-flavor thresholds and thus good candidates of hadronic molecules.

*e-mail: fkguo@itp.ac.cn

 DOI: 10.1051/epjconf/201613001027

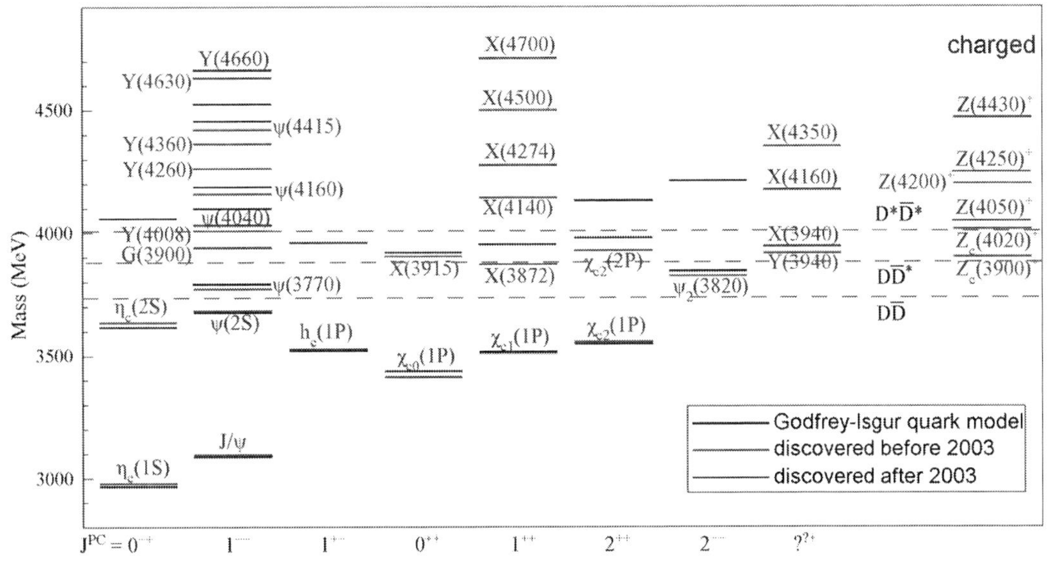

Figure 1. Meson specturm in the charmonium mass region. The black lines are those charmonia predicted in quark model by Godfrey and Isgur [1]. The red lines indicate the states discovered before 2003, and the blue ones correspond to the structures reported after 2003. A few open-charm thresholds are also shown as dashed lines. The quantum numbers of the $X(4140)$, $X(4274)$, $X(4500)$ and $X(4700)$ are according to the LHCb amplitude analysis in Ref. [2]. Note that although the $X(3915)$ is listed with 0^{++}, it is difficult to understand its properties as the $\chi_{c0}(2P)$ [3, 4] and it might be just the same state as the $\chi_{c2}(2P)$ with 2^{++} [5].

2 Model-independent methods for hadronic molecules

As mentioned above, the feature of hadronic molecules that they are located close to thresholds makes the investigation of their properties using model-independent EFT approaches possible. Such approaches can be traced back to 1960's, and Weinberg showed that the deutron could not be an elementary particle [6]. The analysis is based on the observation that the probability of finding a physical state in the two-body continuum in an S-wave, i.e. the probability of such a state being the corresponding two-body bound state, is related to physical observables. This probability is sometimes called compositeness and denoted by $1 - Z$. Its relations to the coupling constant $g_{\rm NR}$ (here the subscript 'NR' indicates that the nonrelativistic normalization is used), scattering length a and effective range r_e are as follows [6–8]:

$$g_{\rm NR}^2 \approx (1-Z)\frac{2\pi}{\mu^2}\sqrt{2\mu E_B}, \quad a \approx -\frac{2(1-Z)}{(2-Z)\sqrt{2\mu E_B}}, \quad r_e \approx \frac{Z}{(1-Z)\sqrt{2\mu E_B}}. \tag{1}$$

These relations receive corrections of order $r\sqrt{2\mu E_B}$ with r, typically $\sim 1/\Lambda_{\rm hadron}$, the range of forces. Therefore, for an S-wave near-threshold state, if the partial decay widths into the modes with a small energy release, which are sensitive to the coupling constants, can be measured precisely,[1] one can then extract the probability for that state to be made of two hadrons. If the probability is large, then the state is predominantly a hadronic molecule.

[1]It would be difficult to measure the scattering parameters experimentally for two heavy hadrons, yet they could be calculated using lattice QCD.

In combination with the heavy quark symmetry to be discussed below, specific EFTs for the near-threshold XYZ states as hadronic molecules exist, such as the ones in [9–13]. The above mentioned model-independent relations are built in these EFTs explicitly or implicitly. It is also clear that the coupling constant is bound from above and the upper bound corresponds to the case with $Z = 0$. Thus, the larger the hadronic molecular component, the larger the coupling constant, or vice versa. It is then natural to expect that transitions between two hadronic molecules with a small mass difference are more easily to take place than those between states dominated by $Q\bar{Q}$. In Ref. [12], using a nonrelativistic EFT, we calculated the transition $Y(4260) \to \gamma X(3872)$ assuming the $Y(4260)$ and $X(3872)$ to be $D_1\bar{D}$ [14] and $D\bar{D}^*$ molecules [15, 16], respectively, and predicted a large partial width. The experimental observation of such a decay was soon reported by the BESIII Collaboration [17].

In the limit of the heavy quark mass $m_Q \to \infty$, the heavy quark spin decouples. As a result, we have heavy quark spin symmetry (HQSS). If we decompose the total spin of a hadronic system into two parts: $\vec{J} = \vec{s}_Q + \vec{s}_\ell$ with s_Q the total spin of all the heavy quarks and s_ℓ the total anglular momentum of the light degrees of freedom, then angular momentum conservation implies that s_ℓ is a good quantum number as well in the heavy quark limit. We thus have spin multiplets, which are degenerate for $m_Q \to \infty$, for both singly heavy hadrons and heavy quarkonia, such as the (B, B^*) and (η_b, Υ), respectively.

Now let us discuss the consequences of HQSS for hadronic molecules with hidden $Q\bar{Q}$. If the hadronic molecule is a loosely bound state of a heavy quarkonium and light hadrons, then because they can only interact through the exchange of at least two gluons, HQSS would lead to a spin multiplet of hadronic molecules [18]. This was used to predict the properties of an $\eta_c' f_0(980)$ bound state with $J^{PC} = 0^{-+}$ [18] as the spin partner of the $Y(4660)$ were it a $\psi' f_0(980)$ bound state [19]. The same analysis in fact can also be applied to hadro-quarkonia as discussed in Ref. [20].

For the hadronic molecules formed by a pair of heavy and anti-heavy mesons, the HQSS structure is more complicated than the above case. As an example, we consider the S-wave interaction between a pair of $s_\ell^P = \frac{1}{2}^-$ (anti-)heavy mesons. We have a basis of six states of meson pairs:

$$\begin{aligned} 0^{++}: &\quad D\bar{D}, \quad D^*\bar{D}^*, & 1^{+-}: &\quad \frac{1}{\sqrt{2}}\left(D\bar{D}^* + D^*\bar{D}\right), \quad D^*\bar{D}^*, \\ 1^{++}: &\quad \frac{1}{\sqrt{2}}\left(D\bar{D}^* - D^*\bar{D}\right), & 2^{++}: &\quad D^*\bar{D}^*. \end{aligned} \tag{2}$$

Because of HQSS, the interaction at leading order is independent of the heavy quark spin, and thus can be described by the matrix elements $\left\langle s'_{\ell 1}, s'_{\ell 2}, s_L \left|\hat{\mathcal{H}}\right| s_{\ell 1}, s_{\ell 2}, s_L \right\rangle$ where $s_{\ell 1,\ell 2}$ and $s'_{\ell 1,\ell 2}$ are for the heavy mesons in the initial and final states, respectively, s_L is the total angular momentum of the light quarks in the meson-pair system. Thus, for the systems under consideration, we have two independent terms for each isospin ($I = 0$ or 1): $\left\langle \frac{1}{2}, \frac{1}{2}, 0 \left|\hat{\mathcal{H}}\right| \frac{1}{2}, \frac{1}{2}, 0 \right\rangle$ and $\left\langle \frac{1}{2}, \frac{1}{2}, 1 \left|\hat{\mathcal{H}}\right| \frac{1}{2}, \frac{1}{2}, 1 \right\rangle$. This simple observation leads to the conclusion that in the strict heavy quark limit the six pairs in Eq. (2) are grouped in two multiplets with $s_L = 0$ and 1, respectively. In the limit $m_c \to \infty$, D and D^* are degenerate, it is convenient to use the basis of states $s_L^{PC} \otimes s_{c\bar{c}}^{PC}$ with $s_{c\bar{c}}$ refers to the total spin of the c and $\bar{c}$ pair. For the case of S-wave interaction, both of s_L^{PC} and $s_{c\bar{c}}^{PC}$ can only be 0^{-+} or 1^{--}. Therefore, the spin multiplet with $s_L = 0$ contains two states with quantum numbers:

$$0_L^{-+} \otimes 0_{c\bar{c}}^{-+} = 0^{++}, \qquad 0_L^{-+} \otimes 1_{c\bar{c}}^{--} = 1^{+-}, \tag{3}$$

and the spin multiplet for $s_L = 1$ has the following four states:

$$1_L^{--} \otimes 0_{c\bar{c}}^{-+} = 1^{+-}, \qquad 1_L^{--} \otimes 1_{c\bar{c}}^{--} = 0^{++} \oplus 1^{++} \oplus 2^{++}. \tag{4}$$

It becomes clear that if the 1^{++} state $X(3872)$ is a $D\bar{D}^*$ molecule, then it is in the multiplet with $s_\ell = 1$ [21], and has three spin partners with $J^{PC} = 0^{++}$, 2^{++} and 1^{+-} in the strict heavy quark limit, as pointed out in Refs. [22, 23]. This is also the reason why it was suggested that the $Z_b(10610)$ and $Z_b(10650)$ might have four more isovector partners $W_{b0}^{(\prime)}$, W_{b1} and W_{b2} [24–26].

It is worthwhile to notice that the two 1^{+-} states are in different multiplets with $s_\ell = 0$ and 1, respectively, and thus cannot be related to each other via HQSS. However, the isovector $Z_b(10610)$ and $Z_b(10650)$ are located with similar distances to the $B\bar{B}^*$ and $B^*\bar{B}^*$ thresholds, respectively. Such an approximate degeneracy implies that the isovector interactions in the $s_\ell = 0$ and $s_\ell = 1$ sectors are approximately the same, and the off-diagonal transition strength in the isovector channel between the two meson pairs with $J^{PC} = 1^{+-}$ in Eq. (2) approximately vanishes.[2] This leads to the 'light quark spin symmetry' hypothesis proposed by Voloshin very recently [28]. A deeper understanding for such a phenomenon is still needed. Meanwhile, it is worthwhile to notice the analysis that channel-coupling effects could be neglected based on power counting arguments [29]. Neglecting channel coupling, the charm partners for the Z_b states were predicted in Ref. [30], which could correspond to the charged $Z_c(3900)$ [31, 32] and $Z_c(4020)$ [33] observed by the BESIII and Belle collaborations.

When the physical nondegenerate masses for the heavy mesons are used, one needs the physical particle basis given in Eq. (2), and the two-multiplet pattern gets changed. The effective Lagrangian for the leading order calculation was given in Ref. [9]. It turns out that the contact interactions for the 1^{++} and 2^{++} channels are the same. One generally expects a 2^{++} $D^*\bar{D}^*$ bound state assuming the $X(3872)$ to be a $D\bar{D}^*$ bound state [22]. Such a X_2 state should decay dominantly into the $D\bar{D}$ in a D-wave, and the width could be up to a few tens of MeV [23, 34]. The complexity on the X_2 due to the coupling of bare quark model $c\bar{c}$ poles was recently examined in Ref. [35].

3 What can we say about productions and decays of hadronic molecules?

Nowadays lots of phenomenological studies of hadron spectrum are concentrated on candidates of exotic hadrons, and what the nature of the observed XYZ states is. Many of such calculations focus on the mass spectrum, and suggest certain assignments to the observations using the methods of quark model, QCD sum rules and so on. However, no assertive statements can be made on the nature of these experimental structures based solely on the mass. On one hand, different models can produce the same mass, in particular if one only study a single state, by adjusting parameters, which is one of the reasons for the uncontrollable inherent uncertainty of such calculations. On the other hand, as discussed in the preceding section, it is the coupling strength that contains important information on the nature. Therefore, for the purpose of understanding the XYZ structures, it is unavoidable to study carefully their decays and productions which depend on the coupling strengths. In this regard, it is worthwhile to make the following comment: in an EFT, it is necessary to include all the relevant low-energy degrees of freedom, unless their couplings are very weak, in order to describe the system under consideration properly. In this spirit, no matter what model is used to describe a structure close to the threshold of a strongly coupled channel, the effect of that channel has to be taken into account. This is also the reason for the nowadays well-known pattern that the observed charmoium spectrum higher than open-charm thresholds does not agree with the quark model predictions, as shown in Fig. 1.

One sees the peculiarity of hadronic molecules as the closeness of their masses to the corresponding thresholds allows for a model-independent nonrelativistic EFT treatment. The EFT treatment applies if all the dynamical momenta in the involved processes are small, i.e. for long-distance processes. Yet, this also means that once we go to the processes happening at short distances which

[2] A fit to the Belle data of the line shapes of the Z_b states with HQSS constraints implemented also leads to nearly vanishing channel coupling [27].

contain dynamics inaccessible in the low-energy EFT, model-independence would be lost without further inputs. We thus conclude that hadronic molecules, as extended objects with a size larger than the normal hadrons, can be best studied in long-distance processes with a small energy release, while the short-distance processes are insensitive to the hadronic molecular component of physical states. Examples of processes dominated by the short-distance physics include productions of the $X(3872)$ state in high-energy hadron colliders at high-p_T, in B decays and the decay of the $X(3872)$ into the $J/\psi\gamma$. Examples of long-distance processes for the $X(3872)$ include the decays of the $X(3872)$ into $D^0\bar{D}^0\pi^0$ and $D^0\bar{D}^0\gamma$. Measuring the latter processes precisely would be very important in understanding the nature of the $X(3872)$. These decay modes have been discussed by various groups, e.g., Refs. [10, 16, 36–41].

The $X(3872) \to D^0\bar{D}^0\pi^0$ decay has been already observed by the Belle Collaboration [42, 43], which triggered the virtual state explanation for the $X(3872)$ [44]. If the final state interaction (FSI) between the charmed mesons can be neglected, the predictions can be reliably predicted [10]. However, the strength of the interaction between the charmed meson pairs in the $s_\ell = 0$ sector in Eq. (3) is not known. It is possible that the interaction is attractive and is strong enough to produce a near-threshold pole in the 0^{++} channel to which the $D^0\bar{D}^0$ couples. In this case, the partial width of $X(3872) \to D^0\bar{D}^0\pi^0$ will be largely modified by the $D\bar{D}$ FSI as discussed in details in Ref. [41].

4 Summary

We have emphasized that hadronic molecules can only be sensibly defined as near-threshold states. This allows them to be treated model-independently using EFT methods. For hadronic molecules with hidden heavy $Q\bar{Q}$, interesting insights can be obtained using HQSS. The hadronic molecular component for a near-threshold state can be directly studied by long-distance decay or production processes instead of the short-distance ones. Therefore, it is important to measure precisely the masses and the long-distance decays for the hadronic molecule candidates such as the $X(3872)$.

I would like to thank M. Albaladejo, C. Hanhart, C. Hidalgo-Duque, Y. Kalashnikova, U. Meißner, A. Nefediev, J. Nieves, A. Ozpineci and M. Pavón Valderrama for the fruitful collaborations and insights on the topics discussed here. This work is supported in part by DFG and NSFC through funds provided to the Sino-German CRC 110 "Symmetries and the Emergence of Structure in QCD" (NSFC Grant No. 11621131001), by the Thousand Talents Plan for Young Professionals, and by the Chinese Academy of Sciences (Grant No. QYZDB-SSW-SYS013).

References

[1] S. Godfrey and N. Isgur, Phys. Rev. D **32**, 189 (1985)
[2] R. Aaij *et al.* [LHCb Collaboration], arXiv:1606.07898 [hep-ex]
[3] F.-K. Guo and U.-G. Meißner, Phys. Rev. D **86**, 091501 (2012)
[4] S. L. Olsen, Phys. Rev. D **91**, 057501 (2015)
[5] Z. Y. Zhou, Z. Xiao and H. Q. Zhou, Phys. Rev. Lett. **115**, 022001 (2015)
[6] S. Weinberg, Phys. Rev. **137**, B672 (1965)
[7] V. Baru, J. Haidenbauer, C. Hanhart, Y. Kalashnikova and A. E. Kudryavtsev, Phys. Lett. B **586**, 53 (2004)
[8] T. Hyodo, Int. J. Mod. Phys. A **28**, 1330045 (2013)
[9] M. T. AlFiky, F. Gabbiani and A. A. Petrov, Phys. Lett. B **640**, 238 (2006)
[10] S. Fleming, M. Kusunoki, T. Mehen and U. van Kolck, Phys. Rev. D **76**, 034006 (2007)

[11] J. Nieves and M. P. Valderrama, Phys. Rev. D **84**, 056015 (2011)
[12] F.-K. Guo, C. Hanhart, U.-G. Meißner, Q. Wang and Q. Zhao, Phys. Lett. B **725**, 127 (2013)
[13] E. Braaten, Phys. Rev. D **91**, 114007 (2015)
[14] Q. Wang, C. Hanhart and Q. Zhao, Phys. Rev. Lett. **111**, 132003 (2013)
[15] N. A. Törnqvist, Phys. Lett. B **590**, 209 (2004)
[16] M. B. Voloshin, Phys. Lett. B **579**, 316 (2004)
[17] M. Ablikim *et al.* [BESIII Collaboration], Phys. Rev. Lett. **112**, 092001 (2014)
[18] F.-K. Guo, C. Hanhart and U.-G. Meißner, Phys. Rev. Lett. **102**, 242004 (2009)
[19] F.-K. Guo, C. Hanhart and U.-G. Meißner, Phys. Lett. B **665**, 26 (2008)
[20] M. Cleven, F.-K. Guo, C. Hanhart, Q. Wang and Q. Zhao, Phys. Rev. D **92**, 014005 (2015)
[21] M. B. Voloshin, Phys. Lett. B **604**, 69 (2004)
[22] C. Hidalgo-Duque, J. Nieves, A. Ozpineci and V. Zamiralov, Phys. Lett. B **727**, 432 (2013)
[23] V. Baru, E. Epelbaum, A. A. Filin, C. Hanhart, U.-G. Meißner and A. V. Nefediev, arXiv:1605.09649 [hep-ph]
[24] A. E. Bondar, A. Garmash, A. I. Milstein, R. Mizuk and M. B. Voloshin, Phys. Rev. D **84**, 054010 (2011)
[25] M. B. Voloshin, Phys. Rev. D **84**, 031502 (2011)
[26] T. Mehen and J. W. Powell, Phys. Rev. D **84**, 114013 (2011)
[27] F.-K. Guo, C. Hanhart, Y. S. Kalashnikova, P. Matuschek, R. V. Mizuk, A. V. Nefediev, Q. Wang and J.-L. Wynen, Phys. Rev. D **93**, 074031 (2016)
[28] M. B. Voloshin, Phys. Rev. D **93**, 074011 (2016)
[29] M. P. Valderrama, Phys. Rev. D **85**, 114037 (2012)
[30] F.-K. Guo, C. Hidalgo-Duque, J. Nieves and M. P. Valderrama, Phys. Rev. D **88**, 054007 (2013)
[31] M. Ablikim *et al.* [BESIII Collaboration], Phys. Rev. Lett. **110**, 252001 (2013)
[32] Z. Q. Liu *et al.* [Belle Collaboration], Phys. Rev. Lett. **110**, 252002 (2013)
[33] M. Ablikim *et al.* [BESIII Collaboration], Phys. Rev. Lett. **111**, 242001 (2013)
[34] M. Albaladejo, F.-K. Guo, C. Hidalgo-Duque, J. Nieves and M. P. Valderrama, Eur. Phys. J. C **75**, 547 (2015)
[35] E. Cincioglu, J. Nieves, A. Ozpineci and A. U. Yilmazer, arXiv:1606.03239 [hep-ph].
[36] E. S. Swanson, Phys. Lett. B **588**, 189 (2004)
[37] M. B. Voloshin, Int. J. Mod. Phys. A **21**, 1239 (2006)
[38] W. H. Liang, R. Molina and E. Oset, Eur. Phys. J. A **44**, 479 (2010)
[39] E. Braaten and J. Stapleton, Phys. Rev. D **81**, 014019 (2010)
[40] V. Baru, A. A. Filin, C. Hanhart, Y. S. Kalashnikova, A. E. Kudryavtsev and A. V. Nefediev, Phys. Rev. D **84**, 074029 (2011)
[41] F.-K. Guo, C. Hidalgo-Duque, J. Nieves, A. Ozpineci and M. P. Valderrama, Eur. Phys. J. C **74**, 2885 (2014)
[42] G. Gokhroo *et al.* [Belle Collaboration], Phys. Rev. Lett. **97**, 162002 (2006)
[43] T. Aushev *et al.* [Belle Collaboration], Phys. Rev. D **81**, 031103 (2010)
[44] C. Hanhart, Y. S. Kalashnikova, A. E. Kudryavtsev and A. V. Nefediev, Phys. Rev. D **76**, 034007 (2007)

Quarkonium and heavy flavour meson production at 13 TeV at ATLAS

Leonid Gladilin[1,★]
for the ATLAS Collaboration

[1] *Lomonosov Moscow State University Skobeltsyn Institute of Nuclear Physics (SINP MSU), Moscow 119991, Russian Federation*

Abstract. First results of the ATLAS experiment at LHC on quarkonium and heavy flavour meson production in proton-proton collisions at 13 TeV are presented. A brief summary of the results obtained at 7-8 TeV is also reported. Comparison of the data cross sections with various theoretical predictions is discussed.

1 Introduction

Measurements of the charmed and beauty mesons production and the non-prompt (from b-hadron decays) production of charmonium states probe production and hadronisation of c and b quarks. Measurements of the bottomonium and prompt charmonium production probe a heavy quark pair production and its subsequent evolution into a bound state. The latter includes non-perturbative effects and can be described with colour-singlet (CS) and colour-octet (CO) contributions in the framework of non-relativistic QCD (NRQCD). In this framework, the non-perturbative evolution is described with long-distance matrix elements (LDME) tuned to experimental results. First results on quarkonium and heavy flavour meson production in proton-proton collisions at 13 TeV and recent results at 7-8 TeV obtained with the ATLAS [1] detector at the LHC are described in this note.

2 Charmonium production

The non-prompt J/ψ production fraction has been measured differentially as a function of the J/ψ transverse momentum and rapidity using 6.4 pb^{-1} of proton-proton collision data at a centre-of-mass energy of 13 TeV [2]. The fraction in intervals of dimuon p_{T} and $|y|$ is measured and summarised in Fig. 1(left). The non-prompt fraction is found to increase steadily from 0.25 at a transverse momentum of 8 GeV to 0.65 at 40 GeV, with no significant variation with rapidity observed within the precision of the measurement. The centre-of-mass energy (and initial-state) dependence of the fraction is studied by comparing these results, in the J/ψ rapidity interval $|y| < 0.75$, to previous ATLAS measurements in the same rapidity region at $\sqrt{s} = 2.76$ TeV and a comparable rapidity interval ($0.25 < |y| < 0.50$) at $\sqrt{s} = 7$ TeV, and CDF measurements in a slightly different rapidity interval ($|y| < 0.6$) at $\sqrt{s} = 1.96$ TeV. The comparison is illustrated in Fig. 1(right). No significant change in the non-prompt

★e-mail: gladilin@mail.cern.ch

fraction is observed from $\sqrt{s} = 7\,\text{TeV}$ to $\sqrt{s} = 13\,\text{TeV}$, contrary to the significant difference observed between the $\sqrt{s} = 7\,\text{TeV}$ measurement and the measurements at lower energies.

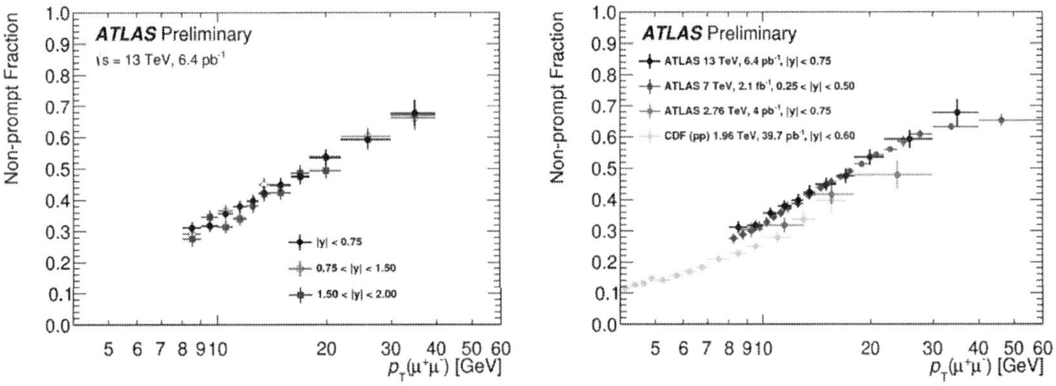

Figure 1. Non-prompt J/ψ production fraction as a function of J/ψ p_T in (left) three intervals of J/ψ rapidity and in (right) the most central rapidity region ($|y| < 0.75$) [2] compared to previous measurements from ATLAS in pp collisions at 2.76 GeV and 7 GeV, and from CDF in $p\bar{p}$ collisions at $\sqrt{s} = 1.96\,\text{GeV}$.

Differential cross sections of the charmonium prompt and non-prompt production have been measured in pp collisions at $\sqrt{s}$ = 7-8 TeV for the $J/\psi \to \mu^+\mu^-$ and $\psi(2S) \to \mu^+\mu^-$ [3], $\psi(2S) \to J/\psi\pi^+\pi^-$ [4, 5], $\chi_{c1/2} \to J/\psi\gamma$ [6] and $X(3872) \to J/\psi\pi^+\pi^-$ [5] final states.

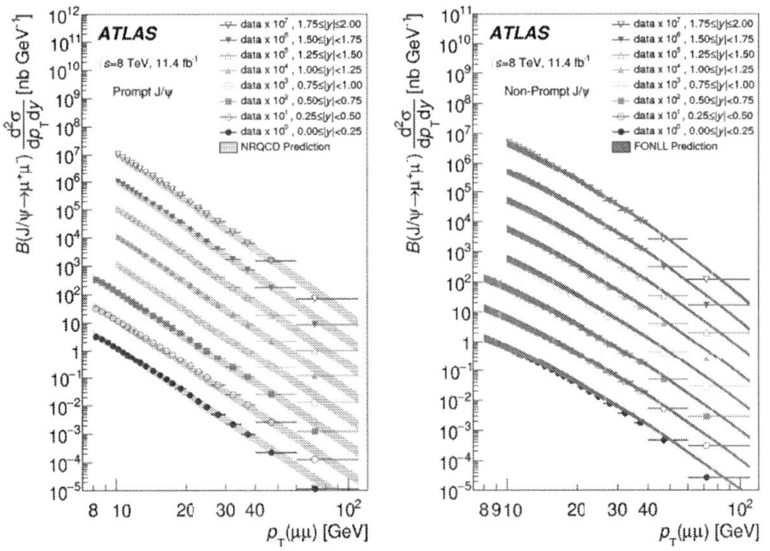

Figure 2. The differential (left) prompt and (right) non-prompt cross section times dimuon branching fraction of J/ψ as a function of J/ψ p_T for eight slices in rapidity [3]. For each increasing rapidity slice, an additional scaling factor of 10 is applied to the plotted points for visual clarity. Theoretical predictions are also shown.

Figure 2 shows the differential prompt and non-prompt cross section times dimuon branching fraction of J/ψ as a function of J/ψ p_T for eight slices in rapidity. For the prompt production, predictions from the NRQCD model, which includes colour-octet contributions with LDMEs tuned to earlier collider data, are in good agreement with the data. For the non-prompt production, the fixed-order next-to-leading-logarithm (FONLL) calculations reproduce the data reasonably well, with a

 DOI: 10.1051/epjconf/201613001028

slight overestimation of the differential cross sections at the highest transverse momenta. Similar conclusions have been done for prompt and non-prompt production of $\psi(2S)$ and $\chi_{c1/2}$.

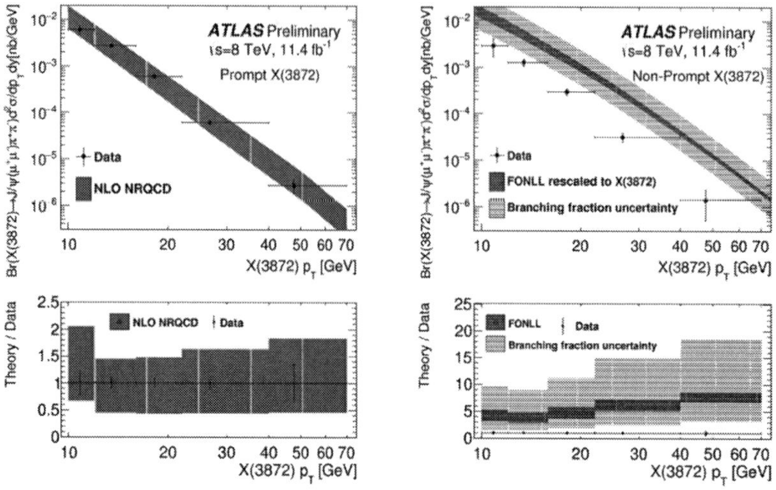

Figure 3. The differential (left) prompt and (right) non-prompt cross section times branching fractions of $X(3872)$ as a function of $X(3872)$ p_T [5]. Theoretical predictions are also shown.

Figure 3 shows differential prompt and non-prompt cross section times branching fractions of $X(3872)$ as a function of $X(3872)$ p_T. For the prompt production, good agreement is found with theoretical predictions within the NLO NRQCD model, which considers $X(3872)$ to be a mixture of $\chi_{c1}(2P)$ and a $D^0\bar{D}^{*0}$ molecular state, with the production being dominated by the $\chi_{c1}(2P)$ component and the normalisation fixed through the fit to the earlier LHC data. For the non-prompt production, the FONLL calculations for $\psi(2S)$, recalculated for X(3872) using the branching fraction $Br(B \to X(3872))Br(X(3872) \to J/\psi\pi^+\pi^-) = (1.9 \pm 0.8) \times 10^{-4}$ estimated in [7] from the Tevatron data, overestimate the data, especially at large transverse momenta.

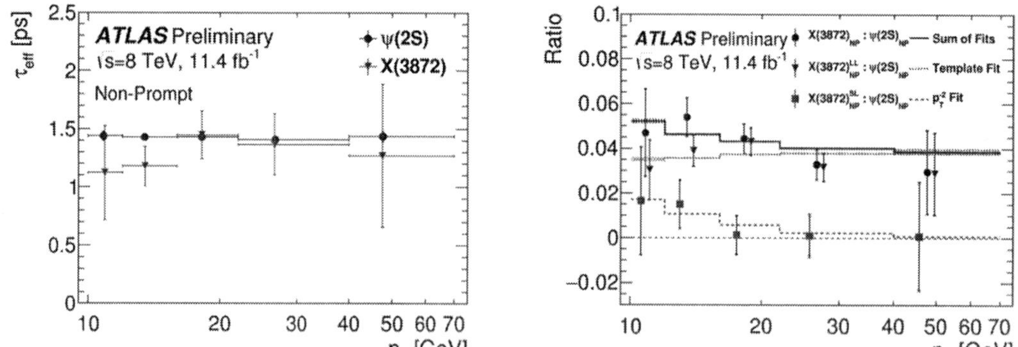

Figure 4. Measured (left) effective pseudo-proper lifetimes for non-prompt $\psi(2S)$ and $X(3872)$ and (right) ratio of cross section times branching fraction between $X(3872)$ and $\psi(2S)$ for non-prompt production [5]. In (right), the total non-prompt ratio (black circles) is separated into short-lived (red squares) and long-lived (blue triangles) components for the $X(3872)$, shown with respective fits described in the text.

Figure 4(left) shows measured effective pseudo-proper lifetimes for non-prompt $\psi(2S)$ and $X(3872)$. While the pseudo-proper lifetime distribution is nearly flat for $\psi(2S)$, the signal from $X(3872)$ at low p_T tends to have shorter lifetimes. The non-prompt production cross section of $X(3872)$ is split into long-lived ($\tau = 1.45 \pm 0.05$ ps) and short-lived ($\tau = 0.40 \pm 0.05$ ps) components. Figure 4(right) shows ratio of cross section times branching fraction between $X(3872)$ and $\psi(2S)$ for the total non-prompt production and for the long-lived and short-lived components. The measured ratio of long-lived $X(3872)$ is well described by the Monte Carlo kinematic template which is nearly flat. The short-lived component can originate from B_c production, which is expected to be dominated by non-fragmentation processes at low transverse momentum [8]. These processes are expected to have p_T dependence $\propto p_T^{-2}$ relative to the fragmentation contribution. So the ratio of short-lived non-prompt $X(3872)$ to non-prompt $\psi(2S)$ is fitted with a function a/p_T^2 to find $a = 2.04 \pm 1.43(stat) \pm 0.34(sys)\,\mathrm{GeV}^2$ [5]. This value is used to determine the fraction of the short-lived component in the non-prompt $X(3872)$ production, for $p_T > 10$ GeV, to be $(25 \pm 13(stat) \pm 2(sys) \pm 5(spin))\%$ [5], where the last uncertainty comes from the variation of the spin alignment of $X(3872)$.

3 Beauty and charmed meson production

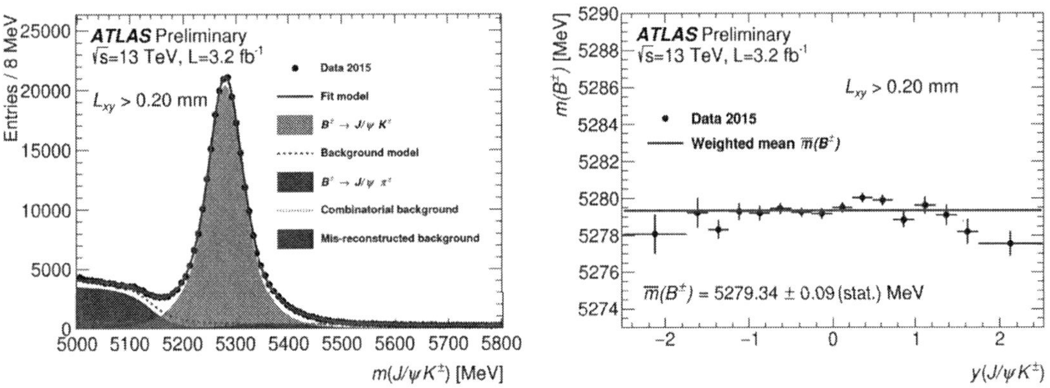

Figure 5. B^+ mass (left) reconstructed for all $B^+ \to J/\psi K^+$ candidates and (right) fitted in several bins of $y(B^+)$ [9].

The performance of B^+ mass reconstruction in $B^+ \to J/\psi K^+$ decay at 13 TeV is verified using 3.2 fb^{-1} of pp collision data [9]. Figure 5 shows the B^+ mass reconstructed for all $B^+ \to J/\psi K^+$ candidates and fitted in several bins of $y(B^+)$. The measured B^+ mass is in good agreement with the world average value.

The strangeness-suppression factor, $\gamma_{s/d}$, in beauty fragmentation is determined, using decays $B_s^0 \to J/\psi\phi$ and $B_d^0 \to J/\psi K^{0*}$, to be $(0.240 \pm 0.004(stat) \pm 0.010(sys) \pm 0.017(th))\%$ [10], where the last uncertainty is due to the uncertainty of the perturbative QCD prediction for $Br(B_s^0 \to J/\psi\phi)/Br(B_d^0 \to J/\psi K^{0*})$ [11].

The differential cross section for B^+ meson production has been studied with 2.4 fb^{-1} of pp collision data at 7 TeV [12]. The FONLL calculations reproduce the data reasonably well.

The production of $D^{*\pm}$, $D^{\pm}$ and $D_s^{\pm}$ charmed mesons has been measured in the kinematic region $3.5 < p_T(D) < 100$ GeV and $|\eta(D)| < 2.1$ in pp collisions at 7 TeV, using an integrated luminosity of

 DOI: 10.1051/epjconf/201613001028

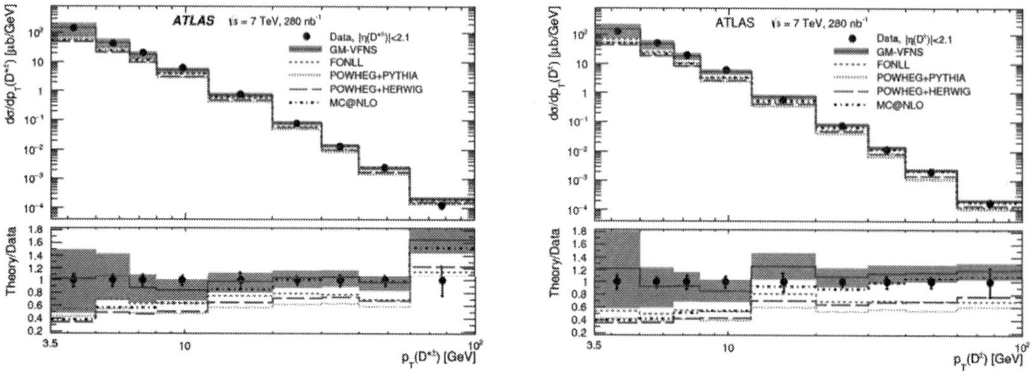

Figure 6. Differential cross sections for (left) $D^{*\pm}$ and (right) $D^{\pm}$ mesons as a function of $p_{\rm T}$ for data (points) [13] compared to the NLO QCD calculations of FONLL, POWHEG+PYTHIA, POWHEG+HERWIG, MC@NLO and GM-VFNS (histograms). The bands show the estimated theoretical uncertainty of the GM-VFNS calculation.

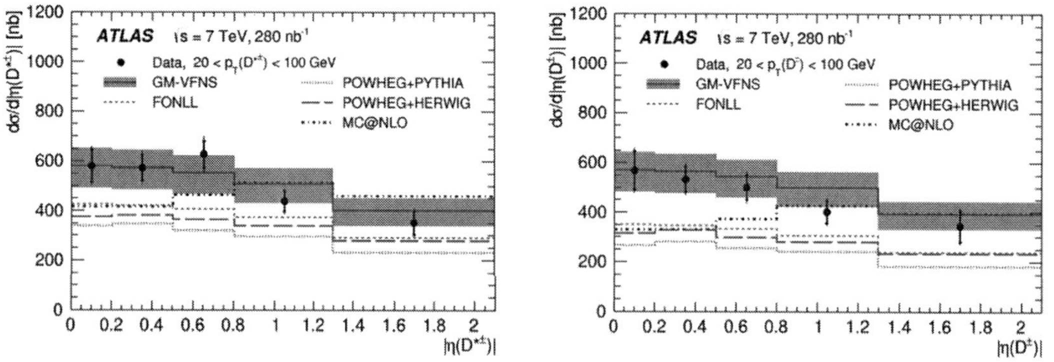

Figure 7. Differential cross sections for (left) $D^{*\pm}$ and (right) $D^{\pm}$ mesons as a function of $|\eta|$ for data (points) [13] compared to the NLO QCD calculations of FONLL, POWHEG+PYTHIA, POWHEG+HERWIG, MC@NLO and GM-VFNS (histograms). The bands show the estimated theoretical uncertainty of the GM-VFNS calculation.

up to $280\,{\rm nb}^{-1}$ [13]. Figures 6-7 show differential cross sections for $D^{*\pm}$ and $D^{\pm}$ mesons as functions of $p_{\rm T}$ and $|\eta(D)|$ for data compared to the NLO QCD calculations. The FONLL, POWHEG+PYTHIA and POWHEG+HERWIG predictions are generally below the data. They are consistent with the data in the measured $p_{\rm T}(D)$ and $|\eta(D)|$ ranges within the large theoretical uncertainties. The FONLL and POWHEG predictions reproduce the shapes of the data distributions. The $p_{\rm T}$ shape of the MC@NLO prediction is harder than that for the data. The $|\eta|$ shape of the MC@NLO prediction in the range $20 < p_{\rm T} < 100\,{\rm GeV}$ differs from the data and all other predictions. The general-mass variable-flavour-number scheme (GM-VFNS) predictions agree with data in both shape and normalisation.

The visible D cross sections are extrapolated to the cross sections in the full kinematic phase space after subtracting the cross-section fractions originating from beauty production. To calculate the total cross section of charm production, the total production cross section of a given D meson should be divided by twice the value of the corresponding charm fragmentation fraction [14]. The weighted mean of the two values calculated from $D^{*\pm}$ and $D^{\pm}$ cross sections is [13]:

$$\sigma_{c\bar{c}}^{\text{tot}} = 8.6 \pm 0.3(stat) \pm 0.7(sys) \pm 0.3(lum) \pm 0.2(ff)^{+3.8}_{-3.4}(extr)\,\text{mb},$$

where the fourth uncertainty is due to the uncertainty of the fragmentation fractions and the last uncertainty is due to the extrapolation procedure. The total cross section of charm production agree with the result of the ALICE collaboration.

The total cross sections for D production are also used to calculate two fragmentation ratios for charged charmed mesons: the strangeness-suppression factor, $\gamma_{s/d}$, and the fraction of charged non-strange D mesons produced in a vector state, P_{v}^{d} [13]:

$$\gamma_{s/d} = 0.26 \pm 0.05(stat) \pm 0.02(sys) \pm 0.02(br) \pm 0.01(extr),$$

$$P_{\text{v}}^{d} = 0.56 \pm 0.03(stat) \pm 0.01(sys) \pm 0.01(br) \pm 0.02(extr).$$

The fragmentation ratios agree with those obtained by the ALICE collaboration at the LHC, and those measured in e^+e^- annihilations at LEP and in $e^{\pm}p$ collisions at HERA. The value of strangeness-suppression in charm fragmentation agrees with that measured in beauty fragmentation.

Acknowledgments

Participation in the conference was supported by the Russian Foundation for Basic Research, grant 15-02-08133.

References

[1] ATLAS Collaboration, JINST **3**, S08003 (2008)
[2] ATLAS Collaboration, ATLAS-CONF-2015-030, http://cds.cern.ch/record/2037967 (2015)
[3] ATLAS Collaboration, Eur. Phys. J. **76**, 283 (2016)
[4] ATLAS Collaboration, JHEP **09**, 079 (2014)
[5] ATLAS Collaboration, ATLAS-CONF-2016-028, http://cds.cern.ch/record/2160169 (2016)
[6] ATLAS Collaboration, JHEP **07**, 154 (2014)
[7] P. Artoisenet and E. Braaten, Phys. Rev. D **81**, 114018 (2010)
[8] A. V. Berezhnoy and A. K. Likhoded, arXiv:1309.1979 (2013)
[9] ATLAS Collaboration, ATLAS-CONF-2015-064, http://cds.cern.ch/record/2114830 (2015)
[10] ATLAS Collaboration, Phys. Rev. Lett. **115**, 262001 (2015)
[11] X. Liu, W. Wang and Y. Xie, Phys. Rev. D **89**, 094010 (2014)
[12] ATLAS Collaboration, JHEP **10**, 042 (2013)
[13] ATLAS Collaboration, Nucl. Phys. B **907**, 717 (2016)
[14] L. Gladilin, Eur. Phys. J. C **75**, 19 (2015)

First CMS results at 13 TeV

P. Ronchese[1,⋆]
on behalf of the CMS Collaboration
[1] *INFN and University of Padova, Italy*

Abstract. Measurements of b-hadron and quarkonium production cross sections provide essential information to understand QCD. This talk will show new measurements of double-differential production cross sections vs. transverse momentum p_T and rapidity y, at center-of-mass energy $\sqrt{s}$ = 13 TeV, performed by CMS experiment with data collected in 2015. Comparisons with results obtained at $\sqrt{s}$ = 7 TeV will be shown.

1 Introduction

The motivations to study heavy flavour physics at CMS [1] include the probe of the underlying QCD processes, obtained by measuring b-hadron production cross-section and quarkonium polarization and by looking for new and exotic states. The measurements of the production cross-section for B^+ mesons and ψ,Υ quarkonia from CMS at a center of mass energy $\sqrt{s}$ = 13 TeV will be shown.

2 The CMS experiment

The main components of the CMS experiment [1] are the tracker with pixels and microstrips, the electromagnetic and hadron calorimeters, and the muon detectors. Three different detectors are dedicated to muon trigger and detection, having the capability to perform a standalone reconstruction of muons. The match between the reconstruction from the tracker and the muon detectors can be conducted inside-out, that is more efficient at low tranverse momentum, or outside-in, that is more efficient at high p_T. The misidentification probability is less than 1% and the momentum resolution ranges from 1% to 6% for muons up to 100 GeV.

Muons play an essential role in triggering events in the harsh LHC environment, characterized by a very high luminosity, while keeping the frequency inside a limited bandwidth. Triggers for B physics require the presence of two muons at least, possibly together with an additional track. The trigger system consists of a first level, based on a dedicated hardware taking fast decision and giving a trigger rate of the order of ~ 100 kHz, followed by an higher level based on software performing full track reconstruction and giving a rate of ~ 1 kHz. Specific triggers have been developed for different analyses, applying cuts on muon and dimuon transverse momentum and rapidity, the presence of secondary vertices and their displacement from the interaction point, the dimuon mass and the angle formed by the dimuon momentum and the flight direction, given by the position of the secondary

⋆e-mail: paolo.ronchese@pd.infn.it

 DOI: 10.1051/epjconf/201613001029

vertex relative to the collision point. In the first year of run at $\sqrt{s}$ = 13 TeV LHC delivered an integrated luminosity $\mathcal{L} \sim 4$ fb^{-1}, and CMS recorded data corresponding to $\mathcal{L} \simeq 3.6$ fb^{-1}. For muon-based analyses an integrated luminosity $\mathcal{L} \simeq 2.7$ fb^{-1} was collected.

3 Double Υ production

In the high parton densities in the proton-proton collisions produced by LHC, quarkonium pairs can be produced in single parton scattering (SPS) or double parton scattering (DPS) processes. SPS is assumed to dominate and produces pairs of strongly correlated mesons with small rapidity difference, SPS leads to the production of multiple heavy flavour quarks and meson pairs with large rapidity difference [2, 3]. The quarkonium production mechanism involves the production of color-singlet states, that is dominant at low transverse momentum, and color-octet states that become more important at higher p_T [4, 5]. Another source of quarkonium pairs is the decay of tetra-quarks [6].

The selection of double-Υ events at CMS [7] required the reconstruction of muons with hits in the strip tracker and pixel detector and the match of the extrapolated track with at least one segment reconstructed in the muon chambers. Cuts on muon (dimuon) p_T and η (y) were then applied to select a region with an uniform acceptance: $p_{T,\mu} > 3.5$ GeV, $|\eta_\mu| < 2.4$, $p_{T,\Upsilon} < 50$ GeV, $|y_\Upsilon| < 2.0$. Events with multiple Υ were discarded and the production of non prompt pairs was neglected.

In the invariant mass distribution the yield of two Υ(1S) or a Υ(1S) and a Υ(2S) were estimated, while no visible signal appeared for Υ(3S). A two-dimensional fit was performed over the mass distribution, with five components corresponding to Υ(1S)Υ(1S), Υ(1S)Υ(2S) pairs, one Υ(1S) or Υ(2S) and one dimuon from combinatorial background and two combinatorial dimuons. The signal was described with a Crystal-Ball function and background with a Chebyshev polynomial; the fit result is shown in Fig. 1.

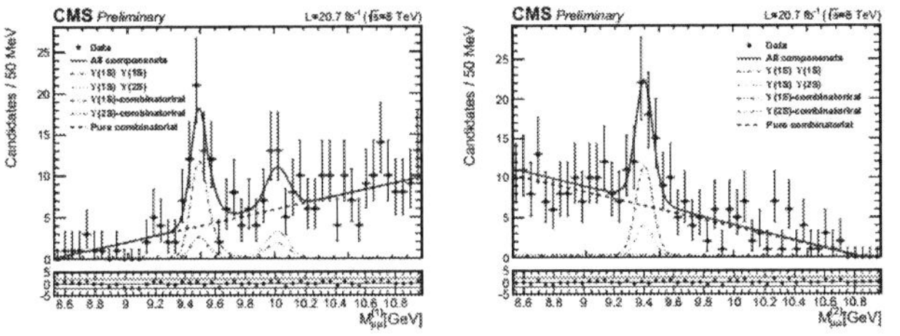

Figure 1. Projections of the 2D fit: invariant mass of higher mass muon pairs, $M^{(1)}_{\mu\mu}$ (left) and lower mass muon pairs, $M^{(2)}_{\mu\mu}$ (right). The lower pads show the difference between the total fit curve and data, with red lines corresponding to a difference of ±2 standard deviations.

The cross section was extracted from the signal yield multiplying the number of fitted signal events n_{sig} for a correction factor $\bar{\omega}$, accounting for acceptance and efficiency, and dividing by the squared branching ratio of the Υ and the integrated luminosity:

$$\sigma_T = \frac{n_{\text{sig}} \cdot \bar{\omega}}{\mathcal{B}(\Upsilon(1S) \rightarrow \mu^+\mu^-)^2 \cdot \mathcal{L}}.$$

Acceptance and efficiency were computed event-by-event on a simulated sample assuming isotropic decays.

The total measured cross-section was [7]

$$\sigma_T = (68.8 \pm 12.7(\text{stat}) \pm 7.4(\text{syst}) \pm 2.8(\text{BR}))\ \text{pb},$$

where the last systematic uncertainty comes from the $\Upsilon \rightarrow \mu^+\mu^-$ branching ratio. Other systematic uncertainties were estimated by changing the functions used to describe signal and background, the estimation of acceptance and efficiency and the integrated luminosity. Υ polarization affects significantly the acceptance, that was estimated assuming unpolarized production: that was compared with the hypotheses of full longitudinal or transverse polarization showing a variation of -38% and +36%.

4 B^+ production

The measurement and study of beauty hadron production cross-section at high energies provide new important tests of theoretical calculations [8, 9]. The B^+ production cross-section was measured at the CMS experiment looking at the decay $B^+ \to J/\psi K^+$, followed by the decay $J/\psi \to \mu^+\mu^-$ [10]; charge conjugation states are implied here and in the following. The differential cross-section versus transverse momentum or rapidity was measured in a rapidity range $|y_B| < 2.4$ and transverse momentum range 10 GeV $< p_{T,B} <$ 100 GeV, using a subsample of data collected in 2015 corresponding to an integrated luminosity $\mathcal{L} = 50.8$ pb^{-1}. The differential cross-section was extracted from the signal yield in each bin dividing by the branching ratios $\mathcal{B}$, acceptance A, efficiency ϵ, integrated luminosity and the bin width Δz, with $z = p_{T,B}, y_B$. An additional factor 2 was included to account for B charge symmetry, while the cross-section for the production of B^+ only was quoted:

$$\frac{d\sigma(pp \to B^+X)}{dz} = \frac{n_{\mathrm{sig}}(z)}{2 \cdot \mathcal{B} \cdot A \cdot \epsilon(z) \cdot \mathcal{L} \cdot \Delta z}.$$

Events were selected applying the requirements on muon quality as described in Sect. 3 ; candidate J/ψ mesons were reconstructed combining pairs of oppositely-charged muons having an invariant mass within ±150 MeV of the nominal J/ψ mass, $p_{T,J/\psi} > 8$ GeV and a χ^2 probability of the vertex-fit larger than 10%. Candidate B^+ mesons were reconstructed combining a J/ψ candidate with a charged track with $p_T > 1$ GeV; the track was assumed to be a kaon and the track fit was required to give $\chi^2/n_{\mathrm{d.o.f.}} < 5$. A kinematic fit was performed, constraining the dimuon mass to the nominal J/ψ mass; the three-track combination was required to be compatible with having a common vertex with a vertex-fit χ^2 probability larger than 10% and a reconstructed invariant mass $M_{J/\psi K}$ in the range $5 \div 6$ GeV. The transverse decay distance divided by its uncertainty was required to exceed 3.5 and the cosine of the angle between B^+ candidate momentum and the vector formed by the interaction point and the $\mu\mu K$ vertex in the transverse plane had to be greater than 0.99.

The B^+ invariant mass distribution was built in bins of transverse momentum and an unbinned maximum likelihood fit was performed; the signal was described with the sum of two gaussians and the background with an exponential plus an error function to describe the contribution of mis-reconstructed decays of B^+ with additional particles in the final state. The invariant mass distribution for the inclusive sample in the total acceptance region is shown in Fig. 2.

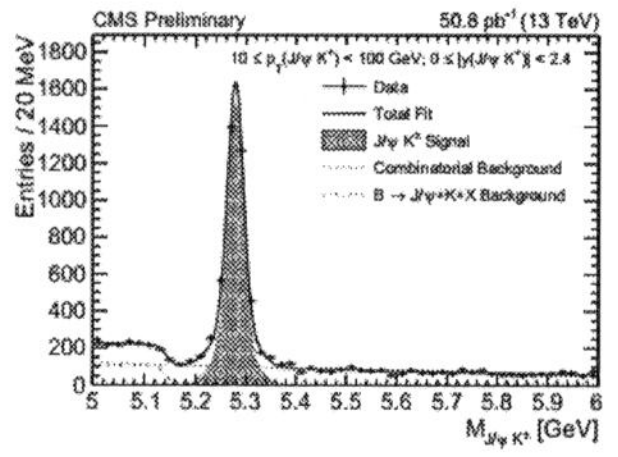

Figure 2. Invariant mass distribution of the full $B^\pm \to J/\psi K^\pm$ candidates sample: the solid curve shows the result of the fit, the shaded area represents the signal component, the dashed and dash-dotted curves represent the combinatorial and mis-reconstructed $B^\pm \to J/\psi K^\pm X$ background components, respectively.

The acceptance and efficiency were estimated with a sample of simulated events; trigger and muon reconstruction efficiency was estimated also in data selecting an inclusive J/ψ sample and applying a tag-and-probe method: the difference with the one obtained from the simulation was included in the systematic uncertainties.

To estimate other systematic uncertainties the mass modeling functions were changed; the rare decay $B^+ \to J/\psi\pi^+$ was included in the fit; the uncertainty coming from bin to bin migration of events, due to finite resolution, was estimated by examining the generator information in the simulated samples. Uncertainties associated to the $p_{T,B}$ and y_B distributions used in the generation were evaluated

DOI: 10.1051/epjconf/201613001029

with event-by-event weights determined from the differences between the distributions generated by PYTHIA and given by the FONLL calculations. The uncertainty on the B^+ lifetime was also included, the error on the integrated luminosity is 4.8% and the uncertainty on the branching ratios is 3.1%.

The differential cross-section versus transverse momentum and rapidity [10] is shown in Fig. 3, together with the results obtained at $\sqrt{s}$ = 7 TeV [11] and the predictions from FONLL and PYTHIA [8, 12–14].

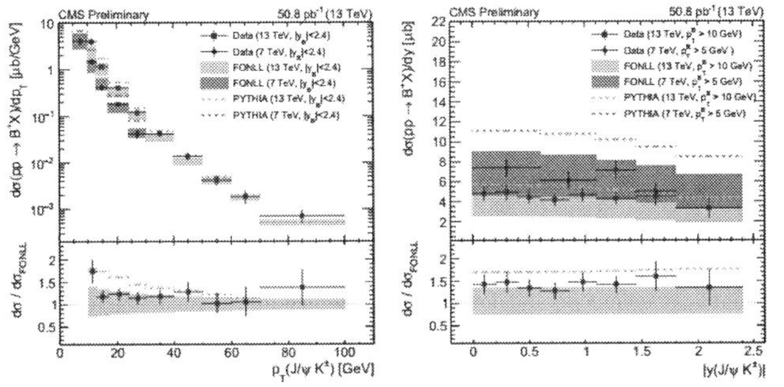

Figure 3. Differential cross-section versus transverse momentum (left) and rapidity (right), compared with predictions from FONLL (boxes) and PYTHIA (dashed lines). The lower pads display the data over FONLL cross section ratios (dots) and the PYTHIA/FONLL ratios (dashed lines).

5 Quarkonium production

Quarkonium production can be described in two phases, with a perturbative generation of a quark-antiquark pair in a singlet or octet state followed by the hadronization producing a bound state, described by a long-distance matrix element [15–18]. Measurements at different energies allow for a test of this factorization; predictions are done performing different perturbative calculations for each different collision energy and using the same matrix element at all energies [19, 20].

Higher energies, with higher cross-sections, allow for a measurement with an extended reach in transverse momentum, up to 120 GeV for rapidity up to 1.2, using the full available data sample collected in 2015 [21]. The product of double-differential cross-section by decay branching ratio was extracted from the signal yield in each bin dividing by acceptance A, efficiency ϵ, integrated luminosity and the bin widths:

$$\mathcal{B}(Q\bar{Q} \to \mu^+\mu^-) \cdot \frac{d^2\sigma(pp \to Q\bar{Q}X)}{dp_T dy} = \frac{N_{Q\bar{Q}}(z)}{\mathcal{L} \cdot \Delta p_T \Delta y} \cdot \langle \frac{1}{A(p_T, y) \cdot \epsilon(p_T, y)} \rangle.$$

The signal yield in each bin was extracted from an unbinned maximum likelihood fit; the signal was described with a Crystal Ball function and the background with an exponential.

Charmonia can be produced in primary proton-proton interactions (prompt) and b-hadron decays (non-prompt); the two components were separated performing a simultaneous fit to the invariant mass and the "pseudo-proper decay length" [22], with a function describing the prompt component with a resolution function, the non prompt component with an exponential convoluted with a resolution function, and the background with a gaussian plus an exponential.

In Fig. 4 the fits to the mass and pseudo-proper decay length are shown for J/ψ in a p_T, y bin.

Acceptance has been estimated in a generated event sample by looking at the fraction of generated events passing the selection; in the generation an unpolarized production of quarkonia has been assumed. Efficiency has been estimated with the tag-and-probe method.

Uncertainties related to the signal yield estimation have been evaluated by changing the parameters in the Crystal Ball function, leaving the masses free in the fit and changing the functions for the background. Uncertainties on the non-prompt fraction have been estimated by calculating the decay

 DOI: 10.1051/epjconf/201613001029

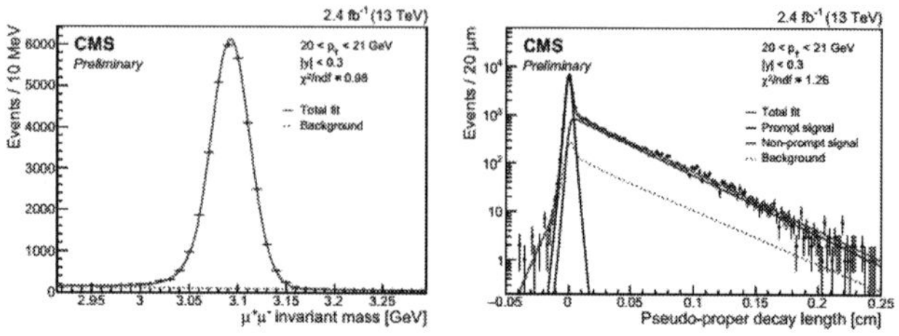

Figure 4. J/ψ fits: mass (left), pseudo-proper decay length (right).

length from the primary vertex in place of the average collision point, using different functions for the background or changing parameters constraints.

The double-differential cross-sections vs. p_T for different y bins are shown in Fig. 5 for J/ψ and ψ(2S) and in Fig. 6 for Υ(1S), Υ(2S) and Υ(3S); the comparisons with the measurements at 7 TeV [23, 24] are shown in Fig. 7.

The non-prompt fractions versus p_T for J/ψ and ψ(2S), compared with the corresponding measurements at 7 TeV [25], are shown in Fig. 8.

6 Conclusions

Differential cross section for B^+ production at $\sqrt{s} = 13$ TeV has been measured for p_T up to 100 GeV; a reasonable agreement with FONLL calculations and with PYTHIA has been found.

The double differential production cross sections at $\sqrt{s} = 13$ TeV for J/ψ, ψ(2S), $\Upsilon(n\text{S})$ has been measured. These results shall contribute to consolidate the underlying hypotheses of NRQCD and provide further input to constrain the theory parameters.

The simultaneous production of two Υ(1S) at $\sqrt{s} = 8$ TeV has been observed and the total cross-section has been measured.

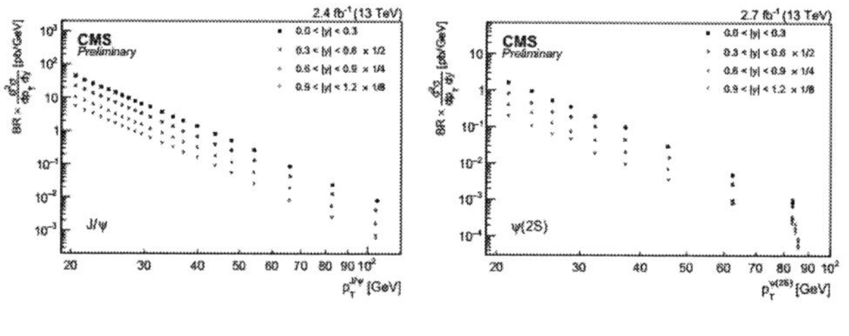

Figure 5. Double-differential cross-sections vs. p_T: J/ψ (left), ψ(2S) (right).

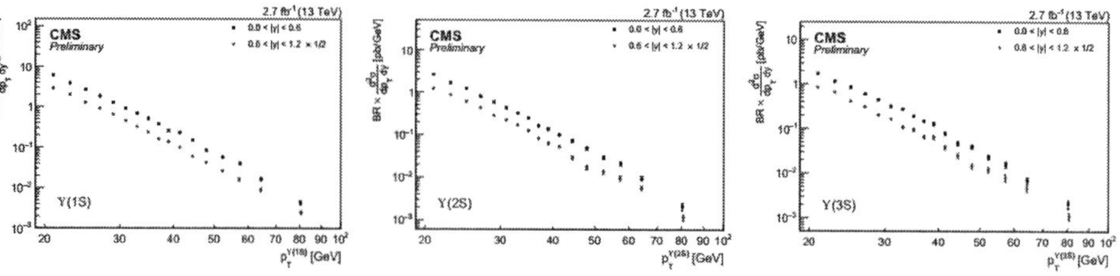

Figure 6. Double-differential cross-sections vs. p_T: Υ(1S) (left), Υ(2S) (middle), Υ(3S) (right).

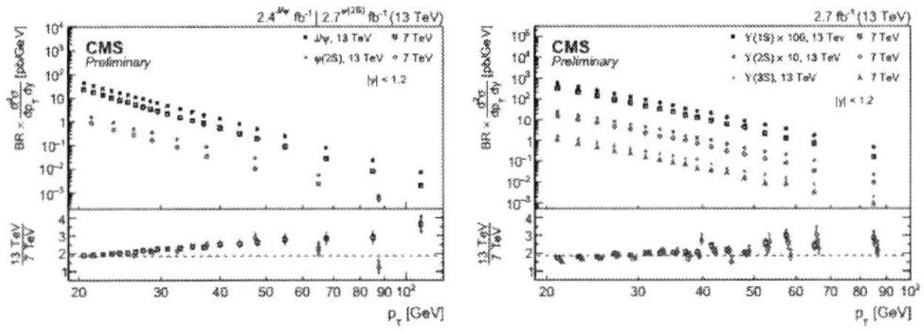

Figure 7. Comparison of the double-differential cross-sections vs. p_T for 13 TeV and 7 TeV: J/ψ and $\psi(2S)$ (left), $\Upsilon(1S)$, $\Upsilon(2S)$ and $\Upsilon(3S)$ (right). The lower pads display the cross section ratios.

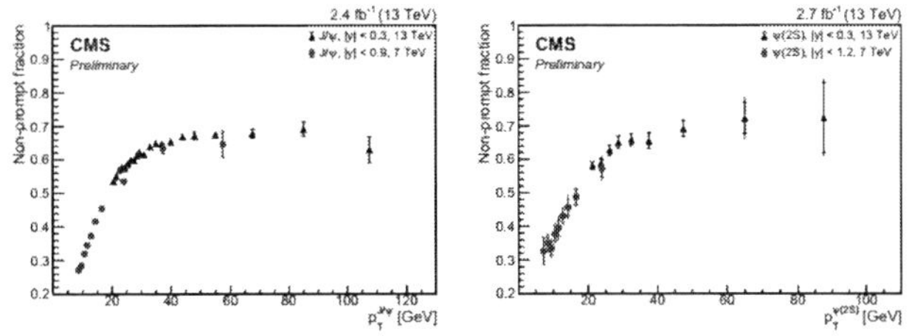

Figure 8. Comparison of the non-prompt fraction of J/ψ (left) and $\psi(2S)$ (right) as a function of p_T for 13 TeV and 7 TeV.

References

[1] CMS Collaboration, J. Inst. **03**, S08004 (2008)
[2] S.P. Baranov *et al.*, Phys. Lett. B **705**, 116 (2011)
[3] C.H. Kom *et al.*, Phys. Rev. Lett. **107**, 082002 (2011)
[4] P. Ko *et al.*, J. High Energy Phys. **01**, 070 (2011)
[5] J. Campbell *et al.*, Phys. Rev. Lett. **98**, 252002 (2007)
[6] A.V. Berezhnoy *et al.*, Phys. Rev. D **86**, 034004 (2012)
[7] CMS Collaboration, CMS-PAS-BPH-14-008 (2016)
[8] M. Cacciari *et al.*, J. High Energy Phys. **05**, 007 (1998)
[9] B.A. Kniehl *et al.*, Phys. Rev. D **77**, 014011 (2008)
[10] CMS Collaboration, CMS-PAS-BPH-15-004 (2016)
[11] CMS Collaboration, Phys. Rev. Lett. **106**, 112001 (2011)
[12] M. Cacciari *et al.*, J. High Energy Phys. **03**, 006 (2001)
[13] M. Cacciari *et al.*, J. High Energy Phys. **10**, 137 (2012)
[14] M. Cacciari *et al.*, arXiv:1507.06197 (2015)
[15] G.T. Bodwin *et al.*, Phys. Rev. D **51**, 1125 (1995)
[16] G.T. Bodwin *et al.*, Phys. Rev. D **55**, 5853 (1997)
[17] P.Cho and A.K.Leibovich, Phys. Rev. D **53**, 150 (1996)
[18] P.Cho and A.K.Leibovich, Phys. Rev. D **53**, 6203 (1996)
[19] P. Faccioli *et al.*, Phys. Lett. B **736**, 98 (2014)
[20] G.T. Bodwin *et al.*, Phys. Rev. Lett. **113**, 022001 (2014)
[21] CMS Collaboration, CMS-PAS-BPH-15-004 (2016)
[22] CMS Collaboration, Eur. Phys. J. C **71**, 1575 (2011)
[23] CMS Collaboration, Phys. Rev. Lett. **114**, 191802 (2015)
[24] CMS Collaboration, Phys. Lett. B **749**, 14 (2015)
[25] CMS Collaboration, J. High Energy Phys. **02**, 011 (2012)

EPJ Web of Conferences **130**, 01030 (2016) DOI: 10.1051/epjconf/201613001030
MESON 2016

MESON2016 – Concluding Remarks

Avraham Gal[1,*]

[1] *Racah Institute of Physics, The Hebrew University, Jerusalem 91904, Israel*

Abstract. Several topics presented and discussed at MESON2016 are highlighted, including pentaquarks, dibaryons and meson-nuclear bound states.

1 Introduction

The scope of topics presented and discussed in MESON2016 is too broad to be covered in one concluding talk. I therefore selected a few central topics discussed in MESON2016 where my personal involvement helped making some meaningful remarks. These topics are Pentaquarks, Dibaryons, and Meson-Nuclear Bound States. I apologize to the many speakers whose presentations were not mentioned in this concluding talk.

2 Exotics: remarks on Pentaquarks

Regarding *pentaquarks*, it is appropriate perhaps to note that the $\mathcal{S} = -1$ $\Lambda(1405)$ resonance, defying a three-quark classification, was predicted in 1959 by Dalitz and Tuan as a $\bar{K}N$ quasibound state [1] five years *before* the term 'quark' was transformed by Gell-Mann from Fiction to Physics. A recent LQCD calculation confirms its $\bar{K}N$ hadronic structure [2] as opposed to a tightly bound genuine pentaquark. Indeed, the $\Lambda(1405)$ emerges naturally below the K^-p threshold in chiral EFT hadronic approaches [3], although as shown in Cieplý's talk [4] the *subthreshold* $\bar{K}N$ scattering amplitudes exhibit appreciable model dependence, with consequences for K^-pp quasibound-state searches.

A $\mathcal{S} = +1$ $\Theta^+(1530)$ pentaquark was claimed more than 10 years ago. Recent dedicated experimental searches have failed to confirm it, placing instead upper limits on its coupling to the KN channel [5]. It was argued that the Θ^+ might be formed copiously in nuclei by absorption on *two* nucleons, e.g. $K^+d \to \Theta^+p$ [6, 7] thereby resolving a long-standing puzzle, discussed in Friedman's talk [8], regarding the size and A dependence of K^+ nuclear cross sections at low energies.

The recent LHCb discovery of hidden-charm structures [9] has led to several serious attempts to interpret these in terms of pentaquark(s). As argued in Karliner's talk the relatively small width of order 40 MeV for $P_c(4550)$ supports a $\Sigma_c\bar{D}^*$ hadronic molecule structure of two quark clusters, rather than a tightly bound pentaquark; see Fig. 1 on next page. This $\Sigma_c\bar{D}^*$ hadronic molecule is apparently the lightest of several other predicted doubly-heavy hadronic molecules [10]. It was emphasized that this calls for a new rich meson-meson, meson-baryon and baryon-baryon heavy-flavor QCD spectroscopy. Other speakers too discussed various aspects of heavy-flavor Exotics, demonstrating that no clear consensus has yet been reached on this topic.

* e-mail: avragal@savion.huji.ac.il

DOI: 10.1051/epjconf/201613001030

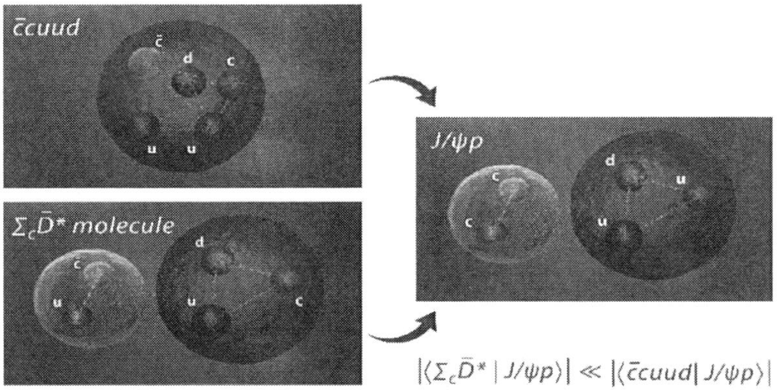

Figure 1. Left: two alternative visual descriptions of the LHCb hidden-charm pentaquark $P_c(4550)$ as a tightly bound $\bar{c}cuud$ pentaquark or as a $\Sigma_c\bar{D}^*$ hadronic molecule. Right: $P_c(4550) \to J/\psi + p$ decay channel which for a $\bar{c}cuud$ pentaquark implies a considerably larger width than reported. Figure adapted from Karliner's talk.

3 Exotics: remarks on Dibaryons

3.1 Nonstrange dibaryons

The only dibaryon for which good experimental evidence exists to date is the nonstrange $I = 0$ $J^P = 3^+$ $\mathcal{D}_{03}(2380)$, peaking ≈85 MeV below the $\Delta\Delta$ threshold. The WASA-at-COSY experiments that established it, see Fig. 2, were ranked in Wilkin's obituary of COSY at MESON2016 a top no. 2 in COSY's impact list. The small width of $\mathcal{D}_{03}(2380)$, $\Gamma \approx 70$ MeV, less than even a single Δ width, was shown (on p. 479 in Ref. [14]) to follow from phase-space and quantum-statistics arguments.

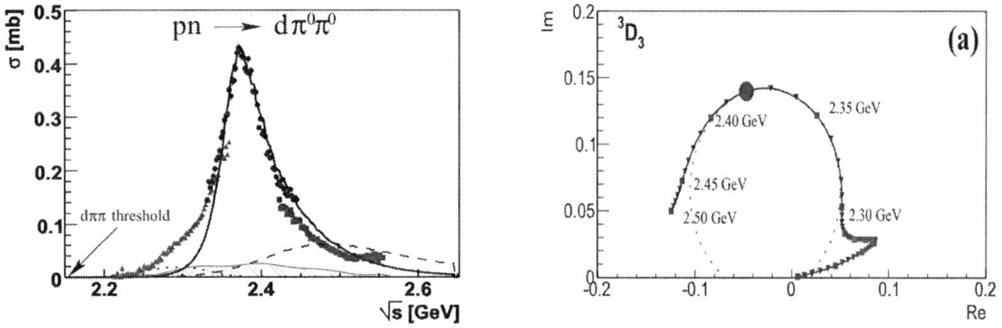

Figure 2. Evidence for the $\mathcal{D}_{03}(2380)$ dibaryon from WASA-at-COSY. Left: from $p + d \to d\pi^0\pi^0 + p_s$ [11]. Right: from the Argand diagram of the $^3D_3$ partial wave in np scattering [12], with full account of the recent measurement of the np analyzing power [13].

The large binding energy of $\mathcal{D}_{03}(2380)$ with respect to $\Delta\Delta$, exceeding by far the scale of nucleon separation energies in nuclei, does not mean it is a deeply bound dibaryon if one recalls the existence of a lower two-body channel, $\pi\mathcal{D}_{12}(2150)$, relative to which $\mathcal{D}_{03}(2380)$ resonates. Here, $\mathcal{D}_{12}(2150)$ stands for a near-threshold $N\Delta$ $I = 1$ $J^P = 2^+$ πNN quasibound state. Both $\mathcal{D}_{12}$ and $\mathcal{D}_{03}$, together with their $I \leftrightarrow S$ twins $\mathcal{D}_{21}$ and $\mathcal{D}_{30}$, were proposed by Dyson and Xuong [15] based on symmetry arguments, and have been considered subsequently in numerous quark-based works; and recently also in terms of 'meson assisted dibaryons' [14] in the hadronic basis [16, 17].

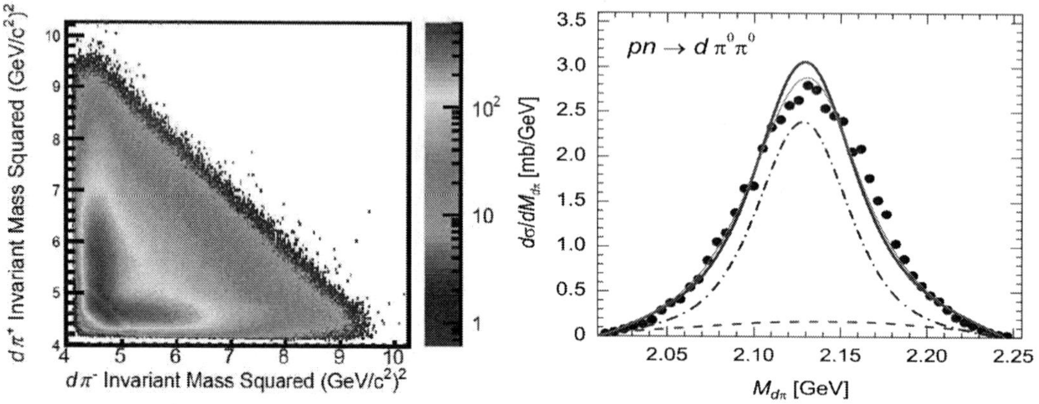

Figure 3. Left: $\mathcal{D}_{12}(2150)$ $N\Delta$ dibaryon signal in the Dalitz plot of $M^2_{d\pi^+}$ vs. $M^2_{d\pi^-}$ from a preliminary report on $\gamma d \to d\pi^+\pi^-$ measurements in the g13 experiment (CLAS Collaboration) at JLab [18]. Right: The $pn \to d\pi^0\pi^0$ WASA-at-COSY $M_{d\pi}$ distribution [11] and, in solid lines, as calculated [19] for two input parametrizations of $\mathcal{D}_{12}(2150)$. The dot-dashed line gives the $\pi\mathcal{D}_{12}(2150)$ contribution to the two-body decay of $\mathcal{D}_{03}(2380)$, and the dashed line gives a scalar-isoscalar σ-meson emission contribution.

The relevance of the $\mathcal{D}_{12}(2150)$ $N\Delta$ dibaryon to the physics of the $\mathcal{D}_{03}(2380)$ $\Delta\Delta$ dibaryon is demonstrated in Fig. 3 by showing, on the left panel, a $d\pi^\pm$ invariant-mass correlation near the $N\Delta$ threshold as deduced from preliminary CLAS data on the $\gamma d \to d\pi^+\pi^-$ reaction [18] and, on the right panel, a $d\pi$ invariant-mass distribution peaking near the $N\Delta$ threshold as deduced from the WASA-at-COSY $pn \to d\pi^0\pi^0$ reaction by which the $\mathcal{D}_{03}(2380)$ was discovered [11]. These preliminary CLAS data for $\gamma d \to d\pi^+\pi^-$ suggest a subthreshold $\mathcal{D}_{12}(2150)$ dibaryon with mass 2115±10 MeV and width 125±25 MeV [18], consistently with past deductions. The peaking of the $d\pi$ invariant-mass distribution in the $pn \to d\pi^0\pi^0$ reaction essentially at this $\mathcal{D}_{12}(2150)$ mass value suggests that the two-body decay modes of $\mathcal{D}_{03}(2380)$ are almost saturated by the $\pi\mathcal{D}_{12}(2150)$ decay mode, as reflected in the calculation [19] depicted in the right panel.

The success of *hadronic* model calculations [16, 17] to reproduce such $\mathcal{D}_{IS}$ dibaryon signals is consistent with the failure of recent quark-based calculations [20] to find tightly bound *hexaquarks* by using realistic color-spin hyperfine and color confinement quark-quark interactions. An hexaquark with quantum numbers identical to those of $\mathcal{D}_{03}$ lies at least 150 MeV above the $\Delta\Delta$ threshold, and this gap gets larger for other hexaquark candidates; a similar conclusion also holds for Jaffe's $\mathcal{S} = -2$ $I(J^P) = 0(0^+)$ $\mathcal{H}$ hexaquark [21]. This means that the proper degrees of freedom in the case of nonstrange dibaryons are nucleons, pions and Δ baryons, and that physical thresholds and p-wave pions must be realistically incorporated in future considerations of such dibaryons.

3.2 Strange dibaryons

Following recent searches for a $\bar{K}NN$ $I(J^P)=\frac{1}{2}(0^-)$ quasibound state (loosely termed K^-pp) in Frascati [22, 23], SPring-8 [24] and GSI [25, 26], Iwasaki reported in MESON2016 on dibaryon candidates

DOI: 10.1051/epjconf/201613001030

from J-PARC Experiments E27 [27, 28] and E15 [29, 30], with binding energies given by

$$\text{deep}: \; B_{K^-pp}(\text{E27}) \approx 95 \text{ MeV}, \qquad \text{shallow}: \; B_{K^-pp}(\text{E15}) \approx 15 \text{ MeV}, \tag{1}$$

relative to the K^-pp threshold at 2370 MeV. To understand the possible origin of such radically different $\mathcal{S} = -1$ dibaryon candidates, it is instructive to look at the E27 $d(\pi^+, K^+)X$ small-angle missing-mass spectrum, Fig. 4(left), which indicates ≈22 MeV attractive shift of the $Y^*(1385+1405)$ unresolved quasi-free peak, consistently with the attraction calculated in the $I(J^P)=\frac{1}{2}(0^-)$ $\Lambda(1405)N$ s-wave channel that overlaps substantially with K^-pp [31]. Chirally motivated K^-pp calculations also suggest binding of order 20 MeV, as reviewed in Ref. [32], in rough agreement with the E15 $^3\text{He}(K^-, n)X$ near-threshold signal but not with the E27 deeply-bound signal shown in Fig. 4(right). The relatively shallow K^-pp binding persists in three-body calculations upon including the $\pi\Lambda N$ and $\pi\Sigma N$ lower-mass channels [33] which play only a secondary role in binding $\bar{K}$ mesons.

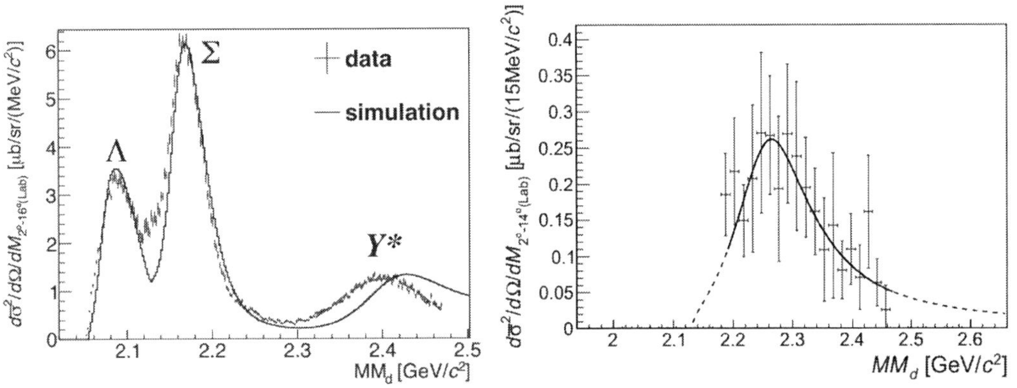

Figure 4. J-PARC E27 $d(\pi^+, K^+)X$ missing-mass spectra at $p_{\pi^+} = 1.69$ GeV/c. Left: small-angle K^+ inclusive quasi-free spectrum [27]. Right: $\Sigma^0 p$ decay branch of a pp coincidence spectrum [28].

The $\pi\Lambda N$–$\pi\Sigma N$ system, however, may benefit from sizable meson-baryon p-wave interactions, in terms of $\Delta(1232) \to \pi N$ and $\Sigma(1385) \to \pi\Lambda$–$\pi\Sigma$ strong-decay form factors, by aligning isospin and angular momentum to total $I(J^P)=\frac{3}{2}(2^+)$. Such a pion assisted dibaryon was studied in Ref. [34] by solving πYN coupled-channel Faddeev equations, thereby predicting a new $\mathcal{S} = -1$ dibaryon resonance $\mathcal{Y}_{\frac{3}{2}2^+}$ slightly below the $\pi\Sigma N$ threshold ($\sqrt{s_{\rm th}} \approx 2270$ MeV). Adding a $\bar{K}NN$ channel hardly matters, since its leading $^3S_1$ NN component is Pauli forbidden. The E27 deeply bound broad signal at $\sqrt{s} \sim 2275$ MeV shown in Fig. 4(right) may then correspond to the production of such $\mathcal{Y}^+_{\frac{3}{2}2^+}$ in $\pi^+ + d \to \mathcal{Y}^+ + K^+$, followed by its decay to $\Sigma^0 + p$ [35]. We note that the $\mathcal{S} = -1$ $\mathcal{Y}_{\frac{3}{2}2^+}(2275)$ dibaryon may have good overlap with $^5S_2$, $I = \frac{3}{2}$ $\Sigma(1385)N$ and $\Delta(1232)Y$ dibaryon configurations, the lowest threshold of which, that of $\Sigma(1385)N$, is only ∼50 MeV above the $\pi\Sigma N$ threshold.

Other possible search reactions that are isospin selective as far as the final $\mathcal{Y} \to \Sigma N$ decay is concerned are

$$\pi^\pm + d \to \mathcal{Y}^{++/-} + K^{0/+}, \qquad p + p \to \mathcal{Y}^{++} + K^0, \tag{2}$$

in which the produced dibaryon $\mathcal{Y}$ decays to a ΣN final charge state which is uniquely $I = \frac{3}{2}$, viz. $\mathcal{Y}^{++/-} \to \Sigma^\pm + p(n)$. The pp reaction has been reported by the HADES Collaboration at GSI [36], finding no $\mathcal{Y}$ dibaryon signal. It is not clear whether the pp experiments were able to resolve as small cross sections as 0.1 μb or less that are expected in production of $\mathcal{Y}$ dibaryon candidates [26].

3.3 Charmed dibaryons

Charmed, $C = +1$ dibaryons have also been predicted: (i) a $I(J^P)=\frac{1}{2}(0^-)$ dynamically generated DNN quasibound state at 3.5 GeV [37] reminiscent of the $\mathcal{S} = -1$ K^-pp; and (ii) a $I(J^P)=\frac{3}{2}(2^+)$ $\pi\Lambda_c N$ quasibound state below 3.4 GeV [38], analogous to the $\mathcal{S} = -1$ pion assisted dibaryon $\mathcal{Y}_{\frac{3}{2}2^+}$. The prediction of this charmed pion assisted dibaryon $C_{\frac{3}{2}2^+}(3370)$ is robust, since it depends little on the unknown $^3S_1$ $\Lambda_c N$ interaction. The $C_{\frac{3}{2}2^+}(3370)$ is likely to be the *lowest lying* charmed dibaryon. It could be searched with proton beams at GSI, and with pion beams in the high-momentum hadron beam line extension approved at J-PARC, viz.

$$(p+p) \text{ or } (\pi^+ + d) \to C^{+++} + D^-, \qquad C^{+++} \to \Sigma_c^{++}(2455) + p. \tag{3}$$

4 Meson-nuclear bound states

No meson-nuclear bound states have been firmly established so far. For K^- mesons, extrapolating from kaonic atoms [39] it is widely accepted that broad quasibound states exist [40]. K^+ mesons, in contrast, experience repulson in dense (nuclear) matter. This is naively explained by a mean-field treatment of $K^- \equiv s\bar{u}$ and $K^+ \equiv \bar{s}u$ mesons, arguing that a nonstrange antiquark/quark induces attraction/repulsion in dense matter. In the charmed sector, one would then expect *attraction* for $D^+ \equiv c\bar{d}$ and repulsion for $D^- \equiv \bar{c}d$ mesons. This is not borne out in a recent QCD sum-rule calculation, showing in Fig. 5(left) a *repulsive* mass shift in dense matter, as a function of the assumed value of the πN σ term, for *both* $D^\pm$ mesons [41]. Fig. 5(right) shows that an attractive D^+ mass shift is possible in principle, but only for unrealistically high values of the heavy-quark mass m_h. This result has also been explained in Ref. [42] using a constituent quark picture.

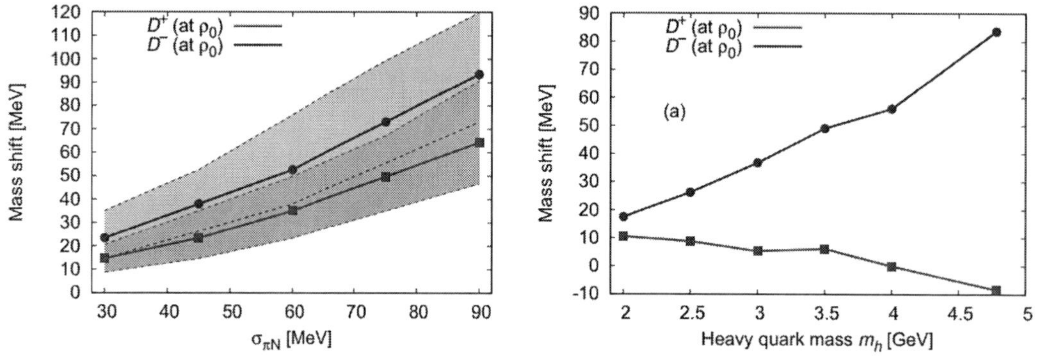

Figure 5. D meson mass shift in nuclear matter [41] vs. $\sigma_{\pi N}$ (left) and vs. the heavy quark mass m_h (right).

Turning to nonstrange and noncharmed meson-nuclear interactions discussed in MESON2016, Nanova reviewed recent ω and η' nuclear photoproduction experiments by the CBELSA/TAPS Collaboration which study the meson momentum dependence of the extracted meson-nucleus optical potential, suggesting that while the ω-nucleus potential is too absorptive to observe distinct quasibound states, the η'-nucleus potential is weakly absorptive [43] and sufficiently attractive [44] to motivate searches for η'-mesic nuclear states. Ongoing searches in ^{12}C(p,d) at GSI were discussed in Tanaka's talk. However, with $p_{\eta'}$ centered about ~1 GeV in the ELSA experiments, the optical potential derived for η' at rest depends on extrapolation from the lowest available value $p_{\eta'} \approx 275$ MeV/c down to

 DOI: 10.1051/epjconf/201613001030

threshold, where COSY-11 data on near-threshold meson production in pp collisions indicate a rather strong ηp attraction that is likely to support η-mesic nuclear states and a much weaker $\eta' p$ interaction [45], for the real part of which only a limit consistent with zero can be placed [46]. Citing from Wilkin's talk: "I would not put any money on bound η' in nuclei!". Within a QCD-inspired $\eta - \eta'$ mixing model [47], η'-nuclear attraction of roughly -40 MeV at saturation density as derived in the ELSA experiment [44] implies about -90 MeV for η attraction in nuclear matter, commensurate with the attraction expected in the ηN interaction model GW considered below. For these and for other reasons specified below, the following discussion is limited to η-nuclear quasibound states.

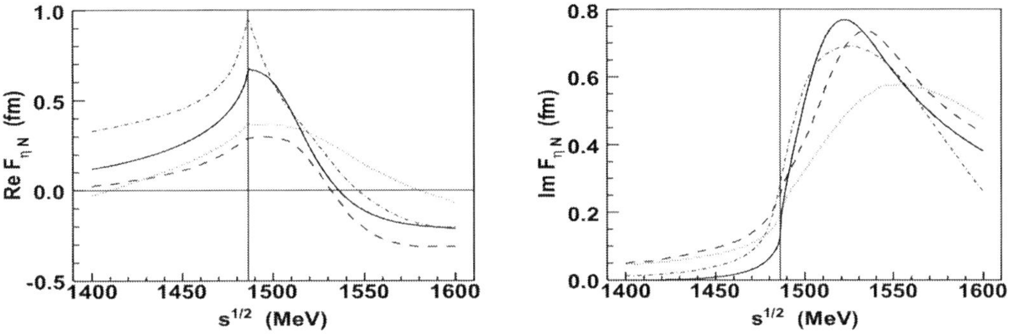

Figure 6. Real (left) and imaginary (right) parts of the ηN s-wave cm scattering amplitude $F_{\eta N}(\sqrt{s})$, compiled in Ref. [48] from several $N^*(1535)$ resonance coupled-channel models, in decreasing order of Re $a_{\eta N}$: GW [49], CS [50], MBM [51] and IOV [52]. The ηN threshold is marked by a thin vertical line.

The ηN near-threshold dynamics is governed by the $N^*(1535)$ resonance, introducing thereby appreciable model dependence in coupled-channel calculations of the s-wave scattering amplitude $F_{\eta N}(\sqrt{s})$, as seen in Fig. 6. Owing to the nearby $N^*(1535)$, both Re $F_{\eta N}$ and Im $F_{\eta N}$ decrease in all models steadily below threshold, which is where bound states are calculated. This decrease persists also in in-medium extensions $F_{\eta N}(\sqrt{s}, \rho)$ of the ηN scattering amplitude, suggesting that η-nuclear states are narrow. I know of no similar mechanism that would suggest as narrow η'-nuclear states.

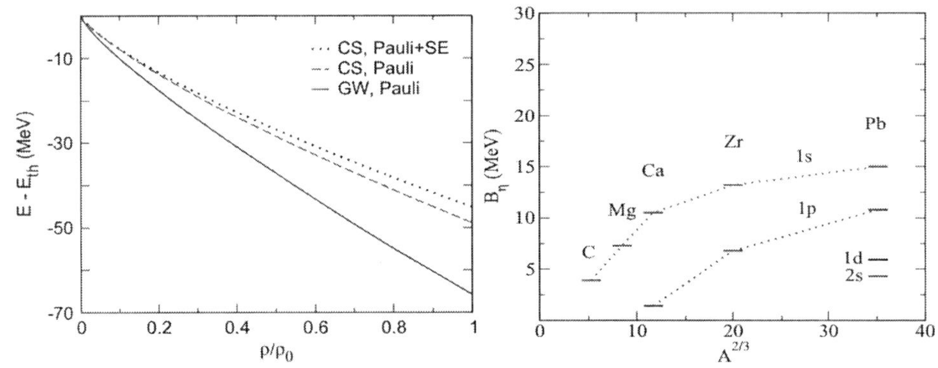

Figure 7. Left: Subthreshold energies probed in the $1s_\eta$-^{40}Ca bound state as a function of nuclear density, calculated self-consistently for in-medium ηN scattering amplitudes in models GW and CS used in Ref. [53]. Right: η-nuclear spectra [54] calculated self-consistently using the in-medium CS model NLO30$_\eta$ [50].

The subthreshold energy $\sqrt{s}$ and the nuclear density ρ, both serving in bound state calculations as arguments of the in-medium meson-nucleon scattering amplitude $F_{mN}(\sqrt{s},\rho)$, are tightly correlated as demonstrated in Fig. 7(left) within a particular η-nucleus calculation. This correlation imposes a self-consistent procedure in bound state calulations [48], as discussed here by Mareš.

A chart of η-nuclear bound states calculated self-consistently in the CS model is shown in Fig. 7(right). Since Im $F_{\eta N}(\sqrt{s})$ is particularly small in model CS below threshold, see Fig. 6(right), the resulting η-nuclear widths are just a few MeV, and only somewhat larger in model GW. Bound states should definitely exist in ^{12}C and beyond, and beginning in ^{6}Li in model GW which according to Fig. 6(left) provides the strongest ηN attraction among the four models exhibited, Few-body calculations have also been reported recently using the ηN interaction models GW and CS. No bound state was found for ηd and for $\eta\,^{3}$He [55]. Calculations are underway for $\eta\,^{4}$He.

5 Summary and outlook

Several topics discussed in MESON2016 were picked up selectively in these Concluding Remarks, the common grounds of which is the rich spectroscopic variety that remains largely to be uncovered in hadronic systems. The impression gained at this Conference is that no consensus has been reached on the hidden-charm structures observed recently for mesons and for baryons in the energy range 2–5 GeV. In the absence of compelling arguments, or calculations classifying these in terms of genuine tetraquarks and pentaquarks, the only logical conclusion is that of hadronic-molecule underlying structure. For dibaryons too, highlighting recent experimental results from COSY (nonstrange) and J-PARC (strange), the evidence points to hadronic structure. Finally, we focused attention to the possible existence of observable η-nuclear quasibound states, which could be explored at GSI using (p,d) and in J-PARC using the (π^{+},p) reaction on nuclear targets.

I would like to thank the Organizers of MESON2016 for trusting me in this unthankful job.

References

[1] R. H. Dalitz, S. F. Tuan, Phys. Rev. Lett. **2**, 425 (1959)
[2] J. M. M. Hall, W. Kamleh, D. B. Leinweber, B. J. Menadue, B. J. Owen, A. W. Thomas, R. D. Young, Phys. Rev. Lett. **114**, 132002 (2015)
[3] T. Hyodo, Nucl. Phys. A **914**, 260 (2013)
[4] A. Cieplý, M. Mai, U.-G. Meißner, J. Smejkal, Nucl. Phys. A **954**, 17 (2016)
[5] M. Moritsu et al. (J-PARC E19 Collaboration), Phys. Rev. C **90**, 035205 (2014)
[6] A. Gal, E. Friedman, Phys. Rev. Lett. **94**, 072301 (2005)
[7] A. Gal, E. Friedman, Phys. Rev. C **73**, 015208 (2006)
[8] E. Friedman, Nucl. Phys. A **954**, 114 (2016)
[9] R. Aaij et al. (LHCb Collaboration), Phys. Rev. Lett. **115**, 072001 (2015)
[10] M. Karliner, J. L. Rosner, Phys. Rev. Lett. **115**, 122001 (2015)
[11] P. Adlarson et al. (WASA-at-COSY Collaboration), Phys. Rev. Lett. **106**, 242302 (2011)
[12] P. Adlarson et al. (WASA-at-COSY Collaboration, SAID Data Analysis Center), Phys. Rev. C **90**, 035204 (2014); see also R. L. Workman, W. J. Briscoe, I. I. Strakovsky, Phys. Rev. C **93**, 045201 (2016)
[13] P. Adlarson et al. (WASA-at-COSY Collaboration, SAID Data Analysis Center), Phys. Rev. Lett. **112**, 2012301 (2014)

[14] A. Gal, Acta Phys. Pol. B **47**, 471 (2016)
[15] F. J. Dyson, N.-H. Xuong, Phys. Rev. Lett. **13**, 815 (1964)
[16] A. Gal, H. Garcilazo, Phys. Rev. Lett. **111**, 172301 (2013)
[17] A. Gal, H. Garcilazo, Nucl. Phys. A **928**, 73 (2014)
[18] R. Schumacher (JLab CLAS Collaboration), private communication (APS 04/2015 meeting)
[19] M. N. Platonova, V. I. Kukulin, Nucl. Phys. A **946**, 117 (2016)
[20] W. Park, A. Park, S. H. Lee, Phys. Rev. D **92**, 014037 (2015)
[21] W. Park, A. Park, S. H. Lee, Phys. Rev. D **93**, 074007 (2016)
[22] M. Agnello et al. (FINUDA Collaboration), Nucl. Phys. A **914**, 310 (2013)
[23] O. Vázquez Doce et al. (data from KLOE Collaboration), Phys. Lett. B **758**, 134 (2016)
[24] A. O. Tokiyasu et al. (LEPS Collaboration), Phys. Lett. B **728**, 616 (2014)
[25] G. Agakishiev et al. (HADES Collaboration), Phys. Lett. B **742**, 242 (2015)
[26] E. Epple, L. Fabbietti, Phys. Rev. C **92**, 044002 (2015)
[27] Y. Ichikawa et al. (J-PARC E27 Experiment), Prog. Theor. Exp. Phys. **2014**, 101D03 (2014)
[28] Y. Ichikawa et al. (J-PARC E27 Experiment), Prog. Theor. Exp. Phys. **2015**, 021D01 (2015)
[29] T. Hashimoto et al. (J-PARC E15 Experiment), Prog. Theor. Exp. Phys. **2015**, 061D01 (2015)
[30] Y. Sada et al. (J-PARC E15 Experiment), Prog. Theor. Exp. Phys. **2016**, 051D01 (2016)
[31] T. Uchino, T. Hyodo, M. Oka, Nucl. Phys. A **868-869**, 53 (2011)
[32] A. Gal, Nucl. Phys. A **914**, 270 (2013)
[33] J. Révai, N. V. Shevchenko, Phys. Rev. C **90**, 034004 (2014)
[34] H. Garcilazo, A. Gal, Nucl. Phys. A **897**, 167 (2013)
[35] T. Nagae, Nucl. Phys. A **954**, 94 (2016)
[36] J. C. Berger-Chen, L. Fabbietti, doi:10.3204/DESY-PROC-2014-04/101 (arXiv:1410.8004)
[37] M. Bayar, C. W. Xiao, T. Hyodo, A. Doté, M. Oka, E. Oset, Phys. Rev. C **86**, 044004 (2012)
[38] A. Gal, H. Garcilazo, A. Valcarce, T. Fernández-Caramés, Phys. Rev. D **90**, 014019 (2014)
[39] E. Friedman, A. Gal, Nucl. Phys. A **881**, 150 (2012), ibid. **899**, 60 (2013)
[40] D. Gazda, J. Mareš, Nucl. Phys. A **881**, 159 (2012)
[41] K. Suzuki, P. Gubler, M. Oka, Phys. Rev. C **93**, 045209 (2016)
[42] A. Park, P. Gubler, M. Harada, S. H. Lee, C. Nonaka, W. Park, Phys. Rev. D **93**, 054035 (2016)
[43] S. Friedrich et al. (CBELSA/TAPS Collaboration), Eur. Phys. J. A **52**, 297 (2016)
[44] M. Nanova et al. (CBELSA/TAPS Collaboration), Phys. Rev. C **94**, 025205 (2016)
[45] P. Moskal et al., Phys. Lett. B **482**, 356 (2000), Phys. Rev. C **69**, 025203 (2004)
[46] E. Czerwiński et al., Phys. Rev. Lett. **113**, 062004 (2014)
[47] S. D. Bass, A. W. Thomas, Phys. Lett. B **634**, 368 (2006), Acta Phys. Pol. B **45**, 627 (2014)
[48] A. Gal, E. Friedman, N. Barnea, A. Cieplý, J. Mareš, D. Gazda, Acta Phys. Pol. B **45**, 673 (2014)
[49] A. M. Green, S. Wycech, Phys. Rev. C **71**, 014001 (2005)
[50] A. Cieplý, J. Smejkal, Nucl. Phys. A **919**, 46 (2013)
[51] M. Mai, P. C. Bruns, U.-G. Meißner, Phys. Rev. D **86**, 094033 (2013)
[52] T. Inoue, E. Oset, M. J. Vicente Vacas, Phys. Rev. C **65**, 035204 (2002)
[53] E. Friedman, A. Gal, J. Mareš, Phys. Lett. B **725**, 334 (2013)
[54] A. Cieplý, E. Friedman, A. Gal, J. Mareš, Nucl. Phys. A **925**, 126 (2014)
[55] N. Barnea, E. Friedman, A. Gal, Phys. Lett. B **747**, 345 (2015)

EPJ Web of Conferences **130**, 02001 (2016) DOI: 10.1051/epjconf/201613002001
MESON 2016

Theoretical approaches to low energy $\bar{K}N$ interactions

Aleš Cieplý[1,⋆] and Maxim Mai[2]

[1] *Nuclear Physics Institute, 250 68 Řež, Czechia*
[2] *The George Washington University, 725 21st St. NW, Washington, DC 20052, USA*

Abstract. We provide a direct comparison of modern theoretical approaches based on the SU(3) chiral dynamics and describing the low energy $\bar{K}N$ data. The model predictions for the $\bar{K}N$ amplitudes and pole content of the models are discussed.

1 Introduction

In our contribution we review the current status of low energy $\bar{K}N$ interactions and concentrate on comparison of the available theoretical approaches derived from the effective SU(3) chiral Lagrangian that describes the interaction of the pseudoscalar meson octet with the ground state baryon octet. In the $S = -1$ sector the involved meson-baryon coupled channels are $\pi\Lambda$, $\pi\Sigma$, $\bar{K}N$, $\eta\Lambda$, $\eta\Sigma$ and $K\Xi$ with threshold energies from about 1250 to 1810 MeV. Since the $\Lambda(1405)$ resonance lies closely below the $\bar{K}N$ threshold the chiral perturbation series does not converge in its vicinity and coupled-channel re-summation techniques are standardly employed to sum the major part of the chiral expansion to obtain the scattering amplitude.

In our recent work [1] we performed a direct comparison of the theoretical approaches that include NLO corrections to the leading order in the chiral expansion and fix the free model parameters, the low energy constants, to reproduce the experimental data on K^-p scattering and reactions including the recent precise measurement of kaonic hydrogen characteristics (the shift and width of the $1s$ level due to strong interaction) by the SIDDHARTA collaboration [2]. The discussed models comprise of

- the Kyoto-Munich [3] and Murcia [4] approaches, which rely on the re-summation of the S-wave projected chiral potential. Both are conceptually identical but differ in their treatment of the experimental data and fitting procedures. In our analysis, we have included the NLO models $\mathrm{KM}_{\mathrm{NLO}}$ from Ref. [3] and the models M_{I} and M_{II} from Ref. [4].
- the Bonn approach [5] that does not rely on partial wave projection of the interaction kernel when solving the Bethe-Salpeter equation. In Ref. [5], two solutions (which we denote as B_2 and B_4 here) of the global fits to the K^-p experimental data were found compatible with the photoproduction data measured by the CLAS collaboration [6].
- the Prague approach [7] that differs from the other considered approaches by relying on effective separable meson-baryon potentials with off-shell form factors that also regularize the intermediate state Green function, which is equivalent to dimensional regularization used in the other approaches. In our analysis we use the NLO model $\mathrm{P}_{\mathrm{NLO}}$, originally denoted as NLO30 in Ref. [7].

⋆e-mail: cieply@ujf.cas.cz

We refer the reader to the original papers [3], [4], [5] and [7] for the specific details of the considered approaches. Without dwelling upon any technicalities we proceed with a presentation of our findings in the next section and conclude the paper with a brief summary.

2 Results and discussion

The considered models represent the current state of theory on low energy meson-baryon interactions in the $S = -1$ sector and describe the K^-p reactions data about equally well. This is demonstrated in the left panel of Fig. 1 which shows the theoretical predictions for the $1s$ level characteristics of the kaonic hydrogen, the energy shift $\Delta E(1s)$ and the absorption width $\Gamma(1s)$, both caused by the strong interaction. The rectangular areas drawn in the figure visualize the experimental progress with the rectangular boxes covering areas within one standard deviation of the experimental data taken from the KEK [8], DEAR [9] and SIDDHARTA [2] measurements. The theoretical approaches reproduce the most recent SIDDHARTA data quite well and are in very close agreement among each other. However, the same cannot be said concerning the positions of the poles assigned to the $\Lambda(1405)$ resonance and shown in the right panel of Fig. 1. All models based on the chiral SU(3) dynamics generate invariably two poles in the $\pi\Sigma$-$\bar{K}N$ coupled channels sector. The models agree on the real part of the complex energy for the pole that couples more strongly to the $\bar{K}N$ channel and is generated at a higher energy of about 1420 MeV. Though, the imaginary part of the pole energy is not established so well and the position of the second pole varies from one model to another, apparently not constrained much by the experimental data. It was already shown in Ref. [5] that the new CLAS data on $\pi\Sigma$ photoproduction off proton [6] provide additional constrains on the pole positions. On the other hand, the theoretical models still find it difficult to explain the peaks in the $\pi\Sigma$ mass spectra observed in the pp collisions by the HADES experiment [10].

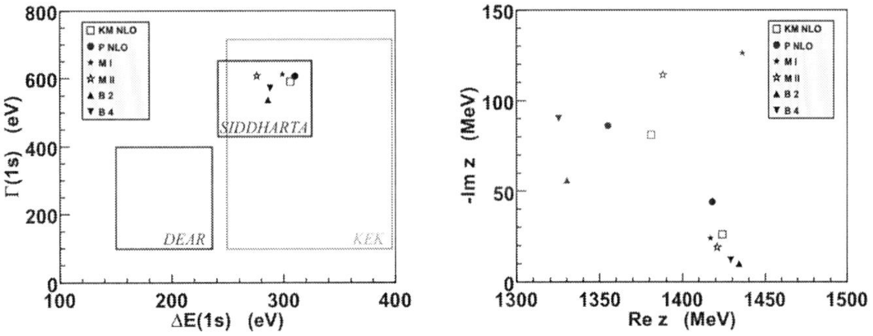

Figure 1. Kaonic hydrogen characteristics (left panel) and positions of the poles assigned to $\Lambda(1405)$ (right panel) as generated by various theoretical approaches.

In Fig. 2 we also demonstrate that the considered approaches lead to very different predictions for the K^-p amplitude extrapolated to sub-threshold energies as well as for the K^-n amplitude. The theoretical ambiguities observed below the $\bar{K}N$ threshold are much larger then those standardly indicated by uncertainty bounds derived from variations of the K^-p scattering length within constraints enforced by the kaonic hydrogen data, see e.g. Ref. [3].

Finally, in Ref. [1] we have also analyzed the origin of the poles of the scattering T-matrix generated by the theoretical models. There, we followed the pole movements to the so-called zero coupling

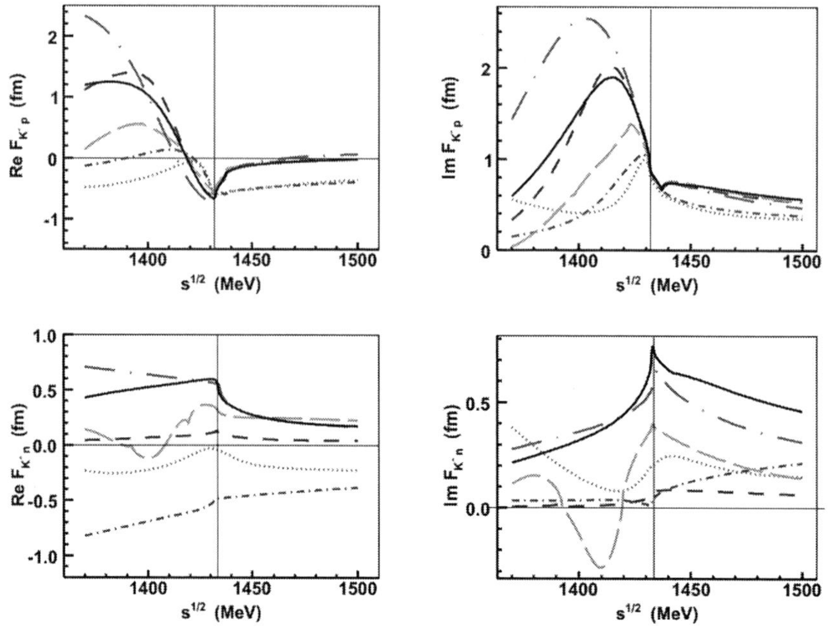

Figure 2. The K^-p (top panels) and K^-n (bottom panels) elastic scattering amplitudes generated by the NLO approaches considered in our work. The various lines refer to the models: B_2 (dotted, purple), B_4 (dot-dashed, red), M_I (dashed, blue), M_{II} (long-dashed, green), P_{NLO} (dot-long-dashed, violet), KM_{NLO} (continuous, black).

limit (ZCL), in which the inter-channel couplings are switched off. Our findings are reviewed in the Table 1 that shows the channels in which a pole assigned to a given resonance persists when the ZCL is reached. As there are two poles assigned to the $\Lambda(1405)$ we present them separately with the indexes 1 and 2. In the complex energy plane the $\Lambda_1(1405)$ pole is usually found at lower energy and further from the real axis than the $\Lambda_2(1405)$ pole. All models have the $\Lambda_1(1405)$ pole in the $\pi\Sigma$ channel when the inter-channel interactions are switched off. The $\Lambda_2(1405)$ pole couples most strongly to the $\bar{K}N$ channel, so it came as a surprise that the pole origin can be traced to the $\eta\Lambda$ channel in the ZCL for the B_2 and M_{II} models. Thus, the other models that have the ZCL pole in the $\bar{K}N$ channel should be preferred if one anticipates a simplified picture of a $\bar{K}N$ bound state submerged in the $\pi\Sigma$ continuum [11]. We have also hinted in Ref. [1] at quite large NLO couplings occurring in the Bonn and Murcia models and being most likely responsible for an appearance of the $\eta\Lambda$ bound state in the ZCL.

In the isoscalar sector the models can also account for the $\Lambda(1670)$ resonance that emerges from the $K\Xi$ pole found in the ZCL. We have argued in Ref. [1] that an appearance of such pole is related to a particular condition imposed on a subtraction constant (or an inverse range in case of the Prague approach). If the condition is not met, the pole is missing as happens for the KM_{NLO} and B_2 models. One should note, however, that with an exception of the Murcia approach the other approaches did not aim at describing the experimental data in the $\Lambda(1670)$ energy region, so it is not surprising that the pole is either completely missing or not at an appropriate position in those models.

Similarly, in the isovector sector the models can provide a pole which can be related to the $\Sigma(1750)$ resonance and the origin of this pole can be traced to the $K\Xi$ virtual (or bound) state in the ZCL. Several of the discussed models also predict an isovector $\bar{K}N$ pole located below the $\bar{K}N$ threshold at the Riemann sheet which is physical in the $\pi\Sigma$ and unphysical in the $\bar{K}N$ channel (it would be the third

Table 1. The origins (channels) of the poles generated by the considered models in which the poles are found when inter-channel couplings are switched off.

Resonances	P_{NLO}	KM_{NLO}	M_I	M_{II}	B_2	B_4
$\Lambda_1(1405)$	$\pi\Sigma$	$\pi\Sigma$	$\pi\Sigma$	$\pi\Sigma$	$\pi\Sigma$	$\pi\Sigma$
$\Lambda_2(1405)$	$\bar{K}N$	$\bar{K}N$	$\bar{K}N$	$\eta\Lambda$	$\eta\Lambda$	$\bar{K}N$
$\Lambda(1670)$	$K\Xi$	—	$K\Xi$	$K\Xi$	—	$K\Xi$
$\bar{K}N(I=1)$	$\bar{K}N$	$\eta\Sigma$	$\bar{K}N$	$\bar{K}N$	—	—
$\Sigma(1750)$	$K\Xi$	—	$K\Xi$	$K\Xi$	—	$K\Xi$

Riemann sheet if only these two channels were coupled). This pole emerges from an isovector $\bar{K}N$ virtual state generated in the ZCL by the Prague and Murcia models, though the Kyoto-Munich model has it in the $\eta\Sigma$ channel. We note that an existence of this pole was already witnessed in Refs. [12], [13] and [14]. It is understood that it relates to the cusp structure in the energy dependence of the elastic K^-n amplitude obtained for both, the P_{NLO} and the KM_{NLO} models as seen in Fig. 2.

3 Summary

In the present work different versions of the modern chiral unitary approaches were compared directly for the first time. Our main observations are as follows:

- We have demonstrated that the available theoretical models lead to very different predictions for the elastic K^-p and K^-n amplitudes at sub-threshold energies.
- The tracking of the poles to the ZCL provides us with new insights related to the appearance of poles in a given approach. The procedure also reveals different concepts of forming the $\Lambda(1405)$.
- Several models predict an existence of an isovector pole close to the $\bar{K}N$ threshold.

The authors acknowledge a collaboration with U.-G. Meißner and J. Smejkal who contributed to Ref. [1] the current presentation is based on.

References

[1] A. Cieplý, M. Mai, U.-G. Meißner and J. Smejkal, Nucl. Phys. A **954**, 17 (2016)
[2] M. Bazzi *et al.* [SIDDHARTA Collaboration], Phys. Lett. B **704**, 113 (2011)
[3] Y. Ikeda, T. Hyodo and W. Weise, Nucl. Phys. A **881**, 98 (2012)
[4] Z. H. Guo and J. A. Oller, Phys. Rev. C **87**, 035202 (2013)
[5] M. Mai and U.-G. Meißner, Eur. Phys. J. A **51**, 30 (2015)
[6] K. Moriya *et al.* [CLAS Collaboration], Phys. Rev. C **87**, 035206 (2013)
[7] A. Cieplý and J. Smejkal, Nucl. Phys. A **881**, 115 (2012)
[8] M. Iwasaki *et al.*, Phys. Rev. Lett. **78**, 3067 (1997)
[9] G. Beer *et al.* [DEAR Collaboration], Phys. Rev. Lett. **94**, 212302 (2005)
[10] G. Agakishiev *et al.* [HADES Collaboration], Phys. Rev. C **87**, 025201 (2013)
[11] T. Hyodo and W. Weise, Phys. Rev. C **77**, 035204 (2008)
[12] J. A. Oller and U.-G. Meißner, Phys. Lett. B **500**, 263 (2001)
[13] D. Jido, J. A. Oller, E. Oset, A. Ramos and U.-G. Meißner, Nucl. Phys. A **725**, 181–200 (2003)
[14] A. Cieplý, E. Friedman, A. Gal, D. Gazda and J. Mareš, Phys. Rev. C **84**, 045206 (2011)

 DOI: 10.1051/epjconf/201613002002

The $B_c \to J/\Psi KD$ weak decay and its relation with the $D^*_{s0}(2317)$ resonance

P. Fernandez-Soler[1,⋆], Zhi-Feng Sun[1], M. Bayar[2], and E. Oset[1]

[1] *Departamento de Física Teórica and IFIC (Centro Mixto Universidad de Valencia-CSIC), Valencia, Spain*
[2] *Department of Physics of Kocaeli University, Izmit, Turkey*

Abstract. We study the influence of the $D^*_{s0}(2317)^+$ resonance in the decay $B_c \to J/\psi DK$. In this process, we consider that the B_c meson decays first into J/ψ and the quark pair $c\bar{s}$, and then the quark pair hadronizes into DK or $D_s\eta$ components. The final state interaction, generating the $D^*_{s0}(2317)^+$ resonance, is described by the chiral heavy meson unitary approach. With the parameters chosen in order to match the pole position of the $D^*_{s0}(2317)^+$, we obtain the DK invariant mass distribution of the decay $B_c \to J/\psi DK$, and also the rate for $B_c \to J/\psi D^*_{s0}(2317)$. We predict the ratio of these two magnitudes. This decay mechanism has not been measured yet, and it would provide insight about the nature of this resonance.

1 Introduction

In this talk we show the results given in Ref. [1] where the role of the unmeasured decay $B_c \to J/\Psi KD$ in the molecular nature of the $D^*_{s0}(2317)^+$ resonance was discussed. Further references and details can be found in this work. The scalar resonance $D^*_{s0}(2317)^+$ was first observed by the BABAR Collaboration [2] and later it was confirmed by several collaborations. Before the BABAR experiment there were some predictions about a similar state. These predictions involved different quark models and potentials, and also some QCD calculations, predicting a meson mass larger than the experimental value. After the BABAR experiment, many theoretical groups performed research on the $D^*_{s0}(2317)^+$ state, see for example Refs. [3, 4]. Since the mass of the $D^*_{s0}(2317)^+$ is close to the threshold of the DK system, being the difference of about 50 MeV, the molecular state interpretation seems natural. The $D^*_{s0}(2317)^+$ state was studied in the frame of molecular, KD mixing with $c\bar{s}$ state, four-quark state, and the mixture of two-meson and four-quark state. It is worth mentioning that the lattice QCD simulations also have provided some hints about this issue after the BABAR discovery. Results for the DK scattering length were extrapolated to physical pion masses with the help of unitarized chiral perturbation theory, and by means of the Weinberg compositeness condition [5] the amount of KD content in the $D^*_{s0}(2317)^+$ was determined, resulting in a sizable fraction of the order of 70% within errors (a more accurate determination of that fraction, also of the order of 70% was done in [6]).

Here we shall give the DK invariant mass distribution in the decay $B_c \to J/\psi DK$, from which information on the internal structure of the $D^*_{s0}(2317)^+$ state will be obtained.

⋆e-mail: pedro.fernandez@ific.uv.es

2 Formalism

As a first step, we write the leading mechanisms describing the weak process by means of which the heavy quark b in the B_c meson decays into three lighter quarks. This is depicted in Fig. 1, and our first assumption will be that the matrix element for this transitions will be factorized and assumed to be constant in the small range of the KD invariant mass that we will study.

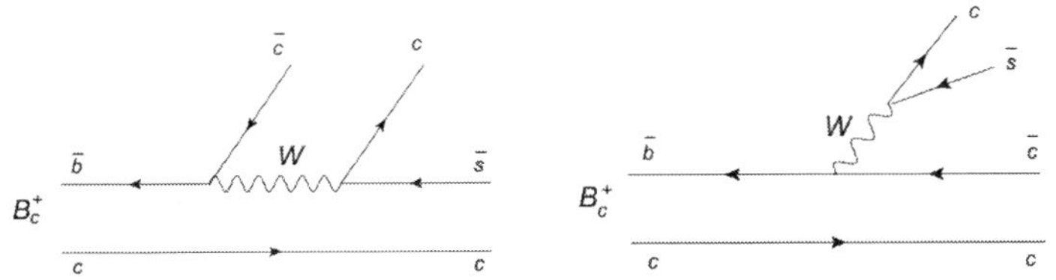

Figure 1. Diagrams for the B_c^+ weak decay mechanism into a final configuration with a $c\bar{s}$ state.

A $c\bar{c}$ pair will combine into a J/ψ, so we need to split the $c\bar{s}$ quark pair into meson states. The interaction of the J/ψ with the pseudoscalars will not be considered. The hadronization of the $c\bar{s}$ pair is achieved introducing a $q\bar{q}$ combination with the quantum numbers of the vacuum: $c\bar{s}(u\bar{u}+d\bar{d}+s\bar{s}+c\bar{c})$. The splitting of this product of quark fields into meson states is done introducing the M matrix, which is nothing but the $q\bar{q}$ flavour matrix with the matrix elements: $M_{i,j} = q_i\bar{q}_j$, $i, j = 1, 2, ..., 4$ where $(q_1, q_2, q_3, q_4) = (u, d, s, c)$. This matrix has the following property $(M^2)_{i,j} = q_i q_j(u\bar{u} + d\bar{d} + s\bar{s} + c\bar{c})$. If we identify the $q\bar{q}$ flavour matrix with the meson's flavour matrix ϕ using the standard $\eta - \eta'$ mixing

$$\phi = \begin{pmatrix} \frac{\eta}{\sqrt{3}} + \frac{\pi^0}{\sqrt{2}} + \frac{\eta'}{\sqrt{6}} & \pi^+ & K^+ & \bar{D}^0 \\ \pi^- & \frac{\eta}{\sqrt{3}} - \frac{\pi^0}{\sqrt{2}} + \frac{\eta'}{\sqrt{6}} & K^0 & D^- \\ K^- & \bar{K}^0 & \frac{\sqrt{2}\eta'}{\sqrt{3}} - \frac{\eta}{\sqrt{3}} & D_s^- \\ D^0 & D^+ & D_s^+ & \eta_c \end{pmatrix}, \tag{1}$$

then our product of quark fields in terms of meson fields is

$$(\phi^2)_{43} = D^0K^+ + D^+K^0 + \eta_c D_s^+ - \frac{1}{\sqrt{3}}\eta D_s^+ + \sqrt{\frac{2}{3}}D_s^+\eta' \equiv \sum_j h_j P_j P'_j, \tag{2}$$

where in Eq. (2) j labels the different channels. We consider only the meson channels whose mass is close to the KD threshold, we keep the D^+K^0, D^0K^+ and ηD_s^+ channel, namely $j = 1, 2, 3$. As shown in Fig. 2, the amplitude for the final $J/\psi K^0 D^+$ state is written in terms of the common V_p constant factor which accounts for the weak production and hadronization mechanisms. There are contributions to the final KD state that involve intermediate meson loops of the different channels. The interaction of the meson channels is denoted by the $t_{i,j}$ vertex, which is obtained after the unitarization of the S-wave leading order amplitudes, using the heavy meson chiral Lagrangian from Ref. [4]. The expression of these tree level amplitudes can be found in Ref. [1], we will denote them here as $V_{i,j}$. The unitarization is achieved solving the on-shell version of the factorized Bethe-Salpeter equation in coupled channels:

$$t^{-1} = V^{-1} - G. \tag{3}$$

The G that appears in Eq. (3) is a diagonal matrix in the coupled channels basis, with the loop function of the different meson channels. The loops are integrated using dimensional regularization,

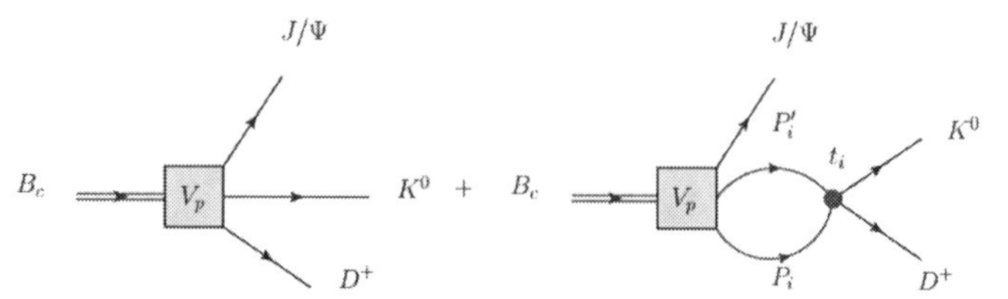

Figure 2. Diagrams for the KD production in the B_c^+ decay. Note the common constant vertex V_p accounting for the hadronization and b decay, and also note the intermediate meson loops contributions. The t_i vertex accounts for the interaction of mesons in coupled channels.

and regularized including a subtraction constant, called a, at some scale μ that we take as 1.5 GeV. The analytic expression can be found in Ref. [1]. The meson dynamics will generate the $D^*_{s0}(2317)$ resonance, manifesting as a pole of the amplitudes in the complex energy plane. In our approach, we fine tune the free parameter a in order to generate a pole in t in the physical Riemann sheet of the complex energy, at the position of the mass of the $D^*_{s0}(2317)$ resonance, below the KD threshold. What we have is a bound state, which is not strictly true but it is a good picture if we recall that the $D^*_{s0}(2317)$ width comes from an isospin violating mode, and is very small. We calculate the coupling g_i of the resonance to the different channels in terms of the residue of t at the pole position s_p, since $t_{i,j} \approx g_i g_j/(s - s_p)$ for values of the invariant s close to the pole. We find a larger coupling to the KD channels $g_{K^+D^0} = g_{K^0D^+} = 7.4$ GeV but the ηD_s coupling is also sizeable, about −6.0 GeV. With the unitarized amplitudes of Eq. (3) we build the decay amplitude which enters in the invariant KD mass distribution of the decay,

$$\frac{d\Gamma}{dM_{\rm inv}} = \frac{A^2 p_{J/\psi} p_{KD}}{(2\pi)^3 m^2_{B_c}} \left| h_1 + \sum_{i=1}^{3} h_i G_i t_{i1} \right|^2 . \quad (4)$$

In order to derive Eq. (4) we consider that the V_p vertex must be a P-wave vertex, $V_p = \sqrt{3}A\cos\theta$, in order to match total spin conservation, since the pseudoscalars interact in S-wave in our approach. The amplitude for the decay reads $V_p\left(h_1 + \sum_{i=1}^{3} h_i G_i t_{i1}\right)$, see Fig. 2. In a similar way we are also able to calculate the coalescence, that is, the width for the direct production of the $D^*_{s0}(2317)$ in the B_c decay: $\Gamma(B_c \to J/\psi D^*_{s0}(2317))$,

$$\Gamma(B_c \to J/\psi D^*_{s0}(2317)) = \frac{A^2 p^3_{J/\psi}}{8\pi m^2_{B_c}} \left| \sum_{i=1}^{3} h_i G_i g_i \right|^2 \Bigg|_{s=s_p} , \quad (5)$$

and as we see in Eq. (5) it is given in terms of the coupling of the resonance to the different channels. The ratio of the KD invariant mass distribution (Eq. (4)) and the coalescence (Eq. (5)) provides a free parameter magnitude tied to the molecular nature of the $D^*_{s0}(2317)$) resonance. We define this ratio removing the phase space and multiplying by the necessary factors in order to make it dimensionless,

$$\frac{d\tilde{\Gamma}}{dM_{inv}} = M^2_{B_c} \frac{(d\Gamma/dM_{\rm inv})/p^3_{J/\psi} p_{DK}}{\Gamma(B_c^+ \to J/\psi D^*_{s0}(2317)))/ \left. p^3_{J/\psi} \right|_{s=s_p}} . \quad (6)$$

3 Results

In the left panel of Fig. 1 we show the results of the KD invariant mass distribution, and on the right panel we find the result of the ratio $\frac{d\tilde{\Gamma}}{dM_{inv}}$ defined in Eq. (6). It is interesting to compare the

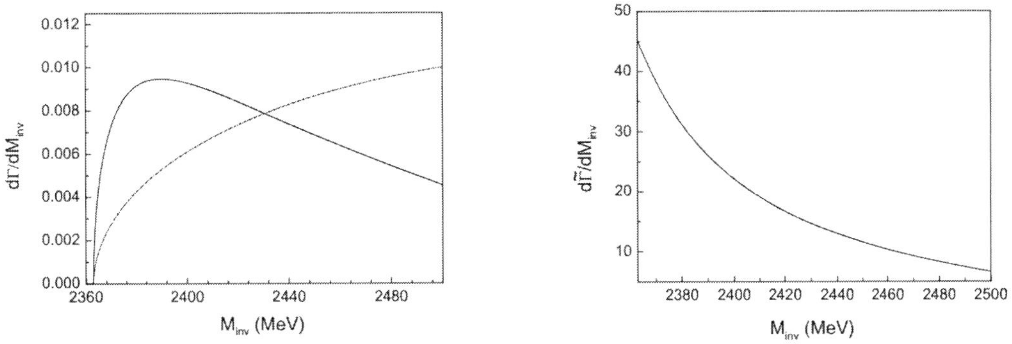

Figure 3. (Left) KD invariant mass distribution for the $B_c \to J/\psi KD$ decay, the dashed line is the phase space. Both curves are normalized in the energy range showed; (Right) The ratio of the KD invariant mass distribution and the coalescence without the phase space factor.

KD invariant mass distribution with the phase space, the solid and dashed lines respectively. The A constant in Eq. (5) is chosen in a way that the invariant mass distribution is normalized in the energy range of the plot. The phase space is also normalized in the same energy range. We clearly see that both curves have a different behaviour near the threshold. Even if the phase space is relatively small, we observe a huge enhancement in the KD invariant mass distribution. This enhancement has been also observed experimentally in several decays of heavy mesons, see for example in Ref. [7] the B's and B_s's decays into KD final states. We present an unmeasured reaction where this feature is also predicted, but in our approach is due to the dynamically generation of the $D^*_{s0}(2317)$. Furthermore, we give the ratio of the KD invariant mass distribution and the $D^*_{s0}(2317)$ production. In the right panel of Fig. 3 we can observe the features of this magnitude. It has a finite value at threshold, and it looks like the tail of a peak located below. Here this is indeed due to the presence of the pole of the $D^*_{s0}(2317)$ below the threshold, because Eq. (6) is nothing but the ratio of the amplitudes of Fig. 2 and the coalescence. Since we are forcing the presence of the pole with the fine tuning of the subtraction constant a, this is what explains the shape of the plot shown in the right panel of Fig. 3. The comparison of our predictions with the measurement results of the process proposed here would provide further information about the molecular nature of the $D^*_{s0}(2317)$ resonance. As a final comment, we mention that possible $q\bar{q}$ small components were considered adding a CDD pole to the meson's potential and using modified Weinberg compositeness rule to match different KD amounts obtained from the lattice studies, see Refs. [1, 5, 6]. This contribution does not change the conclusions and general features shown here.

References

[1] Z. F. Sun, M. Bayar, P. Fernandez-Soler and E. Oset, Phys. Rev. D **93** no.5, 054028 (2016)
[2] B. Aubert *et al.* [BaBar Collaboration], Phys. Rev. Lett. **90**, 242001 (2003)
[3] E. van Beveren and G. Rupp, Phys. Rev. Lett. **91**, 012003 (2003)
[4] D. Gamermann, E. Oset, D. Strottman and M. J. Vicente Vacas, Phys. Rev. D **76**, 074016 (2007)
[5] S. Weinberg, Phys. Rev. **137**, B672 (1965)
[6] A. Martinez Torres, E. Oset, S. Prelovsek and A. Ramos, JHEP **1505**, 153 (2015)
[7] R. Aaij *et al.* [LHCb Collaboration], Phys. Rev. D **90** no.7, 072003 (2014)

EPJ Web of Conferences **130**, 02003 (2016) DOI: 10.1051/epjconf/201613002003
MESON 2016

The $\rho(\omega)/B^*(B)$ system and bound states in the unitary local Hidden Gauge approach

P. Fernandez-Soler[1,⋆], Zhi-Feng Sun[1], J. Nieves[2], and E. Oset[1]

[1] *Departamento de Física Teórica and IFIC (Centro Mixto Universidad de Valencia-CSIC), Valencia, Spain*
[2] *IFIC (Centro Mixto Universidad de Valencia-CSIC), Valencia, Spain*

Abstract. In this work, we study systems composed of a ρ/ω and B^* meson pair. We find three bound states in isospin, spin-parity channels $(1/2, 0^+)$, $(1/2, 1^+)$ and $(1/2, 2^+)$. The state with $J = 2$ can be a good candidate for the $B_2^*(5747)$. We also study the ρB system, and a bound state with mass 5728 MeV and width around 20 MeV is obtained, which can be identified with the $B_1(5721)$ resonance. In the case of $I = 3/2$, one obtains repulsion and thus, no exotic (molecular) mesons in this sector are generated in the approach.

1 Introduction

The present work is the extension to the b sector of the work of Ref. [1], for the $\rho(\omega)D^*$ system. Furhter details and references can be found in Ref. [2]. The interaction of the ρ, ω and D^* mesons was studied using the Local Hidden Gauge (LHG) Lagrangians of Refs. [3], where the vector mesons are considered as gauge bosons of a hidden symmetry transforming inhomogeneously. Choosing the appropriate gauge they can transform as in the non-linear realization of chiral symmetry. The extension to the charm sector, together with a non-perturbative treatment of the LHG amplitudes, provided the set of predictions that is compiled in Table 1. Fitting the $\frac{1}{2}(2^+)$ state to an already existing state, two more states were predicted and identified with other experimental states [4], the $D(2600)$ and $D^*(2640)$. It is remarkable that the $D(2600)$, the first row in Table 1, was measured after the theoretical prediction [5].

Table 1. Results of the unitary LHG approach in the charm sector. The $\rho(\omega)D^*$ interaction.

Model result			PDG state association [4]		
$I(J^{PC})$	Mass	Width	$I(J^{PC})$	Mass	Width
$\frac{1}{2}(0^+)$	2600 MeV	≈ 61 MeV	$D(2600)\ \frac{1}{2}(?^?)$	2612 ± 6 MeV	93 ± 14 MeV
$\frac{1}{2}(1^+)$	2620 MeV	0 MeV	$D^*(2640)\ \frac{1}{2}(?^?)$	2637 ± 6 MeV	< 15 MeV
$\frac{1}{2}(2^+)$	fitted	fitted	$D_2^*(2460)\ \frac{1}{2}(2^+)$	2462.6 ± 0.6	49.90 ± 1.3

⋆e-mail: pedro.fernandez@ific.uv.es

2 Formalism

The meson-meson bound states produced with this model will appear as poles in the scattering t-matrix, T, on the physical Riemann Sheet of the energy, in our case of $s = P^2$, being P the total four-momenta of the meson-meson system. Imposing unitarity on the right-hand cut is equivalent to solve the on-shell version of the factorized Bethe-Salpeter equation:

$$T^{I,J}(s) = V^{I,J}\left(1 - G(s)V^{I,J}(s)\right)^{-1}, \tag{1}$$

which depend on the total isospin I and spin J of the ρB^* and ωB^* two possible channels. Eq. (1) is a matrix equation in the channels space. The $V^{I,J}(s)$ are the tree level Feynman amplitudes associated to the possible $\rho(\omega)B^* \to \rho(\omega)B^*$ transitions, and the $G(s)$ is the two-meson loop function, regularized with a sharp cut-off Λ. Eq. (1) is equivalent to the following re-summation:

$$T^{I,J} = V^{I,J} + V^{I,J}GV^{I,J} + V^{I,J}GV^{I,J}GV^{I,J} + \dots. \tag{2}$$

2.1 The Local Hidden Gauge amplitudes

The Local Hidden Gauge Lagrangian for the vector meson interaction is the following,

$$\mathcal{L}_{III} = -\frac{1}{4}\left\langle V_{\mu\nu}V^{\mu\nu}\right\rangle, \tag{3}$$

where the symbol $\langle\rangle$ represents the trace in SU(4) flavor space (we consider u, d, s and b quarks), with

$$V_{\mu\nu} = \partial_\mu V_\nu - \partial_\nu V_\mu - ig[V_\mu, V_\nu], \tag{4}$$

and V_μ is the following matrix,

$$V_\mu = \begin{pmatrix} \frac{\rho^0}{\sqrt{2}} + \frac{\omega}{\sqrt{2}} & \rho^+ & K^{*+} & B^{*+} \\ \rho^- & -\frac{\rho^0}{\sqrt{2}} + \frac{\omega}{\sqrt{2}} & K^{*0} & B^{*0} \\ K^{*-} & \bar{K}^{*0} & \phi & B_s^{*0} \\ B^{*-} & \bar{B}^{*0} & \bar{B}_s^{*0} & \Upsilon \end{pmatrix}_\mu . \tag{5}$$

As can be seen in Eq. (5), the degrees of freedom are the vector meson fields. The local hidden gauge Lagrangian (3) contains a four vector contact term $\mathcal{L}_c \propto g^2$ and a meson exchange term $\mathcal{L}_{ex} \propto g$. At tree-level, the interaction involves contact diagrams and one meson exchange, as it is depicted in Fig. 1.

- In the Local Hidden Gauge approach the interaction is mainly given by the vector meson exchange, being the exchange of heavy mesons penalized by their large mass.
- We make the approximation of neglecting the three-momenta of the external mesons, with respect to the masses: $\frac{\vec{k}_i^2}{M_i^2} \to 0$. This helps to simplify the formalism and the spin projections.
- We also have considered the large width of the ρ meson, and its two pion decay mechanism in order to give some width to the dynamically generated states.
- It can be shown that the meson exchange term of the Local Hidden Gauge Lagrangian Eq. (3), in the $\frac{\vec{k}_i^2}{M_i^2} \to 0$ limit, has the same structure as the vector-pseudoscalar term, thus the ρB interaction is easily obtained replacing $B^* \to B$ in the ρ exchange amplitudes. In our formalism there are no terms that link the ρB^* and ρB channels, the terms leading to these transitions are anomalous and small. There is another possibility using box diagrams, but this contribution was found also small.

DOI: 10.1051/epjconf/201613002003

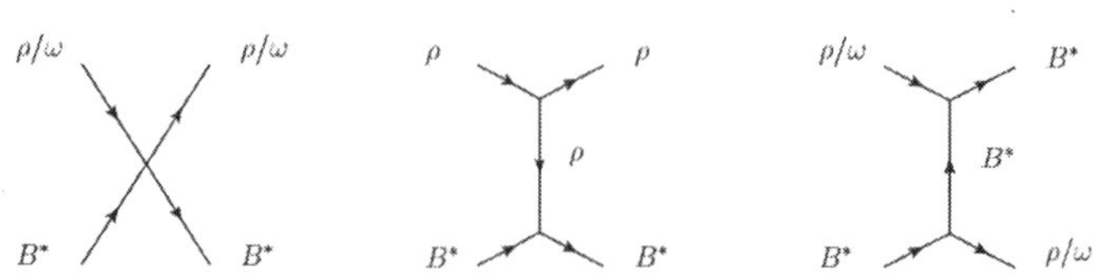

Figure 1. The possible tree-level diagrams for the $\rho(\omega)B^*(B)$ interaction.

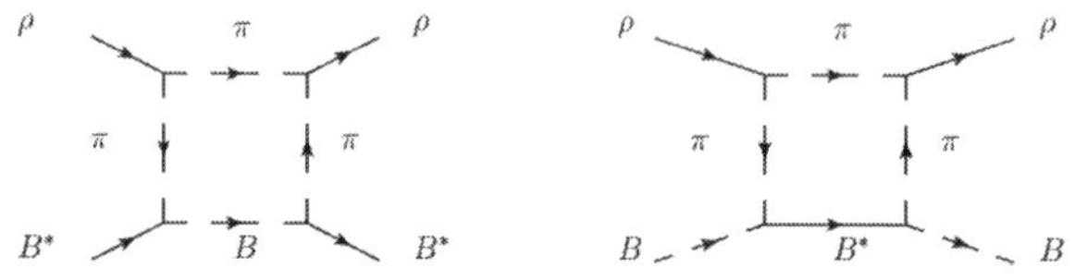

Figure 2. The $\rho \to \pi\pi$ decay is taken into account in the interaction $V^{I,J}$ in order to provide some width to the states.

- When we split the $\rho B^*(B)$ states in terms of the spin of the light quarks, we find that neglecting $\rho B^* \to \rho B$ transitions leads to a spin degeneracy in the mass of the states. Furthermore, it is found that the ρ exchange amplitudes are spin degenerated in the LHG, and not suppressed in the heavy quark mass power counting. The rest of amplitudes do not conserve the heavy quark spin symmetry, but they grow as $1/m_Q$ in the $m_Q \to \infty$ limit.

The $V^{I,J}$ potential employed to solve Eq. (1) is the sum of the meson exchange amplitudes and the contact terms of Fig 1, and the box diagrams contributions of Fig. 2.

3 Results

In Table 2 we find the set of predictions that it is obtained when fitting the $\frac{1}{2}(2^+)$ state, generated by the ρB^* interaction, to an already existing state, the $B_2^*(5747)$. We can appreciate that the mass difference of the states is small, revealing the near spin-degeneracy of the masses. As we have stated, this is due to the larger contribution of the ρ exchange in the interaction, and the fact that this contribution is spin degenerated in our formalism. We observe also that we predict states with spin 0 and 1 generated by the $\rho(\omega)B^*$ interaction. The ρB state, which is also a prediction, can be identified with an already $\frac{1}{2}(1^+)$ existing state, the $B_1(5721)^0$.

Table 2. Summary of the states found in the $\rho(\omega)B^*$ and ρB sectors using the unitary LHG approach.

Main channel	$I(J^P)$	M [Mev]	Γ [MeV]	Main decay channel	Exp (M,Γ) [MeV]
ρB^*	$\frac{1}{2}(0^+)$	5812	$25-45$	πB	
ρB^*	$\frac{1}{2}(1^+)$	5817	0		
ρB^*	$\frac{1}{2}(2^+)$	5745	$25-35$	πB	$(5743 \pm 5\,,\, 23^{+5}_{-11})$
ρB	$\frac{1}{2}(1^+)$	5728	$18-24$	πB^*	$(5726.8 \pm 2\,,\, 49^{+14}_{-23})$

References

[1] R. Molina, H. Nagahiro, A. Hosaka and E. Oset, Phys. Rev. D **80**, 014025 (2009)
[2] P. Fernandez-Soler, Z. F. Sun, J. Nieves and E. Oset, Eur. Phys. J. C **76** (2016) no.2, 82
[3] M. Bando, T. Kugo, S. Uehara, K. Yamawaki and T. Yanagida, Phys. Rev. Lett. **54**, 1215 (1985)
[4] K. A. Olive *et al.* [Particle Data Group Collaboration], Chin. Phys. C **38**, 090001 (2014)
[5] P. del Amo Sanchez *et al.* [BaBar Collaboration], Phys. Rev. D **82**, 111101 (2010)

EPJ Web of Conferences **130**, 02004 (2016) DOI: 10.1051/epjconf/201613002004
MESON 2016

A three body state with J=3 in the $\rho B^* \bar{B}^*$ interaction

P. Fernandez-Soler[1,⋆], M. Bayar[2], Zhi-Feng Sun[1], and E. Oset[1]

[1] *Departamento de Física Teórica and IFIC (Centro Mixto Universidad de Valencia-CSIC), Valencia, Spain*
[2] *Department of Physics of Kocaeli University, Izmit, Turkey*

Abstract. We study the $\rho B^* \bar{B}^*$ system solving the Faddeev equations in the fixed center approximation. The $B^* \bar{B}^*$ system will be considered forming a cluster, and using the two-body ρB^* unitarized scattering amplitudes in the local Hidden Gauge approach we find a new $I(J^{PC}) = 1(3^{--})$ state. The mass of the new state corresponds to a two particle invariant mass of the ρB^* system close to the resonant energy of the $B_2^*(5747)$, indicating that the role of this $J = 2$ resonance is important in the dynamical generation of the new state.

1 Introduction

Here we study the interaction of the three body system composed of a ρ meson and a $B^* \bar{B}^*$ pair, further details and references can be found in Ref. [1]. We use the Fixed Center Approximation (FCA), that is, we consider the pair of b mesons forming a $J = 2$ cluster, and let the ρ meson interact with the cluster as can be seen in Fig. 1. The motivation for study such a system is that in previous works, the ρB^* interaction was found very attractive, specially in $J = 2$ where it was found the most bound state [2]. On the other hand, the $B^* \bar{B}^*$ interaction was studied in [3] and found degenerate states in all spins. We can think that the combination of these two attractive subsystems would generate a new $J = 3$ bound state with our formalism.

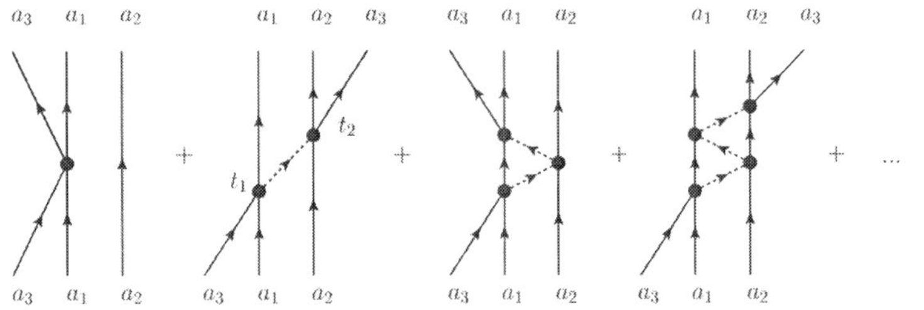

Figure 1. Interaction of three particles a_1, a_2 and a_3, in the FCA. The a_3 particle is the lighter one, in our case the ρ meson.

⋆e-mail: pedro.fernandez@ific.uv.es

2 Formalism

The main assumption is that the heavy B^* and $\bar{B}^*$ mesons will be forming a cluster of spin two. Then we can construct the diagrammatic series of Fig. 1 which account all the possible terms for the scattering of the ρ (particle a_3) and say, the B^* (particle a_1). Thus, denoting this sum as T_1:

$$T_1 = t_1 + t_1 G_0 t_2 + t_1 G_0 t_2 G_0 t_1 + \ldots, \tag{1}$$

$$T_2 = t_2 + t_2 G_0 t_1 + t_2 G_0 t_1 G_0 t_2 + \ldots, \tag{2}$$

where we have also written in Eq. (2) the series for the scattering with the $\bar{B}^*$ (particle a_2). From Eqs. (1) and (2) we deduce the coupled Faddeev equations,

$$T_1 = t_1 + t_1 G_0 T_2, \tag{3}$$

$$T_2 = t_2 + t_2 G_0 T_1, \tag{4}$$

and the full scattering T-matrix is $T = T_1 + T_2$. The G_0 (dashed line in Fig. 1) is the ρ propagator inside the cluster of mass M_c,

$$G_0(q^0) = \frac{1}{2M_c} \int_{\mathbb{R}^3} \frac{d^3q}{(2\pi)^3} F_R(\vec{q}^{\,2}) \frac{1}{(q^0)^2 - \vec{q}^{\,2} - m_{a_3}^2 + i\epsilon}, \tag{5}$$

the F_R function in Eq. (5) is the form factor of the resonance or cluster, which is the Fourier transform of its wave function. There q^0 and m_{a_3} are the energy and the mass of the ρ meson. We will consider the following form factor description for s-wave functions

$$F_R(\vec{q}^{\,2}) = \frac{1}{N} \int_{\Omega} d^3p\, \mathcal{A}(\vec{p}) \mathcal{A}(\vec{p} - \vec{q}), \quad \Omega := \left\{ |\vec{p}|, |\vec{p} - \vec{q}| < \Lambda \right\}, \tag{6}$$

where $\mathcal{A}$ and N are defined by,

$$\mathcal{A}(\vec{p}) = \frac{1}{M_c - \omega_{a_1}(\vec{p}) - \omega_{a_2}(\vec{p})}, \tag{7}$$

$$N = F_R\left(\vec{q}^{\,2} = 0\right), \tag{8}$$

and Λ is a three-momentum cutoff used to regularize the meson-meson loop function in the $B^*\bar{B}^*$ system in order to obtain the $B^*\bar{B}^*$ as a bound state [4].

- It is interesting to note that this is the only information needed from the $B^*\bar{B}^*$ system. One is not taking the $B^*\bar{B}^*$ interaction explicitly, but it is considered implicitly since it leads to the binding of the $B^*\bar{B}^*$ system, and we can determine the wave function and the form factor from this information.
- The t_1 and t_2 are the ρB^* and $\rho\bar{B}^*$ interaction in $J = 2$, studied in [2] with the unitarization of amplitudes given by the local Hidden Gauge approach, that generated the $B_2^*(5747)$. In this formalism we have $t_1 = t_2$.
- The quantum numbers of the cluster are $I(J^{PC}) = 0(2^{++})$, and the ρ meson is an isotriplet, thus it is necessary to write the 3 body $T(I = 1)$ matrix in terms of proper isospin states of the ρB^* and $B^*\bar{B}^*$ subsystems. This provides the result

$$T = \frac{2\tilde{t}}{1 - \tilde{t} G_0}, \tag{9}$$

where $\tilde{t} = \frac{t_1(I_{\rho B^*}=1/2) + 2t_1(I_{\rho B^*}=3/2)}{3} \times \frac{M_c}{m_{B^*}}$, being the last factor included to ensure the correct normalization convention of the amplitudes in the sums of Eqs. (1) and (2), as in Ref. [4], with respect to the convention used in [2].

- The sources of uncertainties of the model are the propagator cut off, Λ, which is translated into uncertainties in the cluster mass, and the way of sharing the total energy between the two subsystems. We have estimated the variation of the results tied to these sources of about 60 MeV in the mass, and 10 MeV in the width of the $J = 3$ bound state found.

3 Results

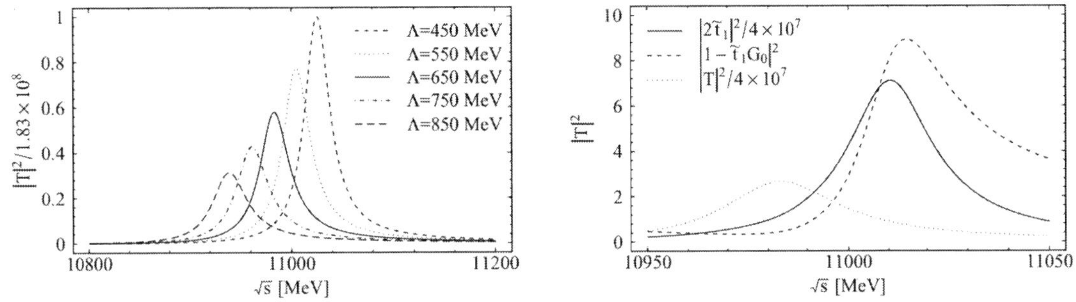

Figure 2. (Left) modulus squared of the total T matrix of Eq. (9). (Right) plot of the numerator and denominator of Eq. (9) together with the total T.

As we can see in the left panel of Fig. 2, we find a $I(J^P) = 1(3^-)$ bound state of the three body system with a mass $M = 10968 \pm 57$ MeV and a width of 36 ± 7 MeV. In the right panel of Fig. 2 we have plotted the modulus squared of the t_1, which is the contribution of the $J = 2$ ρB^* interaction, the solid line. We can see a clear peak, which is directly related with the $B_2^*(5747)$ resonant state found in the unitary Hidden Gauge formalism. Furthermore, we have plotted the denominator of T, $1 - G_0\tilde{t}_1$, dashed line, which is the effect of the ρ orbiting around the heavy mesons in the FCA. This term is the responsible of displacing the peak to lower energies, see the dotted line, and we appreciate that the strong binding energy of the two subsystems plays a fundamental role on the generation of the three-body state.

References

[1] M. Bayar, P. Fernandez-Soler, Z. F. Sun and E. Oset, Eur. Phys. J. A **52** no.4, 106 (2016)
[2] P. Fernandez-Soler, Z. F. Sun, J. Nieves and E. Oset, Eur. Phys. J. C **76** no.2, 82 (2016)
[3] A. Ozpineci, C. W. Xiao and E. Oset, Phys. Rev. D **88** 034018 (2013)
[4] L. Roca and E. Oset, Phys. Rev. D **82** 054013 (2010)

K^+-nucleon amplitudes in the medium below 800 MeV/c

E. Friedman[1,*]

[1]*Racah Institute of Physics, Hebrew University, Jerusalem, Israel*

Abstract. Simple in-medium meson-nucleon kinematics has been applied recently in calculations of strong interaction effects in kaonic atoms, pionic atoms and elastic scattering of low energy pions by nuclei. More sensitive tests of this approach are possible with K^+-nucleus interactions below 800 MeV/c because of the superior penetration of kaons into nuclei. The energy dependence of calculated reaction and total cross sections for K^+ on ^{6}Li, C, Si and Ca agrees better with experiment when this approach is used. Calculated absolute scales are in full agreement with experiment for the very low density nucleus ^{6}Li whereas calculations are 23±4 % too low for C, Si and Ca, thus quantifying phenomenologically the enhancement in the nuclear medium observed before.

1 Introduction

The present note deals with K^+ meson-nucleus interaction at low energies where the total K^+-nucleon cross sections are small and the kaon penetrates well into the nucleus. The prime motivation for this work was to test recent approach to the calculation of optical potentials based on in-medium kinematics, used so far in analyses of pionic and kaonic atoms and of elastic scattering of low energy $\pi^\pm$ by nuclei [1]. With the deep penetration of K^+ mesons into the nucleus at momenta below 800 MeV/c it was possible to test this approach with greater sensitivity than previously done with K^- and pions.

Previous analyses of experimental K^+-nucleus integral cross sections showed that the in-medium K^+-nucleon amplitude is enhanced compared to the free interaction in this energy range. The present work reaffirms quantitatively the enhancement.

2 Analysis

Applying the in-medium approach to K^+ mesons, the kaon-nucleon scattering amplitude is presented in terms of the Mandelstam variable s in the nuclear medium

$$s = (E_K + E_N)^2 - (\vec{p}_K + \vec{p}_N)^2, \tag{1}$$

where $E_K = m_K + E_{\rm lab}$, $E_N = m_N - B_N$. $E_{\rm lab}$ is the laboratory kinetic energy of the kaon and B_N is an average binding energy of a nucleon. In the nuclear medium $\vec{p}_K + \vec{p}_N = 0$ where $\vec{p}_K$ is determined by the beam energy and the optical potential, $\vec{p}_N$ is determined by the nuclear

*e-mail: elifried@cc.huji.ac.il

environment. Averaging over angles, $(\vec{p}_K + \vec{p}_N)^2 \to (p_K)^2 + (p_N)^2$. For the meson $(p_K)^2$ is obtained by substituting locally

$$p_K^2/2m_K \to E_{\rm lab} - {\rm Re}\ V_{\rm opt}^K - V_c, \tag{2}$$

with $V_{\rm opt}^K$ the kaon-nucleus optical potential and V_c the Coulomb potential. For the nucleon $(p_N)^2$ is taken from the Fermi gas model.

Defining $\delta\sqrt{s} = \sqrt{s} - E_{\rm th}$ with $E_{\rm th} = m_K + m_N$, then to first order in $B/E_{\rm th}$ and $(p/E_{\rm th})^2$ one gets [2, 3]

$$\delta\sqrt{s} - \xi_N E_{\rm lab} = -B_N\rho/\bar{\rho} - \xi_N T_N(\rho/\bar{\rho})^{2/3} + \xi_K[{\rm Re}\ V_{\rm opt} + V_c(\rho/\rho_0)^{1/3}], \tag{3}$$

with $\xi_N = m_N/(m_N + m_K)$, $\xi_K = m_K/(m_N + m_K)$ and T_N the average kinetic energy of a bound nucleon. $\rho, \bar{\rho}$ and ρ_0 are the local, average and central nuclear densities, respectively. This value of $\sqrt{s}$ serves as the argument of the in-medium kaon-nucleon amplitude in constructing the kaon-nucleus optical potential. In another version a further Coulomb potential is subtracted from $\delta\sqrt{s}$ due to 'minimal substitution' (MS) arguments [4].

In this approach the energy of the K^+-nucleon interaction in the medium depends on the optical potential (and density) whereas the optical potential depends on the amplitude, within a '$t\rho$' approach. Therefore a self-consistent solution is required. Good convergence is usually achieved after 4-5 iterations. For K^+ interaction with nucleons in the momentum range of 400-800 MeV/c the relevant energies in the medium are lower than the cm beam energy by up to 30 MeV.

3 Results

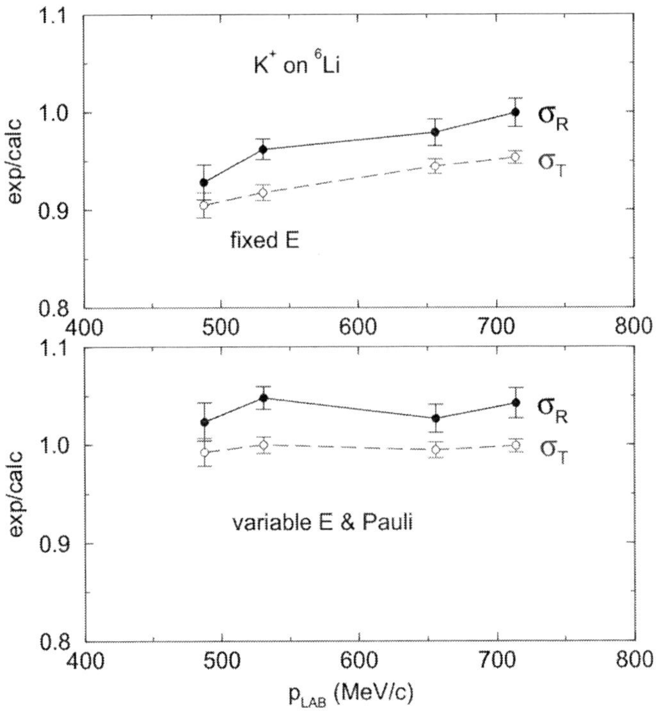

Figure 1. Ratios between experimental and calculated reaction (solid lines) and total cross sections (dashed lines) of K^+ on ^{6}Li. Upper part: using free amplitudes at the relevant cm beam energy. Lower part: using in-medium amplitudes at energies given by Eq. (3), including MS term

Total cross sections for K^+ on D, ^{6}Li, C, Si and Ca were measured in the early 1990s at four momenta below 800 MeV/c [5–8] . The same set of transmission measurements was later reanalysed to determine not only total cross sections but also reaction cross sections, that are less dependent

DOI: 10.1051/epjconf/201613002005

on applied corrections [9]. Comparing total cross sections of K^+ on ^{6}Li with three times the cross sections on deuteron, it was found that within 2% there are no indications for medium effects in the results for ^{6}Li, where the average density is about half of the average density for other nuclei [10, 11]. Consequently in what follows we use ^{6}Li as a reference.

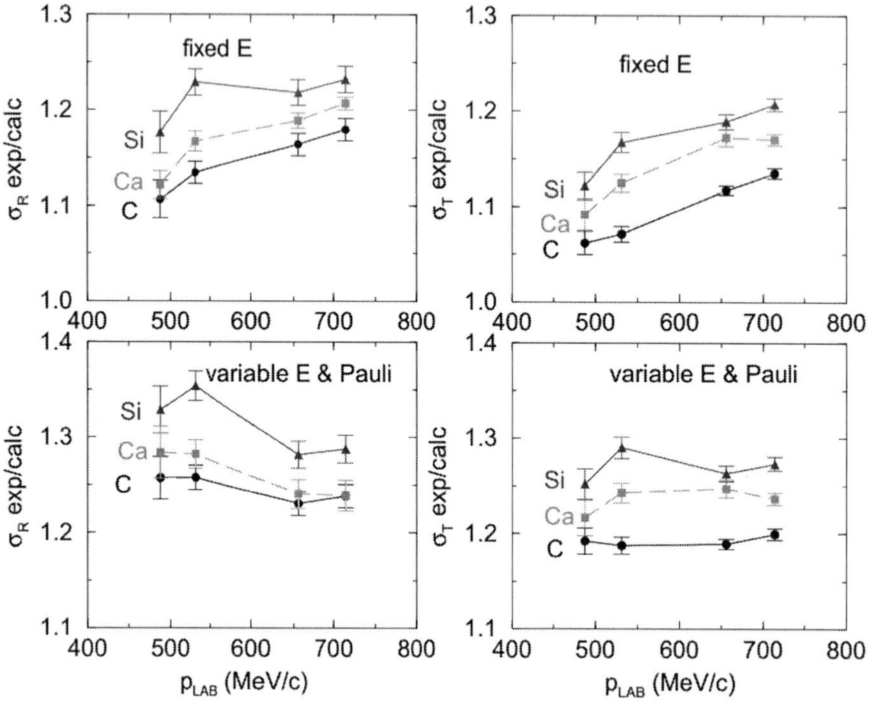

Figure 2. Ratios between experimental and calculated reaction (left panels) and total cross sections (right panels) of K^+ on C, Si Ans Ca. Upper panels are for cm beam energies, lower panels are for energies given by Eq. (3), including MS term

Fig. 1 shows ratios between experimental and calculated reaction (solid lines) and total cross sections (dashed lines) of K^+ on ^{6}Li. Calculations used the isospin-averaged forward scattering amplitudes from the SAID software package [12]. The upper part is for calculations based on using the K^+-nucleon forward scattering amplitudes at the cm beam energy. In the lower part the energy of that amplitude depended on the local nuclear density, as obtained from iterative self-consistent solution of Eq. (3) including MS term. It is evident that the absolute scale is reproduced quite well by calculations in both cases, but the dependence on beam momentum is not reproduced by the calculations at the upper part. In contrast, using Eq. (3) (lower part) the dependence on the beam momentum is reproduced at the level of 2-3% as is also the absolute scale of the cross sections. Including corrections due to Pauli correlations made no difference in the case of ^{6}Li, presumably because of the low density of this nucleus.

Fig. 2 shows ratios between experimental and calculated reaction and total cross sections of K^+ on C (circles), Si (triangles) and Ca (squares). Reaction cross sections are shown on the left panels, total cross sections are shown on the right panels. The upper panels are for calculations using fixed energies for the in-medium K^+-nucleon amplitudes whereas the lower panels are for calculations using the in-medium kinematics as given by Eq. (3), including MS term and corrections due to Pauli correlations [13] which are non-negligible for these normal-density nuclei. Applying only Eq. (3) (not shown) improves the agreement with experiment of the momentum dependence, but further improvement is achieved when Pauli corrections too are included. We note that the results for Si have been scattered more than the results for C and Ca also in previous analyses of the same data. However, the ratios for C and Ca are very close to each other and are essentially independent of momentum to within 2-3%. Averaging these, an enhancement of reaction and total cross sections by 23±4% is established.

4 Discussion

The enhancement of experimental total cross sections on C was already noted in 1985 by Siegel, Kaufmann and Gibbs [14] on the basis of very limited data. They showed that traditional calculations failed to produce the enhancement, thus suggesting more exotic mechanisms, see [3] for additional references. The present data was also analysed in 1997 [10, 11], separately at the different energies, arriving at similar conclusions. Peterson [15] showed in 1999 that the enhancement is reproduced by using in medium an increased average K^+-nucleon S-wave phase-shift compared to free space, suggesting 'swelling' of bound nucleons. A possible link between the additional reactive content and K^+ interactions with two nucleons was explored in [16, 17].

In the present work we were able to handle the four energies together thanks to the energy-dependence of the model Eq. (3). The 23±4% enhancement of cross sections could be reproduced phenomenologically by 40±2% increase of the imaginary part of the in-medium amplitude [3]. This is more than wave function renormalization effects due to partial restoration of chiral symmetry, as suggested recently by Jido [18]. Clearly the observed enhancement is an open problem.

I wish to thank A. Gal for fruitful discussions.

References

[1] A. Gal *et al.*, EPJ Web of Conferences **81**, 01018 (2014) and references therein
[2] E. Friedman, A. Gal, Nucl. Phys. A **899**, 60 (2013)
[3] E. Friedman, Nucl. Phys. A **954**, 114 (2016)
[4] E. E. Kolomeitsev, N. Kaiser, W. Weise, Phys. Rev. Lett. **90**, 092501 (2003)
[5] Y. Mardor *et al.*, Phys. Rev. Lett. **65**, 2110 (1990)
[6] R. A. Krauss *et al.*, Phys. Rev. C **46**, 655 (1992)
[7] R. Sawafta *et al.*, Phys. Lett. B **307**, 293 (1993)
[8] R. Weiss *et al.*, Phys. Rev. C **49**, 2569 (1994)
[9] E. Friedman *et al.*, Phys. Rev. C **55**, 1304 (1997)
[10] E. Friedman, A. Gal, J. Mareš, Phys. Lett. B. **396**, 21 (1997)
[11] E. Friedman, A. Gal, J. Mareš, Nucl. Phys. A **625**, 272 (1997)
[12] http://gwdac.phys.gwu.edu
[13] T. Waas, M. Rho, W. Weise, Nucl. Phys. A **617**, 449 (1997)
[14] P. B. Siegel, W. B. Kaufmann, W. R. Gibbs, Phys. Rev. C **31**, 2184 (1985)
[15] R. J. Peterson, Phys. Rev. C **60**, 022201 (1999)
[16] A. Gal, E. Friedman, Phys. Rev. Lett. **94**, 072301 (2005)
[17] A. Gal, E. Friedman, Phys. Rev. C **73**, 015208 (2006)
[18] D. Jido, arXiv:1603.07083v2 (2016)

EPJ Web of Conferences **130**, 02006 (2016) DOI: 10.1051/epjconf/201613002006
MESON 2016

Calculations of kaonic nuclei based on chiral meson-baryon coupled channel interaction models

J. Hrtánková[1,2,⋆], A. Cieplý[1], and J. Mareš[1]

[1] *Nuclear Physics Institute, 250 68 Řež, Czech Republic*
[2] *Czech Technical University in Prague, Faculty of Nuclear Sciences and Physical Engineering, Břehová 7, 115 19 Prague 1, Czech Republic*

Abstract. We present our latest calculations of K^--nuclear quasi-bound states using a self-consistent scheme for constructing K^--nuclear potentials from various subthreshold chirally inspired $\bar{K}N$ scattering amplitudes. We consider in-medium versions of the scattering amplitudes taking into account Pauli blocking in the intermediate states. The resulting K^- binding energies as well as the widths exhibit the same A dependence, however, the binding energies strongly depend on the model used.

1 Introduction

The aim of the present study is to compare the predictions for K^--nuclear quasi-bound states calculated using different meson-baryon coupled channel interaction models: Prague (P NLO) [1], Kyoto-Munich (KM NLO) [2], Murcia (M1 and M2) [3], and Bonn (B2 and B4) [4]. They capture the physics of the $\Lambda(1405)$ and reproduce low energy K^-N observables, including the $1s$ level shift and width in the K^- hydrogen atom from the SIDDHARTA experiment [5]. However, the corresponding scattering amplitudes differ considerably below threshold, thus in the energy region relevant for K^--nuclear bound-state calculations as shown in Figure 1.

2 Model

The binding energies B_{K^-} and widths Γ_{K^-} of K^--nuclear quasi-bound states are obtained by solving the Klein-Gordon equation

$$\left[\omega_{K^-}^2 + \vec{\nabla}^2 - m_{K^-}^2 - \Pi_{K^-}(\vec{p}_{K^-}, \omega_{K^-}, \rho)\right]\phi_{K^-} = 0\,, \quad (1)$$

where $\omega_{K^-} = m_{K^-} - B_{K^-} - \mathrm{i}\Gamma_{K^-}/2 - V_C = \tilde{\omega}_{K^-} - V_C$, m_{K^-} is the K^- mass, V_C is the Coulomb potential, and $\vec{p}_{K^-}$ represents the kaon momentum. The self-energy operator Π_{K^-} is constructed in a $t\rho$ form:

$$\Pi_{K^-} = 2\mathrm{Re}(\tilde{\omega}_{K^-})V_{K^-} = -4\pi\frac{\sqrt{s}}{m_N}\left(F_0\frac{1}{2}\rho_p + F_1\left(\frac{1}{2}\rho_p + \rho_n\right)\right)\,, \quad (2)$$

where F_0 and F_1 denote the isospin 1 and 0 S-wave in-medium amplitudes, respectively, m_N is the

[⋆] e-mail: hrtankova@ujf.cas.cz

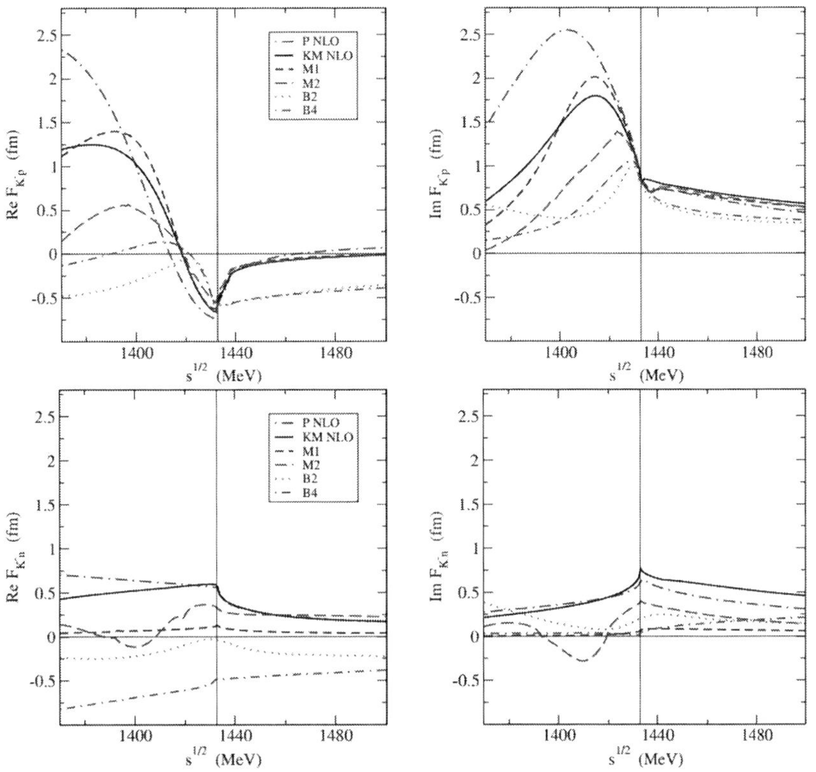

Figure 1. Energy dependence of real (left) and imaginary (right) parts of free-space $K^- p$ (top) and $K^- n$ (bottom) amplitudes in considered models.

nucleon mass and $\sqrt{s}$ is the Mandelstam variable. The proton and neutron density distributions ρ_p and ρ_n are obtained within a relativistic mean-field model. The in-medium amplitudes F_0 and F_1 are obtained from the free-space amplitudes by applying the multiple scattering approach (WRW) [6] which accounts for Pauli correlations:

$$F_1 = \frac{F_{K^-n}(\sqrt{s})}{1 + \frac{1}{4}\xi_k \frac{\sqrt{s}}{m_N} F_{K^-n}(\sqrt{s})\rho}, \quad F_0 = \frac{[2F_{K^-p}(\sqrt{s}) - F_{K^-n}(\sqrt{s})]}{1 + \frac{1}{4}\xi_k \frac{\sqrt{s}}{m_N}[2F_{K^-p}(\sqrt{s}) - F_{K^-n}(\sqrt{s})]\rho}, \tag{3}$$

where ξ_k is adopted from Ref. [6]. In the P NLO model [1], the integration in the underlying Green's function is limited to a certain domain due to the Pauli principle (Pauli) and the in-medium hadron self-energies (Pauli+SE) are considered as well. Figure 2 illustrates that WRW and Pauli approaches yield similar in-medium amplitudes in the subthreshold energy region relevant to our calculations.

The available energy $\sqrt{s}$ in the laboratory frame acquires the form [7] (taking into account non-negligible contribution from particle momenta)

$$\sqrt{s} = m_N + m_{K^-} - B_N - \xi_N B_{K^-} + \xi_{K^-}\mathrm{Re}\mathcal{V}_{K^-}(r) - \xi_N T_N \left(\frac{\rho}{\rho_0}\right)^{2/3}, \tag{4}$$

where B_N is the average binding energy per nucleon, $\xi_{N(K^-)} = m_{N(K^-)}/(m_N + m_{K^-})$, T_N is the nucleon kinetic energy determined from the Fermi Gas model, and $\mathcal{V}_{K^-} = V_{K^-} + V_C$.

 DOI: 10.1051/epjconf/201613002006

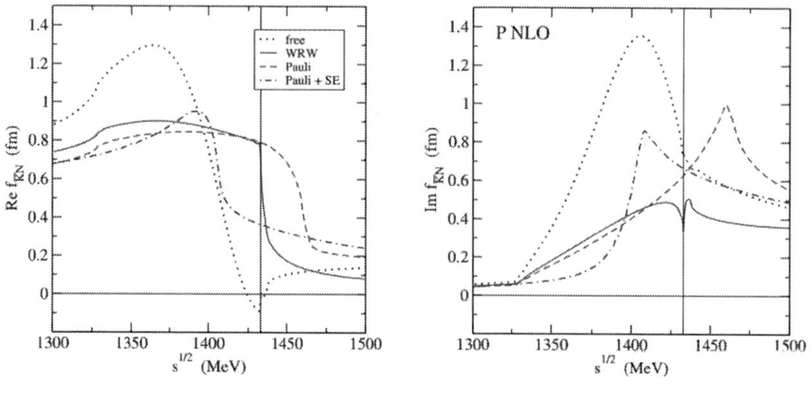

Figure 2. Energy dependence of free-space (dotted line) $f_{\bar{K}N} = \frac{1}{2}(F_{K^-p} + F_{K^-n})$ amplitude compared with WRW modified amplitude (solid line), Pauli (dashed line), and Pauli + SE (dot-dashed line) modified amplitude for $\rho_0 = 0.17$ fm^{-3} in the P NLO model.

3 Results

With the formalism introduced above we performed self-consistent calculations of K^- quasi-bound states in nuclei across the periodic table. In Figure 3, we present the 1s K^- binding energies B_{K^-} and corresponding widths Γ_{K^-} as a function of the mass number calculated for different baryon-meson interaction models. The K^- binding energies exhibit considerable model dependence, nevertheless their A dependence is very similar in all models considered. The K^- widths show rather weak A dependence except for the M2 model. The KM NLO model predicts widths twice as large as the P NLO and M1 models. Within the Bonn models B2 and B4, we did not succeed to obtain any bound states since the real parts of the K^-N amplitudes are weakly attractive or even repulsive below threshold (see Figure 1). In Figure 4, the K^- spectrum in ^{40}Ca calculated for various models is shown. Again, the K^- binding energies strongly depend on the model used. In the P NLO and KM NLO

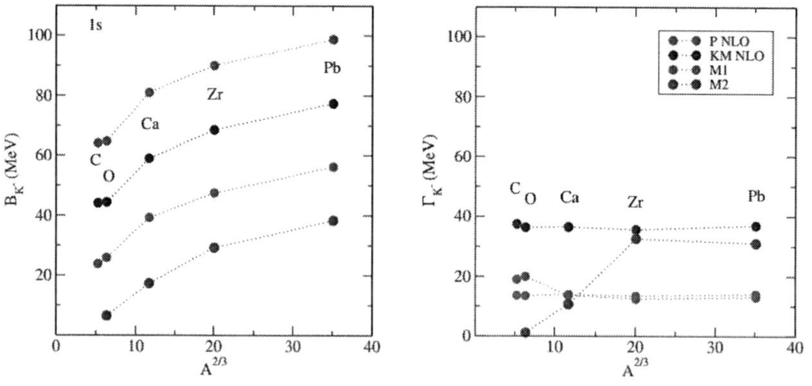

Figure 3. 1s K^- binding energies (left) and corresponding widths (right) in various nuclei calculated self-consistently in the P NLO, KM NLO, M1, and M2 models. $K^-NN \rightarrow YN$ ($Y = \Lambda, \Sigma$) decay modes are not considered.

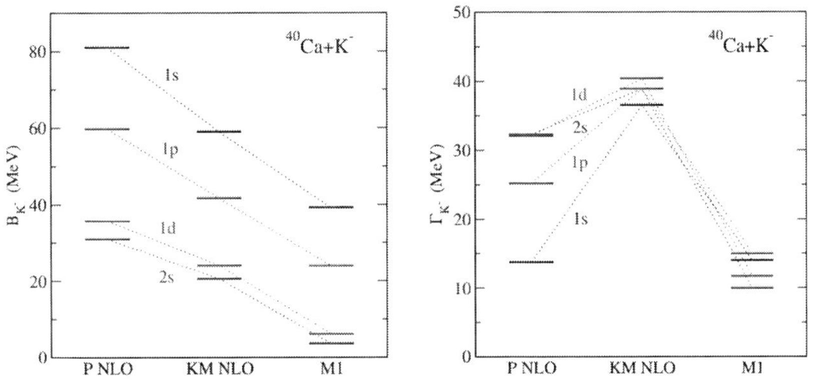

Figure 4. K^- binding energies (left) and widths (right) in s, p and d levels in ^{40}Ca calculated self-consistently in the P NLO, KM NLO, and M1 models. $K^-NN \to YN$ decay modes are not considered.

models, the K^- in the $1s$ state has the smallest $K^-N \to \pi Y$ conversion width due to the considerable energy shift towards the $\pi\Sigma$ threshold. The K^- widths of excited states grow as the $\sqrt{s}$ moves farther from the $\pi\Sigma$ threshold. However, in the M1 model the widths follow the opposite trend. It is due to the fact that $\sqrt{s}$ is much closer to the $\bar{K}N$ threshold where the imaginary part of the K^-p amplitude starts to decrease (see Figure 1).

It is to be noted that the core polarization caused by K^- was not taken into account in the present calculations. Previous studies of kaonic nuclei [8] found this effect quite mild, adding up to 5 MeV to K^- binding energies. Moreover, the annihilation of K^- on 2 nucleons [8] should be considered as well, since it increases the width by almost 50 MeV and, consequently, even the $1s$ state K^- widths become comparable with the binding energies. We intend to include both effects mentioned above in the upcoming calculations.

We wish to thank A. Gal and E. Friedman for fruitful discussion, and M. May for providing us with the free $\bar{K}N$ scattering amplitudes. This work was supported by the GACR Grant No. P203/15/04301S. J. Hrtánková acknowledges financial support from CTU-SGS Grant No. SGS16/243/OHK4/3T/14.

References

[1] A. Cieply, J. Smejkal, Nucl. Phys. **A 881**, 115 (2012)
[2] Y. Ikeda, T. Hyodo and W. Weise, Nucl. Phys. **A 881**, 98 (2012)
[3] Z. H. Guo, J. A. Oller, Phys. Rev. **C 87**, 035202 (2013)
[4] M. Mai and U.-G. Meißner, Nucl. Phys. **A 900**, 51 (2013)
[5] M. Bazzi et al (SIDDHARTA Collaboration), Phys. Lett. **B 704**, 113 (2011)
[6] T. Wass, M. Rho and W. Weise, Nucl. Phys. **A 617**, 449 (1997)
[7] A. Cieplý, E. Friedman, A. Gal, D. Gazda and J. Mareš, Phys. Lett. **B 702**, 402 (2011)
[8] D. Gazda, J. Mareš, Nucl. Phys. **A 881**, 159 (2012)

EPJ Web of Conferences **130**, 02007 (2016) DOI: 10.1051/epjconf/201613002007
MESON 2016

Determination of the ω- and η'-nucleus optical potential

M. Nanova[1,★] and V. Metag[1]
for the CBELSA/TAPS Collaboration

[1] *II. Physikalisches Institut, Justus-Liebig Universität, Giessen, Germany*

Abstract. The ω and η'-nucleus interaction has been studied in photoproduction reactions off C and Nb targets, using the CBELSA/TAPS detector system. Transparency ratio measurements provide information on the inelastic cross section and in-medium width of mesons and thereby on the imaginary part of the meson-nucleus potential. The real part of the optical potential can be deduced from measurements of the excitation function and momentum distribution which are sensitive to the sign and depth of the potential. Data taken on a C and Nb target have been analysed to determine the real and the imaginary part of the ω- and η'-nucleus optical potential. The momentum dependence of the imaginary part of both mesons is presented and discussed. The results are compared to previous experimental results and to model calculations assuming different scenarios. The data are consistent with a weakly attractive potential for both mesons. The relatively small in-medium width of the η' meson encourages the search for η' bound states.

1 Introduction

The in-medium modifications of hadron properties have been identified as one of the key problems in understanding the non-perturbative sector of QCD. Several theoretical papers discuss the possibility of a partial restoration of chiral symmetry in a strongly interacting environment [1–3]. Mesons are considered to be excitations of the QCD vacuum which has a complicated structure with non-vanishing chiral and higher order condensates. These condensates are predicted to change within a strongly interacting medium and, as a consequence, also the mass spectrum of mesons is expected to be modified [2]. This idea inspired hadronic models developed to calculate the in-medium self-energies of hadrons and their spectral functions. Mass shifts and/or in-medium broadening as well as more complex structures in the spectral function due to the coupling of vector mesons to nucleon resonances have been predicted [4, 5].

Recently many studies have focused on the η' meson. Its especially large mass compared to the mass of the other pseudoscalar mesons has to be attributed to chiral and flavor symmetry breaking effects. Due to a reduction of the chiral condensate a drop in the $U_A(1)$ breaking part of the η' mass might be expected [6, 7], causing an η' mass shift of $\approx$-120 MeV at nuclear matter density ρ_0. This prediction is, however, in conflict with earlier calculations within the Nambu-Jona-Lasinio-model which expect almost no change in the η' mass as a function of nuclear density [8]. Further model calculations claim mass shift of the η' of -80, -40 MeV at ρ_0 [9, 10]. It is obvious that these contradictory theoretical predictions call for an experimental clarification.

★e-mail: Mariana.Nanova@exp2.physik.uni-giessen.de

2 Experimental approaches to study mesons in the medium

The meson-nucleus interaction and the in-medium modifications of mesons can be described by an optical potential

$$U(r) = V(r) + iW(r), \tag{1}$$

comprising a real (V(r)) and an imaginary (W(r)) part; r is the distance between the meson and the centre of the nucleus [6]. The strength of the real part of the meson-nucleus potential is connected to the meson in-medium mass shift $\Delta m(\rho_0)$ at saturation density ρ_0 by:

$$V(r) = \Delta m(\rho_0) \cdot c^2 \cdot \frac{\rho(r)}{\rho_0}. \tag{2}$$

The imaginary part of the potential describes the meson absorption in the medium via inelastic channels and is related to the in-medium width Γ_0 of the meson at nuclear saturation density by [11]

$$W(r) = -\frac{1}{2}\Gamma_0 \cdot \frac{\rho(r)}{\rho_0}. \tag{3}$$

The experimental approaches to determine the real part of the potential which were applied for ω and η' mesons and discussed in this paper are: (i) excitation function of the meson and (ii) meson momentum distributions.

The imaginary part of the meson-nucleus potential corresponds to half of the in-medium width: Im$U = \Gamma/2$, which can be extracted from the attenuation of the meson flux deduced from a measurement of the so-called transparency ratio for a number of nuclei [12]. The transparency ratio describes the loss of flux of mesons in nuclei via inelastic processes. Due to these processes the lifetime of mesons in a nuclear medium is reduced leading to an increase in width as compared to the free particle width.

The data on ω and η' photoproduction were taken in a series of experiments in 2009 and 2013/14 with the Crystal Barrel/TAPS detector system at the ELSA accelerator in Bonn [13, 14]. The high segmentation and the almost full solid angle coverage of the detector systems allowed the reconstruction of ω and η' mesons from multi photon final states by invariant mass analysis. The decay modes used in the analyses were $\omega \rightarrow \pi^0\gamma \rightarrow 3\gamma$ and $\eta' \rightarrow \pi^0\pi^0\eta \rightarrow 6\gamma$ with branching ratios of 8.2% and 8.5%, respectively [15].

2.1 The real part of the ω- and η'-nucleus optical potential

Information on the in-medium meson mass and the real part of the meson-nucleus potential can be extracted from the excitation function of the meson as shown in [16, 17]. The experimental results for the ω excitation function in comparison to theoretical calculations are shown in Fig. 1(left). Data were taken on a carbon target with the Crystal Ball and TAPS detector system at MAMI-C. The GiBUU transport model [18] has been applied to calculate the ω excitation function for 6 different scenarios allowing for mass shifts up to -125 MeV at normal nuclear density. It is obvious that the data are not consistent with a strong mass shift scenario, $|V|$>100 MeV as predicted in [2, 19]. A χ^2-fit of the data with the excitation functions calculated for the different scenarios gives a potential depth of $V_{\omega A}(\rho = \rho_0) = -(42 \pm 17(\text{stat}) \pm 20(\text{syst}))$ MeV [20].

The excitation functions for η' meson produced off C and Nb appear to be incompatible with η' mass shifts of -100 MeV and more at normal nuclear matter density, as clearly seen in Fig. 1(middle and right). A χ^2-fit of the data with the excitation functions calculated for the different scenarios over

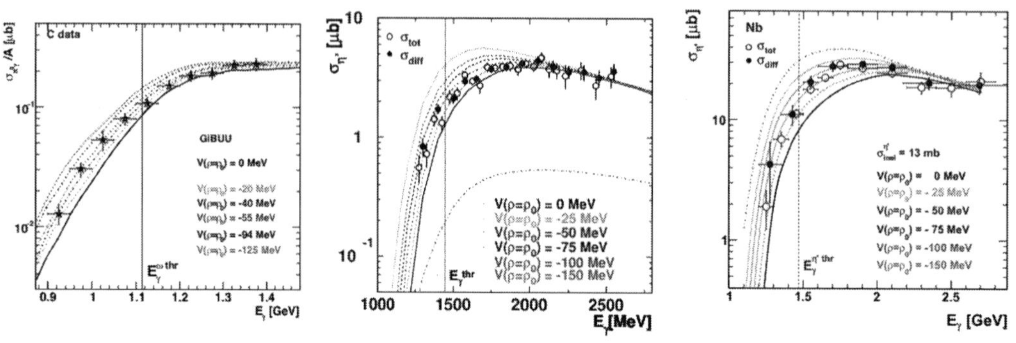

Figure 1. (Left) Measured excitation function for the ω meson in comparison to GiBUU transport calculations for several in-medium modification scenarios [20] assuming mass shifts up to -125 MeV. Measured excitation function for the η' meson off ^{12}C (middle) and ^{93}Nb (right), in comparison to theoretical calculations for different scenarios [14, 21]. The experimental data are extracted by integrating the differential cross sections (full circles) and by direct measurement of the η' yield in incident photon energy bins (open circles). The calculations are for $\sigma_{\eta' N}$=11 mb (for C data) and for $\sigma_{\eta' N}$=13 mb (for Nb data), and for potential depths: V=0 MeV (black line), -25 MeV (green), -50 MeV (blue), -75 MeV (black dashed), -100 MeV (red) and -150 MeV (magenta) at normal nuclear density, respectively, and using the full nucleon spectral function. The dot-dashed blue curve (in the middle spectrum) indicates the fraction of η' mesons produced off correlated nucleon pairs.

the full range of incident energies gives a potential depth of -(40±6) MeV and -(40±12) MeV for C and Nb, respectively [14, 21].

As discussed in [16, 17], the momentum distribution of the mesons is also sensitive to the potential depth and has been investigated as well. A χ^2-fit of the data with the momentum distributions calculated for the different scenarios gives a depth of -(32±11) MeV and -(45±20) MeV for the η'-C and η'-Nb potential, respectively [14, 21].

To study the possible momentum dependence of the real part of the meson-nucleus optical potential an experiment has been performed to produce ω mesons with low momenta (< 500 MeV/c). In the reaction $\gamma C \to \omega p \to \pi^0 \gamma p$ the proton, detected in the TAPS detector ($\Theta_p = 1° - 11°$), takes over most of the momentum of the incoming photon beam. Being so low in energy, the ω-mesons are particularly sensitive to the ω nucleus potential. In case of a strongly repulsive (attractive) interaction one would expect the peak in the kinetic energy distribution to be shifted to higher (lower) energies for the C target. The sensitivity of the peak position in the kinetic energy distribution on the potential depth, studied in [22], can be exploited to deduce the depth of the real part of the ω-nucleus potential. The experimental data for the ω kinetic energy distribution have been compared with the theoretical calculations and a potential depth of -(15±35) MeV has been determined, which appears to be too small to allow for the formation of ω mesic states in view of the large in-medium width of $\approx$ 90 MeV (see Sec. 2.2). Combining the results from the analysis of the excitation function and the peak position from the kinetic energy distribution an average depth of the real part of the ω-C optical potential of V($\rho = \rho_0$) = -(29 ±19(stat)±20(syst)) MeV is obtained Fig. 2 (left).

From the analysis of the excitation functions and the momentum distributions for the η' meson and by proper weighting of the errors a depth of the real part of the η'-C and -Nb optical potential of V($\rho = \rho_0$) = -(39 ±7(stat)±15(syst)) MeV is obtained (see Fig. 2 (right)). This result for V($\rho = \rho_0$) is consistent with predictions of the η'-nucleus potential depth within the Quark-Meson Coupling model [10] and with calculations in [23] but does not support larger mass shifts as discussed in [6, 9].

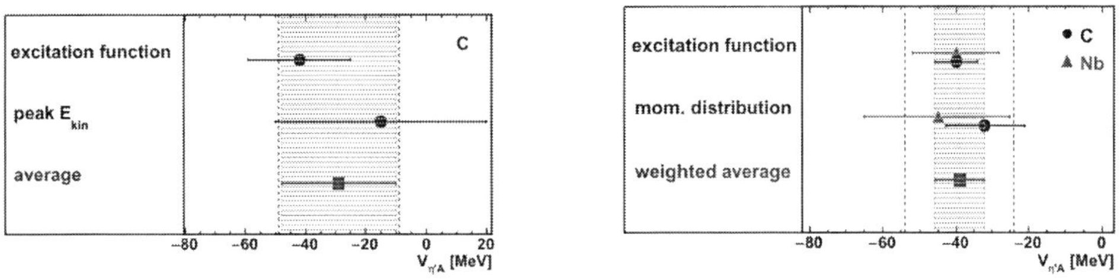

Figure 2. Depths of the real part of the ω (left) and η' (right) -nucleus potential determined by analyzing the excitation function and the momentum distributions for C [20, 21] (full black circles) and for Nb [14] (red triangles). The weighted overall average is indicated by a blue square. The shaded area indicates the statistical error. The vertical hatched lines mark the range of systematic uncertainties [14, 24].

2.2 The imaginary part of the ω- and η'-nucleus optical potential

To determine the meson attenuation in nuclei and the inelastic meson-nucleon cross sections as a function of the meson momentum, the transparency ratio is deduced from the data as defined in Eq.4 [11, 13, 25]. To suppress nuclear effects not related to meson absorption, the transparency ratio is not taken relative to the free nucleon but rather to a nucleon in a light nucleus like carbon. The momentum dependence of the transparency ratio for a meson m is thus obtained by dividing the differential inclusive meson production cross sections for niobium by the one for carbon according to Eq.4

$$T^{\rm m}_{\rm Nb/C}(p_m) = \frac{12 \cdot \frac{d\sigma_{\gamma \rm Nb \to mX}}{dp_m}(p_m)}{93 \cdot \frac{d\sigma_{\gamma \rm C \to mX}}{dp_m}(p_m)}, \tag{4}$$

where 12 and 93 are the nuclear mass numbers of carbon and niobium, respectively. The resulting transparency ratios as a function of the meson momentum are shown in Fig. 3. Consistent with earlier measurements [21, 25], a slight increase with momentum is observed for the ω meson while for the η' meson the transparency ratio is almost independent of momentum. The momentum depen-

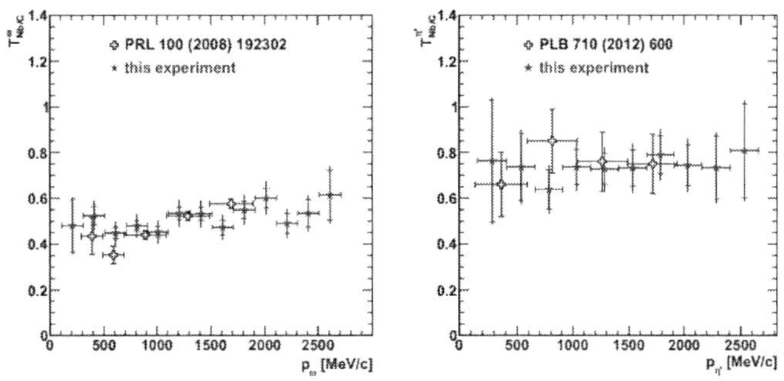

Figure 3. The transparency ratio according to Eq. 4 for (Left) ω and (Right) η' photoproduction off carbon and niobium (red stars) in comparison to earlier measurements (open crosses) [11, 25]. For the present data the thick error bars (red) represent the statistical errors. The thin error bars (black) include the systematic errors added in quadrature [13].

dence of the transparency ratio allows the determination of the in-medium ω and η' widths which can be converted into the dependence of the imaginary part of the ω- and η'-nucleus potential as a function of the available energy in the meson-^{93}Nb system, as shown in Fig. 4 (for more details see [13]). The data have been fitted and extrapolated towards the production threshold. For the ω meson the modulus of the imaginary part of the meson nucleus potential near threshold is found to be (48 ± 12(stat)± 9(syst)) MeV comparable to the modulus of the real part of about 30 MeV, determined in [20, 22, 27]. For the η' meson the extrapolation towards the production threshold yields an imaginary potential of (13 ± 3(stat)±3(syst)) MeV, corresponding to an imaginary part of the η' scattering length $\mathrm{Im}(a_{\eta' N}) = (0.16 \pm 0.05)$ fm. This is about a factor two smaller than obtained in the direct determination of the $\eta' N$ scattering length from an analysis of near threshold η' production in the $pp \rightarrow pp\eta'$ reaction [28], but almost overlaps within the errors.

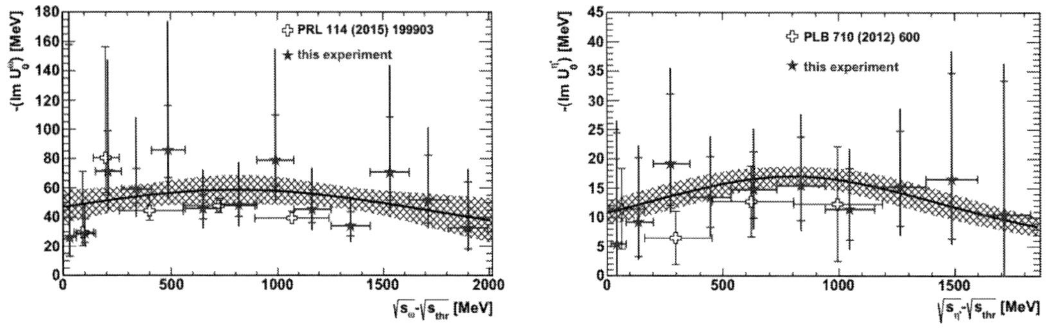

Figure 4. Imaginary part of the (Left) ω-nucleus and (Right) η'-nucleus optical potential as a function of the available energy in the meson-^{93}Nb system (red stars) [13] in comparison to earlier measurements (open crosses) [11, 25, 26]. The solid curves are Breit-Wigner fits to the present data. The shaded areas indicate a confidence level of $\pm 1\sigma$ of the fit curve taking statistical and systematic errors into account (taken from [13]).

3 Conclusion

The photoproduction of ω and η' mesons off nuclei has been studied in the 1-3 GeV energy range to extract information on the in-medium properties of these mesons. The interaction of mesons with nuclei and the in-medium effects have been described with an optical model, comprising a real part, associated with the mass modification in the nuclear medium, and an imaginary part, accounting for the absorption of mesons in nuclei. For the ω- and η'- nucleus optical potential values of $U_{\omega A}(\rho = \rho_0) =$ -((29 ±19(stat)±20(syst)) + i(48±12(stat)±9(syst))) MeV and $U_{\eta' A}(\rho = \rho_0) =$ -((39 ±7(stat)±15(syst)) + i(13±3(stat)±3(syst))) MeV have been obtained (Fig. 5) [24]. The latter measurement establishes an in-medium mass drop of a pseudoscalar meson at normal nuclear matter density. Although not very deep, the real part of the η' - nucleus potential is about three times larger than the imaginary part and may thus allow the existence and observation of η'-nucleus states, only bound by the strong interaction. Corresponding experiments are ongoing.

Acknowledgements

We would like to thank our coworker Stefan Friedrich as well as all other colleagues in the CBELSA/TAPS collaboration for important contributions to this work. This work is supported by the *Deutsche Forschungsgemeinschaft* within SFB/TR16.

DOI: 10.1051/epjconf/201613002007

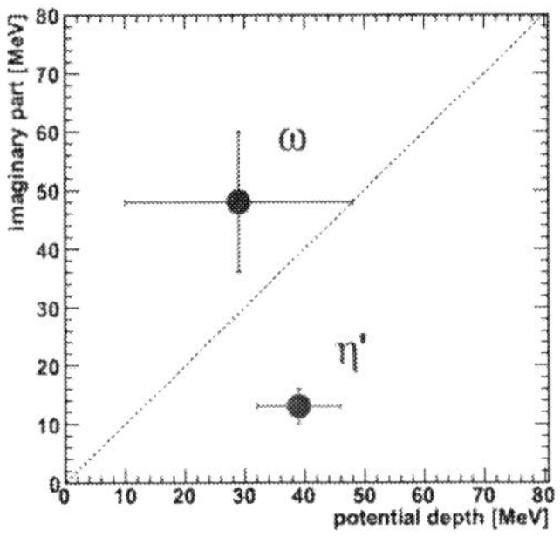

Figure 5. The imaginary part versus the real part of the meson-nucleus potential for ω and η' mesons [24].

References

[1] V. Bernard and U.-G. Meissner, Nucl. Phys. A **489**, 647 (1988)
[2] T. Hatsuda and S. Lee, Phys. Rev. C **46**, R34 (1992)
[3] G. E. Brown and M. Rho, Phys. Rev. Lett. **66**, 2720 (1991)
[4] P. Mühlich et al., Nucl. Phys. A **780**, 187 (2006)
[5] M. Lutz *et al.*, Nucl. Phys. A **706**, 437 (2002)
[6] H. Nagahiro, M. Takizawa, and S. Hirenzaki, Phys. Rev. C **74**, 045203 (2006)
[7] Y. Kwon *et al.*, Phys. Rev. D **86**, 034014 (2012)
[8] V. Bernard and U.-G. Meissner, Phys. Rev. D **38**, 1551 (1988)
[9] S. Sakai and D. Jido, Phys. Rev. C **88**, 064906 (2013)
[10] S. D. Bass and A. W. Thomas, Phys. Lett. B **634** 368 (2006)
[11] M. Nanova *et al.*, CBELSA/TAPS Collaboration, Phys. Lett. B **710**, 600 (2012)
[12] D. Cabrera *et al.*, Nucl. Phys. A **733**, 130 (2004)
[13] S. Friedrich *et al.*, CBELSA/TAPS Collaboration, Eur. Phys. J. A **52**, 297 (2016)
[14] M. Nanova *et al.*, CBELSA/TAPS Collaboration, Phys. Rev. C **94**, 025205 (2016)
[15] K. A. Olive *et al.*, Particle Data Group, Chin. Phys. C **38**, 090001 (2014)
[16] J. Weil, U. Mosel, and V. Metag, Phys. Lett. B **723**, 120 (2013)
[17] E. Ya. Paryev, J. Phys. G: Nucl. Part. Phys. **40**, 025201 (2013)
[18] O. Buss *et al.*, Phys. Rep. **512**, 1 (2012)
[19] F. Klingl *et al.*, Nucl. Phys. A **610**, 297 (1997), Nucl. Phys. A **650**, 299 (1999)
[20] V. Metag *et al.*, A2 Collaboration, Prog. Part. Nucl. Phys. **67**, 530 (2012)
[21] M. Nanova *et al.*, CBELSA/TAPS Collaboration, Phys. Lett. B **727**, 417 (2013)
[22] S. Friedrich *et al.*, CBELSA/TAPS Collaboration, Phys. Lett. B **736**, 26 (2014)
[23] H. Nagahiro, S. Hirenzaki, E. Oset, and A. Ramos, Phys. Lett. B **709**, 87 (2012)
[24] V. Metag, M. Nanova, and K.-T. Brinkmann, CRC 16 Symp., Proceed. June (2016)
[25] M. Kotulla *et al.*, CBELSA/TAPS Collaboration, Phys. Rev. Lett. **100**, 192302 (2008)
[26] M. Kotulla *et al.*, CBELSA/TAPS Collaboration, Phys. Rev. Lett. **114**, 199903 (2015)
[27] V. Metag, Hyp. Int. **234**, 25 (2015)
[28] E. Czerwinski *et al.*, Phys. Rev. Lett. **113**, 062004 (2014)

Searching for η-mesic Helium with WASA-at-COSY facility

Oleksandr Rundel[1,*], Oleksandr Khreptak[1], Magdalena Skurzok[1], Wojciech Krzemien[2], and Pawel Moskal[1]
for the WASA-at-COSY Collaboration

[1] *M. Smoluchowski Institute of Physics, Jagiellonian University, 30-059 Cracow, Poland*
[2] *National Centre for Nuclear Research, 05-400 Otwock, Świerk, Poland*

Abstract. The search for the ^{4}He-η and ^{3}He-η bound states was performed with the WASA-at-COSY facility using a ramped beam technique in 2008, 2010 and 2014, respectively. The 2008 and 2010 data analyses has already been completed and resulted in the upper limits for the total cross sections for the η-mesic ^{4}He production in $dd \to {}^3\text{He}n\pi^0$ and $dd \to {}^3\text{He}p\pi^-$ processes. No narrow structure which could be interpreted as a bound state was observed. The experiment carried out in 2014 collected the largest data sample for ^{3}He-η in the world. The data analysis is in progress. This article presents results of the 2010 data analysis and preliminary results of the 2014 data analysis.

1 Introduction

The existence of η-mesic nuclei in which the η meson is bound with the nucleus with the strong interaction was postulated by Haider and Liu already in 1986 [1], however till now no experiment confirmed it empirically. Recent theoretical studies of hadronic- and photo- production of the η meson, reveal the possibility of the existence of η-mesic bound states for light nuclei like ^{4}He and ^{3}He [2, 3], and give hope of their observation in deuteron-deuteron and proton-deuteron fusion reactions, respectively. Interested readers could find valuable information in the recent reviews on mesic nuclei search [4–12].

This article presents brief description of the search for the η-mesic Helium performed by the WASA-at-COSY collaboration carried out in 2008, 2010 (^{4}He-η nucleus) and 2014 (^{3}He-η nucleus). The results concerning ^{4}He-η bound states have been already published in references [4, 13–15] while analysis of the data taken in 2014 is still in progress.

2 Experiment

The search for η-mesic Helium was carried out using the WASA detector system [16] with a deuteron target of pellet type installed at the COSY synchrotron [17] in the Jülich Research Center in Germany. COSY provided deuteron and proton beam for the search for ^{4}He-η and ^{3}He-η bound systems, respectively. The main advantage of the used experimental setup was the possibility of continuous beam energy changing and the simultaneous registration of all particles taking part in the reaction. The

*e-mail: alexrndl@gmail.com

deuteron beam momentum was varying from 2.127 GeV/c to 2.422 GeV/c, while the proton beam momentum from 1.426 GeV/c to 1.635 GeV/c, crossing the kinematic threshold for η production.

The signatures of the η-mesic nuclei are searched for by studying the excitation function for the chosen decay channels of the ^{4}He-η and ^{3}He-η systems, formed in d-d and p-d collisions, respectively.

3 Results for the search of ^{4}He-η bound states

The first experiment carried out in June 2008, was focused on the measurement of the excitation function of the $dd \to {}^3\mathrm{He}p\pi^-$ process near the η production threshold covering the excess energy range from -51.4 MeV up to 22 MeV. The excitation function does not reveal the resonance structure which could be interpreted as the η-mesic ^{4}He. Therefore, only the upper limit of the total cross-section for the bound state formation and decay in the $dd \to ({}^4\mathrm{He}\text{-}\eta)_{bound} \to {}^3\mathrm{He}p\pi^-$ reaction was determined which varies from 20 nb to 27 nb at the confidence level of 90% [18, 19].

The second experiment, performed in November 2010, brought in about a 10 times higher statistic with respect to the previous measurement. The search for the ^{4}He-η bound state was performed for two reactions $dd \to {}^3\mathrm{He}n\pi^0$ and $dd \to {}^3\mathrm{He}p\pi^-$ via the measurement of the excitation function for each of them in the excess energy range Q∈(-70,30) MeV. Since no narrow peak, which could originate from the mesic nuclei, was observed the upper limit of the total cross section for each process was determined independently at the confidence level of 90%. The obtained upper limits vary from 21 to 36 nb for the $dd \to ({}^4\mathrm{He}\text{-}\eta)_{bound} \to {}^3\mathrm{He}n\pi^0$ process and from 5 to 9 nb for the $dd \to ({}^4\mathrm{He}\text{-}\eta)_{bound} \to {}^3\mathrm{He}p\pi^-$ process for the bound state width ranging from 5 to 50 MeV [4, 14].

A new analysis taking into account the isospin relation between the $n\pi^0$ and $p\pi^-$ pairs outgoing from the N^* decay (probability of $p\pi^-$ pair production is two times higher than in case of $n\pi^0$ production) results in the upper limit varying in the range from 2.5 to 3.5 nb for the $dd \to ({}^4\mathrm{He}\text{-}\eta)_{bound} \to {}^3\mathrm{He}n\pi^0$ process and from 5 to 7 nb for the $dd \to ({}^4\mathrm{He}\text{-}\eta)_{bound} \to {}^3\mathrm{He}p\pi^-$ process (for the fixed binding energy B_s=30 MeV and width range Γ ∈(5,50) MeV). The green area presents the systematics errors.

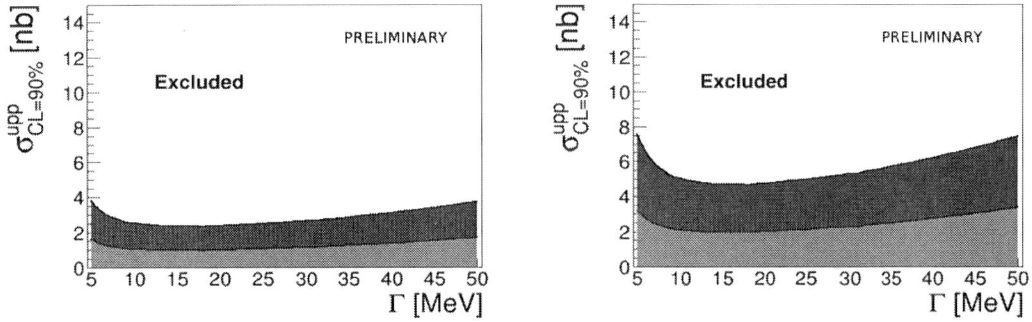

Figure 1. Upper limit of the total cross-section for $dd \to ({}^4\mathrm{He}\text{-}\eta)_{bound} \to {}^3\mathrm{He}n\pi^0$ (left panel) and $dd \to ({}^4\mathrm{He}\text{-}\eta)_{bound} \to {}^3\mathrm{He}p\pi^-$ (right panel) reaction as a function of the width of the bound state. The binding energy was fixed to 30 MeV. The upper limit was determined via the simultaneous fit for both channels. The green area denotes the systematic uncertainties.

4 Status of the search for ^{3}He-η nuclei

The experiment for searching ^{3}He-η mesic nuclei was carried out in May 2014. Three processes corresponding to the three mechanisms were measured: (i) absorption of the η meson by one of the nucleons, which subsequently decays into N-π pair e.g.: $pd \to ({}^3\mathrm{He}\text{-}\eta)_{bound} \to ppp\pi^-$, (ii) decay of the η -meson while it is still "orbiting" around a nucleus e.g.: $pd \rightarrow (^3\text{He-}\eta)_{bound} \rightarrow {}^3\text{He}6\gamma$ or $pd \rightarrow (^3\text{He-}\eta)_{bound} \rightarrow {}^3\text{He}2\gamma$ reactions and (iii) η meson absorption by few nucleons [20, 21] e.g.: $pd \rightarrow (^3\text{He-}\eta)_{bound} \rightarrow ppn$ or $pd \rightarrow (^3\text{He-}\eta)_{bound} \rightarrow pd$. At present, the analysis of data is in progress. There are preliminary results of luminosity estimation obtained based on $pd \rightarrow {}^3\text{He}\eta$ reaction. The estimated average luminosity is equal to about $4.5 \cdot 10^{30}$ cm^{-2}s^{-1}. For the purpose of this research, the luminosity determination is crucial, both above and below the η meson production threshold. This will be determined based on the quasi-free proton-proton scattering reaction using methods described in [22, 23].

5 Conclusion

The upper limits for the cross section for the creation of ^{4}He-η mesic nuclei have been preliminary determined for $dd \rightarrow {}^3\text{He}n\pi^0$ and $dd \rightarrow {}^3\text{He}p\pi^-$ processes after analysing the data obtained in experiments in June 2008 and November 2010. The analysis of data obtained in the experiment for the search of ^{3}He-η mesic nuclei that took place in May 2014 is in progress. Based on preliminary calculations we expect to reach sensitivity of the cross section of about 10 nb which is much lower than in case of COSY-11 result obtained for $pd \rightarrow (^3\text{He-}\eta)_{bound} \rightarrow ppp\pi^-$ reaction (270 nb) [24].

We acknowledge support by the Polish National Science Center through grants No. DEC-2013/11/N/ST2/04152, 2011/01/B/ST2/00431 and 2011/03/B/ST2/01847.

References

[1] Q. Haider, L. C. Liu, Phys. Lett. **B172**, 257 (1986)
[2] S. Wycech, A. M. Green, J. A. Niskanen, Phys. Rev. **C52**, 544 (1995)
[3] C. Wilkin, Phys. Rev. **C47**, 938 (1993)
[4] M. Skurzok, W. Krzemień, O. Rundel and P. Moskal, Acta. Phys. Polon. **B47**, 503 (2016)
[5] A. Gal et al., Acta Phys. Polon. **B45**, 673 (2014)
[6] H. Machner, J. Phys. **G42**, 043001 (2015)
[7] C. Wilkin, Acta Phys. Polon. **B47**, 249 (2016)
[8] B. Krusche, C. Wilkin, Prog. Part. Nucl. Phys. **80**, 43 (2014)
[9] N. G. Kelkar et al., Rept. Prog. Phys. **76**, 066301 (2013); Acta Phys. Polon. B46, 113 (2015)
[10] S. Bass et al., Acta Phys. Polon. **B47**, 373 (2016)
[11] Q. Haider, L. C. Liu, Int. J. Mod. Phys. **E24**, 1530009 (2015)
[12] P. Moskal, Few Body Syst. **55**, 667 (2014); Acta Phys. Polon. **B47**, 97 (2016)
[13] W. Krzemień, P. Moskal, M. Skurzok, Few Body Syst. **55**, 795 (2014)
[14] M. Skurzok, *arXiv:1509:01385*, Ph.D. thesis, Jagiellonian University (2015)
[15] W. Krzemień, *arXiv:1202:5794*, Ph.D. thesis, Jagiellonian University (2011)
[16] WASA-at-COSY Collaboration: H.-H. Adam *et al.*, *arXiv:nucl-ex/0411038* (2004)
[17] R. Maier, Nucl. Instrum. Meth. **A390**, 1 (1997)
[18] W. Krzemien, P. Moskal, M. Skurzok, Acta Phys. Polon. **B45**, 689 (2014)
[19] P. Adlarson *et al.*, Phys. Rev. **C87**, 035204 (2013)
[20] C. Wilkin, Acta Phys. Polon. **B45**, 603 (2014)
[21] S. Wycech, W. Krzemien, Acta Phys. Polon. **B45**, 745 (2014)
[22] P. Moskal and R. Czyżykiewicz, AIP Conf. Proc. **950**, 118 (2007)
[23] P. Moskal et al., Phys. Rev. **C79**, 015208 (2009)
[24] P. Moskal, J. Smyrski, Acta Phys. Polon. **B41**, 2281 (2010)

EPJ Web of Conferences **130**, 02009 (2016) DOI: 10.1051/epjconf/201613002009
MESON 2016

Molecular components in $D_{s0}^*(2317)$ and $D_{s1}(2460)$ mesons

Pablo G. Ortega[1]*, Jorge Segovia[2,]**, David R. Entem[3], and Francisco Fernández[3]

[1] *CERN (European Organization for Nuclear Research), CH-1211 Geneva, Switzerland*
[2] *Technische Universität München, D-85748 Garching bei München, Germany*
[3] *Universidad de Salamanca, E-37008 Salamanca, Spain*

Abstract. Different experiments have confirmed that the $D_{s0}^*(2317)$ and $D_{s1}(2460)$ mesons are very narrow states located, respectively, below the DK and D^*K thresholds. This is markedly in contrast with the expectations of naive quark models and heavy quark symmetry. We address the mass shifts of the $c\bar{s}$ ground states with quantum numbers $J^P = 0^+$ ($D_{s0}^*(2317)$) and $J^P = 1^+$ ($D_{s1}(2460)$) using a nonrelativistic constituent quark model in which quark-antiquark and meson-meson degrees of freedom are incorporated. The quark model has been applied to a wide range of hadronic observables and thus the model parameters are completely constrained. We observe that the coupling of the 0^+ (1^+) meson sector to the DK (D^*K) threshold is a key feature in lowering the masses of the corresponding $D_{s0}^*(2317)$ and $D_{s1}(2460)$ states predicted by the naive quark model, but also in describing the $D_{s1}(2536)$ meson as the 1^+ state of the $j_q^P = 3/2^+$ doublet predicted by heavy quark symmetry and thus reproducing its strong decay properties. Two features of our formalism cannot be address nowadays by other approaches: the coupling of the D-wave D^*K threshold in the $J^P = 1^+$ $c\bar{s}$ channel and the computation of the probabilities associated with different Fock components in the physical state.

1 Introduction

Prior to the discovery in 2003 of the $D_{s0}^*(2317)$ ($J^P = 0^+$) [1] and $D_{s1}(2460)$ (1^+) [2] resonances, the heavy-light meson sectors were reasonably well understood in the $m_Q \to \infty$ limit. In such a limit, heavy quark symmetry (HQS) holds [3]. The heavy quark acts as a static color source, its spin s_Q is decoupled from the total angular momentum of the light quark j_q and they are separately conserved. Then, the heavy-light mesons can be organized in doublets, each one corresponding to a particular value of j_q and parity. For the lowest P-wave charmed-strange mesons, HQS predicts two doublets which are labeled by $j_q^P = 1/2^+$ with $J^P = 0^+, 1^+$ and $j_q^P = 3/2^+$ with $J^P = 1^+, 2^+$. Moreover, the strong decays of the $D_{sJ}(j_q = 3/2)$ proceed only through D-waves, while the $D_{sJ}(j_q = 1/2)$ decays happen only through S-waves [3]. The D-wave decay is suppressed by the barrier factor which behaves as q^{2L+1} where q is the relative momentum of the two decaying mesons. Therefore, states decaying through D-waves are expected to be narrower than those decaying via S-waves.

The $D_{s0}^*(2317)$ and $D_{s1}(2460)$ mesons are considered to be the members of the $j_q^P = 1/2^+$ doublet and thus being almost degenerated and broad due to its S-wave decay. However, neither experimental values of their masses nor their empirical widths accommodate into the theoretical expectations.

*Currently at Instituto de Física Corpuscular (IFIC), CSIC-Universidad de Valencia, E-46071 Valencia, Spain
**speaker, e-mail: jorge.segovia@tum.de

 DOI: 10.1051/epjconf/201613002009

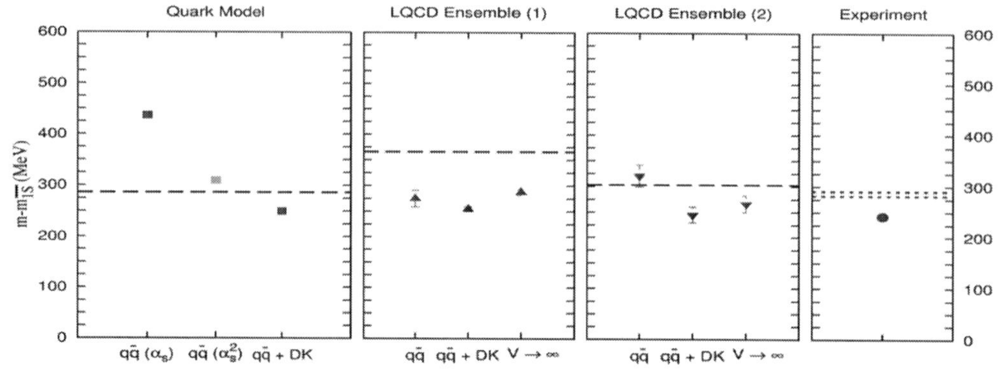

Figure 1. Energy levels from constituent quark model (CQM), from Lattice QCD [16] using Ensemble (1) and Ensemble (2), and from experiment [26]. We show, for CQM results, the quark-antiquark value taking into account the OGE potential (α_s), including its one-loop corrections (α_s^2) and coupling with the DK threshold. For the lattice QCD results, in each ensemble, we show values with just a $q\bar{q}$ interpolator basis and with a combined basis of $q\bar{q}$ and DK interpolating fields. The value of the bound $D_{s0}^*(2317)$ state position in the infinite volume limit $V \longrightarrow \infty$ is obtained by an analytical continuation of the scattering amplitude combined with Lüscher's finite volume method. The dashed lines represent the threshold for DK in each approach and the dotted lines are the thresholds for D^0K^+ and D^+K^0 in experiment.

These results led to many theoretical speculations about the nature of these resonances ranging from conventional $c\bar{s}$ states [4, 5] to molecular or compact tetraquark interpretations [6–12].

Certainly quark models predict $c\bar{s}$ ground states with quantum numbers $J^P = 0^+$ and 1^+ that do not fit the experimental data. As the predictions of the quark models are roughly reasonable for other states in the charmed-strange sector [13], one must expect that the $D_{s0}^*(2317)$ and $D_{s1}(2460)$ resonances should be modifications of the genuine $c\bar{s}$ states rather than new states out of the systematics of the quark model. On this respect, particularly relevant was the suggestion [14, 15] that the coupling of the $J^P = 0^+$ (1^+) $c\bar{s}$ state to the DK (D^*K) threshold plays an important dynamical role in lowering the bare mass to the observed value. Moreover, in a recent lattice study of the $D_{s0}^*(2317)$ and $D_{s1}(2460)$ mesons [16], good agreement with the experimental mass was found when operators for $D^{(*)}K$ scattering states are included.

In this contribution to the proceedings we present the work performed in Ref. [17][1]. Therein, we study the low-lying P-wave charmed-strange mesons using a nonrelativistic constituent quark model in which quark-antiquark and meson-meson degrees of freedom are incorporated. The constituent quark model (CQM) was proposed in Ref. [18] (see references [19] and [20] for reviews). This model successfully describes hadron phenomenology and hadronic reactions and has been recently applied to mesons containing heavy quarks (see, for instance, Refs. [21–25]).

2 Results for the $D_{s0}^*(2317)$ meson

Figure 1 compares our results for the $D_{s0}^*(2317)$ meson with the lattice QCD study of Ref. [16] and with experiment [26]. Instead of the $D_{s0}^*(2317)$ mass itself, following the lattice study, we compare the values of $m_{D_{s0}^*(2317)} - m_{\overline{1S}}$, where $m_{\overline{1S}} = 1/4(m_{D_s} + 3m_{D_s^*})$ is the spin-averaged ground state mass.

The mass of the $D_{s0}^*(2317)$ state obtained using the naive quark model and without the 1-loop corrections to the one-gluon exchange (OGE) potential is much higher than the experimental value. In this case, the $m_{D_{s0}^*(2317)} - m_{\overline{1S}} = 437\,\text{MeV}$ is almost twice the empirical figure. The mass associated

[1]All the details about the computation and the theoretical framework can be found in Ref. [17] and references therein.

 DOI: 10.1051/epjconf/201613002009

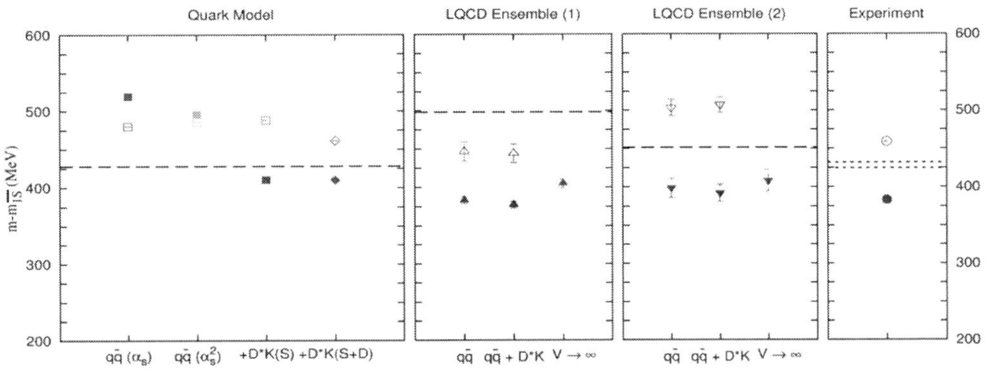

Figure 2. Energy levels from constituent quark model (CQM), from Lattice QCD [16] using Ensemble (1) and Ensemble (2), and from experiment [26]. We show, for CQM results, the quark-antiquark value taking into account the OGE potential (α_s), including its one-loop corrections (α_s^2) and coupling with the D^*K threshold in S- and D-wave. For the lattice QCD results, in each case, we show values with just a $q\bar{q}$ interpolator basis and with a combined basis of $q\bar{q}$ and D^*K interpolating fields. Remember that in the lattice QCD computations the D^*K threshold is coupled only in an S-wave. The value of the bound $D_{s1}(2460)$ state position in the infinite volume limit $V \longrightarrow \infty$ is obtained by an analytical continuation of the scattering amplitude combined with Lüscher's finite volume method. This method has not been used for the $D_{s1}(2536)$ meson. The dashed lines represent the threshold for D^*K in each approach and the dotted lines are the thresholds for $D^{*0}K^+$ and $D^{*+}K^0$ in experiment.

to the $D^*_{s0}(2317)$ state is very sensitive to the α_s^2-corrections of the OGE potential. This effect brings down the $m_{D^*_{s0}(2317)} - m_{\overline{1S}}$ splitting to 309 MeV, which is now only 30% higher than the experimental value. However, as one can see in Fig. 1, the hypothetical $D^*_{s0}(2317)$ state would be above the DK threshold and thus would decay into this final channel in an S-wave making the state wider than the observed one. The mass-shift due to the α_s^2-corrections allows that the 0^+ state be close to the DK threshold. This makes the DK coupling a relevant dynamical mechanism in the formation of the $D^*_{s0}(2317)$ bound state. When we couple the 0^+ $c\bar{s}$ ground state with the DK threshold, the splitting $m_{D^*_{s0}(2317)} - m_{\overline{1S}} = 249.6$ MeV is in good agreement with experiment. Regarding the probabilities of the different Fock components in the physical state, we obtain 66% for $q\bar{q}$ and 34% for DK reflecting that the $D^*_{s0}(2317)$ meson is mostly of quark-antiquark nature in our approach.

3 Results for the $D_{s1}(2460)$ and $D_{s1}(2536)$ mesons

Figure 2 compares our results for the $m_{D_{s1}} - m_{\overline{1S}}$ mass splitting of the first two $J^P = 1^+$ charmed-strange states with the lattice QCD study of Ref. [16] and with experiment [26].

The naive quark model predicts that the states corresponding to the $D_{s1}(2460)$ and $D_{s1}(2536)$ mesons are almost degenerated, with masses close to the experimentally observed mass of the $D_{s1}(2536)$. The inclusion of the 1-loop corrections to the OGE potential does not improve the situation, making the splitting between the two states even smaller. Following lattice criteria, we couple first the D^*K threshold in an S-wave with the two 1^+ $c\bar{s}$ states. One can see in Fig. 2 that the state associated with the $D_{s1}(2460)$ meson goes down in the spectrum and it is located below D^*K threshold with a mass compatible with the experimental value. The state associated with the $D_{s1}(2536)$ meson is almost insensitive to this coupling because it is the $J^P = 1^+$ member of the $j_q = 3/2$ doublet predicted by HQS and thus it couples mostly in a D-wave to the D^*K threshold. Lattice QCD has not yet computed the coupling in D-wave of the D^*K threshold with the 1^+ $c\bar{s}$ sector. This coupling is trivially implemented in our approach. The state associated with the $D_{s1}(2460)$ meson experience a very small modification because it is almost the $|1/2, 1^+\rangle$ eigenstate of HQS, whereas the state associated with $D_{s1}(2536)$ meson suffers a moderate mass-shift approaching to the experimental value.

When the D^*K threshold is coupled, the meson-meson component is around 50% for both $D_{s1}(2460)$ and $D_{s1}(2536)$ mesons. It is also relevant to realize that the quark-antiquark component in the wave function of the $D_{s1}(2536)$ meson holds quite well the ${}^1P_1$ and ${}^3P_1$ composition predicted by HQS, which is crucial in order to have a very narrow state and describe well its decay properties.

4 Summary

We have performed a coupled-channel computation taking into account the $D^*_{s0}(2317)$, $D_{s1}(2460)$ and $D_{s1}(2536)$ mesons and the DK and D^*K thresholds within the framework of a constituent quark model. Our method allows to introduce the coupling with the D-wave D^*K channel and the computation of the probabilities associated with the different Fock components of the physical state.

This work has been partially funded by Ministerio de Ciencia y Tecnología under Contract no. FPA2013-47443-C2-2-P, by the Spanish Excellence Network on Hadronic Physics FIS2014-57026-REDT, and by the Junta de Castilla y León under Contract no. SA041U16. P.G.O. acknowledges the financial support from the Spanish Ministerio de Economia y Competitividad and European FEDER funds under the contract no. FIS2014-51948-C2-1-P. J.S. acknowledges the financial support from Alexander von Humboldt Foundation.

References

[1] B. Aubert et al. (BaBar), Phys. Rev. Lett. **90**, 242001 (2003)
[2] D. Besson et al. (CLEO), Phys. Rev. **D68**, 032002 (2003)
[3] N. Isgur, M.B. Wise, Phys. Rev. Lett. **66**, 1130 (1991)
[4] Fayyazuddin, Riazuddin, Phys. Rev. **D69**, 114008 (2004)
[5] O. Lakhina, E.S. Swanson, Phys. Lett. **B650**, 159 (2007)
[6] T. Barnes, F.E. Close, H.J. Lipkin, Phys. Rev. **D68**, 054006 (2003)
[7] H.J. Lipkin, Phys. Lett. **B580**, 50 (2004)
[8] P. Bicudo, Nucl. Phys. **A748**, 537 (2005)
[9] D. Gamermann, E. Oset, D. Strottman, M.J. Vicente Vacas, Phys. Rev. **D76**, 074016 (2007)
[10] D. Gamermann, E. Oset, Eur. Phys. J. **A33**, 119 (2007)
[11] A. Martínez Torres, E. Oset, S. Prelovsek, A. Ramos, JHEP **05**, 153 (2015)
[12] Z.H. Guo, U.G. Meißner, D.L. Yao, Phys. Rev. **D92**, 094008 (2015)
[13] J. Segovia, D.R. Entem, F. Fernandez, Phys. Rev. **D91**, 094020 (2015)
[14] E. van Beveren, G. Rupp, Phys. Rev. Lett. **91**, 012003 (2003)
[15] E. van Beveren, G. Rupp, Eur. Phys. J. **C32**, 493 (2004)
[16] C.B. Lang et al., Phys. Rev. **D90**, 034510 (2014)
[17] P.G. Ortega, J. Segovia, D.R. Entem, F. Fernández (2016), `arXiv:hep-ph/1603.07000`
[18] J. Vijande, F. Fernández, A. Valcarce, J. Phys. **G31**, 481 (2005)
[19] A. Valcarce, H. Garcilazo, F. Fernández, P. Gonzalez, Rept. Prog. Phys. **68**, 965 (2005)
[20] J. Segovia, D.R. Entem, F. Fernández, E. Hernandez, Int. J. Mod. Phys. **E22**, 1330026 (2013)
[21] P.G. Ortega, J. Segovia, D.R. Entem, F. Fernandez, Phys. Rev. **D81**, 054023 (2010)
[22] J. Segovia, D.R. Entem, F. Fernández, Phys. Rev. **D83**, 114018 (2011)
[23] J. Segovia, D. Entem, F. Fernández, Nucl. Phys. **A915**, 125 (2013)
[24] J. Segovia, E. Hernández, F. Fernández, D.R. Entem, Phys. Rev. **D87**, 114009 (2013)
[25] J. Segovia, D.R. Entem, F. Fernández, Phys. Rev. **D91**, 014002 (2015)
[26] K.A. Olive et al. (Particle Data Group), Chin. Phys. **C38**, 090001 (2014)

Search for η' mesic nuclei by missing-mass spectroscopy of the ^{12}C(p,d) reaction

Yoshiki K. Tanaka[1,2,⋆], Yassid Ayyad[3], Jose Benlliure[4], Kai-Thomas Brinkmann[5], Stefan Friedrich[5], Hiroyuki Fujioka[6], Hans Geissel[1,5], Jnaneswari Gellanki[7], Chenlei Guo[8], Eric Gutz[5], Emma Haettner[1], Muhsin N. Harakeh[7], Ryugo S. Hayano[2], Yuko Higashi[9], Satoru Hirenzaki[9], Christine Hornung[5], Yoichi Igarashi[10], Natsumi Ikeno[11], Kenta Itahashi[12], Masahiko Iwasaki[12], Daisuke Jido[13], Nasser Kalantar-Nayestanaki[7], Rituparna Kanungo[14], Ronja Knoebel[1,5], Nikolaus Kurz[1], Volker Metag[5], Ivan Mukha[1], Tomofumi Nagae[6], Hideko Nagahiro[9], Mariana Nanova[5], Takahiro Nishi[12], Hooi Jin Ong[3], Stephane Pietri[1], Andrej Prochazka[1], Christophe Rappold[1], Moritz P. Reiter[1], Jose Luis Rodríguez Sánchez[4], Christoph Scheidenberger[1,5], Haik Simon[1], Branislav Sitar[15], Peter Strmen[15], Baohua Sun[8], Ken Suzuki[16], Imrich Szarka[15], Maya Takechi[17], Isao Tanihata[3,8], Satoru Terashima[8], Yuni N. Watanabe[2], Helmut Weick[1], Eberhard Widmann[17], John S. Winfield[1], Xiaodong Xu[1], Hiroki Yamakami[6], and Jianwei Zhao[8] for the η-PRiME/Super-FRS Collaboration

[1] *GSI Helmholtzzentrum für Schwerionenforschung GmbH, Darmstadt, Germany*
[2] *The University of Tokyo, Tokyo, Japan*
[3] *RCNP, Osaka University, Osaka, Japan*
[4] *Universidade de Santiago de Compostela, Santiago de Compostela, Spain*
[5] *Universitä Giessen, Giessen, Germany*
[6] *Kyoto University, Kyoto, Japan*
[7] *KVI-CART, University of Groningen, Groningen, the Netherlands*
[8] *Beihang University, Beijing, China*
[9] *Nara Women's University, Nara, Japan*
[10] *KEK, Ibaraki, Japan*
[11] *Tottori University, Tottori, Japan*
[12] *RIKEN Nishina Center, Saitama, Japan*
[13] *Tokyo Metropolitan University, Tokyo, Japan*
[14] *Saint Mary's University, Nova Scotia, Canada*
[15] *Comenius University Bratislava, Bratislava, Slovakia*
[16] *Stefan-Meyer-Institut für subatomare Physik, Vienna, Austria*
[17] *Niigata University, Niigata, Japan*

Abstract. We performed a missing-mass spectroscopy experiment of the ^{12}C(p,d) reaction at 2.5 GeV proton energy in order to search for η' mesic nuclei. An excitation-energy spectrum of ^{11}C was obtained around the η' production threshold. As no significant peak structure was observed in the spectrum, upper limits of the formation cross section of η' mesic nuclei have been determined. Preliminary results of the experiment as well as future prospects are given.

⋆e-mail: y.tanaka@gsi.de

1 Introduction

The η' meson has a mass of 958 MeV/c^2, which is especially large compared with the other pseudoscalar mesons such as π, K, and η. This is theoretically understood as a consequence of the $U_A(1)$ anomaly, whose contribution to the η' mass is present under spontaneous and/or explicit chiral symmetry breaking in the low-energy region of QCD [1, 2]. In the nuclear medium, where chiral symmetry is partially restored, the mass of the η' meson may be reduced. Such a mass reduction would induce an attractive η'-nucleus interaction, and thus existence of η' meson-nucleus bound states (η' mesic nuclei) is theoretically suggested [3–5].

So far, in-medium properties of the η' meson are barely known. Theoretically, mass reductions of the η' meson at normal nuclear density have been predicted by various model calculations [5–8], but their results range from 37 MeV/c^2 to 150 MeV/c^2. From the experimental side, the CBELSA/TAPS collaboration measured η' meson photo-production off carbon nuclei, and indicated the real part of the η'-nucleus potential of $-(37 \pm 10 \pm 10)$ MeV with the imaginary part of $-(10 \pm 2.5)$ MeV at normal nuclear matter density [9, 10]. Such a small imaginary part implies possibility of experimentally observing η' meson-nucleus bound states as distinct narrow structures.

We designed an inclusive measurement of the $^{12}\mathrm{C}(p, d)$ reaction to search for η' mesic nuclei, which can provide a unique opportunity to directly study in-medium properties of the η' meson [11]. We adopted the proton incident energy of 2.5 GeV to populate η' mesic states coupled with neutron hole states via the $^{12}\mathrm{C}(p, d)\eta'\otimes^{11}\mathrm{C}$ reaction. The missing mass of the reaction is then obtained by measuring the momentum of the ejected deuteron. A measurement with good statistics is required, because a large amount of background events due to quasi-free multi-pion production (p+$N \rightarrow d$+π's) are expected in the inclusive spectrum. In the first experiment reported in this article, we aimed at achieving a high statistical sensitivity with a relative error size of $\sim 1\%$ in the spectrum to overcome the expected small signal-to-noise ratio.

2 Experiment

We performed a missing-mass spectroscopy experiment of the $^{12}\mathrm{C}(p, d)$ reaction at GSI, Darmstadt in 2014. A 2.5 GeV proton beam extracted from the synchrotron (SIS-18) impinged onto a 4 g/cm^2-thick carbon target. Deuterons emitted at 0° in the $^{12}\mathrm{C}(p, d)$ reaction were momentum-analyzed by the fragment separator (FRS) [12] used as a high-resolution spectrometer with an experimental setup shown in Fig. 1. We adopted a new ion optics mode which has an achromatic focal plane at F2 and a dispersive focal plane at F4 with a designed dispersion of $\sim$ 3.6 cm/%. Two sets of multi-wire drift chambers (MWDCs) installed at F4 were used to measure the deuteron tracks. Data in the momentum range from 2.77 GeV/c to 2.93 GeV/c were accumulated by scaling the whole magnetic fields in the FRS by factors between 0.980 and 1.020.

Particle identification was provided by a time-of-flight between the two focal planes measured with plastic scintillation counters (SC2H, 2V, 41, and 42). In this experiment, a major experimental background particles at the downstream focal planes were protons emitted by the inelastic (p, p') reaction. Since the background protons have higher velocities ($\beta_p \sim 0.95$) than those of the signal deuterons ($\beta_d \sim 0.84$), they were clearly distinguished by a DAQ trigger based on the time-of-flight between F2 and F4. Furthermore, by analyzing recorded shapes of the signals of the scintillation counters, accidental multi-proton events were also rejected in the analysis. As a result, a contamination of the background protons in the spectrum was reduced to a negligible level with a fraction of $\sim 10^{-4}$ of the deuteron events.

The calibration of the spectrometer was carried out by measuring the $\mathrm{D}(p, d)p$ elastic scattering using a 1.6 GeV proton beam and a CD_2 target. Monochromatic deuterons with the momentum of

 DOI: 10.1051/epjconf/201613002010

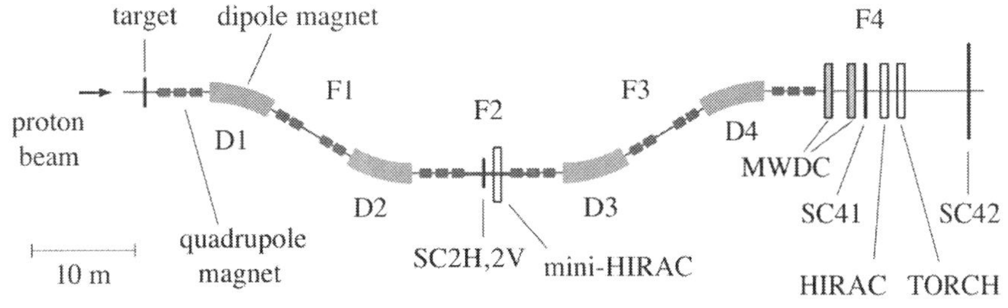

Figure 1. A schematic view of the experimental setup at the FRS. MWDCs were installed at the dispersive focal plane F4 to measure the deuteron tracks. Scintillation counters (SC2H, 2V, 41, and 42) were used for time-of-flight measurements. Čerenkov detectors (mini-HIRAC, HIRAC and TORCH) were used to tune a DAQ trigger condition.

2.8 GeV/c are emitted in this reaction, which is at the middle of the momentum range in the main measurements. By measuring these deuterons with various central momenta of the FRS, we confirmed the proper functionality of the spectrometer system, and calibrated the ion-optical parameters. Such measurements were repeated every $\sim$ 8 hours, and we monitored the stability of the whole system.

3 Preliminary results

The excitation-energy spectrum of ^{11}C was successfully obtained from −91 MeV to +34 MeV with respect to the η' emission threshold, as presented in Ref. [13]. In the spectrum, a high statistical sensitivity with a relative error of $< 1\%$ was achieved owing to the high-intensity proton beam and the thick carbon target. The experimental energy resolution was evaluated to be 2.5 MeV (σ), which is sufficiently small compared with expected widths of η' mesic nuclei.

Since no peak structure corresponding to η' mesic nuclei was observed in the excitation-energy spectrum, we determined upper limits of the formation cross sections of η' mesic nuclei. We tested a Voigtian peak, a Lorentzian peak folded by a Gaussian resolution, for each assumed energy and width of the state by a statistical method. The obtained upper limits near the η' emission threshold are about 0.1–0.2 μb/sr for the assumed Lorentzian width of $\Gamma = 5$ MeV, 0.2–0.4 μb/sr for $\Gamma = 10$ MeV, and 0.3–0.6 μb/sr for $\Gamma = 15$ MeV at 95% confidence level [13]. These upper limits are as small as theoretically-expected cross sections in case of deep η'-nucleus potential of the order of 100 MeV [4]. Furthermore, the obtained spectrum was also directly compared with the theoretically-calculated spectra to discuss a constraint for the η'-nucleus potential parameters [13].

4 Summary and future outlook

We performed the inclusive measurement of the ^{12}C(p,d) reaction at the proton energy of 2.5 GeV to search for η' meson bound states in carbon nuclei. The excitation-energy spectrum of ^{11}C was successfully obtained around the η' production threshold with the high statistical sensitivity and the sufficiently small energy resolution. In the spectrum, no significant peak structure corresponding to η' mesic bound states was observed. Thus, upper limits of the formation cross section of η' mesic states were determined. The experimental spectrum was also compared with theoretically calculated formation spectra to set a constraint on the η'-nucleus potential.

In the future, further experiments to search for η' mesic nuclei are necessary toward a direct study of in-medium properties of the η' meson. In order to achieve an even higher experimental sensitivity, we plan a semi-exclusive measurement of the $^{12}C(p, dp)$ reaction at GSI/FAIR [14]. In this experiment, while ejected deuterons are measured by a forward spectrometer such as FRS/Super-FRS, protons emitted in the decay of the η' mesic nuclei are identified around the reaction target. Tagging such protons is expected to suppress the background processes dominating in the present inclusive spectrum. R&D for the proton tagging system is in progress.

Y. K. T. acknowledges financial support from Grant-in-Aid for JSPS Fellows (No. 258155), and H. F. from Kyoto University Young Scholars Overseas Visit Program. This work is partly supported by a Grant-in-Aid for Scientific Research on Innovative Areas (No. 24105705) from the Ministry of Education, Culture, Sports, Science and Technology (MEXT), a Grant-in-Aid for Young Scientists (A) (No. 25707018) from Japan Society for the Promotion of Science (JSPS) and by the Bundesministerium für Bildung und Forschung.

References

[1] D. Jido *et al.*, Phys. Rev. C **85**, 032201(R) (2012)
[2] S. H. Lee and T. Hatsuda, Phys. Rev. D **54**, 1871 (1996)
[3] H. Nagahiro and S. Hirenzaki, Phys. Rev. Lett. **94**, 232503 (2005)
[4] H. Nagahiro *et al.*, Phys. Rev. C **87**, 045201 (2013)
[5] H. Nagahiro, M. Takizawa, and S. Hirenzaki, Phys. Rev. C **74**, 045203 (2006)
[6] P. Costa *et al.*, Phys. Rev. D **71**, 116002 (2005)
[7] S. Sakai and D. Jido, Phys. Rev. C **88**, 064906 (2013)
[8] S. D. Bass and A. W. Thomas, Phys. Lett. B **634**, 368 (2006)
[9] M. Nanova *et al.*, Phys. Lett. B **710**, 600 (2012)
[10] M. Nanova *et al.*, Phys. Lett. B **727**, 417 (2013)
[11] K. Itahashi *et al.*, Prog. Theor. Phys. **128**, 601 (2012)
[12] H. Geissel *et al.*, Nucl. Instrum. Meth. B **70**, 286 (1992)
[13] Y. K. Tanaka *et al.*, Phys. Rev. Lett. in print (2016)
[14] H. Fujioka *et al.*, EPJ Web of Conferences **66**, 09006 (2014)

EPJ Web of Conferences **130**, 02011 (2016) DOI: 10.1051/epjconf/201613002011
MESON 2016

Trapping of Σ^+ hyperons in nuclei

Sławomir Wycech[1,⋆] and *Kristian* Piscicchia[2,3]

[1] *National Centre for Nuclear Studies, Hoża 69, 00-681 Warsaw, Poland*
[2] *LNF-INFN, Italy*
[3] *Museo Storico della Fisica e Centro Studi e Ricerche Enrico Fermi*

Abstract. The nuclear capture of K^- studied by FINUDA discovered a puzzling low momentum component in the spectrum of final Σ^+ hyperon. This component is interpreted here as the effect of Gamov state formed by the hyperon. Such state is quasi-localized in space with a radius in between the hyper-nuclear and the hyper-atomic radius. Experimental and theoretical consequences of this discovery are studied.

1 Introduction

The formation of the hyperon and meson pairs $\Sigma^\pm, \pi^\mp$ following the K^-p capture in nuclei has been studied by a number of experimental techniques. Early experiments performed in nuclear emulsion and He(D) bubble chambers [1] concentrated on studies of total $P_{\Sigma\pi}$ momenta or invariant masses $M_{\Sigma\pi}$. These experiments allow to test invariant mass of the K^-p pair in the subthreshold region. The purpose of such research has been to learn the structure of $\Lambda(1405)$ resonance located below the K^-p threshold. Properties of the latter may be detected with a simultaneous measurement of : $M_{\Sigma,\pi}$, $P_{\Sigma\pi}$ and the ratio of two formation rates $\sigma(\Sigma^+,\pi^-)/\sigma(\Sigma^+,\pi^-) = R_\pm(M_{\Sigma,\pi})$. This ratio depends strongly on the invariant mass and reflects an interference of the resonant amplitude with a background one. Recent experiments by FINUDA collaboration allowed fairly precise measurements of both, the meson and the hyperon momenta [2]. The reaction in question is

$$K^- \, ^6Li \rightarrow \Sigma^\pm \, \pi^\mp \, R \tag{1}$$

with some residual system R, apparently $R \approx^5$He. In agreement with old measurements it has been found that the spectrum of Σ^- momentum is described by a regular peak centered at $\sim$ 200 MeV/c. On the other hand, the spectrum of Σ^+ momentum indicates two peaks apparently due to two different formation mechanisms. The additional low energy peak is located in the segment <0,50> MeV/c corresponds to centroid kinetic energy of $\sim$ 0.3 MeV, very small on the nuclear scale.

The difference in spectra is apparently related to different hyperon charge which matters at low momenta. The $R_\pm(M_{\Sigma,\pi})$ is also dramatically changed for those events which involve low energy hyperons. This finding has an indirect precedence. The European K^- collaboration studied subthreshold ratio $R_\pm$ in light emulsion nuclei. Results obtained by Keane [3] indicated an anomaly of $R_\pm$ around $M_{\Sigma,\pi} \sim$ 1395 MeV some 35 MeV below the K^-p threshold. This region of $M_{\Sigma,\pi}$ corresponds essentially to $p(\Sigma) \sim$ 0 events. The anomaly itself was studied in analysis [4] motivated by the idea of

⋆e-mail: wycech@fuw.edu.pl

Dalitz [5] that in addition to the baryonic component a sizable fraction of $\Lambda(1405)$ may be due to a direct three quark component. Now, in the light of the FINUDA finding this anomaly may be related to the low momentum peak in the Σ^+ spectrum.

Here, the low momentum peak is attributed to the Σ^+ hyperon trapping in the residual nucleus. A fraction of hyperons is trapped into a Gamov state formed by the attractive nuclear potential and repulsive Coulomb barrier. In section 2 we discuss properties of such $(\Sigma^+,{}^5\text{He})$ state formed in an intermediate state of K^- capture in ^{6}Li. We conclude that the Gamov state may offer explanation of the low momentum peak. However, a good quantitative description requires more data. Next, a new chapter in the physics of Σ-nucleus interactions ma be opened.

2 The origin of anomalous threshold peak in the Σ^+ momentum

The mechanism for the peak formation is sequential : (1) the K meson undergoes capture $K^-p \to \Sigma^+\,\pi^-$. Residual nucleus R is a spectator (2) the Σ^+ hyperon and R are trapped by Coulomb barrier into Gamov state (3) the Gamov state decays into low momentum Σ^+ and R.

If the Gamov state is very large the internal momenta are small and Σ^+ is released with a negligible momentum. In such conditions, the final meson momentum p_π is $\sim$ 180 MeV/c and the momentum taken by the hyperon is well fixed at value $p_\Sigma \sim p_\pi\, m_\Sigma/(m_\Sigma + m_R) \sim 20$ MeV which is the center of the experimental peak in the Li case.

The Gamov state of Σ^+ *hyperon* is a quasi-localised state formed in the continuum by an attractive nuclear potential and repulsive Coulomb barrier. Historically, such states were introduced to describe the α decay of heavy nuclei. Quantum mechanics allows to understand the unstable state of α particle localised within a nucleus and decaying via sub-barrier transition. The technical description requires solution of the corresponding Schroedinger equation with the outgoing wave boundary condition. Such solutions are possible for negative energies and describe bound states. For positive energies such solutions might also exists but with complex energies and imaginary parts of these energies give lifetimes of such systems. There is a difference between Gamov states of α formed in heavy nuclei and Σ^+ formed in light nuclei. In the latter case Coulomb barriers are low, energies of the Gamow states are low and the hyperon has a sizable probability to stay outside the nucleus. Moreover, the hyperon is absorbed by nuclei and the wave equation has to be determined in complex domain anyway, with the help of a complex optical potential. Here we use $V_n = [V_o - iW_o]\,\rho(r)$ and search for S wave quasi-bound states of positive energies

$$-\frac{1}{2M}\frac{dR}{d^2r} + (V_{culomb} + V_n)R = (E_G - i\Gamma/2)R. \tag{2}$$

The three step mechanism specified above generates a product of three factors in the formula for K^- capture amplitude

$$A_\Sigma = C_G\, D(p_\pi)\, \widetilde{\Phi}(q). \tag{3}$$

The first factor G is an amplitude (here it is an unspecified parameter) for the Σ^+ to be captured into Gamov state. Propagation of the Gamov hyper-nuclear system is described by

$$D(p_\pi) = [E_\Sigma(p_\Sigma) + E_R(p_R) - E_G(p_\pi) + i\Gamma_G/2]^{-1} \tag{4}$$

where E_Σ , E_R and E_G are the hyperon residual nucleus and Gamov-system energies. As we are in the continuum there is a pole in D. The last factor $\widetilde{\Phi}$ is the Fourier transform of the wave function R/r expressed in terms of relative momentum $\mathbf{q} = \mathbf{p}_r\, m_\Sigma/(m_\Sigma + m_R)) - \mathbf{p}_\Sigma\, m_R/(m_\Sigma + m_R))$ Profiles of the final $p(\Sigma)$ distribution presented below are given by the last two factors, both are important.

Several values of well depth V_o have been tried to study the effect of nucleus on the Gamov state. Apparently there exist no Σ^+ hypernuclei (other than ${}^4_\Sigma He$ [6]) and we use potentials which do not generate bound states. The mere existence of a Gamov state puts strong limitations on V_o from above as well as from below. The limiting values in ^{5}He are in the bracket (-28 till -26 MeV) depending on absorptive strength tried in the region $W_o < 10$ MeV motivated by the hyperhelium width. Differential equation (2) is integrated numerically to distances of 50 fm For technical simplicity the Coulomb potential is suppressed at distances exceeding 28 fm. These boundary conditions are good enough to obtain stability of the method and a three digit precision of energies and widths. As a consistency check we use the condition $\int ImV_n|\Phi|^2 = 2\Gamma$. The complex energy $E_G - i\Gamma/2$ is obtained with " asymptotic" wave function $(R/r)^2$ which drops to the level of 10^{-8} of the initial one. Under the cut-off condition one can define mean square radii of the Gamov systems. For the cases presented in the figure these are less than 15 fm.

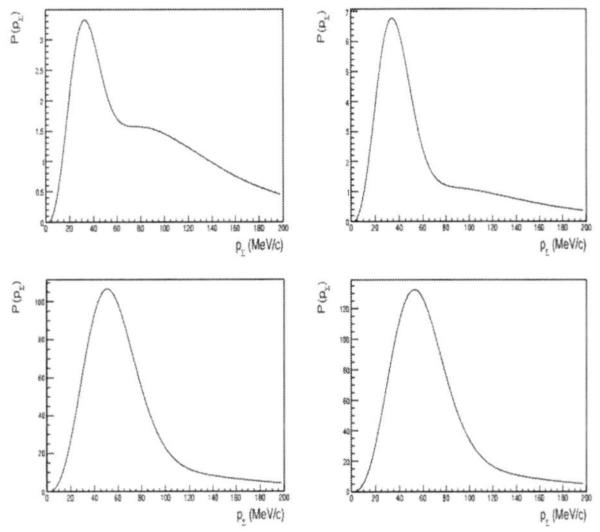

Figure 1. Calculated profiles of the low energy peaks in Σ^+ momentum distribution. Corresponding quantum numbers of Gamov states (E_G, Γ) in MeV are: up left (0.068,1.35), up right (0.03,1.38), down left (0.12,0.19) down right (0.02 ,0)

Amplitude $|A|^2$ averaged over final 3-body phase space generates the shape of low energy momentum peak. Several examples are plotted in the figure. The results indicate existence of threshold peak centered around 50 but not around 20 MeV/c found by FINUDA. There might be several effects which are missing: (1) The residual nucleus (^{5}He is unstable with $\Gamma \sim 0.5$ MeV), its density is calculated [7] but not tested. (2) There may be some momentum dependence of the Gamov state formation amplitude G. (3) Some loss of the final hyperon momentum in the target may take place. These questions may be answered with more data and those are forthcoming [8].

An additional asymmetry in the $\Sigma^\pm, \pi^\mp$ formation may also stem from Σ, N threshold enhancement [9]. This effect we find to be weak and also unlikely to produce a narrow peak.

DOI: 10.1051/epjconf/201613002011

3 Conclusions

The description of the anomalous low energy momentum distribution $P(p_\Sigma)$ in terms of the Gamov state is in a qualitative consistency with the FINUDA data. The mere fact of existence of the low energy peak puts strong limitation on the hyperon nucleus potential. The shape of of the peak may help to determine the strength of Σ absorption in nuclei. These two pieces of information are to some extent equivalent to a detection of (apparently nonexistent) Σ hypernuclei. The FINUDA data is related to a hyperon ^{5}He system. The only, so far, detected sigma hypernucleus has also been found in He. It would be very interesting to check the existence of Gamov states of Σ^+ also in heavier nuclei. Positive answer would open a new branch of hypernuclear spectroscopy.

The full understanding of p_Σ spectrum in particular the weight of the anomalous peak relative to the standard spectrum might be very informative, perhaps more than the Σ hypernuclear spectroscopy would be.

References

[1] E.H.S. Burhop, High Energy Phys. Vol. **3** Acad. Press. N. York, (1967)
[2] M. Angelo for FINUDA, Phys. Lett. B **704**, 474 (2011)
[3] D.F. Keane, Thesis, Univ. College Belfield, Dublin, (1981)
[4] L.R. Staronski and S. Wycech, J. Phys. G **13**, 1361 (1987)
[5] R.H. Dalitz and J.G. McGinley, Low and Intermediate Energy KN Physics ed. E. Ferrari and G. Violini, Springer, 1980, p. 381
[6] T. Nagae, *et al.*, Phys. Rev. Lett. **80**, 1608 (1998)
[7] T. Neff and H. Feldmeier, Nucl. Phys. A. **738**, 357 (2004)
[8] AMADEUS/KLOE project: K^- capture in He,B,C,Al, private information
[9] H. Machner *et al.*, Nucl. Phys. A **901**, 65 (2013)

Search for the C-violating meson decay $\eta \rightarrow \pi^0 e^+ e^-$ with WASA-at-COSY

Kay Demmich[1,⋆], Florian Bergmann[1], Nils Hüsken[1], and Alfons Khoukaz[1]
for the WASA-at-COSY Collaboration

[1] *Westfälische Wilhelms-Universität Münster, Institut für Kernphysik, Münster Germany*

Abstract. The WASA-at-COSY experiment is, besides meson production processes, dedicated to studies on rare and forbidden decays of light mesons. In extensive η-production beam times, a high-statistics data set have been collected by means of proton-proton scattering which opens the door to studies on forbidden decays like the C-parity violating process $\eta \rightarrow \pi^0 e^+ e^-$. In this article, an optimized detector calibration leading to a significantly improved missing mass resolution and a preliminary decay analysis are presented.

1 Introduction

Investigations on fundamental symmetries like the C-parity are a main tool for testing standard model predictions. The decay of the η-meson into $\pi^0 e^+ e^-$ via a virtual photon can be described by a vector meson dominance model but violates the C-parity and, hence, is forbidden in the frame of the standard model. The final state can be achieved by higher order processes fulfilling all conservation laws but these highly suppressed, i.e., a relative branching ratio in the order of 10^{-12} to 10^{-9} is expected by calculations. Thus, this decay of the η-meson is a perfect probe for testing the C-parity conservation within the standard model up to a relative branching ratio of $\approx 10^{-9}$ and may give access to physics beyond the standard model by assuming a dark Z-like U-boson as an intermediate state [1]. Up to now, this decay has not been observed and only an upper limit for the branching ratio of 4×10^{-5} is quoted by the PDG [2].

Studies on this decay rely on data sets with very high statistics and an optimal detector resolution to distinguish between possible signal events and the large number of background reactions which may mimic the same signature in the detector.

The WASA detector [3] loacated at the accelerator facility COSY at the Forschungszentrum Jülich, Germany, was utilized to obtain large data sets in η-production runs. Two different production mechanisms were exploited, namely the fusion reaction $p + d \rightarrow {}^3\text{He} + \eta$ characterized by a low fraction of background reactions, i. e. the direct pion production, and the scattering $p + p \rightarrow p + p + \eta$ with a significantly larger cross section leading to $\approx 5 \times 10^8$ η-mesons recorded. This article presents results obtained with the latter data set.

⋆e-mail: Kay.Demmich@uni-muenster.de

2 Analysis

The data set is analyzed in order to observe the decay $\eta \to \pi^0 e^+ e^-$ or, at least, to improve the existing upper limit for the relative branching ration. A key point is the identification of the signal channel and the separation from background reactions like pion production processes. Several methods are used to optimize the analysis. The protons are identified by their characteristic energy loss bands in the forward detector part, pions and electrons are separated by energy-momentum correlations (see Fig. 1). Utilizing the missing mass method, the reaction $p + p \to p + p + \eta$ can be identified. In the missing mass spectrum of the protons, this reaction forms a narrow peak upon the continuous distribution of background reaction like direct pion production as shown in the left histogram of Fig. 2. The separation efficiency by a missing mass window obvioulsy relies strongly on the energy resolution of the forward detector.

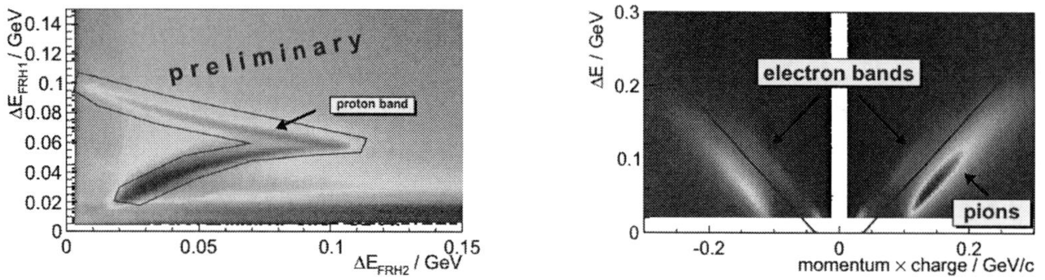

Figure 1. Left: Example of the energy loss correlation in two subsequent scintillator planes of the forward detector. A cut around the characteristic proton band is shown. Right: Separation between electrons and pions by the means of energy-momentum correlation measured in the central detector.

For the calculation of the missing mass, the kinetic energies of the proton ejectiles are used. These are calculated by the energy losses in subsequent scintillator planes. Assuming a perfect calibration of the energy loss detectors, the obtained missing mass peak position is expected to be at the mass of the η-meson of $m_\eta = 547.862\,\text{MeV}$ [2]. The measured data, however, show systematic dependencies of the missing mass position from the polar scattering angle ϑ and the energy deposition E_{dep} in the stopping plane of the protons. To correct for this, a new calibration routine has been developed.

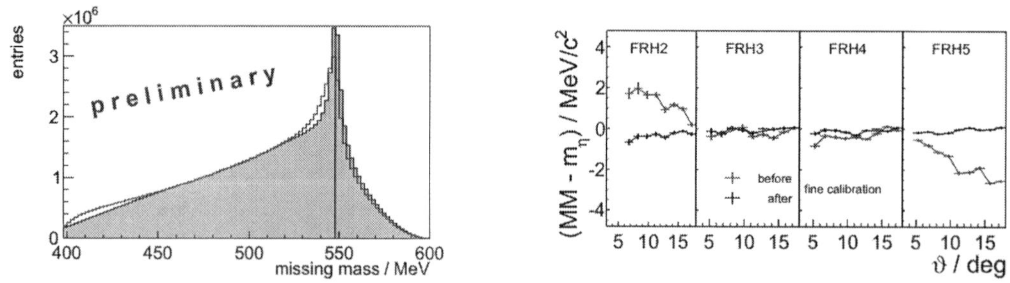

Figure 2. Left: Missing mass spectrum before (red line) and after the fine calibration (shaded filled area). Right: Deviation of the missing mass peak position depending on the polar scattering angle before (red line) and after (black line) the fine calibration.

A scaling factor depending on ϑ and E_{dep} is applied to the reconstructed kinetic energy of each proton. These factors are highly correlated, since the missing mass is always calculated out of two protons, and has to be determined in an iterative way. As a result, the missing mass resolution is improved by 40 %. Fig. 2 shows the shift of the missing mass position before and after this fine calibration for a small data subset.

The data were taken in a beam time lasting for several months. During this time, slight changes in the detector settings are possible, leading to a time depending miscalibration. The effect has been corrected by an additional time depending factor and a steadily high resolution throughout the complete beam time could be realized. Fig. 2 compares the missing mass distribution before and after the fine calibration. The different widths of the η-peak are clearly visible.

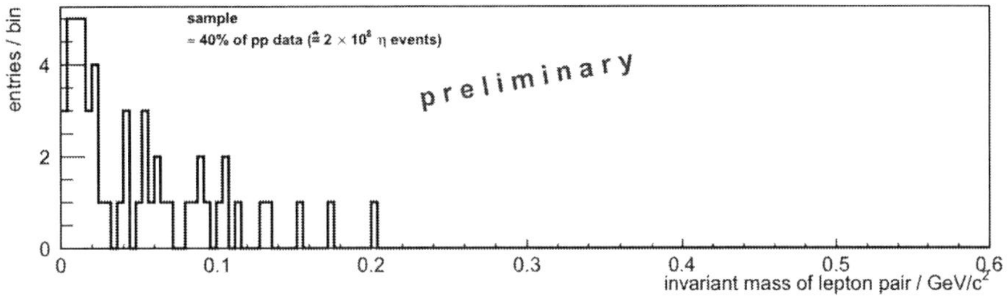

Figure 3. Invariant mass spectrum of the lepton pair after all cuts.

Fig. 3 shows the invariant mass spectrum of the lepton pair for candidates of the decay $\eta \to e^+e^-\pi^0 \to e^+e^-\gamma\gamma$ as a preliminary analysis result for $\approx 40\,\%$ of the data set. Besides a cut on the proton-proton missing mass around the η-mass, cuts on the invariant mass of two photons around the π^0-mass and on the invariant mass of all decay particles around the η-mass are applied. After a final probability cut on a kinematic fit with the hypothesis $pp \to ppe^+e^-\gamma\gamma$ only a few events remain out of $\approx 200 \times 10^6$ η-events.

3 Summary and outlook

The decay $\eta \to \pi^0 e^+e^-$ is a perfect probe for testing C-violation and the search for particles beyond the standard model. A large data set, obtained with the WASA-at-COSY experiment by proton-proton scattering, allows for a very sensitive search for this reaction. The missing mass technique is a key tool to separate between η-productions and background reactions like multi-pion production. A new calibration method has been implemented to improve the missing mass resolution drastically.

The analysis will be further improved by adjusted Monte Carlo settings to the new calibration and cut optimizations based on Monte Carlo cocktails with the goal to significantly improve the existing upper limit as well as to study possible contributions from non-standard particles.

The support from COSY-FFE grants is kindly acknowledged.

References

[1] H. Davoudiasl, H.-S. Lee, and W. J. Marciano, Phys. Rev. D **89**, 095006 (2014)
[2] K. A. Olive et al. (Particle Data Group), Chin. Phys. C, 38, 090001 (2014)
[3] Bo Höistad, James Ritman (WASA-at-COSY Collaboration), arXiv:nucl-ex/0411038

EPJ Web of Conferences **130**, 03002 (2016) DOI: 10.1051/epjconf/201613003002
MESON 2016

Dalitz plot of $\eta' \rightarrow \eta\pi^+\pi^-$

Sudeep Ghosh[1,*] and Ankhi Roy[1]
for the CLAS Collaboration

[1] *Indian Institute of Technology Indore, Khandwa Road, Simrol, M.P - 452020, India*

Abstract. In this talk we present preliminary experimental results of Dalitz plot analysis of the decay $\eta' \rightarrow \eta\, \pi^+\, \pi^-$, based on the CLAS data collected during photoproduction experiment γ p $\rightarrow \eta'$ p for the center-of-mass energy from 1.90 to 2.62 GeV at Jefferson Lab. This experiment will enable us to report precise Dalitz plot parameters of the $\eta' \rightarrow \eta\, \pi^+\, \pi^-$ decay for the first time in a photoproduction experiment.

1 Introduction

The work is on the measurement of Dalitz plot parameters from the $\eta' \rightarrow \eta\, \pi^+\, \pi^-$ decay through the "g12" experiment at Thomas Jefferson National Accelerator Facility (JLab), Newport News,VA, USA. A Dalitz plot is used to study the decay dynamics of mesons. Chiral Perturbation Theory is the effective theory to understand mesons in the low energy regime and Dalitz plot helps to understand the correct input to theoretical distribution of the effective chiral Lagrangian. The $\eta' \rightarrow \eta\, \pi^+\, \pi^-$ decay has the lowest Q-value compared to the other decay modes of the η' meson, because of its heavier decay products. So a study of this channel attributes to understanding effective chiral perturbation theory at a low Q limit.

Three body decay of the η' meson has two degrees of freedom, and has two independent Dalitz plot variables X and Y below, completely describes the decay.

$$X = \frac{\sqrt{3}(T_{\pi^+} - T_{\pi^-})}{Q}, Y = \frac{(m_\eta + 2m_\pi)}{m_\pi} \cdot \frac{T_\eta}{Q} - 1, \tag{1}$$

where T_η, T_{π^+} and T_{π^-} are the kinetic energy of a given particles η, π^+ and π^- in the rest frame of η' and $Q = T_{\pi^+} + T_{\pi^-} + T_\eta$, m_η and m_π are the mass of the η and π meson.

2 Experiment

The Continuous Electron Beam Accelerator Facility (CEBAF) at JLab has an ability to accelerate electrons to a maximum energy of 6 GeV. The accelerated electrons are then passed to the experimental hall B, where the electrons find the gold (Au) radiator of 10^{-4} radiation length. The interaction of electrons with the Au foil produces real photons for the "g12" experiment via the bremsstrahlung process. The Dipole Magnet then separates charged electrons from the photons. The recoiled and

*e-mail: phd12115113@iiti.ac.in

diverted electrons are then passed to set of E-plane and T-plane scintillators to measure the energy and timing respectively. The whole process indirectly tag the photons and measures the photon energies from 20% to 95% of the incident electron beam energy, in "g12" that makes an available photon energies to range from 1.142 - 5.425 GeV. The extracted photons out of radiator are then collimated before it enters inside the CEBAF Large Acceptance Spectrometer (CLAS) and hits the not polarised liquid hydrogen (lH_2) target placed -90 cm from the CLAS centre. The interaction between the incoming photon beam and proton target produces particles that finds the Start Counter with 6 sectors and 4 scintillator paddles each near to the interaction region. This was used in the event trigger and to measure the start time. Then comes the Superconducting Toroidal Magnet which produces the magnetic field to bend the charge particles and then passed through the three layers of drift chamber with each having six sectors. Finally the particles hit the Time of Flight (TOF) detector where the instant of hit time are recorded. Using all the information recorded starting from the Start Counter to Time of Flight detector the particle identification is done in CLAS. All the sub-detectors together make the CLAS detector well optimised for the detection of charged particles and gives a coverage of 6° to 100° of the polar angle [1].

3 Data Analysis

The complete reaction under this study is $\gamma\, p \rightarrow \eta'(\rightarrow \eta\, \pi^+\, \pi^-)\, p$ and η meson is reconstructed using the missing mass information. The calibrated along with CLAS particle identified "g12" corrected data with one p, one π^+, one π^- and any number of neutral particles is selected for analysis. A total of 661 runs with a photon beam energy range from the η' production threshold of 1.455 to 3.2 GeV is analysed. The selected data is then corrected to set of standard correction procedures in "g12" [2]. The correction includes beam energy, removal of bad TOF paddle and geometric fiducial cut. Events surviving the cuts are corrected and made to pass the tuned "g12 Kinematic Fitter" which improves the signal to background ratio of the selection. The fitter used a 1-C constrained fit requiring the four-momentum of [beam] + [target] - ([π^+] [π^-] [p]) to be an [η] meson. The probability of each event are then calculated and finally a cut is made on 1% Pull probability cut to reject events with very low probability to satisfy the constrained condition. To select the events of interest a cut is placed to $M_x(p\, \pi^+\, \pi^-)$ around the η meson mass window 0.547 ± 0.01 GeV. The η' meson going along the most forward and backward region is eliminated with $|\cos\theta_{center-of-mass}$ of $\eta'\, | \leq 0.85$ cut to reject the low yield region. The $M_x(p)$ spectrum can then be explained with Voigt signal and the background with a polynomial of order 3 as shown in Fig. 1. The background subtracted events lying within the 3σ region is the total number of signal events.

PLUTO, an event generator is used for this analysis. It uses ROOT based programmes and it is very commonly used in Hadron Physics experiments to generate hadronic production and decay of mesons. The generated events from PLUTO is also modelled with the bremsstrahlung photon, differential cross-setion of the η' meson [3] and Dalitz plot parameters of $\eta' \rightarrow \eta\, \pi^+\, \pi^-$ decay [4]. The modelled Monte Carlo events are then passed through the standard Geant3 - based CLAS simulation software which finally simulates the experimental conditions reflecting proper detector responses. The reconstructed events are then passed through the same conditions in the data before using it for analysis.

3.1 Dalitz plot

The analysis uses a binned Dalitz plot, and the binning in the Dalitz variables X and Y should be greater than the resolution of these variables. Fig. 1 shows the resolution of variables X and Y from

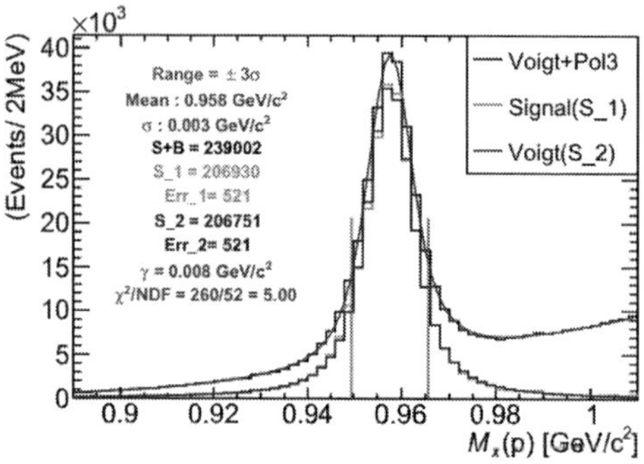

Figure 1: $M_x(p)$ vs the number of events plotted after implementing all correction and conditions. The signal is fitted with the Voigt function and the background fitted with a polynomial of order 3.

the simulation, it is the difference of the true and reconstructed information of the variables. The Dalitz plot analysis is done further with a binning of 15 x 15 bins in X and Y.

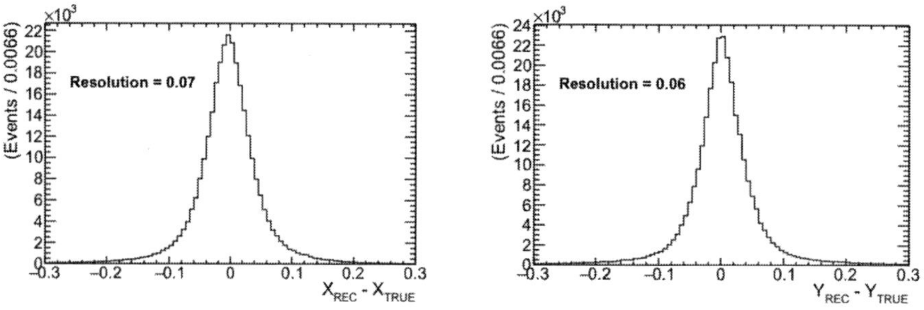

(a) (b)

Figure 2: The figure (a) shows the difference of Reconstructed and True value of the events vs the number of events for variable X and the figure (b) shows the difference of Reconstructed and True value of the events vs the number of events for variable Y.

3.2 Background subtracted Dalitz plot

The background subtraction is performed to each bin of the Dalitz plot as explained in Sect. 3 and the number of η' meson events are calculated in each bin of the Dalitz plot. Bins with less than 100 events and low acceptance ($< 1\%$) are rejected from the analysis along with the events lying outside the Dalitz plot boundary.

3.3 Fitting Procedure to the Dalitz Plot

Once the $\eta' \rightarrow \eta\, \pi^+\, \pi^-$ events from the data is filled in each bin of Dalitz plot, a Dalitz plot fit with the general parametization function in Eq.(2) is performed. The square of the decay amplitude

$$M^2 = A(1 + aY + bY^2 + cX + dX^2), \tag{2}$$

 DOI: 10.1051/epjconf/201613003002

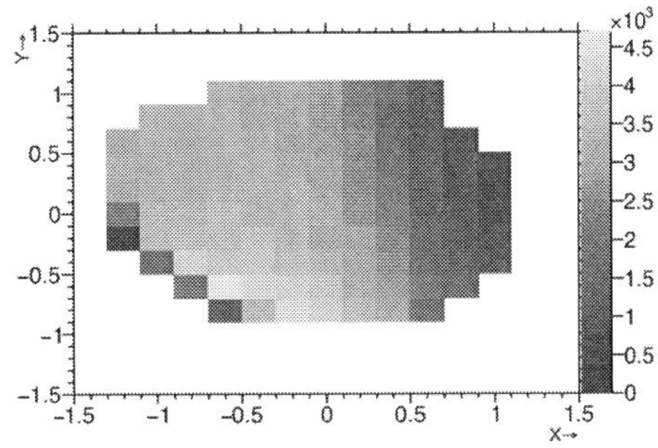

Figure 3: Dalitz variable X vs Y plot for $\eta' \to \eta\,\pi^+\,\pi^-$ events without acceptance correction.

where a, b, c and d are the Dalitz plot parameters of the decay and A is the normalization constant. The fitting is performed with the least square fitting procedure using MINUIT available in ROOT, which minimises the χ^2 using Eq.(3) in each bin of the Dalitz plot [5].

$$\chi^2 = \sum_{i=1}^{Nbins} \left(\frac{N_i - \sum_{j=1}^{Nbins} \epsilon_{i,j} N_{theory,j}}{\sigma_i} \right)^2 \tag{3}$$

Where,

- N_i is the number of $\eta' \to \eta\,\pi^+\,\pi^-$ events in the ith Dalitz plot bin.
- $\epsilon_{i,j}$ is acceptance with smearing matrix, i.e., it gives acceptance of jth bin when events are generated in the ith bin only.
- $N_{theory,j} = \int_{Boundary} A(1 + aY + bY^2 + cX + dX^2)\, dXdY$.
- σ_i is the error associated with the ith Dalitz plot bin.

4 Conclusion

The background subtraction is performed and acceptance with smearing matrix has been calculated for the $\eta' \to \eta\,\pi^+\,\pi^-$ decay. The final Dalitz plot parameters and systematics are under study.

References

[1] B. A. Mecking *et al.* [CLAS Collaboration], Nucl. Instrum. Meth. A **503**, 513 (2003).
[2] Z. Akbar et al. g12 Analysis Procedures, Statistics and Systematics. Technical report, CLAS Technical Note, 2016.
[3] M. Williams *et al.* [CLAS Collaboration], Phys. Rev. C **80**, 045213 (2009)
[4] M. Ablikim *et al.* [BESIII Collaboration], Phys. Rev. D **83**, 012003 (2011)
[5] L. Caldeira Balkestahl [KLOE-2 Collaboration], Acta Phys. Polon. B **46**, no. 1, 31 (2015)

 DOI: 10.1051/epjconf/201613003003

Electromagnetic Transition Form Factor of the η meson with WASA-at-COSY

A. Goswami[1,*]
for the WASA-at-COSY Collaboration

[1] *IIT Indore, Indore, India*

Abstract. In this work we present a study of the Dalitz decay $\eta \to \gamma e^+ e^-$. The aim of this work is to measure the transition form factor of the η meson. The transition form factor of the η meson describes the electromagnetic structure of the meson. The study of the Dalitz decay helps to calculate the transition form factor of the η meson. When a particle is point-like it's decay rate can be calculated within QED. However, the complex structure of the meson modifies its decay rate. The transition form factor is determined by comparing the lepton-antilepton invariant mass distribution with QED. For this study data on proton-proton reaction at a beam energy of 1.4 GeV has been collected with WASA-at-COSY detector at Forschungszentrum Juelich, Germany. In the higher invariant mass region recent theoretical calculations slightly deviate from the fit to the data. We expect better results in the higher invariant mass region than previous measurements. The preliminary results of the analysis will be presented.

1 Introduction

The η meson is a Pseudoscalar meson. The Dalitz decay ($\eta \to \gamma^*\gamma \to e^+e^-\gamma$) of the η meson which has a branching ratio 6.9×10^{-3}, can be used to extract the information on the it's electromagnetic transition form factor. This kind of study has been proved as a good testing ground for the model like VMD. The transition form factor might also contribute to the hadronic light by light scattering of the g-2 of muon [1] which is a limiting contribution of the Standard Model. The main goal of this analysis is to measure transition form factor of the the η meson in the time-like region. Transition form factor of the η meson describes the electromagnetic structure of the meson. The decay rate of a point-like particle can be calculated within QED. However, the complex structure of the meson modifies its decay rate. The transition form factor is determined by comparing the experimentally measured lepton-antilepton invariant mass distribution of the Dalitz decay of the η with QED. Transition form factor of the η can be calculated using the following equation1[2]:

$$\frac{d\Gamma(\eta \to \gamma e^+ e^-)}{dq^2 \cdot \Gamma(\eta \to \gamma\gamma)} = \frac{2\alpha}{3\pi}[1 - \frac{4m_e^2}{q^2}]^{1/2}[1 + \frac{2m_e^2}{q^2}]\frac{1}{q^2}[1 - \frac{q^2}{m_\eta^2}]^3 \mid F_\eta(q^2) \mid^2 \quad (1)$$

Where m_e stands for the lepton mass, m_η is the mass of the eta meson and q^2 is the squared four-momentum of $e^\pm$.

*e-mail: phd1301151001@iiti.ac.in

According to VMD, transition form factor is parameterized using single pole approximation:

$$F_{\eta}(q^2) = \frac{1}{1 - \frac{q^2}{\Lambda^2}} \tag{2}$$

Slope of the form factor is related with the radius of the charge distribution. A recent work from the CB/TAPS at MAMI collaboration reports that the value of the slope of the form factor is $\Lambda^{-2} = (1.95 \pm 0.15_{stat} \pm 0.10_{syst})\ GeV^{-2}$ [3]. Their result significantly agrees with recent theoretical calculations and with the VMD model. We aim to calculate a more precise value of the transition form factor with our high statistics data.

2 Experimental Setup

WASA (Wide Angle Shower Apparatus) is a 4π detector setup [4]. The high intensity proton beam provided by COoler SYnchrotron (COSY) at Forschungzentrum Juelich in combination with a high density pellet target of the WASA detector make possible to achieve high luminosities which is required for the study of rare decays. This detector setup consists of the two parts, the forward detector and the central detector. The recoil protons are detected with the forward detector which covers 3-18 degrees in the polar angle and decay fragments of the η meson are identified with the central part of the detector which covers 20-169 degrees.

3 Analysis

A large data sample of pp→ppη ($10^9\eta$) has been collected in 2008, 2010 and 2012 at WASA-at-COSY experimental setup. However, in this study data collected in 2010 has been used. In the analysis, the recoil protons are identified in the forward part of the detector using the ΔE-E method as shown in Figure 1 (Left), where ΔE is the energy deposited by protons in the layers of Forward Range Hodoscope (FRH).

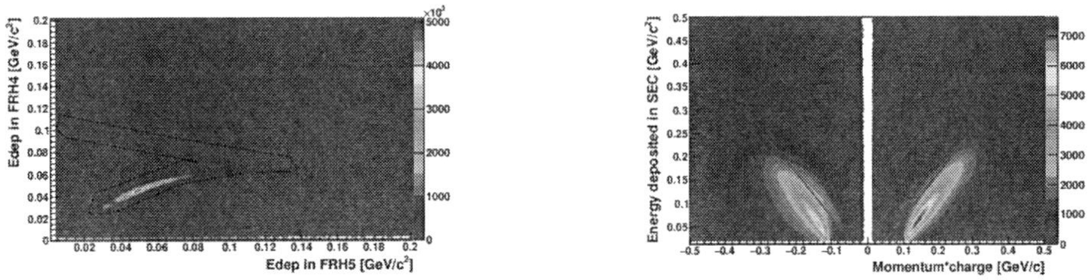

Figure 1: Energy deposited by the charged particles in the FRH4 v/s FRH5 of the forward detector (left) and momentum times charge of the particles v/s energy deposited by the particles in the elctromagnetic calorimeter (Right). The graphical region in the left figure indicates all proton tracks and the lines in the right figure indicate the separation of electrons and protons.

Furthermore, charged decay particles (e^+, e^-) of the produced η meson are identified in the central part of the detector using ΔE-P method as shown in Figure 1 (Right). Here, ΔE is the energy measured by the electromagnetic calorimeter and momentum information is provided by the mini drift chamber in presence of the magnetic field of solenoid. Kinematic conditions have been implemented in the

DOI: 10.1051/epjconf/201613003003

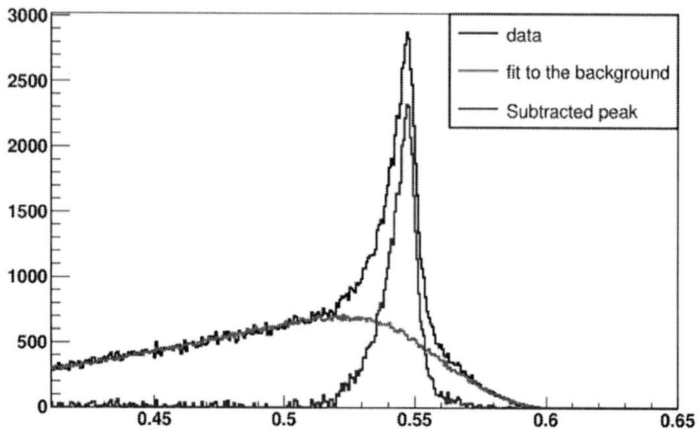

Figure 2: Missing mass of two initial and two final state photons after all kinematic conditions.

data analysis to identify all final state particles and to suppress the background. The missing mass technique is used to tag the η mesons, as shown in Figure 2.

A 4th order polynomial multiplied with the phase space of multipion production pp$\rightarrow$pp$\pi^0\pi^0$ has been used to fit background excluding the peak region (0.53-0.57 GeV). Total 43000 $\eta \rightarrow e^+e^-\gamma$ events have been reconstructed. However, background is still present inside the peak region. Thus, we have performed a detailed study of contributing background channels to identify sources of the background. For that all possible background channels (both phase space and competing) have been simulated which are mentioned in Table 1. The values of cross-section and branching ratios in the table are taken from the Particle Data Group[5].

Table 1: List of simulated background channels.

Channel	Cross-section / Branching ratio
$pp \rightarrow pp\pi^0(e^+e^-\gamma)\pi^0(\gamma\gamma)$	324 μb
$pp \rightarrow pp\pi^+\pi^-\pi^0(e^+e^-\gamma)$	4.6 μb
$pp \rightarrow pp\pi^0(e^+e^-\gamma)\pi^0(\gamma\gamma)\pi^0(\gamma\gamma)$	1.34 μb
$\eta \rightarrow \pi^+\pi^-\pi^0$	22.6%
$\eta \rightarrow \pi^+\pi^-\gamma$	4.68%
$\eta \rightarrow \gamma\gamma$	39%
$\eta \rightarrow \pi^0(e^+e^-\gamma)\pi^0(\gamma\gamma)\pi^0(\gamma\gamma)$	32%

The channel $\eta \rightarrow \gamma\gamma$ contributes as a background if one of the photons interacts with the beampipe material and converts into $e^{\pm}$ pair. Both $\eta \rightarrow \gamma\pi^+\pi^-$ and $\eta \rightarrow \pi^0\pi^+\pi^-$ channels have a similar topology to the signal channel as in both channels there are two oppositely charged tracks and one neutral track. It is seen in the simulation that pp$\rightarrow$pp$\pi^0\pi^0$ is the main source of the background in the phase space region and 8.4% background still may contribute from the competing decay channels $\eta \rightarrow \gamma\gamma$, $\eta \rightarrow \gamma\pi^+\pi^-$, and $\eta \rightarrow \pi^0\pi^+\pi^-$ inside the peak region. We are also planning to look at the Δ-Δ resonances and $\pi^+\pi^-$ correlations as a part of background study. We plan to perform a study of

DOI: 10.1051/epjconf/201613003003

relative branching ratio in order to verify it from the PDG value from the cut based analysis. Finally, we will measure the transition form factor of the η meson.

This work is supported by Forschungszentrum Juelich within the COSY-FFE program and by Indian Institute of Technology Indore.

References

[1] F. Jegerlehner and A. Nyffeler, Phys. Rep. **477**, 1 (2009)
[2] L. G. Landsberg, Phys. Rep. **128**, 301 (1985)
[3] P. Aguar-Bartolomé *et al.* (A2 Collaboration at MAMI), Phys. Rev. C **89**, 044608 (2014)
[4] H. Adam *et al.*, Proposal for the Wide Angle Shower Apparatus (WASA) at COSY-Juelich, arXiv:nucl-ex/0411038v1 (2004)
[5] K. A. Olive *et al.* (Particle Data Group), Chin. Phys. C **38**, 090001 (2014)

Production and interaction of the η meson with nucleons and nuclei

Wojciech Krzemień[1,⋆], Oleksandr Khreptak[2], Paweł Moskal[2], Iryna Ozyrianska-Schätti[2], Oleksandr Rundel[2], Magdalena Skurzok[2], and Marcin Zieliński[2]
for the WASA-at-COSY Collaboration

[1] *National Centre for Nuclear Research, 05-400 Otwock, Świerk, Poland*
[2] *Faculty of Physics, Astronomy and Computer Science, Jagiellonian University, 30-348 Cracow, Poland*

Abstract. We report on the status of the search for η-mesic nuclei and the studies of the interaction of the η meson with nucleons. Recently we have completed the analysis of the new WASA-at-COSY data on the production of the η meson with polarized proton beam. New results on the analyzing power for the $\vec{p}p \to pp\eta$ reaction with more than an order of magnitude improved precision shed a new light on the production mechanism of the η meson in nucleon-nucleon collisions. Also, the latest results of the search for η-mesic nuclei are discussed.

1 Introduction

The η particle together with isoscalar η' and isovector π^0 lay in the origin ($S = 0$, $I_3 = 0$) of the nonet of pseudoscalar mesons representation. However, its behaviour is very different with respect to its interaction with nucleons [1]. In the low energy region, the η meson interaction with nucleons is dominated by the S_{11} resonance, which with its mass of 1535 MeV lays very close to the η-N threshold. This makes the s-wave $\eta-N$ interaction very strong and - as shown the analysis of Bhalerao and Liu - attractive [2]. This can be contrasted with the pion case which, dominated by the p-wave interaction from the $\Delta(1232)$ resonance, is much weaker [3]. Also, the measurement of the η'N scattering length shows that its interaction is rather weak [4]. The large value of the η-N scattering length led to the hypothesis, proposed by the Haider and Liu, who postulated that the total interaction in a nucleus- system is strong enough to form a bound-state - the so called mesic nuclei [5]. The second question raised and not unequivocally answered by the earlier measurements was about the η production mechanism in the nucleon-nucleon collisions.

Due to the short lifetime of the meson ($t \approx 10^{-18}$ s) it is not feasible to create the η beam. Therefore, its interaction with nucleon or nuclei must be studied via the observation of final states of nuclear reactions including the η-nucleon (or η-nuclei) pair. The Final State Interaction between produced particles can strongly influence the production cross-sections and, in this way, can be used for studies of the interaction itself [6].

⋆e-mail: wojciech.krzemien@ncbj.gov.pl

In this contribution we discuss the η production mechanism in the interaction with nucleons and the search for the η-mesic nuclei in the context of the recent experimental results from WASA-at-COSY collaboration.

2 η production mechanism in the interaction with nucleons

The measurements of the large total cross-section of $NN \rightarrow NN\eta$ reaction near the η production threshold [7–18] motivated the two-step η production model proposed in [19]. In this scenario one of the proton is firstly excited through the exchange of a single meson and forms the S_{11} resonance, which in the second step deexcites via the emission of η and nucleon. In principle, the excitation to S_{11} state can occur by exchanging π, η, ω or/and ρ mesons. The measurements of total cross-section isospin dependence by WASA/PROMICE and COSY-11 showed that the η production in the total isosinglet state is much higher that in the isotriplet state [18, 20]. This result strongly suggest the isovector meson exchange, reducing the candidates to π and ρ particles [19, 21, 22]. To further distinguish between the π and ρ meson exchange models, the determination of the analysing power in the polarization measurement was required. The first measurement by COSY-11 gave a indication in favour of the pseudoscalar exchange model, although due to the limited statistics the decisive conclusions could not be done [23–25]. The WASA-at-COSY performed a high-statistics measurement of the $\vec{p}p \rightarrow pp\eta$ with the polarized beam. The spin of the polarization was flipped from cycle to cycle. The data was gathered for two separated beam momenta 2026 MeV/c and 2188 MeV/c, which correspond to excess energy over the η production threshold of 15 MeV and 72 MeV, respectively. More details of the analysis can be found in [26].

3 Search for the η-mesic nuclei

The recent reviews on the search for mesic nuclei can be found in [3, 27–33]. The WASA-at-COSY collaboration [34] performed three dedicated experiments with aim to search for the η-mesic nuclei in ^{4}He and ^{3}He systems in the deuteron-deuteron and proton-deuteron collisions, respectively [35–39]. The choice of the light nuclei was motivated by both theoretical considerations (see e.g. [40, 41]) as well as the earlier measurements by SATURNE, ANKE and COSY-11 [42–46], that provided strong experimental hints for the existence of the bound state in the $^3\text{He} - \eta$ and $^4\text{He} - \eta$ systems. The main experimental idea for the ^{4}He is based on the measurement of the excitation function of the $dd \rightarrow {}^3\text{He}N\pi$ reaction for energies in the vicinity of the η production threshold and on the selection of events with low ^{3}He center-of-mass (CM) momenta. In the case of existence of the $^4\text{He} - \eta$ bound state we expect to observe a resonance-like structure in the excitation function below the threshold for the production of the $^4\text{He} - \eta$ system. The ^{3}He state is investigated in proton on deuteron collisions. The details can be found in [47].

4 Summary

The latest preliminary results from the 2010 experiment in $^4\text{He}-\eta$ system, do not confirm the existence of the η-mesic nuclei. The preliminary value of the upper limit obtained from the simultaneous fit, taking into account the isospin dependence of the $dd \rightarrow (^4\text{He-}\eta)_{bound} \rightarrow {}^3\text{He}n\pi^0$ and $dd \rightarrow (^4\text{He-}\eta)_{bound} \rightarrow {}^3\text{He}p\pi^-$ excitation functions , of order of few nb can be compared to the theoretical estimate of 4 nb [48]. In case of the $^3\text{He} - \eta$ system, the analysis is ongoing. The current experimental upper limit for the production of $pd \rightarrow (^3\text{He-}\eta)_{bound} \rightarrow ppp\pi^-$ comes from the COSY-11 measurement and is equal to about 270 nb [49]. Due to the high statistics the expected sensitivity in current WASA analysis is of order of 10 nb, which, taking into account the theoretical estimate of 80 nb [50], should be sufficient to confirm or rule out the hypothesis of existence of the ^{3}He $- \eta$ mesic nuclei.

The WASA-at-COSY determined the analysing power in the $\vec{p}p \longrightarrow pp\eta$ reaction with two order of magnitude higher precision than the previous COSY-11 measurement. The preliminary results of the angular dependency of the analysing power is in disagreement with the prediction by both the pseudo-scalar and the vector exchange models. For higher energy (Q =72 MeV), the Ps-Pp interference is clearly observed.

We acknowledge support by the Polish National Science Center through grant No. 2011/03/B/ST2/01847,2011/01/B/ST2/00431, 2013/11/N/ST2/04152, by the FFE grants of the Research Center Juelich, by the EU Integrated Infrastructure Initiative HadronPhysics Project under contract number RII3-CT-2004-506078 and by the European Commission under the 7th Framework Programme through the Research Infrastructures action of the Capacities Programme, Call: FP7- INFRASTRUCTURES-2008-1, Grant Agreement N. 227431.

References

[1] P. Moskal *et al.*, Phys. Lett. B **482**, 356 (2000)
[2] R. S. Bhalerao and L. C. Liu, Phys. Rev. Lett. **54**, 865 (1985)
[3] B. Krusche and C. Wilkin, Prog. Part. Nucl. Phys. **80**, 43 (2014)
[4] E. Czerwiński *et al.*, Phys. Rev. Lett. **113**, 062004 (2014)
[5] Q. Haider and L. C. Liu, Phys. Lett. B **172**, 257 (1986)
[6] P. Moskal *et al.*, Prog. Part. Nucl. Phys. **49**, 1 (2002)
[7] E. Chiavassa *et al.*, Phys. Lett. B **322**, 270 (1994)
[8] H. Calén *et al.*, Phys. Lett. B **366**, 39 (1996)
[9] H. Calén *et al.*, Phys. Rev. Lett. **79**, 2642 (1997)
[10] F. Hibou *et al.*, Phys. Lett. B **438**, 41 (1998)
[11] J. Smyrski *et al.*, Phys. Lett. B **474**, 182 (2000)
[12] A M. Bergdolt *et al.*, Phys. Rev. D **48**, 2969 (1993)
[13] M. Abdel-Bary *et al.*, Eur. Phys. J. A **16**, 127 (2003)
[14] P. Moskal *et al.*, Phys. Rev. C **69**, 025203 (2004)
[15] P. Moskal *et al.*, Eur. Phys. J. A **43**, 131 (2010)
[16] H. Petren *et al.*, Phys. Rev. C **82**, 055206 (2010)
[17] H. Calén *et al.*, Phys. Rev. C **58**, 2667 (1998)
[18] P. Moskal *et al.*, Phys. Rev. C **79**, 015208 (2009)
[19] G. Fäldt and C. Wilkin, Phys. Scripta **64**, 427 (2001)
[20] H. Calen *et al.*, Phys. Rev. C **58**, 2667 (1998)
[21] K. Nakayama *et al.*, Phys. Rev. C **68**, 045201 (2003)
[22] K. Nakayama *et al.*, Phys. Rev. C **65**, 045210 (2002)
[23] P. Winter *et al.*, Eur. Phys. J. A **18**, 355 (2003)
[24] P. Winter *et al.*, Phys. Lett. B **544**, 251 (2002)
[25] R. Czyzykiewicz *et al.*, Phys. Rev. Lett. **98**, 122003 (2007)
[26] I. Ozerianska-Schatti arXiv:1607.07261 (2016)
[27] A. Gal *et al.*, Acta Phys. Polon. B **45**, 673 (2014)
[28] H. Machner, J. Phys. G **42**, 043001 (2015)
[29] C. Wilkin, Acta Phys. Polon. B **47**, 249 (2016)

[30] N. G. Kelkar *et al.*, Rept. Prog. Phys. **76**, 066301 (2013); Acta Phys. Polon. B **46**, 113 (2015)
[31] S. Bass *et al.*, Acta Phys. Polon. B **47**, 373 (2016)
[32] Q. Haider, L. C. Liu, Int. J. Mod. Phys. E **24**, 1530009 (2015)
[33] P. Moskal, Few Body Syst. **55**, 667 (2014); Acta Phys. Polon. B **47**, 97 (2016)
[34] WASA-at-COSY Collaboration: H. -H. Adam *et al.*, arXiv:nucl-ex/0411038 (2004)
[35] P. Adlarson *et al.*, Phys. Rev. C **87**, 035204 (2013)
[36] W. Krzemien, P. Moskal, M. Skurzok, Acta Phys. Polon. B **45**, 689 (2014)
[37] M. Skurzok, W. Krzemień, O. Rundel and P. Moskal, Acta. Phys. Polon. B **47**, 503 (2016)
[38] W. Krzemień, *arXiv:1202:5794*, Ph.D. thesis, Jagiellonian University (2011)
[39] M. Skurzok, *arXiv:1509:01385*, Ph.D. thesis, Jagiellonian University (2015)
[40] S. Wycech, A. M. Green, J. A. Niskanen, Phys. Rev. C **52**, 544 (1995)
[41] C. Wilkin, Phys. Rev. C **47**, 938 (1993)
[42] N. Willis *et al.*, Phys.Lett. B **406**, 14 (1997)
[43] P. Moskal, J. Smyrski, Acta. Phys. Pol. B **41**, 2281 (2010)
[44] J. Smyrski *et al.*, Phys. Lett. B **649**, 258 (2007)
[45] T. Mersmann *et al.*, Phys. Rev. Lett. **98**, 242301 (2007)
[46] W. Krzemien *et al.*, Int. J. Mod. Phys. A **24**, 576 (2009)
[47] O. Rundel *et al.* these proceedings, (2016)
[48] S. Wycech, W. Krzemien, Acta Phys. Polon. B **45**, 745 (2014)
[49] P. Moskal, J. Smyrski, Acta Phys. Polon. B **41**, 2281 (2010)
[50] C. Wilkin, Acta Phys. Polon. B **45**, 603 (2014)

Studying η-Meson Decays with WASA-at-COSY

Daniel Lersch[1,*]
for the WASA-at-COSY Collaboration

[1] *Juelich Research Center, Germany*

Abstract. The η-meson is a unique tool in a way that it provides access to (rare) decay processes, which allow to probe symmetry breaking phenomena, determine transition form factors, or to explore the anomalous sector of QCD. Those decays have been studied with the WASA-at-COSY experiment. Two data sets with different η-production mechanisms have been acquired. The analysis of those data sets are restricted to: The investigation of the isospin violating decay $\eta \to \pi^+\pi^-\pi^0$; Exploring the box anomaly in $\eta \to \pi^+\pi^-\gamma$; The determination of the electromagnetic transition form factor via the reactions: $\eta \to e^+e^-\gamma$ and $\eta \to e^+e^-e^+e^-$; Testing the CP-violation in $\eta \to \pi^+\pi^-e^+e^-$. An overview of the analysis of those decays will be given in the following.

1 The WASA-at-COSY Facility

The different η-decays are measured and reconstructed within the 4π wide angle shower apparatus (WASA) (see Fig. 1), which is located at the cooler synchrotron (COSY) at the Jülich Research Center in Germany. A proton beam, provided by COSY, is impinged on a liquid deuterium / hydrogen target producing η-mesons via: $pd \to {}^3\mathrm{He}\eta$ and $pp \to pp\eta$. The scattered projectiles are detected in the forward part of WASA, where the angular information is provided by tracking detectors and the kinetic energy is reconstructed by a range hodoscope (see Fig. 1). The decay products are detected in the central part of WASA. Charged particles are reconstructed within a drift chamber in combination with a $\sim 1\,\mathrm{T}$ magnetic solenoid field. Additional plastic scintillators and an electromagnetic calorimeter allow for the distinction between pions and electrons. Neutral particles are reconstructed within the electromagnetic calorimeter.

2 The Data Sets

A first iteration of measurements was done using the $pd \to {}^3\mathrm{He}\eta$ reaction ($\sim 3 \cdot 10^7$ η-mesons have been detected) for the study of the more abundant η-decay channels (such as $\eta \to \pi^+\pi^-\pi^0$) and to setup the framework for a common analysis. The rare η-decay modes (e.g. $\eta \to e^+e^-e^+e^-$) are studied by using the high-statistics $pp \to pp\eta$ data set ($\sim 10^9$ η-mesons have been produced). The η-mesons are reconstructed via their missing mass, which is defined as: Missing mass $= \sqrt{(P_{\mathrm{in}} - P_{\mathrm{out}})^2}$, where P_{in} is the four momentum of all incoming particles (i.e. beam and target) and P_{out} is the corresponding four momentum of all scattered projectiles. A plot of the missing mass distribution for the two data

DOI: 10.1051/epjconf/201613003005

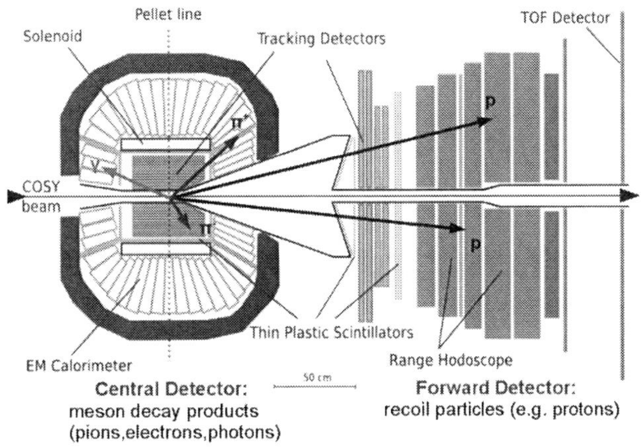

Figure 1. Schematic representation of the 4π WASA detector. The reaction $\eta \to \pi^+\pi^-\gamma$ (see arrows) is shown as an example for the reconstruction of an η-decay.

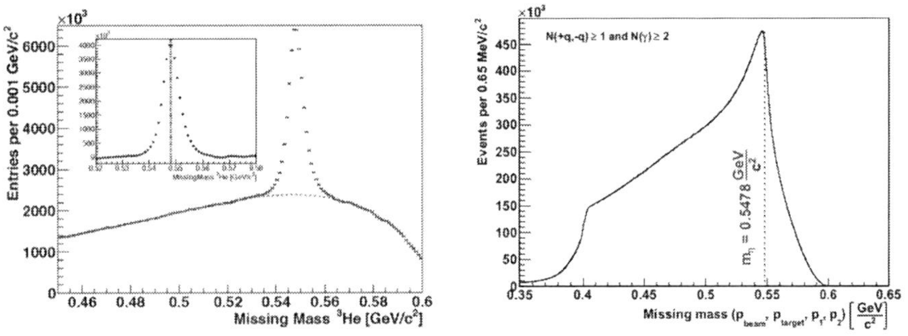

Figure 2. Missing mass distribution of the measured data for the reactions: $pd \to {}^3\text{He}\eta$ (left) and $pp \to pp\eta$ (right). The statistics shown on the right hand side correspond to $\sim$ 50% of the total proton-proton data set.

sets is shown in Fig. 2. The missing mass distribution shown on the left of Fig. 2 is deduced from the $pd \to {}^3\text{He}\eta$ data set. A clear peak is pronounced at the η mass. The continuous background below the peak is related to direct pion production reactions such as: $pd \to {}^3\text{He}\pi^+\pi^-$, $pd \to {}^3\text{He}\pi^0\pi^0$ and $pd \to {}^3\text{He}\pi^+\pi^-\pi^0$. The corresponding distribution for the $pp \to pp\eta$ reaction is shown on the right hand side of Fig. 2. A peak at the η-mass is visible as well, however the non-resonant background is more dominating, compared to the left plot. This is related to the fact that, besides the η-production, the production cross sections for the multi-pion background is increased in the proton-proton case.

★e-mail: d.lersch@fz-juelich.de

3 Measurements and Results

3.1 $\eta \rightarrow \pi^+\pi^-\pi^0$ - The Dalitz Plot

This isospin violating decay allows to probe the quark mass ratio, as the decay amplitude Γ is proportional to Q^{-4} with [1]:

$$Q^2 \equiv \frac{m_s^2 - \hat{m}^2}{m_d^2 - m_u^2} \text{ and } \hat{m} = \frac{1}{2}(m_u + m_d) \quad (1)$$

The decay amplitude is experimentally accessible via the Dalitz Plot analysis:

$$\frac{d^2\Gamma}{dXdY} \propto (1 + aY + bY^2 + dX^2 + fY^3 + ...) \quad (2)$$

where a,b,d and f are fit-parameter and X and Y are the dimensionless Dalitz Plot variables. Table 1 summarizes theoretical approaches such as nonrelativistic effective field theory (NREFT) and chiral pertubation theory for the determination of the Dalitz Plot parameters a,b,d and f. The corresponding measured results are listed in the bottom half of Table 1. The results measured by WASA-at-COSY are based on 120 k $\eta \rightarrow \pi^+\pi^-\pi^0$ events which have been reconstructed from the $pd \rightarrow {}^3\text{He}\eta$ data set. Table 1 shows that the current WASA-at-COSY results for the Dalitz Plot parameters are consistent with the KLOE results. A partial wave analysis (PWA) has been performed with the WASA data including the calculation of Q [4]: $Q = 21.4 \pm 0.4$. In a next step, the Dalitz Plot analysis will be redone using the $pp \rightarrow pp\eta$ data set.

Table 1. Comparison of the WASA-at-COSY results from the Dalitz Plot analysis with theory and recent results from the KLOE collaboration.

	Parameter:	– a	b	d	f
Theor.	ChPT (NNLO) [2]	1.271(75)	0.394(102)	0.055(57)	0.025(160)
	NREFT [3]	1.213(14)	0.308(23)	0.050(3)	0.083(19)
	PWA [4]	1.116(32)	0.188(12)	0.063(4)	0.091(3)
Exp.	KLOE (08) [5]	$1.090(5)(^{+8}_{-19})$	0.124(6)(10)	$0.057(6)(^{+7}_{-16})$	0.14(1)(2)
	WASA [6]	1.144(18)	0.219(19)(47)	0.086(18)(15)	0.115(37)
	KLOE (16) [7]	1.104(3)(2)	$0.142(3)(^{5}_{-4})$	$0.073(3)(^{+4}_{-3})$	$0.154(6)(^{+4}_{-5})$

3.2 $\eta \rightarrow \pi^+\pi^-\gamma$ - The Box Anomaly and $\pi^+\pi^-$ FSI

In the limit of vanishing quark masses the decay amplitude $A_{\eta\rightarrow\pi^+\pi^-\gamma}$ of $\eta \rightarrow \pi^+\pi^-\gamma$ is sensitive to the box anomaly term which is part of the Wess-Zumino-Witten lagrangian [8, 9]. However, this amplitude is dominated by pion-pion final state interactions (FSI) when going to physical quark masses. Therefore $A_{\eta\rightarrow\pi^+\pi^-\gamma}$ has to be modified in terms of the squared pion invariant mass $s_{\pi\pi}$: [10]

$$A_{\eta\rightarrow\pi^+\pi^-\gamma} \mapsto A_{\eta\rightarrow\pi^+\pi^-\gamma} \times [F_{PV}(s_{\pi\pi}) \times (1 + \alpha s_{\pi\pi})] \quad (3)$$

where F_{PV} is the pion vector form factor, which describes all resonant and non-resonant pion-pion interactions. This factor is independent of the underlying reaction mechanism. Reaction specific contributions are described by the linear term $(1 + \alpha s_{\pi\pi})$ with α being a fit parameter. Fig. 3 shows the single photon energy E_γ in the η-rest frame which is proportional to $s_{\pi\pi}$ [11]. The left hand side of Fig. 3 shows the result obtained from the reaction $pd \rightarrow {}^3\text{He}\eta$ [11] where 13 k $\eta \rightarrow \pi^+\pi^-\gamma$ events

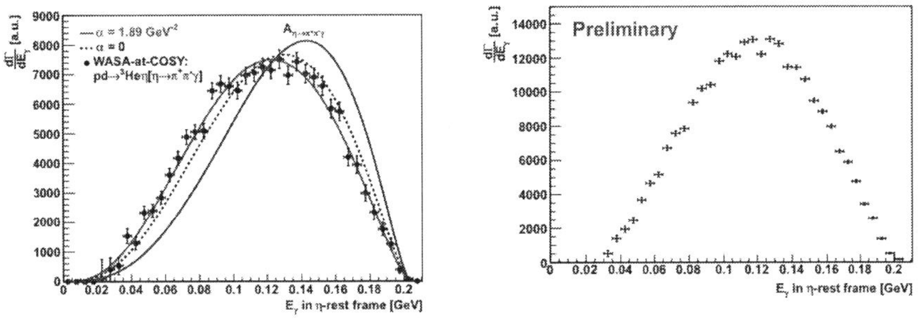

Figure 3. Single photon energy E_γ in the η-rest frame obtained from the $pd \to {}^3\mathrm{He}\eta$ data set [11] (left) and $pp \to pp\eta$ data set (right). The black points represent the measured data. The solid / dashed lines correspond to a fit of Eq. 3.

have been reconstructed. The black data points are fitted by Eq. 3 with α fixed to zero (black dashed curve) and α left as a free parameter (red) curve. The blue curve is a prediction of the E_γ-distribution without including final state interactions at all [14] (i.e. using $A_{\eta\to\pi^+\pi^-\gamma}$ solely). A fit of Eq. 3 to the $pd \to {}^3\mathrm{He}\eta$ data with α left free (red curve in Fig. 3) leads to [11]:

$$\alpha = (1.89 \pm 0.25_{stat} \pm 0.59_{sys} \pm 0.02_{theor.})\,\mathrm{GeV}^{-2} \tag{4}$$

This value is significantly different from zero and indicates, that reaction dependent contributions from the production mechanisms of the two pions have to be included. The result in Eq. 4 is inconsistent with the theoretical predictions [14–16]. However, this value is supported by the latest KLOE result [12] with: $\alpha = (1.32 \pm 0.08_{stat} \pm^{0.1}_{0.09 sys} \pm 0.02_{theor.})\,\mathrm{GeV}^{-2}$. Another observable to investigate the box anomaly and final state interactions is the relative branching ratio $\Gamma(\eta \to \pi^+\pi^-\gamma)/\Gamma(\eta \to \pi^+\pi^-\pi^0)$. The value found for the WASA proton-deuteron data set [13] shows a slight missmatch with the current PDG value [17]. The right hand side of Fig. 3 shows the E_γ-distribution obtained from analyzing $\sim 50\%$ of the $pp \to pp\eta$ data set, resulting in 209 k reconstructed $\eta \to \pi^+\pi^-\gamma$ events. This distribution has already been corrected for multi-pion background. Currently ongoing steps include a correction for misidentified $\eta \to \pi^+\pi^-\pi^0$ events as well as an efficiency correction. The α-parameter as well as the relative branching ratio will be determined for the high statistics proton-proton data set, in order to decrease the statistical uncertainties and to recheck the existing results. Theoretically, the decay $\eta \to \pi^+\pi^-\gamma$ would be suitable to investigate CP-violation via measuring E_1 transitions of the single photon. However, this would require information about the photon polarisation, which is not possible with the WASA setup. An other elegant way to probe the CP-violation is to look at this decay with the photon being virtual and decaying into a dilepton-pair: $\eta \to \pi^+\pi^-\gamma^*[\gamma^* \to e^+e^-]$. This leads to four charged particles in the final state which can be reconstructed with WASA.

3.3 $\eta \to \pi^+\pi^-e^+e^-$ - CP-Violation

The information about a possible CP-Violation is embedded in the asymmetry A_Φ [18]:

$$A_\Phi = \frac{N(\sin[\Phi]\cos[\Phi] > 0) - N(\sin[\Phi]\cos[\Phi] < 0)}{N(\sin[\Phi]\cos[\Phi] > 0) + N(\sin[\Phi]\cos[\Phi] < 0)} \tag{5}$$

where Φ is the angle between the decay planes of the dileptons and the charged pions. The upper limit for Eq. 5 predicted by theory is $\sim 1\%$ [18]. The current measured result for A_Φ has been found by KLOE [19]: $A_\Phi = (-0.6 \pm 2.5_{stat} \pm 1.8_{sys}) \cdot 10^{-2}$. The branching ratio BR for this decay has been measured by KLOE as well and is given by [19]: $\mathrm{BR}(\eta \to \pi^+\pi^-e^+e^-) = (2.68 \pm 0.09_{stat} \pm 0.07_{sys}) \cdot 10^{-4}$. Fig. 4 shows the missing mass as well as the $\sin\Phi\cos\Phi$-distribution after reconstructing 251 ± 17

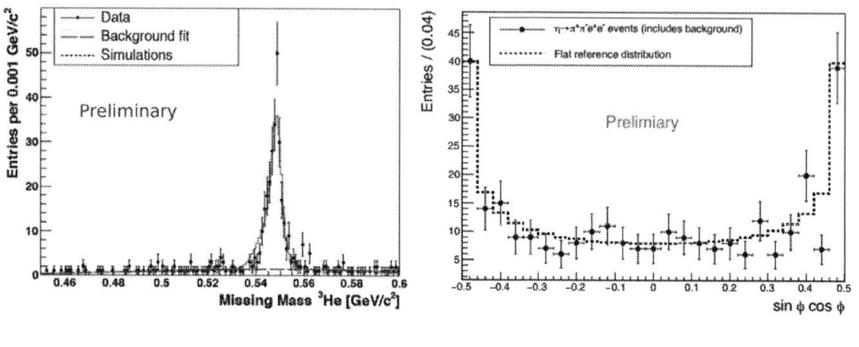

Figure 4. Left: Missing mass after reconstructing $\eta \to \pi^+\pi^-e^+e^-$ from the proton-deuteron data set [13]. The horizontal red line represents a fit of the background. Right: $\sin\Phi\cos\Phi$-distribution for data [13] (black points) and Monte Carlo simulations of $\eta \to \pi^+\pi^-e^+e^-$ (blue dashed line).

$\eta \to \pi^+\pi^-e^+e^-$ events from the proton-deuteron data set. The preliminary results deduced from Fig. 4 are [13]:

$$A_\Phi = (-1.1 \pm 6.6_{stat} \pm 0.2_{sys}) \cdot 10^{-2} \quad (6)$$
$$\mathrm{BR}(\eta \to \pi^+\pi^-e^+e^-) = (2.7 \pm 0.2_{stat} \pm 0.2_{sys}) \cdot 10^{-4} \quad (7)$$

Both values are in agreement with the current KLOE results. However, Fig. 4 already indicates a need for more statistics in order to decrease the statistical uncertainties. Therefore, this measurement will be redone with the proton-proton data set. A first estimation shows that about $\sim 1\,\mathrm{k}\ \eta \to \pi^+\pi^-e^+e^-$ events are expected for the full $pp \to pp\eta$ data set, which is comparable with the statistics from the KLOE experiment.

3.4 $\eta \to e^+e^-\gamma$ and $\eta \to e^+e^-e^+e^-$ - Dalitz Decays

The Dalitz decay $\eta \to e^+e^-\gamma$ allows to investigate the single off-shell transition form factor $F(q^2)$, by measuring the dilepton invariant mass distribution q as well as the relative branching ratio $\frac{\Gamma(\eta \to e^+e^-\gamma)}{\Gamma_\eta}$. If the photon becomes virtual, the Dalitz decay turns into the double Dalitz decay: $\eta \to e^+e^-e^+e^-$ which gives access to the double off-shell transition form factor $F(q_1^2, q_2^2)$. The observable to check theoretical predictions for $F(q_1^2, q_2^2)$ [20] is the branching ratio $\mathrm{BR}(\eta \to e^+e^-e^+e^-)$. Fig. 5 shows the missing masses after reconstructing $\eta \to e^+e^-\gamma$ and $\eta \to e^+e^-e^+e^-$ events. The corresponding preliminary values for the relative branching ratios are given by [13]:

$$\mathrm{BR}(\eta \to e^+e^-\gamma) = (6.72 \pm 0.07_{stat} \pm 0.31_{sys}) \cdot 10^{-3} \quad (8)$$
$$\mathrm{BR}(\eta \to e^+e^-e^+e^-) = (3.2 \pm 0.9_{stat} \pm 0.5_{sys}) \cdot 10^{-5} \quad (9)$$

Both results are in agreement with current measurements [17] and based on: $14,040 \pm 120$ reconstructed $\eta \to e^+e^-\gamma$ events and 18 ± 5 reconstructed $\eta \to e^+e^-e^+e^-$ events. Each reaction is reanalyzed in the proton-proton data set, in order to improve statistics. Additionally the dileption invariant mass for the single Dalitz decay will be determined. The large invariant mass region will profit from the high statistics and allow for a precise determination of $F(q^2)$ at large q-values.

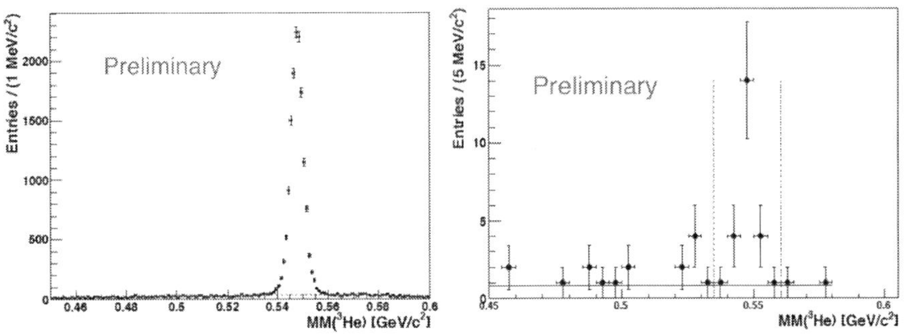

Figure 5. Missing mass after reconstructing $\eta \to e^+e^-\gamma$ (left) and $\eta \to e^+e^-e^+e^-$ (right) from the proton-deuteron data set [13]. The red horizontal lines in each plot represent the fitted background. The vertical lines in the right diagram correspond to the integration window.

4 Summary and Outlook

The η-decays discussed in the previous section have been measured in the $pd \to {}^3\text{He}\eta$ and $pp \to pp\eta$ reaction. The analysis for the proton-deuteron data set is finished and results have been published or are on the way of publishing [6, 11, 13]. The analysis of the proton-proton data set is still ongoing and will profit from the analysis that has been set up for the proton-deuteron data. New results with improved statistics and a more detailed insight into the nature of η-decays are expected in the near future.

References

[1] D. B. Kaplan and A. V. Manohar, Phys. Rev. Lett. **56**, 2004 (1986)
[2] J. Bijnens and K. Ghorbani., J. High Energy Phys. 11, 030 (2007)
[3] S.P. Schneider, B. Kubis and C. Ditsche, J. High Energy Phys. 02, 028 (2007)
[4] Peng Guo *et al.*, Phys. Rev. D **92**, 054016 (2015)
[5] F. Ambrosio *et al.* (KLOE Collaboration), J. High Energy Phys. 05, 006 (2008)
[6] P. Adlarson *et al.* (WASA collaboration), Phys. Rev., C **90**, 045207 (2014)
[7] F. Ambrosio *et al.* (KLOE Collaboration), J. High Energy Phys. 05, 019 (2016)
[8] J. Wess and B. Zumino, Phys. Lett. B **37**, 95 (1971)
[9] E. Witten, Nucl. Phys. B **223**, 422 (1983)
[10] F. Stollenwerk *et al.*, Phys. Lett. B **707**, 184 (2012)
[11] P. Adlarson *et al.* (WASA collaboration), Phys. Lett. B **707**, 243 (2012)
[12] F. Ambrosio *et al.* (KLOE collaboration), Phys. Lett. B **718**, 910 (2013)
[13] P. Adlarson *et al.*, arXiv:1509.06588 [nucl-ex]
[14] B.R. Holstein, Phys. Scripta T **99**, 55 (2002)
[15] M. Benayoun *et al.*, Eur. Phys. J. C **31**, 525 (2003)
[16] J. Bijnens *et al.*, Phys. Lett. B **237**, 488 (1990)
[17] K. Olive *et al.*, Chin. Phys. C **38**, 090001 (2014)
[18] D. Gao, Mod. Phys. Lett. A **17**, 1583 (2002)
[19] F. Ambrosio *et al.*, Phys. Lett. B **675**, 283 (2009)
[20] B. Bijnens and F. Persson, arXiv:hep-ph/0106130

EPJ Web of Conferences **130**, 03006 (2016) DOI: 10.1051/epjconf/201613003006
MESON 2016

ηN interactions in the nuclear medium. η-nuclear bound states

J. Mareš[1,*], N. Barnea[2], A. Cieplý[1], E. Friedman[2], and A. Gal[2]

[1] *Nuclear Physics Institute, 250 68 Řež, Czech Republic*
[2] *Racah Institute of Physics, The Hebrew University, 91904 Jerusalem, Israel*

Abstract. We report on our recent study of in-medium ηN interactions and η-nuclear quasi-bound states. The ηN scattering amplitudes considered in the calculations are constructed within coupled-channel models that incorporate the S_{11} $N^*(1535)$ resonance. The implications of self-consistent treatment and the role played by subthreshold dynamics are discussed.

1 Introduction

The ηN attraction generated by the $N^*(1535)$ resonance near threshold seems to be strong enough to allow binding of the η meson in nuclei. However, in-medium modifications and strong energy dependence of the ηN scattering amplitudes have to be carefully taken into account. This contribution briefly summarizes systematic treatment of energy dependence within self-consistent calculations of η quasi-bound states in selected nuclei (see [1–3] for more details).

2 Methodology

The ηN scattering amplitudes are highly model dependent as illustrated in Fig. 1 for meson-baryon interaction models GW [4], CS [5], M2 [6], and GR [7]. The ηN amplitudes differ below as well as above the ηN threshold, except perhaps common value $\mathrm{Im}F_{\eta N} \approx 0.2 - 0.3$ fm at threshold.

The in-medium amplitudes which serve as an input in our many-body self-consistent calculations are obtained from the free-space amplitudes GW and M2 by applying the multiple scattering approach [8] (see Ref. [2] for details). In the GR and CS models, the Pauli principle restricts integration domain in the Green's function which enters the underlying Lippmann-Schwinger equations [2].

The strong energy dependence of the ηN scattering amplitudes $F_{\eta N}(\sqrt{s})$ has to be treated self-consistently [1, 2]. The argument $\sqrt{s}$ in the scattering amplitudes is given by

$$\sqrt{s} = \sqrt{(\sqrt{s_{\mathrm{th}}} - B_\eta - B_N)^2 - (\vec{p}_\eta + \vec{p}_N)^2} \leq \sqrt{s_{\mathrm{th}}}, \quad (1)$$

where $\sqrt{s_{\mathrm{th}}} \equiv m_\eta + m_N$ and B_η and B_N are η and nucleon binding energies. In the nuclear medium (for $A \gg 1$ approximated by the lab system) the momentum dependent term causes additional downward energy shift, since $(\vec{p}_\eta + \vec{p}_N)^2 \neq 0$, which can be approximated as [2]

$$\delta\sqrt{s} \approx -B_N\frac{\rho}{\bar{\rho}} - \xi_N B_\eta \frac{\rho}{\rho_0} - \xi_N T_N (\frac{\rho}{\rho_0})^{2/3} - \xi_\eta \frac{\sqrt{s}}{\omega_\eta E_N} 2\pi \mathrm{Re}\, F_{\eta N}(\sqrt{s}, \rho)\rho\,, \quad (2)$$

*e-mail: mares@ujf.cas.cz

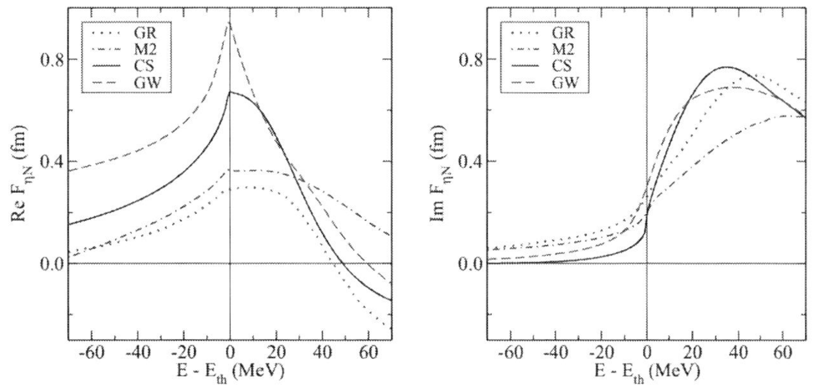

Figure 1. Energy dependence of the real (left panel) and imaginary (right panel) parts of the free ηN scattering amplitude in interaction models GW [4] (dashed), CS [5] (solid), M2 [6] (dot-dashed), and GR [7] (dotted). The vertical line denotes the ηN threshold.

where $\xi_{N(\eta)} = m_{N(\eta)}/(m_N + m_\eta)$, $T_N = 23.0$ MeV at ρ_0, $B_N \approx 8.5$ MeV is the average nucleon binding energy, and $\bar{\rho}$ is the average nuclear density. For attractive scattering amplitudes, all terms in Eq. 2 are negative definite, providing substantial downward energy shift. A variant of Eq. 2 was used in η-nuclear three- and four-body calculations (see Ref. [3] for details).

In few-body ηNN and ηNNN systems, the η-nuclear cluster wave functions were expanded in a hyperspherical basis and the ground-state binding energies were calculated variationally. For the NN interaction, the Minnesota central potential [9] and the Argonne AV4' potential [10] were used. The ηN interaction was described by energy dependent local ηN potentials that reproduce the ηN scattering amplitudes below threshold in considered interaction models [3].

The conversion widths were evaluated through the expression $\Gamma/2 \approx \langle \Psi_{g.s.}| - \mathrm{Im} V_{\eta N} |\Psi_{g.s.}\rangle$, where $V_{\eta N}$ sums overall pairwise ηN interactions [1].

The interaction of the η meson with the nuclear many-body system was described by the Klein–Gordon (KG) equation of the form

$$[\,\nabla^2 + \tilde{\omega}_\eta^2 - m_\eta^2 - \Pi_\eta(\omega_\eta, \rho)\,]\,\psi = 0\,, \tag{3}$$

where $\tilde{\omega}_\eta = \omega_\eta - \mathrm{i}\Gamma_\eta/2$ is complex energy of η, $\omega_\eta = m_\eta - B_\eta$, and Γ_η is the width of the η-nuclear bound state. The self-energy operator $\Pi_\eta(\sqrt{s}, \rho) \equiv 2\omega_\eta V_\eta = -(\sqrt{s}/E_N)\, 4\pi F_{\eta N}(\sqrt{s}, \rho)\rho$ was constructed self-consistently using the RMF density distributions in a core nucleus.

It is to be stressed that $\mathrm{Re}F_{\eta N}(\sqrt{s})$ and B_η appear as arguments in the expression for $\delta\sqrt{s}$ (Eq. 2), which in turn serves as an argument for $F_{\eta N}$ and thus for the self-energy Π_η. Therefore, a self-consistency scheme in terms of both Π_η and B_η is required in calculations.

3 Results

Our few-body calculations of the ηNN system found *no* bound states in the considered coupled-channel models. For ηNNN, a relatively broad and weakly bound state (with η separation energy below 1 MeV) was found for the Minesota NN potential and one particular variant of the ηN potential

[1] We found an error in normalization in Ref. [3], which made the calculated widths about factor of 2 larger.

that reproduced the GW scattering amplitudes (see Ref. [3] for details). *No* ηNNN bound states were found using more realistic NN interaction models.

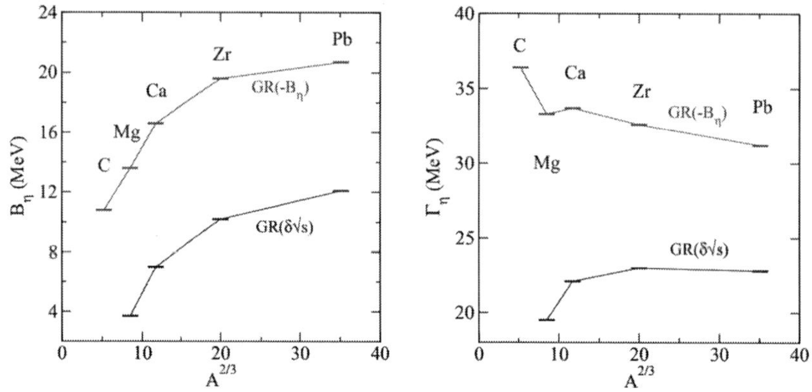

Figure 2. Binding energies (left) and widths (right) of the $1s$ η-nuclear states in selected nuclei calculated using the GR ηN scattering amplitude [7] with different procedures for subthreshold energy shift $\delta\sqrt{s}$.

Figure 2 illustrates the role of the energy dependence of ηN scattering amplitudes in self-consistent evaluations of η nuclear-states in many-body nuclear systems. A comparison is made for the in-medium GR amplitude: our self-consistency scheme based on $\delta\sqrt{s}$ of Eq. 2 (marked $\delta\sqrt{s}$) reduces considerably the GR binding energies and widths with respect to the original calculations of Ref. [7] that used the $\delta\sqrt{s} = -B_\eta$ procedure (marked $-B_\eta$). However, even the reduced GR widths are still quite high, suggesting that η-nuclear states will be extremely difficult to resolve if the GR model is the realistic one.

The model dependence of the ηN scattering amplitudes shown in Fig. 1 manifests itself in the calculations of η-nuclear states. Figure 3 presents binding energies B_η and widths Γ_η calculated for the $1s$ η-nuclear states in selected nuclei using the above ηN amplitudes.

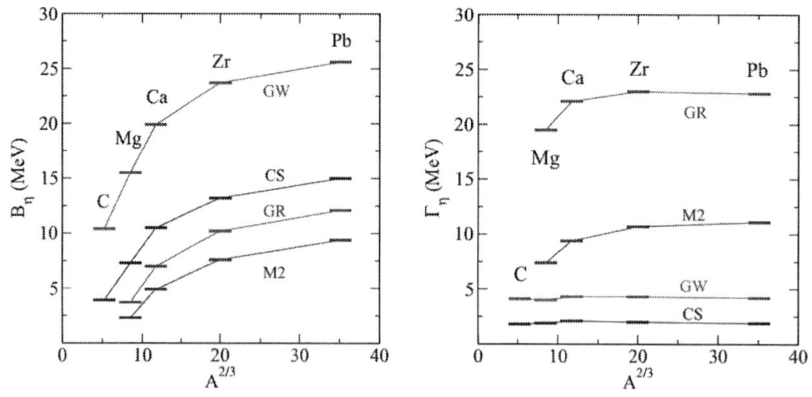

Figure 3. Binding energies (left) and widths (right) of $1s$ η-nuclear states in selected nuclei calculated self-consistently using the M2, GR, CS, and GW ηN scattering amplitudes (see text).

The left panel of Fig. 3 demonstrates that for each of the ηN amplitude models the binding energy increases with A and tends to saturate for large values of A. The hierarchy of the curves reflects the strength of $\mathrm{Re}F_{\eta N}(\sqrt{s})$ in the subthreshold region (see Fig. 1). The M2 amplitude is too weak to produce the $1s$ η bound state in ^{12}C. In contrast, $\mathrm{Re}F_{\eta N}(\sqrt{s})$ of the GW model is strong enough to bind η in ^{12}C and even in lighter nuclei, e.g., it predicts the $1s$ η bound state in ^{4}He with $B_\eta = 1.2$ MeV and $\Gamma_\eta = 2.3$ MeV (calculated using a static ^{4}He density).

The right panel shows substantial differences between the widths Γ_η calculated using the above mentioned models. The CS and GW models yield relatively small uniform widths of order 2 and 4 MeV, respectively. On the other hand, the GR and M2 models predict much larger widths which increase with A. This reflects partly the energy dependence of $\mathrm{Im}F_{\eta N}(\sqrt{s})$ in the subthreshold region and partly the difference in the in-medium renormalization stemming from $\mathrm{Re}F_{\eta N}(\sqrt{s})$. For instance, the large downward energy shift due to the subthreshold amplitude in the GW model (57 MeV at ρ_0) causes a particularly large reduction in the strength of the $\mathrm{Im}F_{\eta N}(\sqrt{s})$ input.

The widths calculated here do not include contributions from two-nucleon processes which are estimated to add a few MeV. We may therefore conclude that η-nuclear states could in principle be observed if the CS and GW models turn out to be realistic ones, provided a suitable production/formation reaction is found. Other models give either too large widths or are too weak to generate η-nuclear bound states in lighter nuclei.

This work was supported by the GACR Grant No. P203/15/04301S, as well as by the EU initiative FP7, HadronPhysics3, under the SPHERE and LEANNIS cooperation programs.

References

[1] E. Friedman, A. Gal, J. Mareš, Phys. Lett. **B 725**, 334 (2013).
[2] A. Cieplý, E. Friedman, A. Gal, J. Mareš, Nucl. Phys. **A 925**, 126 (2014).
[3] N. Barnea, E. Friedman, A. Gal, Phys. Lett. **B 747**, 345 (2015).
[4] A.M. Green, S. Wycech, Phys. Rev. **C 71**, 014001 (2005).
[5] A. Cieplý, J. Smejkal, Nucl. Phys. **A 919**, 46 (2013).
[6] M. Mai, P.C. Bruns, U.-G. Meissner, Phys. Rev. **D 86**, 015201 (2013).
[7] T. Inoue, E. Oset, Nucl. Phys. **A 710**, 354 (2002); C. García-Recio, T. Inoue, J. Nieves, E. Oset, Phys. Lett. **B 550**, 47 (2002).
[8] T. Wass, M. Rho and W. Weise, Nucl. Phys. **A 617**, 449 (1997).
[9] D. R. Thomson, M. LeMere, Y. C. Tang, Nucl. Phys. **A 286**, 53 (1977).
[10] R. B. Wiringa, S. C. Pieper, Phys. Rev. **C 89**, 182501 (2002).

Role of the $a_0(980)$, $f_0(980)$ resonances in $\eta \to 3\pi$ from the Khuri-Treiman formalism

B. Moussallam[1,⋆] and M. Albaladejo[2]

[1] *Groupe de Physique Théorique, IPN (UMR8608), Université Paris-Sud 11, Orsay, France*
[2] *Instituto de Física Corpuscular (IFIC), Centro Mixto CSIC-Universidad de Valencia, Spain*

Abstract. The chiral expansion of the $\eta \to 3\pi$ decay amplitude at NLO fails to reproduce the recent high accuracy measurements of the Dalitz plot parameters. We reconsider the idea of employing the chiral expansion in an unphysical region surrounding the Adler zero and then deducing the amplitude in the physical region using the Khuri-Treiman dispersive formalism. We further present an extension of this formalism which takes into account $K\bar{K}$ inelastic rescattering effects and thus provides a realistic description of the double $a_0(980)$, $f_0(980)$ resonance effect. We evaluate how the influence of these resonances propagates down to the low-energy decay region and show that it is significant, in particular for the $\eta \to 3\pi^0$ decay, and improve the agreement of the predicted Dalitz plot parameters with experiment.

1 Introduction

The fact that the masses of the u, d, s quarks are much smaller than 1 GeV has an essential influence on the physics of light mesons (indeed if these masses where $O(1)$ GeV the lightest meson in QCD would be the 0^{++} glueball). In this context, $\eta \to 3\pi$ decays are key processes for the determination of the $u - d$ quark mass difference. At next-to-leading order (NLO) in the low-energy chiral expansion, the amplitude can be expressed in a very predictive form involving the double quark mass ratio

$$Q^{-2} = \frac{m_d^2 - m_u^2}{m_s^2 - (m_d + m_u)^2/4} \tag{1}$$

as an overall multiplicative factor [1, 2]. Unfortunately, the NLO amplitude fails to reproduce the experimental values of the Dalitz plot parameters which have been measured quite precisely (see [3, 4] and references therein), which indicates that higher order chiral effects must be accounted for.

A calculation of the complete NNLO amplitude has been performed [5] but our present knowledge of the $O(p^6)$ couplings is not sufficient for making model independent predictions. It is plausible, however, that the important $O(p^6)$ effects are those associated with the restoration of unitarity, or final-state interactions (FSI), effects. This has led to the suggestion of relying on the chiral expansion in the subthreshold unphysical region only, where these effects are suppressed, and construct the physical amplitude from an extrapolation based on exact unitarity and analyticity [6, 7]. A dispersive framework for $\eta \to 3\pi$ was developed in refs. [2, 8] based on the classic work of Khuri and Treiman [9]. In

⋆e-mail: moussall@ipno.in2p3.fr

this formalism, $\pi\pi$ rescattering is treated in the elastic approximation. We present here an extension which accounts for the leading inelastic effects in $\pi\pi$ scattering and also includes $\eta\pi$ rescattering such that a realistic description of the role of the two light resonances $a_0(980)$, $f_0(980)$ in the dispersive integrals can be achieved.

2 The Khuri-Treiman equations with elastic unitarity

We restrict ourselves to regions of the Mandelstam plane where the imaginary parts of the $\eta\pi \to \pi\pi$ partial-waves with $J \geq 2$ can be neglected compared to those with $J = 0, 1$ (i.e. $|s|, |t|, |u| \lesssim 1$ GeV^2). It can then be shown [2, 8] that the η decay amplitudes can be expressed in terms of three functions of a single variable,

$$\begin{aligned} \mathcal{T}_{\eta\to\pi^+\pi^-\pi^0}(s,t,u) &= -\epsilon_L \Big[M_0(s) - \tfrac{2}{3}M_2(s) + (s-u)M_1(t) + (s-t)M_1(u) + M_2(t) + M_2(u)\Big] \\ \mathcal{T}_{\eta\to 3\pi^0}(s,t,u) &= -\epsilon_L \Big[M_0(s) + M_0(t) + M_0(u) + \tfrac{4}{3}(M_2(s) + M_2(t) + M_2(u))\Big] \end{aligned} \tag{2}$$

where the Mandelstam variables (in the charged case) are given by $s = (p_{\pi^+} + p_{\pi^-})^2$, $t = (p_{\pi^-} + p_{\pi^0})^2$, $u = (p_{\pi^+} + p_{\pi^0})^2$ and the overall factor $\epsilon_L = Q^{-2} m_K^2(m_K^2 - m_\pi^2)/3\sqrt{3}F_\pi^2 m_\pi^2$. The functions $M_I(w)$, furthermore, are analytic in the whole complex plane except for a cut along the positive real axis $w \geq 4m_\pi^2$. From eq. (2), it is easy to derive the form of the three partial-waves with $J = 0, 1$ corresponding to a definite isospin, I, of one of the $\pi\pi$ pairs

$$\mathcal{M}_J^I(s) = \lambda^{(I)} \frac{\epsilon_L}{16\pi} \kappa^J(s)(M_I(s) + \hat{M}_I(s)) \tag{3}$$

where $\kappa^2(s) = (1 - 4m_\pi^2/s)(s - (m_\eta + m_\pi)^2)(s - (m_\eta - m_\pi)^2)$ and the isospin factors are $\lambda^{(0)} = \sqrt{6}/2$, $\lambda^{(1)} = 1/3$, $\lambda^{(2)} = -1$. The functions $\hat{M}_I$ can be expressed as simple integrals of the functions M_I (see [2]). They have left-hand cuts which extend into the complex plane and approach infinitesimally close to the unitarity cut but remain well separated from it.

Assuming elastic unitarity for the three partial waves

$$\mathrm{Im}\,[\mathcal{M}_J^I(s)] = \sigma_{\pi\pi}(s)(f_J^I(s))^* \mathcal{M}_J^I(s) \tag{4}$$

(where f_J^I is a $\pi\pi \to \pi\pi$ partial-wave) one can derive the discontinuities of the functions M_I and then, using dispersive representations together with the Omnès method[10] arrive at the following set of coupled integral equations for the functions M_I [2]

$$\begin{aligned} M_0(w) &= \Omega_0(w)[\alpha_0 + w\beta_0 + w^2\left(\gamma_0 + \hat{I}_0(w)\right)] \\ M_1(w) &= \Omega_1(w)\, w[\beta_1 + \hat{I}_1(w)] \\ M_2(w) &= \Omega_2(w)\, w^2[\hat{I}_2(w)] \end{aligned} \tag{5}$$

where the functions $\hat{I}_I$ involve integrals over the left-cut functions

$$\hat{I}_I(w) = -\frac{1}{\pi}\int_{4m_\pi^2}^{\infty} ds' \frac{\mathrm{Im}\,[1/\Omega_I(s')]}{(s')^n(s'-w)} \hat{M}_I(s') \tag{6}$$

(with $n = 2$ when $I = 0, 2$ and $n = 1$ when $I = 1$) and $\Omega_I(s)$ are the Omnès functions associated with the elastic $I = 0, 2$, $J = 0$ and $I = J = 1$ $\pi\pi$ phase-shifts.

2.1 Matching with $O(p^4)$ ChPT

Dispersive Omnès representations involve a polynomial arbitrariness. The form used above, following ref. [2], is chosen to depend on four polynomial parameters $\alpha_0, \beta_0, \gamma_0, \beta_1$. This is particularly convenient for the present purpose since all four parameters can be fixed from matching conditions with the NLO amplitude. These conditions must ensure that,

$$\mathcal{T}^{disp}(s,t,u) - \mathcal{T}^{NLO}(s,t,u) = O(p^6)\,. \tag{7}$$

This will be implemented here in a form essentially equivalent to ref. [2], but slightly more symmetric, obtained by expanding the difference in eq. (7) as a polynomial in s, t, u and requiring that the $O(p^4)$ part vanishes.

3 KT equations with inelastic channels

The formalism discussed above does not account properly for the role of the $f_0(980)$ resonance in the dispersive integrals, which induces a strong inelasticity in $I = J = 0$ $\pi\pi$ scattering and also ignores the role of the $a_0(980)$ resonance. In order to improve on that, we must go beyond the elastic approximation and include the $K\bar{K}$ and $\eta\pi$ channels in the unitarity relation. The $I = 0$ S-wave rescattering will then be described by a 2×2 matrix involving the $(\pi\pi)_{I=0}$ and $(K\bar{K})_{I=0}$ channels which we call $\boldsymbol{T}^{(0)}$. Similarly, we will consider a 2×2 matrix related to the $I = 1$ channels $\eta\pi$ and $(K\bar{K})_{I=1}$, which will play a role as an initial-state interaction. We also collect the S-wave isospin violating amplitudes into two matrices

$$\boldsymbol{T}^{(01)} = \begin{pmatrix} (\pi\pi)_0 \to \eta\pi & (K\bar{K})_0 \to \eta\pi \\ (\pi\pi)_0 \to (K\bar{K})_1 & (K\bar{K})_0 \to (K\bar{K})_1 \end{pmatrix}, \quad \boldsymbol{T}^{(12)} = \begin{pmatrix} \eta\pi^+ \to \pi^+\pi^0 \\ K^+\bar{K}^0 \to \pi^+\pi^0 \end{pmatrix} \tag{8}$$

For illustration, the unitarity relation for $\boldsymbol{T}^{(01)}$ reads

$$\mathrm{Im}\,[\boldsymbol{T}^{(01)}] = \boldsymbol{T}^{(0)*}\,\Sigma^0\,\boldsymbol{T}^{(01)} + \boldsymbol{T}^{(01)*}\,\Sigma^1\,\boldsymbol{T}^{(1)} + \boldsymbol{T}^{(0)*}\begin{pmatrix} 0 & 0 \\ 0 & \Delta\sigma_K \end{pmatrix}\boldsymbol{T}^{(1)} \tag{9}$$

and contains three contributions. In the third one, isospin violation is induced by the physical $K^+ - K^0$ mass difference,

$$\Delta\sigma_K(s) = \frac{1}{2}\left(\theta(s-4m_{K^+}^2)\sqrt{1-4m_{K^+}^2/s} - \theta(s-4m_{K^0}^2)\sqrt{1-4m_{K^0}^2/s}\right). \tag{10}$$

Correspondingly to this matrix unitarity relation, one can derive a matrix Omnès generalisation of the equation for the amplitude function M_0,

$$\boldsymbol{M}_0(w) = \boldsymbol{\Omega}_0(w)\left[\boldsymbol{P}_0(w) + w^2\left(\hat{\boldsymbol{I}}_A(w) + \hat{\boldsymbol{I}}_B(w)\right)\right]{}^t\boldsymbol{\Omega}_1(w) \tag{11}$$

and a similar one for the M_2 equation. Here $\boldsymbol{\Omega}_I$ are Omnès matrices corresponding to the scattering matrices $\boldsymbol{T}^{(I)}$ and $\boldsymbol{P}_0$ is a matrix of polynomials which will get completely determined, as in the elastic case, from chiral matching conditions. The matrix $\hat{\boldsymbol{I}}_B$, which corresponds to the third term in the unitarity relation (9) reads

$$\hat{\boldsymbol{I}}_B(w) = \frac{32}{\sqrt{6}\,\epsilon_L}\int_{4m_{K^+}^2}^{\infty}\frac{ds'\,\Delta\sigma_K(s')}{(s')^2(s'-w)}\boldsymbol{\Omega}_0^{-1*}(s')\,\boldsymbol{T}^{(0)*}(s')\begin{pmatrix} 0 & 0 \\ 0 & 1 \end{pmatrix}\boldsymbol{T}^{(1)}(s')\,{}^t\boldsymbol{\Omega}_1^{-1}(s')\,. \tag{12}$$

Finally, the matrix $\hat{\boldsymbol{I}}_A$ collects integrals over the left-cut pieces of the partial-waves (analogous to eq. (6)).

4 Results and conclusions

Concerning the left-cut parts of the partial-wave amplitudes, we know how to relate $\hat{M}_I$ to the right-cut functions M_I (via the representation (2) of the $\eta \to 3\pi$ amplitude). Similar relations can be derived, in principle, for the $K\bar{K}$ amplitudes (which requires to consider also the cross-channel amplitudes). We adopt here the simple approximation of neglecting the left-cut integrals involving $K\bar{K}$ amplitudes. As a related approximation, we perform the chiral matching of the $K\bar{K}$ amplitudes using the LO expressions (which have no left-hand cut).

We have solved numerically the KT equations using an $I = 0$ T-matrix obtained from simple fits to available experimental data and imposing asymptotic conditions which insure a unique solution for the Omnès matrix. For the $I = 1$ T-matrix, we have used a model constrained by chiral symmetry, at low energy, and by experimental information on the isovector resonances (see ref. [11]).

The results for the Dalitz plot parameters are shown in Table 1. One sees that the KT amplitudes matched to the NLO amplitudes provide a considerable improvement over the NLO amplitudes in the physical region. This is in agreement with the findings of ref. [8]. The table shows that including the effects of the 1 GeV scalar resonances further improves the agreement with experiment. Concerning the quark mass ratio, we find $Q \simeq 21.8$ with the elastic KT amplitudes and essentially the same result with the inelastic ones. The main source of uncertainty for Q arises from the NNLO contributions in the matching equations.

Table 1. $\eta \to 3\pi$ Dalitz plot parameters from the NLO chiral expansion and from elastic and inelastic chirally matched Khuri-Treiman amplitudes.

$\eta \to \pi^+\pi^-\pi^0$	$O(p^4)$	single-ch.	coupled-ch.	KLOE	BESIII
a	-1.328	-1.154	-1.142	-1.095(4)	-1.128(15)
b	0.429	0.202	0.171	0.145(6)	0.153(17)
d	0.089	0.094	0.097	0.081(7)	0.085(16)
f	0.016	0.108	0.123	0.141(10)	0.173(28)
g	-0.081	-0.087	-0.088	-0.044(16)	–
$\eta \to \pi^0\pi^0\pi^0$				PDG	
α	+0.0142	-0.0274	-0.0337	-0.0315(15)	

References

[1] J. Gasser, H. Leutwyler, Nucl. Phys. B **250**, 539 (1985)
[2] A. Anisovich, H. Leutwyler, Phys. Lett. B **375**, 335 (1996)
[3] M. Ablikim *et al.* (BESIII), Phys. Rev. D **92**, 012014 (2015)
[4] A. Anastasi *et al.* (KLOE-2) (2016), JHEP **1605**, 019 (2016)
[5] J. Bijnens, K. Ghorbani, JHEP **0711**, 030 (2007)
[6] A. Neveu, J. Scherk, Annals Phys. **57**, 39 (1970)
[7] C. Roiesnel, T.N. Truong, Nucl. Phys. B **187**, 293 (1981)
[8] J. Kambor, C. Wiesendanger, D. Wyler, Nucl. Phys. B **465**, 215 (1996)
[9] N. Khuri, S. Treiman, Phys. Rev. **119**, 1115 (1960)
[10] R. Omnès, Nuovo Cim. **8**, 316 (1958)
[11] M. Albaladejo, B. Moussallam, Eur. Phys. J. C **75**, 488 (2015)

EPJ Web of Conferences **130**, 03008 (2016) DOI: 10.1051/epjconf/201613003008
MESON 2016

Near threshold production of η-mesons in proton neutron collisions at ANKE

Daniel Schröer[1,⋆], Christopher Fritzsch[1], Alfons Khoukaz[1], and Marcel Rump[1]
for the ANKE Collaboration

[1] *Institut für Kernphysik, Westfälische Wilhelms-Universität Münster, Germany*

Abstract. The interaction between hadrons and η-mesons is an intensively studied topic. Due to its stength it might lead to the formation of η-mesic nuclei. In order to study the characteristics of this interaction a measurement of the reaction $p + d \rightarrow d + \eta + p_{\mathrm{sp}}$ has been performed at the ANKE spectrometer at the COSY accelerator of the Forschungszentrum Jülich. The two different beam momenta ($p_1 = 2.09\,\mathrm{GeV/c}$ and $p_2 = 2.25\,\mathrm{GeV/c}$) in combination with the Fermi motion inside the target deuteron grant access to the determination of total and differential cross sections in an excess energy range from threshold up to $Q = 90\,\mathrm{MeV}$. While the course of the total cross section, especially near threshold, will allow to extract the scattering length $a_{\mathrm{d}\eta}$ via an s-wave final state interaction ansatz, the differential cross sections permit to verify the legitimacy of the s-wave assumption.

1 Introduction

The ANKE spectrometer [1] at the accelerator facility COSY is very well suited for the production of light mesons. Particularly the interaction of η-mesons with hadrons is of special concern, as this might lead to the formation of η-mesic nuclei [2]. To search for such η-mesic nuclei there are principally two possibilities. The first way is to look for a decay signal below the production threshold. In this case the η-mesic nucleus would decay e.g. into a nucleus and a pion, which would influence the cross section of pion production, but up to now no clear sign has been encountered yet. The second approach is to examine the production above threshold, where a pole near threshold could cause a steeper rise in the cross section than expected from pure phase space behaviour.

The difference between pure phase space behaviour and the final state interaction ("FSI") effect for a two-particle final state can be described by an s-wave FSI-Ansatz

$$\frac{p_i}{p_f} \cdot \frac{d\sigma}{d\Omega} = |f|^2 = |f_s \cdot \mathrm{FSI}|^2 \tag{1}$$

with f_s being a constant production amplitude

$$\mathrm{FSI} = \frac{1}{1 - i \cdot a \cdot p_f + \frac{1}{2} r_0 a p_f^2} = \frac{1}{(1 - p_f/p_a)(1 - p_f/p_b)}, \tag{2}$$

⋆e-mail: d_schr13@uni-muenster.de

 DOI: 10.1051/epjconf/201613003008

with p_f (p_i) being the final (initial) state momentum, a the scattering length and r_0 the effective range.

Investigations of the reaction $d+p \rightarrow {}^3He+\eta$ have shown a steep rise close to the threshold which could be described by a pole at $|Q_0| \approx 0.3\,\mathrm{MeV}$ [3]. As the sign of this pole is not determined, further studies have to be undertaken. For this purpose the reaction $p+d \rightarrow d+\eta+p_{sp}$ has been measured at ANKE.

2 The reaction $p+n \rightarrow d+\eta$

To measure the reaction $p+n \rightarrow d+\eta$ COSY provided a proton beam with momenta $p_1 = 2.09\,\mathrm{GeV/c}$ and $p_2 = 2.25\,\mathrm{GeV/c}$ while the cluster-jet target at ANKE produced a stream of deuterons, which served as an effective neutron target with the proton being a spectator particle. This combination allows to investigate the cross section and thence the FSI in an excess energy range from threshold up to $Q = 90\,\mathrm{MeV}$. The impact of the FSI is strongest in the vicinity of the threshold so that it is essential to extract a clean signal at low Q-values.

The reaction is identified via the missing mass method. For this purpose both the deuteron and proton have to be reconstructed. Two Silicon Tracking Telescopes ("STT"), placed near the interaction point, are used for the detection of the spectator proton while the deuteron is observed in the Forward system ("Fd") (Fig. 1). As there is a huge proton background, great emphasis was placed on the clean separation between protons and deuterons [4]. Since near threshold the peak in the missing mass spectrum is close to the kinematical limit an elegant description of the multi-pion background is mandatory. For this goal the data taken at the second beam energy analyzed as if they happened at the the first one. This results in a shift of the recorded events so that the kinematical limits are at the same position, whereas the η-peaks are separated. If both spectra are scaled accordingly and subtracted a dip (peak) for the (un-)shifted data sets can be seen (Fig. 2).

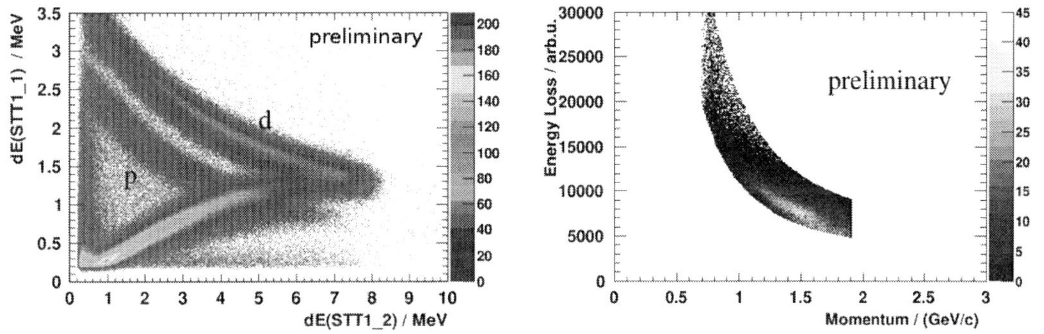

Figure 1. Left: Energy loss in first layer vs energy loss in second layer of one STT; right: Energy Loss in Fd Counter vs momentum of incident particle with cut on deuterons.

The resulting count rates have to be corrected for acceptance and the Fermi motion inside the deuteron as the latter results in different effective luminosities for each Q value. Both has been achieved by executing Monte-Carlo simulations. The resulting, unnormalized cross sections are shown in Fig. 3. A fit with Eq. 1 shows a scattering length of $|a_{d\eta}| \approx 1.2\,\mathrm{fm}$.

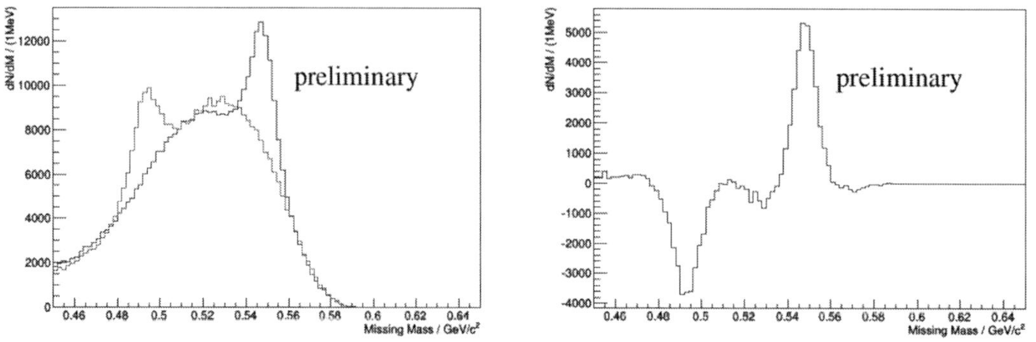

Figure 2. Left: Missing mass for both beam momenta (p_1 = 2.09 GeV/c in black and shifted for p_2 = 2.25 GeV/c in red); right: Difference between the missing mass spectra for both beam momenta.

3 Summary and outlook

Approximately 100k events of the reaction $p + n \rightarrow d + \eta$ have been gathered at ANKE. By detecting and reconstructing the deuteron and the spectator proton a clean signal can be seen in missing mass spectra. A preliminary scattering length could be determined via the course of the total cross section but the differential cross sections will be investigated to determine the limit for the s-wave FSI-ansatz. Additionally the calculation of the luminosity is on its way to allow to value the total cross section.

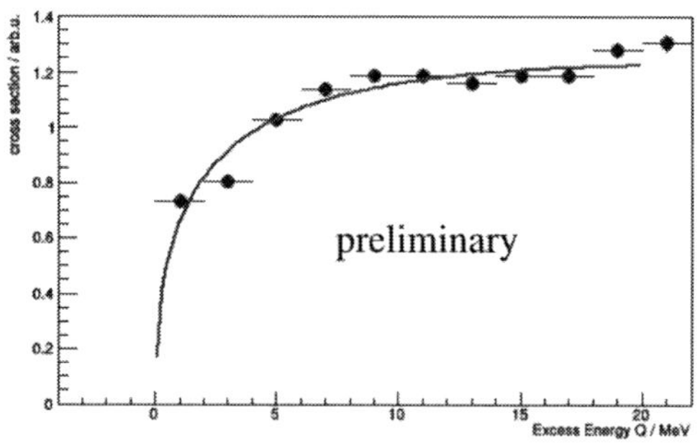

Figure 3. Cross section of the reaction $p + n \rightarrow d + \eta$. In red the fit with Eq. 1 is shown.

Acknowledgements

The support from COSY-FFE grants is kindly acknowledged.

References

[1] S. Barsov *et al.*, Nuclear Inst. and Methods in Physics Research A **462**, 364 - 381 (2001)

[2] Q. Haider and L.C. Liu, Phys. Lett. B **172**, 257 (1986)

[3] T. Mersmann *et al.*, Phys. Rev. Letters **98**, 242301 (2007)

[4] M. Rump, bachelor thesis, Westfälische Wilhelms-Universität Münster, Germany, 2013

Recent Results from Photoproduction of Mesons in A2

Natalie Walford[1,*]
for the A2 Collaboration

[1] *University of Basel (Basel, Switzerland)*

Abstract. Recent experiments using the Crystal Ball/TAPS setup at the MAMI accelerator in Mainz, Germany continue to study the properties and the excitation spectrum of the nucleon with meson photoproduction. Electromagnetic excitations of the proton and neutron are essential for understanding their isospin decomposition. The electromagnetic coupling of photons to protons is different than that of neutrons in certain states. Hence, a complete partial wave analysis (PWA) can assist in yielding more information about any reaction, but gains from polarization observables constraining the fits. Polarization observables play a crucial role as they are essential in disentangling the contributing resonant and non-resonant amplitudes, whereas cross-section data alone is not sufficient for separating resonances. Preliminary results of polarization observables (E, T, and F) of η and double π production off a polarized neutron (dButanol) target are shown with comparison to predictions of recent multipole analyses. These results will allow for developing the world database.

1 Introduction

A complete picture of the spectrum of N^* resonances is essential to understanding the excitation spectrum of the nucleon. For many years now, the study of the nucleon and its excitation spectrum has been studied. Understanding the dynamics and relevant degrees of freedom within hadrons in the non-pertubative regime of quantum chromodynamics (QCD) is crucial, as QCD describes the strong force acting between quarks, which currently can only be solved at very low and high energies where perturbative methods can be used.

Approximation models describe the behaviour of quarks in nucleons at medium energies. Meson photoproduction can specifically allow for unique possibilities to investigate these nucleons and their excited states (or resonances). Unfortunately, many states are overlapping and cannot be easily differentiated from each other and many have been predicted, but not yet detected. [1] The states that have not yet been observed could be missing because they have not been seen experimentally until now or do not exist at all. Most earlier experiments were preformed with pion beams and some resonances might couple less strongly to pions and couple more strongly to rare channels and have circumvented detection. Most results still arise from experiments on the proton, which does not allow for as much information regarding the isospin structure of the electromagnetic transitions. Therefore, advances in

*e-mail: natalie.walford@unibas.ch

DOI: 10.1051/epjconf/201613003009

experiments on the neutron can provide an integral piece to the understanding of the nucleon spectrum. Unfortunately, it is not possible to perform measurements on the free nuetron, but a deuterated butanol target has allowed for the possibility to study spin effects with quasi-free neutrons.

For years, cross section data has been used to learn about the nucleon spectrum. However, cross section data alone is not enough to distinguish the broad overlapping resonances. Polarization observables can provide understanding of these overlapping resonances by discovering more information about the complex helicity amplitudes, which describe the interaction between photon beams and nucleons. These amplitudes are fully determined when a complete set of measurements is performed and give rise to the cross section, complemented by polarization observables including beam, target, and recoil asymmetries. The following observables will be discussed: E (longitudinally polarized target and circularly polarized photon beam), F (transversely polarized target and circularly polarized photon beam), and T (transversely polarized target and an unpolarized photon beam).

The relationship between the cross section and polarization observable E can be written in terms of the helicity of beam and target as follows:

$$E = \frac{\sigma_{1/2} - \sigma_{3/2}}{\sigma_{1/2} + \sigma_{3/2}} = \frac{\sigma_{1/2} - \sigma_{3/2}}{2 * \sigma_0},$$

where $\sigma_{1/2}$ is the cross section of events with anti-parallel beam and target polarizations with total spin $\frac{1}{2}$ and $\sigma_{3/2}$ is the cross section of events with parallel beam with total spin of $\frac{3}{2}$. Due to the page limit, the T and F observable, which have a similar equation, will not be shown, but were discussed during the workshop.

2 Experimental Setup

Data described herein is from experiments performed at the MAMI-C accelerator in Mainz, Germany [2]. A longitudinally polarized electron beam of energy 1557 MeV and polarization degree of 80% is delivered into the A2 experimental Hall. Circularly polarized photons are produced from a radiator and energy tagged using the Glasgow-Mainz photon tagger [3] with energies between 470 and 1450 MeV. The targets consists of a deuterated butanol material (dButanol or D_9C_4OD) cooled to a low temperature with the deuterons either transversally or longitudinally polarized up to 80%. The target is surrounded by a cylindrical particle identification detector (PID) made up of 24 plastic scintillator strips that each cover 15° in the azimuthal angle. The PID is then surrounded by a multi-wire poroportional chamber (MWPC) and the MWPC is surrounded by the spherical Crystal Ball (CB) calorimeter [4]. The CB consists of 672 NaI(Tl) crystals and covers 20° to 160° in the polar angle. In the forward direction, a hexagonal two arm photon spectrometer (TAPS) built from 72 $PbWO_4$ (two innermost rings) and 366 BaF_2 crystals (remaining rings) is present. A veto wall is present in front of TAPS is used for particle identification. The combination of the CB and TAPS provides an almost 4π acceptance in the center of mass fram with a high angular and energy resolution.

3 Analysis and Results

Using information from the detectors, events are collected and then selected based on the number of charged or neutral hits. Neutral mesons are identified via a χ^2 test, which tries to the find the best combination of photon clusters for the meson invariant mass. To eliminate accidental conincident tagger photons, coincidence time cuts are applied and random background subtraction was performed. To separate the background from the signal, kinematic cuts were applied for each W and $\cos\theta$ bin. Relevant events come from the polarized deuterons with cuts applied, but events from carbon and

DOI: 10.1051/epjconf/201613003009

oxygen are subtracted using a separate beam time on a carbon (^{12}C) target with similar experimental conditions. The subtraction of reactions on unpolarized carbon and oxygen nuclei inside the dButanol target was performed. By comparing the missing mass spectra of the dButanol, liquid deuterium, and carbon beam times, allows for further studies of the contributions of reactions on unpolarized nuclei.

Different versions were used to extract the final E asymmetry. *Version 1* refers to where the asymmetry was normalized with twice the unpolarized cross section σ_0, which was measured with a liquid deuterium target, and does not need to utilize carbon background subtraction since the unpolarized background cancels in the difference of the two helicity states ($\sigma_\Delta = \sigma_{1/2} - \sigma_{3/2}$). *Version 2* refers to the normalization of the numerator using the sum of the two spin configurations ($\sigma_\Sigma = \sigma_{1/2} + \sigma_{3/2}$). Here, the background from unpolarized carbon and oxygen nuclei inside the target have to be subtracted, allowing for events only on polarized protons and neutrons. *Version 3* refers to using the sum and difference of the two helicity states to calculate E and the carbon was used to subtract the unpolarized carbon and oxygen nuclei. Finally, *version 4* refers to using a direct calculation of $\sigma_{1/2}$ and $\sigma_{3/2}$. An overview of the different versions are summarized in Table 1.

Table 1. Overview of the versions used to extract the E asymmetry.

Version	E	$\sigma_{1/2}$	$\sigma_{3/2}$
1	$\frac{\sigma_\Delta}{2*\sigma_0}$	$\sigma_0(1 + E)$	$\sigma_0(1 - E)$
2	$\frac{\sigma_\Delta}{\sigma_\Sigma}$	$\sigma_0(1 + E)$	$\sigma_0(1 - E)$
3		$\frac{\sigma_\Sigma+\sigma_\Delta}{2}$	$\frac{\sigma_\Sigma-\sigma_\Delta}{2}$
4	$\frac{\sigma_{1/2}-\sigma_{3/2}}{\sigma_{1/2}+\sigma_{3/2}}$	$\sigma_{1/2}$	$\sigma_{3/2}$

Fig. 1 shows the results for $\gamma p \to p\eta$ and $\gamma n \to n\eta$ for E, $\sigma_{1/2}$, and $\sigma_{3/2}$. For the E asymmetry, it is expected to be at unity close to threshold due to the dominance of the $N(1535)1/2^-$. This is seen in the results off the neutron and the two extraction versions are in good agreement. A deviation occurs in the proton data of E, but is related to the detection efficiency for protons. The results of E are in good agreement with the BnGa model [6], but differ from MAID [5] due to the unrealistic contribution of $N(1675)5/2^-$ in the model. For the $\sigma_{1/2}$ proton results, they are in reasonable agreement with model predictions. An enhancement is seen around $W \approx 1720$ MeV for the $\sigma_{3/2}$ proton component. For the $\sigma_{1/2}$ neutron results, a narrow structure is seen around $W \approx 1670$ MeV, which is related to the $J = 1/2$ partial wave and appears to be a contribution from a narrow P_{11} resonance.

Results for $\gamma p \to p\pi^0$ and $\gamma n \to n\pi^0$ for E, $\sigma_{1/2}$, and $\sigma_{3/2}$ are shown in Fig. 2. Here, it can be seen that the individual contributions to $\sigma_{1/2}$ and $\sigma_{3/2}$ in the second and third resonance region are under- or overestimated by the models. It is worth noting that the MAID [8] model achieves the poorest agreement. SAID [7] and BnGa [9] are, apart from E on the neutron (where SAID achieves the better agreement), are very similar to each other and agree with the shape of the data quite well. Due to page limitations, the double π^0 results have been omitted, but were discussed in the workshop.

4 Conclusion

In conclusion, these measurements of coincident neutrons in neutral final states are crucial for the analysis of N^* properties. Due to the lack of neutron data in the world database, these results will provide insight into constraining the current PWA models and help to discover all the states of the nucleon.

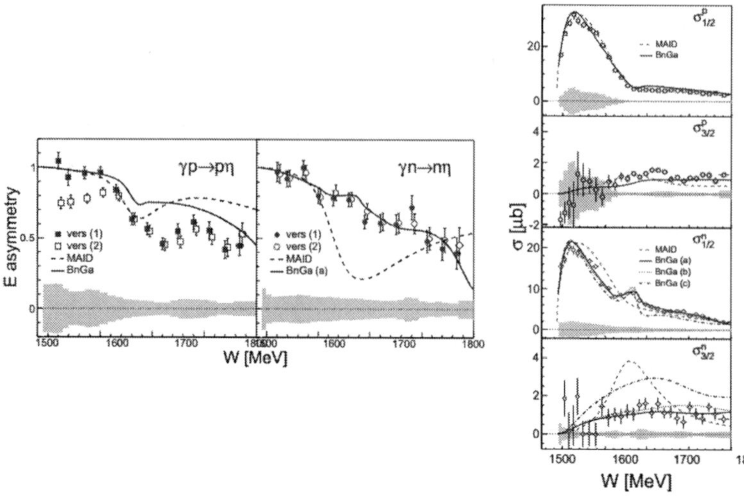

Figure 1. Double polarization observable E for $\gamma p \to p\eta$ (left) and $\gamma n \to n\eta$(right). Solid symbols: *Version 1*, open symbols: *Version 2*, gray shaded areas: systematic uncertainties. Curves: predic- tions from MAID (dashed) [5] and BnGa (model based on S_{11} interference) [6]. Total cross sections are also shown. $\gamma p \to p\eta$ (blue squares) and $\gamma n \to n\eta$ (red circles), gray shaded areas (systematic uncertainties). Curves: model predictions from MAID (dashed green) [5], BnGa(a) (model version with interference in S_{11} wave, solid black) [6], BnGa(b) (model with narrow P_{11} resonance with positive A1/2 coupling, dotted black) [6], and BnGa(c) (narrow P_{11} resonance with negative $A_{\frac{1}{2}}$ coupling, dash-dotted, black) [6].

Acknowledgments

All results presented in this paper have been obtained in the framework of the A2 collaboration at the Mainz MAMI accelerator. This work was supported by Swiss National Fund (SNF) and Deutsche Forschungsgemeinschaft (DFG).

References

[1] R.G. Edwards *et al.*, Phys. Rev. **D84** 074508 (2011)
[2] K. H. Kaiser *et al.*, NIM **A593** 159 (2008)
[3] J. C. George *et al.*, Eur. Phys. J. **A37** 129 (2008)
[4] A. Starostin *et al.*, Phys. Rev. **C64** 055205 (2001)
[5] W.-T. Chiang *et al.*, Nucl. Phys. A **700**, 429 (2002)
[6] A.V. Anisovich *et al.*, Eur. Phys. J. A **51**, 72 (2015)
[7] R.L. Workman *et al.*, Phys. Rev. **C86**, 015202 (2012)
[8] D. Drechsel *et al.*, Nucl. Phys. A **645**, 145 (1998)
[9] E. Gutz *et al.*, Eur. Phys. J. **A50** 74 (2014)

 DOI: 10.1051/epjconf/201613003009

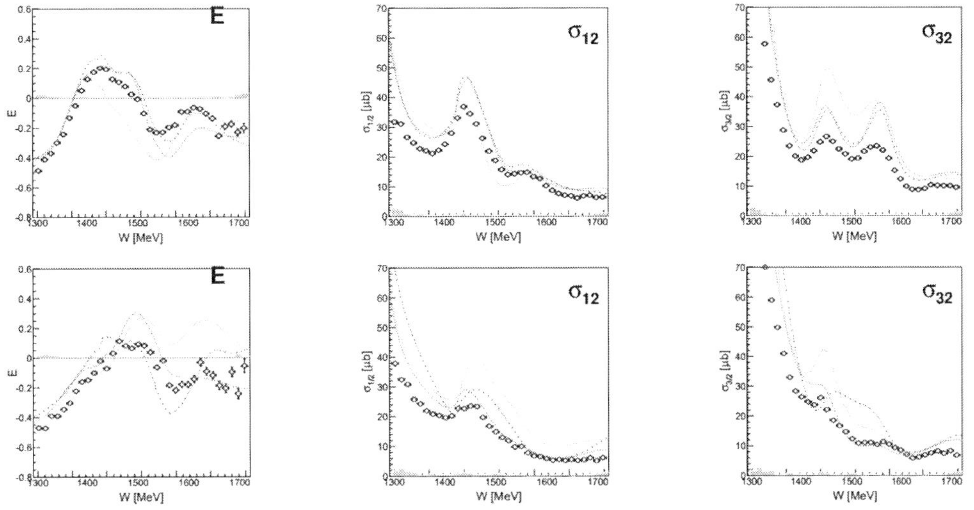

Figure 2. E (first column), $\sigma_{1/2}$ (middle column), and $\sigma_{3/2}$ (last column) asymmetries for single π^0 for quasi-free protons (upper row) and quasi-free neutrons (lower row). The cyan lines are SAID [7] nucleon models, the orange lines are MAID [8] nucleon models, and the magenta lines are Bonn-Gatchina [9] nucleon models.

DOI: 10.1051/epjconf/201613003010

Determination of the analysing power for the $\vec{p}p \rightarrow pp\eta$ reaction using the WASA-at-COSY detector system

I. Schätti-Ozerianska, P. Moskal, and M. Zieliński
for the WASA-at-COSY Collaboration

[1] *Faculty of Physics, Astronomy and Applied Computer Science, Jagiellonian University in Kraków, 30-348 Kraków, Poland*

Abstract. High statistics measurements of the analyzing power for the $\vec{p}p \rightarrow pp\eta$ reaction were performed with the WASA-at-COSY detector at the Cooler Synchrotron COSY. The η meson was produced at two excess energies: Q = 15 MeV and Q = 72 MeV, and it was registered using the missing and invariant mass techniques. The determined analyzing power is consistent with zero for Q = 15 MeV implying the dominance of s wave and an insignificant contribution of the Sd partial wave. However, at Q = 72 MeV the angular distribution of the analyzing power reveals a strong interference of Ps and Pp partial waves. Moreover, the available theoretical predictions for the angular dependence of the analyzing power strongly deviate from our experimental results.

1 Introduction

The aim of the presented studies is the determination of the analyzing power $A_y(\theta)$ for the reaction $\vec{p}p \rightarrow pp\eta$, which can provide the information about the production mechanism of the η meson and meson-nucleon final state interaction. So far the total and differential cross sections for the η meson production in nucleon-nucleon collisions [1] have been determined precisely [2–14]. However A_y has been determined only by the DISTO [15] and COSY-11 [16–18] collaboration with large uncertainties. The presented experiment was conducted using the azimuthally symmetric WASA-at-COSY detector [19] and a polarized proton beam provided by COSY colliding with a proton target at two beam momenta: 2026 MeV/c and 2188 MeV/c corresponding to 15 MeV and 72 MeV excess energies, respectively. The polarization degree was measured using proton-proton elastic scattering reaction, registered simultaneously with the main reaction. The determined value of the polarization amounts to about 65%, and it was stable over the whole experiment [20, 21]. Compared to the previous COSY-11 experiments with about 2000 events only [16], WASA-at-COSY collected nearly 400 000 events with an η meson identified.

2 Analyzing power of the η meson

The analyzing power A_y was determined by measuring the $pp \rightarrow pp\eta$ reaction and two neutral particles originating from the η meson decays: $\eta \rightarrow \gamma\gamma$ and $\eta \rightarrow 3\pi^0$. Values of $A_y(\theta)$ were determined for spin up and spin down mode and both decay channels separately.

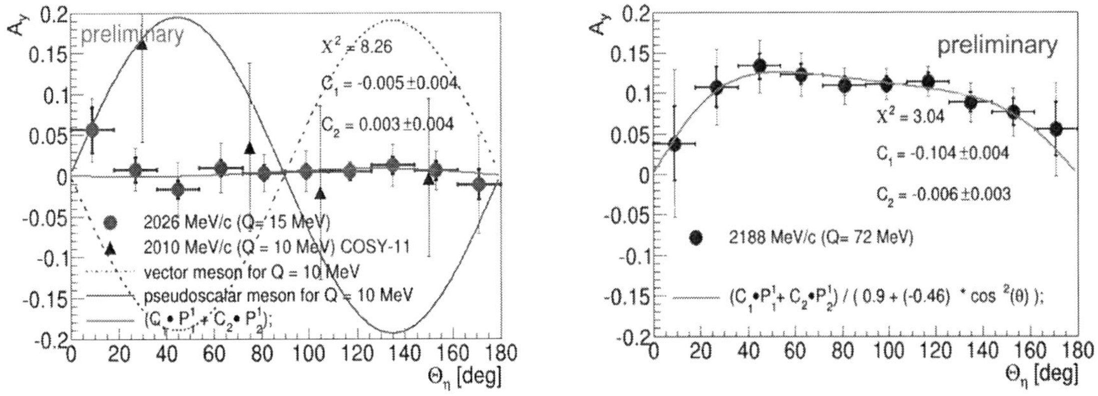

Figure 1. Analyzing power of the η meson as a function of the centre-of-mass emission angle of the η meson θ_η. (Left) Angular distribution of A_y for excess the energy 15 MeV, where superimposed lines indicate theoretical predictions: vector dominance model [22] for 10 MeV - solid red line; pseudoscalar model [23, 24] - dashed blue line. (Right) Angular distribution of A_y for 72MeV, where the superimposed solid pink line denotes the fit of A_y with the sum of the two associated Legendre polynomials P_1^1 and P_2^1.

The analyzing power was determined for ten angular θ intervals together with statistical and systematic uncertainties (Fig. 1). The obtained angular dependence of the analyzing power agrees with the previous experiments. However it differs from the theoretical predictions based on the pseudoscalar or vector meson dominance models [22, 24].

Figure 1 shows the obtained results based on the measured experimental data with superimposed lines corresponding to a fit of the formula:

$$A_y \frac{d\sigma}{d\Omega} = C_1 \cdot sin\theta_\eta + C_2 \cdot cos\theta_\eta sin\theta_\eta \tag{1}$$

with $C_1 = \Im(A_{Ps} A^*_{Pp})$ and $C_2 = \Im(A_{Ss} A^*_{Sd})$ treated as free parameters of the fit.

3 Result

The elaborated angular dependence of the analyzing power agrees with previous results obtained by the COSY-11 experiment but disagrees with the theoretical predictions [23, 24]. The analysis using the associated Legendre polynomials shows that $Ss\ Sd$ and $Pp\ Ps$ interference is not present for Q = 15 MeV. For the higher excess energy (72 MeV), the observed Sd partial wave contribution is also negligible. However, a contribution of $Ps\ Pp$ interference is large which indicates that at Q = 72 MeV the η meson is produced in the s and p wave, but a contribution from the d wave is negligible.

Acknowledgment

We acknowledge support by the Polish National Science Center through grant No. 2011/03/B/ST2/01847, by the FFE grants of the Research Center Juelich, by the EU Integrated Infrastructure Initiative HadronPhysics Project under contract number RII3-CT-2004-506078 and by the European Commission under the 7th Framework Programme through the Research Infrastructures action of the Capacities Programme, Call: FP7- INFRASTRUCTURES-2008-1, Grant Agreement N. 227431.

References

[1] P. Moskal *et al.*, Prog. Part. Nucl. Phys. **49**, 1 (2002)
[2] E. Chiavassa *et al.*, Phys. Lett. B **322**, 270 (1994)
[3] H. Calén *et al.*, Phys. Lett. B **366**, 39 (1996)
[4] H. Calén *et al.*, Phys. Rev. Lett. **79**, 2642 (1997)
[5] F. Hibou *et al.*, Phys. Lett. B **438**, 41 (1998)
[6] J. Smyrski *et al.*, Phys. Lett. B **474**, 182 (2000)
[7] A.M. Bergdolt *et al.*, Phys. Rev. D **48**, 2969 (1993)
[8] M. Abdel-Bary *et al.*, Eur. Phys. J. A **16**, 127 (2003)
[9] P. Moskal *et al.*, Phys. Rev. C **69**, 025203 (2004)
[10] P. Moskal *et al.*, Eur. Phys. J. A **43**, 131 (2010)
[11] H. Petren *et al.*, Phys. Rev. C **82**, 055206 (2010)
[12] H. Calén *et al.*, Phys. Rev. Lett. **79**, 2642 (1997)
[13] H. Calén *et al.*, Phys. Rev. C **58**, 2667 (1998)
[14] P. Moskal *et al.*, Phys. Rev. C **79**, 015208 (2009)
[15] F. Balestra *et al.*, Phys. Rev., C **69**, 064003 (2004)
[16] R. Czyżykiewicz *et al.*, Phys. Rev. Lett. **98**, 122003 (2007)
[17] P. Winter *et al.*, Eur. Phys. J. A **18**, 355 (2003)
[18] P. Winter *et al.*, Phys. Lett. B **544** , 251 (2002)
[19] P. Moskal, M. Hodana, J. Phys. Conf. Ser. 295:012080, 2011
[20] I. Ozerianska *et al.*, EPJ Web Conf. **81**, 02013 (2014)
[21] I. Schätti-Ozerianska *et al.*, Acta Phys. Polon. B **46**, 153 (2015)
[22] K. Nakayama *et al.*, Phys. Rev. C **68**, 045201 (2003)
[23] K. Nakayama *et al.*, Phys. Rev. C **65**, 045210 (2002)
[24] G. Fäldt and C. Wilkin, Phys. Scripta **64**, 427 (2001)

A data-driven approach to η and η' Dalitz decays

Rafel Escribano[1,2,★]

[1] *Grup de Física Teòrica, Departament de Física, Universitat Autònoma de Barcelona, E-08193 Bellaterra (Barcelona), Spain*
[2] *Institut de Física d'Altes Energies (IFAE), The Barcelona Institute of Science and Technology (BIST), Campus UAB, E-08193 Bellaterra (Barcelona), Spain*

Abstract. Data-driven model-independent predictions for the dilepton invariant mass spectra and branching ratios of the single and double Dalitz decays $\eta^{(\prime)} \to \ell^+\ell^-\gamma$ and $\eta^{(\prime)} \to \ell^+\ell^-\ell^+\ell^-$ ($\ell = e$ or μ) are provided by means of an approach based on the use of rational approximants applied to the available experimental data in the space-like region of the η and η' transition form factors.

1 Introduction

In this talk I have provided data-driven model-independent predictions for the dilepton invariant mass spectra and integrated branching ratios (BRs) of the single and double Dalitz decays $\eta^{(\prime)} \to \ell^+\ell^-\gamma$ and $\eta^{(\prime)} \to \ell^+\ell^-\ell^+\ell^-$ with $\ell = e$ or μ. For a detailed and exhaustive analysis of all the processes considered, including the π^0 ones not reported here, I refer to the original work in Ref. [1]. It is worth noticing that the previous analysis is the first complete work including all the possible Dalitz reactions. Our approach is based on the use of rational approximants applied to the available experimental data in the space-like region of the η and η' transition form factors (TFFs) [2]. It is defined as model independent because we are able to ascribe a conservative systematic error to our predictions. In this sense, the traditional VMD description of the TFF in terms of a monopole used commonly by experimental collaborations must be considered only as a first step in this systematic approximation. For the case of the double Dalitz decays, which demand TFFs of double virtuality not yet experimentally available, we have made use of the factorization hypothesis stating that the normalised version of the double-virtual TFFs is nothing else than the product of the two single-virtual TFFs.

2 Transition form factors

The energy dependence of the η and η' TFFs in the time-like region required for describing these Dalitz decays has been predicted after fixing this dependence but from existing data in the space-like region reported by the CELLO [3], CLEO [4], L3 [5], and *BABAR* [6] collaborations. A more precise determination of these TFFs could have been obtained here taking into account in addition the recent data in the time-like region coming from the NA60 [7] ($\eta \to \mu^+\mu^-\gamma$), A2 [8, 9] ($\eta \to e^+e^-\gamma$) and BESIII [10] ($\eta' \to e^+e^-\gamma$) Colls. Nevertheless, the impact of the incorporation of these

★e-mail: rescriba@ifae.es

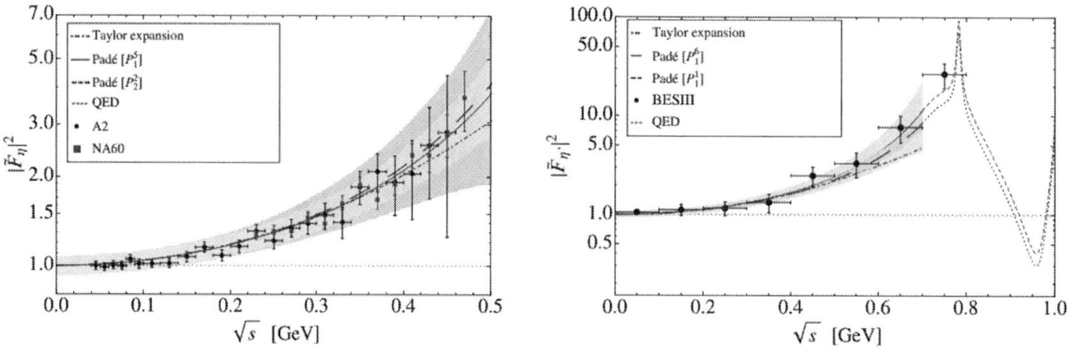

Figure 1. Modulus square of the normalised time-like η *(left)* and η' *(right)* TFF as a function of the invariant dilepton mass, $\sqrt{s} \equiv m_{\ell\ell}$. The predictions are compared to the experimental data from $\eta \to e^+e^-\gamma$ [9] (black circles), $\eta \to \mu^+\mu^-\gamma$ [7] (green squares) and $\eta' \to e^+e^-\gamma$ [10] (black circles).

sets of time-like data was studied in Refs. [11] and [12], for the η and η' cases, respectively. The application of Padé approximants (PAs) in the space-like is safe and reliable since the TFFs do not contain singularities in this region. The TFFs are expected to be well-behaved analytical functions except for a cut starting at the $\pi\pi$ threshold. The PAs, being rational functions and thus containing only poles by construction, are not sensitive to this unitary cut belonging to the time-like region. Therefore, one could think that PAs cannot be applied to this time-like region. However, it was shown in Ref. [12] that the effect of this unitary cut is minor and then one needs only to worry about the appearance of possible poles on the real (energy) axis that would prevent PAs from being used from that singularity on. Such poles could be understood as the projection on the real axis of resonance poles (or a combination of them) in the complex plane. For the η, the first pole encountered is well beyond the region of available phase-space and therefore its TFF is well established. For the η', instead, the first pole appears in the upper part of the phase-space, somewhere inside the resonance region attributed to the lowest-lying vector mesons. For that reason, we match the η' TFF predicted by PAs to the more common VMD description to be able to predict the integrated BRs of the η' Dalitz decays (see Ref. [1] for details). In Fig. 1, our predictions for the the time-like region energy dependence of the η *(left)* and η' *(right)* TFFs are shown. We display the predictions for two different types of PAs, those with a single pole in the denominator, P^L_1, motivated by the likely dominance of resonances, and those diagonal, P^N_N, where the high-energy behaviour of the TFF predicted by perturbative QCD is imposed[1]. The QED prediction (pointlike TFF) and the Taylor expansion including the slope and the curvature of the TFF are also included for comparison in each case. The error bands incorporate not only the statistical but also the systematic error associated to our approach. Higher the order of the PA is, smaller the systematic error is. In this line, the corresponding systematic uncertainty of a VMD description (monopole FF) is much bigger than ours. As seen, the agreement between our predictions for the two TFFs and the experimental data is remarkable, making the whole approach trustworthy.

3 Single Dalitz decays

For the numerical computations, we have employed the PDG fitted values $\Gamma_{\eta\to\gamma\gamma} = 0.516(18)$ keV and $\Gamma_{\eta'\to\gamma\gamma} = 4.35(14)$ keV [13]. In Fig. 2, we display the decay rate distributions for $\eta^{(\prime)} \to e^+e^-\gamma$

[1]The fits in the space-like region were performed to $Q^2|F_{\eta^{(\prime)}}(Q^2)|$ and this behaves as a constant at high energies.

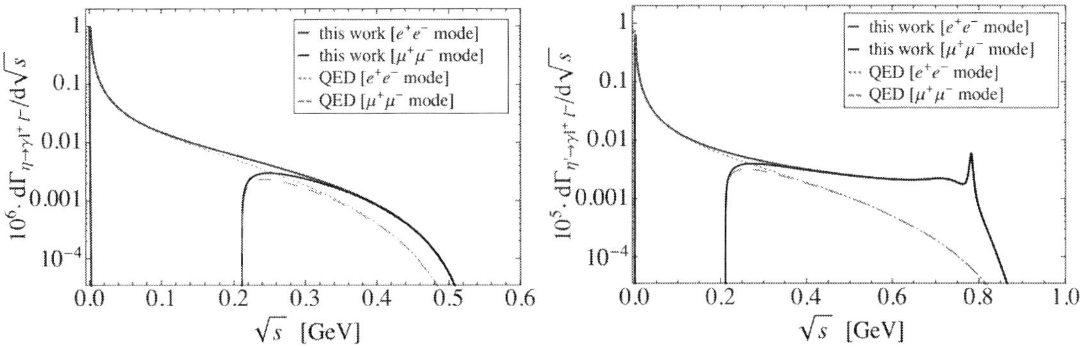

Figure 2. Decay rate distributions for $\eta^{(\prime)} \to e^+e^-\gamma$ (blue solid curves) and $\eta^{(\prime)} \to \mu^+\mu^-\gamma$ (black solid curves). The corresponding QED estimates are also displayed (grey dotted and long-dashed curves, respectively).

Table 1. Branching ratio predictions for $\eta \to \ell^+\ell^-\gamma$ ($\ell = e, \mu$) and their comparisons with experimental measurements and previous theoretical calculations.

Source	$\mathcal{BR}(\eta \to e^+e^-\gamma) \times 10^3$	$\mathcal{BR}(\eta \to \mu^+\mu^-\gamma) \times 10^4$
QED	6.38	2.17
This work (P_1^5)	$6.60^{+0.50}_{-0.47}$	$3.25^{+0.40}_{-0.36}$
This work (P_2^2)	$6.61^{+0.53}_{-0.49}$	$3.30^{+0.65}_{-0.56}$
PDG [13]	6.9 ± 0.4	3.1 ± 0.4
H. Berghauser *et al.* [8]	$6.6 \pm 0.4_{\mathrm{stat}} \pm 0.4_{\mathrm{syst}}$	
WASA-at-COSY Coll. [14]	$6.72 \pm 0.07_{\mathrm{stat}} \pm 0.31_{\mathrm{syst}}$	
QED [16]	6.38	2.18
FF 2,3 [17]	6.57	3.05
FF 4 [17]	6.53	2.87
LFQM [18]	6.95	2.94
Hidden gauge [19]	6.57 ± 0.03	3.05 ± 0.04
Modified VMD [19]	6.55 ± 0.03	2.97 ± 0.05

(blue solid curves) and $\eta^{(\prime)} \to \mu^+\mu^-\gamma$ (black solid curves). The non-pointlike behaviour of the TFFs is notably seen in the upper part of the spectra, being its numerical effect (in the integrated BRs) much bigger in the dimuon case than in the dielectron one for obvious phase-space reasons. Our BR predictions for $\eta^{(\prime)} \to \ell^+\ell^-\gamma$ ($\ell = e, \mu$) and their comparisons with experimental measurements and previous theoretical calculations can be found in Table 1, for the η, and Table 2, for the η', respectively. As seen from the Tables, with the current precision, there is no difference within errors between the predictions from monopole PAs and diagonal ones. Indeed, at the level of integrated BRs, our predictions are incompatible with the QED ones only for the dimuon cases. In any case, the preferred channel for testing the energy dependence of the TFFs would be the radiative dimuon decay of the η'. At the present accuracy, however, there is not yet possible to distinguish among different theoretical frameworks. So far, all of them are compatible with each other and with the data. More refined experimental analyses will be required in the future.

Table 2. Branching ratio predictions for $\eta' \to \ell^+\ell^-\gamma$ ($\ell = e, \mu$) and their comparisons with experimental measurements and previous theoretical calculations.

Source	$\mathcal{BR}(\eta' \to e^+e^-\gamma) \times 10^4$	$\mathcal{BR}(\eta' \to \mu^+\mu^-\gamma) \times 10^4$
QED	3.94	0.38
This work (P^6_1)	$4.42^{+0.39}_{-0.35}$	$0.81^{+0.16}_{-0.13}$
This work (P^1_1)	$4.35^{+0.29}_{-0.27}$	0.74 ± 0.06
PDG [13, 15]		1.08 ± 0.27
BESIII Coll. [10]	$4.69 \pm 0.20_{\rm stat} \pm 0.23_{\rm syst}$	
Hidden gauge [19]	4.62 ± 0.17	0.98 ± 0.05
Modified VMD [19]	4.53 ± 0.17	0.90 ± 0.05

Table 3. Central final branching ratio predictions as a combined weighted average of the results presented. Errors are symmetrised. n_σ stands for the number of standard deviations the measured results are from our predictions.

Decay	This work	Experimental value [13]	n_σ
$\pi^0 \to e^+e^-\gamma$	1.169(1)%	1.174(35)%	0.15
$\eta \to e^+e^-\gamma$	$6.61(50) \times 10^{-3}$	$6.90(40) \times 10^{-3}$	0.45
$\eta \to \mu^+\mu^-\gamma$	$3.26(46) \times 10^{-4}$	$3.1(4) \times 10^{-4}$	0.26
$\eta' \to e^+e^-\gamma$	$4.38(32) \times 10^{-4}$	$4.69(20)(23) \times 10^{-4}$	0.70
$\eta' \to \mu^+\mu^-\gamma$	$0.75(6) \times 10^{-4}$	$1.08(27) \times 10^{-4}$	1.19
$\pi^0 \to e^+e^-e^+e^-$	$3.36689(5) \times 10^{-5}$	$3.34(16) \times 10^{-5}$	0.17
$\eta \to e^+e^-e^+e^-$	$2.71(2) \times 10^{-5}$	$2.4(2)(1) \times 10^{-5}$	1.38
$\eta \to \mu^+\mu^-\mu^+\mu^-$	$3.98(15) \times 10^{-9}$	$< 3.6 \times 10^{-4}$	
$\eta \to e^+e^-\mu^+\mu^-$	$2.39(7) \times 10^{-6}$	$< 1.6 \times 10^{-4}$	
$\eta' \to e^+e^-e^+e^-$	$2.10(45) \times 10^{-6}$	not seen	
$\eta' \to \mu^+\mu^-\mu^+\mu^-$	$1.69(36) \times 10^{-8}$	not seen	
$\eta' \to e^+e^-\mu^+\mu^-$	$6.39(91) \times 10^{-7}$	not seen	

4 Double Dalitz decays

For the double Dalitz decays, $\eta^{(\prime)} \to \ell^+\ell^-\ell^+\ell^-$ with $\ell = e$ or μ, we will mostly use the factorisation ansatz for the energy dependence of the doubly-virtual TFFs. However, bivariate approximants named Chisholm (CAs) can also be used. In these, besides the constraints imposed by perturbative QCD, one can implement OPE constraints as well. In both cases, the numerical predictions are indistinguishable. Again, the effect of the energy dependence of the TFFs is much bigger for the dimuon states, and in particular for the η' channels. Our predictions for the η and η' BRs involving two dielectron, two dimuon, and one dielectron and one dimuon final states are shown in the lower part of Table 3. Only the decay $\eta \to e^+e^-e^+e^-$ is measured at present. The other two η decays have upper bounds while nothing is known for the η' decays. There is some hope to measure in the near future the channels involving dielectron pairs but very unlikely for only dimuon pairs.

5 Conclusions

The detailed and exhaustive analysis performed in Ref. [1] can be summarised in Table 3, where central final branching ratio predictions as a combined weighted average of the results presented there are displayed, not only for those processes involving η and η' but also the π^0 ones for completeness.

This work was supported in part by the Ministerio de Economía y Competitividad under grants CICYT-FEDER-FPA 2014-55613-P and SEV-2012-0234, and the Secretaria d'Universitats i Recerca del Departament d'Economia i Coneixement de la Generalitat de Catalunya under grant 2014 SGR 1450.

References

[1] R. Escribano and S. Gonzàlez-Solís, arXiv:1511.04916 [hep-ph]
[2] R. Escribano, P. Masjuan and P. Sanchez-Puertas, Phys. Rev. D **89**, 034014 (2014)
[3] H. J. Behrend *et al.* [CELLO Collaboration], Z. Phys. C **49**, 401 (1991)
[4] J. Gronberg *et al.* [CLEO Collaboration], Phys. Rev. D **57**, 33 (1998)
[5] M. Acciarri *et al.* [L3 Collaboration], Phys. Lett. B **418**, 399 (1998)
[6] P. del Amo Sanchez *et al.* [BaBar Collaboration], Phys. Rev. D **84**, 052001 (2011)
[7] R. Arnaldi *et al.* [NA60 Collaboration], Phys. Lett. B **677**, 260 (2009)
[8] H. Berghauser *et al.*, Phys. Lett. B **701**, 562 (2011)
[9] P. Aguar-Bartolome *et al.* [A2 Collaboration], Phys. Rev. C **89**, 044608 (2014)
[10] M. Ablikim *et al.* [BESIII Collaboration], Phys. Rev. D **92**, 012001 (2015)
[11] R. Escribano, P. Masjuan and P. Sanchez-Puertas, Eur. Phys. J. C **75**, 414 (2015)
[12] R. Escribano, S. Gonzàlez-Solís, P. Masjuan and P. Sanchez-Puertas, Phys. Rev. D **94**, 054033 (2016)
[13] K. A. Olive *et al.* [Particle Data Group Collaboration], Chin. Phys. C **38**, 090001 (2014)
[14] P. Adlarson *et al.*, arXiv:1509.06588 [nucl-ex]
[15] R. I. Dzhelyadin *et al.*, Sov. J. Nucl. Phys. **32**, 520 (1980) [Yad. Fiz. **32**, 1005 (1980)]
[16] T. Miyazaki and E. Takasugi, Phys. Rev. D **8**, 2051 (1973)
[17] F. Persson, hep-ph/0106130
[18] C. C. Lih, J. Phys. G **38**, 065001 (2011)
[19] T. Petri, arXiv:1010.2378 [nucl-th]

EPJ Web of Conferences **130**, 04002 (2016) DOI: 10.1051/epjconf/201613004002
MESON 2016

Electromagnetic transition form factor and radiative corrections in decays of neutral pions

Tomáš Husek[1,★]

[1] *Charles University, Prague, Czech Republic*

Abstract. We briefly summarize experimental and theoretical results on the rare decay $\pi^0 \to e^+e^-$. The notorious $3.3\,\sigma$ discrepancy between the SM prediction and the experimental value provided by KTeV collaboration is discussed in the view of a complete set of NLO QED radiative corrections. We also present the Two-Hadron Saturation (THS) scenario for the PVV correlator and apply it to the decay under discussion. The discrepancy then reduces down to $1.8\,\sigma$.

1 Introduction

The rare decay of a neutral pion, i.e. the process $\pi^0 \to e^+e^-$, drew attention of theorists during last years due to the precise measurement of its branching ratio done by KTeV experiment at Fermilab [1]

$$B^{\mathrm{KTeV}}(\pi^0 \to e^+e^-(\gamma),\, x_{\mathrm{D}} > 0.95) = (6.44 \pm 0.25 \pm 0.22) \times 10^{-8}\,. \tag{1}$$

Soon after it has been shown [2] that the prediction based on the Standard Model (SM) is in $3.3\,\sigma$ disagreement with this measurement. One can immediately think about the possibility that this is a sign for a new physics. But our strategy is to look first properly for a more conventional solution. In what follows we will investigate in detail the radiative corrections and suggest a new model for the electromagnetic transition form factor of a neutral pion.

2 Rare decay $\pi^0 \to e^+e^-$ and radiative corrections

The pure-QED two-loop virtual radiative corrections were calculated in [3]. Since the compensation of the IR divergences is necessary, also the exact bremsstrahlung contributions were additionally taken into account in [4]. The overall size of the corrections was found to be

$$\delta(0.95) \equiv \delta^{\mathrm{virt.}} + \delta^{\mathrm{BS}}(0.95) = (-5.5 \pm 0.2)\,\%\,, \tag{2}$$

which differs significantly from previous approximate calculations done in [5] and [6], where for $\delta(0.95)$ the authors obtained -13.8 % and -13.3 %, respectively. After taking into account a proper calculation of the radiative corrections in the QED sector, the original discrepancy between SM and KTeV experiment reduces to the $2\,\sigma$ level or less. In what follows we would like to check if using a more sophisticated model for the pion transition form factor $\mathcal{F}_{\pi^0\gamma^*\gamma^*}$ helps the current situation.

★e-mail: husek@ipnp.mff.cuni.cz

3 Two-Hadron Saturation for the PVV correlator

Within a family of large-N_c motivated resonance-saturation models for the PVV correlator in the chiral limit, we consider the following ansatz [7]:

$$\Pi^{\text{THS}}(r^2; p^2, q^2) = \frac{B_0 F^2}{r^2(r^2 - M_P^2)} \frac{P(r^2; p^2, q^2)}{(p^2 - M_{V_1}^2)(p^2 - M_{V_2}^2)(q^2 - M_{V_1}^2)(q^2 - M_{V_2}^2)} . \tag{3}$$

We have taken into account two multiplets in both vector and pseudoscalar channel. To our knowledge such an approach was not used in the literature before. Together with the constraints we apply further we call this approach the Two-Hadron Saturation (THS). In the numerator there stands a general polynomial which is symmetric in p^2 and q^2. Since the correlator must drop at large momenta, we can, in the very beginning, restrict the number of free parameters to 22, so the polynomial looks as follows:

$$P(r^2; p^2, q^2) = c_0 p^2 q^2 + c_1[(p^2)^3 q^2 + (q^2)^3 p^2] + c_2 (r^2)^2 p^2 q^2 + \ldots \tag{4}$$

To minimize the number of unknown parameters, we apply high- as well as low-energy theoretical constraints. First we demand satisfying of the leading-order OPE constraints [8] by our ansatz. Next, we define the πVV correlator:

$$\mathcal{F}_{\pi VV}(p^2, q^2) \equiv \frac{1}{\mathcal{Z}_\pi} \lim_{r^2 \to 0} r^2 \Pi(r^2; p^2, q^2) , \tag{5}$$

where $\mathcal{Z}_\pi = B_0 F$ denotes the overlap between the pion field and the pseudoscalar quark current. Instead of involving subleading orders in the high-energy expansion we in addition apply the Brodsky–Lepage (B-L) constraint [9] on the resulting πVV correlator. Finally, after matching at the photon point to the chiral anomaly we are left with a correlator that appears to have only one free dimensionless parameter κ:

$$\begin{aligned}\mathcal{F}_{\pi VV}^{\text{THS}}(p^2, q^2) = &-\frac{N_c}{8\pi^2 F} \frac{M_{V_1}^4 M_{V_2}^4}{(p^2 - M_{V_1}^2)(p^2 - M_{V_2}^2)(q^2 - M_{V_1}^2)(q^2 - M_{V_2}^2)} \\ &\times \left\{ 1 + \frac{\kappa}{2N_c} \frac{p^2 q^2}{(4\pi F)^4} - \frac{4\pi^2 F^2 (p^2 + q^2)}{N_c M_{V_1}^2 M_{V_2}^2} \left[6 + \frac{p^2 q^2}{M_{V_1}^2 M_{V_2}^2} \right] \right\} .\end{aligned} \tag{6}$$

We determine the value of this parameter from the fit to the ω-π transition form factor measurements with the result $\kappa = 21 \pm 3$. Note that our result is independent of the mass M_P of the first pseudoscalar excitation. This happens due to the fact that at the end of the day we could conveniently rescale the only free parameter left. For the mass of the first vector-meson multiplet M_{V_1} we use the physical masses of ρ or ω mesons. To account for the uncertainty given by taking only two multiples out of the whole tower of resonances, we vary the mass of the second multiplet M_{V_2} within a range given by physical masses of the first and second vector-meson excitations, i.e. $M_{V_2} \in [1400, 1740]$ MeV.

By construction, our approach satisfies all the theoretical LO constraints given by QCD and might be compared with other models, like Vector-Meson Dominance (VMD), which violates OPE, or Lowest-Meson Dominance (LMD), which strongly violates B-L.It is worth to note that in the singly off-shell regime, the correlator does not depend on κ, so we have a full predictive power for associated quantities.

DOI: 10.1051/epjconf/201613004002

4 Results

Taking into account the complete set of NLO radiative corrections and THS model, we get for the theoretical prediction of the quantity measured by KTeV experiment $B^{\mathrm{THS}} = (5.8 \pm 0.2) \times 10^{-8}$. If we recall the experimental value $B^{\mathrm{KTeV}} = (6.44 \pm 0.33) \times 10^{-8}$, we find that the disagreement is only at the level of only 1.8 σ. Thus, if the KTeV result will be confirmed by future experiments, two scenarios are conceivable. Either there are indeed some aspects of the THS approach, which are not well-suited for $\pi^0 \to e^+e^-$, or the beyond-Standard Model physics influences the rare pion decay significantly. But under the present circumstances the current discrepancy is inconclusive. Actually, the quantity measured by KTeV was a ratio of branching ratios, where the Dalitz decay was used as the normalization channel. This is why it is important to have the latter process under control.

5 Dalitz decay of a neutral pion

Corrections to the Dalitz plot in the form of a table of values were calculated in the classical work [10]. The new investigation [11] of this topic was motivated by needs of NA48/NA62 experiments at CERN, the aim of which is, among other goals, to measure the slope a of $\mathcal{F}_{\pi^0\gamma^*\gamma^*}(0, q^2)$. Unlike before, we have added 1γIR contribution, which was for long time considered to be negligible. Additionally, no approximation considering masses of the particles involved was used during the calculation, so the results are applicable for related decays like $\eta \to \ell^+\ell^-\gamma$. Note though that new contributions are needed in the case of η decays. C++ code was developed, which returns the correction for any given x and y and which became a part of the simulation software of NA62 experiment.

6 Conclusions

The complete set of NLO radiative corrections in the QED sector for both discussed decays, the Dalitz decay and the rare decay $\pi^0 \to e^+e^-$, is now available. We showed on the case of KTeV experiment that their use in future experimental analyses should be essential. We also presented THS model for electromagnetic transition form factor $\mathcal{F}_{\pi^0\gamma^*\gamma^*}(p^2, q^2)$, which satisfies all main theoretical constraints and is phenomenologically successful. Taking all the pieces together, we get reasonable SM prediction for the pion rare decay branching ratio, which differs from the KTeV result by inconclusive 1.8 σ.

This contribution is based on collaboration with K. Kampf, J. Novotný and S. Leupold. It was supported by Charles University (grant GAUK 700214) and by the Czech Science Foundation (grant GAČR 15-18080S).

References

[1] E. Abouzaid et al. (KTeV), Phys. Rev. **D75**, 012004 (2007)
[2] A.E. Dorokhov, M.A. Ivanov, Phys. Rev. **D75**, 114007 (2007)
[3] P. Vaško, J. Novotný, JHEP **1110**, 122 (2011)
[4] T. Husek, K. Kampf, J. Novotný, Eur. Phys. J. **C74**, 3010 (2014)
[5] Bergström, L., Z. Phys. **C20**, 135 (1983)
[6] A. Dorokhov, E. Kuraev, Y. Bystritskiy, M. Secansky, Eur. Phys. J. **C55**, 193 (2008)
[7] T. Husek, S. Leupold, Eur. Phys. J. **C75**, 586 (2015)
[8] M. Knecht, A. Nyffeler, Eur. Phys. J. **C21**, 659 (2001)
[9] S.J. Brodsky, G.P. Lepage, Phys. Rev. **D24**, 1808 (1981)
[10] K. Mikaelian, J. Smith, Phys. Rev. **D5**, 1763 (1972)
[11] T. Husek, K. Kampf, J. Novotný, Phys. Rev. **D92**, 054027 (2015)

EPJ Web of Conferences **130**, 04003 (2016) DOI: 10.1051/epjconf/201613004003
MESON 2016

PWA with full rank density matrix of the $\pi^+\pi^-\pi^-$ and $\pi^-\pi^0\pi^0$ systems at VES setup

Igor Kachaev[1] and Dmitry Ryabchikov[1,★] for the VES group
[1] *IHEP, Protvino, Russia*

Abstract. Partial Wave Analysis of the $\pi^+\pi^-\pi^-$ and $\pi^-\pi^0\pi^0$ final states produced by $29\,GeV/c$ π^- beam on beryllium target is presented. About $42 \cdot 10^6$ events for the first system and $22 \cdot 10^6$ events for the second one are collected with VES setup. The statistics for 3π neutral is known to be the largest in the world while statistics for 3π charged is next to largest. The t' range for the analysis is $0 < |t'| < 0.8\ GeV^2/c^2$. The data are analysed using formalism of full rank density matrix. The comparison of the analysis results for two systems is presented.

1 Preface

Large statistics for the reactions $\pi^- A \to \pi^+\pi^-\pi^- A$ and $\pi^- A \to \pi^-\pi^0\pi^0 A$ is collected with VES setup. Brief description of the upgraded VES setup can be found in [1]. Final states for both reactions should have similar resonant content. Still these states are quite different from the point of view of experimental setup, acceptance and the structure of Dalitz plot. So the comparison of these reactions can be very useful. This permits us to test isospin relations between these states, better understand possible resonances, check for issues in the setup, hardware, methods and programs involved in the analysis.

One should stress that partial wave analysis (PWA) is complex and model dependent procedure. Results of PWA can depend on how the amplitudes are constructed, what parametrization of density matrix is used (rank 1, rank 2, full rank), how good is the description of the acceptance of the experimental setup. To test the stability and physical meaning of PWA results it may also be very useful to compare these results for given PWA model and two final states.

In this article we present first attempt to compare PWA results of $\pi^+\pi^-\pi^-$ and $\pi^-\pi^0\pi^0$ final states on large statistics in a wide range of momentum transfer $|t'|$. The statistics used here is somewhat larger than one presented on the conference due to improvements in methods used in offline reconstruction. The same parametrization, namely full density matrix, is used to analyse both states. General properties of the data are shown in Fig. 1 a) – c). Here and below results for $\pi^+\pi^-\pi^-$ final state are drawn in black while results for $\pi^-\pi^0\pi^0$ are drawn in red (grey in b/w figures). The analysis is performed in four t' regions 0–0.03–0.15–0.30–0.80 GeV^2/c^2.

2 Method of the analysis

Our method of the analysis is based on Illinois PWA [2]. Isobar model is used, namely sequential decay of 3π system via $\pi\pi$ subsystems. Partial wave has quantum numbers $J^P LM^\eta$ R where J^P is

★e-mail: Dmitry.Ryabchikov@ihep.ru

DOI: 10.1051/epjconf/201613004003

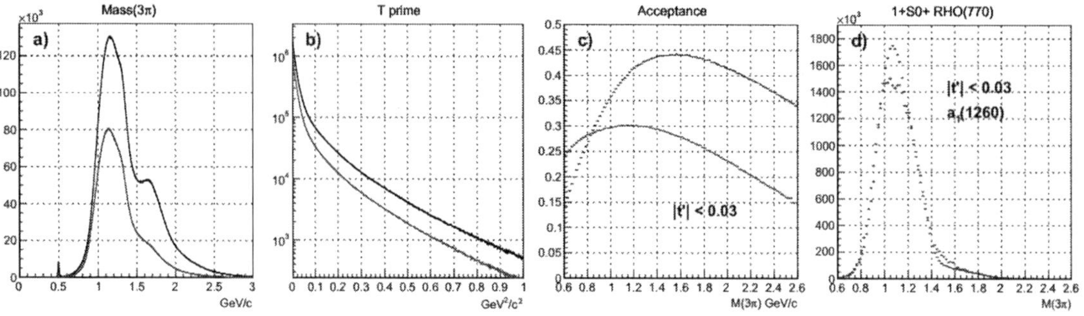

Figure 1. Graph of a) $M(3\pi)$, b) t', c) acceptance, d) intensity of wave $1^+S0^+\rho(770)\pi$ for $\pi^+\pi^-\pi^-$ (black) and $\pi^-\pi^0\pi^0$ (red) final states.

spin and parity for 3π system, M^η is its projection of spin and exchange naturality, R is the known resonance in $\pi\pi$ system, L is orbital momentum in Rπ decay. Isospin and G-parity $I^G = 1^-$ are the same for both 3π final states. Amplitudes are constructed using d-functions. Relativistic corrections are applied according to [3, 4]. Two particle states f_0, ρ, f_2, ρ_3 in $\pi\pi$ subsystems are described by relativistic Breit-Wigners with parameters from PDG [6]. To describe $\pi\pi$ S-wave this is not enough so we use here modified M solution from [5] for broad part of the amplitude and include narrow $f_0(975)$ and $f_0(1500)$ separately. To make M solution broad we drop 4-th order terms and coupling to $K\bar{K}$. We name this pseudo state ε, it should describe $f_0(1370)$ and $f_0(500)$.

We are using positive definite density matrix of unlimited rank and event by event extended log likelihood fit to describe data. No restrictions are imposed on the rank of the matrix. The advantage of this approach is that interference terms can vary freely; the disadvantage is that number of parameters is large which limits total number of waves used. Here we have 37 waves, only some of them are forced to be fully coherent with some others. This is done to reduce total number of parameters, fit has more than 900 of them. GEANT4 based model is used to describe the acceptance of the VES setup.

Let us define coherent part of the density matrix ρ as the largest part of the matrix which has rank one and so behaves like vector of amplitudes. It can be constructed as follows. Let us decompose hermitian matrix ρ with dimension d into its eigenvalues and eigenvectors:

$$\rho_{ij} = \sum_{k=1}^{d} e_k V_k^i V_k^{j+} \quad \text{where} \quad \begin{cases} e_k \text{ is k-}th\text{ eigenvalue} \\ V_k \text{ is k-}th\text{ eigenvector} \end{cases}$$

Let $e_1 > e_2 > \ldots > e_d > 0$. Let us define coherent part of the density matrix as "leading term" with matrix elements $\rho_{ijL} = e_1 V_1^i V_1^{j+}$. If $e_1 \gg e_2$, as is often the case for 3π systems, then matrix elements of this term are stable with respect to small variations of ρ matrix elements. So this decomposition can have physical meaning. Experience shows that resonances tend to concentrate in ρ_L.

3 Fit results

For $\pi^+\pi^-\pi^-$ states f_0, ρ, f_2, ρ_3 are connected to $\pi^+\pi^-$ subsystem while for $\pi^-\pi^0\pi^0$ ρ and ρ_3 are seen in $\pi^-\pi^0$ and f_0, f_2 in $\pi^0\pi^0$. It is known from isospin consideration that if we neglect phase space factors and the difference in Dalitz plot structure for given states we should have

$$R = \frac{\sigma(\pi^-\pi^0\pi^0)}{\sigma(\pi^+\pi^-\pi^-)} = \begin{cases} 1 & \text{for waves with } \rho(770), \rho_3(1690) \\ 1/2 & \text{for waves with } f_0(...), f_2(1270) \end{cases}$$

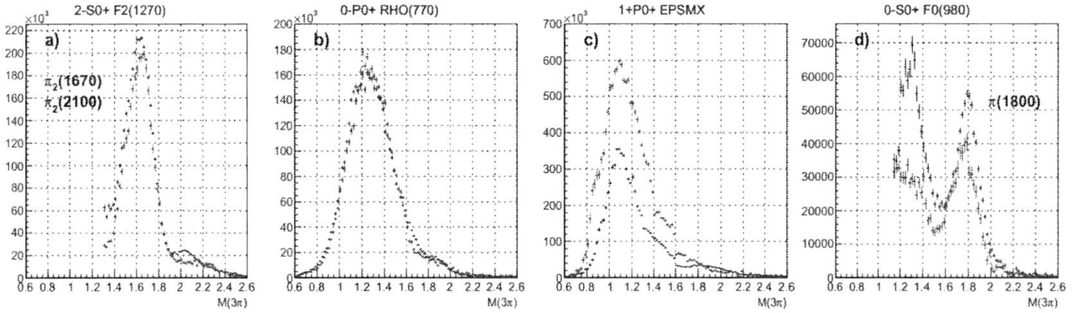

Figure 2. Intensities of waves a) $2^-S0^+f_2(1270)\pi$ b) $0^-P0^+\rho(770)\pi$ c) $1^+P0^+\varepsilon\pi$ d) $0^-S0^+f_0(980)\pi$ for $t' < 0.03$ GeV^2/c^2.

All waves for $\pi^-\pi^0\pi^0$ state coupled to $\pi^0\pi^0$ have factor 1/2 in cross section. To simplify comparison they are scaled by 2 in the following pictures. All intensities produced by PWA program are corrected by acceptance. After this correction and scaling intensities of resonances should be approximately the same for both 3π systems. We apply an additional correction factor 1.10 to $\pi^-\pi^0\pi^0$ intensities to make $2^+D1^+\rho\pi$ signal to be the same in both systems. We choose the $a_2(1320)$ for that as this resonance is narrow and has minimal background. This correction accounts for an error in overall normalization.

We have found that most of large waves are comparable for both final states. This can be seen for $1^+S0^+\rho(770)\pi$, $2^-S0^+f_2(1270)\pi$, $0^-P0^+\rho(770)\pi$, $0^-S0^+f_0(980)\pi$ for $|t'| < 0.030\,GeV^2/c^2$ in Fig. 1 d), Fig. 2 a),b),d). Notable exception is the wave $1^+P0^+\varepsilon\pi$ in Fig. 2 c), which is almost two times larger than expected for $\pi^-\pi^0\pi^0$ system. Excellent coincidence is seen for waves $2^+D1^+\rho(770)\pi$ and $4^+G1^+\rho(770)\pi$ in all t' regions, these waves are shown for $|t'| = 0.03 - 0.15\ GeV^2/c^2$ in Fig. 3 a),b).

For small waves the degree of agreement depends of the wave. For exotic wave with $J^{PC} = 1^{-+}$ and positive exchange naturality, namely the wave $1^-P1^+\rho(770)\pi$, the coincidence is good for $|t'| > 0.03\ GeV^2/c^2$, see Fig. 4 a),b). For this wave in the coherent part of the density matrix, where this wave is 5–20 times smaller than in full density matrix, the coincidence is good in all $|t'|$ regions, see Fig. 5 a),b). But exotic waves with negative exchange naturality, namely $1^-P0^-\rho(770)\pi$ and $1^-P1^-\rho(770)\pi$, are suppressed by 4–5 times in $\pi^-\pi^0\pi^0$ final state, see Fig. 4 c),d).

Some possible resonant states are better observed in $\pi^+\pi^-\pi^-$ final state. It is suggested by some experimental groups that there exists an object $a_1(1420)$ in the wave $1^+P0^+f_0(980)\pi$, see Fig. 3 c). One can see that the bump at $M(3\pi)\sim 1.4\ GeV/c$ is much more significant in $\pi^+\pi^-\pi^-$. The same is true for $\pi_2(2100)$ in Fig. 2 a) and not well established $a_3(1875)$ in the wave $3^+S0^+\rho_3(1690)\pi$ in Fig. 3 d). The same can be stated about possible decay mode $\pi(1800)\to f_0(1500)\pi$ in the wave $0^-S0^+f_0(1500)\pi$, see Fig. 5 c). Possible $a_2(1700)$ object is not seen in waves $2^+D1^+\rho(770)\pi$ and $2^+P1^+f_2(1270)\pi$ in neither 3π final state in neither $|t'|$ region, see Fig. 3 a), Fig. 5 d).

4 Conclusions

Mass-independent PWA is done for $\pi^+\pi^-\pi^-$ and $\pi^-\pi^0\pi^0$ data collected with VES setup. Large waves are mostly consistent with isospin symmetry. Decay modes $\pi(1800) \to f_0(1500)\pi$, $a_3(1875) \to \rho_3(1690)\pi$ are possibly seen in 0^-S0^+ and 3^+S0^+ waves in $\pi^+\pi^-\pi^-$ but are less prominent in $\pi^-\pi^0\pi^0$. The same can be said about the bump in $1^+P0^+f_0(980)\pi$ wave at $M \sim 1.4\ GeV/c^2$. State $a_2(1700)$ is not seen in $2^+D1^+\rho\pi$ and $2^+P1^+f_2(1270)\pi$ in neither final state. The wave $1^-P1^+\rho(770)$ with $J^{PC} = 1^{-+}$ is small in both 3π final states in all $|t'|$ regions. Its coherent part is severely smaller.

This research was supported in part by RFBR grant 16-02-00737.

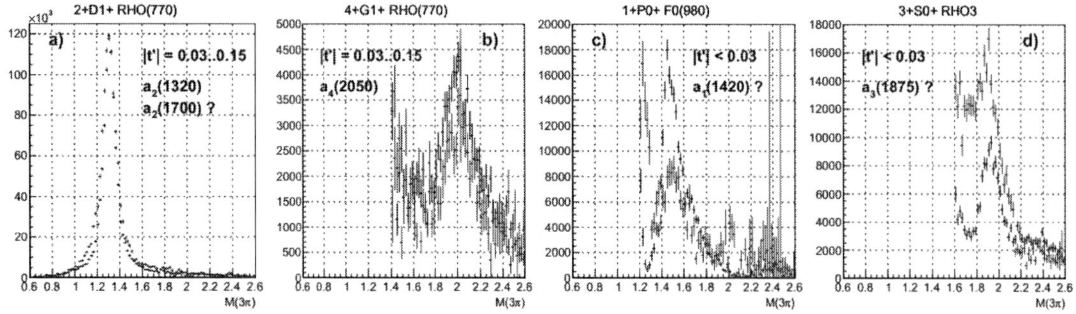

Figure 3. Intensities of waves a) $2^+D1^+\rho(770)\pi$ b) $4^+G1^+\rho(770)\pi$ c) $1^+P0^+f0(980)\pi$ d) $3^+S0^+\rho_3(1690)\pi$.

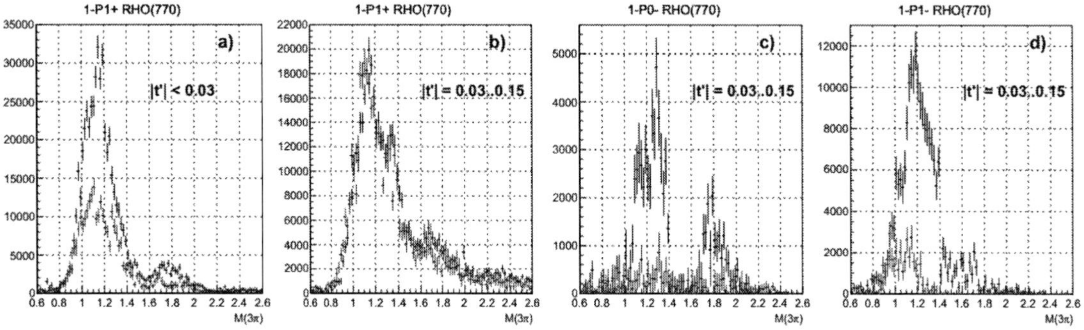

Figure 4. Intensities of waves a), b) $1^-P1^+\rho(770)\pi$, c) $1^-P0^-\rho(770)\pi$, d) $1^-P1^-\rho(770)\pi$.

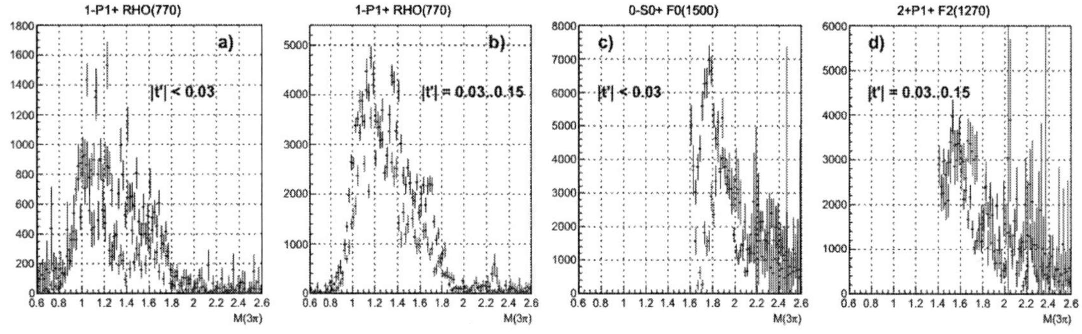

Figure 5. Intensities of waves a), b) $1^-P1^+\rho\pi$, coherent part of the density matrix; c) $0^-S0^+f_0(1500)\pi$, d) $2^+P1^+f_2(1270)\pi$, full density matrix.

References

[1] Yu. Khokhlov *et al.*, EPJ Web of Conferences **37**, 01029 (2012)
[2] J.D. Hansen *et al.*, Nucl. Phys. B **81** 403 (1974)
[3] S.U. Chung *et al.*, Phys. Rev. D **48**, N3 1225 (1993)
[4] F. Filippini *et al.*, Phys. Rev. D **51**, N5 2247 (1995)
[5] Au, Morgan, Pennington, Phys. Rev. D **35** 1633 (1987)
[6] K.A. Olive *et al.* (Particle Data Group), Chin. Phys. C **38**, 090001 (2014)

EPJ Web of Conferences **130**, 04004 (2016) DOI: 10.1051/epjconf/201613004004
MESON 2016

Light meson decays from photon-induced reactions with CLAS

Michael C. Kunkel[1,⋆]
for the CLAS Collaboration

[1] *Forschungszentrum Jülich, Jülich Germany*

Abstract. Photoproduction experiments with the CEBAF Large Acceptance Spectrometer CLAS at the Thomas Jefferson National Facility produce data sets with competitive statistics of light mesons. With these data sets, measurements of transition form factors for η, ω, and η' mesons via conversion decays can be performed using the invariant mass distribution of the final state dileptons. Tests of fundamental symmetries and information on the light quark mass difference can be performed using a Dalitz plot analysis of the meson decay. An overview of preliminary results, from existing CLAS data, and future prospects within the newly upgraded CLAS12 apparatus are given.

1 Introduction

Decays of light mesons provide insight into the structure of the meson. The Light Meson Decay (LMD) group, established at the Thomas Jefferson National Facility with worldwide collaboration, investigates physics pertaining to, but not limited to, transition form factors, anomalous decays and the search for CP violation through Dalitz plot analysis. The presentation given was an overview of the LMD program, recent updates on measurements and an outlook on measurements that can be taken with the CLAS12 detector.

2 Light meson decay program

The light meson group was established in 2013. The goal of the group is to investigate properties of light meson decays using data obtained with the CLAS detector. Since decays of hadrons are independent of production, all CLAS data can be used, however there are two experiments that were chosen as flagships for the program, the g11 and g12 experiment. Both experiments use a photon beam incident on a liquid hydrogen target which created photo-induced reactions with photon beam energies of 800 MeV - 3.8 GeV for g11 and 1.1 GeV - 5.5 GeV for g12. See [1] for a complete list of meson decays the LMD group plans to investigate.

⋆e-mail: m.kunkel@fz-juelich.de

 DOI: 10.1051/epjconf/201613004004

2.1 The radiative decay of the η and η' meson

Radiative decays of η, $\eta' \to \pi^+\pi^-\gamma$ are tools to explore QCD anomalies, i.e. box anomaly, and pion final state interactions. The radiative decay widths of η' and η' can be expressed as [2];

$$\frac{d\Gamma(\eta^{(\prime)} \to \pi^+\pi^-\gamma)}{ds_{\pi\pi}} = A|P(s_{\pi\pi})F_V(s_{\pi\pi})\Gamma_0(s_{\pi\pi})| \tag{1}$$

where $\Gamma_0(s_{\pi\pi})$ is denoted as

$$\Gamma_0(s_{\pi\pi}) = \frac{\kappa\left(M^2_{\eta^{(\prime)}} - s_{\pi\pi}\right)^3 s_{\pi\pi}\left(1 - \frac{4M_\pi^2}{s_{\pi\pi}}\right)^{\frac{3}{2}}}{M^3_{\eta^{(\prime)}}} \tag{2}$$

with κ being a numerical constant. $F_V(s_{\pi\pi})$ is the pion form factor and $P(s_{\pi\pi})$ is written as a polynomial in $s_{\pi\pi}$ as

$$P(s_{\pi\pi}) = 1 + \alpha s_{\pi\pi} + O(s^2_{\pi\pi}) \tag{3}$$

where α is the variable of measurement. Previous results on the radiative decay for the η meson from WASA-at-COSY [3] and KLOE [4] would profit from further measurements. Only one measurement exists for the η' radiative decay [5]. With the CLAS g11 experiment, both the η and η' radiative decay width will be measured. In Fig. 1a the CLAS g11 data is shown for the particle selection of exclusive $\gamma p \to p\pi^+\pi^-\gamma$. Selecting events within a 2.5σ range of η' and applying efficiency corrections, the $s_{\pi\pi}$ distribution is shown in Fig. 1b along with preliminary fit results using Eq. 3 and theory predictions [6].

$$s_{\pi\pi} = m^2 - 2E_\gamma m \tag{4}$$

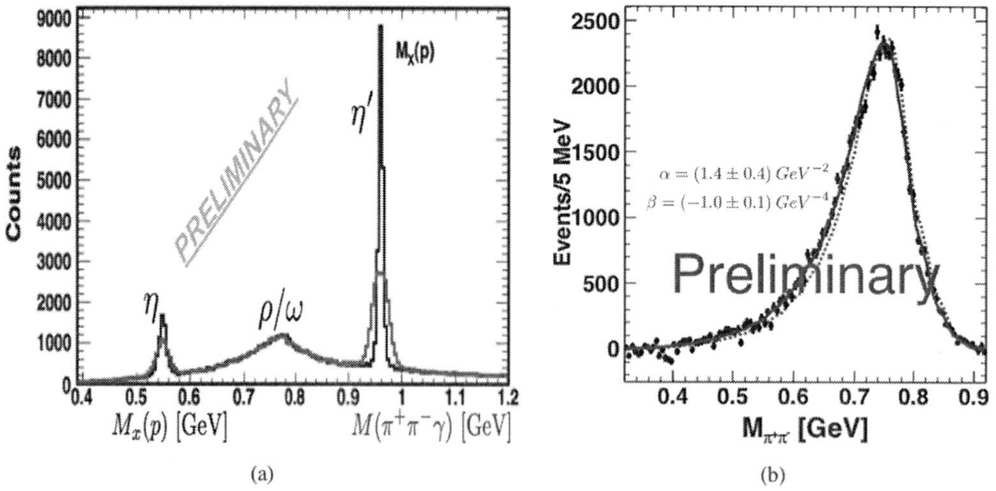

Figure 1: CLAS yield for $\gamma p \to p\eta^{(\prime)} \to p\pi^+\pi^-\gamma$ from the g11 data set (a).(black) CLAS preliminary $s_{\pi\pi}$ distribution for η' with (blue solid line) preliminary fit results using Eq.(3) and (red dashed line) theory predictions [6] (b).

2.2 Update on the dalitz plot analysis of $\eta' \to \pi^+\pi^-\eta$

The Dalitz plot of $\eta' \to \pi^+\pi^-\eta$ provides kinematic information of the decay, enabling the study of low energy dynamics of QCD and heavier mass pseudoscalar mesons. The $\eta' \to \pi^+\pi^-\eta$ decay has a low Q-value due to the decay products being relatively heavy, this helps test and limit the effective chiral Lagrangian theory. The measurable of the Dalitz plot X and Y are projected and fitted to

$$f(X,Y) = A(1 + a(Y) + b(Y^2) + c(X) + d(X^2) \tag{5}$$

to extract the parameters a, b, c, d. This topic was briefly discussed, as a full update was given by later in the session.

2.3 The transition form factor of the ω Meson

Transition form factors characterize modifications of the point-like photon-meson vertex due to the structure of the meson. The virtual photon can interact with quarks, therefore it can be used as a probe for the internal structure of mesons and its electromagnetic interaction is calculable with the Kroll-Wada formula [7];

$$\frac{d\Gamma_{M\to l^+l^-X}}{dq^2 d\Gamma_{M\to X\gamma}} = \frac{\alpha}{3\pi q^2}\left(\left(1+\frac{q^2}{m_M^2-m_X^2}\right)^2 - \frac{4m_M^2 q^2}{(m_M^2-m_X^2)^2}\right)^{\frac{3}{2}}\left(1-\frac{4m_l^2}{q^2}\right)^{1/2}\left(1+\frac{2m_l^2}{q^2}\right)\Bigg|_{\text{Q.E.D}}. \tag{6}$$

where M is the species of meson i.e. π^0, η, ω, η', etc., X is the child particle in the decay, m_M the mass of the meson, m_X the mass of the child particle, m_l the mass of the lepton species in the decay, i.e. $e^\pm$ or $\mu^\pm$ and q being the momentum transfer which is identical to the invariant mass of the dilepton. Deviations of Equation 6 represent the internal structure of the meson for pseudoscalar mesons, while for vector mesons the deviation represents the transition from $M \to X$. These deviations are the transition form factor $|F(q^2)|$ and can be determined by comparing Equation 6 to what is measured experimentally.

$$\frac{d\Gamma_{M\to l^+l^-X}}{dq^2 d\Gamma_{M\to X\gamma}}\Bigg|_{\text{measured}} = \frac{d\Gamma_{M\to l^+l^-X}}{dq^2 d\Gamma_{M\to X\gamma}}\Bigg|_{\text{Q.E.D}} |F(q^2)|^2 . \tag{7}$$

Depending on the decay width of the meson of interest, the transition can be modeled as a simple pole,

$$|F(q^2)| = \frac{1}{1-\frac{q^2}{\Lambda^2}} \tag{8}$$

a complex pole,

$$|F(q^2)|^2 = \frac{\Lambda^2(\Lambda^2+\gamma^2)}{(\Lambda^2-q^2)\Lambda^2\gamma^2}, \tag{9}$$

or some other function that describes the transition. In both Equation 8 and Equation 9, Λ is the mass and width of the virtual vector meson, while γ is the width of the virtual vector meson. Recent measurements of the transition form factor for $\omega \to \mu^+\mu^-\pi^0$ have shown unexpected discrepancies with the Vector Dominance Model [8] and recent models of chiral Lagrangian field theory [9], and dispersion theory [10] attempt to predict the contributions of the virtual vector meson, as seen in Fig. 2.

With the CLAS g12 experiment, lepton ($e^\pm$) identification is done using Cherenkov detectors and electromagnetic calorimetry, providing a $e^+e^-/\pi^+\pi^-$ rejection of 10^6. Using the $e^\pm$ data from CLAS g12, the $\omega \to pe^+e^-\pi^0$ transition form factor can be extracted. Preliminary studies have shown that the number of $\omega \to pe^+e^-\pi^0$ events in CLAS to be $\approx 3 \cdot 10^3$.

 DOI: 10.1051/epjconf/201613004004

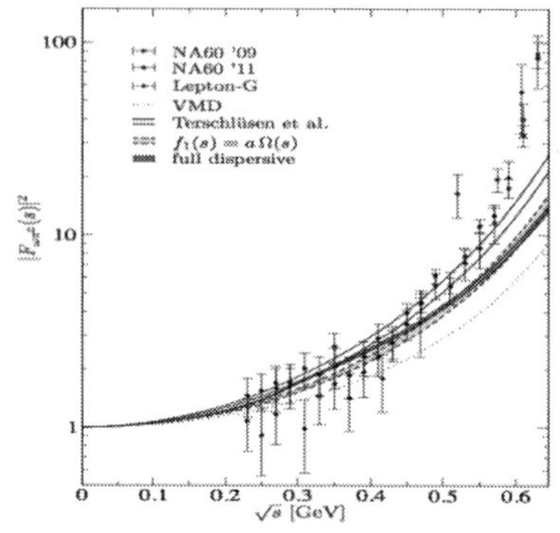

Figure 2: Form factor of the decay $\omega \rightarrow l^+l^-\gamma$ compared to dimuon data by the NA60 Collaboration [8, 11] and LEPS [12]. Shown are VMD (dotted line) Eqs. (6, 9) using the mass of the ρ meson as the virtual vector meson, the results of a chiral Lagrangian treatment with explicit vector mesons [9] (white shaded curve with solid borders), a simplified approximation to the full dispersive solution [10] (gray shaded curve with dashed borders), full dispersive solution [10] (black hatched curve with solid borders). [10]

2.4 Update on the branching ratio measurement of the $\eta' \rightarrow e^+e^-\gamma$ decay

A recent measurement of BESIII reports the branching ratio $\Gamma(\eta' \rightarrow e^+e^-\gamma)/\Gamma(\eta' \rightarrow \gamma\gamma)$ to be $2.13\pm0.09(stat.)\pm0.07(sys.))10^{-2}$ from 864 events [13]. Using the $e^\pm$ data from CLAS g12, preliminarily 172 events of the $\eta' \rightarrow e^+e^-\gamma$ decay were observed and analyzed, using the Q-factor [14] method to suppress background from neighboring $A \rightarrow e^+e^-X$ decays, Figure 3. A preliminary branching

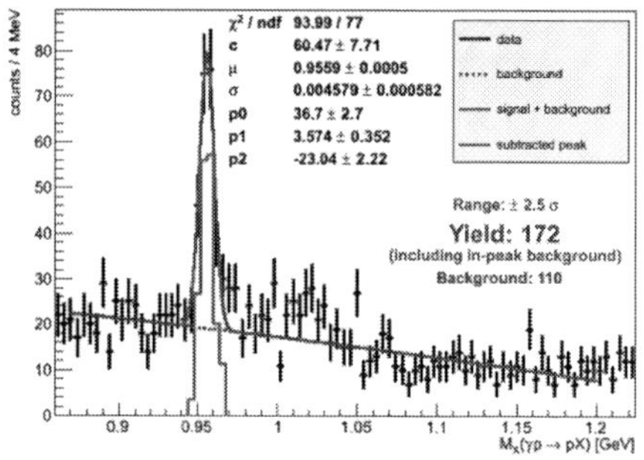

Figure 3: (black)CLAS yield for $\gamma p \rightarrow p\eta' \rightarrow pe^+e^-\gamma$ from g12 data set. (red line) Fit of signal and background using a Voigt signal and 1st order polynomial background function. (magenta line) Background subtracted signal peak, (blue dashed line) represents the background.

ratio $\Gamma(\eta' \rightarrow e^+e^-\gamma)/\Gamma(\eta' \rightarrow \gamma\gamma)$ was measured to be consistent with the BESIII measurement. However, the statistics of either BESIII or CLAS are not ample enough to determine, statistically, which theoretical model best represents the structure of the η' meson [15–17]. Table 1 outlines the current measurements and theoretical predictions.

Charge Radius $\langle r \rangle$	Measurement (M) / Prediction (P)
CLAS ($\eta' \to e^+e^-\gamma$)	TBD
BESIII ($\eta' \to e^+e^-\gamma$)	(M) $1.60 \pm 0.17(stat) \pm 0.08(sys)\text{GeV}^{-2}$ [13]
CELLO ($\eta' \to \mu^+\mu^-\gamma$)	(M) $1.7 \pm 0.4\text{GeV}^{-2}$ [18]
Dispersion	(P) $1.53^{0.15}_{-0.08}\text{GeV}^{-2}$ [17]
ChPT	(P) 1.6GeV^{-2} [16]
VMD	(P) 1.45GeV^{-2} [15]

Table 1: Current Measurements and Theoretical Predictions of the η' charge radius.

2.5 Future measurement of the η' meson transition form factor with CLAS12

With the newly built CLAS12 detector, $e^\pm$ identification can be achieved with a $e^+e^-/\pi^+\pi^-$ rejection of $10^6 - 10^{11}$ while retaining $e^+e^-\gamma$ acceptance $\sim 1\% - 0.1\%$.

Using the CLAS12 FASTMC simulation, a simulation of 10^6 $ep \to e'p\eta' \to pe^+e^-\gamma(e')$ was performed. The generation of events included cross-section information obtained from previous CLAS measurements, the s^n scaling law on the cross-section and the VMD model for the decay of the η' meson to achieve a reasonable model of the production of the η' meson. Estimations of the integrated quasi-real photon rate from an electron beam of luminosity $10^{35}\text{cm}^{-2}\text{s} - 1$ impinged on a 5 cm ℓH_2 target yielded that the production rate for hadrons to be 80 kHz. Therefore, the total rate of η' can be calculated by defining the following ratio $R(W)$:

$$R(W) = \frac{\sigma(W)}{\eta' \text{ integrated } \sigma(W)} \tag{10}$$

and scaling the total rate by $R(W)$. This leads to:

$$\eta' \text{ total rates / 80 Days } (W) = 80\,\text{kHz} \cdot \frac{6.912\text{e} \cdot 10^6 \text{ seconds}}{80 \text{ days}} \cdot \frac{1}{\text{R(W)}} \tag{11}$$

The total $\eta' \to e^+e^-\gamma$ rates per 80 days, and as a function W, is calculated by multiplying Eq.(11) with the product of the average detection efficiency $\epsilon \approx 5\%$ as well as the branching fraction $\mathcal{BR} = 4.69 \cdot 10^{-4}$ [13] for $\eta' \to e^+e^-\gamma$. The total number N_{tot} of expected $\eta' \to e^+e^-\gamma$ events after 80 days of measurement is given by the integral of Eq.(11) over W. This leads to the expected yield:

$$N_{tot} = \int_{1.9\,\text{GeV}}^{2.8\,\text{GeV}} \left[N(W)_{\eta' \to e^+e^-\gamma\,/\,80\text{ Days}}\right] dW = \frac{28,200 \text{ events}}{80\text{Days}} \tag{12}$$

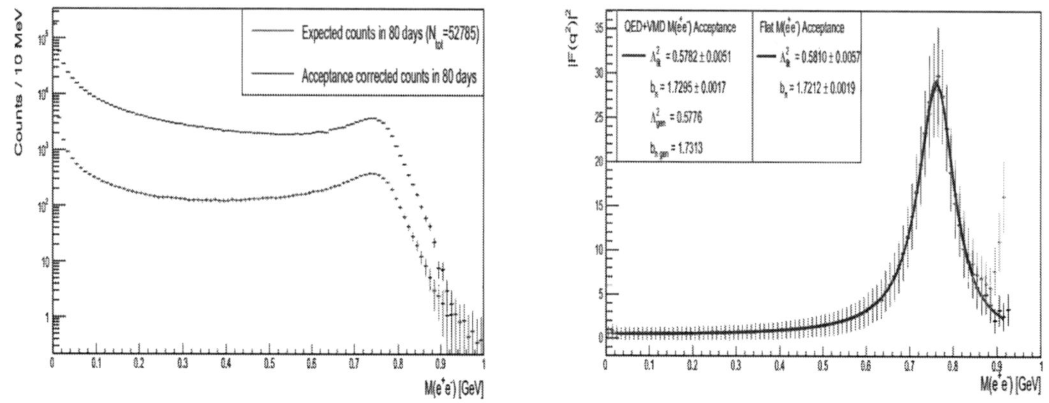

(a) (b)

Figure 4: (Color Online) Expected yield as a function of $M(e^+e^-)$ (a).(Color Online) $\left|F(q^2)\right|^2$ as a function of $M(e^+e^-)$ using two acceptance models. (black points) QED+VMD $M(e^+e^-)$ acceptance model. (green points) Flat $M(e^+e^-)$ acceptance model. The solid lines represent a fit using Eq.(9) to the data points (b).

From Fig. 4a the QED normalized spectrum can be deduced and is shown in Fig. 4b. Both generation models (i.e. flat and QED+VMD) are used to determine the transition form factor. It is shown that there exists a systematic uncertainty depending on the chosen generation model. However, the final calculation on the slope parameter or the TFF shows a negligible impact of this uncertainty. Regardless of the generation model, it is shown in Fig. 4b that the accumulated statistics collected by CLAS12 allow for a precision of each parameter $\lesssim 0.5\%$. Compared to current experimental uncertainties of 10% and differences in theoretical approaches of $\approx 10\%$, Table 1, therefore a measurement by CLAS12 would not only be in the position to decisively discriminate between the theoretical predictions, but also help to further constrain the hadronic light-by-light amplitude to the muon $g-2$ anomaly [19].

3 Summary

The CLAS LMD program covers a variety of meson decay physics. With the data sets collected by the former CLAS detector, transition form factor measurements, investigation of symmetry breaking phenomena and probing QCD anomalies are done with high accuracy. The newly built CLAS12 detector will provide more statistically supported data samples, allowing LMD analyses to probe further into light meson decay physics.

References

[1] M. Amaryan et al., *Photoproduction and Decay of Light Mesons in CLAS*
[2] F. Stollenwerk et al., Physics Letters B **707**, 184 (2012)
[3] P. Adlarson et al., Physics Letters B **707**, 243 (2012)
[4] D. Babusci et al., Physics Letters B **718**, 910 (2013)

[5] A. Abele et al., Physics Letters B **402**, 195 (1997)
[6] B. Kubis, J. Plenter, The European Physical Journal C **75**, 1 (2015)
[7] N.M. Kroll, W. Wada, Phys. Rev. **98**, 1355 (1955)
[8] R. Arnaldi et al., Physics Letters B **677**, 260 (2009)
[9] C. Terschlusen, S. Leupold, Physics Letters B **691**, 191 (2010)
[10] S.P. Schneider et al., Phys. Rev. D **86**, 054013 (2012)
[11] G. Usai, Nuclear Physics A **855**, 189 (2011)
[12] R. Dzhelyadin et al., Physics Letters B **102**, 296 (1981)
[13] M. Ablikim et al. (BESIII), Phys. Rev. **D92**, 012001 (2015)
[14] M. Williams et al., JINST **4**, P10003 (2009)
[15] A. Bramon, E. Massó, Physics Letters B **104**, 311 (1981)
[16] L. Ametller et al., Phys. Rev. D **45**, 986 (1992)
[17] C. Hanhart et al., Eur. Phys. J. C **73**, 2668 (2013)
[18] R. Dzhelyadin et al., Physics Letters B **88**, 379 (1979)
[19] T. Blum et al. (2013), `1311.2198`

The interference effects of multi-channel pion-pion scattering contributions to the final states of Ψ- and Υ-meson family decays

Yu.S. Surovtsev[1,a], P. Bydžovský[2], T. Gutsche[3], R. Kamiński[4], V.E. Lyubovitskij[3,5,6], and M. Nagy[7]

[1] *Bogoliubov Lab of Theoretical Physics, JINR, 141 980 Dubna, Russia*
[2] *Nuclear Physics Institute, AS CR, 25068 Řež/Prague, Czech Republic*
[3] *Institut für Theoretische Physik, Universität Tübingen, D–72076 Tübingen, Germany*
[4] *Institute of Nuclear Physics, PAN, Krakow 31342, Poland*
[5] *Department of Physics, Tomsk State University, 634050 Tomsk, Russia*
[6] *Mathematical Physics Department, Tomsk Polytechnic University, Lenin ave. 30, 634050 Tomsk, Russia*
[7] *Institute of Physics, SAS, Bratislava 84511, Slovak Republic*

Abstract. It is shown that the basic shape of dipion mass distributions in the two-pion transitions of both charmonia and bottomonia states are explained by an unified mechanism based on the contribution of the $\pi\pi$, $K\overline{K}$ and $\eta\eta$ coupled channels including their interference.

1 Introduction

In the analysis of practically all available data on two-pion transitions of the Υ mesons from the ARGUS, CLEO, CUSB, Crystal Ball, Belle, and *BaBar* Collaborations — $\Upsilon(mS) \to \Upsilon(nS)\pi\pi$ ($m > n$, $m = 2,3,4,5$, $n = 1,2,3$) — the contribution of multi-channel $\pi\pi$ scattering in the final-state interactions is considered. The analysis, which is aimed at studying the scalar mesons, is performed jointly considering the above bottomonia decays, the isoscalar S-wave processes $\pi\pi \to \pi\pi, K\overline{K}, \eta\eta$, which are described in our model-independent approach based on analyticity and unitarity and using an uniformization procedure, and the charmonium decay processes — $J/\psi \to \phi\pi\pi$, $\psi(2S) \to J/\psi\pi\pi$ — with data from the Crystal Ball, DM2, Mark II, Mark III, and BES II Collaborations.

We show that the experimentally observed interesting (even mysterious) behavior of the $\pi\pi$ spectra of the Υ-family decays, beginning from the second radial excitation and higher, — a bell-shaped form in the near-$\pi\pi$-threshold region, smooth dips about 0.6 GeV in the $\Upsilon(4S,5S) \to \Upsilon(1S)\pi^+\pi^-$, about 0.45 GeV in the $\Upsilon(4S,5S) \to \Upsilon(2S)\pi^+\pi^-$, and about 0.7 GeV in the $\Upsilon(3S) \to \Upsilon(1S)(\pi^+\pi^-,\pi^0\pi^0)$, and also sharp dips about 1 GeV in the $\Upsilon(4S,5S) \to \Upsilon(1S)\pi^+\pi^-$ — is explained by the interference between the $\pi\pi$ scattering, $K\overline{K} \to \pi\pi$ and $\eta\eta \to \pi\pi$ contributions to the final states of these decays (by the constructive one in the near-$\pi\pi$-threshold region and by the destructive one in the dip regions).

[a] e-mail: surovcev@theor.jinr.ru

2 The effect of multi-channel $\pi\pi$ scattering in decays of the ψ- and Υ-meson families

Considering multi-channel $\pi\pi$ scattering, we shall deal with the 3-channel case ($\pi\pi \to \pi\pi, K\overline{K}, \eta\eta$) because it was shown [1] that this is a minimal number of coupled channels needed for obtaining correct values of f_0-resonance parameters. When performing our combined analysis data for the multi-channel $\pi\pi$ scattering were taken from many papers (see Refs. in our paper [2]). For the decay $J/\psi \to \phi\pi^+\pi^-$ data were taken from Mark III, DM2 and BES II Collaborations; for $\psi(2S) \to J/\psi(\pi^+\pi^-$ and $\pi^0\pi^0)$ — from Mark II and Crystal Ball(80) (see Refs. also in [2]). For $\Upsilon(2S) \to \Upsilon(1S)(\pi^+\pi^-$ and $\pi^0\pi^0)$ data were used from ARGUS [3], CLEO [4], CUSB [5], and Crystal Ball [6] Collaborations; for $\Upsilon(3S) \to \Upsilon(1S)(\pi^+\pi^-, \pi^0\pi^0)$ and $\Upsilon(3S) \to \Upsilon(2S)(\pi^+\pi^-, \pi^0\pi^0)$ — from CLEO [7, 8]; for $\Upsilon(4S) \to \Upsilon(1S, 2S)\pi^+\pi^-$ — from *BaBar* [9] and Belle [10]; for $\Upsilon(5S) \to \Upsilon(1S, 2S, 3S)\pi^+\pi^-$ — from Belle Collaboration [10].

The used formalism for calculating the di-meson mass distributions in the quarkonia decays is analogous to the one proposed in Ref. [11] for the decays $J/\psi \to \phi(\pi\pi, K\overline{K})$ and $V' \to V\pi\pi$ ($V = \psi, \Upsilon$) but with allowing for also amplitudes of transitions between the $\pi\pi$, $K\overline{K}$ and $\eta\eta$ channels in decay formulas. There was assumed that the pion pairs in the final state have zero isospin and spin. Only these pairs of pions undergo final state interactions whereas the final $\Upsilon(nS)$ meson ($n < m$) remains a spectator. The amplitudes of decays are related with the scattering amplitudes T_{ij} ($i, j = 1 - \pi\pi, 2 - K\overline{K}, 3 - \eta\eta$) as follows

$$F(J/\psi \to \phi\pi\pi) = c_1(s)T_{11} + \Big(\frac{\alpha_2}{s-\beta_2} + c_2(s)\Big)T_{21} + c_3(s)T_{31}, \quad (1)$$

$$F(\psi(2S) \to \psi(1S)\pi\pi) = d_1(s)T_{11} + d_2(s)T_{21} + d_3(s)T_{31}, \quad (2)$$

$$F(\Upsilon(mS) \to \Upsilon(nS)\pi\pi) = e_1^{(mn)}T_{11} + e_2^{(mn)}T_{21} + e_3^{(mn)}T_{31}, \quad (3)$$

$$m > n,\ m = 2, 3, 4, 5,\ n = 1, 2, 3$$

where $c_i = \gamma_{i0} + \gamma_{i1}s$, $d_i = \delta_{i0} + \delta_{i1}s$ and $e_i^{(mn)} = \rho_{i0}^{(mn)} + \rho_{i1}^{(mn)}s$; indices m and n correspond to $\Upsilon(mS)$ and $\Upsilon(nS)$, respectively. The free parameters α_2, β_2, γ_{i0}, γ_{i1}, δ_{i0}, δ_{i1}, $\rho_{i0}^{(mn)}$ and $\rho_{i1}^{(mn)}$ depend on the couplings of J/ψ, $\psi(2S)$ and the $\Upsilon(mS)$ to the channels $\pi\pi$, $K\overline{K}$ and $\eta\eta$. The pole term in eq.(1) in front of T_{21} is an approximation of possible ϕK states, not forbidden by OZI rules.

The amplitudes T_{ij} are expressed through the S-matrix elements $S_{ij} = \delta_{ij} + 2i\sqrt{\rho_1\rho_2}T_{ij}$ where $\rho_i = \sqrt{1 - s_i/s}$ and s_i is the reaction threshold. The S-matrix elements are parameterized on the uniformization plane of the $\pi\pi$ scattering amplitude by poles and zeros which represent resonances. The uniformization plane is obtained by a conformal map of the 8-sheeted Riemann surface, on which the three-channel S matrix is determined, onto the plane. In the uniformizing variable used we have neglected the $\pi\pi$-threshold branch point and allowed for the $K\overline{K}$- and $\eta\eta$-threshold branch points and left-hand branch point at $s = 0$ related to the crossed channels. The background is introduced to the amplitudes in a natural way: on the threshold of each important channel there appears generally speaking a complex phase shift. It is important that we have obtained practically zero background of the $\pi\pi$ scattering in the scalar-isoscalar channel. It confirms well our representation of resonances.

The expression $N|F|^2\sqrt{(s-s_1)[m_\psi^2 - (\sqrt{s} - m_\phi)^2][m_\psi^2 - (\sqrt{s} + m_\phi)^2]}$ for decay $J/\psi \to \phi\pi\pi$ and the analogues relations for $\psi(2S) \to \psi(1S)\pi\pi$ and $\Upsilon(mS) \to \Upsilon(nS)\pi\pi$ give the di-meson mass distributions. N (normalization to experiment) is: for $J/\psi \to \phi\pi\pi$ 0.5172 (Mark III), 0.1746 (DM 2) and 3.8 (BES II); for $\psi(2S) \to J/\psi\pi^+\pi^-$ 1.746 (Mark II); for $\psi(2S) \to J/\psi\pi^0\pi^0$ 1.6891 (Crystal Ball(80)); for $\Upsilon(2S) \to \Upsilon(1S)\pi^+\pi^-$ 4.1758 (ARGUS), 2.0445 (CLEO(94)) and 1.0782 (CUSB); for $\Upsilon(2S) \to \Upsilon(1S)\pi^0\pi^0$ 0.0761 (Crystal Ball(85)); for $\Upsilon(3S) \to \Upsilon(1S)(\pi^+\pi^-$ and $\pi^0\pi^0)$ 19.8825

 DOI: 10.1051/epjconf/201613004005

and 4.622 (CLEO(07)); for $\Upsilon(3S) \to \Upsilon(2S)(\pi^+\pi^-$ and $\pi^0\pi^0)$ 1.6987 and 1.1803 (CLEO(94)); for $\Upsilon(4S) \to \Upsilon(1S)\pi^+\pi^-$ 4.6827 (*BaBar*(06)) and 0.3636 (Belle(07)); for $\Upsilon(4S) \to \Upsilon(2S)\pi^+\pi^-$, 37.9877 (*BaBar*(06)); for $\Upsilon(5S) \to \Upsilon(1S)\pi^+\pi^-$, $\Upsilon(5S) \to \Upsilon(2S)\pi^+\pi^-$ and $\Upsilon(5S) \to \Upsilon(3S)\pi^+\pi^-$ respectively 0.2047, 2.8376 and 6.9251 (Belle(12)).

Satisfactory combined description of all considered processes (including $\pi\pi \to \pi\pi, K\overline{K}, \eta\eta$) is obtained with the total $\chi^2/\text{ndf} = 736.457/(710 - 118) \approx 1.24$; for the $\pi\pi$ scattering, $\chi^2/\text{ndf} \approx 1.15$.

Studying the decays of charmonia and bottomonia, we investigated the role of the individual f_0 resonances in contributing to the shape of the dipion mass distributions. In this case we switched off only those resonances [$f_0(500)$, $f_0(1370)$, $f_0(1500)$ and $f_0(1710)$], removal of which can be somehow compensated by correcting the background to have the more-or-less acceptable description of the multichannel $\pi\pi$ scattering. First, when leaving out before-mentioned resonances, a minimal set of the f_0 mesons consisting of the $f_0(500)$, $f_0(980)$, and $f_0'(1500)$ is sufficient to achieve a description of the processes $\pi\pi \to \pi\pi, K\overline{K}, \eta\eta$ with a total $\chi^2/\text{ndf} \approx 1.20$. Second, from these three mesons only the $f_0(500)$ can be switched off while still obtaining a reasonable description of multi-channel $\pi\pi$-scattering (though with an appearance of the pseudo-background) with a total $\chi^2/\text{ndf} \approx 1.43$. In figure 1 we show here those two-pion spectra of the charmonia and bottomonia decays which are most typical from the point of view of interference effects; the solid lines correspond to contribution of all relevant f_0-resonances; the dotted, of the $f_0(500)$, $f_0(980)$, and $f_0'(1500)$; the dashed, of the $f_0(980)$ and $f_0'(1500)$.

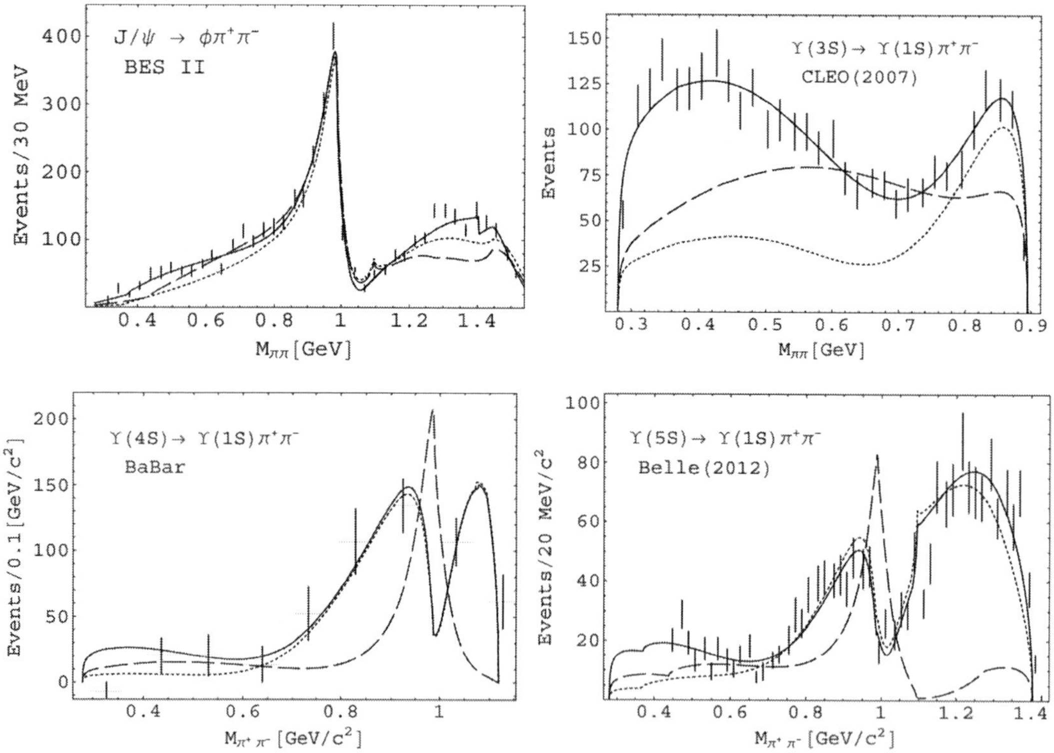

Figure 1. The decays $J/\psi \to \phi\pi\pi$ and $\Upsilon(3S, 4S, 5S) \to \Upsilon(1S)\pi^+\pi^-$. The solid lines correspond to contribution of all relevant f_0-resonances; the dotted, of the $f_0(500)$, $f_0(980)$, and $f_0'(1500)$; the dashed, of the $f_0(980)$ and $f_0'(1500)$.

3 Conclusions

The combined analysis was performed for data on isoscalar S-wave processes $\pi\pi \to \pi\pi, K\overline{K}, \eta\eta$ and on the decays of the charmonia — $J/\psi \to \phi\pi\pi$, $\psi(2S) \to J/\psi\,\pi\pi$ — and of the bottomonia — $\Upsilon(mS) \to \Upsilon(nS)\pi\pi$ ($m > n$, $m = 2,3,4,5$, $n = 1,2,3$) from the ARGUS, Crystal Ball, CLEO, CUSB, DM2, Mark II, Mark III, BES II, *BaBar*, and Belle Collaborations.

It is shown that the dipion mass spectra in the above-indicated decays of charmonia and bottomonia are explained by the unified mechanism which is based on our previous conclusions on wide resonances [1, 2] and is related to contributions of the $\pi\pi$, $K\overline{K}$ and $\eta\eta$ coupled channels including their interference. It is shown that in the final states of these decays (except $\pi\pi$ scattering) the contribution of coupled processes, e.g., $K\overline{K}, \eta\eta \to \pi\pi$, is important even if these processes are energetically forbidden.

The role of the individual f_0 resonances in making up the shape of the dipion mass distributions in the charmonia and bottomonia decays is considered. Note the unexpected result — a considerable contribution of the $f_0(1370)$ to the bell-shaped form of the dipion mass spectra of bottomonia decays in the near-$\pi\pi$-threshold region.

Since describing the bottomonia decays, we did not change resonance parameters in comparison with the ones obtained in the combined analysis of the processes $\pi\pi \to \pi\pi, K\overline{K}, \eta\eta$ and charmonia decays, the results of this analysis confirm all of our earlier conclusions on the scalar mesons [1].

This work was supported in part by the Heisenberg-Landau Program, by the Votruba-Blokhintsev Program for Cooperation of Czech Republic with JINR, by the Grant Agency of the Czech Republic (grant No. P203/15/04301), by the Grant Program of Plenipotentiary of Slovak Republic at JINR, by the Bogoliubov-Infeld Program for Cooperation of Poland with JINR, by the BMBF (Project 05P2015, BMBF-FSP 202), by Tomsk State University Competitiveness Improvement Program, the Russian Federation program "Nauka" (Contract No. 0.1526.2015, 3854), by Slovak Grant Agency VEGA under contract No.2/0197/14, and by the Polish National Science Center (NCN) grant DEC-2013/09/B/ST2/04382.

References

[1] Yu.S. Surovtsev *et al.*, J. Phys. G Nucl. Part. Phys. **41**, 025006 (2014); Phys. Rev. D **86**, 116002 (2012)
[2] Yu.S. Surovtsev *et al.*, Phys. Rev. D **89**, 036010 (2014)
[3] H. Albrecht *et al.* (ARGUS Collaboration), Phys. Lett. B**134**, 137 (1984)
[4] D. Besson *et al.* (CLEO Collaboration), Phys. Rev. D **30**, 1433 (1984)
[5] V. Fonseca *et al.* (CUSB Collaboration), Nucl. Phys. B **242**, 31 (1984)
[6] D. Gelphman *et al.* (Crystal Ball Collaboration), Phys. Rev. D **32**, 2893 (1985)
[7] D. Cronin-Hennessy *et al.* (CLEO Collaboration), Phys. Rev. D **76**, 072001 (2007)
[8] F. Butler *et al.* (CLEO Collaboration), Phys. Rev. D **49**, 40 (1994)
[9] B. Aubert *et al.* (BaBar Collaboration), Phys. Rev. Lett. **96**, 232001 (2006)
[10] A. Sokolov *et al.* (Belle Collaboration), Phys. Rev. D **75**, 071103 (2007); A. Bondar *et al.* (Belle Collaboration), Phys. Rev. Lett. **108**, 122001 (2012)
[11] D. Morgan and M.R. Pennington, Phys. Rev. D **48**, 1185 (1993)

DOI: 10.1051/epjconf/201613005001

$a_0(980)$ photoproduction in the coupled channel model

Ł. Bibrzycki[1,⋆] and R.Kamiński[2]

[1] *Institute of Computer Science, Pedagogical University of Cracow, Podchorążych 2, 30-084 Kraków, Poland*
[2] *H. Niewodniczański Institute of Nuclear Physics, Polish Academy of Sciences, Radzikowskiego 152, 31-342 Kraków, Poland*

Abstract. We calculated the coupled channel photoproduction amplitudes of the scalar isovector resonace $a_0(980)$, which include the $\pi\eta$, K^+K^- and $K^0\overline{K^0}$ intermediate states. Based on them we obtained the mass distribution in the $\pi\eta$ channel at photon energy E_γ=7 GeV, corresponding to energy accissible in new JLab experiments CLAS12 and GlueX. We also analyzed the shape of mass distribution as a function of the phase of the background amplitude.

1 Introduction

As dominating decay channels of the $a_0(980)$ are $\pi\eta$ and $K\overline{K}$, the construction of the amplitudes of the resonant $\pi\eta$ photoproduction is inevitably the coupled channel problem. The model described in this paper assumes that the isovector $a_0(980)$ resonance observed in the $\pi\eta$ channel results from the $\pi\eta$ and $K\overline{K}$ interactions in the final state [1]. We describe the $a_0(980)$ photoproduction as a two stage process shown schematically in Fig. 1. First, a pair of pseudoscalar mesons $\pi\eta$, $K^0\overline{K^0}$ or

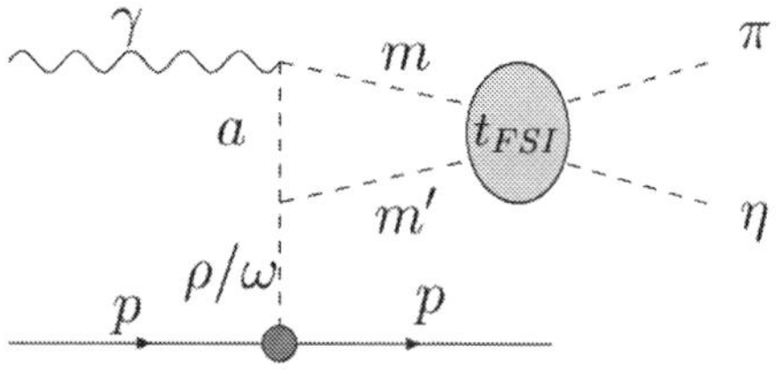

Figure 1. Diagram for the photoproduction process with final state interactions.

K^+K^- is photoproduced, which we describe in terms of Born amplitudes. The final state interactions are described in terms of the coupled channel and unitary amplitudes. The general form of the $\pi\eta$ photoproduction amplitude, which takes into account the final state interactions is

$$\langle\lambda'|A_{\pi\eta}|\lambda_\gamma\lambda\rangle = \langle\lambda'|V_{\pi\eta}|\lambda_\gamma\lambda\rangle + 4\pi\sum_{\{mm'\}}\int_0^\infty \frac{k'^2dk'}{(2\pi)^3}F(k')t_{\pi\eta;mm'}G_{mm'}\langle\lambda'|V_{mm'}|\lambda_\gamma\lambda\rangle \tag{1}$$

⋆e-mail: lukasz.bibrzycki@ifj.edu.pl

 DOI: 10.1051/epjconf/201613005001

where $V_{\pi\eta}$ ($V_{mm'}$) is the Born amplitude of the $\pi\eta$ (mm') pair photoproduction, $t_{\pi\eta;mm'}$ is the coupled channel scattering amplitude, λ, λ' and λ_γ are respectively the helicities of the initial and final proton, and photon helicity. $G_{mm'}$ is the propagator of the intermediate mm' pair defined as

$$G_{mm'} = \frac{-1}{M_{mm'} - M'_{mm'}(k') + i\epsilon} \tag{2}$$

and $F(k')$ is the form-factor needed to regularize the divergent mesonic loop of the diagram shown in Fig. 1. In present calculations we limit ourselves to the on-shell part of the amplitude and are concerned with the S-wave, so projecting out the higher partial waves, we arrive at the following coupled channel S-wave $\pi\eta$ photoproduction amplitude:

$$A_{\pi\eta} = \left[1 + ir_{\pi\eta}t^{I=1}_{\pi\eta}\right] V_{\pi\eta} + ir_{K\overline{K}}t^{I=1}_{\pi\eta;K\overline{K}}\frac{1}{\sqrt{2}}\left(V_{K^+K^-} + V_{K^0\overline{K^0}}\right), \tag{3}$$

where $r_{\pi\eta} = -k_{\pi\eta}M_{\pi\eta}/8\pi$, $r_{K\overline{K}} = -k_{K\overline{K}}M_{\pi\eta}/8\pi$, $M_{\pi\eta}$ is effective mass of the $\pi\eta$ system, $t^{I=1}_{\pi\eta}$ and $t^{I=1}_{K\overline{K};\pi\eta}$ are isovector elastic $\pi\eta$ scattering amplitude and $K\overline{K} \to \pi\eta$ transition amplitude respectively. The $k_{\pi\eta}$ and $k_{K\overline{K}}$ are CM momenta in the $\pi\eta$ and $K\overline{K}$ channels defined as

$$k_{\pi\eta} = \frac{\sqrt{[M^2_{\pi\eta} - (m_\eta + m_\pi)^2][M^2_{\pi\eta} - (m_\eta - m_\pi)^2]}}{2M_{\pi\eta}} \quad \text{and} \quad k_{K\overline{K}} = \sqrt{\frac{M^2_{\pi\eta}}{4} - m^2_K}. \tag{4}$$

Further we use numerical channel indices for the elements of the final state scattering amplitude and denote $t_{\pi\eta}$ as T_{11} and $t_{\pi\eta;K\overline{K}}$ as T_{12} respectively.

2 Born amplitudes

In Fig. 2 we show the general structure of Feynman diagrams representing Born amplitudes. In these diagrams a corresponds to ρ or ω in the $\pi\eta$ channel and K or K^* in the $K\overline{K}$ channel. b corresponds to the ρ and ω exchange. Inclusion of diagram c) is necessary for electromagnetic current conservation when charged kaon pair is produced in the intermediate state. The amplitudes corresponding to

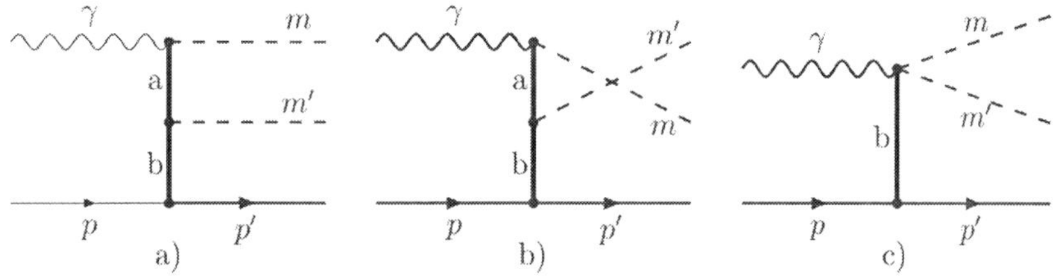

Figure 2. Diagrammatic representation of the Born amplitudes for $\pi\eta$, $K^0\overline{K^0}$ and K^+K^- intermediate states.

diagrams of Fig. 2 have a general form of:

$$V_{mm'} = \sum_{\{mm'\}} \overline{u}(p', s')J_{r,mm'} \cdot \varepsilon(q, \lambda^\gamma)u(p, s), \tag{5}$$

where the sum is over intermediate states mm', $J_{r,mm'}$ is the hadronic current, $u(p, s)$ and $\overline{u}(p', s')$ - wave functions of the initial and final proton respectively and ε - polarisation 4-vector of the incident photon. Details of the partial wave expansion of the Born amplitudes are discussed in [1, 2].

3 Final state interaction amplitudes

To obtain the final state rescattering amplitudes one solves the coupled channel Lippman-Schwinger equations which in momentum space read [3–5]

$$\langle q|\hat{T}|k\rangle = \langle q|\hat{V}|k\rangle + \int \frac{d^3p}{(2\pi)^3}\langle q|\hat{V}|p\rangle\langle p|\hat{G}|p\rangle\langle p|\hat{T}|k\rangle, \tag{6}$$

where $\hat{V}$, $\hat{G}$ and $\hat{T}$ are 2×2 matrices in channel space. $\hat{G}$ denotes the diagonal propagator matrix, with the elements defined as

$$G_i(p) = \frac{1}{E - E_i(p) + i\epsilon}. \tag{7}$$

$\hat{V}$ is the interaction matrix which is assumed in a separable form

$$\langle q|V_{ij}|k\rangle = \lambda_{ij} g_i(q) g_j(k), \tag{8}$$

where λ_{ij} is a 2×2 coupling constant matrix and the form factor is defined as

$$g_i(p) = \sqrt{\frac{4\pi}{m_i}}\frac{1}{p^2 + \beta_i^2}, \tag{9}$$

with m_i being the reduced mass of mesons in i–th channel. The on-shell scattering matrix elements $T_{ij}(k_i, k_j)$ can be expressed in terms of the S-matrix elements

$$S_{ij} = \delta_{ij} - \frac{i}{\pi}\sqrt{k_i\alpha_i k_j\alpha_j}\, T_{ij}(k_i, k_j) \tag{10}$$

where i and j enumerate channels, k_i, k_j are center of mass momenta in respective channels and α_i are defined as:

$$\alpha_1 = \frac{E_\pi E_\eta}{E_\pi + E_\eta} \tag{11}$$

and

$$\alpha_2 = \frac{E_K}{2}. \tag{12}$$

Here E_π, E_η and E_K denote the π, η and K energies in the center of mass of respective channels. For the 2-channel case, the elements of the S–matrix can be related to the Jost function with the following expressions [4]:

$$S_{11} = \frac{D(-k_1, k_2)}{D(k_1, k_2)}, \tag{13}$$

$$S_{22} = \frac{D(k_1, -k_2)}{D(k_1, k_2)}, \tag{14}$$

$$S_{12}^2 = S_{11}S_{22} - \frac{D(-k_1, -k_2)}{D(k_1, k_2)}. \tag{15}$$

Information on the location of amplitude poles in the complex momenta planes, which correspond to zeroes of Jost function are then used to obtain 2 complex (or 4 real) equations:

$$D(k_1^r, k_2^r) = 0, \qquad D(k_1^R, k_2^R) = 0. \tag{16}$$

The superscripts r and R correspond to resonances $a_0(980)$ and $a_0(1450)$ which are included in the amplitude parameterization. Eqs. (16) along with information on $a_0(980)$ and $a_0(1450)$ branching fractions to $\pi\eta$ and $K\overline{K}$ channels are used to fix 5 parameters of the FSI amplitude [5].

4 Mass distributions and phase dependence

We define the double differential cross section as

$$\frac{d\sigma}{dtdM_{\pi\eta}} = \frac{1}{4}\frac{1}{(2\pi)^3}\frac{|k|}{32m^2E_\gamma{}^2}\sum_{\lambda_\gamma,\lambda,\lambda'}|\langle\lambda'|A_{\pi\eta}|\lambda_\gamma\lambda\rangle|^2, \tag{17}$$

where m and E_γ are the proton mass and photon energy in the laboratory frame respectively. For the calculation of the mass distribution we integrate Eq. (17) over the physical region of t. The calculation was performed at the photon energy E_γ=7 GeV which roughly corresponds to energies accessible in new and upgraded JLab experiments. As we are interested in the high energy regime we use the reggeised version of the photoproduction amplitude defined in Eq. (1). The striking feature of mass distribution in Fig. 3 is that interference of the real Born amplitude and the final state interaction terms results in the resonance being represented by the minimum (double dot-dash line). To develop a maximum, the background term needs to acquire the imaginary part. Similar feature was observed for the $a_0(980)$ production in the $\gamma\gamma$ fusion [6]. Dot-dashed, dashed and dotted curves in Fig. 3 represent mass distributions resulting from interference of the final state interactions terms with the background amplitude "corrected" by additional phases of $\frac{\pi}{2}$, π and $\frac{3}{2}\pi$ respectively. Detailed study of the size and origin of the imaginary part of the background will be possible when high statistics data on $\pi\eta$ photoproduction from CLAS12 and GlueX experiments will be accessible.

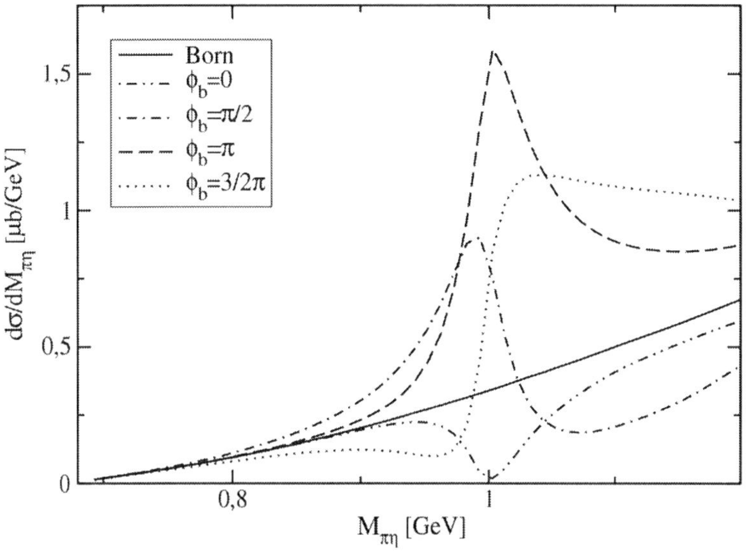

Figure 3. Dependence of the mass distribution on the phase of the background amplitude.

5 Summary

We calculated the $M_{\pi\eta}$ mass spectrum in the region corresponding to isovector resonance $a_0(980)$ at energy of 7 GeV which will be measured in upgraded JLab facilities. We have analyzed the interference pattern of resonance and background amplitude and found that for purely real Born background the $a_0(980)$ resonance is represented by the minimum in the mass distribution. The maximum can be developed only when the background acquires the imaginary part whose size can be estimated by experiment.

Acknowledgments

This research has been funded by the Polish National Science Center (NCN) grant No. DEC-2013/09/B/ST2/04382

References

[1] Ł. Bibrzycki, R. Kamiński, Int. J. Mod. Phys. A **31**, 1650139 (2016)
[2] Ł. Bibrzycki, R. Kamiński, Phys. Rev. D **87**, 114010 (2013)
[3] R. Kamiński, L. Leśniak and J. P. Maillet, Phys. Rev. D **50**, 3145 (1994)
[4] L. Leśniak, Acta Phys. Polon. B **27**, 1835 (1996)
[5] A. Furman and L. Leśniak, Phys. Lett. B **538**, 266 (2002)
[6] N. N. Achasov, G. N. Shestakov, JETP Lett. 92, 1 (2010)

Measurement of the cross section ratio $\sigma_{\psi(2S)}/\sigma_{J/\psi(1S)}$ in deep inelastic exclusive ep scattering at HERA

Jacek Ciborowski[1,⋆]
for the ZEUS Collaboration

[1] *University of Warsaw, Faculty of Physics, L. Pasteura 5, 02-093 Warsaw, Poland*

Abstract. The exclusive deep inelastic electroproduction of $\psi(2S)$ and $J/\psi(1S)$ at an ep centre-of-mass energy of 317 GeV has been studied with the ZEUS detector at HERA in the kinematic range $2 < Q^2 < 80\,\mathrm{GeV}^2$, $30 < W < 210\,\mathrm{GeV}$ and $|t| < 1\,\mathrm{GeV}^2$, where Q^2 is the photon virtuality, W is the photon–proton centre-of-mass energy and t is the squared four-momentum transfer at the proton vertex. The data for $2 < Q^2 < 5\,\mathrm{GeV}^2$ were taken in the HERA I running period and correspond to an integrated luminosity of $114\,\mathrm{pb}^{-1}$. The data for $5 < Q^2 < 80\,\mathrm{GeV}^2$ are from both HERA I and HERA II periods and correspond to an integrated luminosity of $468\,\mathrm{pb}^{-1}$. The decay modes analysed were $\mu^+\mu^-$ and $J/\psi(1S)\pi^+\pi^-$ for the $\psi(2S)$ and $\mu^+\mu^-$ for the $J/\psi(1S)$. The cross-section ratio $\sigma_{\psi(2S)}/\sigma_{J/\psi(1S)}$ has been measured as a function of Q^2, W and t. The results are compared to predictions of QCD-inspired models of exclusive vector-meson production.

Exclusive electroproduction of vector mesons in deep inelastic scattering at high energies, $ep \to eVp$, where V denotes a vector meson, may be described as a multi-step process. The electron emits a virtual photon, γ^*, with virtuality Q^2 and $\gamma^* p$ centre-of-mass energy W, the virtual photon fluctuates into a $q\bar{q}$ pair and which subsequently interacts with the proton via a colour-neutral exchange, e.g. through a two-gluon ladder, and then hadronises into the vector meson. The $\psi(2S)$ and the $J/\psi(1S)$ have the same quark content, different radial distributions of the wave functions, and their mass difference is small compared to the HERA centre-of-mass energy. Therefore, the ratio of their electroproduction cross sections allows checking perturbative QCD predictions regarding their wave functions[1].

The luminosity used for this analysis was 468 pb^{-1} in total, which consists of data from 1996-2000 and 2002-2007 running periods (HERA I and HERA II, respectively). The kinematic range was $2 < Q^2 < 80$ GeV^2, $30 < W < 210$ GeV and $|t| < 1$ GeV^2. Events were selected with no activity in the central ZEUS detector in addition to signals from the scattered electron and the decay products of the studied mesons. The sample contained exclusive and a small fraction of proton-dissociative events with diffractive masses $M_Y < 4$ GeV which was assumed to cancel in the cross section ratio. The decay channels were: $J/\psi(1S), \psi(2S) \to \mu^+\mu^-$, and $\psi(2S) \to J/\psi(1S)\pi^+\pi^-$ with the subsequent decay $J/\psi(1S) \to \mu^+\mu^-$. The analysis was based on data collected with the ZEUS detector when 920(820) GeV protons collided with 27.5 GeV electrons or positrons. The DIFFVM [2] Monte Carlo (MC) was used for simulating the studied process. Exclusive and diffractive Bethe-Heitler dimuon production were simulated using the GRAPE package [3]. The results for the three cross-section

⋆e-mail: cib@fuw.edu.pl

ratios $\sigma_{\psi(2S)}/\sigma_{J/\psi(1S)}$: $R_{\mu\mu}$ for $\psi(2S) \to \mu^+\mu^-$, $R_{J/\psi\pi\pi}$ for $\psi(2S) \to J/\psi(1S)\pi^+\pi^-$ and R_{comb} for the combination, for the kinematic range $5 < Q^2 < 80\,\text{GeV}^2$, $30 < W < 210\,\text{GeV}$ and $|t| < 1\,\text{GeV}^2$ and for the total integrated luminosity of 468 pb^{-1}, are: $R_{J/\psi\pi\pi} = 0.26 \pm 0.03^{+0.01}_{-0.01}$, $R_{\mu\mu} = 0.24 \pm 0.05^{+0.02}_{-0.03}$, $R_{\text{comb}} = 0.26 \pm 0.02^{+0.01}_{-0.01}$ and $R_{\psi(2S)} = 1.1 \pm 0.2^{+0.2}_{-0.1}$.

Fig. 1 shows the values of R_{comb} as a function of W and $|t|$ for $5 < Q^2 < 80\,\text{GeV}^2$. In Fig. 2, the values of R_{comb} as a function of Q^2 are shown (including $2 < Q^2 < 5\,\text{GeV}^2$ from the HERA I data) with H1 measurements [5]. The H1 collaboration has also measured $R = \sigma_{\psi(2S)}/\sigma_{J/\psi(1S)}$ in photoproduction ($Q^2 \approx 0$), and found a value of $R = 0.150 \pm 0.035$ [4], which is consistent with the observed trend. The results are compared to model predictions from six different groups labelled: HIKT, KNNPZZ, AR, LP, FFJS and KMW. In order to calculate the exclusive production of vector

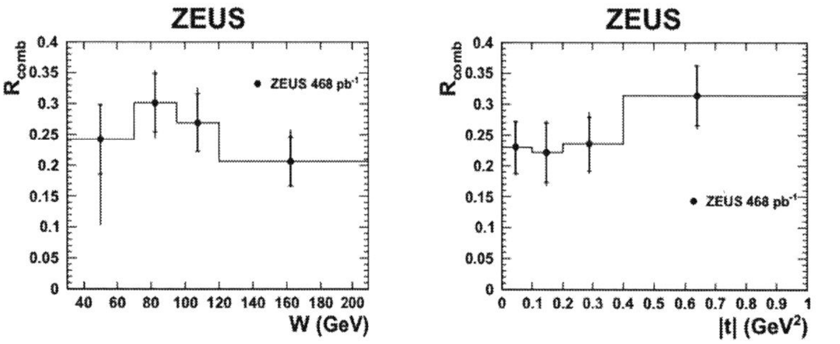

Figure 1. Cross-section ratio $R_{comb} = \sigma_{\psi(2S)}/\sigma_{J/\psi(1S)}$ for the combined $psi(2S)$ decay modes as a function of W and $|t|$. The horizontal lines show the bin widths. The inner error bars show the statistical and the outer error bars show the quadratic sum of statistical and systematic uncertainties.

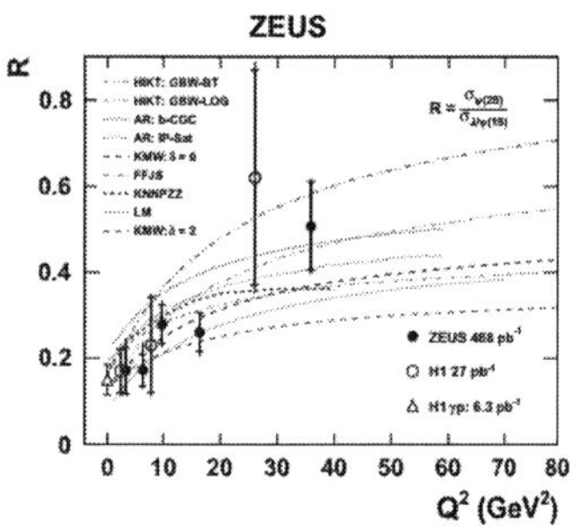

Figure 2. Comparison of the ZEUS measurement of the $R = \sigma_{\psi(2S)}/\sigma_{J/\psi(1S)}$ ratio and the previous H1 results as a function of photon virtuality Q^2 together with theoretical predictions.

charmonium states one has to determine: the probability of finding a $c\bar{c}$-dipole of transverse size r and impact parameter b in the photon in the infinite momentum frame; the $c\bar{c}$-dipole scattering amplitude or cross section of the proton as a function of r, b and $x_{\text{Bj}} \approx (M_V + Q^2)/(W^2 + Q^2)$; the probability that the $c\bar{c}$-dipole forms the vector state V in the infinite momentum frame. The probability distribution of $c\bar{c}$-dipoles in the photon can be calculated in QED [6, 7]. For the probability that the $c\bar{c}$-dipole forms the vector state, its centre-of-mass wave function has to be boosted into the infinite momentum frame, which can be done using the boosted Gaussian model [8] calculations are used. All models predict a Q^2 dependent suppression of exclusive $\psi(2S)$ relative to $J/\psi(1S)$ production. For those, which explicitly use the wave functions of the vector mesons, this is caused by the node of the radial $\psi(2S)$ wave function, which leads to a destructive interference of the contributions to the production amplitude from small and large dipoles. Hüfner et al. [9] (HIKT) use two phenomenological parameterisations of the $c\bar{c}$-dipole cross section, GBW[10] and KST[11], which both describe the low-x inclusive DIS data from HERA. For the centre-of-mass wave functions of the $J/\psi(1S)$ and $\psi(2S)$, they use four different phenomenological potentials, BT, LOG, COR and POW, and c-quark masses between 1.48 and 1.84 GeV. However, only the models with c-quark masses around 1.5 GeV, GBW–BT and GBW–

LOG, are able to describe the cross sections of exclusive $J/\psi(1S)$ production measured at HERA. For the boost of the charmonium wave functions into the infinite momentum frame, they find the wave function from the Schrödinger equation and then boost the result. A major progress made [9] is the inclusion of the Melosh spin rotation into the boosting procedure, which enhances the $\psi(2S)$ to $J/\psi(1S)$ cross-section ratio by a factor of two to three. The model of Kopeliovich et al. [12–15] (KNNPZZ) uses the running gBFKL approach for the $c\bar{c}$-dipole cross section and the diffractive slope for its t dependence. The parameterisation of the $c\bar{c}$-dipole cross section used gives a quantitative description of the rise of the proton structure function at small x values as well as of the Q^2 and W dependence of diffractive $J/\psi(1S)$ production. KNNPZZ use parametrisations of the vector meson wave functions, inspired by the conventional spectroscopic models and short-distance behaviour driven by hard QCD gluon exchange. Armesto and Rezaeian [16] (AR) calculate the $c\bar{c}$-dipole cross section using the Impact-Parameter-dependent Color Glass Condensate model (b-CGC) [17] as well as the Saturation (IP-Sat) [18] dipole model, recently updated with fits to the HERA combined data [19, 20]. In the b-CGC model, which is restricted to the gluon sector, saturation is driven by the BFKL evolution, and its validity is therefore limited to $x_{\mathrm{Bj}} \lesssim 10^{-2}$. The IP-Sat model uses DGLAP evolution and smoothly matches the perturbative QCD limit at high values of Q^2. For the calculation of the light-cone $J/\psi(1S)$ and $\psi(2S)$ wave functions, the boosted Gaussian model and the leptonic decay widths $\Gamma_{ee}^{J/\psi(1S)}$ and $\Gamma_{ee}^{\psi(2S)}$ are used. Lappi and Mäntysaari [21, 22] (LM) use the BFKL evolution as well as the IP-Sat model to predict vector-meson production in ep and electron–ion collisions in the dipole picture. The wave functions of the $J/\psi(1S)$ and $\psi(2S)$ have been calculated according to the procedure developed previously [23, 24] and the low-x inclusive HERA data have been used to constrain the $c\bar{c}$-dipole cross section. Fazio et al. [25] (FFJS) use a two-component Pomeron model to predict the cross sections for vector-meson production. A normalisation factor of $f_{\psi(2S)}^{-1} = 0.45$ ensures that the value of the $\psi(2S)$ cross section is the same as for the other vector mesons at the same values of W, t and $Q_V^2 = M_V^2 + Q^2$ (i.e. $f_{\psi(2S)}\sigma_{\psi(2S)} = \sigma_{J/\psi}$). In this model the Q^2 dependence of the $\psi(2S)$ to $J/\psi(1S)$ suppression is caused by the difference of Q_V^2 at a fixed Q^2 due to the $\psi(2S) - J/\psi(1S)$ mass difference. Kowalski, Motyka and Watt [23] (KMW) assume the universality of the production of vector quarkonia states in the scaling variable Q_V^2 in their calculation of R. With the assumptions that the $c\bar{c} \to V$ transition is proportional to the leptonic decay width Γ_{ee}^V and that the interaction is mediated by two-gluon exchange and therefore proportional to $(\alpha_s(Q_V)\,x_{\mathrm{Bj}}\,g(x_{\mathrm{Bj}}, Q_V^2))^2$, R is given in the leading-logarithmic approximation [26–28] by

$$R = \left(\frac{\alpha_s(Q_{\psi(2S)})}{\alpha_s(Q_{J/\psi(1S)})}\right)^2 \frac{\Gamma_{\psi(2S)} M_{\psi(2S)}^{1-\delta}}{\Gamma_{J/\psi(1S)} M_{J/\psi(1S)}^{1-\delta}} \left(\frac{Q_{\psi(2S)}}{Q_{J/\psi(1S)}}\right)^{-6-4\bar{\lambda}+\delta}. \quad (1)$$

The running strong coupling constant, $\alpha_s(Q)$, and the gluon density, $g(x, Q^2)$, are evaluated at Q_V and x_{Bj}. For small x_{Bj} values, the gluon density can be parameterised as $x_{\mathrm{Bj}}\,g(x_{\mathrm{Bj}}, Q_V^2) \propto x_{\mathrm{Bj}}^{-\lambda(Q_V)}$ with $\lambda(Q_V) \simeq \bar{\lambda} = 0.25$ in the Q_V region of this measurement, while δ depends on the choice of the charmonium wave functions. For the non-relativistic wave functions $\delta = 0$ [26–28] and for the relativistic boosted Gaussian model $\delta \approx 2$ [23]. The Q^2 dependence of the ratio R in this approach is driven by kinematic factors and not by the form of the charmonia wave functions. All models predict only a weak W and $|t|$ dependence of R, as shown in Fig. 1. Therefore, only the comparison of the model calculations with the measurements as a function of Q^2 is presented in Fig. 2. All models predict an increase of R with Q^2, as observed in the data. The models are discussed below in the sequence from higher to lower predicted R values at high Q^2. From the HIKT calculations, the results for R for the two charmonium potentials BT and LOG with c-quark mass around 1.5 GeV and the GBW model for the $c\bar{c}$-dipole cross section are shown. The difference of the results when using the KST dipole cross sections are small. For Q^2 values below 24 GeV^2, the predicted R values for the

BT model are significantly larger than the measured values, whereas the values predicted by the LOG model agree with the data. For the AR calculations, the results for the b-CGC and IP-Sat models of the dipole cross sections are shown in the figure. For Q^2 values below 24 GeV^2, the b-CGC prediction for R is significantly higher than the data, whereas the IP-Sat model gives a good description of the data for the entire Q^2 range. The KMW model with $\delta = 0$ provides a good description of the observed Q^2 dependence of R, whereas the prediction for $\delta = 2$ is about 2 standard deviations below the R value measured for $Q^2 > 24$ GeV^2. The predictions of the models FFJS, KNNPZZ and LM also provide fair descriptions of the measurements. The material described above has recently been published [29].

References

[1] L. Frankfurt, W. Koepf and M. Strikman, Phys. Rev. D **54**, 3194 (1996)
[2] B. List, A. Mastroberardino, in: A.T. Doyle, G. Grindhammer, G. Ingelman, H. Jung (Eds.), Proc. Workshop on Monte Carlo Generators for HERA Physics, DESY, Hamburg, Germany, 1999, p.396, also in preprint DESY-PROC-1999-02
[3] T. Abe, A generator for dilepton production in *ep* collisions, Comput. Phys. Commun. **136**, 126 (2001)
[4] H1 Collaboration, C. Adloff *et al.*, Phys. Lett. B **421**, 385 (1998)
[5] H1 Collaboration, C. Adloff *et al.*, Eur. Phys. J. C **10**, 373 (1999)
[6] J.B. Kogut, D.E. Soper, Phys. Rev. D **1**, 2901 (1970)
[7] J.D. Bjorken, J.B. Kogut, D.E. Soper, Phys. Rev. D **3**, 1382 (1971)
[8] M.V. Terentev, Sov. J. Nucl. Phys. **24**, 106 (1976)
[9] J. Hüfner, et al., Phys. Rev. D **62**, 094022 (2000)
[10] K.J. Golec-Biernat, M. Wüsthoff, Phys. Rev. D **59**, 014017 (1998)
[11] B.Z. Kopeliovich, A. Schäfer, A.V. Tarasov, Phys. Rev. D **62**, 054022 (2000)
[12] B.Z. Kopeliovich, B.G. Zakharov, Phys. Rev. D **44**, 3466 (1991)
[13] B.Z. Kopeliovich, Phys. Lett. B **324**, 469 (1994)
[14] J. Nemchik, N.N. Nikolaev, B.G. Zakharov, Phys. Lett. B **341**, 228 (1994)
[15] J. Nemchik, J. Exp. Theor. Phys. **86**, 1054 (1998)
[16] N. Armesto, A.H. Reazeian, Phys. Rev. D **90**, 054003 (2014)
[17] A.H. Reazeian, I. Schmidt, Phys. Rev. D **88**, 074016 (2013)
[18] A.H. Reazeian *et al.*, Phys. Rev. D **87**, 034002 (2014)
[19] H1 Collaboration, F.D. Aaron *et al.*, J. High Energy Phys. **01**, 109 (2010)
[20] H1 Collaboration, ZEUS Collaboration, H. Abramowicz *et al.*, Eur. Phys. J. C **73**, 2311 (2013)
[21] T. Lappi, H. Mäntysaari, Phys. Rev. C **83**, 065202 (2011)
[22] T. Lappi, H. Mäntysaari, PoS DIS2014, 069 (2014)
[23] H. Kowalski, L. Motyka, G. Watt, Phys. Rev. D **74**, 074016 (2006)
[24] B. Cox, J. Forshaw, R. Sandapen, J. High Energy Phys. **0906**, 034 (2009)
[25] S. Fazio *et al.*, Phys. Rev. D **90**, 016007 (2014)
[26] M.G. Ryskin, Z. Phys. C **57**, 89 (1993)
[27] S.J. Brodsky *et al.*, Phys. Rev. D **50**, 3134 (1994)
[28] S.P. Jones *et al.*, J. High Energy Phys. **1311**, 085 (2013)
[29] ZEUS Collaboration, H. Abramowicz *et al.*, Nucl. Phys. B **909**, 934 (2016)

Inclusive production of J/ψ and ψ' mesons at the LHC

Anna Cisek[1,★] and Antoni Szczurek[1,2]

[1] *University of Rzeszów, PL-35-959 Rzeszów, Poland*
[2] *Institute of Nuclear Physics PAN, PL-31-342 Cracow, Poland*

Abstract. We discuss the prompt production of J/ψ mesons in proton-proton collisions at the LHC within a NRQCD k_t-factorization approach using Kimber-Martin-Ryskin (KMR) unintegrated gluon distributions (UGDF). We include both direct color-singlet production ($gg \to J/\psi g$) as well as a feed-down from $\chi_c \to J/\psi\gamma$ and $\psi' \to J/\psi X$. The production of the decaying mesons (χ_c or ψ')is also calculated within NRQCD k_t-factorization. The corresponding matrix elements for $gg \to J/\psi$, $gg \to \psi'$ and $gg \to \chi_c$ include parameters of the nonrelativistic spatial wave functions of quarkonia at $r = 0$, which are taken from potential models from the literature. We get the ratio of the corresponding of the cross sections for $\chi_c(2)$-to-$\chi_c(1)$ much closer to experimental data than obtained in recent analyses. Differential distributions in rapidity of J/ψ and ψ' are calculated and compared to experimental data of the ALICE and LHCb collaborations. We discuss possible onset of gluon saturation effects at forward/backward rapidities. One can describe the experimental data for J/ψ production within model uncertainties with color-singlet component only. Therefore our theoretical results leave only a relatively small room for the color-octet contributions.

1 Introduction

For a long time there are discrepancies among authors about the production mechanism of J/ψ quarkonia in proton-proton and proton-antiproton collisions. Some authors think that the cross section is dominated by the color-octet contribution. Some authors believe that the color-singlet contribution dominates. The color-octet contribution cannot be calculated from first principle and is rather fitted to the experimental data. Different fits from the literature give different magnitudes of the color-octet contributions. Therefore we concentrate on the color-singlet contribution. In the present paper we wish to calculate the color-singlet contribution as well as possible in the NRQCD k_t-factorization and see how much room is left for the more difficult color-octet contribution. In the present approach we concentrate rather on small transverse momenta of J/ψ or ψ' relevant for ALICE and LHCb data [1–5]. We expect that color-singlet contributions may dominate in this region of the phase space. Finally ψ' quarkonium also has a sizable branching fraction into $J/\psi X$ [6]. Fortunately this contribution is much smaller than the direct one as will be discussed in [8]. It was considered recently in an almost identical approach in [9].

★e-mail: acisek@ur.edu.pl

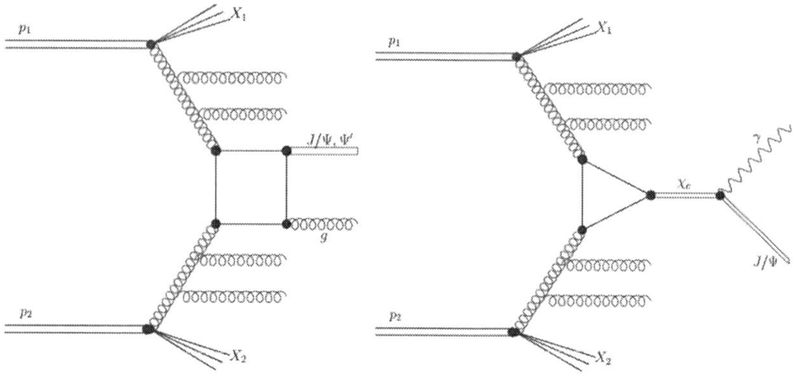

Figure 1. The leading-order diagram for prompt J/ψ (ψ') meson production in the k_t-factorization approach.

2 Inclusive production of J/ψ and ψ' mesons in the NRQCD k_t-factorization approach

The main color-singlet mechanism for the production of J/ψ and ψ' mesons is shown in Fig.1 (left panel). We restrict ourselves to the gluon-gluon fusion mechanism. In the NLO the differential cross section in the k_t-factorization can be written as:

$$\frac{d\sigma(pp \to J/\psi gX)}{dy_{J/\psi}dy_g d^2p_{J/\psi,t}d^2p_{g,t}} = \frac{1}{16\pi^2\hat{s}^2}\int\frac{d^2q_{1t}}{\pi}\frac{d^2q_{2t}}{\pi}\overline{|\mathcal{M}^{off-shell}_{g^*g^*\to J/\psi g}|^2} \times \delta^2\left(\vec{q}_{1t}+\vec{q}_{2t}-\vec{p}_{H,t}-\vec{p}_{g,t}\right)\mathcal{F}_g(x_1,q_{1t}^2,\mu_F^2)\mathcal{F}_g(x_2,q_{2t}^2,\mu_F^2)\,. \tag{1}$$

We calculate the dominant color-single $gg \to J/\psi g$ contribution taking into account transverse momenta of initial gluons. The corresponding matrix element squared for the $gg \to J/\psi g$ is

$$|\mathcal{M}_{gg\to J/\psi g}|^2 \propto \alpha_s^3|R(0)|^2\,. \tag{2}$$

The matrix element is taken from [10]. In our calculation we choose the scale of the running coupling constant as:

$$\alpha_s^3 \to \alpha_s(\mu_1^2)\alpha_s(\mu_2^2)\alpha_s(\mu_3^2)\,, \tag{3}$$

where $\mu_1^2 = max(q_{1t}^2, m_t^2)$, $\mu_2^2 = max(q_{2t}^2, m_t^2)$ and $\mu_3^2 = m_t^2$, where here m_t is the J/ψ transverse mass. The factorization scale in the calculation was taken as $\mu_F^2 = (m_t^2 + p_{t,g}^2)/2$.

Similarly we calculate the P-wave χ_c meson production. Here the lowest-order subprocess $gg \to \chi_c$ is allowed by positive C-parity of χ_c mesons.

In the k_t-factorization approach the leading-order cross section for the χ_c meson production can be written as:

$$\sigma_{pp\to\chi_c} = \int dyd^2p_td^2q_t\frac{1}{sx_1x_2}\frac{1}{m_{t,\chi_c}^2}\overline{|\mathcal{M}_{g^*g^*\to\chi_c}|^2}\mathcal{F}_g(x_1,q_{1t}^2,\mu_F^2)\mathcal{F}_g(x_2,q_{2t}^2,\mu_F^2)/4\,, \tag{4}$$

which can also be used to calculate rapidity and transverse momentum distributions of the χ_c mesons. In the last equation $\mathcal{F}_g$ are unintegrated gluon distributions and $\sigma_{gg\to\chi_c}$ is $gg \to \chi_c$ (off-shell) cross section. The situation is illustrated diagrammatically in Fig.1 (right panel).

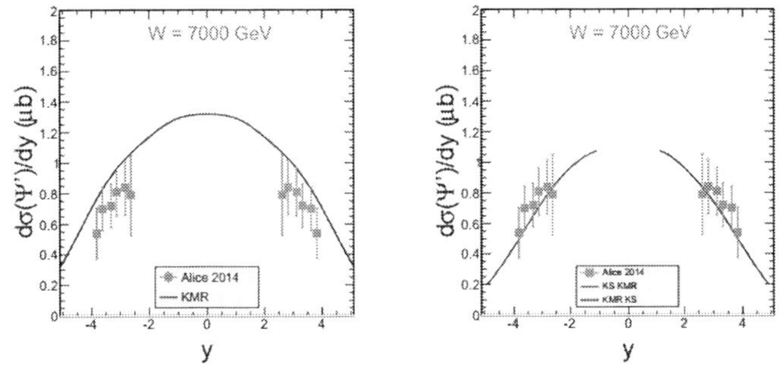

Figure 2. Rapidity distribution of ψ' meson with KMR (left plots) and mixed UGDFs (KS and KMR, right plots). The ALICE data [4] are shown for comparison.

The matrix element squared for the $gg \to \chi_c$ subprocess is

$$|\mathcal{M}_{gg\to\chi_c}|^2 \propto \alpha_s^2 |R'(0)|^2 \,. \tag{5}$$

We used the matrix element taken from the Kniehl, Vasin and Saleev paper [7].

For this subprocess the best choice for running coupling constant is:

$$\alpha_s^2 \to \alpha_s(\mu_1^2)\alpha_s(\mu_2^2) \,, \tag{6}$$

where $\mu_1^2 = max(q_{1t}^2, m_t^2)$ and $\mu_2^2 = max(q_{2t}^2, m_t^2)$. Above m_t is transverse mass of the χ_c meson.

The factorization scale for the χ_c meson production is fixed as $\mu_F^2 = m_t^2$.

3 Results

In Fig.2 we show differential cross section in rapidity for ψ' production at 7 TeV. Our results are compared with ALICE experimental data [4]. In the left panel we present results for Kimber-Martin-Ryskin (KMR) UGDF and in the right panel for mixed Kimber-Martin-Ryskin (KMR) and Kutak-Stasto (KS) UGDFs. Because KMR alone overshoot experimental data for rapidity distribution the best solution is to take the KMR distribution for large x and KS for small x. For ψ' meson we have to include only the direct diagram so it's easy to compare our result with experimental data.

For J/ψ meson we have to include both diagrams. Below we present results for these two subprocesses. In Fig.3 we show rapidity distribution for direct J/ψ meson production. We present results for three different values of energy: W = 2.76 TeV (left), W = 7 TeV (middle) and W = 13 TeV (right). Our results are compared with ALICE and LHCb experimental data [1–5].

In Fig.4 we present results for three different values of energy: W = 2.76 TeV (left), W = 7 TeV (middle) and W = 13 TeV (right) panel. The dotted lines are for χ_{c1} meson contribution, the dot-dashed lines are for χ_{c2} meson contributions ant the solid lines are sum of these two components. The presented here results are calculated with mixed UGDFs (KMR and KS).

4 Conclusion

We have calculated the color-singlet contribution in the NRQCD k_t-factorization and compared our results with ALICE and LHCb data. Our results in rapidity are almost consistent or even exceed

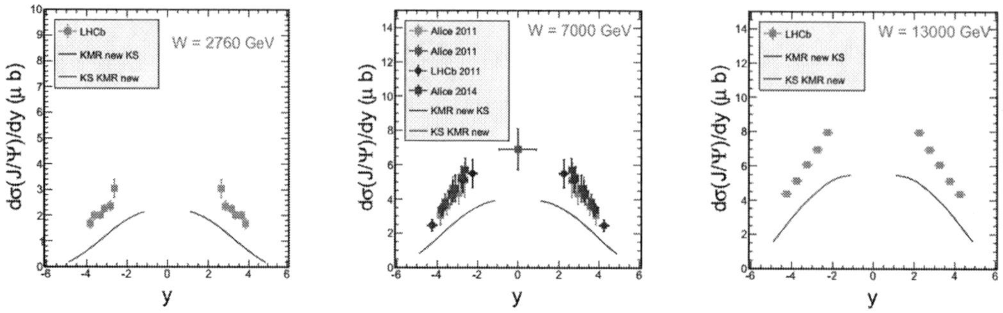

Figure 3. Rapidity distribution of J/ψ meson with KMR (upper plots) and mixed UGDFs (Kutak-Stasto and KMR). The ALICE and LHCb data points [1–5] are shown for comparison.

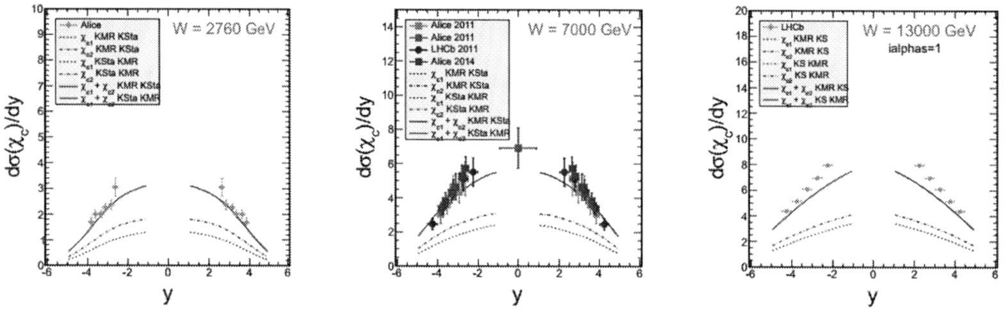

Figure 4. Rapidity distribution of χ_c meson with mixed UGDFs (Kutak-Stasto and KMR). The ALICE and LHCb data points [1–5] for J/ψ are shown for comparison.

experimental data. Cross section strongly depends on UGDF and we think the best solution is to use mixed UGDFs (KMR-KS). In our approach only small room is left for color-octet contribution.

References

[1] B. Abelev *et al.* (ALICE collaboration), Phys. Let. B **718**, 295 (2012)
[2] R. Aaij *et al.* (LHCb collaboration), Eur. Phys. J. C **71**, 1645 (2011)
[3] K. Aamodt *et al.* (ALICE collaboration), Phys. Let. B **704**, 442 (2011)
[4] B.Abelev *et al.* (ALICE collaboration), Eur. Phys. J. C **74**, 2974 (2014)
[5] R. Aaij *et al.* (LHCb collaboration), JHEP **1510**, 172 (2015)
[6] K.A. Olive *et al.* (PDG Collaboration), Chin. Phys. C **38**, 090001 (2014)
[7] B.A. Kniehl, D.V. Vasin and V.A. Saleev, Phys. Rev, D **73**, 074022 (2006)
[8] A. Cisek and A. Szczurek, a paper in preparation
[9] S.P. Baranov, A.V. Lipatov and N.P. Zotov, Eur. Phys. J. C **75**, 128 (2015)
[10] S.P. Baranov, Phys. Rev. D **66**, 114003 (2002)

Measurement of hadron cross sections with the SND detector

V. P. Druzhinin[1,2,⋆], M. N. Achasov[1,2], A. Yu. Barnyakov[1,2], K. I. Beloborodov[1,2], A. V. Berdyugin[1,2], A. G. Bogdanchikov[1], A. A. Botov[1], T. V. Dimova[1,2], V. B. Golubev[1,2], L. V. Kardapoltsev[1,2], A. G. Kharlamov[1,2], A. A. Korol[1,2], S. V. Koshuba[1], D. P. Kovrizhin[1,2], A. S. Kupich[1], K. A. Martin[1,2], A. E. Obrazovsky[1], E. V. Pakhtusova[1], S. I. Serednyakov[1,2], D. A. Shtol[1,2], Z. K. Silagadze[1,2], I. K. Surin[1,2], Yu. V. Usov[1,2], and A. V. Vasiljev[1,2]
The SND Collaboration

[1] *Budker Institute of Nuclear Physics, Novosibirsk, 630090, Russia*
[2] *Novosibirsk State University, Novosibirsk, 630090, Russia*

Abstract. New results on exclusive hadron production in e^+e^- annihilation obtained in experiments with the SND detector at the VEPP-2M and VEPP-2000 e^+e^- colliders are presented.

1 Detector and experiment

The SND [1–4] is the general purpose nonmagnetic detector. Its main part is a spherical three-layer electromagnetic calorimeter containing 1640 NaI(Tl) crystals. Directions of charged particles are measured by a tracking system based on a nine-layer drift chamber. The particle identification is provided by dE/dx measurements in the tracking system and a system of aerogel Cherenkov counters. Outside the calorimeter a muon detector is located.

SND collected data at the VEPP-2M [5] and VEPP2000 [6] e^+e^- colliders. At VEPP-2M, data with an integrated luminosity of about 30 pb^{-1} were recorded in 1996-2000 in the energy range 0.4–1.4 GeV. At VEPP2000, a wider energy interval, 0.3–2.0 GeV, is studied. A 69 pb^{-1} data sample was collected during 2010-2013. Currently the VEPP-2000 accelerator complex is under upgrade. Experiments with increased luminosity are expected to be started in the end of 2016.

The main goal of the SND experiments is careful measurement of exclusive hadronic cross sections below 2 GeV, which are, in particular, needed for the Standard Model calculation of the muon $(g-2)$ and running α_{QED}. Here we present new results on measurement of four exclusive cross sections.

2 Precise cross-section measurements

The $e^+e^- \to \pi^0\gamma$ cross section is the third largest cross section (after $e^+e^- \to \pi^+\pi^-$ and $\pi^+\pi^-\pi^0$) below 1 GeV. From analysis of the $e^+e^- \to \pi^0\gamma$ data in the vector meson dominance (VMD) model, the widths of vector-meson radiative decays are extracted, which are widely used in phenomenological

⋆e-mail: druzhinin@inp.nsk.su

models. The most accurate data on this process were obtained in experiments at the VEPP-2M e^+e^- collider with the SND [7, 8] and CMD-2 [9] detectors. The SND results [7, 8] are based on about 25% of data collected at VEPP-2M. Here we present a new analysis [10] using the full SND@VEPP-2M data sample.

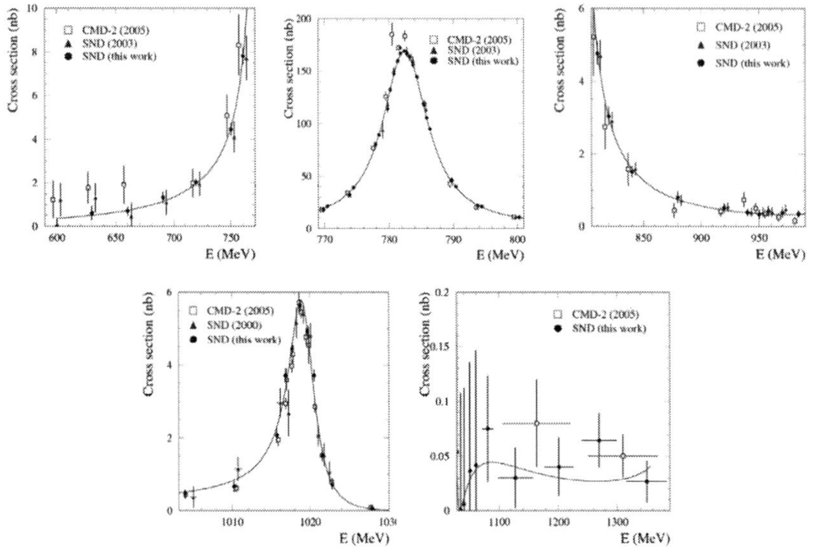

Figure 1. The $e^+e^- \to \pi^0\gamma$ cross section measured by SND using the full VEPP-2M data sample in comparison with the previous most accurate measurements. The curve is the result of the VMD fit. Only statistical errors are shown. The systematic errors are 3.2%, 3%, and 6% for SND (2000), SND (2003), and CMD-2 (2005) data, respectively. Our systematic uncertainty at the ω and ϕ peaks is 1.4%.

The measured $e^+e^- \to \pi^0\gamma$ cross section shown in Fig. 1 agrees with previous SND [7, 8] and CMD-2 [9] measurements within the systematic uncertainties, but is significantly more precise. From the fit to cross section data we obtain the products of branching fractions:

$$\begin{aligned} B(\rho \to \pi^0\gamma)B(\rho \to e^+e^-) &= (1.98 \pm 0.22 \pm 0.10) \times 10^{-8}, \\ B(\omega \to \pi^0\gamma)B(\omega \to e^+e^-) &= (6.336 \pm 0.056 \pm 0.089) \times 10^{-6}, \\ B(\phi \to \pi^0\gamma)B(\phi \to e^+e^-) &= (4.04 \pm 0.09 \pm 0.19) \times 10^{-7}, \end{aligned} \tag{1}$$

For the ω meson, we improve accuracy of the product by a factor of 1.6. There is a tension between the value of the ratio $B(\omega \to \pi^0\gamma)/B(\omega \to \pi^+\pi^-\pi^0)$ measured by KLOE [11] and the same value obtained using VEPP-2M data. After our $\pi^0\gamma$ measurement this tension increases up to 3.4σ. Our $\rho \to \pi^0\gamma$ measurement is most precise. The obtained branching fraction $B(\rho \to \pi^0\gamma) = (4.20 \pm 0.52) \times 10^{-4}$ is lower than the PDG value by 1.8σ, but agrees with the branching fraction for the charged ρ decay. The result for the $\phi \to \pi^0\gamma$ has an accuracy comparable with that of the PDG value. It is obtained in the fit, in which the relative phase between the ϕ and ω amplitudes ($\varphi_{\phi\omega}$) is taken from data on the $e^+e^- \to \pi^+\pi^-\pi^0$ process. In the fit with free $\varphi_{\phi\omega}$, $B(\phi \to \pi^0\gamma)$ is determined with about 20% uncertainty.

We also improve accuracy of $e^+e^- \to K^+K^-$ cross section measurement in the energy range 1.05–2.00 GeV [13]. Our result in comparison with most precise previous measurement by BABAR [12] is shown in Fig. 2.

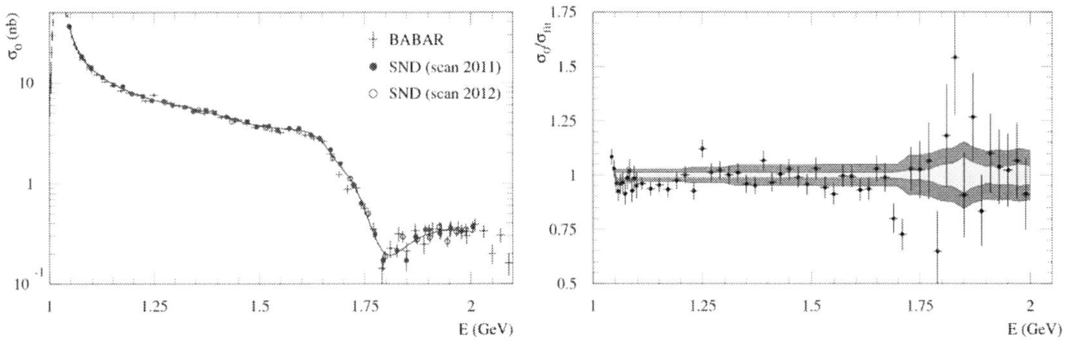

Figure 2. Left panel: The $e^+e^- \to K^+K^-$ cross section measured by SND at VEPP2000 and in the BABAR experiment [12]. The curve is the result of the VMD fit. Right panel: The relative difference between the $e^+e^- \to K^+K^-$ cross sections measured by BABAR and the fit to the SND data. The SND and BABAR systematic uncertainties are shown by the light and dark shaded bands, respectively.

3 Previously unmeasured cross sections

Relatively precise inclusive data on e^+e^- annihilation to hadrons exist above 2 GeV. Below the total hadronic cross section is calculated as a sum of exclusive cross sections. However, in the energy region between 1.5-2.0 GeV the exclusive data are incomplete. In this section we discuss two previously unstudied process.

The process $e^+e^- \to \pi^+\pi^-\pi^0\eta$ has complex internal structure. Our preliminary study show that there are at least four mechanisms for this reaction: $\omega\eta$, $\phi\eta$, $a_0(980)\rho$, and structureless $\pi^+\pi^-\pi^0\eta$. The known $\omega\eta$ and $\phi\eta$ contributions explain about 50-60% of the cross section below 1.8 GeV. Above 1.8 GeV the dominant mechanism is $a_0\rho$. The preliminary result on the $e^+e^- \to \pi^+\pi^-\pi^0\eta$ cross section is shown in Fig. 3 (left). The cross section for the subprocess $e^+e^- \to \omega\eta$ is measured separately [14] and shown in Fig. 3 (right) in comparison with the BABAR measurement [15]. Our results have better accuracy and disagree with the BABAR data at $E > 1.6$ GeV.

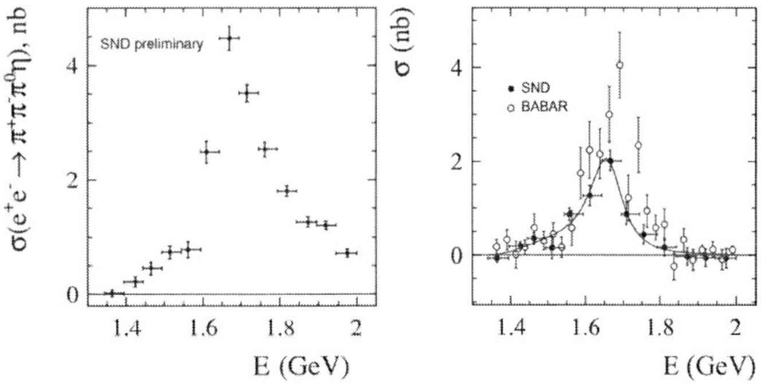

Figure 3. Left panel: The $e^+e^- \to \pi^+\pi^-\pi^0\eta$ cross section measured by SND. Right panel: The $e^+e^- \to \omega\eta$ cross section measured by SND in comparison with BABAR data [15]. The curve is the result of the VMD fit.

The process $e^+e^- \to \omega\pi^0\eta$ is studied in the seven-photon final state [16]. Events of the $e^+e^- \to \pi^0\pi^0\eta\gamma$ process are selected. The analysis of their $\pi^0\gamma$ invariant-mass distribution shows

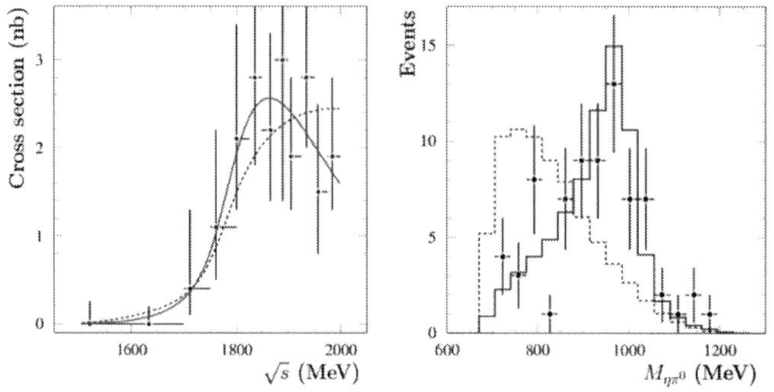

Figure 4. Left panel: The $e^+e^- \to \omega\eta\pi^0$ cross section measured by SND. The solid (dashed) curve shows the result of the fit in the model of $\omega a_0(980)$ intermediate state with (without) a resonance contribution. Right panel: The $\eta\pi^0$ invariant mass spectrum for selected $e^+e^- \to \omega\eta\pi^0$ events. The solid histogram represents $e^+e^- \to \omega a_0(980)$ simulation, while the dashed histogram represents $\omega\eta\pi^0$ phase-space simulation.

the dominance of the $\omega\pi^0\eta$ intermediate state. The measured $e^+e^- \to \omega\pi^0\eta$ cross section is shown in Fig. 4 (left). Figure 4 (right) shows the $\pi^0\eta$ mass distribution for selected $\omega\pi^0\eta$ events, which is well described by the model the of $\omega a_0(980)$ intermediate state.

Both the previously unmeasured cross sections discussed in this section give a sizable contribution ($\sim 5\%$) to the total hadronic cross section.

Acknowledgments: This work is supported in part by the RFBR grants 16-02-00327, 16-02-00014, 15-02-01037, 16-32-00542, and 14-02-00129. Part of this work related to the photon reconstruction algorithm in the electromagnetic calorimeter is supported by the Russian Science Foundation (project No. 14-50-00080).

References

[1] M. N. Achasov *et al.*, Nucl. Instrum. Methods Phys. Res., Sect. A **598**, 31 (2009)
[2] V. M. Aulchenko *et al.*, Nucl. Instrum. Methods Phys. Res., Sect. A **598**, 102 (2009)
[3] A. Y. Barnyakov *et al.*, JINST **9**, C09023 (2014)
[4] V. M. Aulchenko *et al.*, Nucl. Instrum. Methods Phys. Res., Sect. A **598**, 340 (2009)
[5] I. A. Koop *et al.*, in *Proceedings of the Workshop on Physics and Detectors for DAPHNE*, Frascati, Italy, 1999 (Frascati, 1999), p. 393
[6] A. Romanov *et al.*, in *Proceedings of Particle Accelerator Conference PAC 2013, Pasadena, CA USA, 2013*, p. 14
[7] M. N. Achasov *et al.* (SND Collaboration), Eur. Phys. J. C **12**, 25 (2000)
[8] M. N. Achasov *et al.* (SND Collaboration), Phys. Lett. B **559**, 171 (2003)
[9] R. R. Akhmetshin *et al.* (CMD-2 Collaboration), Phys. Lett. B **605**, 26 (2005)
[10] M. N. Achasov *et al.* (SND Collaboration), Phys. Rev. D **93**, 092001 (2016)
[11] F. Ambrosino *et al.* (KLOE Collaboration), Phys. Lett. B **669**, 223 (2008)
[12] J. P. Lees *et al.* (BaBar Collaboration), Phys. Rev. D **88**, 032013 (2013)
[13] M. N. Achasov *et al.* (SND Collaboration), arXiv:1608.08757 [hep-ex]
[14] M. N. Achasov *et al.* (SND Collaboration), arXiv:1607.00371 [hep-ex]
[15] B. Aubert *et al.* (BABAR Collaboration), Phys. Rev. D **73**, 052003 (2006)
[16] M. N. Achasov *et al.* (SND Collaboration), Phys. Rev. D **94**, 032010 (2016)

EPJ Web of Conferences **130**, 05005 (2016) DOI: 10.1051/epjconf/201613005005
MESON 2016

Study of coherent pion production in proton-deuteron collisions with polarized beams and target at ANKE-COSY

S. Dymov[1,2,⋆] and V. Shmakova[2]
for the ANKE Collaboration

[1] *Institut für Kernphysik, Forschungszentrum Jülich, D-52425 Jülich, Germany*
[2] *Laboratory of Nuclear Problems, Joint Institute for Nuclear Research, 141980 Dubna, Russia*

Abstract. Values of the proton analysing power, deuteron vector analysing power, and the deuteron-proton spin correlations in the $pd \to {}^3\text{He}\,\pi^0/{}^3\text{H}\,\pi^+$ reactions at 350-360 MeV per nucleon were obtained by using polarized beams and target at ANKE-COSY. Data were also obtained on the deuteron-proton spin correlation and proton analysing power at small angles at 600 MeV per nucleon. By combining the extrapolated values of the spin correlations to the forward or backward directions with published measurements of the deuteron tensor analysing powers, the relative phases between the two non-vanishing amplitudes were evaluated.

1 Introduction

Two-body pion production in the interaction of protons with few-nucleon systems is of interest, both from the point of view of studying the reaction mechanism, and from that of determining the structure of light nuclei. The success of microscopic models with explicit Δ-excitation for two-nucleon systems suggests that these models should be tested in the three-nucleon case, where production of Δ is intimately linked to 3N forces. The phenomenological approach, using impulse approximation with $pp \to d\pi^+$ cross section, as input was successful near the reaction threshold but only partial progress has been achieved at higher energies [1].

In general six invariant amplitudes are required to describe the $pd \to {}^3\text{He}\,\pi^0$ reaction and these amplitudes will be functions of the angle between the incident proton and outgoing pion in the c.m. frame. The number of independent functions reduce to two at threshold or in the forward/backward directions. These may be written as [2]

$$F(dp \to {}^3\text{He}\pi^0) = \sqrt{\frac{1}{2}}\bar{u}_\tau\, \vec{p}\cdot(A\vec{\epsilon} + iB\vec{\epsilon}\times\vec{\sigma})u_p. \quad (1)$$

Here $\vec{\epsilon}$ is the deuteron polarisation vector, $\vec{p}$ and $\vec{k}$ the proton and pion centre-of-mass momenta and u_p and u_τ the initial and final fermion spinors. The amplitude should be multiplied by a $\sqrt{2}$ factor if the $pd \to {}^3\text{H}\,\pi^+$ reaction is being considered.

⋆e-mail: s.dymov@fz-juelich.de

 DOI: 10.1051/epjconf/201613005005

If only the two amplitudes A and B are retained, the unpolarised c.m. differential cross section, deuteron tensor analysing power, and vector transverse spin correlation become

$$\begin{aligned}\frac{d\sigma}{d\Omega} &= \frac{kp}{3}(|A|^2 + 2|B|^2),\\ T_{20} &= \sqrt{2}\frac{|B|^2 - |A|^2}{|A|^2 + 2|B|^2},\\ C_{y,y} &= -\frac{2Re(A^*B)}{|A|^2 + 2|B|^2},\end{aligned} \tag{2}$$

whereas iT_{11} and T_{22} , as well as the proton analysing power A_y, must all vanish.

2 Spin correlation coefficients

The $pd \to {}^3\text{He}\,\pi^0$ and $pd \to {}^3\text{H}\,\pi^+$ reactions have been studied experimentally over many decades and a wealth of data on the differential cross sections and analysing powers has been collected for these processes. However, the double polarisation observables have been explored far less and information on the spin correlations is still very scarce. The ANKE spectrometer [3] equipped with an internal polarised target together with the polarised deuteron beam of COSY synchrotron (Jülich, Germany) offer a unique opportunity to conduct measurements of the transverse spin correlation coefficients in these reactions.

Two double-polarisation experiments have been performed at ANKE with a polarised deuteron beam and a polarised hydrogen target, at the beam energies of 363 and 600 & 1115 MeV per nucleon. The data at the two lower energies were analysed to obtain the spin correlations in the $pd \to {}^3\text{He}\,\pi^0$ reaction, and the results have recently been published [4].

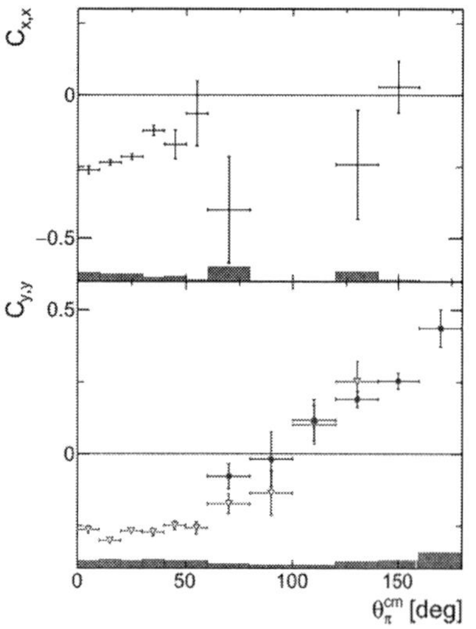

Figure 1. Transverse spin correlation coefficients $C_{x,x}$ and $C_{y,y}$ in the $\vec{d}\vec{p} \to {}^3\text{He}\,\pi^0$ and $\vec{d}\vec{p} \to {}^3\text{H}\,\pi^+$ reactions at 363 MeV per nucleon [4]. In the $C_{y,y}$ case the (red) inverted triangles were obtained through ^{3}He detection and the (blue) circles through ^{3}H detection. The shaded area indicates the systematic uncertainties in the measurement.

The spin correlations at 363 MeV/n are presented in Fig. 1. Both ^{3}He and ^{3}H production processes could be investigated at this energy and the results in the angular range of 60 – 140° covered by the both reactions are completely compatible. The ANKE detector acceptance is limited by the size of the gap in the analysing magnet D2 and is concentrated around $\phi = 0°$ and 180° regions. This leads to a better definition of the $C_{y,y}$ coefficient as compared to $C_{x,x}$.

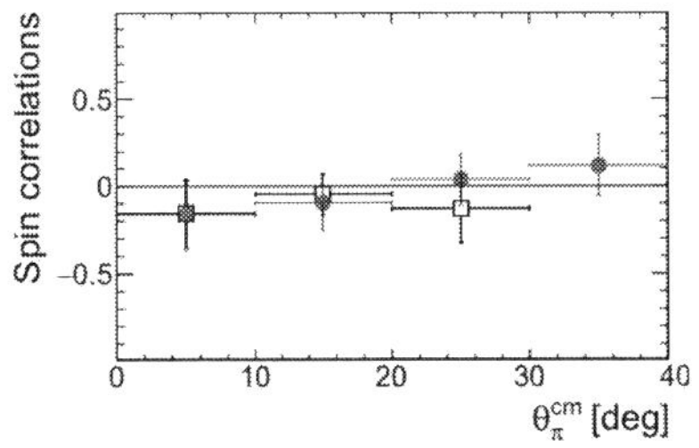

Figure 2. Transverse spin correlation coefficients $C_{y,y}$ (magenta circles) and $C_{x,x}$ (open squares) measured in the $\vec{d}\vec{p} \to {}^3\text{He}\,\pi^0$ reaction at 600 MeV per nucleon [4]. The systematic errors are below 0.03.

Since a clean selection of tritium by the energy loss in the scintillation hodoscope is no longer feasible at 600 MeV per nucleon, only data on the $pd \to {}^3\text{He}\,\pi^0$ reaction are shown at this higher energy. The $C_{x,x}$ and $C_{y,y}$ coefficients measured at small angles are presented in Fig. 2.

The deuteron tensor analysing power T_{20} was measured in the forward/backward directions at Saclay [5] and the values interpolated at 726 MeV are $T_{20}(0^\circ) = -1.01 \pm 0.01$ and $T_{20}(180^\circ) = -1.10 \pm 0.06$. However, it is easily shown from Eq. (2) that the corresponding spin correlation in the forward or backward direction is bounded by

$$\left(C_{y,y}\right)^2 \leq \tfrac{4}{9}\left[1 - T_{20}/\sqrt{2} - (T_{20})^2\right], \tag{3}$$

from which one sees quite generally that $|C_{y,y}| \leq 1/\sqrt{2}$.

Using the forward value of T_{20} measured at Saclay, Eq. (3) shows that $|C_{y,y}(0^\circ)| \leq 0.56 \pm 0.01$ to be compared to the value -0.27 ± 0.03 deduced from extrapolating the combined data of Fig. 1 to the forward direction. The error bars are larger in the backward direction where one finds from the data of Fig. 1 that $C_{y,y}(180^\circ) = +0.46 \pm 0.04$ compared to the upper bound from Eq. (3) of 0.50 ± 0.04.

The change in sign of $C_{y,y}$ between the backward and forward directions is significant. If the phase angle is defined by $\phi = \arg(B/A)$ then, in the forward direction, $\cos\phi = 0.49 \pm 0.05$ whereas in the backward direction $\cos\phi = -0.90 \pm 0.10$. The change in sign of $Re(A^*B)$ could be due to structure in either amplitude. The Saclay data indicate that this is likely to be caused by the B amplitude because $B(180^\circ)$ seems to have a zero in the vicinity of $T_d = 650$ MeV [5].

The deuteron tensor analysing power bound of Eq. (3) provides little real constraint at 1.2 GeV. Using the Saclay value of $T_{20}(0^\circ) = -0.66 \pm 0.02$ one finds that $|C_{y,y}(0^\circ)| < 0.68 \pm 0.01$ to be compared with the extrapolated value of Fig. 2, $C_{y,y}(0^\circ) = -0.10 \pm 0.08$. These mean that in the forward direction the A and B amplitudes at 600 MeV per nucleon are almost completely out of phase, with $\cos\phi = 0.14 \pm 0.12$.

3 Vector analysing powers

In addition to the spin correlation parameters, the proton and deuteron vector analysing powers can also be extracted from these data [4]. At 363 MeV/n, A_y^p can be compared to the data of Ref. [6] and to the high statistics ANKE measurement with a polarised proton beam and a deuterium cluster-jet target (see Fig. 3). Despite the smooth behaviour of the differential cross sections with angle, the rich structure in A_y^p indicates that many partial waves with different phases contribute actively at this energy.

Since the tensor polarisation of the deuteron beam was vanishingly small, the deuteron vector analysing power A_y^d could also be extracted from the data by looking at the dependence of the counting rates on the vector polarisation of the beam. The results obtained from the combined $\vec{d}p \to {}^3\text{He}\,\pi^0/{}^3\text{H}\,\pi^+$ data are shown in Fig. 4. It is perhaps significant that the abrupt change in A_y^d occurs at $\theta_\pi^{cm} \approx 80^\circ$, which is close to the deep minimum in A_y^p.

The available statistics in the 600 MeV/n case is very limited. The uncertainties in the corresponding deuteron vector analysing powers are very large and these data are not shown. However, values of the proton analysing powers could be extracted for small angles by exploiting the polarisation of the hydrogen in the target and these data are presented in Fig. 5. These results for $\theta_\pi^{cm} < 40°$ are significantly larger than those of the lower energy data shown in Fig. 3.

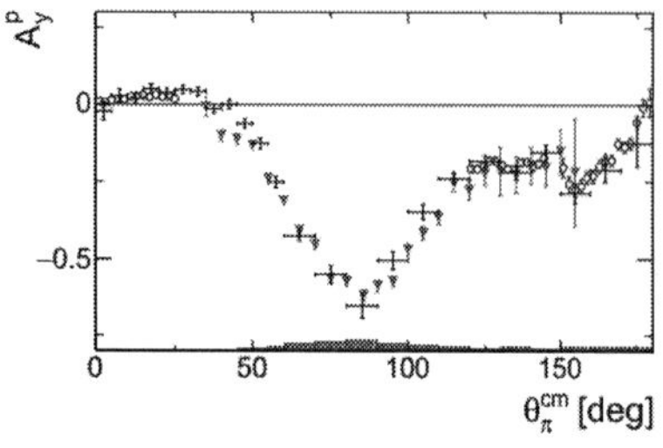

Figure 3. TRIUMF data on the proton analysing power A_y^p in the $\vec{p}d \to {}^3\text{He}\,\pi^0$ reaction at 350 MeV [6] (magenta triangles) are compared to ANKE results at 353 MeV [4] (blue open circles). We do not include here the 9% systematic error that is mainly associated with uncertainties in the COSY beam polarisation. Also shown are the ANKE values at 363 MeV per nucleon [4] deduced from measurements with a deuteron beam incident on a long cell filled with polarised hydrogen gas (black crosses). The shaded area indicates the systematic uncertainties in the deuteron beam measurements.

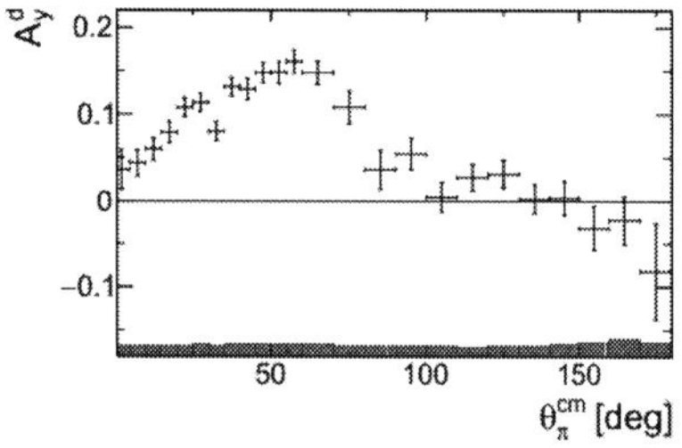

Figure 4. Deuteron vector analysing power A_y^d measured in the $\vec{d}p \to {}^3\text{He}\,\pi^0/{}^3\text{H}\,\pi^+$ reactions at 726 MeV [4] The data are presented in terms of the c.m. angle between the proton and pion. The shaded area indicates the systematic uncertainties in the measurement.

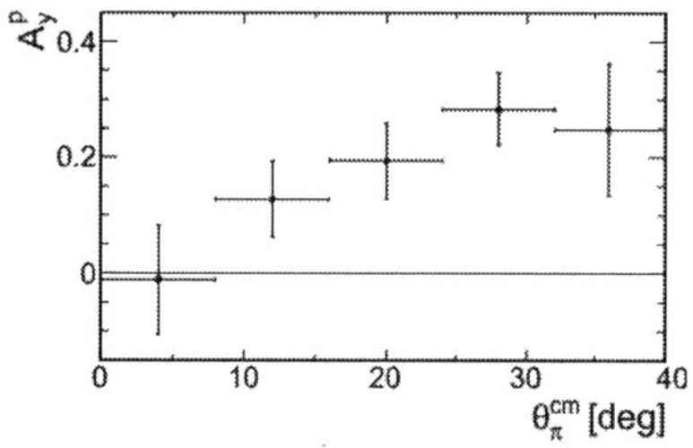

Figure 5. The proton analysing power A_y^p for the $d\vec{p} \to {}^3\text{He}\,\pi^0$ reaction at 600 MeV per nucleon [4] extracted from data obtained with a polarised hydrogen target. Systematic uncertainties, which were dominated by those in the target polarisation, were below 5%.

We are grateful to the COSY crew for providing such good working conditions, especially of the polarised beam. This work has been partially supported by the the COSY FFE programme and the Shota Rustaveli National Science Foundation.

References

[1] W. R. Falk, Phys. Rev. C **50**, 1574 (1994)
[2] J.-F. Germond, C. Wilkin, J. Phys. G **16**, 381 (1990)
[3] S. Barsov *et al.*, Nucl. Instrum. Methods A **462**, 364 (1997)
[4] S. Dymov *et al.*, Phys. Lett. B **762**, 102 (2016)
[5] C. Kerboul *et al.*, Phys. Lett. B **181**, 28 (1986)
[6] J. M. Cameron *et al.*, Nucl. Phys. A **472**, 718 (1987)

Exclusive ρ^0 meson photoproduction with a leading neutron at HERA

Lidia Goerlich[1,*]
for the H1 Collaboration

[1] *Institute of Nuclear Physics Polish Academy of Sciences, Krakow, Poland*

Abstract. A first measurement is presented of exclusive photoproduction of ρ^0 mesons associated with leading neutrons at HERA. The data were taken with the H1 detector in the years 2006 and 2007 at a centre-of-mass energy of $\sqrt{s} = 319\,\text{GeV}$ and correspond to an integrated luminosity of $1.16\,\text{pb}^{-1}$. The ρ^0 mesons with transverse momenta $p_T < 1\,\text{GeV}$ are reconstructed from their decays to charged pions, while leading neutrons carrying a large fraction of the incoming proton momentum, $x_L > 0.35$, are detected in the Forward Neutron Calorimeter. The phase space of the measurement is defined by the photon virtuality $Q^2 < 2\,\text{GeV}^2$, the total energy of the photon-proton system $20 < W_{\gamma p} < 100\,\text{GeV}$ and the polar angle of the leading neutron $\theta_n < 0.75$ mrad. The cross section of the reaction $\gamma p \to \rho^0 n \pi^+$ is measured as a function of several variables. The data are interpreted in terms of a double peripheral process, involving pion exchange at the proton vertex followed by elastic photoproduction of a ρ^0 meson on the virtual pion. In the framework of one-pion-exchange dominance the elastic cross section of photon-pion scattering, $\sigma^{el}(\gamma\pi^+ \to \rho^\circ\pi^+)$, is extracted. The value of this cross section indicates significant absorptive corrections for the exclusive reaction $\gamma p \to \rho^0 n\pi^+$.

1 Introduction

The H1 experiment at HERA has measured exclusive photoproduction of ρ^0 mesons associated with leading neutrons, $\gamma p \to \rho^0 n\pi^+$, with the aim to extract for the first time experimentally, the elastic $\gamma\pi$ cross section. Here, for the ρ meson photoproduction at a soft scale, Regge phenomenology is most appropriate to describe the reaction. In the Regge framework the process is explained by the diagram shown in Fig. 1a which involves an exchange of two Regge trajectories in a process $2 \to 3$, known as Double Peripheral Process (DPP). In this picture the pion exchange at the proton vertex is followed by elastic scattering of this pion on the virtual photon emitted from the beam electron, $\gamma\pi^+ \to \rho^0\pi^+$. The successful description of experimental data in similar reactions at lower energies has shown that in addition to the pion exchange (Fig. 1a) two further contributions (Fig. 1a, 1b) have to be included and interference between the graphs a,b, and c in Fig. 1 plays an important role [1].

The data sample corresponds to an integrated luminosity of $1.16\,\text{pb}^{-1}$ and was collected at $\sqrt{s_{ep}} = 319\,\text{GeV}$ using a special low multiplicity trigger. Exclusive events are selected containing two oppositely charged pions from the ρ^0 decay, the leading neutron and nothing else above noise

*e-mail: lidia.goerlich@ifj.edu.pl

level in the detector. The pion from the proton vertex is emitted under very small angle with respect to the proton beam and escapes detection.

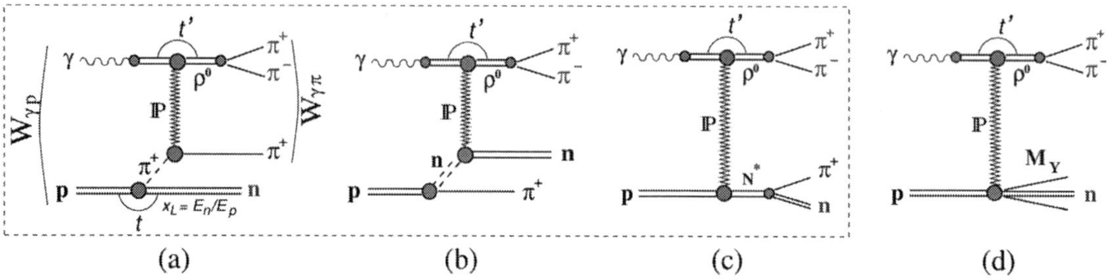

Figure 1. Generic diagrams for processes contributing to exclusive photoproduction of ρ^0 mesons associated with leading neutrons at HERA. The signal corrsponds to the Drell-Hiida-Deck model graphs for the pion exchange (a), neutron exchange (b) and direct pole (c). Background from diffractive ρ^0 production with proton dissociation (d). The N^* in (c) denotes both resonant (via N^+) and possible non-resonant $n + \pi^+$ production.

Signal events (Fig. 1a) are modelled by the Monte Carlo generator POMPYT [2], in which the virtual pion is produced at the proton vertex according to one of the available pion flux parametrisations and then the quasi-elastic scattering process, $\gamma\pi^+ \to \rho^0\pi^+$, is generated. Diffractive ρ^0 production with proton dissociation into a system containing a neutron (Fig. 1d) contributes as background to the DPP and is modelled by the DIFFVM generator [3]. This background is subtracted from the data and its fraction in the final data sample is estimated to be 0.34 ± 0.05. More details of the analysis can be found in [4].

2 Results

Fig. 2a shows the invariant mass distribution of two charged pions corrected for the mass dependent detector efficiency. The shape of the ρ^0 mass deviates from a pure Breit-Wigner resonance and its distortion due to the interference between the resonant and non-resonant $\pi\pi$ production (dashed curve in Fig. 2a) is characterised by the Ross-Stodolsky skewing paramter n_{RS}. The dependence of n_{RS} on p_T^2 of the $\pi\pi$ system measured in this analysis, as shown in Fig. 2b, is in agreement with results for elastic photoproduction of ρ^0 mesons, $\gamma p \to \rho^0 p$, obtained previously by the ZEUS Collaboration.

The measured cross section of the exclusive reaction $ep \to e\rho^0 n\pi^+$ is converted into the γp cross section using the effective photon flux of the Vector Dominance model [5]. The integrated γp cross section in the kinematic range $20 < W_{\gamma p} < 100\,\text{GeV}$, $0.35 < x_L < 0.95$ and $\theta_n < 0.75$ mrad is determined with 2% statistical and 14.6% systematic precision:

$$\sigma(\gamma p \to \rho^0 n\pi^+) = (310 \pm 6_{\text{stat}} \pm 45_{\text{sys}})\,\text{nb}. \quad (1)$$

The differential cross section $d\sigma_{\gamma p}/dx_L$ compared to the predictions based on different models for the pion flux is shown in Fig. 3. The shape of the x_L distribution is well reproduced by most of the pion flux parametrisations (Fig. 3 right). However, some models can be ruled out even on the basis of a shape comparison alone (Fig 3 left). References to pion flux models can be found in [4].

The energy dependence of the reaction $\gamma p \to \rho^0 n\pi^+$ is presented in Fig. 4 (left). The cross section drops with $W_{\gamma p}$ in contrast to the POMPYT MC expectation, where the energy dependence is driven by Pomeron exchange alone. A Regge motivated power law fit to the data, $\sigma_{\gamma p}(W_{\gamma p}) \propto W_{\gamma p}^{\delta}$, yields $\delta = -0.26 \pm 0.06_{stat} \pm 0.07_{sys}$.

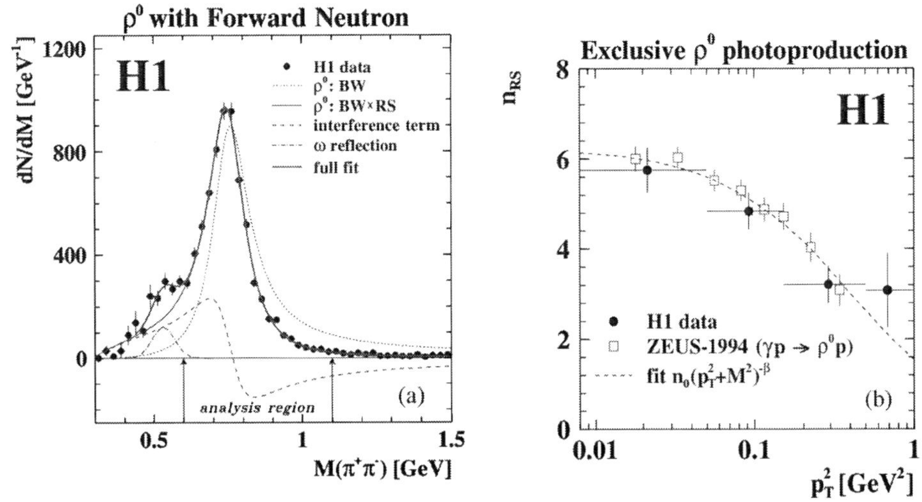

Figure 2. (a) Invariant mass distribution of two oppositely charged pions with $p_T > 0.2\,\text{GeV}$ and polar angles within $20° < \theta < 160°$. (b) Ross-Stodolsky skewing parameter, n_{RS}, as a function of p_T^2 of the $\pi^+\pi^-$ system.

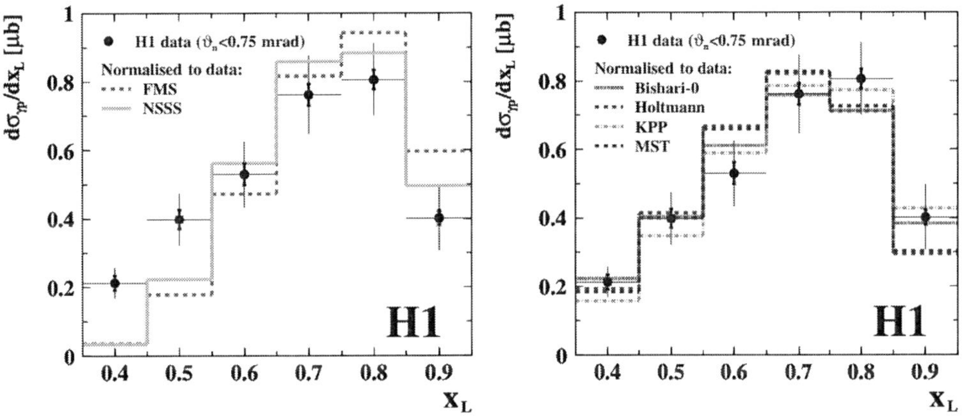

Figure 3. Differential cross section $d\sigma_{\gamma p}/dx_L$ in the range $20 < W_{\gamma p} < 100\,\text{GeV}$ compared to the predictions based on different models for the pion flux. The data points are shown with statistical (inner error bars) and total (outer error bars) uncertainties, excluding an overall normalisation error of 4.4%. All predictions are normalised to the data.

The cross section as a function of the four-momentum transfer squared of the ρ^0 meson, t', is shown in Fig. 4 (right). The t' distribution is fitted with the sum of two exponential functions with distinctly different slopes between the low-t' and the high-t' regions as expected for double peripheral processes. In a geometric picture, the large value of b_1 suggests that for a signifant part of the data ρ^0 mesons are produced at large impact parameter values of order $< r^2 > = 2b_1 \cdot (\hbar c)^2 \simeq 2\text{fm}^2 \approx (1.6\text{R}_\text{p})^2$. In this picture, photons find pions in a cloud which extends far beyond the proton radius. The small value of b_2 corresponds to a target size of 0.5 fm.

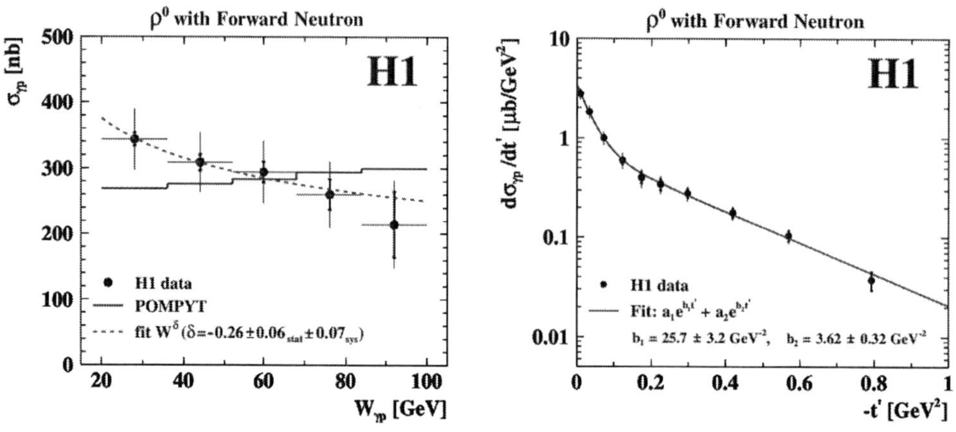

Figure 4. (a) Cross section of the reaction $\gamma p \to \rho^0 n\pi^+$ as a function of $W_{\gamma p}$ compared to the prediction from the POMPYT MC program, which is normalised to the data. (b) Differential cross section $d\sigma_{\gamma p}/dt'$ of ρ^0 mesons.

At small mometum transfer squared at the proton vertex, t, the pion exchange diagram (Fig. 1a) dominates the photoproduction cross section for exclusive ρ^0 production with a leading neutron. Thus the elastic photon-pion cross section is extracted from $d\sigma_{\gamma p}/dx_L$ in the one-pion-exchange (OPE) approximation [6], using the pion flux parametrisation of the Holtmann model [7] compatible with the data. Using the pion flux integrated in x_L and $p_{T,n}$, the elastic $\gamma\pi$ cross section is determined at an average energy $< W_{\gamma\pi} > \simeq 24\,\text{GeV}$:

$$\sigma(\gamma\pi^+ \to \rho^0\pi^+) = (2.33 \pm 34(\text{exp})^{+0.47}_{-0.40}(\text{model}))\,\mu\text{b}. \qquad (2)$$

The estimated cross section ratio for the elastic photoproduction of ρ^0 mesons on the pion and on the proton, $r_{el} = \sigma_{el}^{\gamma\pi}/\sigma_{el}^{\gamma p} = 0.25 \pm 0.06$, indicates an absorption factor of $K_{abs} = 0.44 \pm 0.11$ for the exclusive reaction $\gamma p \to \rho^0 n\pi^+$.

3 Summary

The photoproduction cross section for exclusive ρ^0 production associated with a leading neutron is measured for the first time at HERA. Differential cross sections for the reaction $\gamma p \to \rho^0 n\pi^+$ show behaviour typical for exclusive double peripheral exchange process. The elastic photon-pion cross section, $\sigma(\gamma\pi^+ \to \rho^0\pi^+)$, is extracted in the one-pion-exchange approximation. The value of this cross section indicates large absorptive corrections for the exclusive reaction $\gamma p \to \rho^0 n\pi^+$.

References

[1] S. Drell and K. Hiida, Phys. Rev. Lett. **7**, 199 (1961); R. Deck, Phys. Rev. Lett. **13**, 169 (1964)
[2] P. Bruni and G. Ingelman, *Proc. of the Europhysics Conference* (C93-07-22, Marseille) 595
[3] B. List and A. Mastroberardino, DESY-PROC-1999-02, 396 (1999)
[4] V. Andreev *et al.* [H1 Collaboration], Eur. Phys. J. C**76**, 41 (2016)
[5] T. H. Bauer *et al.*, Rev. Mod. Phys. **50**, 261 (1978)
[6] J. D. Sullivan, Phys. Rev. D**5**, 1732 (1972); V. Pelosi, Lett. Nuovo. Cim. **4**, 502 (1972)
[7] H. Holtmann *et al.*, Phys. Lett. B**338**, 363 (1994)

EPJ Web of Conferences **130**, 05007 (2016) DOI: 10.1051/epjconf/201613005007
MESON 2016

Pion and eta production in elementary and heavy-ion collisions at SIS energies

M. Gumberidze[1,★] for the HADES Collaboration

[1] *Technische Universität Darmstadt, 64289 Darmstadt, Germany*

Abstract. A main emphasis of the HADES program is on the dielectron signal from the early phase of heavy-ion collisions. We discuss the data with respect to the emission of charged hadrons, specifically the production of π^+ and π^- mesons, which are related to neutral pions representing a dominant contribution to the dielectron yield. In addition, inclusive pion production obtained with the photon-conversion method is disscused. Preliminary results from Au+Au at 1.23 AGeV are presented.

1 Introduction

Pion production is the dominating inelastic process in nucleus-nucleus collisions. At beam energies in the range of 1 - 2 GeV per nucleon pions are predominantly created via the excitation and the decay of hadron resonances. The Δ(1232) baryon resonance is most copiously produced, but as the incident energy increases, higher-lying resonances, such as N*(1440), N*(1520), N*(1535), N*(1600), etc., also contribute to pion production in a wide energy range and to production close to the threshold (pions contribute to the production of heavier mesons close to their respective thresholds). These contributions stand out clearly in the dielectron invariant mass distributions measured in elementary and heavy-ion collisions at SIS18 energies. A detailed description of the resonance and meson production in elementary hadronic collisions is indeed a crucial ingredient of transport model calculations used to 2frvb describe particle production at these energies.

2 The HADES experiment

The High Acceptance DiElecton Spectrometer (HADES), installed at SIS18 at the GSI Helmholtzzentrum für Schwerionenforschung (Germany), is a unique apparatus to study dielectron (e^+e^-) and hadron production in heavy-ion collisions, as well as in proton- and pion-induced reactions in the energy range of 1 - 4 GeV. HADES is a fixed-target experiment with large angular acceptance (18° - 85° in polar and almost full coverage in azimuthal angles). The setup comprises a Ring Imaging Cerenkov Detector (RICH) in the field-free region around the target, a low-mass tracking system, a superconducting toroidal magnet, and additional detectors for particle identification. A detailed description of the setup can be found in [1]. As of now, HADES has measured rare and penetrating probes in C+C, Ar+KCl, p+p, d+p, p+Nb, and Au+Au collisions. Measurements in p+p collisions

★e-mail: m.gumberidze@gsi.de

at various beam energies (1.25, 2.2, 3.5 GeV) and d+p (at 1.25 GeV) provided valuable constraints on the various sources contributing to the dielectron signal and allowed to obtain model-independent reference spectra for studies of cold (p+A collisions), and hot and dense matter effects (A+A collisions).

3 Charged pion reconstruction

The interpretation of the dilepton spectra relies on a precise knowledge of the underlying hadronic cocktail composed of components from various sources contributing to the differential yields. The neutral mesons, and foremost the π^0 constitute the source of the bulk of the detected dielectron pairs, namely via the π^0 Dalitz and $\gamma\gamma$ decays. The number of π^0 produced in the reaction can be estimated as an average of π^+ and π^- yields, due to isospin symmetry. It serves as an input for the normalization of the dielectron data obtained in the same experiment.

Identification of charged pions is done based on velocity of particles detected in the time-of-flight detectors as a function of momentum. Reconstructed yields of pions were corrected for efficiency and purity. Preliminary transverse-mass spectra of negatively charged pions in different intervals of rapidity are shown in Fig. 1. Experimental data (colored points) have been fitted in each rapidity bin using the sum of two exponential functions (colored lines) in order to extrapolate down to transverse mass 0. Statistical errors are neglegible thanks to our large data sample (plotted spectra correspond to only 4% of the collected data !). Systematic errors are estimated to be of the order of 5%.

Thanks to the wide acceptance of the present experiment we will be able to do a direct comparison of our Au+Au at 1.23 AGeV data with the results of previous experiments e.g. FOPI [2].

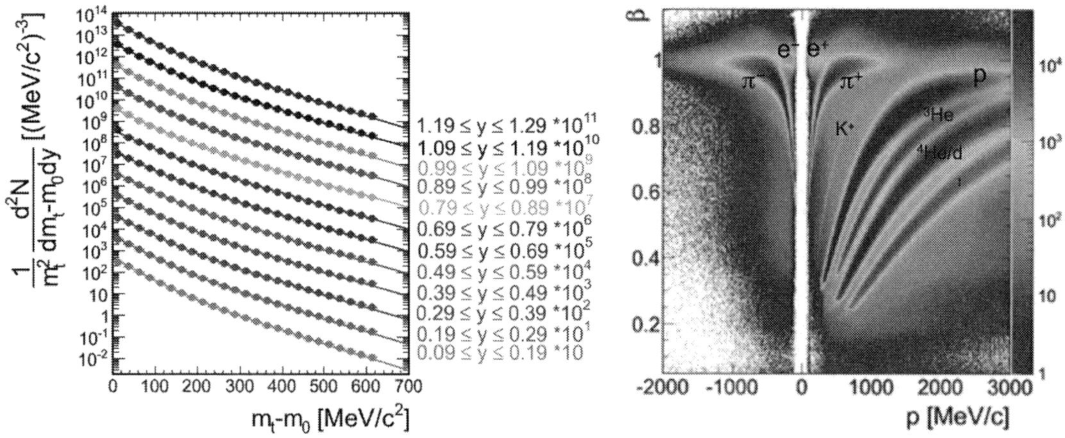

Figure 1. (Color online) Transverse-mass distributions for negatively charged pions in different slices of rapidity (left hand-side). Velocity vs. charge x momentum of charged particles as seen by the HADES detector (right hand-side).

4 Full conversion method

The contribution of the neutral mesons to the dilepton mass spectra is given by their Dalitz decays. However full reconstruction of the meson requires measuring also the γ from Dalitz decays. Furthermore, π^0 and η mesons decay dominantly into two photons, and this channel can be studied as well. As HADES is presently not equipped with an electromagnetic calorimeter direct measurement of photons is not possible. As an alternative, a γ-conversion technique in which meson decay photons are detected via their external conversion into an e^+e^- pair has been used (for details see [3]). In this case meson reconstruction is done by identifing lepton tracks and joining them into opposite-sign pairs (e^+e^-), and by further combining those dilepton pairs into $e^+e^-e^+e^-$ multiplets.

In order to suppress combinations of uncorrelated leptons, several topological cuts are applied on the opening angle between the leptons and the reconstructed pairs. Calculating the invariant mass $M_{e^+e^-e^+e^-}$ and setting appropriate mass cuts allows to select the π^0 and η mesons, respectively. Phase space coverage of π^0 as a function of transverse momentum and rapidity is shown in Fig. 2. As one can see, in rapidity we are covering the mid-rapidity region. The cutoff of the π^0 yield at $p_\perp < 0.2$ GeVc is caused mostly by the strong bending of low-momentum tracks in the HADES magnetic field.

The resulting $M_{e^+e^-e^+e^-}$ distribution from Au+Au at 1.23 AGeV is shown in Fig. 3. A prominent peak appears at the mass of π^0 meson on top of a combinatorial background, which can be well described by event-mixed continuum (see Fig. 3).

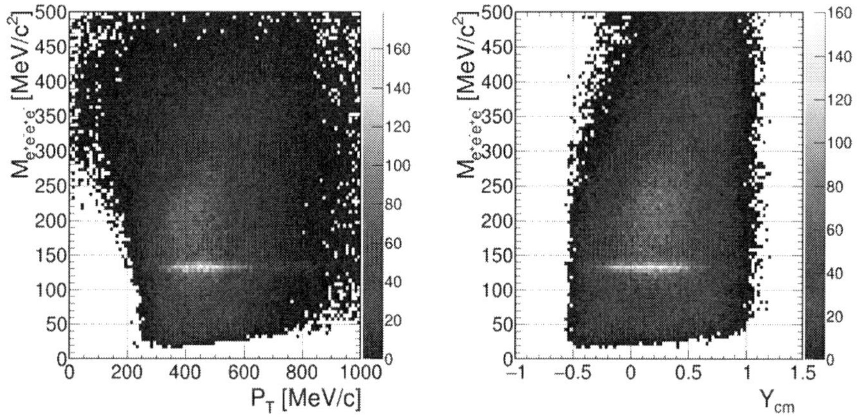

Figure 2. Phase space coverage of four-lepton $e^+e^-e^+e^-$ multiplets.

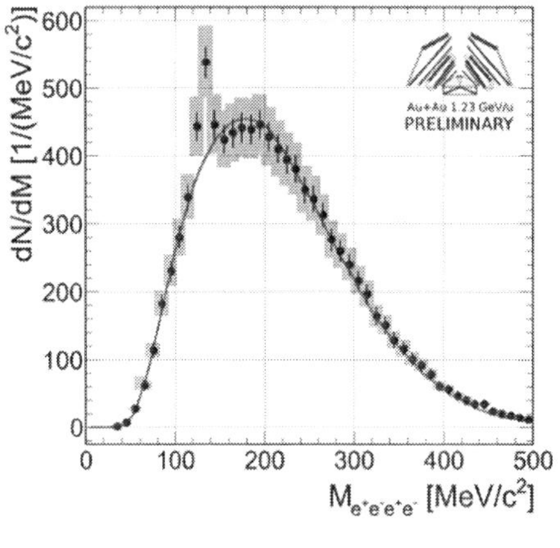

Figure 3. Invariant mass distribution of all $e^+e^-e^+e^-$ multiplets measured with HADES in the Au+Au at 1.23 AGeV reaction. The background of uncorrelated combinations obtained with event mixing is shown as well (red points).

5 Summary

HADES has measured charged pion yields in Au+Au collisions at 1.23 AGeV with unprecedented statistics. The external conversion method allowed to also measure neutral mesons, despite the lack of dedicated photon detector. This provides a necessary baseline for further analysis of the electromagnetic radiation originating from the hot and dense hadronic medium formed in these reactions. Furthermore, the high-quality charged pion data will allow to provide results on collective and correlation effects in the future.

In the future an electromagnetic calorimeter will be installed at HADES and it will largely improve the reconstruction of photons and meutral mesons.

This work has been supported by TU Darmstadt: VHNG823, Helmholtz Alliance HA216/EMMI and GSI.

References

[1] G. Agakishiev *et al.* (HADES Collaboration), Eur. Phys. J. A **41**, 243 (2009)
[2] W. Reisdorf *et al.* (FOPI Collaboration) Nucl. Phys. A **781** 459-508 (2007)
[3] G. Agakishiev *et al.* (HADES Collaboration), Phys. Rev. C **88**, 024904 (2013)

EPJ Web of Conferences **130**, 05008 (2016) DOI: 10.1051/epjconf/201613005008
MESON 2016

Amplitude analysis of resonant production in three pions

Andrew Jackura[1,⋆], Mikhail Mikhasenko[2], and Adam Szczepaniak[1,3]
for the JPAC Collaboration

[1] *Indiana University, Bloomington IN, U.S.A.*
[2] *Universität Bonn, Helmholtz-Institut für Strahlen- und Kernphysik, Bonn, Germany*
[3] *Thomas Jefferson National Accelerator Facility, Newport News VA, U.S.A.*

Abstract. We present some results on the analysis of three pion resonances. The analyses are motivated by the recent release of the largest data set on diffractively produced three pions by the COMPASS collaboration. We construct reaction amplitudes that satisfy fundamental S-matrix principles, which allows the use of models that have physical constraints to be used in fitting data. The models are motivated by the isobar model that satisfy unitarity constraints. The model consist of a Deck production amplitude with which final state interactions are constrained by unitarity. We employ the isobar model where two of the pions form a quasi-stable particle. The analysis is performed in the high-energy, single Regge limit. We specifically discuss the examples of the three pion $J^{PC} = 2^{-+}$ resonance in the $\rho\pi$ and $f_2\pi$ channels.

1 Introduction

We aim to study the peripheral resonance production of 3π systems. Motivated by the large data set at COMPASS, we construct a reaction amplitude model for $\pi^- p \to \pi^-\pi^-\pi^+ p$ at high energies. The model will be used in a mass-dependent analysis of the COMPASS partial wave intensities in [1]. The COMPASS results provide high statistics, high purity data which allows for a detailed analysis. The Joint Physics Analysis Center (JPAC) is affiliated with COMPASS to perform these analysis.

The model is constructed to satisfy fundamental principles of S-matrix theory. We impose unitarity on the reaction amplitudes, and ensure that the model obeys the proper analytic constraints required by causality. We emphasize the production process as an important role in determining physics. The $\pi^-\pi^-\pi^+ \equiv (3\pi)^-$ system is produced by a particle exchange in between the incident pion and proton, as well as an exchange between the incident pion and the reggeon that is exchanged in the full reaction. This production process is known as the Deck production process, and has been studied as possible explanations for various resonances [2–4].

Peripheral production is advantageous as it allows for the factorization of the amplitude into two parts, one being the effective $2 \to n$ meson-Reggeon scattering amplitude, where n is the number of produced mesons. We develop the model for any J^{PC} quantum number, but focus first on the $J^{PC} = 2^{-+}$ sector. The 2^{-+} sector gives us the opportunity to study the long standing puzzle of the $\pi_2(1670) - \pi_2(1880)$ interplay. These mesons have the exact same quantum numbers, but are only ~ 200 MeV apart, which disagrees with Regge trajectories of the mesons.

⋆e-mail: ajackura@indiana.edu,

2 Formalism

The reaction model exhibits numerous features. We want to model the $\pi^- p \to (3\pi)^- p$ reaction in the high energy regime to compare to COMPASS. At high energies, the reaction is dominated by pomeron exchange. At high energies, we can use factorization and effectively separate the nucleon-pomeron-nucleon vertex from the the $\pi^- \mathbb{P} \to \pi^- \pi^- \pi^+$ amplitude. We work in the quasi-two body limit, where the $\pi^- \pi^+$ mesons form a quasi-stable neutral isobar of spin S denoted I_S^0. This leads to the factorization $\pi^- \mathbb{P} \to I_S^0 \pi^-$ and $I_S^0 \to \pi^- \pi^+$. We model the pomeron phenomenologically as having a vector coupling since the Regge pole for the pomeron is $\alpha_{\mathbb{P}} \approx 1$. The amplitude for the total reaction is denoted by $A(s, \Omega, \Omega')$, where s is the invariant mass-squared of the $(3\pi)^-$ system, Ω and Ω' are the orientations of the isobar and one of the pions in the isobar respectively, and all other kinematic variables are suppressed for notational convenience. The partial wave decomposition is

$$A(s, \Omega, \Omega') \sim \sum_i F_i(s) Z_i(\Omega, \Omega'), \tag{1}$$

where $F_i(s)$ are the partial wave amplitudes for the ith channel, and $Z(\Omega, \Omega')$ contains all the rotational dependence of the system for each channel. The channel index i indicates the channel with $(3\pi)^-$ quantum numbers $I^G(J^{PC})M^\epsilon \; I_S^0 \pi \; L$, where I is the total isospin, G is the G-parity, J is the total angular momentum, P is the parity, C is the C-parity, M is the total spin projection, ϵ is the reflectivity, and L is the angular momentum between the isobar I_S^0 and the spectator pion. For the $(3\pi)^-$ system, $I = 1, G = -$, and $C = +$ always, and for pomeron exchange $\epsilon = +$.

2.1 Unitarity and analyticity

Unitarity of the S-matrix puts constraints on the form of amplitudes. We impose these constraints by requiring the partial wave amplitudes satisfy

$$\Delta F_i(s) = 2i \sum_j t_{ij}^*(s) \rho_j(s) F_J(s) \tag{2}$$

$$\Delta t_{ij}(s) = 2i \sum_k t_{ik}^*(s) \rho_k(s) t_{kj}(s), \tag{3}$$

where $t_{ij}(s)$ is the rescattering amplitude of the isobar and spectator pion, $\rho_j(s)$ is the quasi-two body phase space, and Δ indicates the discontinuity of the amplitude across the unitarity cut. It is assumed the $F_i(s)$ has two cuts, the unitarity cut from the 3π threshold to ∞, and a left hand cut generated by the production process. We can write a dispersive integral for Eq (2) which has a solution

$$F_i(s) = b_i(s) + \sum_j t_{ij}(s) c_j + \sum_j t_i(s) \int_{s_{th}}^{\infty} ds' \, \frac{\rho_j(s') b_j(s')}{s' - s}. \tag{4}$$

Here, $b_i(s)$ is the projection of the Deck production amplitude, c_j is a vector of couplings related to the short range production of the three pions, and the last term is needed to satisfy unitarity, often called the unitarized Deck amplitude. Note that in our amplitude, $b_i(s)$ is a model, but parameters c_j will be fit parameters.

We must parameterize the rescattering amplitude in such a way to satisfy Eq. (3). We choose to use the K-matrix parameterization

$$[t^{-1}]_{ij}(s) = [K^{-1}]_{ij}(s) + I_i(s)\delta_{ij}, \tag{5}$$

where $K_{ij}(s)$ is a real-symmetric matrix, and $I_i(s)$ is the Chew-Mandelstam phase space factor satisfying $\text{Im}\, I_i(s) = -\rho_i(s)$. In this model, it is assumed that $t_{ij}(s)$ has only the unitarity cut. It is also assumed that all resonance content comes from this amplitude. therefore, we must parameterize the K-matrix such that we generate poles on the unphysical Riemann sheets. We try various parameterizations of the K-matrix, and look for the corresponding resonance in systematic studies. The fit parameters are the parameters in the K-matrix, as well as the production vector c_j.

2.2 Production mechanism

The closest left hand cut to the physical region comes from π-exchange in the $\pi^-\mathbb{P} \to I_S^0\pi^-$ amplitude. The production process is modeled by the amplitude

$$A_{Deck}(s,\Omega) = \frac{g_{I\pi\pi}g_{\mathbb{P}\pi\pi}}{t(s,\theta) - m_\pi^2}\epsilon_\lambda \cdot p_2\epsilon_{\lambda'}^{S*} \cdot \{p_a\}, \tag{6}$$

where $g_{I\pi\pi}$ and $g_{\mathbb{P}\pi\pi}$ are couplings for $\pi\pi$ to the isobar and pomeron respectively, $t(s\theta)$ is the invariant momentum transfer between the incident pion and the pomeron, and ϵ_λ and $\epsilon_{\lambda'}^S$ are the polarization tensors of the pomeron and isobar respectively. The four-momenta p_a and p_2 are of the incident and spectator pion respectively. Note the symbol $\{p_a\}$ indicates that there will be S contractions of the momentum to the polarization tensor. The couplings are taken from known data on various reactions involving these processes. Therefore, the Deck amplitude is completely determined, and is used as an input for our model. The partial wave projection of the Deck amplitude $b_i(s)$ is given by projecting out A_{Deck} with the appropriate rotational dependencies.

3 Current progress and future work

At the current state, the formalism has been completed and systematic tests and preliminary fits are underway. It is proposed that the model is first fitted to the partial wave intensities of the 2^{-+} sector from the data gathered by the COMPASS collaboration. Once final fits are obtained, the resonance content will be extracted from the analytically continued amplitudes. We construct intensity distributions

$$I_i(s) = \rho_i(s)|F_i(s)|^2, \tag{7}$$

which are fit to the partial wave intensity distributions found in [1]. We also fit relative phase information between different channels.

3.1 Resonance pole extraction

Resonance content from amplitudes is determined by looking for poles of $t_{ij}(s)$ on the unphysical Riemann sheets. Once the parameters are fitted to the COMPASS partial waves, the rescattering amplitude is analytically continued to the second sheet and we look for resonances. The amplitude on the second sheet is $[t_{\text{II}}^{-1}]_{ij}(s) = [t^{-1}]_{ij}(s) - 2i\rho_i(s)\delta_{ij}$. We need to discuss some complications due to the quasi two-body phase space factor. The quasi two-body phase space factor contains the isobar information, which leads to cuts on the unphysical sheet known as Woolly cuts. Resonance may appear under these cuts, which requires the knowledge of how to continue the amplitude below these cuts. Fig. 1 illustrates this with a simple resonance example. The typical two-body phase space factor is given by $\rho \sim 2k/\sqrt{s}$, where k is the relative momentum between the isobar and the spectator pion. If the isobar decays, the phase space is modified to the quasi two-body phase space ρ_Q,

$$\rho_Q \sim \int ds'\rho(s')\,\text{Im}\, f(s'), \tag{8}$$

where ρ is the typical two-body phase space, and $f(s)$ is the isobar shape.

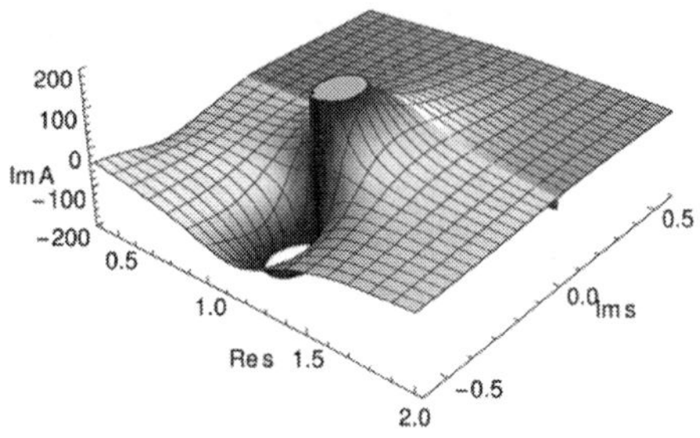

(a) Two-body phase space

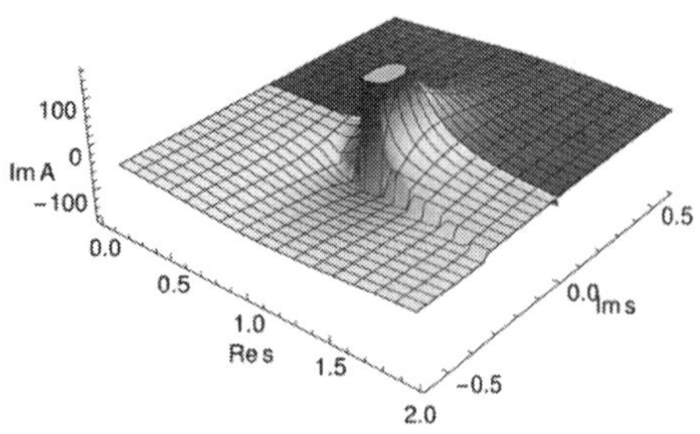

(b) Quasi two-body phase space

Figure 1. Complex s-plane of $\mathrm{Im}\,A$, where A is an amplitude with some resonance. The green section is the first sheet while the yellow section is the second sheet. In (a), we show what the typical pole structure is in the case of a simple two-body phase space factor. In (b), additional cuts appear due to the unstable isobar. Notice that the resonance pole is hidden underneath the cut.

4 Summary

We have developed the formalism to analyze 3π system for peripheral reactions. Our model satisfies the unitarity and analyticity requirements of S-matrix principles, allowing for more rigorous extraction of resonances. We are applying the formalism to the COMPASS results on $\pi^- p \to (3\pi)^- p$ in the 2^{-+} sector. As of the writing of this document, the formalism has been completed, and systematic fits are ongoing. Once fits are complete, we will search for resonances of the $(3\pi)^-$ system. In future studies, we can apply the model to other J^{PC} sectors, and it is of interest to extend the model for photon beams to be used for the JLab12 upgrade.

This work was supported in part by U.S. Department of Energy, Office of Science, Office of Nuclear Physics under contract DE-AC05-06OR23177 and by the U.S. Department of Energy under Grant No. DE-FG0287ER40365.

References

[1] C. Adolph *et al.* (COMPASS Collaboration) [arXiv:1509.00992]
[2] G. Ascoli, L. M. Jones, B. Weinstein and H. W. Wyld, Phys. Rev. D **8**, 3894 (1973)
[3] J. L. Basdevant and E. L. Berger, Phys. Rev. Lett. **114**, no. 19, 192001 (2015)
[4] J. L. Basdevant and E. L. Berger, Phys. Rev. D **16**, 657 (1977)

EPJ Web of Conferences **130**, 05009 (2016) DOI: 10.1051/epjconf/201613005009
MESON 2016

Photoproduction of vector mesons: from ultraperipheral to semi-central heavy ion collisions

Mariola Kłusek-Gawenda[1,⋆] and Antoni Szczurek[1,2]

[1] *Institute of Nuclear Physics, Polish Academy of Sciences, Kraków, Poland*
[2] *University of Rzeszów, Rzeszów, Poland*

Abstract. We discuss nuclear cross sections for $AA \to AAV$ and $AA \to AAVV$ reactions with one or two vector mesons in the final state. Our analysis is done in the impact parameter space equivalent photon approximation. This approach allows to consider the above processes taking into account distance between colliding nuclei. We consider both ultraperipheral and semi-central collisions. We are a first group which undertook a study of single J/ψ photoproduction for different centrality bins. We show that one can describe new ALICE experimental data by including geometrical effects of collisions in the flux factor.
Next, total and differential cross section for double-scattering mechanism in the exclusive $AA \to AAVV$ reaction in ultrarelativistic ultraperipheral heavy ion collisions is presented. In this context we consider double photoproduction and photon-photon processes. Simultaneously, we get very good agreement of our results with STAR (RHIC), CMS and ALICE (LHC) experimental data for single ρ^0 and J/ψ vector meson production. A comparison of our predictions for exclusive four charged pions production is also presented.

1 Introduction

In last years there was an interest in calculating cross sections for exclusive production of vector mesons in ultrarelativistic heavy ion collisions [1–10]. Purely ultraperipheral collisions (UPC) of heavy ions mean the situation when the nuclei do not collide, however, an emission of extra neutrons caused by additional purely electromagnetic interactions could be included. In theoretical calculations UPC means in practice $b > R_A + R_B$ (sum of nuclear radii). All the calculations must be therefore performed in the impact parameter space to include this condition.

We discuss an importance of photoproduction process in the context of J/ψ production in less peripheral collisions. Then, at high energies, nuclei can collide and break apart producing quark-gluon plasma. At low transverse momenta and small multiplicities the ALICE Collaboration [11] observed an enhancement of nuclear modification factor at small transverse momenta $R_{AA} > 1$.

Nuclear cross sections are calculated with the help of equivalent photon approximation (EPA) in the impact parameter space. We include realistic form factor of nuclei which is Fourier transform of the charge distribution in nucleus. The differences of the cross sections with realistic and monopole form factor can be found e.g. in Ref. [12].

⋆e-mail: mariola.klusek@ifj.edu.pl,

2 Some theoretical aspect

Fig. 1 depicts the difference between ultraperipheral and semi-central collisions of heavy ions. The left panel shows UPC for production of vector mesons. The two-dimensional vector $\vec{b}$ is a distance between colliding nuclei, $\vec{b_1}$ is a distance between the photon position and the middle of the first (emitter) nucleus and $\vec{b_2}$ is a distance between the photon position and the middle of the second (medium) nucleus. The production of the vector meson may occur provided the photon (hadronic fluctuation) hits the second nucleus. The hatched area of overlapping nuclei in the right panel (semi-central collision) of the figure represents the area in the impact parameter space for which some absorption may be expected. We include the effect of the "absorption" by modifying effective photon fluxes in the impact parameter space by imposing several additional geometrical conditions on impact parameters (between photon and nuclei and/or between colliding nuclei).

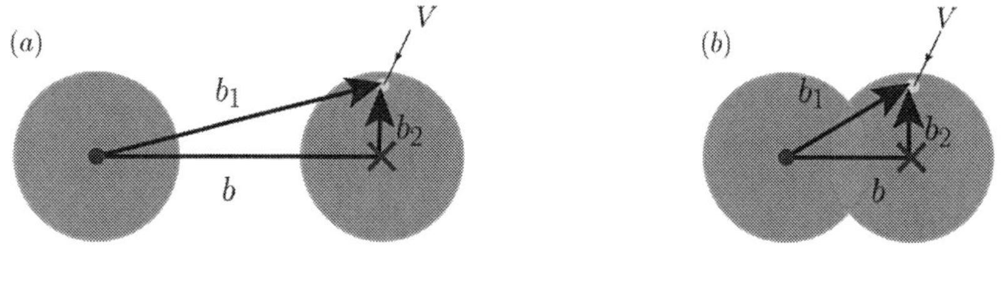

Figure 1. Impact parameter picture of the collision and the production of the vector meson for ultraperipheral (left panel) and for semi-central (right panel) collisions. It is assumed here that the first nucleus is the emitter of the photon which rescatters then in the second nucleus being a rescattering medium.

The left panel of Fig. 2 shows a single vector meson production mechanism. Photon emitted from a nucleus fluctuates into hadronic or quark-antiquark components and converts into an on-shell meson. The differential cross section for this mechanism can be written as

$$\frac{d\sigma_{A_1A_2\to A_1A_2V}}{d^2b dy} = \frac{dP_{\gamma\mathbb{P}}(b,y)}{dy} + \frac{dP_{\mathbb{P}\gamma}(b,y)}{dy} . \quad (1)$$

$P_{\gamma\mathbb{P}}(y,b)/P_{\mathbb{P}\gamma}(y,b)$ are the probability densities for producing a vector meson at rapidity y for fixed impact parameter b (distance between two colliding nuclei) of the heavy ion collision. Each probability is the convolution of a flux of equivalent photon and the $\gamma A \to VA$ cross section.

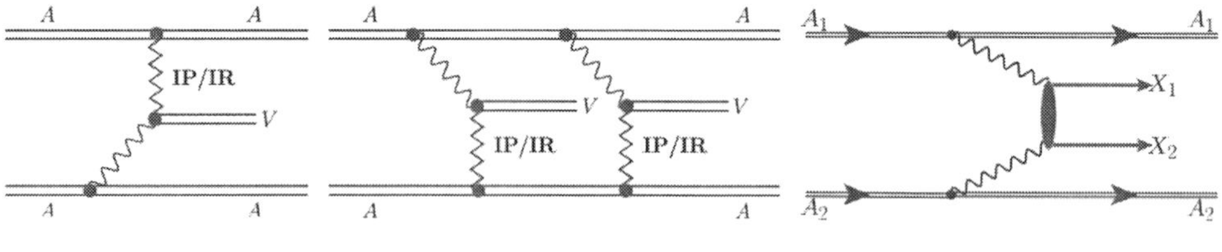

Figure 2. Single vector meson production by photon-Pomeron (or Pomeron-photon) fusion (left panel) and double vector meson production for double-scattering mechanism (middle panel) and for $\gamma\gamma$ fusion (right panel).

The middle panel of Fig. 2 illustrates one of four mechanisms of the double-scattering mechanism (in fact, we take into account all four combinations of $\gamma\mathbb{P}$ exchanges: $\gamma\mathbb{P}-\gamma\mathbb{P}$, $\gamma\mathbb{P}-\mathbb{P}\gamma$, $\mathbb{P}\gamma-\gamma\mathbb{P}$, $\mathbb{P}\gamma-\mathbb{P}\gamma$). Having a formalism for the calculation of single-vector-meson production, one can use

 DOI: 10.1051/epjconf/201613005009

it to calculate cross section for double-scattering mechanisms of two-vector-meson production. The cross section for the double vector meson photoproduction is expressed with the help of probability density of single meson production

$$\frac{d\sigma_{A_1A_2\to A_1A_2VV}}{dy_1 dy_2} = \frac{1}{2}\int d^2b \left[\left(\frac{dP_{\gamma\mathbb{P}}(b,y_1)}{dy_1} + \frac{dP_{\mathbb{P}\gamma}(b,y_1)}{dy_1}\right)\times\left(\frac{dP_{\gamma\mathbb{P}}(b,y_2)}{dy_2} + \frac{dP_{\mathbb{P}\gamma}(b,y_2)}{dy_2}\right)\right] . \quad (2)$$

For completeness, in the right panel of Fig.2, we show Feynman diagram for pair of vector meson production in the $\gamma\gamma$ fusion. A formula for the total cross section for this mechanism can be written in a somewhat simplified way as

$$\sigma_{A_1A_2\to A_1A_2VV} = \int d\omega_1 d\omega_2 n(\omega_1)n(\omega_2)\sigma_{\gamma\gamma\to VV} , \quad (3)$$

where $\omega_{1/2}$ is the photon energy from first/second nucleus. More details referring to Eq. 3 can be found in Ref. [8].

3 Theoretical results versus experimental data

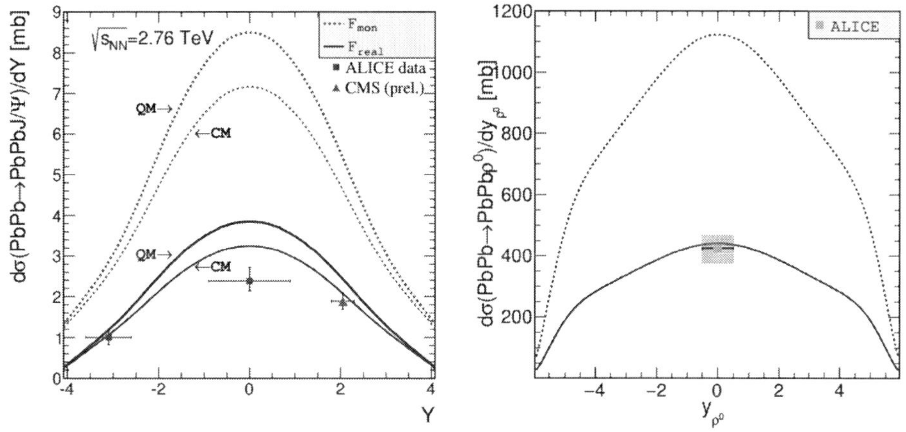

Figure 3. Left panel - differential cross section for coherent production of J/ψ meson in UPC as a function of rapidity of the J/ψ meson compared with the ALICE and CMS data points. Results are shown for both realistic and monopole form factor. Right panel - differential cross section for single ρ^0 photoproduction in UPC as a function of meson rapidity together with the ALICE experimental data point.

We start from purely ultraperipheral collisions. Fig. 3 shows results for coherent J/ψ (left panel) and ρ^0 (right panel) photoproduction in the lead-lead UPC at the LHC energy. We show our results for classical and quantal rescattering (more details in Ref. [10]) in the nucleus-medium and for realistic and monopole nuclear form factors. The result with the monopole form factor overestimates the ALICE [14, 15] and CMS [13] data. The result with the quantal rescattering is a little larger than that for the classical rescattering. However, this difference is much smaller for J/ψ production than for the photoproduction of ρ^0 meson. the ALICE data point (Ref. [16]) for ρ^0 meson production (right panel) is much better described by the classical rescattering calculations (lower line).

Fig. 4 shows differential cross section for the $AA \to AA\rho^0\rho^0$ reaction as a function of one ρ^0 meson rapidity and a comparison of the double-scattering and $\gamma\gamma$ fusion contributions at RHIC (left panel) and at LHC (right panel) energies. The double-scattering component dominates over the $\gamma\gamma$ one. In addition, the distribution in rapidity for the RHIC energy is much narrower than that for LHC energy. The high-energy (VDM-Regge) component of the elementary cross section dominates over the low-energy bump contribution at the LHC energy. This is caused by the fact that at the higher center-of-mass energy the higher values of two-meson invariant mass becomes more important which corresponds to larger values of particle rapidity. Both at the RHIC and LHC energy, the contribution of the $\gamma\gamma$ fusion is one order of magnitude smaller than that of the double-scattering mechanism.

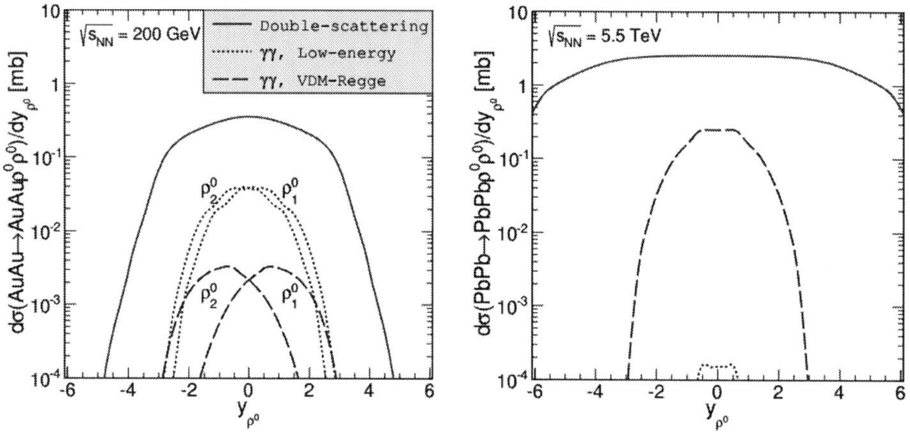

Figure 4. Rapidity distribution of one of the ρ^0 meson produced in the double-scattering mechanism and in the $\gamma\gamma$ fusion at the RHIC (left panel) and at the LHC (right panel) energy.

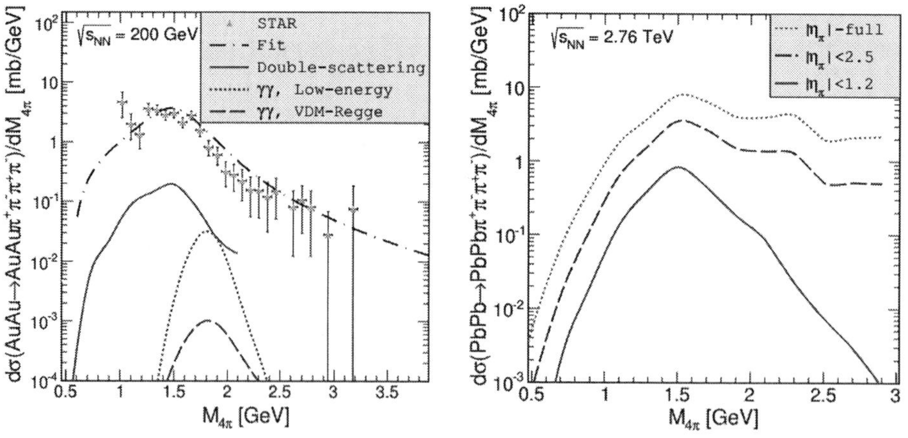

Figure 5. Invariant mass of four charged pions produced in the double-scattering mechanism and in the $\gamma\gamma$ fusion of two ρ^0 mesons at the RHIC (left panel) and at the LHC (right panel) energy.

As in previous figures, Fig. 5 shows contributions from double-scattering, the low-energy bump and the high-energy VDM-Regge $\gamma\gamma$ fusion contributions [6] but now as a function of four-pion invariant mass. The left panel of Fig. 5 shows results for the limited acceptance of the STAR experiment ($|\eta_\pi| < 1$) [17]. The dash-dotted line represents a fit of the STAR Collaboration. The double-scattering contribution accounts only for 20% of the cross section measured by the STAR Collaboration. The STAR experimental data have been corrected by experimental acceptance [17]. Wu suppose that the production of the $\rho^0(1450)$ or $\rho^0(1700)$ resonance and their subsequent decay into the four-pion final state is the dominant effect for the limited STAR acceptance. However, the production mechanism of this two broad resonances and their decay into four charged pions are not yet understood. A model production of the resonances and their decay has to be study in the future. The right panel of Fig. 5 shows four-pion invariant mass distribution for double-scattering mechanism for different limited range of pion pseudorapidity at the LHC energy. The ALICE group collected the data for four-charged-pion production with the limitation $|\eta_\pi| < 1.2$. We cannot compare this distribution with the ALICE data, because those data are not yet absolutely normalized.

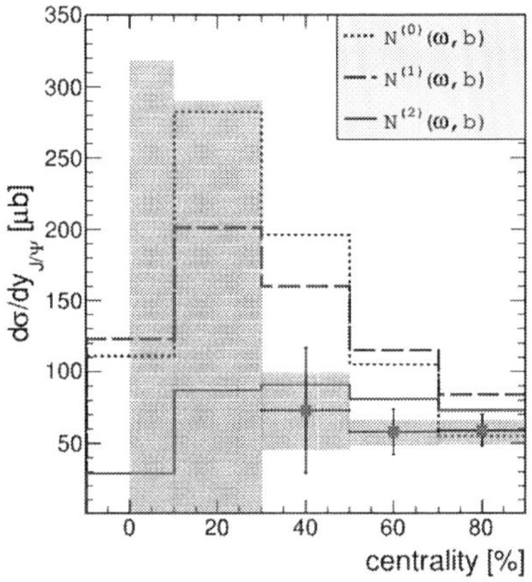

Figure 6. $d\sigma/dy$ cross sections for $PbPb \to PbPbJ/\psi$ reaction for different centrality bins. Theoretical results for three forms of photon fluxes [10] are compared with the ALICE data [18]. The shaded area represents the experimental uncertainties.

Going from peripheral to semi-central collisions we assume that the whole nucleus produces photons and the photon (or hadronic photon fluctuation) must hit the other nucleus to produce the vector meson. In Fig. 6 we show differential nuclear cross section for single J/ψ production as a function of centrality for its small values ($b < R_A + R_B$) i.e. for the semi-central collisions. We present new experimental ALICE data [18] with statistical and systematic error bars (shaded area). The ALICE Collaboration could not extract actual values of the cross section for the two lowest centrality bins. The results for standard photon flux (see Eq. (2.3) in Ref. [10]) exceed the ALICE data. Rather good agreement with the data is achieved for the $N^{(2)}$ photon flux (see Eq. (2.5) in Ref. [10]) obtained with the realistic nucleus form factor. In this approximation we integrate the photon flux of the first (emitter) nucleus only over this part of the second (medium) nucleus which does not collide with the nucleus-emitter (see right panel of Fig. 1).

4 Conclusion

We have studied one and two vector meson production as well as four-pion production in exclusive ultrarelativistic UPC of heavy ions. In addition, we have explained theoretical transition from ultraperipheral to peripheral or semi-central cases. In our calculations we have used equivalent photon approximation in the impact parameter space. We have obtained good description of STAR and ALICE UPC data for $\rho^0(770)$ production and CMS and ALICE data for J/ψ production. For more central nucleus-nucleus collisions we have calculated, for the first time, e.g. the differential cross section as a function of J/ψ quarkonium rapidity and simultaneously we have obtained very good agreement with ALICE experimental data. We think that correct interpretation of the ALICE data suggests that the "coherent" (assumed by the formula used for the $\gamma A \to J/\psi A$ process) scattering of the hadronic fluctuation happens before the quark-gluon plasma is created. In addition, we have compared contribution of four-pion production via $\rho^0\rho^0$ production (double scattering and $\gamma\gamma$ fusion) with experimental STAR Collaboration data. The theoretical predictions have a similar shape as the distribution measured by the STAR Collaboration, but missing contribution probably come from decays of excited states of ρ^0 resonances into four charged pions.

Acknowledgement

This work was partially supported by the Polish grant No. DEC-2014/15/B/ST2/02528 (OPUS) as well as by the Centre for Innovation and Transfer of Natural Sciences and Engineering Knowledge in Rzeszów.

References

[1] L. Frankfurt, M. Strikman, M. Zhalov, Phys. Lett. **B537**, 51 (2002)
[2] C.A. Bertulani, S.R. Klein, J. Nystrand, Ann. Rev. Nucl. Part. Sci. **55**, 271 (2005)
[3] V.P. Goncalves, M.V.T. Machado, J. Phys. **G32**, 295 (2006)
[4] K. Hencken, G. Baur, D. Trautmann. Phys. Rev. Lett. **96**, 012303 (2006)
[5] A.J. Baltz *et al.*, Phys. Rept. **458**, 1 (2008)
[6] M. Kłusek, W. Schäfer, A. Szczurek, Phys. Lett. **B674**, 92 (2009)
[7] S. Baranov, A. Cisek, M. Kłusek-Gawenda, W. Schäfer, A. Szczurek, Eur. Phys. J. **C73**, 2335 (2013)
[8] M. Kłusek-Gawenda, A. Szczurek, Phys. Rev. **C89**, 024912 (2014)
[9] L. Frankfurt, V. Guzey, M. Strikman, M. Zhalov, Phys. Lett. **B752**, 51 (2016)
[10] M. Kłusek-Gawenda, A. Szczurek, Phys. Rev. **C93**, 044912 (2016)
[11] ALICE Collaboration, B. Abelev *et al.*, Phys. Rev. Lett. **109**, 072301 (2012)
[12] M. Kłusek-Gawenda, A. Szczurek, Phys. Rev. **C82**, 014904 (2010)
[13] CMS Collaboration, *CMS-PAS-HIN-12-009* (2014) 11 pp
[14] ALICE Collaboration, B. Abelev *et al.*, Phys. Lett. **B718**, 1273 (2013)
[15] ALICE Collaboration, E. Abbas *et al.*, Eur. Phys. J. **C73**, 2617 (2013)
[16] ALICE Collaboration, J. Adam *et al.*, JHEP 1509, 095 (2015)
[17] STAR Collaboration, B. Abelev *et al.*, Phys. Rev. **C81**, 044901 (2010)
[18] ALICE Collaboration, J. Adam *et al.*, Phys. Rev. Lett. **116**, 222301 (2016)

Central exclusive production of $\rho\rho$ pairs in proton collision

R.A.Kycia[1,⋆] and J.Turnau

[1] *Tadeusz Kościuszko Cracow University of Technology, Kraków, Poland*

Abstract. Model for $pp \to 4\pi$ via 2ρ was outlined. The prediction for the cross section for this process was given for ATLAS and ALICE experimental cuts. This process has sufficient cross section to be observed in these experiments.

1 Introduction

We investigate the central exclusive production of $\rho\rho$ pairs in pp collisions at LHC energies. At low central invariant mass this process can be considered as a possible non-resonant background for glueball searches in four-body final states. Above the resonance region non perturbative production mechanism can be studied. All calculations have been performed using generator GENEX [3], specially designed for effective integration of exclusive processes in restricted phase space.

2 General description of the model

The Lebiedowicz & Szczurek model [1] was used in the form

$$\begin{aligned} ME_{LS2\pi}(p_1,p_2,p_3,p_4) = M_{13}(t_1,s_{13})F(t_a)\tfrac{1}{t_a-m_\pi^2}F(t_a)M_{24}(t_2,s_{24}) + \\ M_{14}(t_1,s_{14})F(t_b)\tfrac{1}{t_b-m_\pi^2}F(t_b)M_{23}(t_2,s_{23}), \end{aligned} \tag{1}$$

where M_{ij} is the elastic $p\pi$ scattering amplitude and hadronic form-factor

$$F(t) = exp\left(\frac{t-m_\pi^2}{\Lambda^2}\right), \tag{2}$$

depends on parameter Λ. The following replacements were made:

- Replace M_{ij} for $p\pi$ amplitude with the amplitude for $p\rho$ scattering according to the Vector-Dominance model, see section 5.3 of [2],
- In (1) and (2) change m_π to m_ρ.

The above described matrix element for $pp \to pp + \rho\rho$ after addition of ρ decay amplitude and symmetrization for the identical particles in final state has the following form[1]:

$$ME = \frac{1}{\sqrt{2!}}(ME(p_1,p_2,p_3,p_4,p_5,p_6) + ME(p_1,p_2,p_5,p_4,p_3,p_6)), \tag{3}$$

⋆e-mail: kycia.radoslaw@gmail.com

[1]The factor $\frac{1}{\sqrt{2!}}$ is the symmetry factor which usually appears in the formula for the cross section.

DOI: 10.1051/epjconf/201613005010

$$ME(p_1, p_2, p_3, p_4, p_5, p_6) = ME_{LS2\rho}(p_1, p_2, p_3, p_4, p_5, p_6) \times \frac{1}{3}(\sum_{\lambda \in \{-1,0,+1\}} ME_{\rho \to \pi^+ + \pi^-}(p_3, p_4, \lambda)) \times (\sum_{\lambda' \in \{-1,0,+1\}} ME_{\rho \to \pi^+ + \pi^-}(p_5, p_6, \lambda')), \tag{4}$$

where $ME_{LS\rho}$ is the modified Lebiedowicz&Szczurek model [1] with Vector-Dominance model (1) and $\rho \to 2\pi$ is given by the standard Zemach formalism:

$$ME_{\rho \to \pi^+ + \pi^-}(1', 2', m) = N \frac{i\varepsilon_\mu(p_r, m)(p_{1'} - p_{2'})^\mu}{p_r^2 - M_\rho^2 + i\Gamma_\rho M_\rho}. \tag{5}$$

where ε is a ρ vector polarisation vector and $p_r = p_{1'} + p_{2'}$.

This simplified model gives similar cross section(to the order of magnitude) as the one recently presented in [5].

3 Results

Cross sections $\sigma_{pp \to pp + \rho\rho}$ are calculated for full and restricted phase space corresponding to ATLAS and ALICE detectors at $\sqrt{(s)} = 7\ TeV$.

For ATLAS we assume:

- Diffractively scattered protons detected in the ATLAS FORWARD DETECTOR (AFD),
- Pions detected in Silicon Central Tracker SCT $|\eta_\pi| < 2.5$ and $p_{t\pi} > 0.1\ GeV$ (enhanced track reconstruction [4]).

For ALICE we assume:

- Pions detected in the Central Barrel (CB) $|\eta_\pi| < 0.9$ and $|p_{t\pi}| > 0.017\ GeV$,
- particles in the rapidity range $-3.7 < \eta_\pi < -0.9$ or $0.9 < \eta_\pi < 5.1$ can be anti-selected in the trigger detectors veto system (TDV) VZERO+FMD+SPD.

W define five phase space regions corresponding to the following event selection setups:

- ATLAS-I: Protons in AFD with $|t_p| < 1\ GeV^2$ and 4π in SCT,
- ATLAS-II: Protons in AFD with $|t_p| < 1\ GeV^2$ and two opposite-sign pions in SCT,
- ATLAS-III: Protons in AFD with $|t_p| < 1\ GeV^2$ and two same-sign pions in SCT,
- ALICE: 4π in CB.

Table 1. Cross section estimation.

Phase Space	$\Lambda[GeV]$	σ
Full	0.5	6.027 nb
Full	1.0	0.41 μb
ATLAS-I	1.0	49.2 nb
ATLAS-II	1.0	60.4 nb
ATLAS-III	1.0	5.97 nb
ALICE	1.0	3.05 nb

The results are presented in Table 1 and in Figs. 1 and 2. In the figures the cross section ratio $R = \sigma_{pp \to pp + \rho\rho} / \sigma_{pp \to pp + \pi\pi}$ was drawn as absorption is to large extent cancelled using this variable. It should be noted that recently presented model calculations for the same process [5] give results differing from ours by more than order of magnitude. This model is based on the tensor pomeron exchange in $p\rho \to p\rho$. The source of so large differences will be investigated.

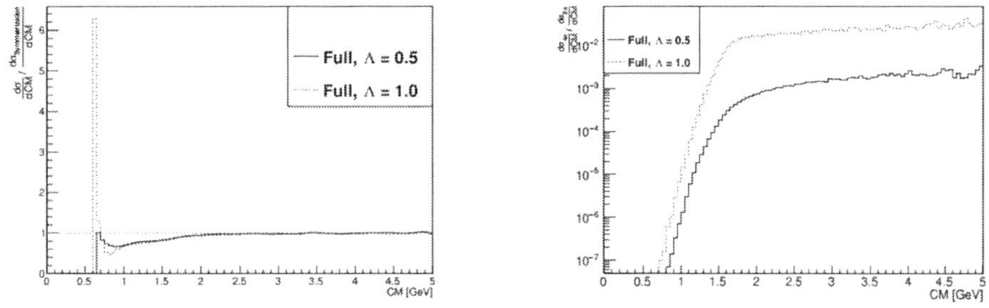

Figure 1. Left: Symmetrization effect full phase space. Right: Cross section ratio full phase space.

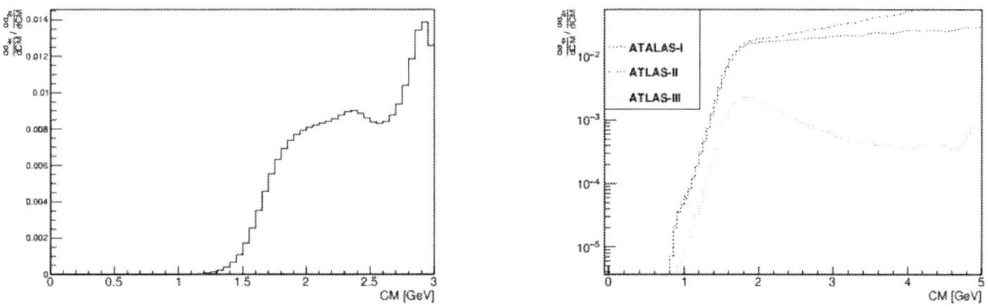

Figure 2. Left: Cross section ratio ALICE. Right: Cross section ratio ATLAS-I/II/III.

4 Conclusions

The cross section for central exclusive production of $\rho\rho$ pairs at LHC energies is rather small in comparison with central exclusive production of pion pairs and depends strongly on parameter Λ. The process $pp \rightarrow pp + \rho\rho \rightarrow pp + 4\pi$ does not contribute to the background of the exclusive production of pion pairs, neither in ALICE nor in the ATLAS experiment. In the resonance region of the central invariant mass $M_{central} < 2.0\ GeV$ the effect of the symmetrisation for identical particles ($\rho\rho$ mixing) is significant, making the signal identification event more difficult.

References

[1] P.Lebiedowicz, A.Szczurek, Phys. Rev. D **81** :036003 (2010)
[2] S. Donnachie et al. *Pomeron Physics and QCD*, Cambridge University Press 2005
[3] R.A.Kycia et al. arXiv:1411.6035 [hep-ph]
[4] R.Staszewski et al., Acta Phys. Polon. B **42** 1861 (2011)
[5] P. Lebiedowicz, O. Nachtmann, A. Szczurek, arXiv:1606.05126 [hep-ph]

EPJ Web of Conferences **130**, 05011 (2016) DOI: 10.1051/epjconf/201613005011
MESON 2016

Exclusive diffractive production of $\pi^+\pi^-$ continuum and resonances within tensor pomeron approach

Piotr Lebiedowicz[1,⋆], Otto Nachtmann[2], and Antoni Szczurek[1,⋆⋆]

[1] *Institute of Nuclear Physics Polish Academy of Sciences, PL-31342 Kraków, Poland*
[2] *Institut für Theoretische Physik, Universität Heidelberg, D-69120 Heidelberg, Germany*

Abstract. We discuss exclusive central diffractive dipion production in proton-(anti)proton collisions at high energies. The calculation is based on a tensor pomeron model and the amplitudes for the processes are formulated in an effective field-theoretic approach. We include the purely diffractive dipion continuum, and the scalar and tensor resonances decaying into the $\pi^+\pi^-$ pairs as well as the photoproduction contributions (ρ^0, Drell-Söding). We discuss how two pomerons couple to tensor meson $f_2(1270)$ and the interference effects of resonance and dipion continuum. The theoretical results are compared with CDF and CMS experimental data. We show the influence of the experimental cuts on the integrated cross section and on various differential distributions for outgoing particles. We find that the relative contribution of resonant $f_2(1270)$ and dipion continuum strongly depends on the cut on proton transverse momenta (or four-momentum transfer squared $t_{1,2}$) which may explain some controversial observations made by different ISR experiments in the past. The cuts may play then the role of a $\pi\pi$ resonance filter. We suggest some experimental analyses to fix model parameters related to the pomeron-pomeron-meson coupling.

1 Introduction

Central production mediated by the "fusion" of two exchanged pomerons [1, 2] is an important diffractive process for the investigation of properties of di-pion resonances, in particular, for search of gluonic bound states (glueballs). The experimental groups such as the COMPASS [3, 4], ISR [5, 6], STAR [7, 8], CDF Tevatron [9], ALICE [10], and CMS [11] all show visible structures in the $\pi^+\pi^-$ invariant mass. Some time ago two of us have formulated a Regge-type model of the dipion continuum for the exclusive reaction $pp \to pp\pi^+\pi^-$ with parameters fixed from phenomenological analysis of total and elastic NN and πN scattering [13]. The number of free model parameters is then limited to a parameter of form factor describing off-shellness of the exchanged pion. The model was extended to include rescattering corrections due to pp nonperturbative interaction [12, 14]. The exclusive reaction $pp \to pp\pi^+\pi^-$ constitutes an irreducible background to the scalar χ_{c0} meson production [14]. These model studies were extended also to exclusive K^+K^- production [15]. The largest uncertainties in the model are due to the unknown off-shell pion form factor and the absorption effects; see Ref. [16].

⋆e-mail: Piotr.Lebiedowicz@ifj.edu.pl
⋆⋆Also at University of Rzeszów, PL-35959 Rzeszów, Poland.

Such an approach gives correct order of magnitude cross sections, however, does not include resonance contributions which interfere with the continuum contribution.

First calculations of central exclusive diffractive production of $\pi^+\pi^-$ continuum together with the dominant scalar $f_0(500)$, $f_0(980)$, and tensor $f_2(1270)$ resonances was performed in Ref. [1]. Here we use the tensor-pomeron model formulated in [17]; see also [18]. In this model pomeron exchange is effectively treated as the exchange of a rank-2 symmetric tensor. In [19] we show that the tensor pomeron is consistent with the STAR pp elastic scattering data [20]. In Ref. [21] the model was applied to the diffractive production of several scalar and pseudoscalar mesons in the reaction $pp \to ppM$. The corresponding pomeron-pomeron-meson coupling constants are not known and have been fitted to existing WA102 experimental data. In most cases one has to add coherently amplitudes for two pomeron-pomeron-meson couplings with different orbital angular momentum and spin of two "pomeron particles". [1] In [22] an extensive study of the photoproduction reaction $\gamma p \to \pi^+\pi^- p$ was presented. The resonant ($\rho^0 \to \pi^+\pi^-$) and non-resonant (Drell-Söding) photon-pomeron/reggeon $\pi^+\pi^-$ production in pp collisions was studied in [23].

The identification of glueballs can be very difficult; see e.g. [24]. The partial wave analyses of future experimental data could be used in this context. Also the studies of different decay channels in central exclusive production would be very valuable. One of the possibilities is the $pp \to pp\pi^+\pi^-\pi^+\pi^-$ reaction being analysed at the LHC. In Ref. [25] we analysed the exclusive diffractive production of four-pion via the intermediate $\sigma\sigma$ and $\rho\rho$ states (for a related work see [26]).

2 Sketch of formalism

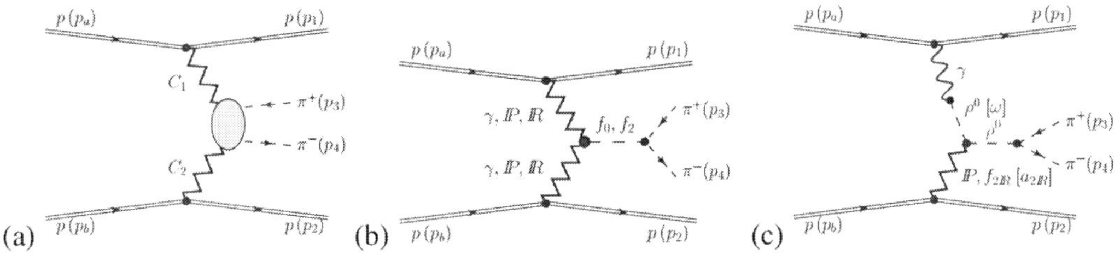

Figure 1. Diagram (a): Generic Born level diagram for central exclusive continuum $\pi^+\pi^-$ production in proton-(anti)proton collisions. Here we labelled the exchange objects by their charge conjugation numbers C_1, $C_2 \in \{+1, -1\}$. Diagram (b): The double-pomeron/reggeon and photon mediated central exclusive scalar and tensor resonance production and their subsequent decays into $\pi^+\pi^-$. Diagram (c): The ρ^0 photoproduction mechanism.

The Born level diagrams for the continuum and resonant $\pi^+\pi^-$ production are shown in Fig. 1. The purely diffractive amplitude is a sum of continuum amplitude and the amplitudes with the s-channel resonances with very restrictive quantum numbers $I^G J^{PC} = 0^+(\text{even})^{++}$:

$$\mathcal{M}_{pp\to pp\pi^+\pi^-} = \mathcal{M}^{\pi\pi-\text{continuum}}_{pp\to pp\pi^+\pi^-} + \mathcal{M}^{(\mathbb{PP}\to f_0\to\pi^+\pi^-)}_{\lambda_a\lambda_b\to\lambda_1\lambda_2\pi^+\pi^-} + \mathcal{M}^{(\mathbb{PP}\to f_2\to\pi^+\pi^-)}_{\lambda_a\lambda_b\to\lambda_1\lambda_2\pi^+\pi^-}\,. \tag{1}$$

[1] We wish to emphasize that the tensorial pomeron can, at least, equally well describe the WA102 experimental data on the exclusive meson production as the less theoretically justified vectorial pomeron frequently used in the literature. The existing low-energy experimental data do not allow to clearly distinguish between the two models as the presence of subleading reggeon exchanges is at low energies very probable for many $pp \to ppM$ reactions.

For instance, the Born (no absorption effects) amplitude for the process $pp \to pp(\mathbb{P}\mathbb{P} \to f_2 \to \pi^+\pi^-)$ can be written in the effective field-theoretic approach as

$$\begin{aligned} \mathcal{M}^{(\mathbb{P}\mathbb{P}\to f_2\to\pi^+\pi^-)}_{\lambda_a\lambda_b\to\lambda_1\lambda_2\pi^+\pi^-} = & (-i)\,\bar{u}(p_1,\lambda_1) i\Gamma^{(\mathbb{P}pp)}_{\mu_1\nu_1}(p_1,p_a)u(p_a,\lambda_a)\; i\Delta^{(\mathbb{P})\mu_1\nu_1,\alpha_1\beta_1}(s_1,t_1) \\ & \times i\Gamma^{(\mathbb{P}\mathbb{P}f_2)}_{\alpha_1\beta_1,\alpha_2\beta_2,\rho\sigma}(q_1,q_2)\; i\Delta^{(f_2)\rho\sigma,\alpha\beta}(p_{34})\; i\Gamma^{(f_2\pi\pi)}_{\alpha\beta}(p_3,p_4) \\ & \times i\Delta^{(\mathbb{P})\alpha_2\beta_2,\mu_2\nu_2}(s_2,t_2)\; \bar{u}(p_2,\lambda_2) i\Gamma^{(\mathbb{P}pp)}_{\mu_2\nu_2}(p_2,p_b)u(p_b,\lambda_b)\,, \end{aligned} \tag{2}$$

where $t_1 = (p_1 - p_a)^2$, $t_2 = (p_2 - p_b)^2$, $s_1 = (p_a + q_2)^2 = (p_1 + p_{34})^2$, $s_2 = (p_b + q_1)^2 = (p_2 + p_{34})^2$, $p_{34} = p_3 + p_4$. $\Delta^{(\mathbb{P})}$ and $\Gamma^{(\mathbb{P}pp)}$ denote the effective propagator and proton vertex function, respectively, for the tensorial pomeron. For the explicit expressions, see Sec. 3 of [17]. The pomeron-pomeron-f_2 vertex is the most complicated element of our amplitude (2). In Ref. [1] we have considered all possible tensorial structures for the $\mathbb{P}\mathbb{P}f_2$ coupling (see Appendix A of [1])

$$i\Gamma^{(\mathbb{P}\mathbb{P}f_2)}_{\mu\nu,\kappa\lambda,\rho\sigma}(q_1,q_2) = \left(i\Gamma^{(\mathbb{P}\mathbb{P}f_2)(1)}_{\mu\nu,\kappa\lambda,\rho\sigma}\mid_{bare} + \sum_{j=2}^{7} i\Gamma^{(\mathbb{P}\mathbb{P}f_2)(j)}_{\mu\nu,\kappa\lambda,\rho\sigma}(q_1,q_2)\mid_{bare} \right) \tilde{F}^{(\mathbb{P}\mathbb{P}f_2)}(q_1^2,q_2^2,p_{34}^2)\,. \tag{3}$$

Other details as form of form factors, the tensor-meson propagator $\Delta^{(f_2)}$ and the $f_2\pi\pi$ vertex are given in Refs. [1, 17]. The production of the f_2 meson via $\mathbb{P}f_{2\mathbb{R}}$, $f_{2\mathbb{R}}\mathbb{P}$, and $f_{2\mathbb{R}}f_{2\mathbb{R}}$ fusion can be treated in an analogous way to the $\mathbb{P}\mathbb{P}$ fusion.

We consider also the production of $\rho(770)$ resonance produced by photon-pomeron/reggeon mechanism studied in detail in [23], see the panel (c) in Fig. 1. In the amplitude for the $\gamma p \to \rho^0 p$ subprocess we included both pomeron and $f_{2\mathbb{R}}$ exchanges. The $\mathbb{P}\rho\rho$ vertex is given in [17] by formula (3.47). The coupling parameters of tensor pomeron/reggeon exchanges was fixed based on the HERA experimental data for the $\gamma p \to \rho^0 p$ reaction. In [23] we showed that the resonant contribution interfere with the non-resonant (Drell-Söding) $\pi^+\pi^-$ continuum and produces a skewing of the $\rho(770)$-meson line shape. Due to the photon propagators occurring in diagrams we expect these processes to be most important when at least one of the protons undergoes only a very small $|t_{1,2}|$.

3 Selected results

We start from a discussion of some dependences for the central exclusive production of the $f_2(1270)$ meson. For detailed study of $f_2(1270)$ production see Ref. [1]. In Fig. 2 we present results for individual pomeron-pomeron-f_2 coupling terms (there are 7 possible terms [1]) at $\sqrt{s} = 200$ GeV and $|\eta_\pi| < 1$. The different predictions differ considerably which could be checked experimentally. We show that only in two cases ($j = 2$ and 5) the cross section $d\sigma/d|t_{1,2}|$ vanishes when $|t_{1,2}| \to 0$. In [1] we tried to understand whether one can approximately describe the dipion invariant mass distribution observed by different experiments assuming only one of the seven possible $\mathbb{P}\mathbb{P}f_2$ tensorial couplings. We found that the feature of the $\pi^+\pi^-$ distribution depends on the cuts used in a particular experiment (usually the t cuts are different for different experiments). As can be clearly seen from Fig. 3 different $\mathbb{P}\mathbb{P}f_2$ couplings generate different interference patterns around $M_{\pi\pi} \sim 1.27$ GeV. A sharp drop around $M_{\pi\pi} \sim 1$ GeV is attributed to the interference of $f_0(980)$ and continuum. We can observe that the $j = 2$ coupling gives results close to those observed by the CDF Collaboration [9]. In this preliminary study we did not try to fit the existing data [9] by mixing different couplings because the CDF data are not fully exclusive (the outgoing p and $\bar{p}$ were not measured). The calculations were done at Born level and the absorption corrections were taken into account by multiplying the cross section by a common factor $\langle S^2 \rangle$ obtained from [16]. The two-pion continuum was fixed by choosing a form factor for the off-shell pion $\hat{F}_\pi(k^2) = \frac{\Lambda^2_{off,M} - m_\pi^2}{\Lambda^2_{off,M} - k^2}$ and $\Lambda_{off,M} = 0.7$ GeV.

DOI: 10.1051/epjconf/201613005011

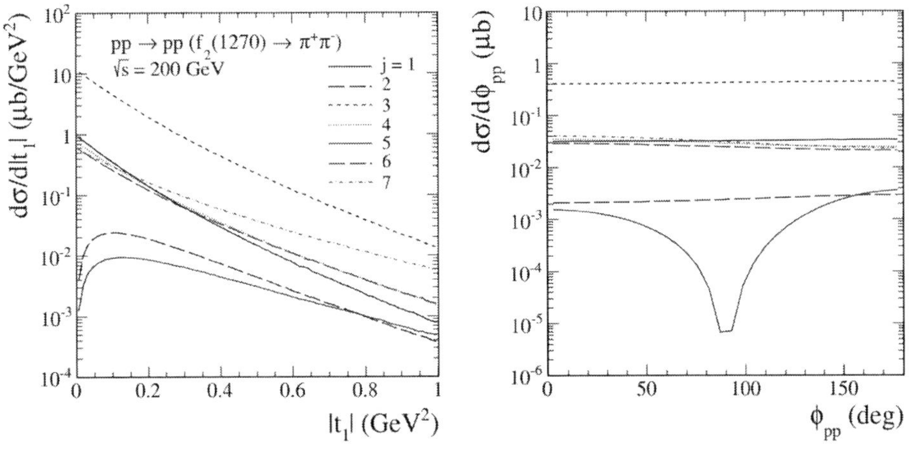

Figure 2. The distribution in transferred four-momentum squared between the initial and final protons (left panel) and the distribution in azimuthal angle between the outgoing protons (right panel) at $\sqrt{s} = 200$ GeV and $|\eta_\pi| < 1$. We show the individual contributions of the different pomeron-pomeron-$f_2(1270)$ couplings: $j = 1$ (the black solid line), $j = 2$ (the black long-dashed line), $j = 3$ (the black dashed line), $j = 4$ (the black dotted line), $j = 5$ (the blue solid line), $j = 6$ (the blue long-dashed line), and $j = 7$ (the red dot-dashed line). For illustration the results have been obtained with coupling constants $g^{(j)}_{\mathbb{P}\mathbb{P}f_2} = 1.0$. No absorption effects were included here.

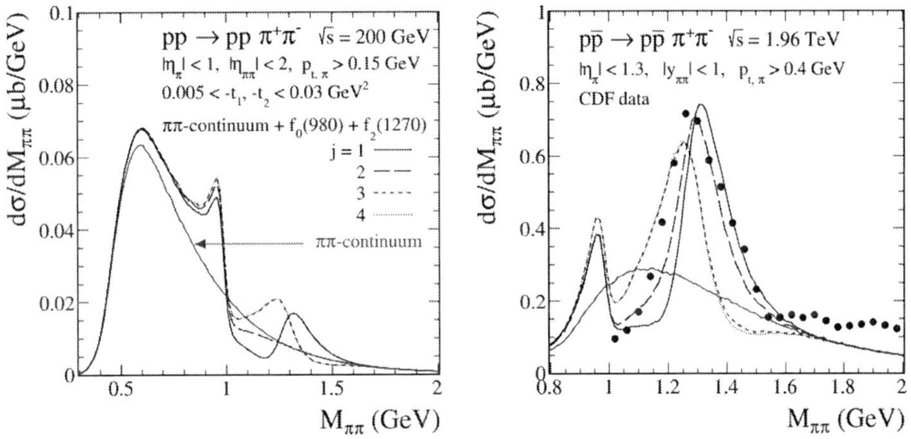

Figure 3. Two-pion invariant mass distribution for the STAR [7] (left) and CDF [9] (right) kinematics. The individual contributions of different $\mathbb{P}\mathbb{P}f_2$ couplings ($j = 1, ..., 4$) are compared with the CDF data [9]. The Born calculations for $\sqrt{s} = 200$ GeV and $\sqrt{s} = 1.96$ TeV were multiplied by the gap survival factors $\langle S^2 \rangle = 0.2$ and $\langle S^2 \rangle = 0.1$, respectively. The blue solid lines represent the non-resonant continuum contribution only ($\Lambda_{off,M} = 0.7$ GeV) while the black lines represent a coherent sum of non-resonant continuum, $f_0(980)$ and $f_2(1270)$ resonant terms.

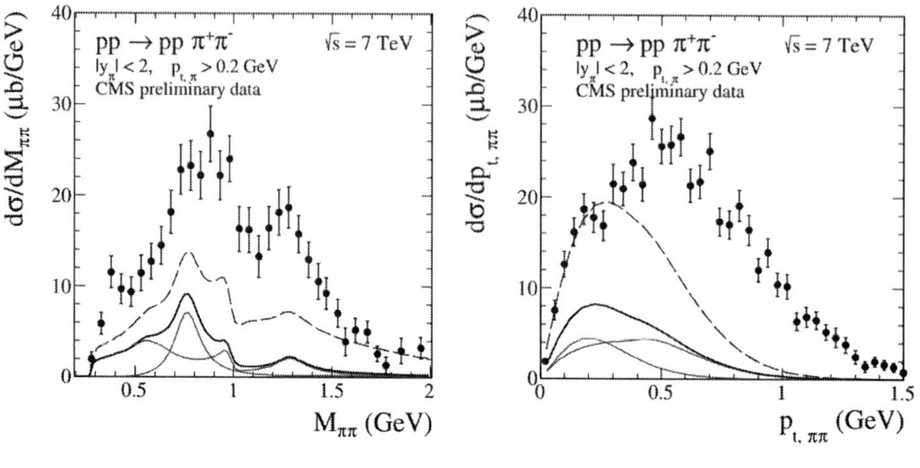

Figure 4. The distributions for two-pion invariant mass (left panel) and transverse momentum of the pion pair (right panel) for the CMS kinematics at $\sqrt{s}$ = 7 TeV. Both photoproduction (red line) and purely diffractive (blue line) contributions multiplied by the factors $\langle S^2 \rangle = 0.9$ and $\langle S^2 \rangle = 0.1$, respectively, are included. The complete results correspond to the black solid line ($\Lambda_{off,M}$ = 0.7 GeV) and the dashed line ($\Lambda_{off,M}$ = 1.2 GeV). The CMS preliminary data scanned from [11] are shown for comparison.

In Fig. 4 we show results including in addition to the non-resonant $\pi^+\pi^-$ continuum, the $f_2(1270)$ and the $f_0(980)$ resonances, the contribution from photoproduction ($\rho^0 \to \pi^+\pi^-$, Drell-Söding mechanism), as well as the $f_0(500)$ resonant contribution. Our predictions are compared with the CMS preliminary data [11]. Here the absorption effects lead to huge damping of the cross section for the purely diffractive term (the blue lines) and relatively small reduction of the cross section for the photoproduction term (the red lines). Therefore we expect one could observe the photoproduction contribution. The CMS measurement [11] is not fully exclusive and, inter alia, the $M_{\pi\pi}$ and $p_{t,\pi\pi}$ spectra contain contributions associated from other processes, e.g., when one or both protons undergoing dissociation. In addition we show results with $\Lambda_{off,M}$ = 1.2 GeV better adjusted to the new CMS data (see the dashed line). If we used this set for the STAR or CDF measurements our results there would be above the preliminary STAR data [7] at $M_{\pi\pi} > 1$ GeV and in complete disagreement with the CDF data from [9]. Only purely central exclusive data expected from CMS-TOTEM and ATLAS-ALFA will allow to draw definite conclusions. The absorption effects due to pp and πp interactions, discussed in [16], lead to a significant modification of the shape of the distributions in ϕ_{pp}, $p_{t,p}$, $t_{1,2}$ and could also be tested by these experimental groups.

4 Conclusions

In our recent paper [1] we have analysed the exclusive central production of dipion continuum and resonances contributing to the $\pi^+\pi^-$ pair production in proton-(anti)proton collisions in an effective field-theoretic approach with tensor pomerons and reggeons proposed in [17]. We have included the scalar $f_0(500)$ and $f_0(980)$ resonances, the tensor $f_2(1270)$ resonance and the vector $\rho(770)$ resonance in a consistent way. In the case of $f_2(1270)$-meson production via "fusion" of two tensor pomerons we have found [1] all (seven) possible $\mathbb{P}\mathbb{P}f_2$ tensorial couplings. The different couplings give different results due to different interference effects of the f_2 resonance and the dipion continuum contributions. We have shown that the resonance structures in the measured two-pion invariant mass spectra depend on the cut on proton transverse momenta and/or on four-momentum transfer squared $t_{1,2}$ used in an experiment. The model parameters of the optimal $\mathbb{P}\mathbb{P}f_2$ coupling ($j = 2$) have been roughly adjusted to the recent CDF and preliminary STAR experimental data and then used for the predictions for the ALICE, and CMS experiments. We have made estimates of cross sections for both the diffractive and photoproduction contributions. We have shown some differential distributions related to produced pions as well as some observables related to final state protons, e.g., different dependence on proton transverse momenta and azimuthal angle correlations between outgoing protons could be used to separate the photoproduction term, see [1]. Future experimental data on exclusive meson production at high energies should thus provide good information on the spin structure of the pomeron and on its couplings to the nucleon and the mesons.

This work was partially supported by the MNiSW Grant No. IP2014 025173 (Iuventus Plus) and the Polish National Science Centre grants DEC-2014/15/B/ST2/02528 and DEC-2015/17/D/ST2/03530.

References

[1] P. Lebiedowicz, O. Nachtmann, A. Szczurek, Phys. Rev. D **93**, 054015 (2016)
[2] R. Fiore, L. Jenkovszky, R. Schicker, Eur. Phys. J. C **76**, 38 (2016)
[3] A. Austregesilo (COMPASS Collaboration), AIP Conf. Proc. **1735**, 030012 (2016)
[4] B. Grube, this Proceedings
[5] T. Åkesson *et al.*, (AFS Collaboration), Nucl. Phys. B **264**, 154 (1986)
[6] A. Breakstone *et al.* (ABCDHW Collaboration), Z. Phys. C **31**, 185 (1986); A. Breakstone *et al.* (ABCDHW Collaboration), Z. Phys. C **42**, 387 (1989); Erratum: Z. Phys. C **43**, 522 (1989); A. Breakstone *et al.* (ABCDHW Collaboration), Z. Phys. C **48**, 569 (1990)
[7] L. Adamczyk, W. Guryn, J. Turnau, Int. J. Mod. Phys. A **29**, 1446010 (2014)
[8] W. Guryn, this Proceedings
[9] T. Aaltonen *et al.* (CDF Collaboration), Phys. Rev. D **91**, 091101 (2015)
[10] R. Schicker (ALICE Collaboration), arXiv:1205.2588 [hep-ex].
[11] CMS Collaboration, Report No. CMS-PAS-FSQ-12-004.
[12] R. Staszewski, P. Lebiedowicz, M. Trzebiński, J. Chwastowski, A. Szczurek, Acta Phys. Polon. B **42**, 1861 (2011)
[13] P. Lebiedowicz and A. Szczurek, Phys. Rev. D **81**, 036003 (2010)
[14] P. Lebiedowicz, R. Pasechnik, A. Szczurek, Phys. Lett. B **701**, 434 (2011)
[15] P. Lebiedowicz and A. Szczurek, Phys. Rev. D **85**, 014026 (2012)
[16] P. Lebiedowicz and A. Szczurek, Phys. Rev. D **92**, 054001 (2015)
[17] C. Ewerz, M. Maniatis, O. Nachtmann, Annals Phys. **342**, 31 (2014)
[18] O. Nachtmann, Annals Phys. **209**, 436 (1991)
[19] C. Ewerz, P. Lebiedowicz, O. Nachtmann, A. Szczurek, arXiv:1606.08067 [hep-ph].
[20] L. Adamczyk *et al.* (STAR Collaboration), Phys. Lett. B **719**, 62 (2013)
[21] P. Lebiedowicz, O. Nachtmann, A. Szczurek, Annals Phys. **344**, 301 (2014)
[22] A. Bolz, C. Ewerz, M. Maniatis, O. Nachtmann, M. Sauter, A. Schöning, JHEP **1501**, 151 (2015)
[23] P. Lebiedowicz, O. Nachtmann, A. Szczurek, Phys. Rev. D **91**, 074023 (2015)
[24] F. Giacosa, this Proceedings
[25] P. Lebiedowicz, O. Nachtmann, A. Szczurek, Phys. Rev. D **94**, 034017 (2016)
[26] R. A. Kycia and J. Turnau, this Proceedings

EPJ Web of Conferences **130**, 05012 (2016) DOI: 10.1051/epjconf/201613005012
MESON 2016

Diffractive production of heavy mesons at the LHC

Marta Łuszczak[1,⋆] and Antoni Szczurek[1,2]

[1] *University of Rzeszów, PL-35-959 Rzeszów, Poland*
[2] *Institute of Nuclear Physics, Polish Academy of Sciences, Kraków, Poland*

Abstract. We discuss diffractive production of heavy mesons at the LHC. In addition to standard collinear approach, for a first time we propose a k_t-factorization approach to the diffractive processes. The unintegrated (transverse momentum dependent) diffractive parton distributions in proton are calculated with the help of the Kimber-Martin-Ryskin prescription where collinear diffractive PDFs are used as input. Some correlation observables, like azimuthal angle correlation between c and $\bar{c}$, and $c\bar{c}$ pair transverse momentum distribution were obtained for the first time. The results of the new approach are compared with those of the standard collinear one. Significantly larger cross sections are obtained in the k_t-factorization approach where some part of higher-order effects is effectively included.

1 Introduction

Diffractive hadronic processes were studied theoretically in the so-called resolved pomeron model [1]. This model, previously used to describe deep-inelastic diffractive processes must be corrected for absorption effects related to hadron-hadron interactions. In theoretical models this effect is taken into account approximately by multiplying the diffractive cross section calculated using HERA diffractive PDFs by a kinematics independent factor called the gap survival probability – S_G. Two theoretical groups specialize in calculating such probabilities [2, 3].

In this study we consider diffractive production of charm for which rather large cross section at the LHC are expected, even within the leading-order (LO) collinear approach [4]. On the other hand, it was shown that for the inclusive non-diffractive charm production the LO collinear approach is a rather poor approximation and higher-order corrections are crucial. Contrary, the k_t-factorization approach, which effectively includes higher-order effects, gives a good description of the LHC data for inclusive charm production at $\sqrt{s}$ = 7 TeV (see *e.g.* Ref. [5]). This strongly suggests that application of the k_t-factorization approach to diffractive charm production is useful. This presentation is based on our recent study presented in [6]. Here we present only results at the quark/antiquark level.

2 Formalism

A sketch of the theoretical formalism is shown in Fig. 1. Here, extension of the standard resolved pomeron model based on the LO collinear approach by adopting a framework of the k_t-factorization is proposed as an effective way to include higher-order corrections.

⋆e-mail: luszczak@ur.edu.pl,

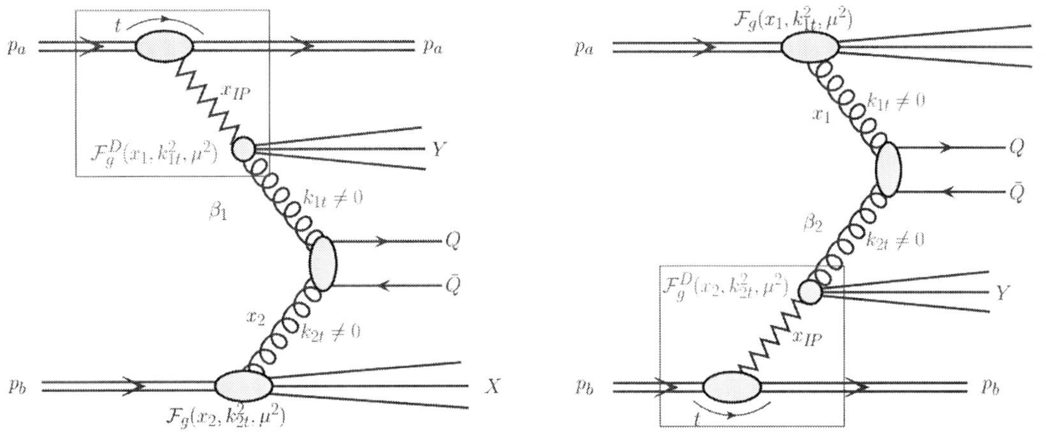

Figure 1. A diagrammatic representation for single-diffractive production of heavy quark pairs within the k_t-factorization resolved pomeron approach.

Details of our new calculations can be found in Ref. [6].

3 Results

In Fig. 2 we show rapidity (left panel) and transverse momentum (right panel) distribution of c quarks (antiquarks) for single diffractive production at $\sqrt{s}$ = 13 TeV. Distributions calculated within the LO collinear factorization (black long-dashed lines) and for the k_t-factorization approach (red solid lines) are shown separately. We see significant differences between the both approaches, that are consistent with the conclusions from similar studies of standard non-diffractive charm production (see *e.g.* Ref. [5]). Here we confirm that the higher-order corrections are very important also for the diffractive production of charm quarks.

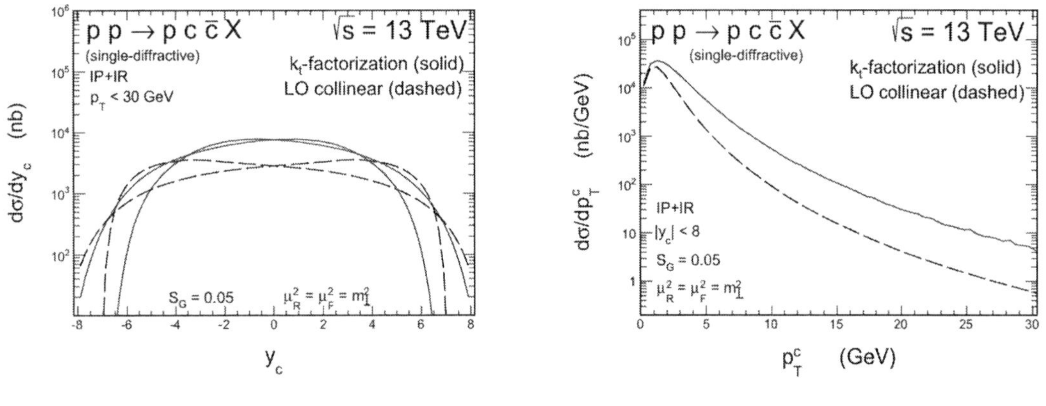

Figure 2. Rapidity (left panel) and transverse momentum (right panel) distributions of c quarks (antiquarks) for a single-diffractive production at $\sqrt{s}$ = 13 TeV. Components of the $g(I\!P)$ - $g(p)$, $g(p)$ - $g(I\!P)$, $g(I\!R)$ - $g(p)$, $g(p)$ - $g(I\!R)$ mechanisms are shown.

The correlation observables cannot be calculated within the LO collinear factorization but can be directly obtained in the k_t-factorization approach. The distribution of azimuthal angle $\varphi_{c\bar{c}}$ between c quarks and $\bar{c}$ antiquarks is shown in the left panel of Fig. 3. The $c\bar{c}$ pair transverse momentum distribution $p_T^{c\bar{c}} = |\vec{p_t^c} + \vec{p_t^{\bar{c}}}|$ is shown in the right panel. Results of the full phase-space calculations illustrate that the quarks and antiquarks in the $c\bar{c}$ pair are almost uncorrelated in the azimuthal angle between them and are often produced in the configuration with quite large pair transverse momenta.

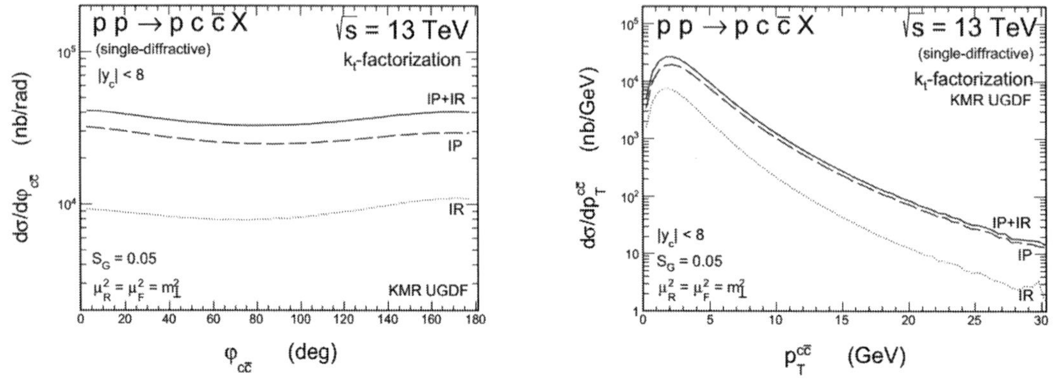

Figure 3. The distribution in $\phi_{c\bar{c}}$ (left panel) and distribution in $p_T^{c\bar{c}}$ (right panel) in the k_t-factorization approach at $\sqrt{s} = 13$ TeV.

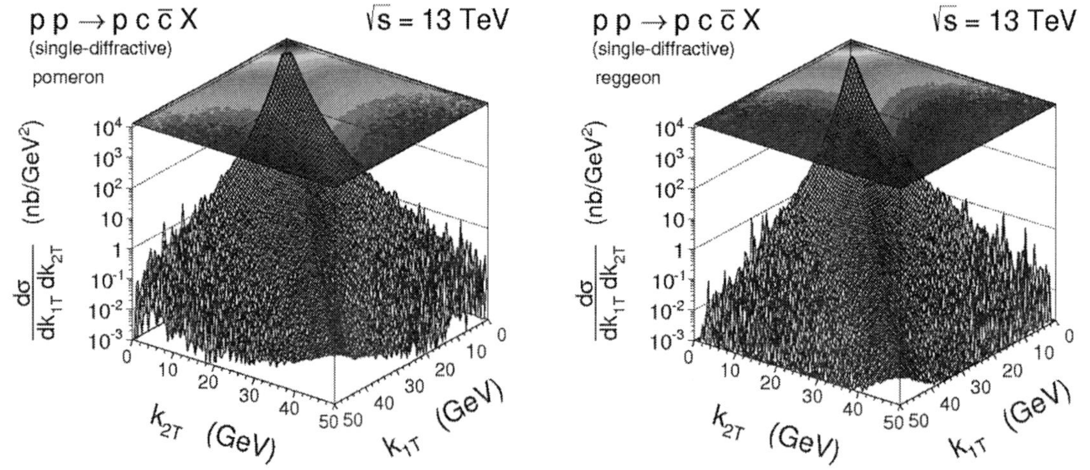

Figure 4. Double differential cross sections as a function of initial gluons transverse momenta k_{1T} and k_{2T} for single-diffractive production of charm at $\sqrt{s} = 13$ TeV. The left and right panels correspond to the pomeron and reggeon exchange mechanisms, respectively.

Figures 4 and 5 show the double differential cross sections as a functions of transverse momenta of incoming gluons (k_{1T} and k_{2T}) and transverse momenta of outgoing c and $\bar{c}$ quarks (p_{1T} and p_{2T}), respectively. We observe quite large incident gluon transverse momenta. The major part of the cross

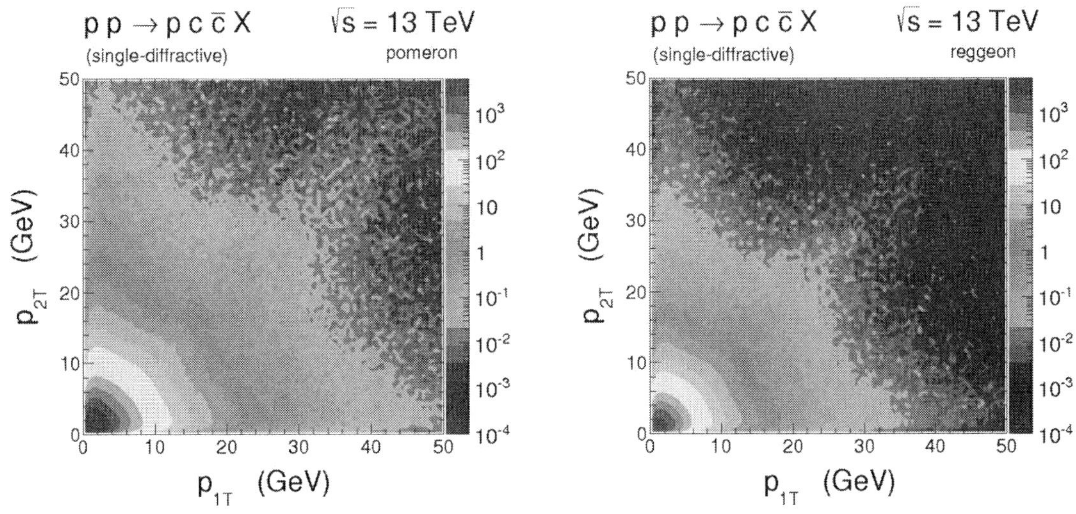

Figure 5. Double differential cross sections as a function of transverse momenta of outgoing c quark p_{1T} and outgoing $\bar{c}$ antiquark p_{2T} for single-diffractive production of charm at $\sqrt{s}$ = 13 TeV. The left and right panels correspond to the pomeron and reggeon exchange mechanisms, respectively.

section is concentrated in the region of small k_t's of both gluons but long tails are present. Transverse momenta of the outgoing particles are not balanced as they were in the case of the LO collinear approximation.

4 Conclusions

Charm production is a good example where the higher-order effects are very important. For the inclusive charm production we have shown that these effects can be effectively included in the k_t-factorization approach [5]. In our approach we decided to use the so-called KMR method to calculate unintegrated diffractive gluon distribution (UDGD). Having obtained the UDGD we have performed calculations of several single-particle and correlation distributions. In general, the k_t-factorization approach leads to larger cross sections. However, the K-factor is strongly dependent on phase space point. Some correlation observables, like azimuthal angle correlation between c and $\bar{c}$, and $c\bar{c}$ pair transverse momentum distributions were obtained in [6] for the first time.

Acknowledgement

This work was partially supported by the Polish National Science Centre grant DEC-2013/09/D/ST2/03724 as well as by the Centre for Innovation and Transfer of Natural Sciences and Engineering Knowledge in Rzeszów.

References

[1] G. Ingelman, P.E. Schlein, Phys. Lett. B **152**, 256 (1985)
[2] V. A. Khoze, A. D. Martin and M. G. Ryskin, Eur. Phys. J. C **18**, 167 (2000)
[3] U. Maor, AIP Conf. Proc. **1105**, 248 (2009)
[4] M. Luszczak, R. Maciula and A. Szczurek, Phys. Rev. D **91**, 054024 (2015)

[5] R. Maciula and A. Szczurek, Phys. Rev. D **87**, 094022 (2013)
[6] M. Luszczak, R. Maciula, A. Szczurek and M. Trzebinski, arXiv:1606.06528 [hep-ph]

EPJ Web of Conferences **130**, 05013 (2016) DOI: 10.1051/epjconf/201613005013
MESON 2016

Double open charm meson production at the LHC: new single- and double-parton scattering mechanisms

Rafał Maciuła[1,*] and Antoni Szczurek[1,2]

[1] *Institute of Nuclear Physics, Polish Academy of Sciences, Radzikowskiego 152, PL-31-342 Kraków, Poland*
[2] *University of Rzeszów, PL-35-959 Rzeszów, Poland*

Abstract. We discuss D^0D^0 meson-meson pair production related to the recent LHCb experimental results obtained at $\sqrt{s} = 7$ TeV. We consider double-parton scattering (DPS) mechanisms of double $c\bar{c}$ production and subsequent standard $cc \to D^0D^0$ scale-independent hadronization as well as double g and mixed $gc\bar{c}$ production with $gg \to D^0D^0$ and $gc \to D^0D^0$ scale-dependent hadronization. The latter scenario with $g \to D^0$ fragmentation components results in a new single-parton scattering (SPS) mechanism of digluon production which is also taken here into account. Results of the numerical calculations are compared with the LHCb data for correlation observables in azimuthal angle $\varphi_{D^0D^0}$ between D^0 mesons, in dimeson invariant mass $M_{D^0D^0}$ and in transverse momentum of the meson from the D^0D^0 pair. The inclusion of the new mechanisms with $g \to D^0$ fragmentation leads to a larger cross sections and to slightly different shapes of the calculated correlation observables.

1 Introduction

Some time ago we have predicted that at large energies, relevant for the Large Hadron Collider, production of double charm should be dominated by the double-parton scattering (DPS) mechanism [1]. In the next step those predictions based on leading-order (LO) collinear approach were improved by application of the k_t-factorization approach which effectively includes higher-order QCD effects [2, 3]. The extended studies provide a relatively good description of the LHCb experimental data [4] for both the total cross section and the dimeson correlation observables. Moreover, within these analyses the single-parton scattering (SPS) $gg \to c\bar{c}c\bar{c}$ contribution was also carefully discussed in both collinear [3] and k_t-factorization [5] approaches. The SPS contribution to the double $c\bar{c}$ cross section was found to be rather small, much smaller than the DPS one, and cannot explain the total and differential cross sections measured by the LHCb.

The studies introduced above were based on the standard $c \to D$ hadronization scenario with scale-independent Peterson fragmentation function (FF) [6]. An alternative approach for hadronization effects is to apply scale-dependent FFs of a parton (gluon, $u, d, s, \bar{u}, \bar{d}, \bar{s}, c, \bar{c}$) to D mesons proposed by Kniehl et al. [7, 8], that undergo DGLAP evolution equations. Both prescriptions were found to provide a very good description of the LHC data on inclusive D meson production (see e.g. Refs. [9, 10]). In the latter approach, a dominant contribution comes from $g \to D$ fragmentation that

*e-mail: rafal.maciula@ifj.edu.pl.pl, supported by the Polish National Science Center grant DEC-2014/15/B/ST2/02528

appears in the evolution of the scale-dependent FFs and the $c \to D$ component is damped with respect to the scale-independent fragmentation scheme.

Here we wish to discuss how the presence of the gluonic components change the overall picture for the double D-meson production. In the scale-dependent (new) scenario for hadronization with $g \to D$ fragmentation the number of contributing DPS processes grows compared to the standard (old) scenario with $c \to D$ fragmentation only. Moreover, a new single-parton scattering mechanism SPS $gg \to DD$ appears. In the new scenario we have more processes for single D meson production (two components) and as a consequence many more processes for the pair production in double-parton scattering. Now there are three classes of DPS contributions. In addition to the coventional DPS $cc \to DD$, considered in Refs. [2, 3, 5] there is a double $g \to D$ fragmentation mechanism, called here DPS $gg \to DD$ as well as the mixed DPS $gc \to DD$ contribution.

Here the gluon and digluon production is considered in the k_t-factorization approach with Reggeized gluons in the t-channel [11] via subprocesses $RR \to g$ and $RR \to gg$, where R is the Reggeized gluon. We use scale-dependent fragmentation functions of Kneesch-Kniehl-Kramer-Schienbein (KKKS08) [12] as implemented in the code available on the Web [13].

All details, relevant diagrams, and the theoretical formalism behind the calculations presented here can be found in our original paper [14] and is not repeated here. There, one can also find a revisited comparison of the theoretical predictions based on the two hadronization scenarios for single D meson production, calculated with the same set of α_S, scales, unPDFs and other details which are relevant for the discussion of double D meson production.

2 Comparison to the LHCb data

We compare now results of our theoretical predictions for double charm production with the LHCb experimental data for D^0D^0 pair production. In Fig. 1 we compare results of our calculation with experimental distribution in transverse momentum of one of the meson from the D^0D^0 (or $\bar{D}^0\bar{D}^0$) pair. We show results for the first scenario when standard Peterson FF is used for the $c \to D^0$ (or $\bar{c} \to \bar{D}^0$) fragmentation (left panel) as well as the result for the second scenario when the KKKS08 FFs with DGLAP evolution for $c \to D^0$ (or $\bar{c} \to \bar{D}^0$) and $g \to D^0$ (or $g \to \bar{D}^0$) are used. One can observe that the DPS $cc \to D^0D^0$ contribution in the new scenario is much smaller than in the old scenario. In addition, the slope of the distribution in transverse momentum changes. Both the effects are due to evolution of corresponding fragmentation function in the second scenario, compared to lack of such an effect in the first scenario. The different new mechanisms give contributions of similar size. We can obtain an agreement in the second case provided σ_{eff} parameter is increased from conventional 15 mb to 30 mb. Even then we overestimate the LHCb data for $3 < p_T < 5$ GeV. Possible effects that may result in larger value of σ_{eff} and in its transverse momentum dependence are discussed in our original paper [14].

In Fig. 2 we show dimeson invariant mass distribution $M_{D^0D^0}$ again for the two cases considered. In the first scenario we get a good agreement only for small invariant masses while in the second scenario we get a good agreement only for large invariant masses. The large invariant masses are strongly correlated with large transverse momenta, so the situation here (for the invariant mass distribution) is quite similar as in Fig. 1 for the transverse momentum distribution.

In Fig. 3 we show azimuthal angle correlation $\varphi_{D^0D^0}$ between D^0 and D^0 (or $\bar{D}^0$ and $\bar{D}^0$ mesons). While the correlation function in the first scenario is completely flat, the correlation function in the second scenario shows some tendency similar as in the experimental data. The increase at small $\Delta\varphi$ for the SPS $gg \to D^0D^0$ contribution is due to s-channel pole in the amplitude for $RR \to gg$ which we regularize by $\hat{s} > 4m_c^2$ condition. In the k_t-factorization, initial partons have transverse momenta, but final gluons may have equal rapidities even when $\Delta\varphi$ is far from π.

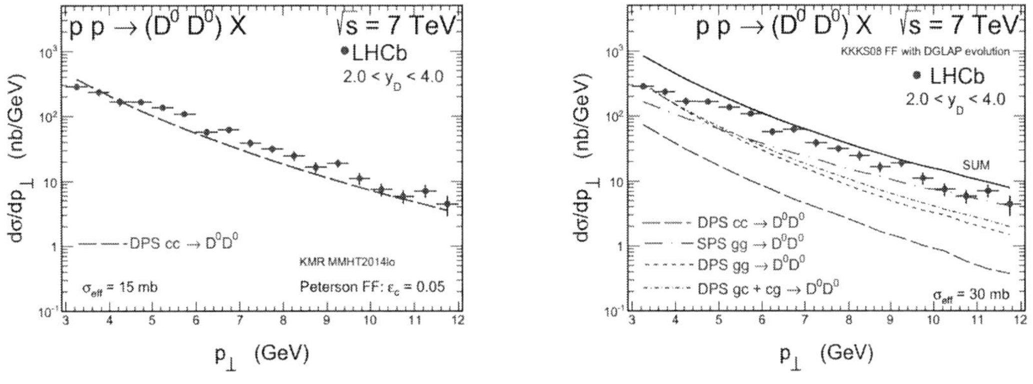

Figure 1. D^0 meson transverse momentum distribution within the LHCb acceptance region. The left panel is for the first scenario and for the Peterson $c \to D$ fragmentation function while the right panel is for the second scenario and for the fragmentation function that undergo DGLAP evolution equation.

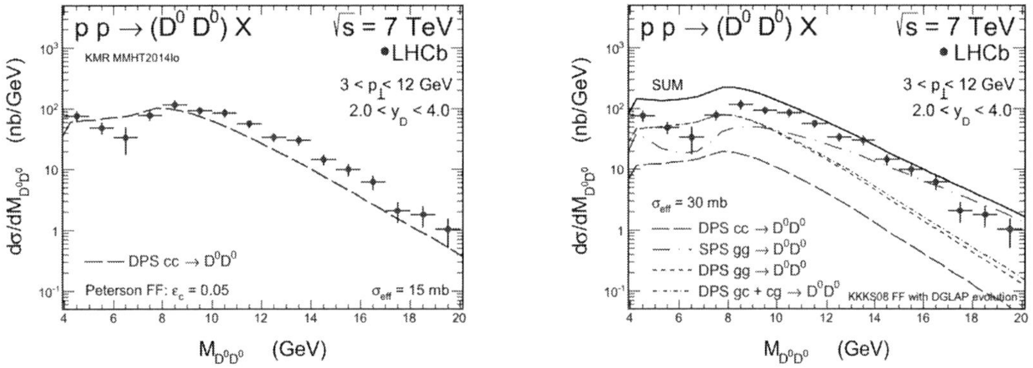

Figure 2. $M_{D^0 D^0}$ dimeson invariant mass distribution within the LHCb acceptance region. The left panel is for the first scenario and for the Peterson $c \to D$ fragmentation function while the right panel is for the second scenario and for the fragmentation function that undergo DGLAP evolution equation.

3 Conclusions

The new multi-component scenario for double D meson production give similar result as the first scenario with one subprocess ($cc \to DD$) and fixed (scale-independent) fragmentation function. However, correlation observables, such as dimeson invariant mass or azimuthal correlations between D mesons, are slightly better described in the second scenario as long as we consider only their shapes. To get the proper normalization of the cross sections calculated within the second scenario a much larger value of σ_{eff} is needed.

The observed overestimation of the correlation observables in the second scenario comes from the region of small transverse momenta. It may be related to the fact that the fragmentation function used in the new scenario were obtained in the DGLAP formalism with massless c quarks and $\bar{c}$ antiquarks which may be a too severe approximation, especially for low factorization scales (i.e. low trans-

 DOI: 10.1051/epjconf/201613005013

verse momenta) for fragmentation functions. On the other hand, the situation can be also improved when a proper transverse momentum dependence of σ_{eff} and/or when perturbative-parton-splitting mechanisms will be included, but this needs further studies.

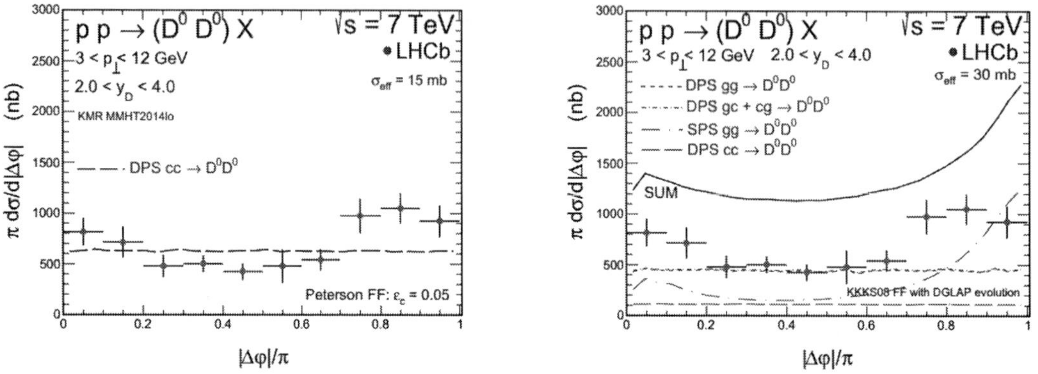

Figure 3. Distribution in azimuthal angle $\varphi_{D^0 D^0}$ between the two D^0 mesons within the LHCb acceptance region. The left panel is for the first scenario and for the Peterson $c \to D$ fragmentation function while the right panel is for the second scenario and for the fragmentation function that undergo DGLAP evolution equation.

We thank V. A. Saleev and A. V. Shipilova for collaboration in obtaining results presented here.

References

[1] M. Łuszczak, R. Maciuła and A. Szczurek, Phys. Rev. D **85**, 094034 (2012)
[2] R. Maciuła and A. Szczurek, Phys. Rev. D **87**, 074039 (2013)
[3] A. van Hameren, R. Maciuła and A. Szczurek, Phys. Rev. D **89**, 094019 (2014)
[4] R. Aaij *et al.* [LHCb Collaboration], J. High Energy Phys. **06**, 141 (2012); [**03**, 108 (2014)]
[5] A. van Hameren, R. Maciuła and A. Szczurek, Phys. Lett. B **748**, 167 (2015)
[6] C. Peterson, D. Schlatter, I. Schmitt and P. M. Zerwas, Phys. Rev. D **27**, 105 (1983)
[7] B. A. Kniehl and G. Kramer, Phys. Rev. D **71**, 094013 (2005)
[8] B. A. Kniehl and G. Kramer, Phys. Rev. D **74**, 037502 (2006)
[9] R. Maciula and A. Szczurek, Phys. Rev. D **87**, 094022 (2013)
[10] A. Karpishkov, M. Nefedov, V. A. Saleev and A. V. Shipilova, Phys. Rev. D **91**, 054009 (2015)
[11] M. A. Nefedov, V. A. Saleev and A. V. Shipilova, Phys. Rev. D **87**, 094030 (2013)
[12] T. Kneesch, B. A. Kniehl, G. Kramer and I. Schienbein, Nucl. Phys. B **799**, 34 (2008)
[13] KKKS08-package: http://lapth.cnrs.fr/ffgenerator/
[14] R. Maciuła, V. A. Saleev, A. V. Shipilova and A. Szczurek, Phys. Lett. B **758**, 458 (2016)
[15] J. R. Gaunt, R. Maciula and A. Szczurek, Phys. Rev. D **90**, 054017 (2014)

Exclusive ω meson production at COMPASS

Wolf-Dieter Nowak[1,*] and Paweł Sznajder[2**]
for the COMPASS Collaboration

[1]*Johannes Gutenberg University, Mainz, Germany*
[2]*National Centre for Nuclear Research, Warsaw, Poland*

Abstract. Exclusive ω meson production is studied by the COMPASS Collaboration using the CERN 160 GeV/c muon beam and a transversely polarised proton target. Single-spin and double-spin asymmetries are measured, some of which are sensitive to the Generalised Parton Distributions E that are related to quark orbital angular momenta. The results, which are sensitive also to the pion-pole contribution to the production mechanism, are compared to the predictions of a phenomenological model.

1 Introduction

Hard exclusive electro and muoproduction of mesons on nucleons has been playing an important role in studies of the hadron structure. Renewed interest arose when it was shown that it allows access to Generalised Parton Distributions (GPDs), which provide a comprehensive description of the partonic structure of the nucleon and contain a wealth of new information. For example, using GPDs the nucleon can be described as an extended object, which is referred to as 3-dimensional nucleon tomography, and access to quark orbital angular momentum becomes possible in principle.

At leading twist, the chiral-even GPDs H and E are sufficient to describe exclusive vector meson production on a spin-$\frac{1}{2}$ target. These GPDs are of special interest as they are related to the total angular momentum carried by partons in the nucleon [1]. Presently, the GPDs H are already well constrained in the accessible x_{Bj} range covered by HERA, HERMES and JLAB data. The constraints on the GPDs E are weak and come mainly from measurements of the nucleon Pauli form factor. The interpretation of the COMPASS results presented in this article is done in the framework of the GPD-based model developed by Goloskokov and Kroll [2, 3]. The model also includes some estimation of the pion-pole contribution to the mechanism of exclusive ω production [2], which is sizeable at COMPASS kinematics. A full account of the content of this article is given in Ref. [4].

2 Formalism

For exclusive ω meson production by a virtual photon, $\gamma^* p \to \omega p$, the total unpolarised cross section can be written as

$$\sigma_0 = \frac{1}{2}(\sigma_{++}^{++} + \sigma_{++}^{--}) + \epsilon\sigma_{00}^{++} = \sigma_T + \epsilon\sigma_L\,. \tag{1}$$

*e-mail: wolf-dieter.nowak@cern.ch
**Supported by the Polish NCN Grant DEC-2011/01/M/ST2/02350.

It is given by the sum of the cross sections for longitudinally and transversely polarised virtual photons, σ_L and σ_T, respectively. The virtual-photon polarisation parameter can be approximated by $\epsilon \simeq (1-y)/(1-y+y^2/2)$.

For the various possible combinations of the helicity of the virtual photon, denoted by $m, n = -1, 0, +1$, and the helicity of the initial-state proton, denoted by $i, j = -\frac{1}{2}, +\frac{1}{2}$, the photoabsorption cross sections or the interference terms are proportional to bilinear combinations of helicity amplitudes $\mathcal{M}$:

$$\sigma_{mn}^{ij} \propto \sum \mathcal{M}^{*}_{m'i',mi} \mathcal{M}_{m'i',nj}. \tag{2}$$

The sum runs over all spin combinations for the final state, given by the spin of the meson, $m' = -1, 0, +1$, and the spin of the final-state proton, $i' = -\frac{1}{2}, +\frac{1}{2}$. For brevity, a dependence on the kinematic variables is omitted here and in the following the helicities are labelled by only their signs or zero.

For exclusive meson production off a transversely polarised target, five single-spin asymmetries $A_{\rm UT}$ and three double-spin asymmetries $A_{\rm LT}$ can be defined:

$$\begin{aligned}
A_{\rm UT}^{\sin(\phi-\phi_s)} &= -\frac{\mathrm{Im}\left(\sigma_{++}^{+-} + \epsilon\sigma_{00}^{+-}\right)}{\sigma_0}, & A_{\rm LT}^{\cos(\phi-\phi_s)} &= \frac{\mathrm{Re}\sigma_{++}^{+-}}{\sigma_0}, & A_{\rm UT}^{\sin(\phi+\phi_s)} &= -\frac{\mathrm{Im}\sigma_{+-}^{+-}}{\sigma_0}, \\
A_{\rm UT}^{\sin(2\phi-\phi_s)} &= -\frac{\mathrm{Im}\sigma_{+0}^{-+}}{\sigma_0}, & A_{\rm LT}^{\cos(2\phi-\phi_s)} &= -\frac{\mathrm{Re}\sigma_{+0}^{-+}}{\sigma_0}, & A_{\rm UT}^{\sin(3\phi-\phi_s)} &= -\frac{\mathrm{Im}\sigma_{+-}^{-+}}{\sigma_0}, \\
A_{\rm UT}^{\sin\phi_s} &= -\frac{\mathrm{Im}\sigma_{+0}^{+-}}{\sigma_0}, & A_{\rm LT}^{\cos\phi_s} &= -\frac{\mathrm{Re}\sigma_{+0}^{+-}}{\sigma_0}.
\end{aligned} \tag{3}$$

Here, U and L denote unpolarised and longitudinally polarised beam, respectively, and T denotes transverse target polarisation. Each asymmetry is related to a specific modulation of the cross section in the angles ϕ and ϕ_s, as indicated by the superscript. The angle ϕ is the azimuthal angle between the lepton plane spanned by incoming and scattered lepton momenta, and the hadron plane spanned by virtual-photon and meson momenta. The angle ϕ_s is the azimuthal angle about the virtual-photon momentum, between the lepton plane and the direction of the transverse component of the polarisation of the target nucleon. The complete formula for the cross section can be found in Ref. [5].

3 COMPASS apparatus

The COMPASS experiment uses the high-intensity M2 muon beam line of the CERN SPS. During data taking, the nominal momentum of the μ^+ beam was 160 GeV/c with a spread of 5% and a longitudinal polarisation of about −80%. The average beam intensity was $3.5 \cdot 10^8$ μ^+/spill with a spill length of about 10 s every 40 s. Trajectory and momentum of each incoming beam particle were measured upstream the target.

The solid-state ammonia (NH_3) target contained transversely polarised protons with a polarisation of about 90%, which was generated by Dynamic Nuclear Polarisation. The target dilution factor, which is defined as cross-section-weighted fraction of polarisable to total material, had a value of about 25%. The direction of the target polarisation was reversed periodically, and three consecutive target cells with alternating polarisation direction (+-+ or -+-) were used to minimise systematic effects due to possible spectrometer instabilities and/or acceptance variation.

The COMPASS spectrometer is a 50 m long two-stage spectrometer with excellent tracking and particle identification capabilities. Each stage is defined by a dipole magnet with tracking detectors before and after it. High redundancy for track reconstruction is accomplished by about 300 tracking-detector planes in total. For muon identification, both stages are equipped with hadron absorbers that are situated between hodoscope stations. Each spectrometer stage contains an electromagnetic and a hadron calorimeter. The former are crucial to identify exclusive ω production, as they allow for the reconstruction of π^0 mesons originating from $\omega \to \pi^+\pi^-\pi^0$ decays. A detailed description of the apparatus can be found in Ref. [6].

4 Event selection

Data taken in the year 2010 are used for the analysis of exclusive ω production. Each selected event contains a primary vertex with exactly one incoming and one outgoing muon track and with exactly two outgoing hadron tracks of opposite charges, which are assumed to be pions. In the electromagnetic calorimeters, exactly two clusters have to be reconstructed that are correlated in time with the vertex and not associated to any charged particle. These clusters are assumed to originate from a pair of photons that was emitted from the primary vertex. The ω resonance is selected by constraining the reconstructed invariant mass as $|M_{\pi^+\pi^-\pi^0} - M_\omega^{\mathrm{PDG}}| < 0.07$ GeV$/c^2$, where $M_\omega^{\mathrm{PDG}} \approx 0.783$ GeV$/c^2$.

The recoiling target particle was not detected during 2010 data taking. The "exclusivity" of the reaction is instead accomplished by constraining the missing energy $E_{miss} = ((p+q-v)^2 - p^2)/2M_p$, where M_p is the proton mass and p, q and v are the four-momenta of proton, photon and meson, respectively. For exclusive events, the reconstructed value of E_{miss} has to be close to zero, and the constraint $|E_{miss}| < 3.0$ GeV is used to account for the experimental resolution in this measured variable. The squared transverse momentum with respect to the virtual-photon direction is required to be in the interval 0.05 (GeV$/c)^2$ $< p_T^2 <$ 0.5 (GeV$/c)^2$, where the lower limit suppresses contributions from coherent production on target nuclei and the upper one provides further reduction of non-exclusive background.

The kinematic region is defined by requiring 1 (GeV$/c)^2$ $< Q^2 <$ 10 (GeV$/c)^2$ on the negative invariant mass squared of the virtual photon, $Q^2 = -q^2$; $0.1 < y < 0.9$ on the fraction of lepton energy lost in the laboratory system, $y = (p\cdot q)/(p\cdot k)$; $0.003 < x_{Bj} < 0.35$ on the Bjorken-x variable; $W > 5$ GeV$/c^2$ on the invariant mass of the system of virtual photon and nucleon, $W = \sqrt{(p+q)^2}$. In the analysis of ω meson the signal and background asymmetries are extracted simultaneously using the unbinned maximum-likelihood method.

5 Results and interpretation

All eight transverse target asymmetries for exclusive ω meson production are shown in Fig. 1 (left). They are given for the mean values of kinematic variables, $\langle Q^2\rangle = 2.2$ (GeV$/c)^2$, $\langle x_{Bj}\rangle = 0.049$, $\langle p_T^2\rangle = 0.17$ (GeV$/c)^2$, $\langle W\rangle = 7.1$ GeV$/c^2$ and $\langle y\rangle = 0.18$. The single spin asymmetries measured as a function of Q^2, x_{Bj} or p_T^2 are shown in Fig. 1 (right).

In Fig. 1 (right) the measured asymmetries are compared to the predictions of the Goloskokov-Kroll model [2, 7]. The authors have provided three sets of predictions: one without the pion pole contribution and two with the contribution that differ by the sign of $\pi - \omega$ transition form factor. While the results for $A_{UT}^{\sin(\phi-\phi_s)}$ and $A_{UT}^{\sin(\phi-\phi_s)}$ asymmetries prefer the positive sign of the $\pi-\omega$ transition form factor, the results for $A_{UT}^{\sin\phi_s}$ asymmetry prefer the negative sign. At present the reason for this discrepancy between the model and the experimental data is unknown. The remaining asymmetries are not sensitive to the sign of $\pi - \omega$ transition form factor.

References

[1] X. Ji, Phys. Rev. Lett. **78** (1997) 610

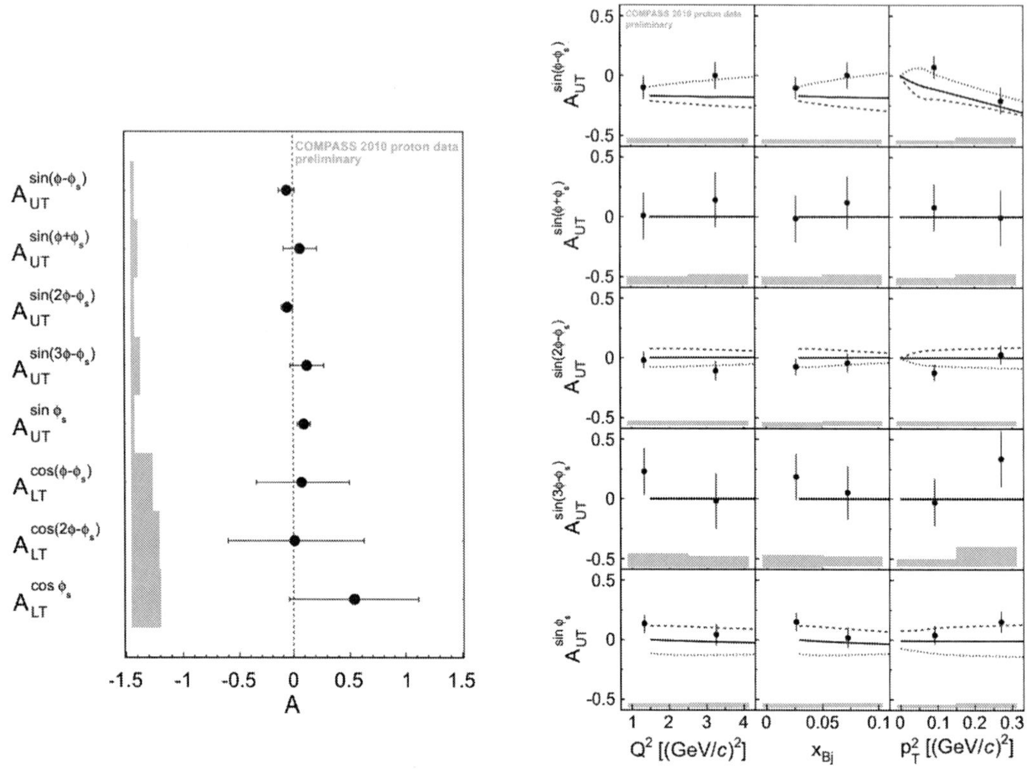

Figure 1. Left: The azimuthal asymmetries for exclusive ω meson production on transversely polarised protons. Error bars and left bands represent statistical and systematic uncertainties, respectively. Mean values of kinematic variables are indicated in the text. Right: Single spin azimuthal asymmetries for exclusive ω meson production on transversely polarised protons. The curves show the predictions of the GPD-based model [2, 7]. The dashed red and dotted blue curves represent the predictions with positive and negative $\pi\omega$ form factor, respectively, while the solid black curves represents the predictions without pion-pole contribution.

[2] S.V. Goloskokov and P. Kroll, Eur. Phys. J. **A50** (2014) 146
[3] S.V. Goloskokov and P. Kroll, Eur. Phys. J. **C74** (2014) 2725
[4] COMPASS Collaboration, C. Adolph *et al.*, arXiv:1606.03725 [hep-ex], subm. to Nucl. Phys. **B**
[5] M. Diehl and S. Sapeta, Eur. Phys. J. **C41** (2005) 515
[6] COMPASS Collaboration, P. Abbon *et al.*, Nucl. Instrum. Meth. **A577** (2007) 455
[7] S.V. Goloskokov and P. Kroll, private communication

EPJ Web of Conferences **130**, 05015 (2016) DOI: 10.1051/epjconf/201613005015
MESON 2016

Constraints on atmospheric charmed-meson production from IceCube

Tomasz Jan Palczewski[1,2,*] for the IceCube Collaboration [**]

[1] *University of California, Berkeley, USA*
[2] *Lawrence Berkeley National Laboratory, Berkeley, USA*

Abstract.
At very-high energies (100 TeV - 1 PeV), the small value of Bjorken-x ($\leq 10^{-3} - 10^{-7}$) at which the parton distribution functions are evaluated makes the calculation of charm quark production very difficult. The charm quark has mass ($\sim$1.5±0.2 GeV) significantly above the Λ_{QCD} scale ($\sim$200 MeV), and therefore its production is perturbatively calculable. However, the uncertainty in the data and the calculations cannot exclude some smaller non-perturbative contribution. To evaluate the prompt neutrino flux, one needs to know the charm production cross-section in pN -> $c\bar{c}$ X, and hadronization of charm particles. This contribution briefly discusses computation of prompt neutrino flux and presents the strongest limit on prompt neutrino flux from IceCube.

1 Introduction

IceCube's detection of ultra-high energy neutrino events heralds the beginning of neutrino astronomy. The flux of cosmic-rays incident on the atmosphere has been measured by a variety of experiments [1–3]. The dominant contribution to the neutrino flux at 10 GeV-1 TeV energies comes from the decays of charged pions and from leptonic and semi-leptonic decays of kaons [4]. The neutrino flux from pions and kaons, commonly referred to as the conventional neutrino flux, is well understood (steep energy spectrum $\sim E^{-3.7}$ below $\sim$100 TeV). Prompt neutrinos, produced in the atmosphere by decays of charmed hadrons (e.g. $D^0, D^\pm \rightarrow l\, \nu_{e,\mu}$ X) that come from cosmic-ray interactions with atmospheric nuclei (pN $\rightarrow c\bar{c}$ X), contribute to the atmospheric neutrino flux at high energies. The long-lived high energy pions and kaons interact before decaying into neutrinos. In contrast charmed particles have short lifetimes and decay into neutrinos independent of their energy up to $\sim 10^{17}$ eV, and arrival direction. Therefore, the prompt neutrino energy spectrum should follow the spectrum of primary cosmic-rays (energy spectrum of $\sim E^{-2.7}$ and an isotropic zenith distribution). The energy spectrum of astrophysical neutrinos is expected to follow the production spectrum at the cosmic-ray accelerator (in a simplistic approach: if Fermi shock acceleration is the responsible mechanism, a power law spectrum $\sim E^{-2}$ is expected).

*e-mail: tpalczewski@lbl.gov
**http://icecube.wisc.edu

2 IceCube detector and detection principle

IceCube is the world's largest neutrino detector, located near the geographic South Pole, instrumenting more than a cubic-kilometer of glacial ice. The detector is composed of Digital Optical Modules (DOMs) which are buried in ice, between depths of 1450m and 2450m, on 86 cables called "strings" [5]. On each string there are 60 DOMs. Each DOM consists of a 10-inch photomultiplier tube (PMT), calibration light sources, and digitizing electronics [6].

3 Estimation of prompt neutrino flux

IceCube observations [7] indicate that the neutrino flux is dominated by conventional atmospheric neutrinos and by cosmic neutrinos at low energies and high energies, respectively. The precise knowledge of the prompt flux is of a key importance for the cosmic neutrino flux measurements.

Many calculations of the prompt neutrino flux have been presented [8–13]. The estimation of the prompt neutrino flux requires knowledge of the charm production cross-section ($\sigma_{c\bar{c}}$) in the process pN $\to$ $c\bar{c}$ X and of the hadronization of charm particles. Theoretical predictions at high energies are uncertain due to uncertainties in charm mass, contributions from next-to-leading order (NLO) corrections, the factorization and renormalization scales, and the choice of parton distribution functions [14, 15]. The charm quark is produced with high fraction of the momentum of the incoming cosmic-ray projectile. The cross-section is sensitive to the domain of parton densities at very small values of Bjorken-x ($\leq 10^{-3} - 10^{-7}$). This region is poorly constrained by the experimental data. The small value of Bjorken-x at which the parton distribution functions are evaluated makes the calculation of charm production very difficult. The computation of the prompt neutrino flux also requires folding of the charmed hadronproduction cross-section with the flux of incoming cosmic-rays. This introduces uncertainty connected with the limited knowledge of the extremely high-energy cosmic ray composition. Recently, new updated parameterisations of the cosmic ray flux have been released [1, 16, 17] with improved description above the 'knee' and more sophisticated composition description. To obtain the prompt neutrino flux at the detector level the propagation of high energy particles and their decay products through the atmosphere needs to be described. The cascade formalism in the framework of the Z-momenta [18] is commonly used for this evaluation. This approach requires, among others, as an input the total inelastic proton-Air cross-section which has large uncertainties at the high energy region [19]. In addition, the assumption that the charm production cross-section scales with the mean atomic number of air as compared to the corresponding pp cross-sections needs to be done. However, small contribution from the nuclear modification effects cannot be fully rejected.

4 Validation of charm hadroproduction using LHC data

The LHC collider $\sqrt{s}$=7 TeV mode corresponds to a fixed target beam energy in pp collisions of 26 PeV. Therefore, the recent results from LHC [20–23] can constrain pQCD parameters (such as factorization and renormalization scales, parton distribution functions, estimates of cosmic ray spectra) which are crucial for the precise calculations of the charm hadroproduction. The charmed particle production cross-sections in pp collisions at a centre-of-mass energy of $\sqrt{s}$=7 and 13 TeV have been measured. As an example Fig. 1 shows measurements and predictions for absolute prompt D^0 cross-sections from pp at $\sqrt{s}$=13 TeV (LHCb experiment). The shapes of differential cross-sections for D^0, D^+, D^{*+}, D_s^+ are in agreement with NLO predictions (the predicted central values are generally below the data but within the uncertainty). These precise LHC results are crucial to increase our knowledge in this field. Although, modern collider experiments are limited due to the lack of the coverage in the very large rapidity region.

 DOI: 10.1051/epjconf/201613005015

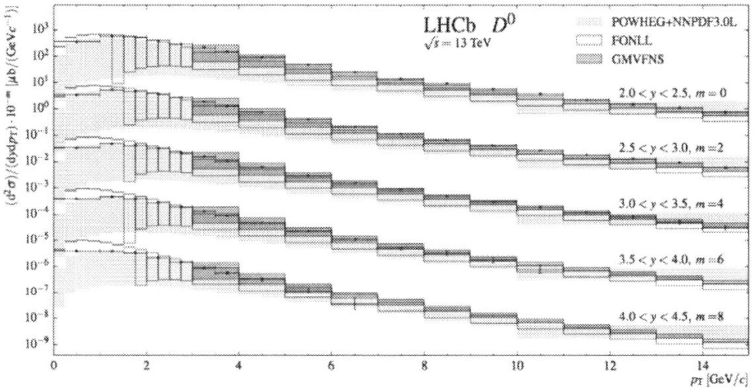

Figure 1. Measurements and predictions for the absolute prompt D^0 cross-section at $\sqrt{s}$=13 TeV. The boxes indicate the $\pm 1\sigma$ uncertainty band on the theory predictions. In cases where this band spans more than two orders of magnitude only its upper edge is indicated. Picture taken from [20].

5 Physics in the forward region

As it was pointed out in the Sect. 4, modern collider experiments, like experiments at RHIC and LHC, have limited coverage in the very large rapidity region. Consequently, the forward charmed particles are produced inside the beampipe. These particles dominate the high-energy atmospheric neutrino flux in underground experiments. In the literature, the forward charm production has been described in many ways and with different complexity, from intrinsic charm [24] to the inclusion of diffractive components in perturbative QCD [25]. In Ref. [26] authors showed that different forward physics assumptions modify the predictions of the prompt neutrino flux greatly (see Fig. 2).

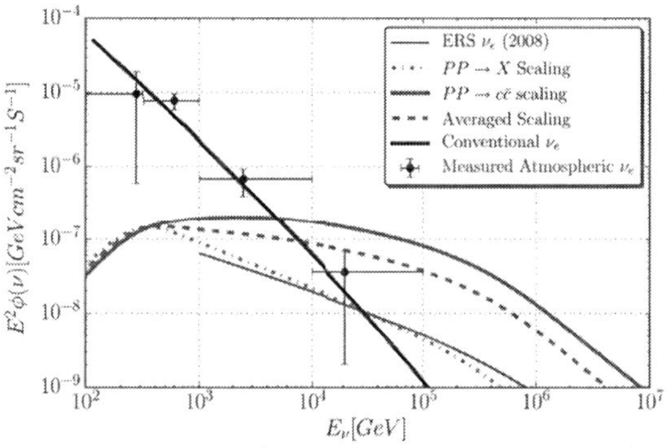

Figure 2. The prompt electron neutrino spectrum from forward charm is shown for extreme assumptions of the energy dependence. Also shown is the result for an intermediate dependence that exceeds the measured flux [27] at the 1σ level at the highest energy of 20 TeV. An estimate of the contribution from centrally produced charm particles by [27] is shown for comparison. The shown conventional atmospheric ν_e flux does not contain contribution from semileptonic decays of K_S which becomes important, regardless of small branching ratios for these decays, at high energies [28]. For more details see [26]. Picture taken from [26].

6 Prompt neutrino flux at IceCube

One of the key IceCube physics goals is searching for astrophysical neutrinos and prompt atmospheric neutrinos in the measured data. The main background for this type of analyses consists of atmospheric muons and conventional atmospheric neutrinos. These backgrounds can be highly suppressed by selecting high-quality upward-going events. In the down-going event region the so called self-veto procedure [29] (an atmospheric neutrino is vetoed when accompanied by atmospheric muons) and high deposited charge cut can be used to suppress backgrounds as well as a selection of the starting events inside the detector (part of the detector is used as a veto). Taking into account that neutrinos from charm have a harder spectrum than atmospheric neutrinos and that they are isotropic this provides an unique opportunity to confront the predicted prompt neutrino fluxes with the observed data. The recent binned likelihood analysis, with assumption of unbroken single power law astrophysical flux (described by two parameters: normalization Φ_{astro} at 100 TeV neutrino energy and the spectral index γ_{astro}) and per-bin expectations given by the sum of three components: conventional, prompt, and astrophysical, of up-going events in 6 years of data [30] gives to date the strongest prompt neutrino flux constraints by IceCube. The prompt normalization as a function of the normalization and spectral index is shown in Fig. 3 (left). It is clearly seen that non-zero prompt normalization appears only for strong deviations from the best fit astrophysical spectrum. Fig. 3 (middle) shows the joint three-dimensional 90% confidence region for the prompt flux and the astrophysical parameters. The maximum prompt flux in the three-dimensional confidence region is 1.06 × prediction taken from [31] (ERS) and can be treated as conservative upper limit. Recent selected perturbative QCD calculations [32–34] are shown in Fig. 3 (right) with the IceCube upper limit from the discussed likelihood 6 year data analysis. The GRSST (H3p) pQCD calculation is based on the same set of tools, POWHEG and NNPDF3.0L, used to calculate predictions of the absolute prompt particle cross-sections at LHC [32] (example shown in Fig. 1)

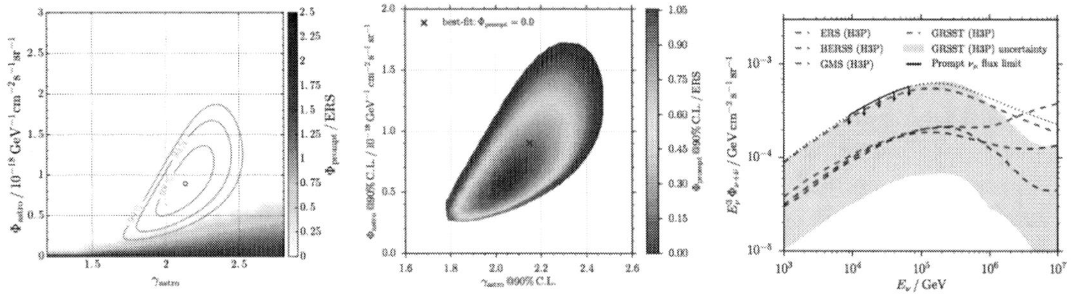

Figure 3. Left: Best-fit prompt normalization in ERS units for each scan point Φ_{astro} and γ_{astro}. Additionally, the two-dimensional contours for Φ_{astro} and γ_{astro} are shown; Middle: 90% CL contour assuming Wilks? theorem based on a three dimensional profile likelihood scans of the astrophysical parameters Φ_{astro} and γ_{astro} and the prompt normalization in ERS units; Right: Prompt atmospheric muon neutrino flux predictions from selected perturbative QCD calculations (dashed lines) in comparison to the upper limit from IceCube likelihood 6 year data analysis based on the up-going data sample. For more details see [30]. Pictures taken from [30].

IceCube has developed techniques for the extraction of physical measurements from atmospheric muon events that can be used to estimate the prompt component [35]. These techniques will not be discussed in this contribution. However, this is an important work as it gives independent verification of the prompt measurement. Currently, a definite measurement of the prompt flux using muon data is not yet possible. Depending on which assumption is chosen for the systematic error, the final result varies considerably (for more details see [35]). In contrast to neutrinos, for muons additional decays of neutral vector mesons contribute to the prompt neutrino flux at an uncertain level [36, 37]. Once systematic effects are fully understood and controlled, it will be possible for the IceCube detector to measure the prompt contribution using muon data.

7 Summary

In the low-energy regime, atmospheric neutrino flux, which mainly comes from pion and kaon intermediate states, is well known. Proper calculation of the prompt neutrino flux is complex and depends on many theoretical aspects briefly discussed in this contribution. Atmospheric lepton flux is a background to cosmic neutrino flux searches but it is interesting in its own right as it probes low-x QCD.

DOI: 10.1051/epjconf/201613005015

A precise measurement of prompt atmospheric muons and neutrinos could therefore complement the knowledge of differential cross sections for parton interactions and structure functions in accelerator physics.

References

[1] T. Stanev, T. K. Gaisser, S. Tilav, Nucl. Instrum. Meth. A **742**, 42 (2014)
[2] E. S. Seo, Astropart. Phys. **39**, 76 (2012)
[3] K. H. Kampert, M. Unger, Astropart. Phys. **35**, 660 (2012)
[4] T. K. Gaisser, Earth Planets Space **62**, 195 (2010)
[5] R. Abbasi *et al.* (IceCube Collaboration), Nucl. Instrum. Meth. A **601**, 294 (2009)
[6] M. G. Aartsen *et al.* (IceCube Collaboration), JINST **9**, P03009 (2014)
[7] M. G. Aartsen *et al.* (IceCube Collaboration), Phys. Rev. D **91**, 022001 (2015)
[8] P. Lipari, Astropart. Phys. **1**, 195 (1993)
[9] L. Pasquali, M. Reno, I. Sarcevic, Phys. Rev. D **59**, 034020 (1999)
[10] R. Enberg, M. H. Reno, I. Sarcevic, Phys. Rev. D**78**, 043005 (2008)
[11] P. Gondolo, G. Ingelman, M. Thunman, Astropart. Phys. **5**, 309 (1996)
[12] A. Martin, M. Ryskin, A. Stasto, Acta Phys. Polon. B **34**, 3273 (2003)
[13] G. Gelmini, P. Gondolo, G. Varieschi, Phys. Rev. D **61**, 036005 (2000)
[14] I. V. Rakobolskaya, T. M. Roganova, L. G. Sveshnikovaa, Nucl. Phys. B **122**, 353 (2003)
[15] A. Bhattacharya, R. Enberg, M. H. Reno, I. Sarcevica, A. Stasto, JHEP **06**, 110 (2015)
[16] T. K. Gaisser, Astropart. Phys. **35**, 801 (2012)
[17] T. Stanev, T. K. Gaisser, S. Tilav, Front. Phys. China **8**, 748 (2014)
[18] P. Lipari, Astropart.Phys. **1**, 195 (1993)
[19] P. Abreu *et al.* (Pierre Auger Observatory), Phys. Rev. Lett. **109**, 062002 (2012)
[20] R. Aaij *et al.* (LHCb Collaboration), JHEP **03**, 159 (2016)
[21] B. Abelev *et al.* (ALICE Collaboration), JHEP **07**, 191 (2012)
[22] B. Abelev *et al.* (ALICE Collaboration), Phys. Lett. B **718**, 279 (2012)
[23] B. Abelev *et al.* (ALICE Collaboration), JHEP **01**, 128 (2012)
[24] S. J. Brodsky, A. Kusina, F. Lyonnet, I. Schienbein, H. Spiesberger, R. Vogt, Adv. High Energy Phys. **2015**, 231547 (2015)
[25] V. D. Barger, F. Halzen, W. Y. Keung, Phys. Rev. D **25**, 112 (1982)
[26] F. Halzen, L. Wille, Phys. Rev. D **94**, 014014 (2016)
[27] R. Enberg, M. H. Reno, I. Sarcevic, Phys. Rev. D **78**, 043005 (2008)
[28] T. Gaisser, S. R. Klein, Astro. Phys. **64**, 13 (2015)
[29] T. K. Gaisser, K. Jero, A. Karle, J. V. Santen, Phys. Rev. **D 90**, 023009 (2014)
[30] M. G. Aartsen *et al.* (IceCube Collaboration), arXiv: **1607.08006** [astro-ph.HE], (2016)
[31] R. Enberg, M. H. Reno, I. Sarcevic, Phys.Rev. D **78**, 043005 (2008)
[32] R. Gauld, J. Rojo, L. Rottoli, S. Sarkar, J. Talbert, JHEP **02**, 130 (2016)
[33] M. V. Garzelli, S. Moch, G. Sigl, JHEP **10**, 115 (2015)
[34] A. Bhattacharya, R. Enberg, M. H. Reno, I. Sarcevic, A. Stasto, JHEP **06**, 110 (2015)
[35] M. G. Aartsen *et al.* (IceCube Collaboration), arXiv: **1506.07981** [astro-ph.HE], (2016)
[36] J. I. Illana, P. Lipari, M. Masip, D. Meloni, Astropart. Phys. **34**, 663 (2011)
[37] J. I. Illana, M. Masip, D. Meloni, JCAP **0909**, 008 (2009)

Electromagnetic effects on meson production: a new tool for studying the space-time evolution of heavy ion collisions

Andrzej Rybicki[1,⋆], Antoni Szczurek[1,2], Mariola Kłusek-Gawenda[1], Nikolaos Davis[1], Vitalii Ozvenchuk[1], and Mirosław Kiełbowicz[1]

[1] *H. Niewodniczański Institute of Nuclear Physics, Polish Academy of Sciences, Radzikowskiego 152, 31-342 Kraków, Poland*
[2] *University of Rzeszów, Rejtana 16, 35-959 Rzeszów, Poland*

Abstract. We review our studies of spectator-induced electromagnetic (EM) effects on the emission of charged mesons in the final state of ultrarelativistic heavy ion collisions. We argue that these effects offer sensitivity to the distance d_E between the charged meson formation zone at freeze-out and the spectator system. As such, they can serve as an independent, new tool to probe the space-time and longitudinal evolution of the system created in the collision. As a phenomenological application for this tool in the context of resonance production and decay, we obtain a first estimate of the time of pion emission from EM effects. This we compare to existing HBT data.

1 Introduction

In the following we address the issue of final state electromagnetic (EM) interactions between the charged mesons emitted in high energy heavy ion collisions and the positively charged nuclear remnant ("spectator system") which does not participate directly in the collision. We will argue that the latter effects can be used as a new tool to study the space-time features of the reaction, applicable over a wide range of collision centrality.

2 EM effects on charged meson spectra and directed flow

An overview of the influence of the EM field from the spectator system over the phase space available for π^+ and π^- meson production is presented in Fig. 1. The measurements come from the NA49 experiment at the CERN SPS, for peripheral and intermediate centrality Pb+Pb collisions at $\sqrt{s_{NN}} = 17.3$ GeV. The invariant double differential spectra of positive (negative) pions (panels a,b) display an evident strong depletion (enhancement) at low transverse momenta in the vicinity of $x_F = 0.15 = m_\pi/m_p$, which corresponds to pions moving parallel to the spectator system and with the same velocity (rapidity). This apparent repulsion of positive (attraction of negative) pions from spectator vicinity results in a very large, isospin symmetry-breaking distortion of π^+/π^- ratios as a function of pion transverse momentum as apparent in panel (c). What may, at first glance, appear surprising is that

⋆e-mail: andrzej.rybicki@ifj.edu.pl

the centrality dependence of this effect is quite weak at least between peripheral and intermediate collisions. However, this can be understood from simple geometrical considerations on the interplay between the decrease of total spectator electric charge and that of the collision impact parameter b with increasing centrality. Analogous EM-induced effects are present for collective azimuthal anisotropies (directed flow) for charged pion emission w.r.t. the reaction plane as described in [1].

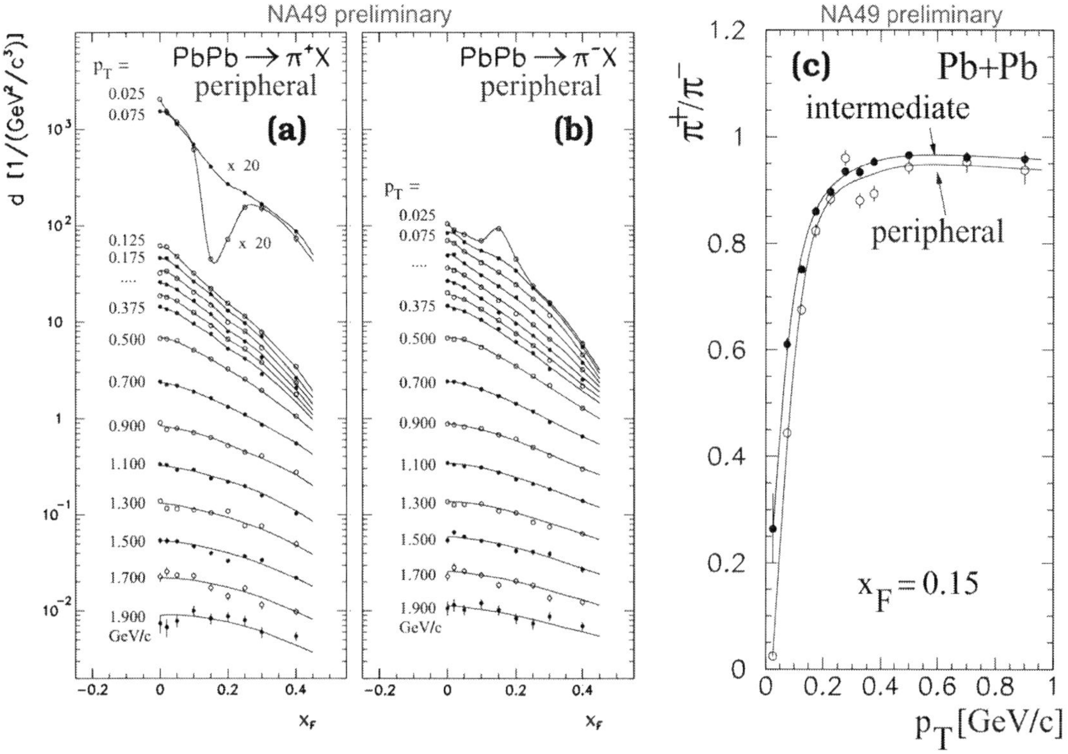

Figure 1. Spectator-induced EM effects on final state charged π meson spectra and π^+/π^- ratios in Pb+Pb collisions at $\sqrt{s_{NN}} = 17.3$ GeV. Double differential $d = E\frac{d^3N}{dp^3}$ distributions of (a) positive and (b) negative pions drawn as a function of $x_F = \frac{2p_L}{\sqrt{s_{NN}}}$ and transverse momentum p_T (both quantities in the collision c.m.s.), and (c) ratio of π^+ over π^- distributions at $x_F = 0.15$.

3 Sensitivity of EM effects to the space-time evolution of the system

As the electromagnetically modified trajectory connects the charged meson produced at freeze-out with its final state measured in the detector, it is probably not very surprising that the EM-induced distortion of the latter appears sensitive to the space-time evolution of the particle production process. As this was demonstrated in [2, 3], the overall shape of the electromagnetic distortion of charged pion spectra, and of charge-dependent effects on directed flow, is directly sensitive to the distance d_E between the longitudinal position of the pion created at freeze-out, and the longitudinal position of the spectator system at the moment (in the collision c.m.s. time) at which the pion is created as shown in Fig. 2 (a). Our subsequent studies based on a simulation of the relativistic motion of charged particles in the EM field indicate that the distortion of π^+ and π^- spectra shown in Fig. 1

 DOI: 10.1051/epjconf/201613005016

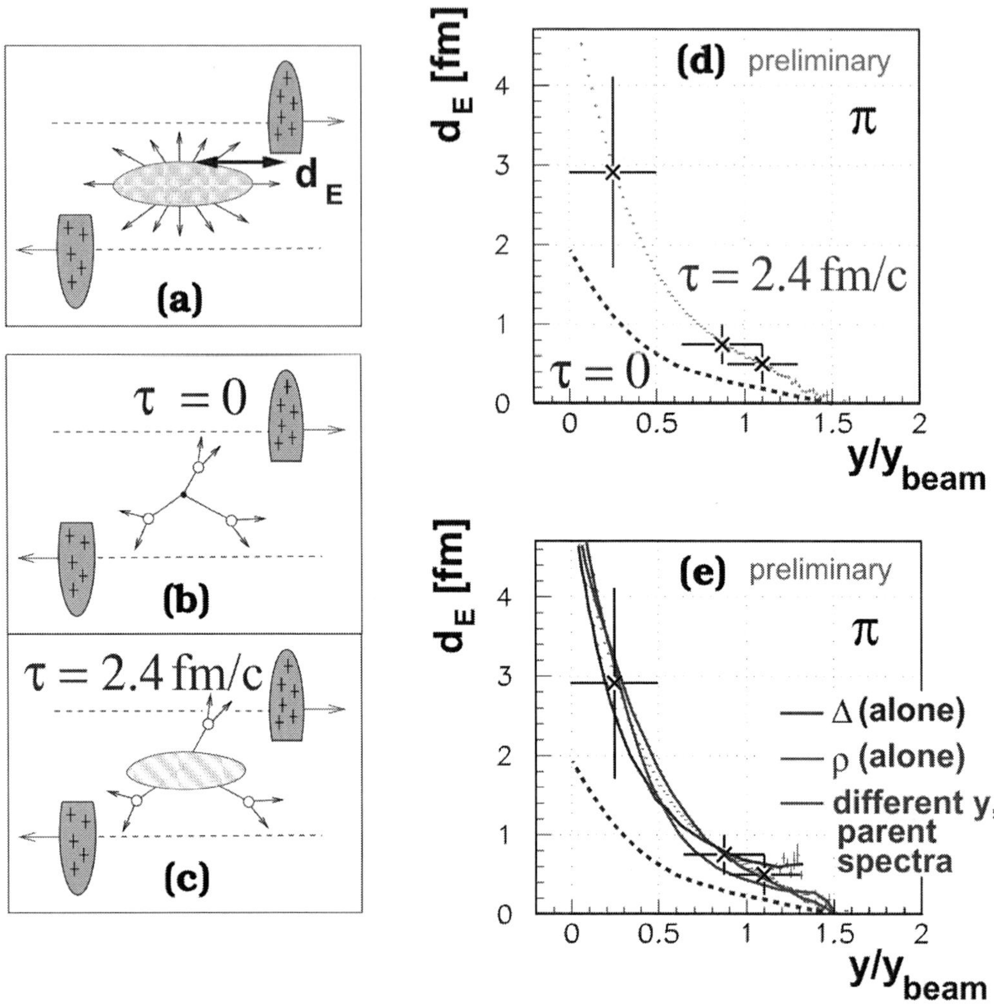

Figure 2. Information obtained on the basis of EM effects described in section 2: (a) schematic definition of d_E, (b) simulation of resonances produced instantly at the moment of the collision ($\tau = 0$) and (c) produced from an intermediate system characterized by a given proper lifetime τ, (d) dependence of d_E on pion rapidity deduced from EM effects (data points), compared to the results of our Monte Carlo simulation illustrated in panels (b) and (c), respectively. (e) uncertainty of the present analysis induced by our incomplete knowledge of resonance production, estimated by assuming different characteristics for the input mix of parent Δ, ρ resonances. All the panels originate from [4].

corresponds to an average longitudinal position of the formation zone of fast ($y \approx y_{beam}$) pions being about $d_E \approx 0.75$ fm behind the spectator system. Simultaneously, the charge splitting of directed flow seen by the STAR Collaboration for Au+Au collisions at $\sqrt{s_{NN}} = 7.7$ GeV [5] would correspond to $d_E \approx 3$ fm for pions produced at low c.m.s. rapidities, while the very large values of π^+ directed flow seen in target vicinity in top SPS energy Pb+Pb collisions by the WA98 experiment [6] would suggest $d_E \approx 0.5$ fm. The situation is summarized in Fig. 2 (d), where a decrease of d_E with increasing pion rapidity is evident. As we described in [4], a Monte Carlo model based on a compilation of knowledge existing on baryonic (Δ) and mesonic (ρ) resonance production and decay can serve as a simple tool for a further investigation of this result. A tentative simulation assuming *immediate* production of these low-lying resonances at the interaction point – $\tau = 0$, as shown in Fig. 2(b) – fails to reproduce the values of d_E obtained from EM effects. This is no longer the case for the space-time evolution simulated as shown in Fig. 2(c), which indicates that the space-time properties and proper lifetime τ of the intermediate system of hot and dense matter created in the collision can be independently studied, and estimated, by means of spectator-induced EM interactions. A tentative evaluation of uncertainties induced by our incomplete knowledge of resonance production is presented in Fig. 2(e). This suggests that the latter uncertainties will in fact remain below the tentative error bars associated at present to our d_E extraction procedures mentioned above. As the latter can definitely be reduced by future experimental and theoretical effort, it can be expected that an overall improvement in knowledge of space-time evolution of the system up to very high rapidities can be obtained from EM effects.

At the present moment, the space-time characteristics of this system, tentatively established from these effects in the collision energy range $\sqrt{s_{NN}} = 7.7 - 17.3$ GeV discussed here and shown in Figs 2(d) and 2(c), can give an independent estimate for the time of pion emission as a function of its rapidity. At midrapidity this time being equal to $d_E(y/y_{\mathrm{beam}} = 0)/\beta$, where β is the spectator velocity, our preliminary estimate gives 5.3 ± 2.2 fm/c for the time of pion emission in this region. For comparison, the pion decoupling times compiled in the published HBT analysis by the ALICE Collaboration [7] show values of approximately $\tau_f \approx 6$ fm/c for the same range of collision energies[1].

4 Conclusions

We conclude at present that spectator-induced electromagnetic effects are a promising field which can be used to obtain new information of the space-time evolution of the system created in heavy ion collisions. There are good reasons to believe that the applicability of these effects for the latter purpose would extend over a broad range of centralities and not remain confined to peripheral reactions, as shown in Fig. 1(c). A further development of these studies, including in particular new measurements at SPS energies [8], will reinforce the present relatively limited database on these phenomena at high energies. Consequently, this can be a chance to significantly deepen our present understanding of the space-time evolution and longitudinal expansion of hot and dense matter created in the course of the collision.

The authors warmly thank the organizers of the MESON 2016 conference, Kraków, Poland. This work was supported by the National Science Centre, Poland (grant no. 2014/14/E/ST2/00018).

References

[1] A. Rybicki, A. Szczurek and M. Kłusek-Gawenda, *Acta Phys. Polon. B* **46**, no. 3, 737 (2015), and references therein
[2] A. Rybicki, A. Szczurek, *Phys. Rev. C* **75**, 054903 (2007)
[3] A. Rybicki and A. Szczurek, *Phys. Rev. C* **87**, 054909 (2013)
[4] A. Rybicki, A. Szczurek, M. Kiełbowicz, N. Davis and V. Ozvenchuk, *Acta Phys. Polon. Supp.* **9**, 303 (2016), and references therein
[5] STAR Collab., L. Adamczyk *et al.*, *Phys. Rev. Lett.* **112**, 162301 (2014)
[6] WA98 Collab., H. Schlagheck, *Nucl. Phys. A* **663**, 725 (2000)
[7] ALICE Collab., K. Aamodt *et al.*, *Phys. Lett. B* **696**, 328 (2011)
[8] NA61 Collab., N. Abgrall *et al.*, CERN-SPSC-2015-038; SPSC-P-330-ADD-8

[1]The corresponding values of τ_f were directly read off the plot published in [7].

EPJ Web of Conferences **130**, 05017 (2016) DOI: 10.1051/epjconf/201613005017
MESON 2016

Photoproduction of J/ψ and Υ in exclusive and proton dissociative diffractive events

Wolfgang Schäfer[1,⋆], Anna Cisek[2], and Antoni Szczurek[1,2]

[1] *Institute of Nuclear Physics Polish Academy of Sciences, PL-31-412 Kraków, Poland*
[2] *University of Rzeszów, Rzeszów, Poland*

Abstract. We use a perturbative QCD based k_T-factorization approach, to calculate the amplitude for the diffractive $\gamma p \to Vp$ processes, where V is a J/ψ or Υ ground state or excited vector meson. Using these amplitudes, we evaluate the cross section for exclusive photoproduction of $J/\psi, \psi', \Upsilon$ mesons in proton-proton collisions. Calculations are performed for a variety of unintegrated gluon distributions, and we compare our results to LHCb data. Absorption effects are taken into account at the amplitude level. We also discuss the related diffractive production in proton dissociative events. Here we concentrate on electromagnetic dissociation , which is calculable without additional free parameters. Besides being of interest in their own right, dissociative events constitute an important experimental background to exclusive production.

1 Introduction

Exclusive processes with large rapidity gaps of the type $pp \to pph$, where h is a hadron produced in the central rapidity region, are described by t-channel exchange mechanisms. If the rapidity gaps are required to be larger than $2 \div 3$ units, the exchange must have a "running spin" $J(t) \geq 1$. If we wish to study the production of vector mesons, the product of C-parities of the exchanges must be negative. All these rquirements are fulfilled for the photon-Pomeron fusion.

In this way, pp interactions allow us to study high-energy photoproduction of vector mesons at previously unaccessible energies [1]. A rich body of experimental data from ep collisions taken at HERA, and a large amount of theoretical works, have convincingly established the diffractive vector meson production as a testing ground of the transition between soft and hard QCD Pomeron physics (for a review, see [2]).

Differently from ep collision, in the pp case, due to the electromagnetic form factors (finite size) of the proton, virtualities of photons are strongly bounded (photons are quasireal) in the exclusive $pp \to Vpp$ processes. This means that photon virtuality Q^2 cannot serve as a hard scale. In order to study the perturbative QCD (pQCD) driven hard Pomeron, we have to restrict ourselves to heavy quarkonia. Secondly, both of the incoming protons can be a source of photons, so that two amplitudes have to be added coherently. Thirdly, protons are strongly interacting, and the many open inelastic channels require the evaluation of absorptive corrections to the naive γ–Pomeron fusion amplitude. For a discussion of these issues and explicit formulas, see e.g. [3, 4].

⋆e-mail: wolfgang.schafer@ifj.edu.pl

2 Exclusive photoproduction of J/ψ, $\psi(2S)$, Υ

The cross section of the $pp \to ppV$ reaction needs as an input the photoproduction amplitude for the $\gamma p \to Vp$ subprocess. For heavy quarkonia the Pomeron exchange can be modelled by a pQCD gluon ladder. We write the full amplitude at finite transverse momentum transfer $\Delta_\perp$ as:

$$\mathcal{M}(W, \Delta_\perp^2) = (i + \rho)\, \Im m \mathcal{M}(W, \Delta_\perp^2 = 0) \cdot f(-\Delta_\perp^2, W)\,. \tag{1}$$

The imaginary amplitude of the forward amplitude can be calculated in terms of the unintegrated gluon distribution in the target proton and the light-cone wave function of the vector meson. For the formalism used, see [4].

The real part of the amplitude is restored from analyticity,

$$\rho = \frac{\Re e \mathcal{M}}{\Im m \mathcal{M}} = \tan\Big(\frac{\pi}{2}\, \frac{\partial \log\big(\Im m \mathcal{M}/W^2\big)}{\partial \log W^2}\Big).$$

The dependence on momentum transfer $t = -\Delta_\perp^2$ is parametrized by the function $f(t, W)$, which dependence on energy derives from the Regge slope

$$B(W) = b_0 + 2\alpha'_{eff} \log\Big(\frac{W^2}{W_0^2}\Big), \tag{2}$$

where we use: $b_0 = 4.88$, $\alpha'_{eff} = 0.164\ \mathrm{GeV}^{-2}$ and $W_0 = 90$ GeV.

Within the diffraction cone it is customary to use the exponential parametrization:

$$f(t, W) = \exp\Big(\frac{1}{2} B(W) t\Big). \tag{3}$$

For our purposes however an extension to larger $|t| \sim 1 \div 2\,\mathrm{GeV}^2$ is desirable. A comparison with ZEUS data [5] (see [4]) shows that a "stretched exponential" parametrization

$$f(t, W) = \exp(\mu^2 B(W)) \exp\Big(-\mu^2 B(W) \sqrt{1 - t/\mu^2}\Big)\,, \tag{4}$$

does a good job with only one new parameter. Different physical process can give rise to the stretched-exponential behaviour, such as a contamination of data by dissociative events, multiple scatterings, or – somewhat unlikely – "hard" scattering.

The exclusive photoproduction of J/ψ and $\psi'(2S)$ charmonia has first been observed by the CDF collaboration [6] and has recently been measured by the LHCb collaboration [7, 8]. In Fig. 1, we compare our calculations of the J/ψ rapidity distributions to the LHCb data, a similar picture is observed for $\psi(2S)$ production (see [4]).

In the leftmost panels we show the result using the Ivanov-Nikolaev glue [9]. While it gives a reasonable account of the shape of the rapidity distribution, it overestimates the cross section. A similar job is done by a gluon distribution of [10], which solves a BFKL-type linear evolution equation. A quite good agreement is found for the unintegrated glue of Kutak and Staśto [10] from the nonlinear evolution equation (rightmost panels). We stress that all the unintegrated gluon distributions reproduce the Tevatron data [6], see [4]. We wish to point out, that also a straight-line extrapolation of a H1-fit used previously in [3] gives a good fit of the cross section (see the left panel of Fig. 2). This would correspond to the exchange of a single Pomeron trajectory and casts some doubts on the necessity of saturation effects in the unintegrated glue. In the right panel of Fig. 2 we show a comparison of our results for Υ production compared to recent LHCb data [11]. The large error bars do not allow us to distinguish different choices for the diffractive slope.

DOI: 10.1051/epjconf/201613005017

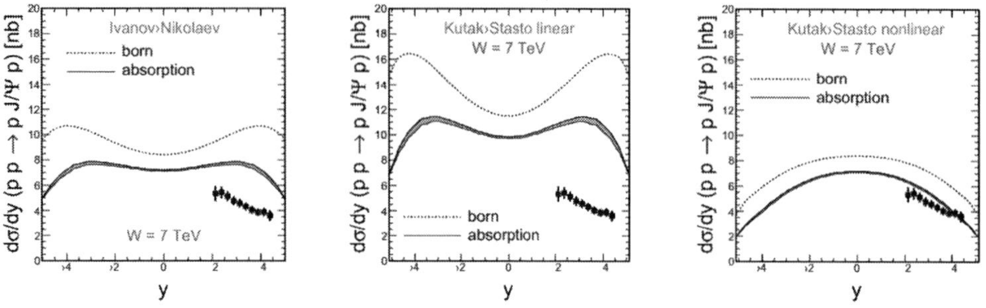

Figure 1. J/ψ rapidity distribution calculated with inclusion of absorption effects (solid lines), compared with the Born result (dashed line) for $\sqrt{s} = 7$ TeV. The LHCb data points [8] are shown for comparison.

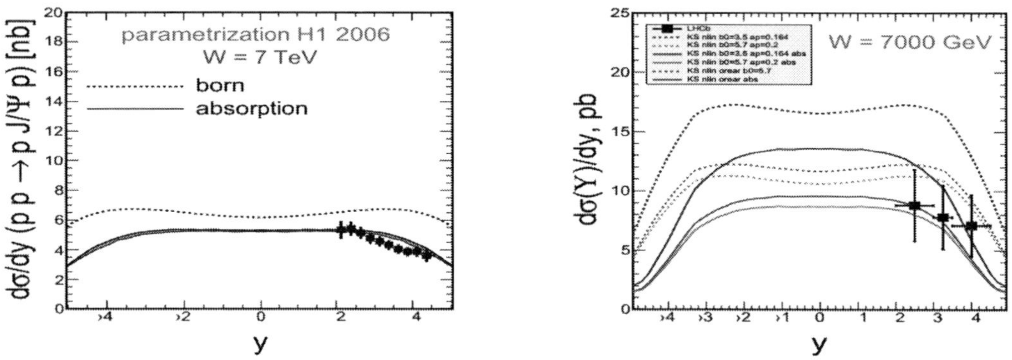

Figure 2. Left panel: J/ψ rapidity distribution calculated with a parametrization of the photoproduction amplitude previously used in [3]. Right panel: Rapidity distribution of exclusive Υ's. Data are taken from [11].

3 Diffractive photoproduction with electromagnetic dissociation

Proton dissociative events are of great practical importance as an often difficult background to exclusive events. Here we concentrate on a simple example relevant to exclusive vector meson production: the events with electromagnetic dissociation of one of the protons. The cross section for such processes is:

$$\frac{d\sigma(pp \to XVp; s)}{dy d^2\mathbf{p}} = \int \frac{d^2\mathbf{q}}{\pi \mathbf{q}^2} \mathcal{F}^{(\mathrm{in})}_{\gamma/p}(z_+, \mathbf{q}^2) \frac{1}{\pi} \frac{d\sigma^{\gamma^* p \to Vp}}{dt}(z_+ s, t = -(\mathbf{q} - \mathbf{p})^2) + (z_+ \leftrightarrow z_-), \qquad (5)$$

where $z_\pm = e^{\pm y} \sqrt{(\mathbf{p}^2 + m_V^2)/s}$. Here the flux of photons associated with the breakup of the proton is calculable in terms of the structure function F_2 of a proton, see [12]: Of special interest is the region of small invariant masses, $M_X < 2$ GeV, where resonance excitation is important. Our recent study (second ref. in [12]) showed, that here it is best to use a parametrization of F_2 given in [13]. In the left panel of Fig. 3 we show a distribution in invariant mass of the excited system for the $pp \to XJ/\psi p$ reaction. In the right panel of Fig. 3 we show the rapidity distribution of J/ψ for elastic and e.m. dissociative events. The e.m. dissociation can reach $10 \div 15\%$ of the exclusive cross section.

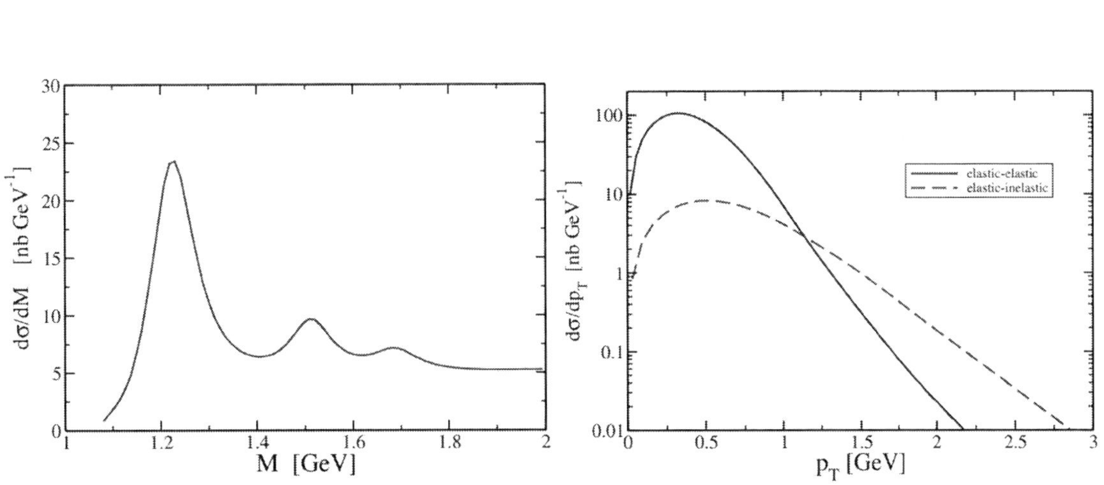

Figure 3. Left panel: distribution in the mass of the electromagnetically excited system, M_X for the $pp \to XJ/\psi p$ process. Right panel: Rapidity distribution of J/ψ for the same process (dashed line), for $M_X < 2$ GeV as well as for the exclusive diffractive production.

4 Conclusions

We have compared our $k_\perp$-factorization results with recent LHCb ($pp \longrightarrow p\ V\ p$) data, for $V = J/\psi, \psi(2S), \Upsilon$. The best description is obtained for a glue which does contain gluon saturation effects. Proton dissociation is a background to exclusive processes. Electromagnetic dissociation is calculable from F_2, excited states $M_X < 2$ GeV make a contribution of $10 \div 15\%$ of the exclusive cross section for J/ψ production.

The work reported here was partially supported by the Polish National Science Centre grant DEC-2014/15/B/ST2/02528.

References

[1] S. R. Klein and J. Nystrand, Phys. Rev. Lett. **92** (2004) 142003.
[2] I. P. Ivanov, N. N. Nikolaev and A. A. Savin, Phys. Part. Nucl. **37** (2006) 1.
[3] W. Schäfer and A. Szczurek, Phys. Rev. D **76** (2007) 094014.
[4] A. Cisek, W. Schäfer and A. Szczurek, JHEP **1504** (2015) 159.
[5] S. Chekanov *et al.* [ZEUS Collaboration], Eur. Phys. J. C **24** (2002) 345.
[6] T. Aaltonen *et al.* [CDF Collaboration], Phys. Rev. Lett. **102** (2009) 242001.
[7] R. Aaij *et al.* [LHCb Collaboration], J. Phys. G **40** (2013) 045001.
[8] R. Aaij *et al.* [LHCb Collaboration], J. Phys. G **41** (2014) 055002.
[9] I. P. Ivanov and N. N. Nikolaev, Phys. Rev. D **65** (2002) 054004.
[10] K. Kutak and A. M. Staśto, Eur. Phys. J. C **41** (2005) 343.
[11] R. Aaij *et al.* [LHCb Collaboration], JHEP **1509** (2015) 084.
[12] G. G. da Silveira, L. Forthomme, K. Piotrzkowski, W. Schäfer and A. Szczurek, JHEP **1502** (2015) 159; M. Łuszczak, W. Schäfer and A. Szczurek, Phys. Rev. D **93** (2016) no.7, 074018.
[13] R. Fiore, A. Flachi, L. L. Jenkovszky, A. I. Lengyel and V. K. Magas, Eur. Phys. J. A **15** (2002) 505.

Kaon photoproduction off proton

Dalibor Skoupil[1,2,*] and Petr Bydžovský[1]

[1]*Nuclear Physics Institute, Czech Academy of Sciencies, Řež near Prague, Czech Republic*
[2]*Czech Technical University in Prague, Czech Republic*

Abstract. We have recently constructed our version of the Regge-plus-resonance (RPR) model and two variants of an isobar model for photoproduction of kaons on the proton, utilizing new experimental data from CLAS, LEPS, and GRAAL collaborations for adjusting free parameters of the models. Higher-spin nucleon (3/2 and 5/2) and hyperon (3/2) resonances were included using the consistent formalism by Pascalutsa and found to play an important role in data description. The set of chosen nucleon resonances in our new isobar models agrees well with the set of the most probable contributing states determined in the Bayesian analysis with the RPR model whilst only 6 out of 10 N^*'s selected in the RPR fit of ours overlap with the nucleon resonant states in the Bayesian analysis. Results of two versions of the isobar model are compared to the new version of the RPR model and experimental data in the third-resonance region and their properties are discussed. We place an emphasis on the choice of resonances, the predictions in the forward- and backward-angle region as well as the choice of the hadron form factor.

1 Introduction and methodology

The study of kaon-hyperon electro- and photoproduction from nucleons in the nucleon resonance region supplies important information about the spectrum of baryon resonances and interactions in hyperon-nucleon systems which arise from Quantum Chromodynamics. A right description of the elementary production process of a Λ hyperon is crucial for obtaining dependable predictions of the excitation spectra for production of Λ hypernuclei. What is more, the "missing" resonances, which were predicted by the quark models [5] but are still elusive in the pion production or πN scattering processes, can be studied by investigating the kaon-hyperon production. In the last decade, experimental data of good quality from the CLAS, LEPS, GRAAL, and MAMI collaborations were produced, which has attracted the attention of many groups of theoretical physicists to this subject.

There are many models of $p(\gamma, K^+)\Lambda$ process at hand, which provide acceptable description of the current experimental data. In our work, we focus on single-channel isobar models and neglect all higher-order effects, *e.g.* rescattering or an interaction in the final state, assuming that their influence on results is included, to some extent, by means of effective values of the coupling constants fitted to experimental data. This simplifying assumption was adopted in numerous single-channel isobar models, *e.g.* Saclay-Lyon (SL) [7], Kaon-MAID (KM) [11] and Ghent Isobar model. The short-range physics, manifesting itself via a spatial structure of hadrons, can be simulated by a hadron form factor

*e-mail: skoupil@ujf.cas.cz

(hff) introduced in the strong interaction vertex. The role of the hff consists also in suppressing the overly strong contributions of Born terms and reducing the high-momentum dependence in exchanges of high-spin resonances [14].

In the isobar model, the amplitude is constructed from effective Lagrangians as a sum of tree-level Feynman diagrams representing the s-, t-, and u-channel exchanges of hadrons in their ground state (Born terms) and various resonances (non Born terms). It is only the exchanges of nucleon resonances in the s channel that make a resonant structure in the observables; the other diagrams constitute the background part of the amplitude as the corresponding poles of theirs are far from the physical region.

The hybrid Regge-plus-resonance (RPR) model, constructed by the Ghent group [6, 8], allows to describe the kaon-hyperon photo- and electroproduction from the threshold up to energies well above the resonance region ($E_\gamma \approx 16\,\mathrm{GeV}$). This model combines merits of the Regge model [9], which is suitable for description above the resonance region, with the isobar model suited for description of the resonance region. The non resonant part of the amplitude, which is a smooth function of energy, is constructed by exchanges of two degenerate kaon trajectories with only three free parameters. The resonant part of the amplitude is then modelled by means of the s-channel exchanges of nucleon resonances whose contributions are made to fade away beyond the resonance region by a strong hff. Thanks to the economical description of the non resonant part of the amplitude, the number of free parameters is significantly reduced in comparison with the isobar model.

Since the kaon production takes place in the third-resonance region, where many nucleon and hyperon higher-spin states might couple to kaon-hyperon channels, the contributions of higher-spin baryon resonances are particularly important in the isobar models. The theoretical description of the interacting baryon fields with spin higher than 1/2 causes problems because of the presence of non physical lower-spin components in the Rarita-Schwinger (RS) field [2]. Provided the RS field is off its mass shell, the non physical lower-spin components of the RS field may participate in the interaction which is then entitled "inconsistent". These problems were eliminated by Pascalutsa who formulated a "consistent" interaction theory for massive spin-3/2 fields demanding invariance of the spin-3/2 interaction vertices under the local U(1) gauge transformation of the RS field [13]. In the consistent theory, the interaction vertices are transverse, which leads to removal of all non physical contributions. This formalism was generalized to arbitrary high spin by the Ghent group [15] and is exploited in this work. The gauge invariance of the interaction leads to a relatively strong momentum dependence of the amplitude, which increases with rising spin of the RS field and regularizes the amplitude. However, the strong dependence on the momentum creates some non physical structures in the energy dependence of the cross section, which then need to be cut off with help of strong hff particularly above the resonance region.

2 Results

The free parameters of the isobar and RPR models, namely the coupling constants and cutoff parameters of the hff, have been fitted to differential-cross-section and hyperon-polarization data. In the fitting procedure of isobar-model parameters, the CLAS data sets were limited in energy to $W < 2.4\,\mathrm{GeV}$ and $W < 2.2\,\mathrm{GeV}$ for differential cross section and hyperon polarization, respectively, whereas for the adjustment of RPR free parameters these data sets were used in their entirety. In comparison to the isobar model, we therefore had less free parameters to optimize and more data available when we went through the fitting procedure of the RPR model.

For the isobar model, many combinations of nucleon and hyperon resonances have been considered and the best solutions have been found: these are the BS1 and BS2 models both with $\chi^2/\mathrm{n.d.f.} = 1.64$. In these models, the set of chosen nucleon resonances largely overlaps with the set found in the Bayesian analysis with the RPR model [8], which suggests that the present data can already fix the resonant part of the amplitude well. Besides others, we confirm the importance of the missing resonances $P_{13}(1900)$ and $D_{13}(1875)$ for the data description in the $K^+\Lambda$ channel but we have replaced the $P_{11}(1880)$ resonance with a $F_{15}(1860)$ resonance which was recently included in the Particle Data Group tables [10]. The BS models differ from one another considerably in the choice of hyperon resonances, which contribute to the non resonant part, and in a form of the hff. While the BS1 model was gained using a multidipole hff, the BS2 model was obtained exploiting a dipole shape of the form factor. The main benefit of the presence of hyperon resonances with spin 3/2 lies in the reduction of the coupling parameters of spin-1/2 hyperon resonances which generally tend to be large. The contributions of spin-3/2 hyperon resonances is perceptible particularly at backward kaon angles.

The uncertainty in the description of the non resonant part, which is quite complicated in the isobar model, has motivated us to fit the data with the RPR model as well, since in this model the background is given in a much simpler form. In the best fit found, which we refer to simply as RPR fit, the resonant content overlaps only partly with that in either BS models or in the RPR-2011A [8] model (only 6 out of 10 nucleon resonances are the same in both cases), the multidipole shape of the form factor is employed and the χ^2/n.d.f. acquires a value of 2.32. Besides other resonances, we have included also a $D_{13}(2120)$ resonance, since its presence leads to a moderate decrease of the χ^2 value.

As shown in the subsequent figures, we have achieved a reasonable description of data with the isobar model as well as with the RPR model. Since predictions of both BS models are nearly identical (see Figs. in Ref. [14]), we show only results of the BS1 model and compare them with predictions of the RPR fit, the RPR-2011A [8] model and the well-known isobar KM [11] and SL [7] models.

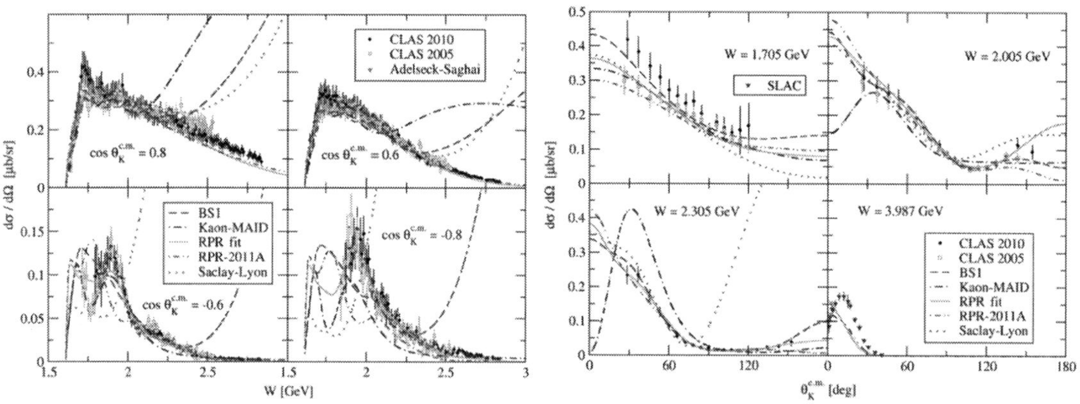

Figure 1. Differential cross section in dependence on the center-of-mass energy W (left) and on the kaon angle $\theta_K^{c.m.}$ (right) is shown as predicted by isobar and RPR models. The data stem from CLAS 2005 [4], CLAS 2010 [12], and SLAC [3] collaborations and from the publication of Adelseck and Saghai [1]. Note that the SLAC data may be a subject of reanalysis in the future.

In Fig. 1, we show predictions of differential cross section in dependence either on the energy W or on the kaon angle $\theta_K^{c.m.}$ computed by several models. There is an ambiguity in the resonance pattern around 1.7 GeV at forward angles (see the left part of Fig. 1): CLAS 2010 data set gives a sharper peak than the CLAS 2005 data, which none of the models displayed captures. The RPR fit is apparently the only model capable of describing the peak around 1.9 GeV at forward angles while both BS1 and BS2 models predict the peak around 1.9 GeV only in the central-angle region. In the forward-angle region, some strength is also apparent around 2.1 GeV, which is modelled by

 DOI: 10.1051/epjconf/201613005018

the high-mass resonances $D_{13}(1875)$, $P_{13}(1900)$, and $F_{15}(2000)$. The RPR fit agrees well with the cross-section data from the threshold up to energies of several GeV where contributions of nucleon resonances vanish and the contribution of kaon trajectories prevails while the isobar models describe data well only up to 2.4 GeV. In the region of backward angles, where the data are not plentiful, there are two broad peaks apparent in the cross section; the one around 2 GeV is quite well captured by the RPR fit. We reckon that this behaviour is achieved by the inclusion of a number of N^*'s and we regard the $P_{11}(1880)$ and $D_{13}(1875)$ resonances as the most important ones in this region.

The RPR fit and the BS models give very similar predictions of the angular dependence of the cross section, which differ mainly in the forward- and backward-angle regions (see the right part of Fig. 1). For $\theta_K^{c.m.} < 40°$, the new models predict decreasing angular dependence like the SL and RPR-2011A models, in contrast to the KM which predicts a very suppressed cross section for $W \geq 2\,\text{GeV}$. Note that the consistency of the cross sections in the region of very small kaon angles with the results of the SL model and the fact that these cross sections are dominated by the spin-flip part of the amplitude could predetermine the new models for successful predictions of the cross sections in the production of hypernuclei, similarly to the SL model.

3 Summary

The isobar models BS1 and BS2 were constructed using a consistent description for spin-3/2 and spin-5/2 resonances. Spin-3/2 hyperon resonances, whose presence is a novel feature of the isobar model, were found to play an important role in depiction of the background part of the amplitude. The set of nucleon resonances selected in our analysis agrees well with the one selected in the analysis with the RPR model done by the Ghent group. A fit with the RPR model, exploiting the consistent high-spin formalism as well, provides a reliable description of data in the resonant and high-spin region. However, the predictions of various models for the cross section at small kaon angles differ and we, therefore, conclude that the data still cannot fix the models fully.

This work was supported by the Grant Agency of the Czech Republic, Grant No. P203/15/04301, and the Grant Agency of the Czech Technical University in Prague, Grant No. SGS16/243/OHK4/3T/14.

References

[1] R. A. Adelseck, B. Saghai, Phys. Rev. C **42**, 108 (1990)
[2] M. Benmerrouche, R. M. Davidson, and N. C. Mukhopadhyay, Phys. Rev. C **39**, 2339 (1989)
[3] A. Boyarski *et al.* (SLAC collaboration), Phys. Rev. Lett **22**, 11311133 (1969)
[4] R. Bradford *et al.* (CLAS collaboration), Phys. Rev. C **73**, 035202 (2006)
[5] S. Capstick and W. Roberts, Prog. Part. Nucl. Phys. **45**, S241 (2000)
[6] T. Corthals, J. Ryckebusch, and T. Van Cauteren, Phys. Rev. C **73**, 045207 (2006)
[7] J. C. David, C. Fayard, G.-H. Lamot, and B. Saghai, Phys. Rev. C **53**, 2613 (1996)
[8] L. De Cruz, T. Vrancx, P. Vancraeyveld, and J. Ryckebusch, Phys. Rev. Lett. **108**, 182002 (2012)
[9] M. Guidal, J.-M. Laget, and M. Vanderhaeghen, Nucl. Phys. A **627**, 645 (1997)
[10] K. A. Olive *et al.* (Particle Data Group), Chin. Phys. C **38**, 090001 (2014)
[11] T. Mart and C. Bennhold, Phys. Rev. C **61**, 012201(R) (1999); T. Mart, *ibid.* **62**, 038201 (2000)
[12] M. E. McCracken *et al.* (CLAS collaboration), Phys. Rev. C **81**, 025201 (2010)
[13] V. Pascalutsa, Phys. Rev. D **58**, 096002 (1998)
[14] D. Skoupil, P. Bydžovský, Phys. Rev. C **93**, 025204 (2016)
[15] T. Vrancx, L. De Cruz, J. Ryckebusch, and P. Vancraeyveld, Phys. Rev. C **84**, 045201 (2011)

Threshold π^- production on the deuteron

B. Strandberg[1,*], J. R. M. Annand[1], W. Briscoe[2], G. Feldman[2], K. Fissum[3], K. Hansen[4], L. Isaksson[4], L. Myers[5], and G. V. O'Reilly[6]
for the MAXTagg@Lund Collaboration

[1] *The University Of Glasgow, Glasgow, United Kingdom*
[2] *The George Washington University, Washington, D.C., USA*
[3] *Lund University, Lund, Sweden*
[4] *MAX IV Laboratory, Lund University, Lund, Sweden*
[5] *Bluffton University, Bluffton, USA*
[6] *University of Massachusetts Dartmouth, Dartmouth, USA*

Abstract. An experiment that aims to measure the π^- photoproduction cross-section $\sigma(E)$ on the deuteron $\gamma + {}^2\mathrm{H} \rightarrow \pi^- + 2\mathrm{p}$ near threshold is discussed. The main concepts of the experimental technique are presented alongside some preliminary signals.

1 Introduction

Pion photoproduction on the nucleus is described by various theoretical frameworks and experimental data is vital to test the accuracy of the models. Numerous experiments have been performed to study the π^0 channel, but data on the charged channels is scarce. To the best of the authors knowledge no data exists for π^- photoproduction below incident photon energy of 158 MeV. In this proceedings paper an experiment that aims to measure the π^- photoproduction

$$\gamma + {}^2\mathrm{H} \rightarrow \pi^- + 2\mathrm{p} \tag{1}$$

cross-section on the deuteron from threshold to $E_\gamma = 160$ MeV is discussed. Sect. 2 outlines the basics of the experimental setup. Sect. 3 is dedicated to our event counting technique and Sect. 4 will present some aspects of the current analysis status.

2 Experimental setup

The experiment was performed at MAXLab in Lund, Sweden. The MAXLab facility provided for the experiment a pulse-stretched electron beam with an energy of $\sim$ 190 MeV. The bremsstrahlung-tagging technique [1, 2] was used to convert the electron beam into a tagged photon beam with an energy range from 140 – 160 MeV. The tagged photon beam was directed on a liquid deuterium target. Three large Sodium-Iodide (NaI) crystal detectors (named BUNI [3], CATS [4] and DIANA [5]) were set up at three theta angles $\theta = 60^\circ, \theta = 120^\circ$ and $\theta = 150^\circ$ to detect photons originating from the liquid deuterium target volume. The floor plan of the experiment is depicted in Fig. 1.

*e-mail: b.strandberg.1@research.gla.ac.uk

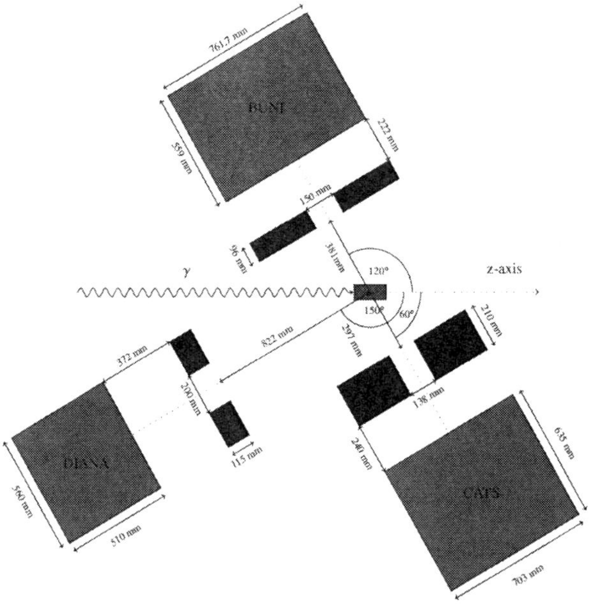

Figure 1. A drawing of the experimental setup in Lund, Sweden. The green-blue colour indicates the NaI detector crystals, the black colour depicts the front collimators and the blue colour represents the liquid deuterium target. There were also thin veto paddles in front of each collimator to veto charged particles (not shown).

3 Event counting

As the pions in reaction (1) are created near threshold their kinetic energy is low and with a high probability they are captured on the deuteron. The capture can go through two dominant channels

$$\pi^- + {}^2\text{H} \to \gamma + 2\text{n}, \tag{2}$$

$$\pi^- + {}^2\text{H} \to 2\text{n}. \tag{3}$$

For our analysis we are especially interested in the radiative capture reaction Eq. (2). The energy spectrum of the emitted photon has a distinct peak at $\sim$ 130 MeV and its shape is well known from before [6–8]. Our plan is to count the π^- that have been produced on the deuteron by counting the π^- that are captured on the deuteron. For this method to succeed one needs to know:

1. The percentage of π^- that are captured.
2. The branching ratio of non-radiative capture (Eq. (3)) to radiative capture (Eq. (2)).
3. The percentage of π^- that undergo some other process - decay, charge exchange, pair production or double-radiative capture.

We have successfully used the Geant4 [10] simulation package to assess how many π^- are captured and how many decay. The branching ratio of reaction (3) relative to (2) has been measured by Highland to be 2.83 ± 0.04 [9]. All other processes have a total probability of less that 1% [11–13]. Therefore the information to estimate the number of produced π^- using the measured number of radiative captures is available.

4 Analysis

4.1 Dominant backgrounds

Multiple background sources contaminate the photon spectrum of reaction (2). There are two channels that produce energetic neutrons which can mimic the detection of a photon, reaction (3) and the

DOI: 10.1051/epjconf/201613005019

photodisintegration reaction

$$\gamma + {}^{2}\mathrm{H} \rightarrow \mathrm{n} + \mathrm{p}. \tag{4}$$

The preliminary simulations of neutron backgrounds and the capture photon spectrum are shown in Fig. 2. However, the π^- capture occurs almost instantaneously and the capture γ's are separable from the neutron backgrounds by analysing the relative arrival times of the event triggers.

In addition to the neutron contamination there are backgrounds from π^0 production and from the Kapton vessel holding the liquid deuterium. The background from Kapton can come from various processes and would thus be difficult to simulate, therefore an empty vessel run was taken to estimate Kapton background contributions. The simulation of the π^0 background alongside the measured Kapton background (preliminary) is shown in Fig. 3. In overall the background contamination sources are well understood, but the analysis has not yet been finalised.

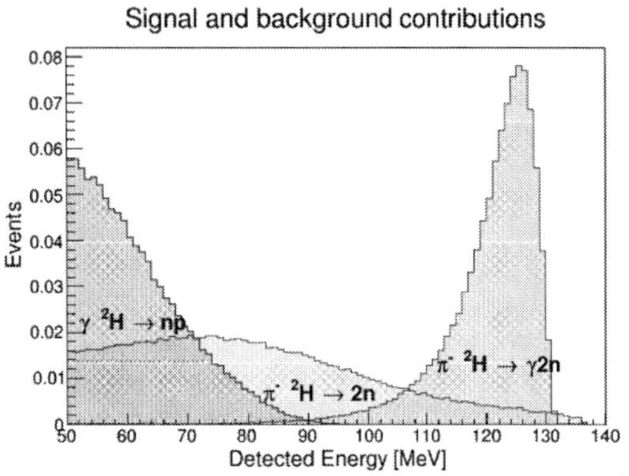

Figure 2. Photon spectrum model of reaction (2) and the neutron background simulations of reactions (3), (4).

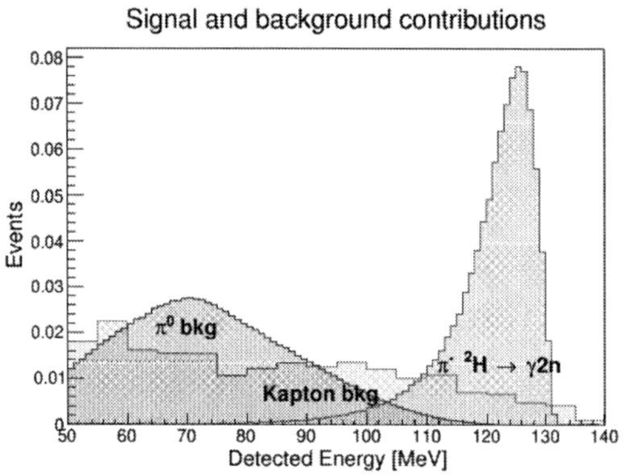

Figure 3. Photon spectrum model of reaction (2), the π^0 background model and the Kapton background.

4.2 Preliminary signals

The preliminary signals in the CATS detector are displayed in Fig. 4. The data has been binned with respect to the incident photon energy to eight bins from 140 – 160 MeV. Some of the neutron background has been successfully separated from the photon signal, but all background has not yet been accounted for. The radiative capture spectrum can be seen to grow significantly with increasing incident photon energy, indicating the increased production of π^-. The yield is dropping in the highest energy bins, but this is due to a reduced tagged photon flux and an increasing number of π^- exiting the target volume without being captured. These preliminary results suggest that it will be possible to extract a π^- photoproduction cross section.

5 Summary

We have provided an overview of an experiment that aims to measure the cross-section $\sigma(E)$ of the reaction $\gamma + {}^{2}\mathrm{H} \rightarrow \pi^- + 2\mathrm{p}$ near the threshold region. A brief summary of the experimental setup and event counting has been presented. In the current status of the analysis the main backgrounds have been identified and the preliminary signals have been extracted.

This work was supported by the United Kingdom's Science and Technology Facilities Council (STFC) from grant number ST/L005719/1. B. Strandberg acknowledges support from The Scottish Universities Physics Alliance (SUPA).

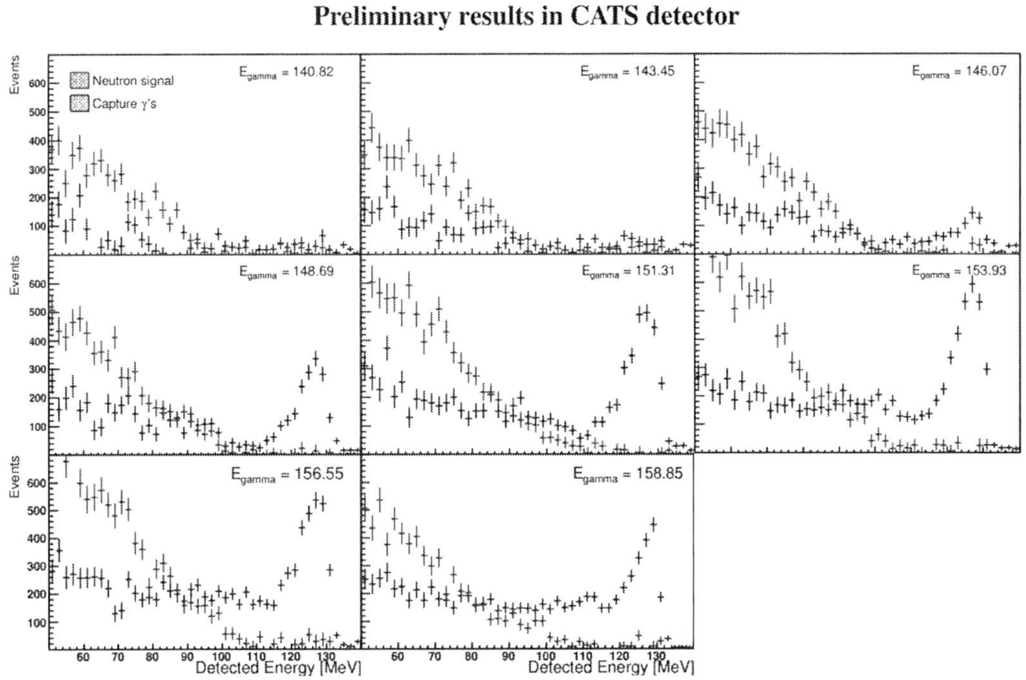

Figure 4. Signals in CATS detector. The photon signal (blue) from the capture reaction (2), the neutron background (red) from reactions (2), (4).

References

[1] J-O. Adler *et al.*, Nucl. Instrum. Meth. A **294**, 15 (1990)
[2] J-O. Adler *et al.*, Nucl. Instrum. Meth. A **388**, 17 (1997)
[3] J. P. Miller *et al.*, Nucl. Instrum. Meth. A **270**, 431 (1988)
[4] A. Hünger *et al.*, Nucl. Phys. A **620**, 385 (1997)
[5] L. S. Myers, Ph.D. thesis, University of Illinois at Urbana-Champaign, Champaign, IL, USA, 2010
[6] B. Gabioud *et al.*, Phys. Rev. Lett. **42**, 1508 (1979)
[7] G. F. de Téramond *et al.*, Phys. Rev. C **36**, 691 (1987)
[8] W. Gibbs *et al.*, Phys. Rev. C **11**, 90 (1975)
[9] V. L. Highland *et al.*, Nucl. Phys. A **365**, 333 (1981)
[10] J. Allison *et al.*, IEEE Trans. Nucl. Sci. **53**, 270 (2006)
[11] S. Tripathi *et al.*, Phys. Rev. C **75**, 064603 (2007)
[12] R. MacDonald *et al.*, Phys. Rev. Lett. **38**, 746 (1977)
[13] D. W. Joseph *et al.*, Phys. Rev. **119**, 805 (1960)

Photoproduction of η and $\eta\prime$ mesons with EtaMAID

Lothar Tiator[1,*], Viktor Kashevarov[1], and Michael Ostrick[1]

[1] *Institut für Kernphysik, Johannes Gutenberg-Universität Mainz, Mainz, Germany*

Abstract. The unitary isobar model EtaMAID has been updated with an extended list of nucleon resonances, fitted to recent and new data for differential cross sections and polarization observables. The nonresonant background is described by Regge trajectories of ω, ρ and a_1, b_1 mesons and in addition Regge cuts, where vector and axial vector mesons are exchanged together with Pomeron and f_2 mesons.

1 Introduction

The isobar model EtaMAID is an online program [1] of the MAID collaboration in Mainz for calculations of observables, amplitudes and multipoles for η and $\eta\prime$ photo- and electroproduction on the nucleon. It was introduced in 2001 [2] and updated in 2003 [3]. Here we will present a new update EtaMAID2016, in progress, taking into account very recent high-precision data of differential cross sections for $p(\gamma,\eta)p$ and $p(\gamma,\eta\prime)p$ from MAMI [4] and polarization observables with beam and target polarization from MAMI [5], ELSA [6], JLab [7] and GRAAL [8]. The high-energy region $W > 2$ GeV is set up with Regge trajectories and Regge cuts and can well describe all high-energy data including polarization observables for γ, π^0, γ, η and $\gamma, \eta\prime$. In the resonance region below $W \approx 2$ GeV we investigated more than 20 N^* resonances and found significant contributions for 18 of them. A very good description has been obtained for all existing photoproduction data.

2 EtaMAID2016

In the spirit of the unitary isobar model MAID for pion photo- and electroproduction [1], the EtaMAID model is set up in the following way

$$t_{\gamma\eta}(W) = t^{bg}_{\gamma\eta}(W) + t^{R}_{\gamma\eta}(W)\,, \tag{1}$$

with a background and a resonance t-matrix term. The background is described in a field-theoretical approach with Feynman diagrams for Born s- and u-channels and vector-meson t-channel exchanges. The resonance part is described with resonance excitations for individual partial waves, where in our current work partial waves up to G waves are considered. Due to isospin zero of η and $\eta\prime$ mesons, Δ resonances can not be excited.

Of course, this ansatz is not unique. However, it is a very important prerequisite to clearly separate resonance and background amplitudes within a Breit-Wigner concept also for higher and overlapping resonances.

*e-mail: tiator@kph.uni-mainz.de

2.1 Regge approach for background

In the asymptotic region, $E > 3$ GeV, the Regge regime, the t-channel vector meson exchange contributions can be replaced by Regge trajectories. This is easily been done by replacing the meson propagator in the t-channel Feynman diagrams by the Regge trajectories, Fig. 1, e.g. for vector-meson exchanges of ρ and ω [3]

$$\frac{1}{t-M_\rho^2} \Rightarrow D_V = \left(\frac{s}{s_0}\right)^{\alpha_V(t)-1} \frac{\pi\,\alpha'_V(e^{-i\pi\alpha_V(t)}-1)}{2\sin[\pi\alpha_V(t)]\,\Gamma[\alpha_V(t)]}\,, \tag{2}$$

where we have already applied the signatures $\mathcal{S} = -1$ of the Regge trajectories for ρ and ω. For b_1 and a_1 axial vector exchange, the signatures of the Regge trajectories are also -1 and the propagators D_A have the same form.

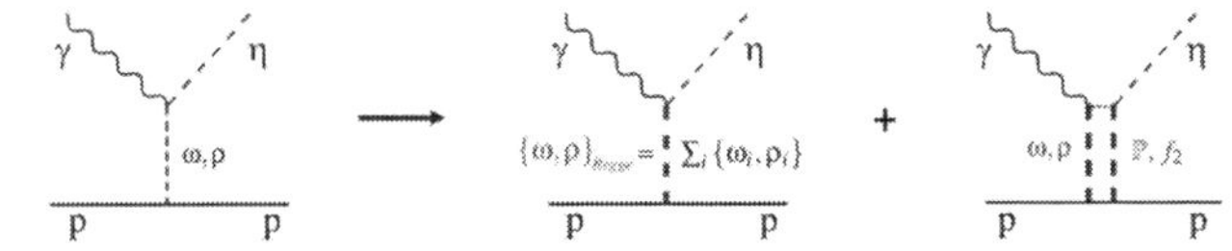

Figure 1. t-channel vector meson exchange diagrams for Regge contributions in pseudoscalar meson photoproduction. The standard Feynman diagram for single poles is replaced by Regge trajectories and Regge cuts with the simultaneous exchange of two Reggeons.

As Donnachie and Kalashnikova [9] suggested, in addition to Regge trajectories, also Regge cuts, Fig. 1, play an important role and can even dominate. These Regge cuts can be considered as a box diagram, where always 2 particles are exchanged, $\rho\mathcal{P}$, ρf_2 and $\omega\mathcal{P}$, ωf_2, where $\mathcal{P}$ is the Pomeron with quantum numbers of the vacuum and f_2 are tensor mesons with quantum numbers $0^+(2^{++})$. All four Regge cuts can contribute to vector and axial vector exchanges and can be written in the following form

$$D_{cut} = \left(\frac{s}{s_0}\right)^{\alpha_c(t)-1} e^{-i\pi\alpha_c(t)/2}\, e^{d_c t}\,. \tag{3}$$

In total, the vector meson propagators are replaced by

$$D_V = D_V + c_{VP}\, D_{VP} + c_{Vf}\, D_{Vf}\,, \quad V = \rho, \omega \tag{4}$$

and the axial vector meson propagators are replaced accordingly. The coefficients c_α for natural and unnatural parity cuts are obtained by a fit to the data. Born terms are not required in the fit.

In Fig. 2 we show the forward cross sections at small t for high energy η photoproduction. The dip around $t \approx -0.6$ GeV^2 shows the typical Regge behavior for trajectories with signature -1. This dip region is filled by the Regge-cut contributions and describe the data very well. No Born terms and N^* resonances are included here.

2.2 Breit-Wigner ansatz for resonance excitations

Due to isospin zero of η and $\eta\prime$ only N^* resonances with isospin 1/2 can be excited. Whereas in the original EtaMAID2001 and also in 2003 only 8 resonances were found to contribute (dominant contributions from: $S_{11}(1535), D_{13}(1520), S_{11}(1650), D_{15}(1675)$, sub-dominant from

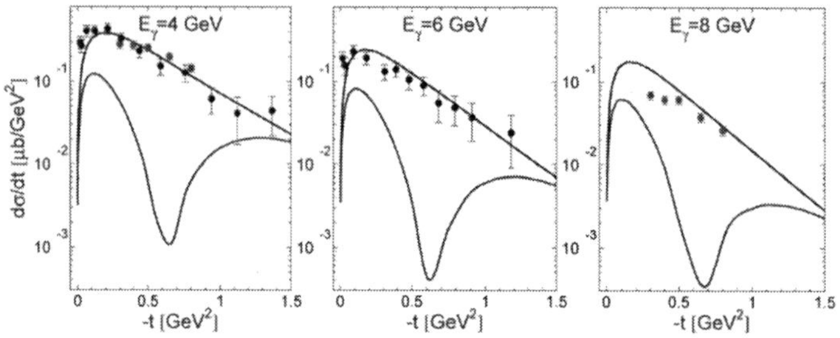

Figure 2. Forward (small t) differential cross sections for η photoproduction described by Regge contributions. The blue lines show the contribution with ρ, ω Regge trajectories alone, giving rise to a pronounced dip, whereas the black lines give the total Regge contribution including the cuts. The data are from [10] (black) and [11] (red).

$F_{15}(1680), D_{13}(1700), P_{11}(1710), P_{13}(1720)$), we have now investigated the sensitivity of up to 20 N^* resonances, practically all known resonances of the Particle Data Tables [12].

The relevant multipoles $\mathcal{M}_{\ell\pm}$ ($E_{\ell\pm}$, $M_{\ell\pm}$) are assumed to have a Breit-Wigner energy dependence of the following form

$$\mathcal{M}_{\ell\pm}(W) = \bar{\mathcal{M}}_{\ell\pm}\, f_{\gamma N}(W)\, \frac{M_R}{M_R^2 - W^2 - iM_R\Gamma_{\text{total}}(W)}\, f_{\eta N}(W)\, C_{\eta N}\,, \tag{5}$$

where $f_{\gamma N}(W)$ and $f_{\eta N}(W)$ are Breit-Wigner factors describing the γN and ηN vertex of an N^* resonance with spin J and partial width $\Gamma_{\eta N}(W)$

$$f_{\gamma N}(W) = \left(\frac{k_\gamma}{k_{\gamma,R}}\right)^2 \left(\frac{X^2 + k_{\gamma,R}^2}{X^2 + k_\gamma^2}\right)^2 \quad \text{and} \quad f_{\eta N}(W) = \zeta_{\eta N}\left[\frac{k_\gamma}{(2J+1)\pi}\,\frac{M_N}{M_R}\,\frac{\Gamma_{\eta N}}{q_\eta}\right]^{1/2}, \tag{6}$$

with k_γ and q_η the photon and η meson momenta in the c.m. system, and $\zeta_{\eta N} = \pm 1$ a relative sign between the $N^* \to \eta N$ and $N^* \to \pi N$ couplings. $C_{\eta N}$ is an isospin factor, which is -1 for ηN and $\eta\prime N$ final states.

The total width is a sum over 7 possible decay channels

$$\Gamma_{total}(W) = \Gamma_{\pi N}(W) + \Gamma_{\pi\pi N}(W) + \Gamma_{\eta N}(W) + \Gamma_{K\Lambda}(W) + \Gamma_{K\Sigma}(W) + \Gamma_{\omega N}(W) + \Gamma_{\eta\prime N}(W)\,. \tag{7}$$

The partial decay widths are typically parameterized in the following way, e.g. for ηN,

$$\Gamma_{\eta N}(W) = \beta_{\eta N}\,\Gamma_R \left(\frac{q_\eta}{q_{\eta,R}}\right)^{2\ell+1} \left(\frac{X^2 + q_{\eta,R}^2}{X^2 + q_\eta^2}\right)^{\ell}, \tag{8}$$

where X is a damping parameter. In EtaMAID2003 it was assumed to be constant 500 MeV for all resonances and all partial widths, but in our current work we are using X as a fit parameter. The c.m. momenta, taken at the resonance position, $W = M_R$, are denoted by an additional index R. In the EtaMAID update we also included further 2-body decay channels, namely $K\Lambda$, $K\Sigma$, ωN. Branching ratios for ηN and $\eta\prime N$ are fitted to the data of η and $\eta\prime$ photoproduction, for all other channels the pure sensitivity of the branching ratios in the total width is not enough to get reliable fits, therefore, these numbers were taken from PDG [12] and from a coupled-channels partial wave analysis [13].

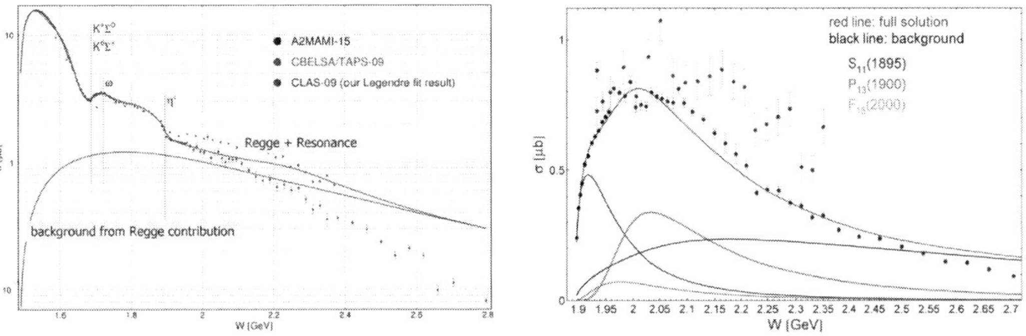

Figure 3. Total cross section for η (left) and $\eta\prime$ (right) photoproduction on the proton. In both figurs the Regge background and the total contributions are shown. For $\eta\prime$ photoproduction we also give the dominant contributions from individual resonances, $N(1895)1/2^-$ (blue), $N(1900)3/2^+$ (magenta) and $N(2000)5/2^+$ (green). The data are from Mainz [4], Bonn [6] and JLab [7], however, the total cross sections of the CLAS data were obtained from a Legendre fit to the angular distributions.

In Fig. 3 the new fit is shown for total cross sections of η and $\eta\prime$ photoproduction on the proton target. In both cases also the flat Regge contribution is plotted, which is rather small for energies below $W \approx 2$ GeV. However, it also interferes with same partial waves for the total cross sections. For angular distributions, the interferences among different resonance contributions and the Regge background become very important. In our fit to the Mainz, Bonn, GRAAL and JLab data we have found significant sensitivity to 18 N^* resonances: $P_{11}(1440), D_{13}(1520), S_{11}(1535), S_{11}(1650), D_{15}(1675), F_{15}(1680), D_{13}(1700), P_{11}(1710), P_{13}(1720), F_{15}(1860), D_{13}(1875), P_{11}(1880), S_{11}(1895), P_{13}(1900), F_{15}(2000), D_{15}(2060), D_{13}(2120), G_{17}(2190)$. Overall, we obtained a very good fit to all data on diff. cross sections and polarization observables. The total cross sections themselves are not fitted, they are the result of the partial wave analysis.

References

[1] D. Drechsel, O. Hanstein, S. S. Kamalov and L. Tiator, Nucl. Phys. A **645**, 145 (1999) http://www.kph.uni-mainz.de/MAID/
[2] W. T. Chiang, S. N. Yang, L. Tiator and D. Drechsel, Nucl. Phys. A **700**, 429 (2002)
[3] W. T. Chiang et al., Phys. Rev. C **68**, 045202 (2003)
[4] M.Ostrick, Proc. 10th Int. Workshop on the Physics of Excited Nucleons (NSTAR2015) JPS Conf. Proc. 10, 010004 (2016)
[5] J. Akondi *et al.* (A2 Collaboration at MAMI), Phys. Rev. Lett. **113**, 102001 (2014)
[6] V. Crede *et al.* (CBELSA/TAPS Collaboration), Phys. Rev. C **80**, 055202 (2009)
[7] M. Williams *et al.* (CLAS Collaboration), Phys. Rev. C **80**, 045213 (2009)
[8] O. Bartalini *et al.* (The GRAAL Collaboration), Eur. Phys. J. A **33**, 169 (2007)
[9] A. Donnachie and Y. S. Kalashnikova, Phys. Rev. C **93**, 025203 (2016)
[10] W. Braunschweig et al., Phys. Lett. B **33**, 236 (1970)
[11] J. Dewire et al., Phys. Lett. B **37**, 326 (1971)
[12] K. A. Olive *et al.* (Particle Data Group), Chin. Phys. C **38**, 090001 (2014)
[13] A. V. Anisovich et al., Eur. Phys. J. A **48**, 15 (2012)

Photo-production of tensor mesons

Ju-Jun Xie[1,*], Li-Sheng Geng[2], and Eulogio Oset[3]

[1] *Institute of Modern Physics, Chinese Academy of Sciences, Lanzhou 730000, China*
[2] *School of Physics and Nuclear Energy Engineering and International Research Center for Nuclei and Particles in the Cosmos, Beihang University, Beijing 100191, China*
[3] *Departamento de Física Teórica and IFIC, Centro Mixto Universidad de Valencia-CSIC Institutos de Investigación de Paterna, Aptdo. 22085, 46071 Valencia, Spain*

Abstract. Assuming that the $f_2(1270)$, $f_2'(1525)$, $a_2(1320)$, and $K_2^*(1430)$ resonances are dynamically generated states from the vector meson-vector meson interactions in $L = 0$ and spin 2, we study the $\gamma p \to f_2(1270)[f_2'(1525)]p$, $\gamma p \to a_2^0(1320)p$, and $\gamma p \to K_2^*(1430)\Lambda(\Sigma)$ reactions. For the $\gamma p \to f_2(1270)p$ reaction, we find that the theoretical results for the differential cross sections are in good agreement with the experimental measurements and provide support for the molecular picture of the $f_2(1270)$ in the first baryonic reaction where it has been tested. Furthermore, we predict also the total and differential cross sections for other reactions. The results can be tested in future experiments and therefore offer new clues on the nature of these tensor states.

1 Introduction

Within quark models, mesons are made of quark-antiquark pairs and baryons are composed of three quarks in the low energy region. However, recent observations of the new baryonic P_c states [1] and the mesonic XYZ states [2–5] by various collaborations have challenged this conventional wisdom. There are many different explanations for the nature of these new observed states, but none of them has been accepted unanimously.

It seems clear that Nature is richer than it is preferred to be. Many low-lying states, even those long believed to be conventional $q\bar{q}$ (or qqq) states, may have large components of other nature. Indeed, it has been shown that many of the low-lying mesonic states can be understood not as of $q\bar{q}$ states but as meson-meson molecules, dynamically generated in the so-called unitary approaches. In Refs. [6–8], the $f_2(1270)$, $f_2'(1525)$, $a_2(1320)$, and $K_2^*(1430)$ are found dynamically generated from the vector meson - vector meson (VV) interaction. The molecular nature of these tensor states has been extensively tested in a large number of processes: the two-photon decay of the $f_2(1270)$ [9], the two-photon and one photon-one vector decays of the $f_2(1270)$, $f_2'(1525)$ and $K_2^*(1430)$ [10], the $J/\psi \to \phi(\omega)f_2(1270)$, $f_2'(1525)$ and $J/\psi \to K^{*0}(892)\bar{K}_2^{*0}(1430)$ decays [11], the radiative decay of J/ψ into $f_2(1270)$ and $f_2'(1525)$ [12], the $\psi(2S)$ decays into $\omega(\phi)f_2(1270)$, $\omega(\phi)f_2'(1525)$, $K^{*0}(892)\bar{K}_2^{*0}(1430)$ and the radiative decays of $\Upsilon(1S)$, $\Upsilon(2S)$, $\psi(2S)$ into $\gamma f_2(1270)$, $\gamma f_2'(1525)$, $\gamma f_0(1370)$, and $\gamma f_0(1710)$ [13, 14], and the ratio of the decay widths of $\bar{B}_s^0 \to J/\psi f_2(1270)$ to $\bar{B}_s^0 \to J/\psi f_2'(1525)$ [15]. The

*e-mail: xiejujun@impcas.ac.cn

agreement with experimental data turns out to be quite good, providing support to the underlying assumption that these states contain large meson-meson components.

Along this line, we study the photo-production of those tensor mesons, assuming that the $f_2(1270)$, $f_2'(1525)$, $a_2(1320)$, and $K_2^*(1430)$ are dynamically generated states from the VV interaction.

2 Formalism and ingredients

From the perspective that the $f_2(1270)$, $f_2'(1525)$, $a_2(1320)$, and $K_2^*(1430)$ are dynamically generated from the VV interaction, the photoproduction of those tensor mesons proceeds via the creation of two vector mesons by the γp initial state in a primary step and the following interaction of the two vector mesons, thus generating the resonance. The corresponding Feynman diagrams are shown in Fig. 1.

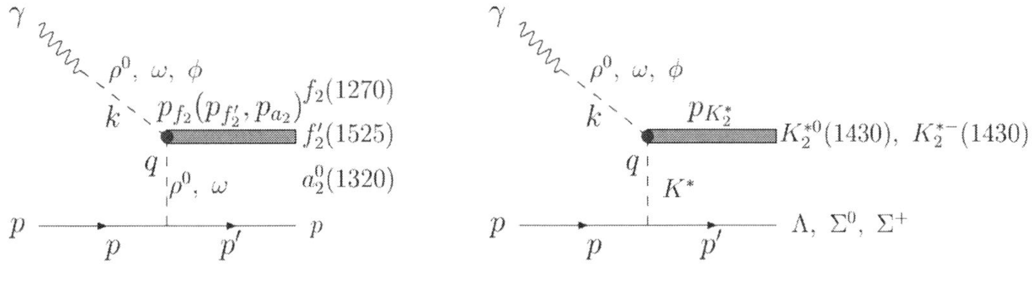

Figure 1. Diagrammatic representation of the photoproduction of tensor mesons.

To evaluate the Feynman amplitudes for the diagrams shown in Fig. 1, we need the coupling of the tensor meson to the respective vector mesons, g_T^{VV}, the γ-V coupling and the VNN coupling. The values for g_T^{VV} are taken from Refs. [7, 18] and shown in Table 1.

Table 1. The coupling constants of tensor mesons to VV.

Resonance	Channel	g_T^{VV} (MeV)
$f_2(1270)$	$\rho\rho$	$(10889, -i99)$
	$\omega\omega$	$(-440, i7)$
	$\phi\omega$	$(777, -i13)$
$f_2'(1525)$	$\rho\rho$	$(-2443, i649)$
	$\omega\omega$	$(-2709, i8)$
	$\phi\omega$	$(5016, -i17)$
$K_2^*(1430)$	ρK^*	$(10901, -i71)$
	ωK^*	$(2267, -i13)$
	ϕK^*	$(-2898, i17)$
$a_2(1320)$	$\rho\omega$	$(0, -i8402)$
	$\rho\phi$	$(0, -i1912)$

The γ-V conversion vertex can be obtained from the local hidden gauge Lagrangian, and the vector-nucleon-nucleon vertex is obtained from the $SU(3)$ baryon-baryon-vector interaction Lagrangian (see more details in Refs. [16, 17]). Then one can easily calculate the scattering amplitude

T and $\overline{\sum}\sum|T|^2$. Here we give explicitly the case of $\gamma p \to f_2'(1525)p$ reaction, as an example,

$$T = e(-\frac{g^{\rho\rho}_{f_2'}}{\sqrt{6}} + \frac{g^{\omega\omega}_{f_2'}}{\sqrt{2}} - \frac{g^{\phi\omega}_{f_2'}}{\sqrt{2}})\frac{1}{q^2 - m_V^2}[\frac{1}{2}\epsilon_i(\gamma)(-g_{j\mu} + \frac{q_j q_\mu}{m_V^2}) + \frac{1}{2}\epsilon_j(\gamma)(-g_{i\mu} + \frac{q_i q_\mu}{m_V^2})$$
$$-\frac{1}{3}\epsilon_m(\gamma)\delta_{ij}(-g_{m\mu} + \frac{q_m q_\mu}{m_V^2})] < p(M')|\gamma^\mu|p(M) >, \qquad (1)$$

with M and M' the third spin component of the initial and final proton. The V stands for the exchanged ρ^0 or ω. Following the same procedure, we can obtain the transition amplitudes for $f_2(1270)$, $a_2(1320)$, and $K_2^*(1430)$ productions as in Refs. [16, 17].

The differential cross section for those reactions are given by $d\sigma/dt = \frac{m_i^2}{16\pi s|\vec{k}|^2}\overline{\sum}\sum|T|^2$ with s the invariant mass squared of the γp system, $m_i^2 = m_p^2$ for $\gamma p \to f_2(1270)$, $[f_2'(1525), a_2^0(1320)]p$ reaction, $m_i^2 = m_p m_\Lambda$ for $\gamma p \to K_2^{*+}(1430)\Lambda$ reaction, and $m_i^2 = m_p m_\Sigma$ for $\gamma p \to K_2^*(1430)\Sigma$ reaction. The variable $\vec{k}$ is the three momenta of photon in the center of mass frame, and $t = q^2 = (p - p')^2$.

3 Numerical results

We show the results of $d\sigma/dt$ at $E_\gamma = 3.4$ GeV for the six reactions in Fig. 2 (left), where the experimental data are taken from Ref. [19] for $\gamma p \to f_2(1270)p$ reaction. It is found that our results are in good agreement with experiment for $f_2(1270)$ production. Besides, we see that the slopes for the six reactions are quite similar.

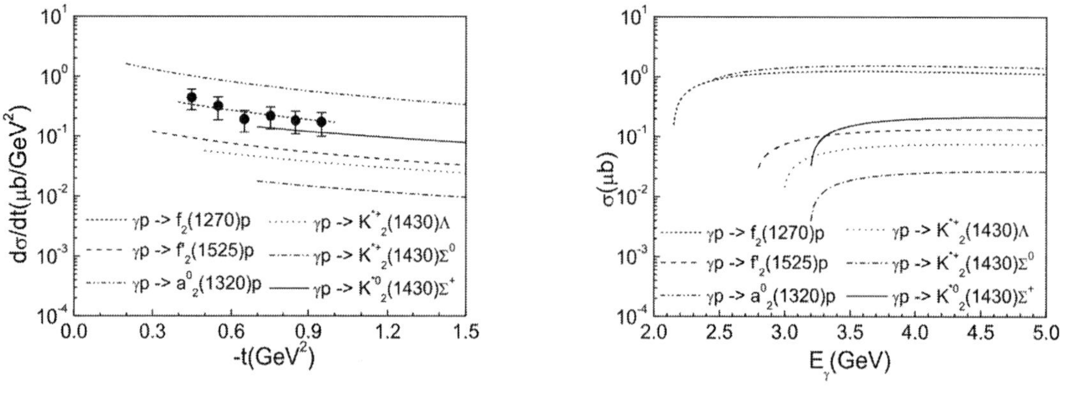

Figure 2. Left: differential cross section $\frac{d\sigma}{dt}$ as a function of t at $E_\gamma = 3.4$ GeV. Right: total cross section for the photo-production of tensor mesons as a function of E_γ.

In addition to the differential cross section, we calculate also the total cross section for the six reactions as a function of the photon beam energy E_γ. The results are shown in Fig. 2 (right). The total cross sections increase rapidly away from threshold and soon become almost constant at higher photon energies. The results for the $f_2(1270)$ and $a_2(1320)$ are larger than for the other productions.

It is worth mentioning that the reaction formalism advocated here involves parameters fixed by previous works, which allow us to make predictions for total cross sections. The differential and total cross sections can be checked in future experiments, such as those at CLAS. In this sense, the reaction mechanism can be easily tested.

4 Conclusions

Recently, it is found that the $f_2(1270)$, $f_2'(1525)$, $a_2(1320)$, and $K_2^*(1430)$ resonances, though long been accepted as ordinary $q\bar{q}$ states, can be dynamically generated from the vector meson-vector meson interaction. Many studies adopting such a scenario have been performed in mesonic reactions and all yield positive results. In this work, we have extended the molecular picture to the γp reaction. The elements needed for the test are very simple, which makes particularly transparent the interpretation of the results. On the one side those tensor mesons couple to VV and the coupling have been fixed before in the unitary approach that generates the $f_2(1270)$, $f_2'(1525)$, $a_2(1320)$, and $K_2^*(1430)$ states as VV molecules based on the local hidden gauge formalism for the interaction of vector mesons. On the other side, with these couplings and the vector meson dominance hypothesis, incorporated in the local hidden gauge approach, the photon gets converted into one of the vector mesons, which interacts with the vector meson emitted by the incoming proton to generate the $f_2(1270)$, $f_2'(1525)$, $a_2(1320)$, and $K_2^*(1430)$ states. With this simple picture we predict both the differential and total cross sections, which could be tested by future experiments, such as those at CLAS.

Acknowledgement

One of us, E. O., wishes to acknowledge support from the Chinese Academy of Science (CAS) in the Program of Visiting Professorship for Senior International Scientists (Grant No. 2013T2J0012). This work is partly supported by the Spanish Ministerio de Economia y Competitividad and European FEDER funds under the contract number FIS2011-28853-C02-01 and FIS2011-28853-C02-02, and the Generalitat Valenciana in the program Prometeo II-2014/068. We acknowledge the support of the European Community-Research Infrastructure Integrating Activity Study of Strongly Interacting Matter (acronym HadronPhysics3, Grant Agreement n. 283286) under the Seventh Framework Programme of EU. This work is also partly supported by the National Natural Science Foundation of China under Grant Nos. 11475227,11375024, and 11522539.

References

[1] R. Aaij *et al.* [LHCb Collaboration], Phys. Rev. Lett. **115**, 072001 (2015).
[2] S. L. Olsen, Front. Phys. China. **10**, 121 (2015).
[3] X. Liu, Chin. Sci. Bull. **59**, 3815 (2014).
[4] G. T. Bodwin, E. Braaten, E. Eichten, S. L. Olsen, T. K. Pedlar and J. Russ, arXiv:1307.7425.
[5] N. Brambilla *et al.*, Eur. Phys. J. C **71**, 1534 (2011).
[6] R. Molina, D. Nicmorus and E. Oset, Phys. Rev. D **78**, 114018 (2008).
[7] L. S. Geng and E. Oset, Phys. Rev. D **79**, 074009 (2009).
[8] L. S. Geng, E. Oset, R. Molina and D. Nicmorus, PoS EFT **09**, 040 (2009).
[9] H. Nagahiro *et al.*, Phys. Rev. D **79**, 114023 (2009).
[10] T. Branz, L. S. Geng and E. Oset, Phys. Rev. D **81**, 054037 (2010).
[11] A. Martinez Torres *et al.*, Phys. Lett. B **680**, 310 (2009).
[12] L. S. Geng *et al.*, Eur. Phys. J. A **44**, 305 (2010).
[13] L. Dai and E. Oset, Eur. Phys. J. A **49**, 130 (2013).
[14] L. R. Dai, J. J. Xie and E. Oset, Phys. Rev. D **91**, 094013 (2015).
[15] J. J. Xie and E. Oset, Phys. Rev. D **90**, 094006 (2014).
[16] J. J. Xie and E. Oset, Eur. Phys. J. A **51**, 111 (2015).

[17] J. J. Xie, E. Oset and L. S. Geng, Phys. Rev. C **93**, 025202 (2016).
[18] C. García-Recio *et al.*, Phys. Rev. D **87**, 096006 (2013).
[19] M. Battaglieri *et al.* [CLAS Collaboration], Phys. Rev. D **80**, 072005 (2009).

EPJ Web of Conferences **130**, 06001 (2016) DOI: 10.1051/epjconf/201613006001
MESON 2016

Search for $K^+ \to \pi^+ \nu\bar{\nu}$ at the NA62 experiment

Viacheslav Duk[1,*] for the NA62 collaboration **

[1] *INFN Perugia, Italy*

Abstract. Looking for phenomena beyond the Standard Model (SM) in rare decays is a complementary approach to direct searches for New Physics (NP) at colliders. One of the theoretically cleanest processes is the ultra rare decay $K^+ \to \pi^+ \nu\bar{\nu}$. The goal of the NA62 experiment at CERN SPS is to measure the branching ratio (BR) of this decay with 10% precision. The experiment has been launched in 2014. In 2015, the detector was commissioned at a low intensity beam. The experimental setup is described and performances achieved in 2015 are discussed in view of the final measurement.

1 $K \to \pi\nu\bar{\nu}$ decays

The ultra rare decays $K^+ \to \pi^+ \nu\bar{\nu}$ and $K_L \to \pi^0 \nu\bar{\nu}$ proceed via flavor changing neutral currents (FCNC) and are highly suppressed in the SM due to the GIM mechanism. The Feynman diagrams of box and penguin contributions are shown in Fig. 1.

*e-mail: Viacheslav.Duk@cern.ch

**G. Aglieri Rinella, R. Aliberti, F. Ambrosino, R. Ammendola, B. Angelucci, A. Antonelli, G. Anzivino, R. Arcidiacono, I. Azhinenko, S. Balev, M. Barbanera, J. Bendotti, A. Biagioni, L. Bician, C. Biino, A. Bizzeti, T. Blazek, A. Blik, B. Bloch-Devaux, V. Bolotov, V. Bonaiuto, M. Boretto, M. Bragadireanu, D. Britton, G. Britvich, M.B. Brunetti, D. Bryman, F. Bucci, F. Butin, E. Capitolo, C. Capoccia, T. Capussela, A. Cassese, A. Catinaccio, A. Cecchetti, A. Ceccucci, P. Cenci, V. Cerny, C. Cerri, B. Checcucci, O. Chikilev, S. Chiozzi, R. Ciaranfi, G. Collazuol, A. Conovaloff, P. Cooke, P. Cooper, G. Corradi, E. Cortina Gil, F. Costantini, F. Cotorobai, A. Cotta Ramusino, D. Coward, G. D'Agostini, J. Dainton, P. Dalpiaz, H. Danielsson, J. Degrange, N. De Simone, D. Di Filippo, L. Di Lella, S. Di Lorenzo, N. Dixon, N. Doble, B. Dobrich, V. Duk, V. Elsha, J. Engelfried, T. Enik, N. Estrada, V. Falaleev, R. Fantechi, V. Fascianelli, L. Federici, S. Fedotov, M. Fiorini, J. Fry, J. Fu, A. Fucci, L. Fulton, S. Gallorini, S. Galeotti, E. Gamberini, L. Gatignon, G. Georgiev, A. Gianoli, M. Giorgi, S. Giudici, L. Glonti, A. Goncalves Martins, F. Gonnella, E. Goudzovski, R. Guida, E. Gushchin, F. Hahn, B. Hallgren, H. Heath, F. Herman, T. Husek, O. Hutanu, D. Hutchcroft, L. Iacobuzio, E. Iacopini, E. Imbergamo, O. Jamet, P. Jarron, E. Jones, T. Jones K. Kampf, J. Kaplon, V. Kekelidze, S. Kholodenko, G. Khoriauli, A. Khotyantsev, A. Khudyakov, Yu. Kiryushin, A. Kleimenova, K. Kleinknecht, A. Kluge, M. Koval, V. Kozhuharov, M. Krivda, Z. Kucerova, Yu. Kudenko, J. Kunze, G. Lamanna, G. Latino, C. Lazzeroni, G. Lehmann-Miotto, R. Lenci, M. Lenti, E. Leonardi, P. Lichard, R. Lietava, L. Litov, R. Lollini, D. Lomidze, A. Lonardo, M. Lupi, N. Lurkin, K. McCormick, D. Madigozhin, G. Maire, C. Mandeiro, I. Mannelli, G. Mannocchi, A. Mapelli, F. Marchetto, R. Marchevski, S. Martellotti, P. Massarotti, K. Massri, P. Matak, E. Maurice, A. Mefodev, E. Menichetti, E. Minucci, M. Mirra, M. Misheva, N. Molokanova, J. Morant, M. Morel, M. Moulson, S. Movchan, D. Munday, M. Napolitano, I. Neri, F. Newson, A. Norton, M. Noy, G. Nuessle, T. Numao, V. Obraztsov, A. Ostankov, S. Padolski, R. Page, V. Palladino, G. Paoluzzi, C. Parkinson, E. Pedreschi, M. Pepe, F. Perez Gomez, M. Perrin-Terrin, L. Peruzzo, P. Petrov, F. Petrucci, R. Piandani, M. Piccini, D. Pietreanu, J. Pinzino, I. Polenkevich, L. Pontisso, Yu. Potrebenikov, D. Protopopescu, F. Raffaelli, M. Raggi, P. Riedler, A. Romano, P. Rubin, G. Ruggiero, V. Russo, V. Ryjov, A. Salamon, G. Salina, V. Samsonov, C. Santoni, G. Saracino, F. Sargeni, V. Semenov, A. Sergi, M. Serra, A. Shaikhiev, S. Shkarovskiy, I. Skillicorn, D. Soldi, A. Sotnikov, V. Sugonyaev, M. Sozzi, T. Spadaro, F. Spinella, R. Staley, A. Sturgess, P. Sutcliffe, N. Szilasi, D. Tagnani, S. Trilov, M. Valdata-Nappi, P. Valente, M. Vasile, T. Vassilieva, B. Velghe, M. Veltri, S. Venditti, P. Vicini, R. Volpe, M. Vormstein, H. Wahl, R. Wanke, P. Wertelaers, A. Winhart, R. Winston, B. Wrona, O. Yushchenko, M. Zamkovsky, A. Zinchenko.

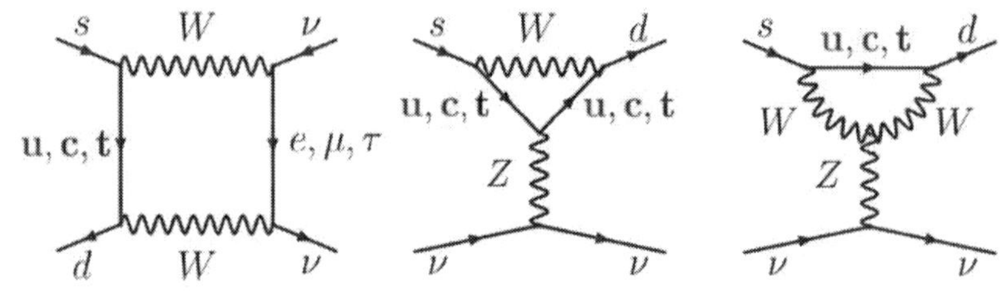

Figure 1. Box and penguin diagrams contributing to $K \to \pi\nu\bar{\nu}$ decays.

Within the SM, BR theoretical values contain CKM matrix elements as external inputs. These elements give the largest uncertainty to the BR. Using tree-level observables, one obtains [1, 2]:

$$BR(K^+ \to \pi^+\nu\bar{\nu}) = (8.4 \pm 1.0) \times 10^{-11} ,$$

$$BR(K_L \to \pi^0\nu\bar{\nu}) = (3.4 \pm 0.6) \times 10^{-11} .$$

These values are not polluted by contributions from the NP and hence are suitable for a NP search as a deviation from the SM BR value. More precise calculation with the loop-level measurements of the CKM elements assumes the absence of NP and provides the best accuracy for further recalculation of CKM elements from the comparison of theoretical and experimental BR values:

$$BR(K^+ \to \pi^+\nu\bar{\nu}) = (9.1 \pm 0.7) \times 10^{-11} ,$$

$$BR(K_L \to \pi^0\nu\bar{\nu}) = (3.0 \pm 0.3) \times 10^{-11} .$$

$K \to \pi\nu\bar{\nu}$ decays are very sensitive to NP. The largest contribution to the BR could come from models with new sources of flavor violation [3, 4]. The 10% precision measurement of the BR will allow to probe NP at mass scales up to 100 TeV.

The only experimental measurement in the charged mode based on the observation of 7 events and decay-at-rest technique has been performed by the E787 and E949 experiments [5], while for the neutral mode there is only an upper limit on the BR from the E391 collaboration [6]:

$$BR(K^+ \to \pi^+\nu\bar{\nu}) = (17.3^{+11.5}_{-10.5}) \times 10^{-11} ,$$

$$BR(K_L \to \pi^0\nu\bar{\nu}) < 2.6 \times 10^{-8} .$$

2 NA62 experiment

The NA62 experiment plans to collect 10^{13} kaon decays in the fiducial volume which corresponds to ~100 SM events of $K^+ \to \pi^+\nu\bar{\nu}$ with ~10% acceptance. The detailed description of the NA62 apparatus can be found in [7]. The schematic view of the setup is shown in Fig. 2.

A proton beam from CERN SPS (400 GeV/c) impinging on a Beryllium target produces a secondary hadron beam of 75 GeV/c with ~6% of K^+. Kaons are identified by the CEDAR detector (**KTAG**), a differential Cherenkov counter which combines high identification efficiency (at least

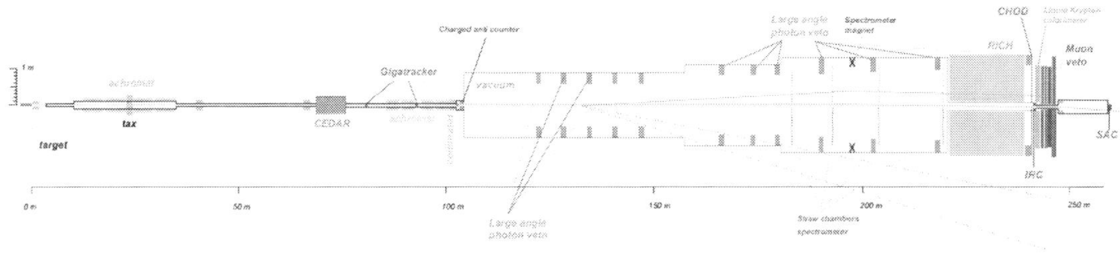

Figure 2. NA62 detector layout.

95%) and excellent time resolution (~100 ps), as needed for the time matching of a primary and a secondary track. The momentum of beam particles is measured by the beam tracker (Gigatracker, or **GTK**). Each of the three stations of GTK is made of 200 μm thick silicon sensors (300×300 μm size, 0.5 X_0 thickness per station, time resolution better than 200 ps, momentum resolution 0.2%). The detector is operated at ~750 MHz rate.

The momentum of secondary particles is measured by the Straw chamber spectrometer (**STRAW**). Each of the four STRAW stations comprises 1792 straw tubes arranged in four views (X, Y, U, V) and is operated in vacuum. The achieved momentum resolution is better than 1%.

The **RICH** detector aims at separating secondary pions from muons at the level of 1% (muon contamination in the pion sample) in the momentum range 15–35 GeV/c. A good time resolution (below 100 ps) is necessary both for time matching of two tracks (primary and secondary) and for providing a signal for the L0/L1 trigger. The charged particle hodoscope (**CHOD**) consisting of 128 scintillator slabs arranged in 2 planes (horizontal and vertical) measures the track crossing time with ~200 ps resolution and also contributes to the L0 trigger.

The photon veto system covers the angular range up to 50 mrad and includes 4 detectors: **LAV** (12 stations operated in vacuum, made of lead glass blocks and covering the angular region from 8.5 to 50 mrad), **LKr** (liquid Krypton calorimeter from the NA48/2 experiment, vetoing the region from 1 to 8.5 mrad), **IRC** and **SAC** (small angle calorimeters covering the angular region <1 mrad). The rejection power of the veto system should be ~10^8 for decays with π^0 in the final state. The **CHANTI** detector placed after the GTK3 identifies upstream inelastic interactions and muon halo. The iron/scintillator calorimeters (**MUV1**, **MUV2**) provide further pion/muon separation. Together with the RICH, this separation (muon mis-ID) should be ~10^{-7}. A fast muon veto **MUV3** (iron wall followed by 5 cm thick scintillator, the light is detected by PMs) identifies muons and provides L0 trigger signals.

3 Analysis strategy

The signal detection principles are driven by the low BR of the decay (and hence a huge background), weak experimental signature (one primary and one secondary track) and the required background level (~10%). This leads to the following strategy:

- high beam intensity and fast timing;
- decays-in-flight technique;
- kinematic reconstruction and efficient particle identification (PID);
- efficient veto system.

The main kinematic variable used in the analysis is the missing mass squared:
$m^2_{miss} = (P_K - P_\pi)^2 \simeq m^2_K(1 - \frac{|p_\pi|}{|p_K|}) + m^2_\pi(1 - \frac{|p_K|}{|p_\pi|}) - |p_K||p_\pi|\theta^2_{\pi K}$. Here P_K and P_π are kaon and

DOI: 10.1051/epjconf/201613006001

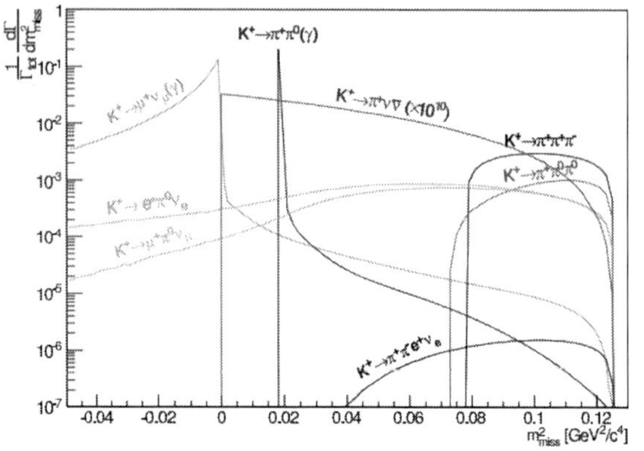

Figure 3. $m^2_{miss} = (P_K - P_\pi)^2$ for the signal and the most frequent background decay modes. The signal is multiplied by 10^{10}, background modes are normalized according to their BR.

pion 4-momenta, p_K and p_π are their 3-momenta and $\theta_{\pi K}$ is the angle between two particles in the laboratory frame.

Fig. 3 shows the distribution of this variable for the signal and backgrounds. Two signal regions are constrained by the main background from the decays $K^+ \to \mu^+\nu_\mu$ (on the left), $K^+ \to \pi\pi\pi$ (on the right) and $K^+ \to \pi^+\pi^0$ (a peak from this decay splits the signal region into 2 halves). Kinematic cuts on the m^2_{miss} allow to reject ~90% of background events. Further suppression is done by the veto system and PID.

4 Preliminary results of the 2015 run

The main goal of the 2015 run was to check the detector performances and the data quality. The data were collected at low beam intensity (~10% of the nominal), the single-track selection criteria were close to those of the final measurement of the BR($K^+ \to \pi^+\nu\bar{\nu}$):

- one primary track in GTK;
- one secondary track in STRAW;
- reconstructed vertex within the decay region;
- energy deposit in LKr;
- hits in the RICH and CHOD;
- spatial matching (STRAW-CHOD, STRAW-LKr);
- time matching (GTK-RICH-CHOD).

The m^2_{miss} distributions for the 2015 data are shown in Fig. 4. The first distribution is done with a kaon-like signal in the KTAG and demonstrates main backgrounds, while the second one illustrates the main sources of downstream tracks not related to the kaon decays: beam elastic scattering in the material along the beamline (KTAG, GTK) and inelastic scattering in the last GTK station.

Time resolutions of the KTAG and GTK are found to match the design values (100 and 200 ps respectively). The resolution of the m^2_{miss} is measured using the peak from the $K^+ \to \pi^+\pi^0$ decay and

DOI: 10.1051/epjconf/201613006001

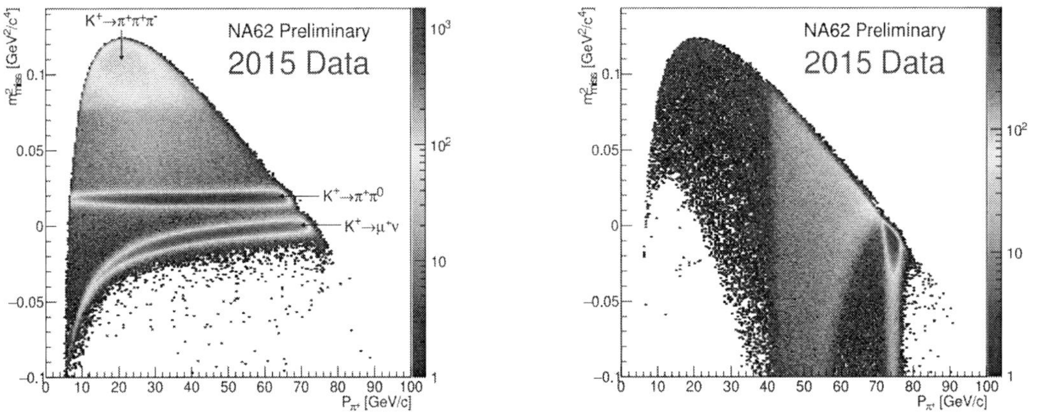

Figure 4. $m^2_{miss} = (P_K - P_\pi)^2$ vs momentum after single-track event selection for kaon decays (left) and without a kaon tag (right).

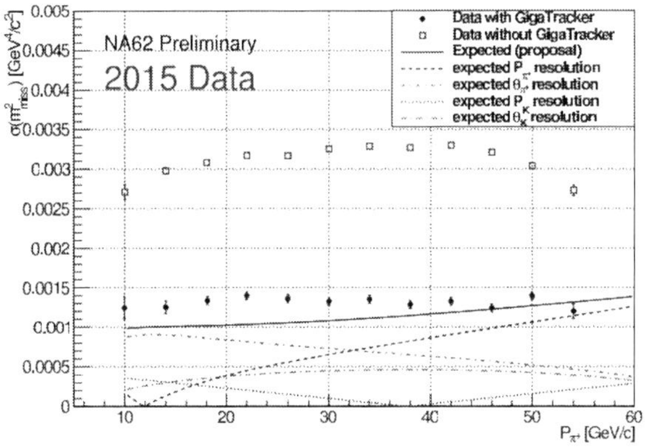

Figure 5. Resolution of the m^2_{miss} vs momentum. Empty squares correspond to the values obtained with the nominal kaon momentum, black points - with the kaon momentum measured by GTK.

found to be 1.2×10^{-3} close to the design value 10^{-3}. The resolution as a function of momentum is shown in Fig. 5.

The identification of secondary charged particles is done in the RICH detector. Fig. 6 illustrates the dependence of the Cherenkov ring radius on the momentum for the single-track selection. The pion-muon separation is based on the difference in ring radii for π and μ. It was measured by selecting a pure muon sample from $K^+ \to \mu^+\nu_\mu$ decay and a pion sample from $K^+ \to \pi^+\pi^0$ decay. In the momentum range between 15 and 35 GeV/c the required muon contamination of 1% was achieved at the 80% pion ID efficiency.

The photon veto system is designed to suppress decays with photons in the final state. For photons from π^0 decays the rejection power provided by LAV, LKr, IRC and SAC detectors should be at least

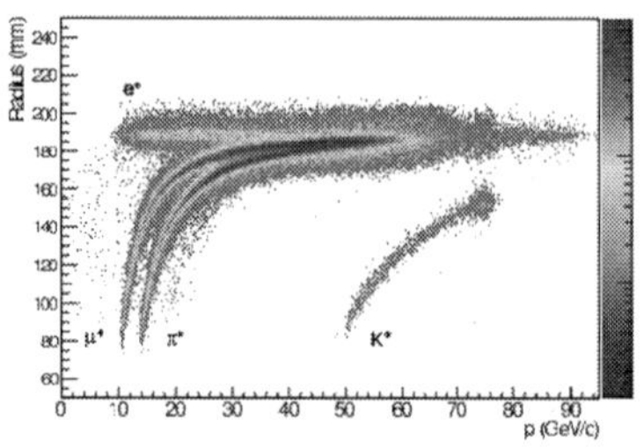

Figure 6. RICH Cherenkov ring radius vs momentum.

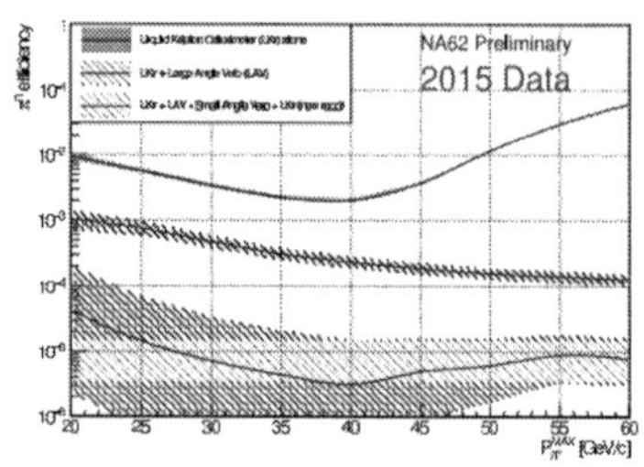

Figure 7. π^0 veto inefficiency for various calorimetric combinations vs momentum.

10^8, or, in other words, the veto inefficiency for photons with energy above 10 GeV should not exceed 10^{-5}. The π^0 veto inefficiency measured from the $K^+ \to \pi^+\pi^0$ decay is shown in Fig. 7. The amount of statistics recorded in 2015 limits this measurement to 10^{-6} (90% CL).

5 Conclusion

The NA62 experiment plans to collect ~100 events of the ultra rare decay $K^+ \to \pi^+\nu\bar{\nu}$ in 2016–2018. In 2015 run the setup performance was studied using the data collected at low beam intensity. The results demonstrate that the experiment is approaching the design sensitivity.

References

[1] J. Brod, M. Gorbahn and E. Stamou, PRD **83**, 034030 (2011)
[2] A.J. Buras, D. Buttazzo, J. Girrbach-Noe and R. Knegjens, JHEP **1511**, 33 (2015)
[3] M. Blanke, A.J. Buras and S. Recksiegel, Eur. Phys. J. **C76** no.4, 182 (2016)
[4] M. Blanke, A.J. Buras, B. Duiling, K. Gemmler and S. Gori, JHEP **903**, 108 (2009)
[5] A. V. Artamonov *et al.*, PRD **79**, 092004 (2009)
[6] J.K. Ahn *et al.*, PRD **81**, 072004 (2010)
[7] The NA62 collaboration, NA62 Technical Design Document, **NA62-10-07** (2010), https://cds.cern.ch/record/1404985

Near-threshold charged kaon pair production in two protons collisions

Damian Gil[1,⋆] and Jerzy Smyrski[1]
for the COSY-11 Collaboration

[1] *Jagiellonian University, Cracow, Poland*

Abstract. The $pp \to ppK^+K^-$ reaction was measured near threshold at excess energy $Q = 4.5$ MeV using the COSY-11 detection system installed at the COSY-Juelich accelerator. Preliminary value of upper limit of the total cross section was determined. This limit indicates that an enhancement of the total cross section close to the threshold, caused by the final state interaction, is smaller than suggested by extrapolation of data measured at higher energies.

1 Physics motivation

The energy dependence of the $pp \to ppK^+K^-$ total cross section close to threshold is very sensitive to the interactions between particles in the final state, and therefore is well suited for studies of these interactions. There are many interesting topics connected with the interactions of the final state kaons and protons. For example, the K^+K^- interaction is crucial for understanding the structure of the scalar mesons $f_0(980)$ and $a_0(980)$, which according Ref. [1, 2] are $K\bar{K}$ molecules. In turn, the K^-p interaction is interesting from the point of view of study the nature of the $\Lambda(1405)$ hyperon, which is considered as a K^-p molecular state [3]. There are also some astrophysical aspects of studying kaons production reactions, like description of neutron stars structure [4]. Elementary reactions of kaon pair production in pp collisions are also important for better understanding of the properties of kaons inside dense baryonic matter [5]. The existing data for the kaon pair production in pp collisions were measured for excess energies higher than $Q = 6$ MeV, with exception of one point at 3 MeV, for which only upper limit of the total cross section was determined. On the other hand, data very close to threshold (for $Q < 6$ MeV) are crucial for distinguishing between various models of the final state interaction (FSI) in the ppK^+K^- system. Therefore, we performed a high luminosity measurement at $Q = 4.5$ MeV.

2 Experiment

The measurement was performed with the COSY-11 detection system, described in details in Ref. [6], at the COSY-Juelich accelerator. A principle of the measurement is presented in Fig. 1. Internal proton beam of the COSY was scattered on the COSY-11 H_2 cluster target. The beam momentum

⋆e-mail: damian.gil@uj.edu.pl

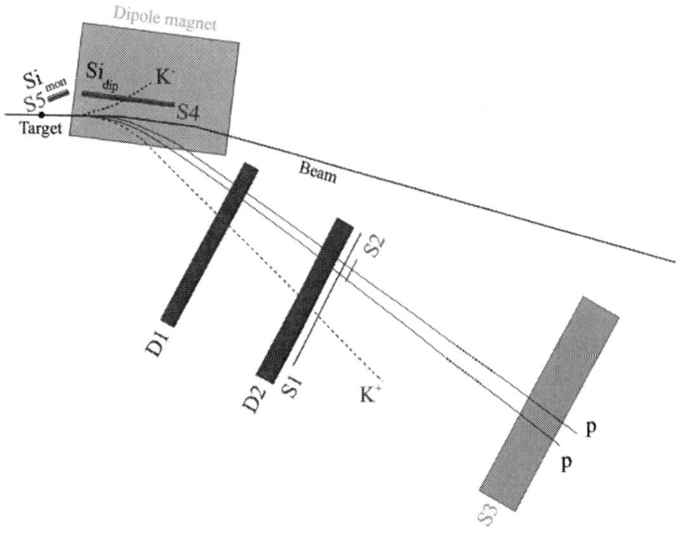

Figure 1. Scheme of the COSY-11 experimental system. Indicated tracks correspond to a simulated ppK^+K^- event.

was 3.316 GeV/c, corresponding to an excess energy in the $pp \to ppK^+K^-$ reaction of 4.5 MeV. With a beam intensity of about 10^{10} stored protons and a target density of about 10^{14} atoms per cm^2, an average luminosity was in the order of 10^{30} cm^{-2}s^{-1}. Positively charged reaction products (ppK^+) were deflected outside the COSY ring in one of regular COSY dipole magnets. Their tracks, registered in a set of two drift chambers (D1 and D2 in Fig. 1), were traced back in the magnetic field of the dipole magnet to the target point in order to determine their momentum vectors. Identification of protons was based on the time-of-flight (TOF) measurement on a path of about 9.1 m between scintillation hodoscopes S1 and S3. For resolving two close proton tracks, a fine granularity hodoscope S2, placed behind the S1 detector, was also used in the TOF measurements. The positively charged kaons were not registered in S3, and therefore they were identified using TOF on a path between the target and the S1 hodoscope. The luminosity was monitored using a coincidence measurement of the elastically scattered protons. The forward scattered protons were measured in the drift chambers D1 and D2, and the recoil protons were registered in a position sensitive silicon pad detector Si_{mon}, placed near the target. For identification of the negatively charged kaons, the ppK^+ missing mass was calculated.

3 Data analysis and results

Figure 2 presents invariant mass squared of the third positively charged particle registered in the S1 detector, besides the two protons, as a function of the missing mass squared of the ppK^+ system. In the spectrum corresponding to simulations of the $pp \to ppK^+K^-$ process, a clear signal from the ppK^+K^- events is visible at the invariant mass and the missing mass equal to the K^+ and K^- mass, respectively. The signal is almost fully confined within 3σ limits indicated with red lines in the figure. In the experimental plot, no events are observed within these limits, and therefore only upper limit for the $pp \to ppK^+K^-$ total cross section can be determined. The background events, visible in the experimental plot, originate mainly from the $pp \to pK^+\Lambda(1405)$ and $pp \to pK^+\Sigma(1385)^0$ reactions. The integrated luminosity was determined on the basis of the registered elastic proton-proton counts, normalized to the high precision pp elastic cross section data measured by the EDDA collaboration [7]. In order to validate obtained integrated luminosity, total cross sections for the $pp \to pp\omega$ and $pp \to pp\eta'$ reactions were determined, based on registered events of the type $pp \to ppX$. For these events, missing mass spectrum of the pp system contains clear signals corresponding to the ω

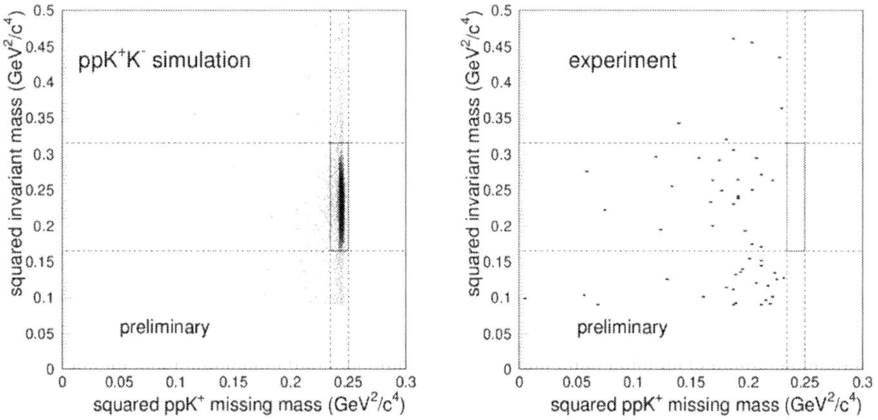

Figure 2. Simulated (left) and measured (right) two-dimensional plots of invariant mass squared of the third (non-proton) particle registered in S1, versus missing mass squared of the ppK^+ system. Red lines present the 3σ limits for distribution of the ppK^+K^- events.

and η' meson production. Obtained values for the ω and η' production cross sections are in a good agreement with results of previous experiments, as shown in Fig. 3. The COSY-11 acceptance for

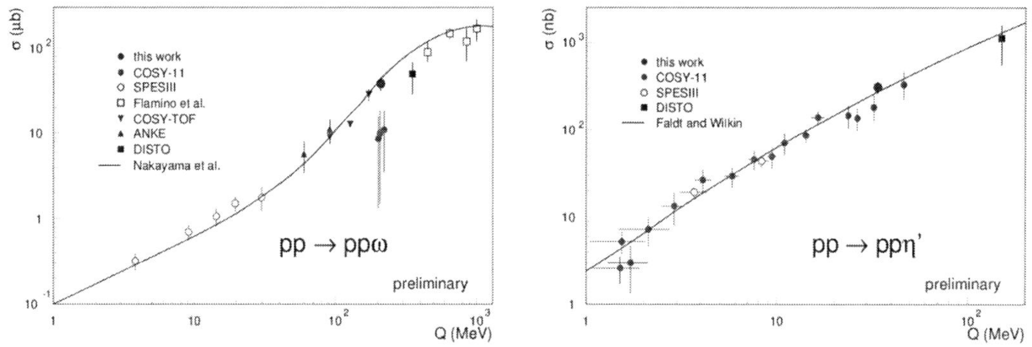

Figure 3. Experimental total cross sections for the ω production [8–13] and η' production [11, 14–19], together with the present data points. Solid lines present theoretical predictions for the ω and η' production cross sections from Ref. [20] and [21], respectively.

the ppK^+K^- events, determined using Monte Carlo simulations taking into account the pp-FSI, is equal to 2.83%. Preliminary upper limit for the $pp \to ppK^+K^-$ total cross section at $Q = 4.5$ MeV, calculated using the obtained integrated luminosity and the acceptance, is presented in Fig. 4, together with existing experimental data. As shown in the figure, energy dependence of the experimental data points is not reproduced by the four-body phase space integral. Inclusion of the pp-FSI still underestimates the experimental values. Also inclusion of the pK^--FSI, parameterized by the scattering length approximation with $a_{pK^-} = (0+1.5i)$ fm, according Ref. [22], underestimates the experimental results, especially in the low energy region. Only after including an extra K^+K^--FSI factor, parameterized with the effective range approximation with $a_{K^+K^-} = (8+0i)$ fm and $b_{K^+K^-} = (-0.1+1.2i)$ fm [23], the previous data points are well described, but the present upper limit of the cross section is overestimated.

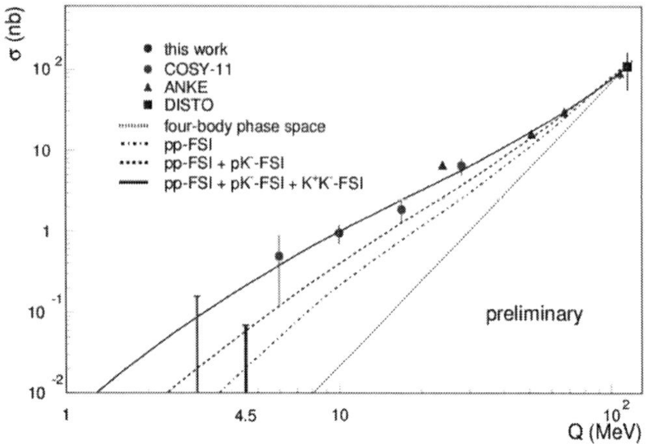

Figure 4. The $pp \to ppK^+K^-$ excitation function obtained from previous experiments [11, 24–28], together with the present upper limit of the total cross section. The dotted line presents an energy dependence according the phase space volume, the dotted-dashed line incorporates the pp-FSI, the dashed line includes additionally the pK^--FSI and solid line is obtained by taking into account also the K^+K^- interaction. All theoretical calculations are normalized to the DISTO point at $Q = 114$ MeV.

4 Conclusions

We measured the $pp \to ppK^+K^-$ reaction at $Q = 4.5$ MeV using the COSY-11 detection system. With no ppK^+K^- observed events, we determined only upper limit of the total cross section. The present limit lies below theoretical calculations, including FSI factors for pp, pK^- and K^+K^- systems, reproducing well experimental data measured at higher energies. This deviation can result from the Coulomb interaction, which was not fully included in the theoretical calculations. It can be also caused by the strong K^+K^- absorption due to formation of hypothetical $K\bar{K}$ molecule [29].

References

[1] M. Bargiotti *et al.*, Eur. Phys. J. C **26**, 371 (2003)
[2] N.N. Achasov and G.N. Shestakov, Phys. Rev. D **58**, 054011 (1998)
[3] J.M.M. Hall *et al.*, Phys. Rev. Lett. **114**, 132002 (2015)
[4] Y. Lim *et al.*, Phys. Rev. C **89**, 055804 (2014)
[5] P. Moskal *et al.*, J. Phys. G **28**, 1777 (2002)
[6] S. Brauksiepe *et al.*, Nucl. Instr. and Meth. A **376**, 397 (1996)
[7] D. Albers *et al.*, Eur. Phys. J. A **22**, 125 (2004)
[8] V. Flaminio *et al.*, Compilation of Cross-Sections, CERN-HERA, 84 (1984)
[9] F. Hibou *et al.*, Phys. Rev. Lett. **83**, 492 (1999)
[10] F. Balestra *et al.*, Phys. Rev. C **63**, 024004 (2001)
[11] C. Quentmeier *et al.*, Phys. Lett. B **515**, 276 (2001)
[12] S. Abd El-Samad *et al.*, Phys. Lett. B **522**, 16 (2001)
[13] M. Abdel-Bary *et al.*, Phys. Lett. B **647**, 351 (2007)
[14] P. Moskal *et al.*, Phys. Rev. Lett. **80**, 3202 (1998)
[15] F. Hibou *et al.*, Phys. Lett. B **438**, 41 (1998)

[16] P. Moskal *et al.*, Phys. Lett. B **474**, 416 (2000)
[17] F. Balestra *et al.*, Phys. Lett. B **491**, 29 (2000)
[18] A. Khoukaz *et al.*, Eur. Phys. J. A **20**, 345 (2004)
[19] P. Klaja *et al.*, Phys. Lett. B **684**, 11 (2010)
[20] K. Nakayama *et al.*, Phys. Rev. C **60**, 055209 (1999)
[21] G. Fäldt and C. Wilkin, Phys. Rev. C **56**, 2067 (1997)
[22] Y. Yan, arXiv:0905.4818
[23] M. Silarski *et al.*, Phys. Rev. C **88**, 025205 (2013)
[24] M. Wolke, Ph.D. thesis, Rheinische Friedrich-Wilhelms-Universität Bonn (1997)
[25] F. Balestra *et al.*, Phys. Lett. B **468**, 7 (1999)
[26] P. Winter *et al.*, Phys. Lett. B **635**, 23 (2006)
[27] Y. Maeda *et al.*, Phys. Rev. C **77**, 01524 (2008)
[28] Q.J. Ye *et al.*, Phys. Rev. C **87**, 065203 (2013)
[29] C. Wilkin, private communication (2016)

EPJ Web of Conferences **130**, 06003 (2016) DOI: 10.1051/epjconf/201613006003
MESON 2016

Theoretical studies of the reaction $e^+e^- \to K^+K^-\gamma$

Leonard Leśniak[1,*] and Michał Silarski[2]

[1] *Institute of Physics, Jagiellonian University, 30-348 Kraków, Poland*
[2] *Laboratori Nazionali di Frascati dell'INFN, Frascati, Italy*

Abstract. We present studies of the $e^+e^- \to K^+K^-\gamma$ reaction for the e^+e^- center-of-mass energies close to the $\phi(1020)$ meson mass. Different mechanisms leading to the final state are considered. The strong interaction amplitudes of the K^+K^- pairs in the S-wave are taken into account. In addition, the photon emission in the initial state, the final state radiation effects and all the possible interference terms are included in the transition matrix elements. The K^+K^- effective mass distributions and the angular dependence of the reaction cross-section are calculated.

1 Introduction

A measurement of the $e^+e^- \to K^+K^-\gamma$ transition could provide a new information about the $K\bar{K}$ strong interactions near threshold as well as about the structure of the $f_0(980)$ and $a_0(980)$ scalar mesons. The kaon-antikaon interaction was studied in several proceses, e.g. in the $pp \to ppK^+K^-$ reaction near the kinematical threshold but the strong interactions of protons in the final state made the interpretation of the results very difficult [1–3]. There are no such obstacles in the $K\bar{K}\gamma$ system. For the $e^+e^- \to K^0\bar{K^0}\gamma$ branching fraction only the upper limit $1.9 \cdot 10^{-8}$ is known [4] and the branching fraction for the $\phi(1020)$ meson decay into the $K^+K^-\gamma$ channel is yet unknown. The $\phi(1020) \to \pi^+\pi^-\gamma$ branching fraction has been measured: $\Gamma(\pi^+\pi^-\gamma)/\Gamma_{total} = (4.1 \pm 1.3) \cdot 10^{-5}$ [5] and the fits for the branching fractions of the $\phi(1020)$ meson decay into the two scalar resonances plus photon give the following results: $\Gamma(f_0(980)\gamma)/\Gamma_{total} = (3.22 \pm 0.19) \cdot 10^{-4}$, $\Gamma(a_0(980)\gamma)/\Gamma_{total} = (7.6 \pm 0.6) \cdot 10^{-5}$ [6]. Both f_0 and a_0 decay to the kaon pairs with low relative momenta, thus studies of the $e^+e^- \to K^+K^-\gamma$ transition at the ϕ mass peak could provide a new information about the K^+K^- strong interactions near threshold as well as about the structure of the scalar mesons.

2 K^+K^- effective mass distributions and angular distributions

The differential cross section for the reaction $e^+(p_{e^+})e^-(p_{e^-}) \to K^+(p_{K^+})K^-(p_{K^-})\gamma(p_\gamma)$ reads

$$\frac{d\sigma}{dm^2 dm^2_{K^-\gamma} dt dt_1} = \frac{|M|^2}{(2\pi)^4 16 s(s-4m_e^2)(s-m^2)r}. \tag{1}$$

The matrix element M can be expressed in terms of five invariants depending on the particle momenta p: $s = (p_{e^+} + p_{e^-})^2$, two effective masses squared $m^2 = (p_{K^+} + p_{K^-})^2$ and $m^2_{K^-\gamma} = (p_{K^-} + p_\gamma)^2$,

[*] e-mail: Leonard.Lesniak@ifj.edu.pl

and two momentum transfers squared $t = (p_{e^-} - p_\gamma)^2$ and $t_1 = (p_{e^-} - p_{K^-})^2$. In Eq. (1) $r = \sqrt{-(t_1 - t_{1min})(t_1 - t_{1max})}$, where t_{1min} and t_{1max} are the kinematically allowed minimum and maximum values of the momentum transfer squared t_1, respectively. The squares of the momentum transfers t_1 and t are related to the K^- and photon polar angles θ_1 and θ_γ, defined in the e^+e^- center-of-mass frame with the z-axis chosen along the e^- momentum:

$$t_1 \approx m_K^2 - \sqrt{s}\, E_1^l (1 - v_1 \cos\theta_1), \qquad t \approx -\sqrt{s}\, \omega^l (1 - \cos\theta_\gamma). \tag{2}$$

In the above equations, where a small value of the electron mass is neglected, v_1 is the K^- velocity, E_1^l and ω^l are the K^- and photon energies given by the following expressions:

$$E_1^l = \frac{m^2 + m_{K^-\gamma}^2 - m_K^2}{2\sqrt{s}}, \qquad \omega^l = \frac{s - m^2}{2\sqrt{s}}. \tag{3}$$

The $K^-\gamma$ effective mass is related to K^- polar angle θ_1^* defined with respect to the direction of the photon momentum in the K^+K^- center-of-mass frame (v being the K^- velocity):

$$m_{K^-\gamma}^2 = m_K^2 + \frac{1}{2}(s - m^2)(1 - v\,z), \quad v = \sqrt{1 - 4m_K^2/m^2}, \quad z = \cos\theta_1^*. \tag{4}$$

In Ref. [7] two theoretical models have been exploited in detailed fits to the KLOE data of the $\phi \to f_0(980)\gamma \to \pi^+\pi^-\gamma$ decay. The first model labeled NS (no-structure model) has been formulated by Isidori, Maiani, Nicolaci and Pacetti in Ref. [8]. The second one called the kaon loop model has been developed by Achasov and his collaborators (see for example [9]). In this study we present results corresponding to the NS-model extended to the $e^+e^- \to K^+K^-\gamma$ reaction. The parameters of the model have been taken from Table 1 of [8].

In Fig. 1 the K^+K^- effective mass distributions and K^- angular distributions with respect to the photon momentum axis in the K^+K^- center-of-mass frame are jointly shown. The K^- angular distributions in the e^+e^- c.m. frame are shown in Fig. 2 for two K^+K^- effective masses m equal to 900 and 998 MeV. In Fig. 3 the photon angular distributions are presented for two values of m.

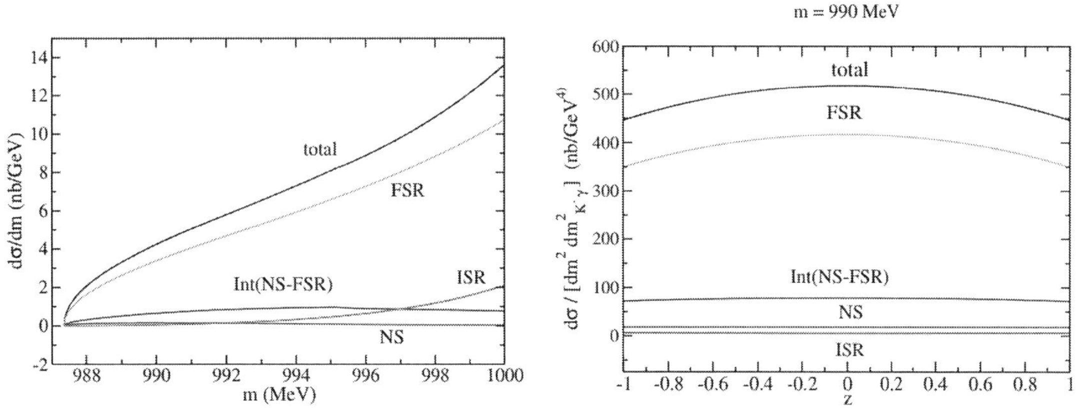

Figure 1. Left panel: K^+K^- effective mass distributions, right panel: double differential cross sections at m=990 MeV versus the cosine of the K^- polar angle defined in Eq. (4). In both cases $45^0 < \theta_\gamma < 135^0$. Four contributions to the total result are presented. They are labeled by FSR (final state radiation), ISR (initial state radiation), NS (NS model) and Int(NS-FSR) (interference of the NS and the FSR amplitudes).

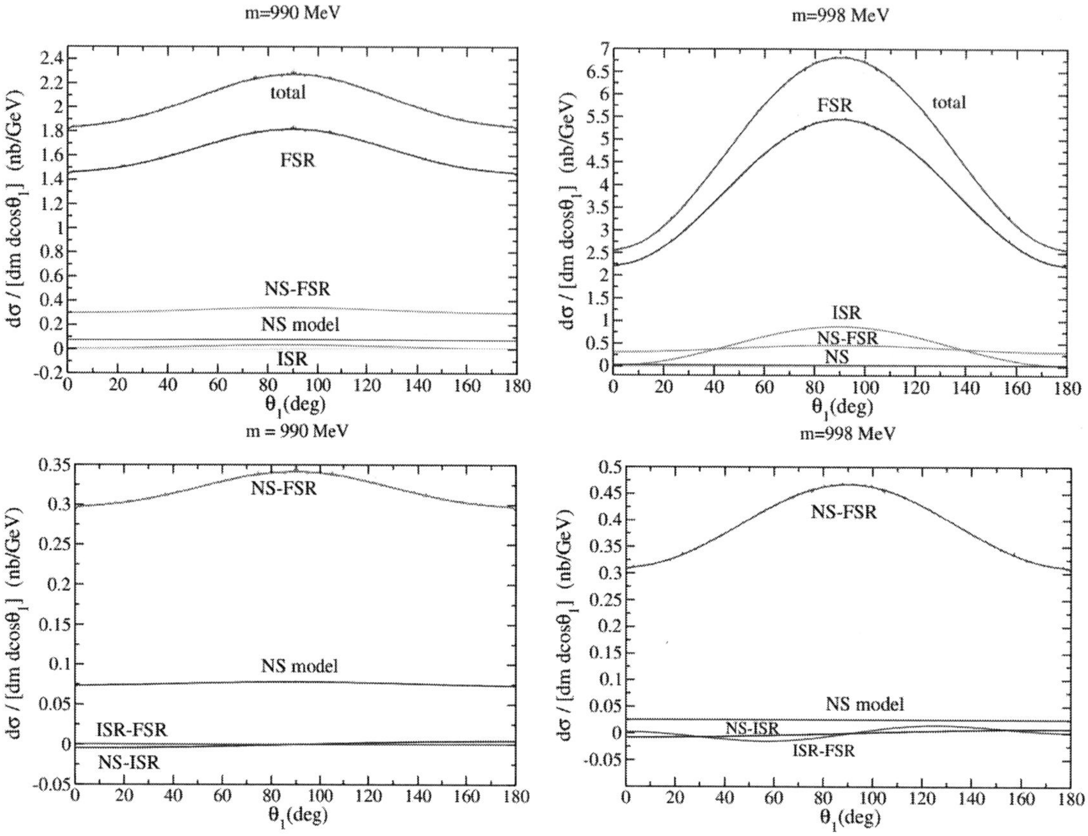

Figure 2. K^- angular distributions at fixed m for $45^0 < \theta_\gamma < 135^0$. The meaning of the labels total, FSR, ISR and NS is the same as in Fig. 1. By NS-FSR, ISR-FSR and NS-ISR we denote the interference terms of the NS and FSR amplitudes, the ISR and FSR amplitudes, and the NS and ISR amplitudes, respectively.

Table 1. The cross sections integrated over the K^+K^- effective mass from the threshold up to 1009 MeV for two ranges of the photon emission angle θ_γ.

reaction mechanism	$24^0 < \theta_\gamma < 156^0$	$45^0 < \theta_\gamma < 135^0$
FSR	0.330 nb	0.238 nb
NS	0.0020 nb	0.0014 nb
Int(NS-FSR)	0.021 nb	0.015 nb
ISR	0.183 nb	0.104 nb
total	0.536 nb	0.358 nb

3 Prospects of the $e^+e^- \to K^+K^-\gamma$ reaction studies with the KLOE data

One of the best data sets which could be used to study the $e^+e^- \to K^+K^- \to K^+K^-\gamma$ reaction was collected by the KLOE experiment operating at the DAΦNE ϕ-factory in Frascati, Italy [10]. Its advantage is a very good kaon momenta determination and the high statistics. However, one needs to deal with a detection of low energy photons ($E_\gamma < 32$ MeV in the e^+e^- c.m. system) and the slow kaon momentum tracks at the low K^+K^- effective masses. Based on the values of the cross sections given in Table 1 we show in Table 2 the estimated numbers of events for the integrated lumonisty of 1.7 fb^{-1}. Corrections for the detector acceptance and efficiency are not yet done.

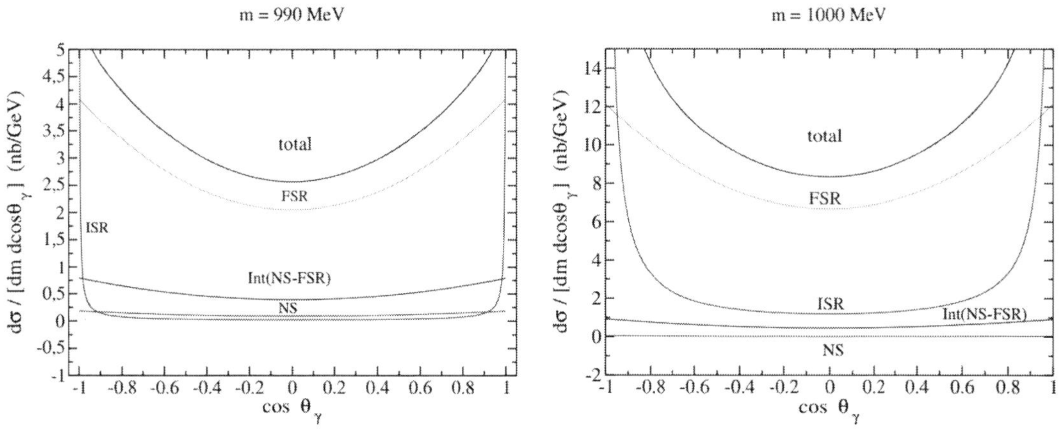

Figure 3. Photon angular distributions at fixed m=990 and 1000 MeV.

Table 2. Numbers of events for the K^+K^- effective mass up to 1009 MeV and for two ranges of θ_γ.

reaction mechanism	$24^0 < \theta_\gamma < 156^0$	$45^0 < \theta_\gamma < 135^0$
FSR	$5.6 \cdot 10^5$	$4.0 \cdot 10^5$
NS	$3.4 \cdot 10^3$	$2.4 \cdot 10^3$
Int(NS-FSR)	$3.6 \cdot 10^4$	$2.5 \cdot 10^4$
ISR	$3.1 \cdot 10^5$	$1.8 \cdot 10^5$
total	$9.1 \cdot 10^5$	$6.1 \cdot 10^5$

These results of theoretical calculations can be used in a future experimental measurement of the unknown branching fraction of the $\phi(1020)$ meson into the $K^+K^-\gamma$ channel.

This work has been supported by the Polish National Science Centre (grant no 2013/11/B/ST2/04245).

References

[1] M. Silarski *et al.*, Phys. Rev. **C 80**, 045202 (2009)
[2] M. Silarski and P. Moskal, Phys. Rev. **C 88**, 025205 (2013)
[3] Q. J. Ye *et al.*, Phys. Rev. **C 87**, 065203 (2013)
[4] F. Ambrosino *et al.* (KLOE Collaboration), Phys. Lett. **B 679**, 10 (2009)
[5] R. R. Akhmetshin *et al.* (CMD-2 Collaboration), Phys. Lett. **B 462**, 371 (1999)
[6] K. A. Olive *et al.* (Particle Data Group), Chin. Phys. **C 38**, 090001 (2014)
[7] F. Ambrosino *et al.* (KLOE Collaboration), Phys. Lett. **B 634**, 148 (2006)
[8] G. Isidori, L. Maiani, M. Nicolaci, S. Pacetti, JHEP **0605**, 049 (2006)
[9] N. N. Achasov, V. V. Gubin and V. I. Shevchenko, Phys. Rev. **D 56**, 203 (1997)
[10] F. Bossi *et al.* , Riv. Nuovo Cim. **31**, 531 (2008)

EPJ Web of Conferences **130**, 06004 (2016) DOI: 10.1051/epjconf/201613006004
MESON 2016

Predictions for pentaquark states of hidden charm molecular nature and comparison with experiment

Eulogio Oset[1,2,⋆], Hua-Xing Chen[3], Albert Feijoo[4], Li-Sheng Geng[3], Wei-Hong Liang[5], De-Min Li[6], Jun-Xu Lu[3], Vladymir K. Magas[4], Juan Nieves[7], Angels Ramos[4], Luis Roca[8], En Wang[6], and Ju-Jun Xie[2]

[1]*Departamento de Física Teórica and IFIC, Centro Mixto Universidad de Valencia-CSIC Institutos de Investigación de Paterna, Aptdo. 22085, 46071 Valencia, Spain*
[2]*Institute of Modern Physics, Chinese Academy of Sciences, Lanzhou 730000, China*
[3]*School of Physics and Nuclear Energy Engineering and International Research Center for Nuclei and Particles in the Cosmos, Beihang University, Beijing 100191, China*
[4]*Departament d'Estructura i Constituents de la Materia and Institut de Ciencies del Cosmos (ICCUB), Universitat deBarcelona, Marti i Franques 1, 08028 Barcelona, Spain*
[5]*Department of Physics, Guangxi Normal University, Guilin 541004, China*
[6]*Department of Physics, Zhengzhou University, Zhengzhou, Henan 450001, China*
[7]*IFIC, Centro Mixto Universidad de Valencia-CSIC, Institutos de Investigacion de Paterna, Aptdo. 22085, 46071 Valencia, Spain*
[8]*Departamento de Fisica, Universidad de Murcia, E-30100 Murcia, Spain*

Abstract. Predictions for hidden charm molecules, with and without strangeness, were made prior to the LHCb experiment. We discuss these issues and how these states can be observed in the reactions, $\Lambda_b \to J/\psi K^- p$, $\Lambda_b \to J/\psi \eta \Lambda$, $\Lambda_b \to J/\psi \pi^- p$, $\Lambda_b \to J/\psi K^0 \Lambda$ and $\Xi_b^- \to J/\psi K^- \Lambda$ by looking at the invariant mass distributions for $J/\psi p$ or $J/\psi \Lambda$.

1 Introduction

Peaks observed recently in the $J/\psi p$ mass distribution in the $\Lambda_b \to J/\psi p K^-$ decay [1, 2], were interpreted in terms of two pentaquark states, $P_c(4380)$, $P_c(4450)$, with widths around 200 MeV and 40 MeV respectively. Summaries of the theoretical and experimental work concerning these findings can be seen in Refs. [3, 4], and a thorough review on the subject has been presented in [5].

Here we concentrate on the hypothesis that the states observed are of molecular type. The molecular nature is quite appealing, since predictions for states of hidden charm, that could be naturally associated to these observed states, had been done before. Indeed in [6, 7] baryon states of hidden charm were found in the study of the interaction of the $\bar{D}\Sigma_c$-$\bar{D}\Lambda_c$, $\bar{D}^*\Sigma_c$-$\bar{D}^*\Lambda_c$ coupled channels, as main building blocks, together with decay channels in the light sector and the $\eta_c N$ and $J/\psi N$ states. Related studies followed and in [8] an admixture of spin flavor symmetry in the light sector and Heavy Quark Spin symmetry, HQSS, was used and states of hidden charm similar to those predicted in [6, 7] were found. Similar results to those of [6, 7] were found in [9] using HQSS and the local hidden

⋆e-mail: oset@ific.uv.es

gauge approach to evaluate the matrix elements of HQSS. More work on this is done in [10], where a mixture of Vector-Baryon, VB, and Pseudoscalar-Baryon (PB), states is allowed in coupled channels such that one can have a better control on the decay width of the states.

In [6, 7] it was also found that in the strange sector there were also some resonances with hidden charm and strangeness S=-1, related to those found in the hidden charm S=0 sector. With this perspective we shall discuss in the present work several promising reactions, some of which are already under analysis by the LHCb collaboration.

2 The $\Lambda_b^0 \to J/\psi p K^-$ reaction and the $P_c(4380)$ and $P_c(4450)$ states

The pentaquarks were found in the $\Lambda_b^0 \to J/\psi p K^-$ reaction, by looking at the $J/\psi p$ mass distribution. At the same time, the pK^- mass distribution was also discussed. It is interesting to mention that this reaction had been studied before theoretically [11] and predictions were made for the pK^- and $\pi\Sigma$ mass distributions, following similar steps as done in the B^0 and B_s^0 decays into $J/\psi\pi^+\pi^-$ in [12] (see also recent review in [13]). It is interesting to recall that the predicctions made for the pK^- mass distribution were in agreement with the $\Lambda(1405)$ contribution deduced from the partial wave analyis of [1, 2]. The consistency of the results of [6] with those of the LHCb experiment was shown in [14].

A recent reanalysis of the $J/\psi p$ and K^-p mass distributions has been done recently, by constructing the suited amplitudes to the possible quantum numbers of the $J/\psi p$ system and taking into account explicitly the K^-p interaction within the chiral unitary approach [15]. What is found there is that the mass distributions by themselves do not favour any particular spin parity for the $J/\psi p$ and there is no compelling need for the broad pentaquark state. Thus, it must be other experimental information what favours one or another assignment.

In the chiral unitary scheme of [16] the $\Lambda(1405)$ appears dynamically in the highly non-linear dynamics involved in the unitarization procedure showing up in the t_{ij} amplitudes. These amplitudes contain two poles for the $\Lambda(1405)$ resonance at $1352-48i$ MeV and $1419-29i$ MeV [16]. The highest mass $\Lambda(1405)$, couples mostly to $\bar{K}N$, and is the one relevant in the present work.

On the other hand, in refs. [6, 9], it was shown that the $J/\psi N$ final state interaction in coupled channels, considering the $\bar{D}^*\Lambda_c$, $\bar{D}^*\Sigma_c$, $\bar{D}\Sigma_c^*$ and $\bar{D}^*\Sigma_c^*$ channels, produces poles in the $J^P = 3/2^-$, $I = 1/2$, sector at $4334 - 19i$ MeV, $4417 - 4i$ MeV and $4481 - 17i$ MeV, which couple sizeably to $J/\psi\, p$ (see table II in ref. [9]). Thus, the $J/\psi\, p$ invariant mass distribution in the $\Lambda_b \to J/\psi\, K^- p$ decay, should show the shape of these resonances, as indeed has been shown in [1].

3 A hidden-charm $S = -1$ pentaquark from the decay of Λ_b into $J/\psi\eta\Lambda$ states

The same mechanism that drives the $\Lambda_b^0 \to J/\psi p K^-$ reaction can lead to other combinations of meson baryon in the final state. Indeed, the $\eta\Lambda$ is one of the channels which is already produced at the tree level in the first step after the hadronization. If we look at the $J/\psi\eta\Lambda$ final state, this channel will be produced with a strength similar to the one of $J/\psi K^- p$, which is already observed. Certainly there will be final state interaction of the meson baryons produced at the first step to give finally the $\eta\Lambda$. And certainly, we can also have rescattering of J/ψ with the Λ. The question now is whether there is some partner pentaquark of those observed in the $J/\psi p$ mass distribution in the $\Lambda_b \to J/\psi K^- p$ reaction, which could show up in the $J/\psi\Lambda$ distribution. This partner state was found in [6] and it couples to $\bar{D}^*\Xi_c$, $\bar{D}^*\Xi_c'$ and $J/\psi\Lambda$.

In [17] predictions are made stating that as far as the coupling of the resonance to $J/\psi\Lambda$ is of the order of 0.5, which is found in [6], one should see clear signals in this reaction.

4 The hidden-charm pentaquark state in $\Lambda_b^0 \to J/\psi p \pi^-$ decay

In the present section we shall discuss the $\Lambda_b^0 \to J/\psi p \pi^-$ reaction which was measured in [18]. The results have a smaller statistics than those of Ref. [1], but in retrospective they are very valuable. In [19] the reaction was studied and $p\pi^-$ as well as $J/\psi p$ invariant mass distributions were calculated. It was found that the peak observed was consistent with the description given before for the $\Lambda_b^0 \to J/\psi p K^-$ decay.

5 The $\Lambda_b \to J/\psi K^0 \Lambda$ reaction and the hidden-charm pentaquark state with strangeness

In this section we report on the study of the $\Lambda_b \to J/\psi K^0 \Lambda$ reaction in [20]. The mechanism is now the same as in the former sections, but instead of looking for $\pi^- p$ in the final state, we look now at its coupled channel $K^0 \Lambda$.

A clear signal was also seen in the $J/\psi \Lambda$ mass distribution. The signal can depend on the precise values of the couplings and the width of the states, but the message of [20] is that the peaks are visible over a wide variation of the parameters around the values chosen according to Refs. [6, 7].

6 The hidden-charm pentaquark state with strangeness $S = -1$ from Ξ_b^- decay into $J/\psi K^- \Lambda$

Finally we report on the work of [21] for the $\Xi_b^- \to J/\psi K^- \Lambda$ reaction, where again the hidden charm strange state discussed in the former section shows up again.

We observe a clear structure around 4650 MeV on top of the background when the $J/\psi \Lambda$ interaction is taken into account. By changing M_R the peak position changes accordingly, but a clear signal is observed in all cases as long as the width is smaller than 100 MeV [21]. We stress however that the strength of the signal will depend strongly on the coupling of the hidden charm state to the $J/\psi \Lambda$, i.e., $g_{J/\psi \Lambda}$, but, again, it was shown in [21] that the signal would still be visible with large changes of this coupling.

7 Further remarks

Considerations are also made about indirect ways of producing the J/ψ- baryon resonance in [17], as the direct production of $\bar{D}^{*0} \Xi_c' \eta$ followed by $\bar{D}^{*0} \Xi_c' \to J/\psi \Lambda$. Once again, the conclusion is that the observation of the predicted strange hidden charm state should be rather clean.

8 Conclusions

From the perspective that the observed hidden charm pentaquark states without strangeness in Ref. [1] might be molecular states of $\bar{D}^* \Sigma_c$-$\bar{D}^* \Sigma_c^*$ nature, as was predicted before the experiment was performed, and the fact that hidden charm states with strangeness were simultaneously predicted, we have looked into some reactions where these states could be seen. We reported on predictions for five different reactions, and neat peaks in the $J/\psi p$ or $J/\psi \Lambda$ mass distributions show up in the calculations. It would be most interesting to perform such experiments, and some of them are already under analysis by the LHCb collaboration. A recent review of all these reactions is available in [22].

One of us, E. Oset, wishes to acknowledge support from the Chinese Academy of Science in the Program of Visiting Professorship for Senior International Scientists (Grant No. 2013T2J0012). This work is partly supported by the National Natural Science Foundation of China under Grant Nos. 11375024, 11522539, 11505158, 11475015, and 11475227, the Spanish Ministerio de Economia y Competitividad and European FEDER funds under the contract number FIS2014-51948-C2-1-P and SEV-2014-0398, and the Generalitat Valenciana in the program Prometeo II-2014/068. This work is also supported by the China Postdoctoral Science Foundation (No. 2015M582197), and the Open Project Program of State Key Laboratory of Theoretical Physics, Institute of Theoretical Physics, Chinese Academy of Sciences, China (No.Y5KF151CJ1).

References

[1] R. Aaij *et al.* [LHCb Collaboration], Phys. Rev. Lett. **115**, 072001 (2015)
[2] R. Aaij *et al.* [LHCb Collaboration], Chinese Phys. C **40**, 011001 (2016)
[3] T. J. Burns, Eur. Phys. J. A **51**, 152 (2015)
[4] R. Chen, X. Liu and S. L. Zhu, arXiv:1601.03233 [hep-ph]
[5] H. X. Chen, W. Chen, X. Liu and S. L. Zhu, arXiv:1601.02092 [hep-ph]
[6] J. J. Wu, R. Molina, E. Oset and B. S. Zou, Phys. Rev. C **84**, 015202 (2011)
[7] J. J. Wu, R. Molina, E. Oset and B. S. Zou, Phys. Rev. Lett. **105**, 232001 (2010)
[8] C. Garcia-Recio, J. Nieves, O. Romanets, L. L. Salcedo and L. Tolos, Phys. Rev. D **87**, 074034 (2013)
[9] C. W. Xiao, J. Nieves and E. Oset, Phys. Rev. D **88**, 056012 (2013)
[10] T. Uchino, W. H. Liang and E. Oset, Eur. Phys. J. A **52**, 43 (2016)
[11] L. Roca, M. Mai, E. Oset and U. G. Meißner, Eur. Phys. J. C **75**, 218 (2015)
[12] W. H. Liang and E. Oset, Phys. Lett. B **737**, 70 (2014)
[13] E. Oset *et al.*, Int. J. Mod. Phys. E **25**, 1630001 (2016)
[14] L. Roca, J. Nieves and E. Oset, Phys. Rev. D **92**, 094003 (2015)
[15] L. Roca and E. Oset, arXiv:1602.06791 [hep-ph]
[16] L. Roca and E. Oset, Phys. Rev. C **88**, 055206 (2013)
[17] A. Feijoo, V. K. Magas, A. Ramos and E. Oset, Eur. Phys. J. C **76**, 446 (2016)
[18] R. Aaij *et al.* [LHCb Collaboration], JHEP **1407**, 103 (2014)
[19] E. Wang, H. X. Chen, L. S. Geng, D. M. Li and E. Oset, Phys. Rev. D **93**, 094001 (2016)
[20] J. X. Lu, E. Wang, J. J. Xie, L. S. Geng and E. Oset, Phys. Rev. D **93**, 094009 (2016)
[21] H. X. Chen, L. S. Geng, W. H. Liang, E. Oset, E. Wang and J. J. Xie, Phys. Rev. C **93**, 065203 (2016)
[22] E. Oset *et al.*, Nucl. Phys. A **954**, 371 (2016)

 DOI: 10.1051/epjconf/201613006005

K-Long Facility for JLab and its Scientific Potential

Igor I. Strakovsky[1,*]

[1] *Institute for Nuclear Studies, Department of Physics, The George Washington University, Washington, D.C. 20052, U.S.A.*

Abstract. Our main interest in creating a secondary high-quality KL-beam is to investigate hyperon spectroscopy through both formation and production processes. We propose to study two-body reactions induced by the KL-beam on the proton target. The experiment should measure both differential cross sections and self-analyzed polarizations of the produced Λ-, Σ-, and Ξ-hyperons using the GlueX detector at the Jefferson Lab Hall D. New data will greatly constrain partial-wave analysis and reduce model-dependent uncertainties in the extraction of strange resonance properties, providing a new benchmark for comparisons with QCD-inspired models and LQCD calculations. The measurements will span c.m. $\cos\theta$ from -0.95 to 0.95 in c.m. range above W = 1490 MeV and up to 4000 MeV.

1 Introduction

At the beginning of February of 2016, Jefferson Lab hosted a Workshop *Physics with Neutral Kaon beam at JLab*. It was dedicated to the physics of hyperons produced by the neutral Kaon beam on both unpolarized and polarized targets [1]. The workshop follows our LoI–12–15–001 [2] (JLab KLF Project) to help to address the comments made by the JLab PAC43 and to prepare the full proposal for PAC45. The emphasis is on the hyperon spectroscopy. Mini-Proceedings of the KL2016 are available at arXiv [3].

The *Excited Hyperons in QCD Thermodynamics at Freeze-Out* (YSTAR2016) Workshop [4] is a successor to the recent KL2016. This workshop will discuss the influence of possible "missing" hyperon resonances on QCD thermodynamics, on freeze-out in heavy ion collisions and in the early universe, and in spectroscopy. Recent studies that compare LQCD calculations of thermodynamic calculations, statistical hadron resonance gas models, and ratios between measured yields of different hadron species in heavy ion collisions provide indirect evidence for the presence of "missing" resonances in all of these contexts. The aim of the workshop is to sharpen these comparisons, advance our understanding of the formation of baryons from quarks and gluons microseconds after the Big Bang and in today's experiments, and to connect these developments to experimental searches for direct, spectroscopic, evidence for these resonances.

The JLab12 energy upgrade, with the new Hall D, is an ideal tool for extensive studies of non-strange and, specifically, strange baryon resonances [5]. Our plan is evolving to take advantage of the existing high quality photon beam line and experimental area in the Hall D complex at Jefferson

*e-mail: igor@gwu.edu

Lab to deliver a beam of K_L particles onto a liquid hydrogen cryotarget within the GlueX detector. The recently constructed GlueX detector in Hall D is a large acceptance spectrometer with good coverage for both charged and neutral particles that can be adapted to this purpose. Obviously, Kaon beam facility (KLF) with good momentum resolution is crucial to provide the data needed to identify and characterize the properties of hyperon resonances. The masses and widths of the lowest Λ and Σ baryons were determined mainly with Kaon-beam experiments in the 1970s [6]. Pole position in complex energy plane for hyperons has began to be studied only recently, first of all for $\Lambda(1520)\frac{3}{2}^-$ [7].

2 Scope of the Proposed KLF Program

A comparison of recent coupled-channel analyses [8–11] comes to the conclusion that, for most cases, it is only the first excited state in each partial wave whose detailed properties [branching reaction (BRs), helicity amplitudes] are known. Different analyses may agree on the existence of the second state (in each partial wave) but not on their decay properties, while there is no agreement even on the existence of a third state in a particular partial wave. Given the arduous nature of the task involved in obtaining high-quality data that enter these multichannel analyses it is reasonable to address the question as to the final scope of this effort. In other words: How many resonances do we need to identify in order to convince ourselves that we have achieved a solid understanding of the baryon spectrum from QCD? As examples, we examine this question from the viewpoint of lattice gauge and constituent quark model (QM) calculations.

Our current "experimental" knowledge of Λ^*, Σ^*, Ξ^*, and Ω^* resonances is far worse than our knowledge of N^* and Δ^* resonances; though they are equally fundamental. Specifically, the properties of multi-strange baryons (Ξ^* and Ω^* states) are poorly known. For instance the *Review of Particle Physics* lists only two states with BR to $K\Xi$, namely, $\Lambda(2100)\frac{7}{2}^-$ (BR < 3%) and $\Sigma(2030)\frac{7}{2}^+$ (BR < 2%) [6]. Clearly, complete understanding of three-quark bound states requires to learn more about baryon resonances in "strange sector" as well.

Reviewing analyses decades worth of data, from both hadronic and EM experiments, we have found numerous baryon resonances, and determined their masses, widths, and quantum numbers. There are 112 baryons in PDG2014 Listings [6] (including both non-strange and strange states) and only 58 of them are 4^* and 3^*. Many more states have been predicted by QMs. For example in case of $SU(6) \times O(3)$, it would be required 434 resonances, if all revealed multiplets were completed (three 70^- and four 56^-).

Three light quarks can be arranged in 6 baryonic families, N^*, Δ^*, Λ^*, Σ^*, Ξ^*, and Ω^*. Number of members in a family that can exist is not arbitrary [12]. If $SU(3)_F$ symmetry of QCD is controlling, then for the octet: N^*, Λ^*, and Σ^*, and for the decuplet: Δ^*, Σ^*, Ξ^*, and Ω^*. Number of experimentally identified resonances of each baryon family in PDG2014 summary tables is 17 N^*, 24 Δ^*, 14 Λ^*, 12 Σ^*, 7 Ξ^*, and 2 Ω^*. Constituent QMs, for instance, predict existence of no less than 64 N^* and 22 Δ^* states with mass less than 3 GeV. Seriousness of "missing-states" problem [13] is obvious from these numbers. To complete $SU(3)_F$ multiplets, one needs no less than 17 Λ^*, 41 Σ^*, 41 Ξ^*, and 24 Ω^*.

3 Observables

There are two particles in the reactions $K_L p \to \pi Y$ and KY that can carry polarization: the target and recoil nucleon/hyperon. Hence, there are two possible double-polarization experiments: target/recoil. While a formally complete experiment requires the measurement, at each energy and angle, of at least three independent observables, the current database for $K_L p \to \pi Y$ and KY is populated mainly by unpolarized cross sections. Figure 1 illustrates this quite clearly.

DOI: 10.1051/epjconf/201613006005

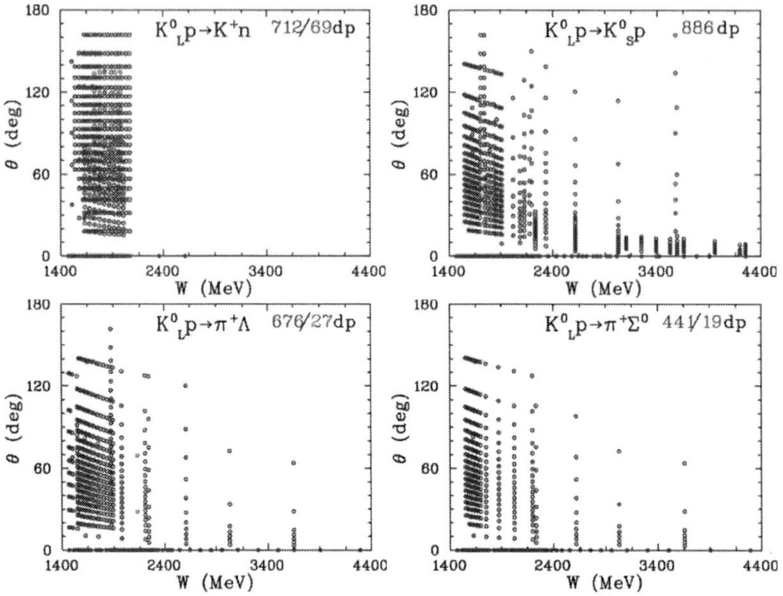

Figure 1. Experimental data available for $K_L p \to K^+ n$, $K_L p \to K_S p$, $K_L p \to \pi^+\Lambda$, and $K_L p \to \pi^+\Sigma^0$ as a function of c.m. energy W [14]. The number of data points (dp) is given in the upper righthand side of each subplot [blue (red) shows amount of unpolarized (polarized) observables]. Total cross sections are plotted at zero degrees.

The experiments using unpolarized LD_2 (to get "neutron" data) and polarized target (aka FROST) for both hydrogen and deuteriun components, we will leave for the following proposals. Obviosly, it will open up a new avenue to the complete experiment. Note that the "neutron" data are critical to determine parameters of neutral Λ^*s and Σ^*s hyperons which were considered recently [15].

4 Phenomenology / Partial-Wave Aanalysis

Following Höhler [16], the differential cross section and polarization for $K_L p \to \pi Y$ and KY are given by

$$\frac{d\sigma}{d\Omega} = \lambda\!\!\!^{-2}(|f(W,\theta)|^2 + |g(W,\theta)|^2), \quad P\frac{d\sigma}{d\Omega} = 2\lambda\!\!\!^{-2}\mathrm{Im}(f(W,\theta)\, g(W,\theta)^*), \tag{1}$$

where $\lambda\!\!\!^- = \hbar/k$, with k the magnitude of c.m. momentum for the incoming meson. Here $f(W,\theta)$ and $g(W,\theta)$ are the usual spin-nonflip and spin-flip amplitudes at c.m. energy W and meson c.m. scattering angle θ. In terms of partial waves, $f(W,\theta)$ and $g(W,\theta)$ can be expanded as

$$f(W,\theta) = \sum_{l=0}^{\infty}[(l+1)T_{l+} + lT_{l-}]P_l(\cos\theta), \quad g(W,\theta) = \sum_{l=1}^{\infty}[T_{l+} - T_{l-}]P_l^1(\cos\theta), \tag{2}$$

where l is the initial orbital angular momentum, $P_l(\cos\theta)$ is a Legendre polynomial, and $P_l^1(\cos\theta)$ is an associated Legendre function. The total angular momentum for the amplitude T_{l+} is $J = l + \frac{1}{2}$, while that for the amplitude T_{l-} is $J = l - \frac{1}{2}$. For hadronic scattering reactions, we may ignore small

CP-violating terms and write

$$K_L^0 = \frac{1}{\sqrt{2}}(K^0 - \overline{K^0}), \quad K_S^0 = \frac{1}{\sqrt{2}}(K^0 + \overline{K^0}). \tag{3}$$

We may generally have both $I = 0$ and $I = 1$ amplitudes for KN and $\overline{K}N$ scattering, so that the amplitudes $T_{l\pm}$ can be expanded in terms of isospin amplitudes as

$$T_{l\pm} = C_0 T^0_{l\pm} + C_1 T^1_{l\pm}, \tag{4}$$

where $T^I_{l\pm}$ are partial-wave amplitudes with isospin I and total angular momentum $J = l \pm \frac{1}{2}$, with the appropriate isospin Clebsch-Gordan coefficients C_I.

We plan to do a coupled-channel PWA with new GlueX KLF data in combination with available and new J-PARC K^-p measurements when they will be available. Then the best fit will allow to determine data driven (model independent) partial-wave amplitudes and associated resonance parameters as the SAID group does, for instance, for analysis of πN-elastic, charge-exchange, and $\pi^-p \to \eta n$ data [17]. With the new GlueX KLF data, the quantitative significance of resonance signals can be determined. Additionally, new PWA with new GlueX data will allow to look for "missing" hyperons via looking for new poles in complex plane positions. It will provide a new benchmark for comparisons with QCD-inspired models and LQCD calculations.

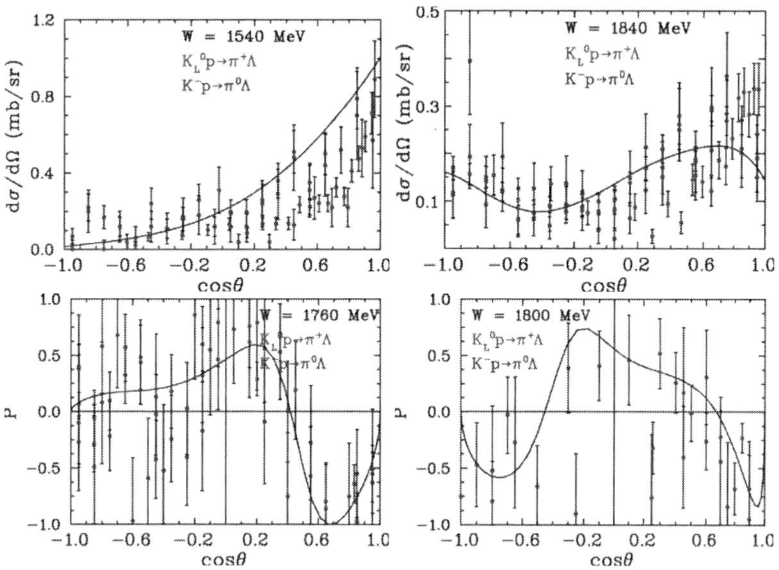

Figure 2. Comparison of selected differential cross section data for $K^-p \to \pi^0\Lambda$ and $K_Lp \to \pi^+\Lambda$ at W = 1540 MeV and 1840 MeV and selected polarization data at W = 1760 MeV and 1800 MeV from previous measurements are those data points within 20 MeV of the Kaon c.m. energy indicated on each panel [14]. The curves are from the recent PWA of $K^-p \to \pi^0\Lambda$ data [8].

The $K^-p \to \pi^0\Lambda$ and $K_Lp \to \pi^+\Lambda$ amplitudes imply that observables for these reactions measured at the same energy should be the same except for small differences due to the isospin-violating mass differences in the hadrons. No differential cross section data for $K^-p \to \pi^0\Lambda$ are available at c.m.

energies $W < 1540$ MeV, although data for $K_Lp \to \pi^+\Lambda$ are available at such energies due to longer K_L life time. At 1540 MeV and higher energies, differential cross section and polarization data for both reactions are in fair agreement, as shown in Fig. 2. Meanwhile, the quality of avilable P measurements do not have a sensitivity to the fit.

5 Proposed Measurements

We propose to use a Hall D Facility with the GlueX spectrometer, to perform precision measurements of two-body reactions induced by the K_L-beam on the liquid hydrogen cryotarget in the resonance region, W = 1490 – 4000 MeV and c.m. $\cos\theta$ from -0.95 to 0.95. This ability of the GlueX provides an ideal environment for these experiment.

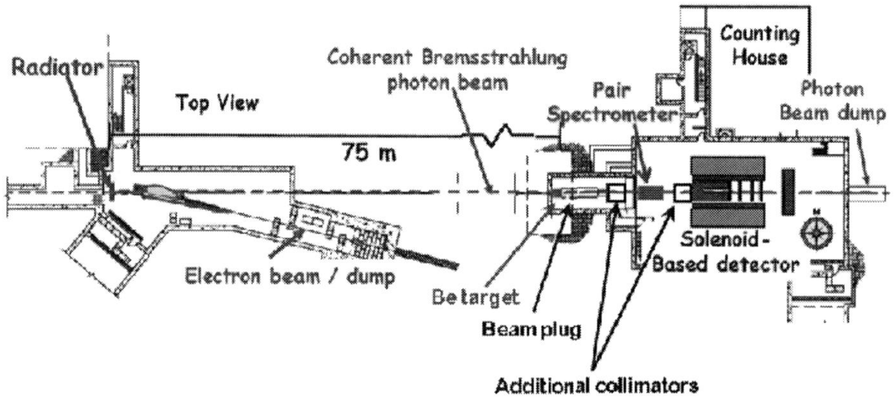

Figure 3. Schematic view of Hall D beamline on the way $e \to \gamma \to K_L$.

There is an advantage factor for K_Lp vs. K^-p experiment. The mean lifetime of the K_L is 51.16 ns ($c\tau$ = 15.3 m) whereas the mean lifetime of the K^- is 12.38 ns ($c\tau$ = 3.7 m) [6]. For this reason, it is much easier to perform measurements of K_Lp scattering at low beam energies compared with K^-p scattering [18].

The recently constructed GlueX detector in Hall-D is a large acceptance spectrometer with good coverage for both charged and neutral particles that can be adapted to this purpose [5]. Schematic view of the Hall D beamline is presented in Fig. 3. At the first stage, E_e = 12 GeV electrons produced at the CEBAF will scatter in a radiator in the target vault, generating intensive beam of bremsstrahlung

 DOI: 10.1051/epjconf/201613006005

photons (we will not need in the Hall D Broadband Tagging Hodoscope). At the second stage, bremsstrahlung photons, created by electrons, hit the Be-target and produce K_L-mesons along with neutron and charged particles. Finaly, K_L will reach the LH_2 cryogenic target within GlueX settings.

We estimated the flux of K_L beam on the GlueX LH_2 target is about 10^5 K_L/s, to be compared to about 10^3 K_L/s used at NINA [19] and SLAC [20], almost comparable to charged Kaon rates obtained at AGS and elsewhere in the past and expected for J-PARC [18]. Momenta of neutral Kaons at JLab will be measured applying the time-of-flight technique using a time structure of 60 ns. The count rate estimates carried out assuming 100 days of data taking are presented in Fig. 4.

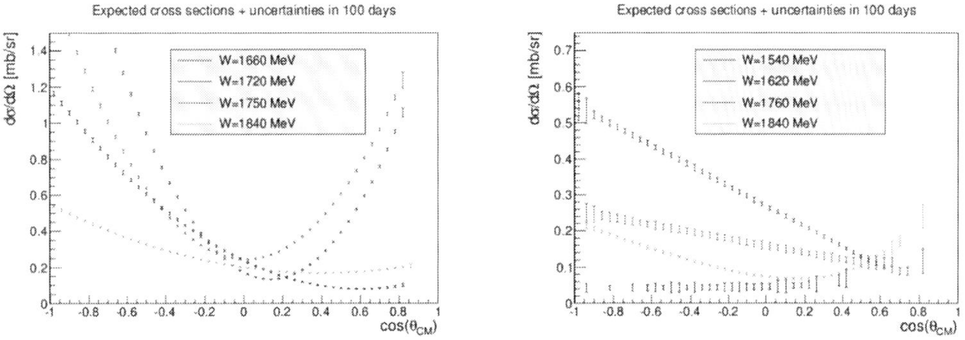

Figure 4. The cross section uncertainty estimates (statistics only) for $K_L p \to p K_S$ (left) and for $K_L p \to \pi^+ \Lambda$ (right).

6 Conclusion and Perspectives

Precise new data (both differential cross section and recoil polarization of hyperons) for $K_L p$ scattering with good kinematic coverage could significantly improve our knowledge on Λ^* and Σ^* resonances. Clearly, complete understanding of three-quark bound states requires to learn more about baryon resonances in "strange sector". Polarization data are very important to be measured in addition to differential cross sections to help remove ambiguities in PWAs.

Unfortunately, the current database for $K_L p$ scattering includes very few polarization data. As noted here, several $K_L p$ reactions are isospin (I=1) selective, which would provide a useful constraint for a combined PWA of $K_L p$ and $K^- p$ reactions. Finally, the long lifetime of the K_L compared with the K^- would allow $K_L p$ measurements to be made easier at lower energies compared with K^- beams. It would be advantageous to combine all $K_L p$ data in a new coupled-channel PWA with available and new J-PARC $K^- p$ data when they will be available. The proposed KL facility potentially may unravel many "missing" hyperons. To complete $SU(3)_F$ multiplets, one needs no less than 17 Λ^*, 41 Σ^*, 41 Ξ^*, and 24 Ω^*.

Measurements of "missing" hyperon states with their spin-parity assignments along with the "missing" non-strange baryons will provide very important ingredients to test QM and LQCD predictions thereby improving our understanding of QCD in a non-perturbative regime.

I thank Moskov Amaryan, Yakov Azimov, William Briscoe, Eugene Chudakov, Ilya Larin, Mark Manley, James Ritman, and, Simon Taylor for comments on the feasibility of future measurements. This work is supported, in part, by the U.S. Department of Energy, Office of Science, Office of Nuclear Physics, under Award Number DE–SC0014133.

References

[1] Workshop on *Physics with Neutral Kaon Beam at JLab*, JLab, VA, USA, Feb. 2016.
[2] *Physics opportunities with secondary K_L beam at JLab*, Spokesperson: M. Amaryan (GlueX Collaborations), JLab LoI12–15–001, Newport News, VA, USA, 2015.
[3] M. Albrow *et al. Mini-Proceedings of the Workshop on Physics with Neutral Kaon Beam at JLab* (KL2016), JLab, VA, USA, Feb. 2016, edited by M. Amaryan *et al.*; arXiv:1604.02141 [hep–ph].
[4] Workshop on *Excited Hyperons in QCD Thermodynamic at Freeze-Out*, JLab, VA, USA, Nov. 2016.
[5] H. Al Ghoul *et al.* (GlueX Collaboration), AIP Conf. Proc. **1735**, 020001 (2016), Proceedings of the *16th International Conference on Hadron Spectroscopy* (Hadron2015), Newport News, VA, Sept. 2015, edited by M. Pennington.
[6] K.A. Olive *et al.* (Particle Data Group), Chin. Phys. C **38**, 090001 (2014).
[7] Y. Qiang *et al.*, Phys. Lett. B **694**, 123 (2010).
[8] H. Zhang *et al.*, Phys. Rev. C **88**, 035204 (2013); *ibid.* **88**, 035205 (2013).
[9] H. Kamano *et al.*, Phys. Rev. C **90**, 065204 (2014); *ibid.* **92**, 025205 (2015).
[10] B. C. Jackson *et al.*, Phys. Rev. C **91**, 065208 (2015).
[11] C. Fernandez-Ramirez *et al.*, Phys. Rev. D **93**, 034029 (2016).
[12] B.M.K Nefkens, πN Newsletter, **14**, 150 (1997).
[13] R. Koniuk and N. Isgur, Phys. Rev. Lett. **44**, 845 (1980).
[14] The Durham HEP Reaction Data Databases (UK) (Durham HepData): http://durpdg.dur.ac.uk/hepdata/reac.html .
[15] B.S. Zou, *Mini-Proceedings of the Workshop on Physics with Neutral Kaon Beam at JLab*, see Ref. [3].
[16] G. Höhler, *Pion-Nucleon Scattering*, Landoldt-Börnstein, Vol. **I/9b2**, edited by H. Schopper (Springer-Verlab, Berlin, 1983).
[17] R. A. Arndt *et al.*, Phys. Rev. C **74**, 045205 (2006).
[18] H. Noumi, *Mini-Proceedings of the Workshop on Physics with Neutral Kaon Beam at JLab*, see Ref. [3].
[19] M.G. Albrow *et al.*, Nucl. Phys. B **23**, 509 (1970).
[20] G.W. Brandenburg *et al.*, Phys. Rev. D **7**, 708 (1973).

EPJ Web of Conferences **130**, 07001 (2016) DOI: 10.1051/epjconf/201613007001
MESON 2016

Charge symmetry breaking in $A = 4$ hypernuclei

P. Achenbach[1,⋆]

A1 Collaboration

[1] *Institut für Kernphysik, Johannes Gutenberg-Universität, 55099 Mainz, Germany*

Abstract. Charge symmetry breaking in the $A = 4$ hypernuclear system is reviewed. The data on binding energies of the mirror nuclei and hypernuclei are examined. At the Mainz Microtron MAMI the high-resolution spectroscopy of decay-pions in strangeness electro-production is used to extract the Λ hyperon ground state binding energy in $^4_\Lambda$H. This binding energy is used together with the $^4_\Lambda$He ground state binding energy from nuclear emulsion experiments and with energy levels of the 1^+ excited state for both hypernuclei from γ-ray spectroscopy to address the charge symmetry breaking in the strong interaction. The binding energy difference of the ground states in the mirror pair is reduced from its long accepted value $\Delta B^4_\Lambda(0^+_{\text{g.s.}}) \approx 0.35$ MeV to ≈ 0.24 MeV. The energy difference of the excited states becomes $\Delta B^4_\Lambda(1^+_{\text{exc}}) \approx -0.08$ MeV, for the first time with opposite sign. These values were not reproduced by theoretical calculations with the exception of very recent approaches, although with a large systematic dependence. The full understanding of the charge symmetry breaking in the $A = 4$ hypernuclei still remains one of the open issues of hypernuclear physics.

1 Charge symmetry and charge independence breaking

The nuclear interactions have a small charge-dependent component breaking the near symmetry between protons and neutrons in their interactions and their contributions to nuclear properties. However, the concept of charge symmetry is quite useful in describing many facets of nuclear physics, e.g. the observation of nearly identical levels and spin-parity assignments of excited nuclear states in mirror nuclei. The fundamental cause of the charge dependence of nuclear forces, i.e. charge-symmetry breaking (CSB) and charge-independence breaking (CIB), is due to the differences in the up and down quark masses and due to electromagnetic effects.

A very important consequence of CSB is the fact that neutrons are heavier than protons (by only approximately $\Delta M/M \sim 0.1\,\%$) despite the larger electrostatic repulsion of the quarks inside the proton that would make the proton heavier. The decay of free neutrons (allowed by their larger mass) during the primordial nucleosynthesis left a large fraction of protons unbound, now existing mainly in the stars and providing a slow-burning fuel for the universe. The very reverse, protons being heavier than neutrons, would be a disaster for life as we know it.

⋆e-mail: achenbach@uni-mainz.de

Differences between the nucleon forces are experimentally mostly clearly observed in the a^N_{pp} and a^N_{nn} proton–proton and neutron–neutron $^1S_0$ singlet scattering lengths and the value for neutron–proton scattering, when electromagnetic effects have been removed. Recent values for the nuclear parts scattering lengths are

$$\begin{aligned} a^N_{pp} &= -17.3 \pm 0.4\,\text{fm} \\ a^N_{nn} &= -18.8 \pm 0.3\,\text{fm} \\ 1/2\,(a^N_{pp} + a^N_{nn}) &= -18.05 \pm 0.5\,\text{fm} \\ a^N_{np} &= -23.77 \pm 0.09\,\text{fm} \end{aligned}$$

following the review of [1]. The np interaction is significantly more attractive than the averaged nn and pp interactions and is a manifestation of CIB. A smaller, but sizable CSB effect is seen in the difference between a^N_{pp} and a^N_{nn} scattering lengths.

In nuclear physics often the meson-exchange theory of nuclear forces is employed. Charged-dependent effects arise through mass splittings and mixing of exchanged vector mesons of different isospin but same spin and parity. CIB in the scattering lengths can be explained quantitatively with the charged and neutral pion mass difference in the one-pion and the two-pion exchange potential [2]. The dominant CSB effect in the scattering lengths can be explained in terms of $\rho^0-\omega$ mixing [3]. Evidence for neutral meson mixing also come from nuclear interactions and cross-section measurements.

2 Charge symmetry breaking in the lightest mirror nuclei and hypernuclei

The CSB effect is manifest in the differences between mirror nuclei and mirror hypernuclei. The differences in these heavier systems (compared to the nucleon) are extensions of the neutron–proton difference.

The lightest mirror pair is the $A = 3$ isodoublet (^{3}H,^{3}He). The measured masses of the $A = 3$ pair are $M(^3\text{H}) = 2808.921\,\text{MeV}/c^2$ and $M(^3\text{He}) = 2808.391\,\text{MeV}/c^2$ with negligibly small errors of $\sim 2\,\text{eV}$ [4]. The corresponding nuclear binding energies are $B(^3\text{H}) = 8.482\,\text{MeV}$ and $B(^3\text{He}) = 7.718\,\text{MeV}$, signifying that the neutron richer nucleus ^{3}H is more deeply bound. The difference in binding energies is $\Delta B^3 = B(^3\text{He}) - B(^3\text{H}) = -0.764\,\text{MeV}$. The repulsive Coulomb interaction in ^{3}He and other electromagnetic contributions need to be removed from the binding energy to determine the strong-interaction CSB. This system is considered best for CSB studies because theoretical studies are well advanced in these few-body cases. The mass and energy values of this mirror pair and of the corresponding $A = 4$ hypernuclei are given in Table 1.

There have been several studies to calculate CSB effects from Coulomb interaction for $A = 3$ [7–9]. A perturbative estimate of the Coulomb contribution can be made from nuclear form factors measured in electron scattering. Reported results include contributions of $\Delta B^3_C = -0.69\,\text{MeV}$ due to the static Coulomb effect, $\Delta B^3_{\text{size}} = +0.04\,\text{MeV}$ due to the finite size effect, and $\Delta B^3_{\text{other}} = -0.04\,\text{MeV}$ from other electromagnetic effects such as the magnetic interaction. Bodmer and Usmani performed a variational calculation and obtained $\Delta B^3_C = -0.67 \pm 0.01\,\text{MeV}$ [10]. In summary, the electromagnetic interaction contributes approximately $\Delta B^3_{\text{em}} = -0.69\,\text{MeV}$ to the binding energy difference. The remaining difference of only $\Delta B^3_{\text{CSB}} = -0.07\,\text{MeV}$ of the original $\sim 0.7\,\text{MeV}$ is attributed to the CSB contribution from the strong interaction.

Brandenburg and Wu et al. estimated the CSB effect with $\pi^0 - \eta$ and $\rho^0 - \omega$ mixing [11, 12] calculating an additional contribution to the binding energy difference of $\Delta B^3_{\text{CSB}} = -0.07\,\text{MeV}$. Adding this contribution to the electromagnetic effects the binding energy difference between ^{3}H and ^{3}He can

Table 1: Known nuclear masses (M in MeV/c^2) and nuclear binding energies (B in MeV) in the system of $A = 4$ mirror hypernuclei from nuclear emulsion measurements [5] including the nuclear core masses from the tabulated mass excess values [4] and the Λ hyperon mass from the Particle Data Group compilation [6]. No electron masses or binding energies were included and differences ΔM and ΔB were calculated by subtracting the H value from the He value, respectively. Note the neutron–proton mass difference of 1.293 MeV/c^2 contributing to the mass differences of the isospin pairs.

$M(^3\text{H}) =$	2808.921	$B(^3\text{H}) =$	8.482
$M(^3\text{He}) =$	2808.391	$B(^3\text{He}) =$	7.718
$\Delta M^3 =$	−0.530	$\Delta B^3 =$	−0.764
$M(^4_\Lambda\text{H}) =$	3922.56	$B_\Lambda(^4_\Lambda\text{H}) =$	2.04
$M(^4_\Lambda\text{He}) =$	3921.68	$B_\Lambda(^4_\Lambda\text{He}) =$	2.39
$\Delta M^4_\Lambda =$	−0.88	$\Delta B^4_\Lambda =$	0.35
$M(^3\text{H} +^3\text{He} + \Lambda) =$	6732.99		
$M(^4_\Lambda\text{H}+^3\text{He}) =$	6730.95		
$M(^3\text{H}+^4_\Lambda\text{He}) =$	6730.60		

be fully explained. Thus, the mechanisms and the effect of CSB in the NN interaction is understood in light nuclei on the keV level.

If the $A = 3$ system is expanded with a bound Λ hyperon to the ($^4_\Lambda$H,$^4_\Lambda$He) mirror pair of hypernuclei, information on the CSB in the ΛN interaction can be extracted. Charge symmetry predicts that the Λp and Λn interactions and consequently their contributions to the binding energies of mirror hypernuclei would be identical. Since the Λ binding (or separation) energy is defined by subtracting the binding energy of the core $B_\Lambda(^4_\Lambda\text{He}) = B(^4_\Lambda\text{He}) - B(^3\text{He})$, where B is the total binding energy, the repulsive Coulomb energy in ^{3}He and $^4_\Lambda$He cancels to first order.

However, the presence of the bound Λ hyperon tends to compress the core nucleus. For $^4_\Lambda$He this compression increases the Coulomb repulsion and lowers the binding energy. It is expected that the change in B_Λ is not more than 10 % of the total Coulomb energy [13]. To leading order in the Coulomb interaction the difference in binding energies is $\Delta E_C = -\Delta B_C = 0.05 \pm 0.02$ MeV and 0.025 ± 0.015 MeV for the ground and excited states, where $\Delta E_C = E_C(^4_\Lambda\text{He}) - E_C(^3\text{He})$ is the difference between the Coulomb energies of $^4_\Lambda$He and ^{3}He [10].

The corresponding differences of B_Λ between $^4_\Lambda$He and $^4_\Lambda$H to be attributed to CSB effects were then found by Bodmer and Usmani to be $\Delta B^{\text{CSB}}_\Lambda = \Delta B^{\text{exp}}_\Lambda - \Delta B_C = (0.35 \pm 0.06)\,\text{MeV} + (0.05 \pm 0.02)\,\text{MeV} = 0.40 \pm 0.06\,\text{MeV}$ for the ground state and $\Delta B^{\text{CSB}}_\Lambda = (0.24 \pm 0.06)\,\text{MeV} + (0.025 \pm 0.015)\,\text{MeV} = 0.27 \pm 0.06\,\text{MeV}$. From these values a phenomenological CSB potential was derived which is effectively spin independent.

Recent experimental and theoretical developments

In 2015 measurements with HPGe detectors on $^4_\Lambda$He hypernuclei produced by (K^-,π^-) reactions on a helium target at the Hadron Experimental Facility of J-PARC found that the transition energy from the exited 1^+ state to the 0^+ ground state is 1.406 ± 0.002 (stat.) ± 0.002 (syst.) MeV [17], falsifying

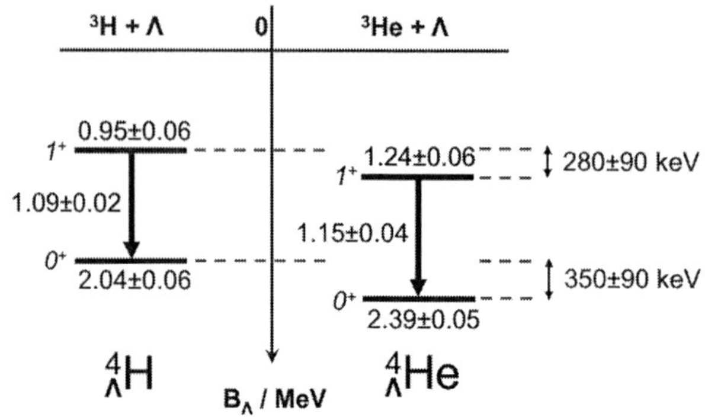

(a) Knowledge before 2015

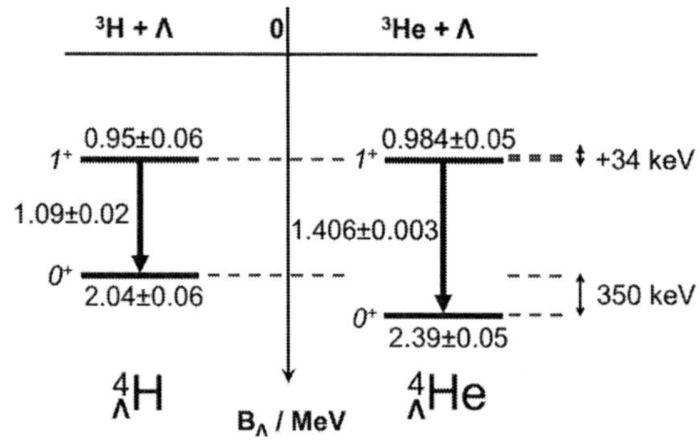

(b) Knowledge in 2015

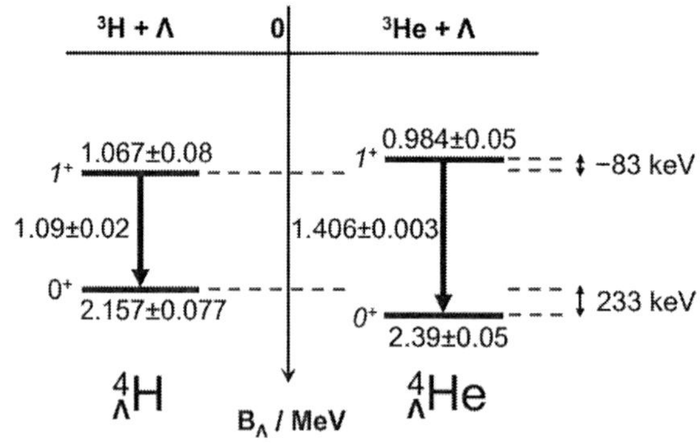

(c) Knowledge after 2015

Figure 1: Advancement of the knowledge on the level diagrams of the ${}^4_\Lambda$H and ${}^4_\Lambda$He mirror hypernuclei. CSB splittings are shown to the right of the ${}^4_\Lambda$He levels, γ-ray transitions are shown by arrows, with γ-ray energies to the left. Diagram (a) shows binding energies, B_Λ, for the ground states from nuclear emulsion measurements [5] in the 1960s and 1970s and for the excited states from low-resolution, low-efficiency NaI detector measurements of γ-ray transitions in the 1970s [14–16]. Diagram (b) includes the ${}^4_\Lambda$He transition energy re-measured recently at J-PARC [17], falsifying the earlier values. Finally, diagram (c) uses the ${}^4_\Lambda$H ground state binding energy recently re-measured at MAMI [18].

earlier results from low-resolution, low-efficiency NaI detector measurements in the 1970s [14]. The excitation energy of ${}^{4}_{\Lambda}$H was known to be only 1.09 ± 0.02 MeV [14–16]. The use of emulsion data for the ground state binding energies lead to a binding energy difference of $\Delta B^4_\Lambda(1^+_{\rm exc}) \approx 0.03 \pm 0.05$ MeV for the excited states. In summary, the breaking of the charge symmetry in $A = 4$ hypernuclei was found to be large and differing between the 0^+ ground state and the 1^+ excited state.

At the Mainz Microtron MAMI the high-resolution spectroscopy of decay-pions in strangeness electro-production is used to extract the Λ hyperon ground state binding energy in ${}^{4}_{\Lambda}$H [19]. A very important result was the confirmation of the B_Λ value for ${}^{4}_{\Lambda}$H independently from the experimental technique. Recently an updated value of $B_\Lambda = 2.157 \pm 0.005$ (stat.) ± 0.077 (syst.) MeV has been published [18] including a detailed error and bias analysis.

This binding energy can be used together with the ${}^{4}_{\Lambda}$He ground state binding energy from nuclear emulsion experiments and with energy levels of the 1^+ excited state for both hypernuclei from γ-ray spectroscopy to arrive at the latest level diagrams and CSB splittings of the ${}^{4}_{\Lambda}$H and ${}^{4}_{\Lambda}$He mirror hypernuclei. Fig. 1 shows the advancement of the knowledge during the last years.

The updated ground state binding energy difference $\Delta B^4_\Lambda(0^+_{\rm g.s.}) = B_\Lambda({}^4_\Lambda{\rm He}(0^+_{\rm g.s.})) - B_\Lambda({}^4_\Lambda{\rm H}(0^+_{\rm g.s.})) = 233 \pm 92$ keV is smaller as measured by the emulsion technique but still supports a sizable CSB effect in the ΛN interaction. The binding energy difference between the excited states of $\Delta B^4_\Lambda(1^+_{\rm exc}) = B_\Lambda({}^4_\Lambda{\rm He}(1^+_{\rm exc})) - B_\Lambda({}^4_\Lambda{\rm H}(1^+_{\rm exc})) = -83 \pm 94$ keV is negative.

These values were not reproduced by theoretical calculations [20, 21] with the exception of very recent approaches [22, 23]. In these approaches a $\Lambda - \Sigma^0$ mixing CSB mechanism due to Dalitz and von Hippel [13] reproduces the large value of ΔB_Λ in the ground state and an opposite sign difference in the first excited state, although with large systematic dependence. In this model the isospin zero singlet Λ hyperon mixes electromagnetically with the $\Delta M_{\Lambda\Sigma} \sim 80\,{\rm MeV}/c^2$ more massive isospin triplet Σ hyperon. The effect of admixture breaks charge symmetry and leads to a binding energy difference in mirror hypernuclei.

Conclusion and outlook

The updated results for the ${}^{4}_{\Lambda}$H binding energy measured at MAMI lead to a difference of $\Delta B^4_\Lambda(0^+_{\rm g.s.}) \approx 0.233 \pm 0.09$ MeV between the binding energies for the ground states in the $A = 4$ system when combined with emulsion data. The data suggest a negative binding energy difference between the excited states of $\Delta B^4_\Lambda(1^+_{\rm exc}) = -0.083 \pm 0.09$ MeV when combined with the known excitation energies supporting a large spin-dependent CSB effect. At J-PARC a precise measurement of the γ-transition energy of ${}^{4}_{\Lambda}$H is planned [24] to confirm the earlier results from the 1970s.

In the mirror pair (${}^{4}_{\Lambda}$He,${}^{4}_{\Lambda}$H) CSB effects appear to be considerably stronger than in any other nuclei or heavier hypernuclei. From the theoretical studies of CSB in $A = 4$ hypernuclei it can be concluded that $\Lambda N - \Sigma N$ coupling and 3-body forces in the hyperon–nucleon interaction are essential ingredients to the ΛN interaction. This also means that the interactions of Λ hyperons in symmetric nuclear matter and neutron rich nuclei would be largely different. At Jefferson Lab a series of measurements on the isospin dependence of the ΛN interaction was proposed and approved [25] to address this issue. The outcome of those and all other ongoing or planned hypernuclear physics endeavors will be a key to understand CSB in light hypernuclei.

This work was supported in part by Deutsche Forschungsgemeinschaft (SFB 1044), the Carl Zeiss Foundation, European Community Research Infrastructure Integrating Activity FP7, U.S.-DOE Contract No. DEFG02-97ER41047, the Strategic Young Researchers Overseas Visits Program for Accelerating Brain Circulation (R2201), and the Core-to-Core program (21002) of JSPS.

References

[1] G.A. Miller, A.K. Opper, E.J. Stephenson, Annu. Rev. Nucl. Part. Sci. **56**, 253 (2006)
[2] G.Q. Li, R. Machleidt, Phys. Rev. C **58**, 3153 (1998)
[3] G.A. Miller, B.M.K. Nefkens, I. Šlaus, Phys. Rep. **194**, 1 (1990)
[4] G. Audi, A.H. Wapstra, C. Thibault, Nucl. Phys. A **729**, 337 (2003)
[5] M. Jurić et al., Nucl. Phys. B **52**, 1 (1973)
[6] K.A. Olive et al. (Particle Data Group), Chin. Phys. C **38**, 090001 (2014)
[7] J.L. Friar, Nucl. Phys. A **156**, 43 (1970)
[8] R.A. Brandenburg, S.A. Coon, P.U. Sauer, Nucl. Phys. A **294**, 305 (1978)
[9] G.A. Miller, Chin. J. Phys. **32**, 1075 (1994)
[10] A. Bodmer, Q. Usmani, Phys. Rev. C **31**, 1400 (1985)
[11] R.A. Brandenburg, S.A. Coon, P.U. Sauer, Phys. Rev. C **37**, 781 (1988)
[12] Y. Wu, S. Ishikawa, T. Sasakawa, Phys. Rev. Lett. **64**, 1875 (1990)
[13] R.H. Dalitz, F. von Hippel, Phys. Lett. **10**, 153 (1964)
[14] M. Bedjidian, E. Descroix, J. Grossiord, A. Guichard, M. Gusakow, M. Jacquin, M. Kudła, H. Piekarz, J. Piekarz, J. Pizzi et al. (CERN-Lyon-Warsaw Collaboration), Phys. Lett. B **83**, 252 (1979)
[15] M. Bedjidian, A. Filipkowski, J. Grossiord, A. Guichard, M. Gusakow, S. Majewski, H. Piekarz, J. Piekarz, J. Pizzi (CERN-Lyon-Warsaw Collaboration), Phys. Lett. B **62**, 467 (1976)
[16] A. Kawachi, Ph.D. thesis, Department of Physics, School of Science, University of Tokyo, Japan (1997)
[17] T.O. Yamamoto et al. (J-PARC E13 Collaboration), Phys. Rev. Lett. **115**, 222501 (2015)
[18] F. Schulz et al. (A1 Collaboration), Nucl. Phys. A **954**, 149 (2016)
[19] A. Esser et al. (A1 Collaboration), Phys. Rev. Lett. **114**, 232501 (2015)
[20] A. Nogga, H. Kamada, W. Glöckle, Phys. Rev. Lett. **88**, 172501 (2002)
[21] A. Nogga, Nucl. Phys. A **914**, 140 (2013)
[22] A. Gal, Phys. Lett. B **744**, 352 (2015)
[23] D. Gazda, A. Gal, Phys. Rev. Lett. **116**, 122501 (2016)
[24] H. Tamura et al. (Hypergamma Collaboration), *Gamma-ray spectroscopy of light* Λ *hypernuclei II (Exp. Proposal E13-2)* (J-PARC, 2016)
[25] F. Garibaldi, P.E.C. Markowitz, S.N. Nakamura, J. Reinhold, L. Tang, G.M. Urciuoli et al. (JLab Hypernuclear Collaboration), *An iso-spin dependence study of the* ΛN *interaction through the high precision spectroscopy of* Λ*-hypernuclei with electron beam (update of Exp. Proposal C12-15-008)* (Jefferson Lab, 2016)

EPJ Web of Conferences **130**, 07002 (2016) DOI: 10.1051/epjconf/201613007002
MESON 2016

Status of the analysis for the search of polarization in the antiproton production process

D. Alfs[1,⋆], A. Asaturyan[2], M. Carmignotto[3], M. Diermaier[4], W. Eyrich[5], B. Głowacz[1], D. Grzonka[6], F. Hauenstein[6], T. Horn[3], K. Kilian[6], S. Malbrunot-Ettenauer[7], A. Mkrtchyan[2], H. Mkrtchyan[2], P. Moskal[1], P. Nadel-Turonski[8], W. Oelert[9], J. Ritman[6], T. Sefzick[6], V. Tadevosyan[2], E. Widmann[4], M. Wolke[10], S. Zhamkochyan[2], M. Zieliński[1], A. Zink[11], and J. Zmeskal[4]

[1]*M. Smoluchowski Institute of Physics, Jagiellonian University, Kraków, Poland*
[2]*A.I. Alikhanyan Science Laboratory, Yerevan, Armenia*
[3]*Physics Department, The Catholic University of America, Washington DC, USA*
[4]*Stefan-Meyer-Institut für subatomare Physik, Wien, Austria*
[5]*Universität Erlangen, Erlangen, Germany*
[6]*Institut für Kernphysik, Forschungszentrum Jülich, Jülich, Germany*
[7]*Physics Department, CERN, Geneve, Switzerland*
[8]*Thomas Jefferson National Accelerator Facility, Newport News, Virginia, USA*
[9]*Johannes Gutenberg-Universität Mainz, Mainz, Germany*
[10]*Department of Physics and Astronomy, Uppsala University, Uppsala, Sweden*
[11]*Erlangen Centre for Astroparticle Physics (ECAP), Erlangen, Germany*

Abstract. The P-349 experiment aims to test whether for antiprotons the production process itself can be a source of polarization in view of the preparation of a polarized antiproton beam. In this article we present the details of performed measurements and report on the status of the ongoing analysis.

1 Motivation

A wide range of fundamental effects in the hadronic sector can be investigated only by means of controlling the spin degrees of freedom. Although polarized proton beams and targets are produced routinely, the preparation of a polarized antiproton beam is still a challenge. A variety of methods to prepare such a beam were proposed [1], however most of them were already discarded due to the low degree of polarization or the expected small number of produced antiprotons.

The only experiment with polarized antiprotons was performed in FERMILAB where a proton beam of 800 GeV/c momentum produced $\bar{\Lambda}$ antihyperons. The observed mean polarization of antiprotons from the decay $\bar{\Lambda} \rightarrow \bar{p}\pi^+$ was equal to 45%. However, it is difficult to generate a well-defined beam by this method especially at lower energies [2].

Therefore, it would be desirable if antiprotons itself would have a non zero polarization degree when produced [3]. However, until now there were no experimental studies performed in this direction for antiprotons. The goal of the P-349 experiment is to test wether antiprotons are produced with polarization.

⋆e-mail: dominika.alfs@uj.edu.pl

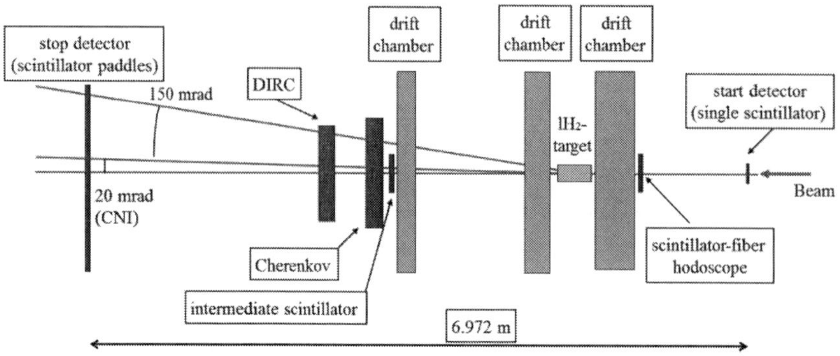

Figure 1. The scheme of the detector arrangement in the horizontal plane. The beam comes from the right side. The total angular range covered by the system is about 150 mrad (horizontally as well as vertically) and the relevant scattering angle for the asymmetry measurement is about 20 mrad. Further explanation is given in the text.

2 Analyzing reaction

In order to measure the polarization of antiprotons the elastic $\bar{p}p$ scattering is used and the scattering asymmetry has to be determined in a region with known analyzing power. The Coulomb-Nuclear Interference (CNI) region is well suited for this kind of measurements because the analyzing power is well known and calculable at high energies. It was shown that in proton-proton and proton-antiproton elastic scattering the value is the same but with opposite sign. The maximum analyzing power is about 4.5% at a four momentum transfer equal to t = -0.0037 GeV/c [4].

In the P-349 experiment the antiprotons have a momentum spectrum peaked around 3.5 GeV/c for which the high energy approximation is not valid anymore. However, according to preliminary calculations for lower antiproton momenta down to 5 GeV/c, the maximum analyzing power is still about 4.5% for a four momentum transfer t = -0.0025 GeV/c [5].

3 Experimental setup

The experiment was performed at the T11 beamline of the CERN/PS complex. Antiprotons were produced in the reaction $pp \to \bar{p}ppp$ by bombarding the solid target with an unpolarized proton beam of momenta equal to about 24 GeV/c. This corresponds to the typical conditions of the antiproton beam production in the existent and planned facilities.

The system was operated in the air. It consisted of (see Fig. 1): scintillators for triggering purposes (start and stop) and TOF measurements (start, intermediate, stop), a scintillator fiber hodoscope for the beam profile control, a set of three drift chambers with vertical and inclined wires for precise 3d track reconstruction [6], a Cherenkov detector for online pion background reduction (it served as trigger veto), and a DIRC for offline particle identification. The beam was scattered on a 15 cm long liquid hydrogen target.

4 Particle identification and track finding

In the first step of the analysis the ability of the online reduction of the pion background and the particle identification by the DIRC detector were verified.

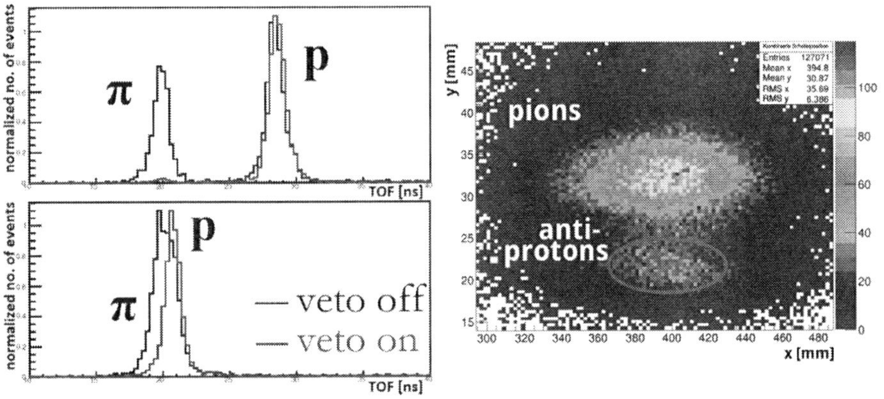

Figure 2. Left: Results of test measurements with the beamline set for positively charged particles of momenta 1 GeV/c (left upper part) and 3 GeV/c (left lower part). Right: Reconstruction of the Cherenkov arcs for each event in one spill in the measurement at 3.5 GeV/c with the bemline set to negatively charged particles. A clear separation of pions and antiprotons is visible.

The ability to suppress the pion background was confirmed in dedicated measurements with the beamline set to positively charged particles of different momenta. Results for momenta of 1 GeV/c and 3 GeV/c are presented in Fig. 2.

For the track finding the method of Hough transformation is applied which is often used in pattern recognition algorithms. In the P-349 experiment we use it to separate the track signals from background and electronic noise in the drift chambers. In the first step only the hits from vertical wires of the drift chambers are used. Each hit with coordinates $P(x, y)$ in 2-dimensional $X - Y$ space corresponds to the sinusoidal line $r = x\cos\theta + y\sin\theta$ in the Hough space (the space parametrized by r and θ). The intersection of the lines in the Hough space provides the parameters of the track candidate.

5 Outlook

The work on the calibration of the drift chambers and the track reconstruction is ongoing.

The majority of events are unscattered primary particles which pass trough the whole detector setup and are used to improve the drift chamber positioning and calibration. The next step is the extraction of antiproton scattering events and the reconstruction of their angular distribution.

This work was supported by the Polish Ministry of Science and Higher Education and by DAAD from resources of Bundesministerium für Bildung und Forschung (BMBF) and Marian Smoluchowski Kraków Research Consortium "Matter–Energy–Future" (KNOW).

References

[1] E. Steffens, AIP Conf. Proc. **1149**, 80 (2009)
[2] A. Bravar *et al.*, Phys. Rev. Lett. **77**, 2626 (1996)
[3] K. Kilian *et al.*, Int. J. Mod. Phys. A **26**, 757 (2011)
[4] H. Okada *et al.*, Phys. Lett. A **638**, 450 (2006)
[5] D. Grzonka, *et. al.*, Acta Phys. Polon. B **46**, 191 (2015)
[6] J. Smyrski, *et. al.*, Nucl. Instrum. Meth. A **541**, 574 (2005)

EPJ Web of Conferences **130**, 07003 (2016) DOI: 10.1051/epjconf/201613007003
MESON 2016

Di-electron production in $d-p$ collisions at E_{kin} =2.5 GeV

Jacek Biernat[1,★] for the HADES Collaboration
[1] *Jagiellonian University Cracow*

Abstract.
Identification of dielectron pair production in the exclusive npe^+e^- final state from quasi-free $n-p$ interactions using deuterium beams on proton target at kinetic beam energy of 1.25 GeV/u with the HADES spectrometer is presented. Comparison of the dielectron invariant mass distribution to the one measured in the $p-p$ reaction shows a prominent enhancement in the high-mass region.

1 Introduction

The main motivation of this work is to study dielectron production from non-mesonic sources in neutron-proton collisions below the η meson production threshold. According to various microscopic calculation in [1, 2] a dominant process leading to the emission of a virtual photon with a mass larger than the pion mass is bremsstrahlung. The understanding of this process is important for the description of the pair production in heavy ion reactions at these energies and also offers a possibility to study electromagnetic form-factors of nucleon in so called "unphysical" region as described in [3]. Studies of the exclusive channel presented in this work extends former investigations of the inclusive pair production published in [4]. A dedicated experimental run was performed with the High Acceptance Di-Electron Spectrometer (HADES) in [5] placed at the GSI Helmholtzzentrum fur Schwerionenforschung, Germany. Deuteron beams of 10^7 particles/s with kinetic energies of 1.25 GeV/u were incident on a liquid hydrogen target cell with a total length of 5 cm, corresponding to a total thickness of 0.35 g/cm^2. Quasi-free $n-p$ reactions were selected on trigger level by detection of fast spectator protons from the deuterium break-up in a dedicated Forward hodoscope Wall (FW) described in [5]. It was placed 7 m downstream of the target and covered polar angles between 0.3^o and 7°. Charged particles ($p, \pi^{\pm}, e^{\pm}$) were detected in the spectrometer as described in [6].

★e-mail: jacek.b.biernat@gmail.com

 DOI: 10.1051/epjconf/201613007003

2 Analysis

Electron identification, track reconstruction, and electron pair (unlike- and like-sign) reconstruction were performed as described in detail in [4–6]. The combinatorial background (CB) was estimated as a function of the pair invariant mass using like sign pairs reconstructed in the same event as follows:

$$\frac{dN_{CB}}{dM_{e^+e^-}} = 2 * \sqrt{\frac{dN}{dM_{e^+e^+}} + \frac{dN}{dM_{e^-e^-}}} \tag{1}$$

This method accounts as well for correlated background from a double conversion of photons from π^0 decay in 2 photons or conversion of a photon accompanied by a pair from the π^0 Dalitz decay as for uncorrelated e^+e^- stemming from multi-pion decays. The signal distribution is then obtained as the difference between unlike sign pairs and CB distributions.

In order to normalize the measured yields, $p-p$ elastic scattering from the quasi-free scattering $d+p \rightarrow ppn_{spec}$ were measured during the experimental run. Since the differential cross section of the proton-proton scattering is well known from the EDDA experiment presented in [7], the measured pair yield can be converted to cross sections.

The exclusive npe^+e^- channel was selected by identification of the proton spectator particle in the FW and a proton-, an electron- and a positron-tracks inside inside the HADES spectrometer acceptance. Finally, the neutron was reconstructed using missing mass technique.

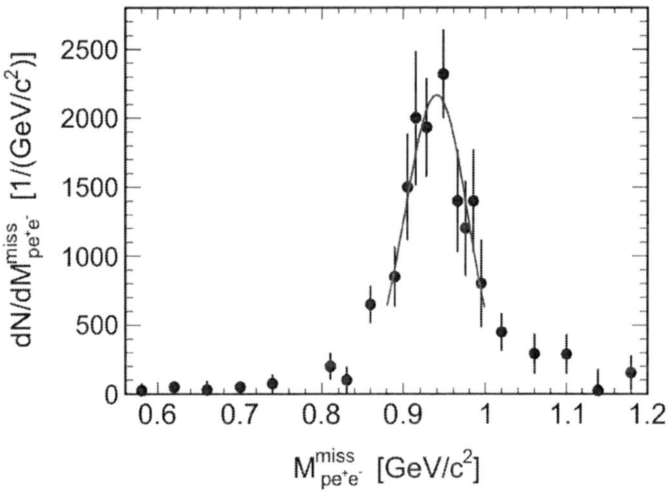

Figure 1. Missing mass distributions for the p, e^+e^- events, with dielectron mass larger than 150 MeV/c^2. The experimental data distribution (black dots) was fitted with a Gaussian distribution (red line).

In Fig. 1 the pe^+e^- missing mass for the dielectron pairs with invariant mass larger than the pion mass and pair opening angle $\alpha > 9^o$ is shown. The obtained distribution shows a clear peak with a mean value of 0.944(3) GeV/c^2 and a sigma of 0.037(4) GeV/c^2 , as expected from simulations. This result shows that the selected events originate dominantly from the $np \rightarrow npe^+e^-$ exclusive process. It is important to stress that the observed width of the distribution is in agreement with the expected combined effect of the experimental resolution and of the momentum smearing of the neutron inside the deuterium.

DOI: 10.1051/epjconf/201613007003

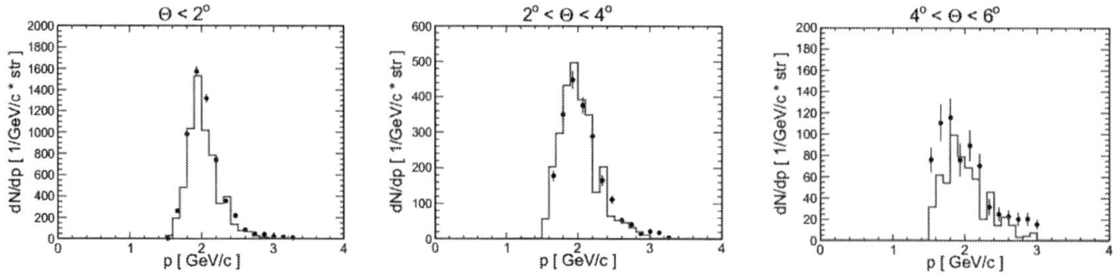

Figure 2. Momentum distributions of the spectator measured by time of flight in FW. Experimental data (black dots) are compared to the simulation (red solid line) for the indicated emission angles and dielectron pairs with $M_{ee} < 0.14$ GeV/c^2. Simulated spectra are scaled to the same global experimental yield.

In Fig. 2 momentum distributions of the proton spectator, determined by the time of flight in FW, for three spectator emission angles and dielectron pairs with $M_{e^+e^-} < 0.14$ GeV/c^2 is presented, where π^0 Dalitz decay is the dominant process. Although, the finite time resolution of the FW (500 ps) dominates the measured width, a broadening of the distributions for larger angles is visible, and is in agreement with the dedicated Monte Carlo simulation performed with the PLUTO event generator described in [8]. The event generator assumes a spectator model with a realistic neutron momentum distribution. For the dielectron production, π^0 Dalitz decay has been simulated, details are described in [9].

Finally in Fig. 3 the distribution of the e^+e^- invariant mass for npe^+e^- events is presented, CB has been subtracted. The signal to CB ratio in the interesting mass region (larger than the pion mass) is better than 10. The distribution is compared to the one measured in the $p-p$ (exclusive $pp \to ppe^+e^-$ reaction) collisions at the same energy, normalized to the same yield in the π^0 region. As one can clearly see, the distribution measured in the $n-p$ reactions has a very different shape, with an apparent excess in the high mass region. The obtained experimental data will be compared to the recent models of Shyam and Model in [2] and Clement and Bashkanov proposed for inclusive data in [10].

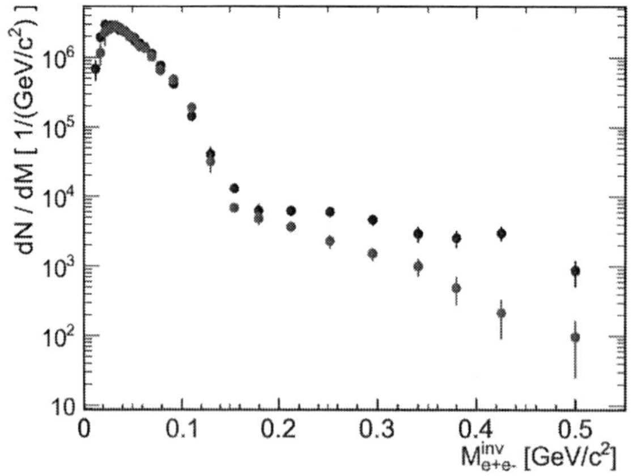

Figure 3. Number of counts as a function of the di- electron invariant mass for the npe^+e^- events measured in quasi-free $n-p$ collisions (black) compared with a similar spectrum obtained for $p-p$ collisions (red).

3 Conclusions

The exclusive pne^+e^- final state in the $d-p$ collisions at beam energy of 1.25 GeV/u has been identified. In particular, it has been demonstrated that the dielectron pairs with a mass larger than the pion mass can be assigned uniquely to the exclusive final state. These pairs show different mass dependence as compared to the one measured in $p-p$ reactions at the same energy. Hence, the measured distributions can be compared to the recent model predictions to identify the underlying process . Furthermore, it was verified that the spectator model applied for $d-p$ reaction describes the shape of the proton spectator momentum distributions for the $M_{ee} < 0.14$ GeV/c^2 very well.

Acknowledgment

The collaboration gratefully acknowledges the support by PTDC/FIS/113339/2009 LIP Coimbra, NCN grant 2013/10/M/ST2/00042 SIP JUC Cracow, Helmholtz Alliance HA216/EMMI GSI Darmstadt, VH-NG-823, Helmholtz Alliance HA216/EMMI TU Darmstadt, 283286, 05P12CRGHE HZDR Dresden, BMBF 06MT7180 TU München, Garching BMBF:05P12RGGHM JLU Giessen, Giessen UCY/3411-23100, University Cyprus CNRS/IN2P3, CNRS/IN2P3 (France), MSMT LG 12007, AS CR M100481202, GACR 13- 06759S NPI AS CR, Rez, EU Contract No. HP3-283286.

References

[1] L.P. Kaptari and B.Kampfer. Nucl. Phys. A **764** 338-370 (2006)
[2] U. Mosel, Annu. Rev. Nucle. Part. Sci **41** 29 (1991)
[3] G. Eichmann Acta Phys. Pol. B Proc. Suppl. **7** 597-606 (2014)
[4] G. Agakichiev *et al.*, Phys. Lett. B **690** 118-122 (2010)
[5] G. Agakichiev *et al.*, Eur. Phys. J. A **41** 243 (2009)
[6] K. Lapidus *et al.*, arXiv:0904.1128 [nucl-ex].
[7] M. Altmeier *et al.*, Eur. Phys. J. A **23**, 351-364 (2005)
[8] I. Froelich *et al.*, J. Phys. Conf. Ser. **219**, 032039 (2010)
[9] I. Frohlich *et al.*, Eur. Phys. J. A **45**, 401-411 (2010)
[10] M. Bashkanov, H. Clement, Eur. Phys. J. A **50**, 107 (2014)

Constraints on hadronic models in extensive air showers with the Pierre Auger Observatory

João Espadanal[1,2,⋆] for the Pierre Auger Collaboration[2]

[1] *LIP, Av. Elias Garcia 14-1, 1000-149 Lisboa, Portugal*
[2] *Pierre Auger Observatory, Malargue, Argentina* ⋆⋆

Abstract. Extensive air showers initiated by ultra-high energy cosmic rays are sensitive to the details of hadronic interactions models, so we present the main results obtained using the data of the Pierre Auger Observatory. The depth at which the maximum of the electromagnetic development takes place is the most sensitive parameter to infer the nature of the cosmic rays. However, the hadronic models cannot describe consistently the maximum and the muon measurements at energies higher than those reached at the LHC.

1 Introduction

Particle physics at energies beyond the reach of accelerators can be studied with air showers induced by ultra high energy cosmic rays (UHECR). Data from the Pierre Auger Observatory [1, 2], indicate that the hadronic interaction models EPOS-LHC and QGSJetII-04 have a muon deficit of around 30% to 80% [3], respectively, to match the primary composition inferred from the depth of the electromagnetic (EM) shower maximum. A UHECR collides with an atmospheric nucleus and initiates an extensive air shower (EAS) of secondary particles. After the first interaction $\sim$ 80% of the particles produced are pions and $\sim$ 8% are kaons. In each hadronic interaction, the produced π^0 carry around 25% of the energy, and they immediately decay into photons, feeding the electromagnetic cascade. The rest of the energy is carried mainly by charged pions (and also other mesons and baryons), which continue the hadronic cascade. The mesons propagate until their energy drops below their critical energy, $O(100\text{ GeV})$, and then they decay into muons among other things. The electromagnetic development and muonic content depend on the primary composition and the hadronic interaction properties, such as the fraction of EM energy per interaction, multiplicities, hadronic cross sections, etc. So, assuming that the composition can be properly understood, a discrepancy between observed and predicted muon numbers, must indicate an inaccurate description of the hadronic models.

The Pierre Auger Observatory is the largest *hybrid* cosmic-ray detector. It has two main detection techniques: the surface detectors (SD) which consists of an array of 1660 water-Cherenkov stations extended over $\sim$ 3000 km^2 with an almost 100% duty cycle, and the fluorescence detectors (FD) composed of 27 telescopes operating with a $\sim$ 15% duty cycle. To understand the muon puzzle, an upgrade is planned for the Observatory[4], to be able to separate the muonic and electromagnetic

⋆e-mail: jespada@lip.pt
⋆⋆Full author list: http://www.auger.org/archive/authors_2016_08.html

DOI: 10.1051/epjconf/201613007004

components of the shower at the ground. Due to the limited space available here, we will briefly discuss the main features of muon studies and electromagnetic profiles.

2 The EM cascade

The longitudinal development of the energy deposit in air by the showers can be measured in the FD, as a function of the atmospheric slant depth. The profile is dominated by the electrons belonging to the EM cascade and the profile maximum is denoted by X_{max}. The depth of shower maximum is mostly sensitive to the first interaction position and the primary particle type, i.e. showers induced by heavy primaries will develop higher (at a shallower depth), faster and with less shower-to-shower fluctuations than those induced by lighter nuclei (due to a larger interaction cross-section and higher number of nucleons). The primary composition can be inferred statistically from the distribution of shower maxima, due to the fluctuations in the first few hadronic interactions in the cascade. The X_{max} distributions can be a superposition of different nuclei, so four component (proton, iron, helium and nitrogen) were considered to fit the data distributions. For both models considered, QGSJetII-04 and EPOS-LHC, the fraction of protons is decreasing and the fraction of helium is increasing for energies above a few EeV, with no significant fraction of iron for all energies (see [5]).

In fig. 1 left, the energy evolution of the average $\langle X_{max}\rangle$ is shown for the data and predictions of the hadronic interaction models. Different selection cuts are applied to ensure good data-taking conditions[6]. The elongation rate ($D_{10} = \mathrm{d}\langle X_{max}\rangle/\mathrm{d}\log_{10} E$) would be constant if the composition were pure. The best fit to our data comprises two linear fits (in units of $\log_{10}(E/\mathrm{eV})$) with a breaking point at $\log_{10}(E/\mathrm{eV}) = 18.27{\pm}04(stat)^{+0.06}_{-0.07}(sys)$. Considering the model predictions, above the breaking point, the observed rate of change of $\langle X_{max}\rangle$ becomes significantly smaller ($\sim$ 26 g.cm$^{-2}/decade$) indicating that the composition is becoming heavier. The observed width of the X_{max} distribution is corrected by subtracting the detector resolution in quadrature to obtain $\sigma(X_{max})$ (fig. 1 right). It also shows that above the breaking point the composition changes from light elements to heavier ones.

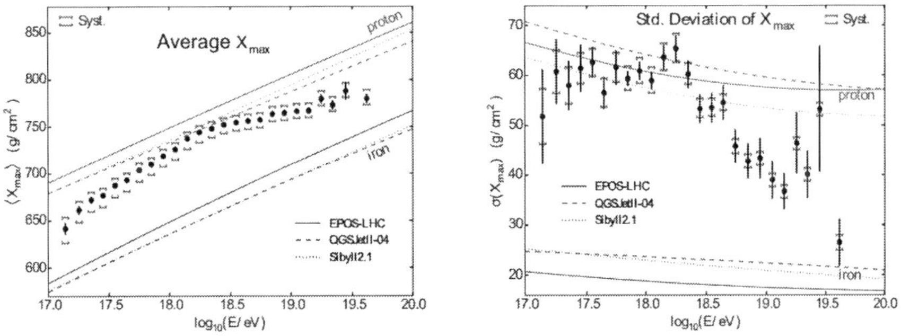

Figure 1. The mean (left) and the standard deviation (right) of the measured X_{max} distributions as a function of energy compared to air-shower simulations for proton and iron primaries [6, 7].

3 The hadronic cascade

In highly-inclined events, the electromagnetic component of a shower is largely absorbed in the atmosphere, while the muons still survive to the ground, which is suitable for direct studies of the EAS muon content. Hybrid events have independent energy and muon measurements, and with zenith angles in the range $62° < \theta < 80°$, the SD signals can be considered as muons. The whole procedure can be seen in [3], schematically: for each event, the muon density at the ground ρ_{μ}^{rec} is recorded with the

SD and adjusted to the shape ρ_μ^{map} and relative muon content R_μ, as $\rho_\mu^{rec} = R_\mu\, \rho_\mu^{map}$. The footprint ρ_μ^{map} is derived from simulations of protons with QGSJETII-03 at an energy of $E = 10^{19}$ eV. A power-law dependence of the mean muon content with energy is expected as $\langle R_\mu \rangle = a(E/10^{19}\mathrm{eV})^b$. The data shown in fig. 2 left, gives $a = 1.84 \pm 0.03(stat) \pm 0.32(sys)$ and $b = 1.03 \pm 0.02(stat) \pm 0.03(sys)$. Even with the large systematic uncertainty (square brackets) data are not compatible with a pure proton composition, and, at higher energy, are marginally compatible with iron. The energy evolution slope also suggests a transition from a lighter to a heavier composition.

The problems with the hadronic models becomes evident when the muon results are compared to the depth of shower maximum (electromagnetic). In fig. 2 right, the average $\langle X_{max} \rangle$ and $\langle R_\mu \rangle$ are shown for the data and simulations at $E = 10^{19}$ eV, where the Auger data fall completely out of the model's phase-space. The average values of $\langle R_\mu \rangle$ and the logarithmic slope $\frac{d\langle \ln R_\mu \rangle}{d \ln E}$ are shown in fig. 3 and compared to the models predictions using the EM X_{max} composition. In all cases the models fail to match both measurements, requiring a substantial increase in muon production. Additionally, the large value of the measured slope favors a changing composition.

The EM component is highly absorbed in the atmosphere, so under certain conditions (large zenith angles, large distances to the shower core), the SD signal is dominated by muons. In those cases, the temporal structure and distance to the shower axis of muons arriving at the ground can be measured.

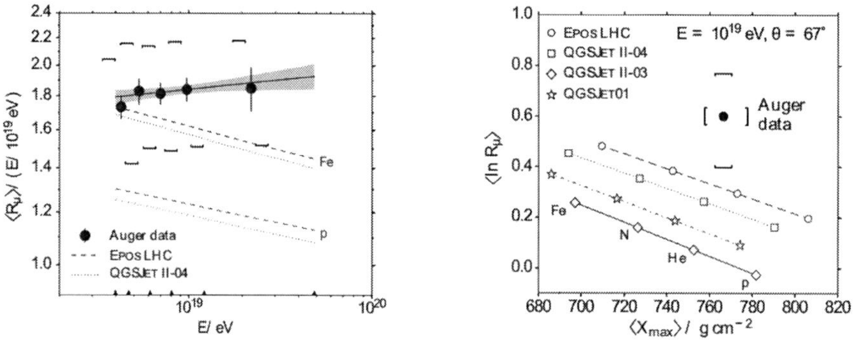

Figure 2. (Left) $\langle R_\mu \rangle$ vs. primary energy, compared to air-shower simulations. (Right) $\langle \ln R_\mu \rangle$ vs. $\langle X_{max} \rangle$. Representative primary masses are indicated by open symbols [3, 7].

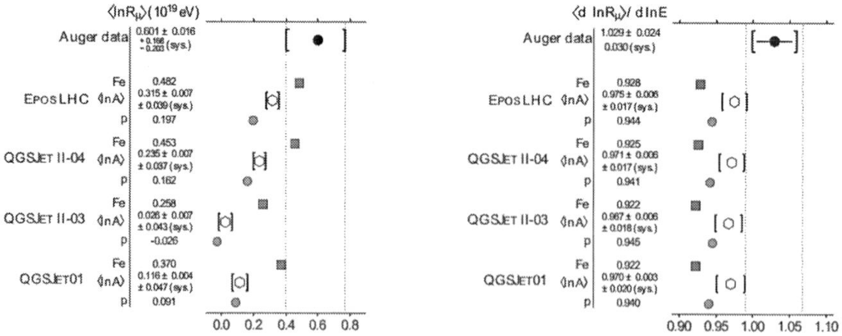

Figure 3. Comparison of $\langle \ln R_\mu \rangle$ (left) and $\frac{d\langle \ln R_\mu \rangle}{d \ln E}$ (right) between 4×10^{18} eV and 5×10^{19} eV with predictions for air-shower simulation models for a pure proton, a pure iron and a mixed composition compatible with the FD measurements (labeled as $\langle \ln A \rangle$) [3, 7].

Using a set of simple assumptions[8], they can be used to obtain the production point of the muons along the shower axis. Summing all muons, the longitudinal development of muon production depth (MPD) can be measured on an event-by-event basis. The evolution with energy of the corresponding maximum X^{μ}_{max} is shown in fig. 4 left. Both muonic $\langle X^{\mu}_{max} \rangle$ and electromagnetic $\langle X_{max} \rangle$ parameters can be translated into $\langle \ln A \rangle$ as shown in fig. 4 right, for two interaction models, allowing a direct comparison of both observables. The QGSJetII-04 model and the EPOS-LHC model give incompatible mass values, from the electromagnetic and muonic components.

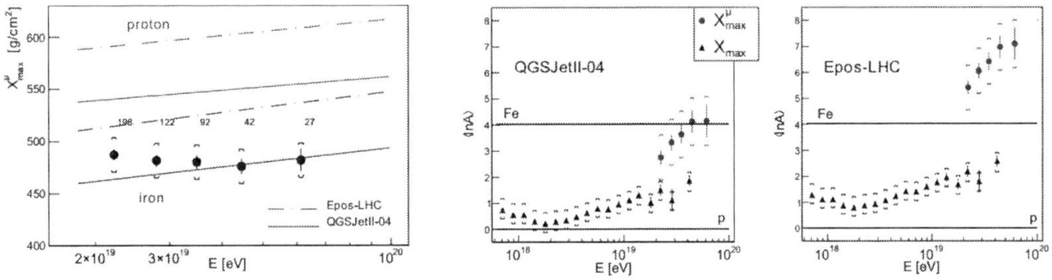

Figure 4. Evolution of X^{μ}_{max} with energy[8](left), the number of events is indicated in each energy bin and brackets representing the systematic uncertainty. $\langle \ln A \rangle$ from $\langle X^{\mu}_{max} \rangle$ and X^{μ}_{max} as a function of energy[7, 8](right).

4 Conclusions

Using data from the Auger Observatory we have found that the average primary composition from the X_{max} observable is not consistent with a pure composition, assuming the overall correctness of the hadronic interaction models considered. Under the assumption that there is no new physics affecting the air-shower development trans-LHC energies, we can see a transition from light to heavy UHECR composition. Nevertheless, muon measurements are inconsistent with hadronic interaction model predictions. The models yield a deficit in the produced number of muons and the interpretation of the longitudinal maxima $\langle X^{\mu}_{max} \rangle$ and $\langle X_{max} \rangle$ leads to an inconsistent average primary mass $\langle \ln A \rangle$. Neither electromagnetic or muonic measurements can be described consistently by the hadronic interaction models, so they can be used to constrain them at energies beyond existing terrestrial accelerators.

Acknowledgments

Thanks to LIP and FCT for a Postdoc grant; this work would not have been possible without the strong commitment and effort of the technical and administrative staff in Malargüe.

References

[1] A. Aab, et al. (Pierre Auger Collaboration), Nucl. Instrum. Meth. A **798**, 172 (2015)
[2] J. Abraham, et al. (Pierre Auger Collaboration), Nucl. Instrum. Meth. A **523**, 50 (2004)
[3] A. Aab, et al. (Pierre Auger Collaboration), Phys. Rev. D **91**, 032003 (2015)
[4] A. Aab, et al. (Pierre Auger Collaboration), arXiv (2016), 1604.03637
[5] A. Aab, et al. (Pierre Auger Collaboration), Phys. Rev. D **90**, 122006 (2014)
[6] A. Aab, et al. (Pierre Auger Collaboration), Phys. Rev. D **90**, 122005 (2014)
[7] A. Aab, et al. (Pierre Auger Collaboration), Proc. of the 34th ICRC (2015)
[8] A. Aab, et al. (Pierre Auger Collaboration), Phys. Rev. D **90**, 012012 (2014)

EPJ Web of Conferences **130**, 07005 (2016) DOI: 10.1051/epjconf/201613007005
MESON 2016

Search for dark forces with KLOE

Aleksander Gajos[1,★]
for the KLOE and KLOE-2 Collaborations

[1] *Institute of Physics, Jagiellonian University, ul. Łojasiewicza 11, 30-348 Cracow, Poland*

Abstract. During the last years several Dark Sector Models have been proposed in order to address striking astrophysical observations which failed standard interpretations. In the minimal case a new vector particle, the so called dark photon, or A' or U boson, is introduced, with small coupling with Standard Model particles. Also, the existence of a dark Higgs boson h' is postulated, in analogy with the Standard Model, to give mass to the U boson through the Spontaneous Symmetry Breaking mechanism. The KLOE experiment working at DAΦNE, an e^+e^- collider operating in Frascati (near Rome), searched for the existence of the U boson investigating three different processes and six different final states: (1) in Dalitz decays of the ϕ meson $\phi \to \eta U$, with $U \to e^+e^-$ and $\eta \to \pi^+\pi^-\pi^0$ and $\pi^0\pi^0\pi^0$, (2) in $e^+e^- \to U\gamma$ events, with U decaying into lepton and pion pairs, (3) in the dark Higgsstrahlung process, $e^+e^- \to Uh'$, $U \to \mu^+\mu^-$, h' invisible. Limits on the model parameters have been set at 90 % CL for U boson masses below 1 GeV. Further improvements are expected in terms of sensitivity and discovery potential with the new KLOE-2 detector.

1 Introduction

Presently, dark matter investigations are one of the most active fronts of searches for physics beyond the Standard Model (SM). The introduction of Weakly Interacting Massive Particles as dark matter candidates would imply the presence of new interaction called the dark force and, consequently, a new gauge vector boson — the U boson, also referred to as the dark photon or A'. It would couple to the SM particles through a kinetic mixing parameter ε equal to the ratio of dark and electromagnetic coupling constants. As ε is expected to be in the range 10^{-4}–10^{-2}, the dark photon should produce observable effects at $O(1\ \text{GeV})$ collider experiments such as KLOE at DAΦNE. Moreover, search for the U boson in such energy range is of special interest as it could explain a number of recent puzzing astrophysical observations such as the excess of positrons in cosmic rays observed by AMS and PAMELA [1, 2], 511 keV gamma rays from the galactic centre seen by INTEGRAL [3] or the annual modulation signal measured by DAMA/LIBRA [4]. The KLOE experiment has searched evidence of the U boson existence as a sharp resonance in the invariant mass spectrum of lepton or pion pairs produced in radiative processes $e^+e^- \to \ell^+\ell^-(\gamma)$ [5, 6], $e^+e^- \to \pi^+\pi^-(\gamma)$ [7] and in decays of the ϕ meson [8, 9], as well as in the dark Higgsstrahlunng process [10].

★e-mail: aleksander.gajos@doctoral.uj.edu.pl

2 The KLOE detector at the DAΦNE collider

DAΦNE is an e^+e^- collider located at the accelerator complex of LNF-INFN in Frascati, Italy. The center-of-mass energy of the colliding beams is typically set at the mass of the ϕ meson resonance ($\sqrt{s}$ ≈1019 MeV) although measurements with off-peak energy are performed as well. The KLOE detector is shaped as a barrel around the e^+e^- collision point, constituted by a cylindrical drift chamber surrounded by an electromagnetic calorimeter and immersed in 0.52 T magnetic field. The drift chamber with a gas mixture of helium (90 %) and isobutane (10 %) provides position resolution of 150 μm in the transverse plane and 2 mm in the z direction whereas momentum of charged particles is reconstructed with a relative resolution of 0.4 % [11]. The electromagnetic calorimeter is a sampling device based on scintillating fibers and lead which provides high efficiency for detection of 20–500 MeV photons with an energy resolution of $5.7\%E/\sqrt{E[GeV]}$ and time resolution of $57\ ps/\sqrt{E[GeV]} \oplus 100\ ps$ [12]. The detector has been recently upgraded and the new KLOE-2 experiment is collecting data with the aim to reach an integrated luminosity of about 5 fb^{-1}.

3 Search for the U boson in Dalitz decays of ϕ

A search for the U boson in the $\phi \to \eta U$ process, with U decaying into an e^+e^- pair, has been performed at KLOE by analysing a data sample of 1.5 fb^{-1} in the case of η decaying into $\pi^+\pi^-\pi^0$ [9] and 1.7 fb^{-1} in the case of $\eta \to \pi^0\pi^0\pi^0$ [8]. No evidence in the e^+e^- invariant mass spectrum was found for both searches [8, 9]. A combined and more stringent upper limit [8] at 90 % CL on the ratio between the coupling constant and the fine structure constant of $\alpha'/\alpha < 1.7 \times 10^{-5}$ for $30 < m_U <$ 400 MeV and $\alpha'/\alpha < 8 \times 10^{-6}$ for the sub-region $50 < m_U < 210$ MeV was evaluated. The result assumes the Vector Meson Dominance expectations for the $\phi\eta\gamma*$ transition form factor.

4 Search for $e^+e^- \to U\gamma$ with U decaying into lepton or pion pairs

Evidence for the $e^+e^- \to U\gamma$ with $U \to \mu^+\mu^-$ was searched in a sample of 239 pb^{-1} in the $e^+e^- \to \mu^+\mu^-\gamma$ process with photon emitted in the initial state (ISR). The presence of the dark photon would be manifested as a peak in the recorded $m_{\mu\mu}$ mass distribution. Events were identified by two charged tracks at a large polar angle ($50° < \theta < 130°$) and an undetected photon at a small angle ($\theta < 15°$ or $\theta > 165°$) which resulted in a strong suppression of the resonant $\phi \to \pi^+\pi^-\pi^0$ and FSR background. The residual background components were simulated and fitted to the di-muon mass spectrum. The experimental and simulated with PHOKHARA [13] $m_{\mu\mu}$ distributions have shown perfect agreement [6]. As no peak was observed in the spectrum, the CL$_S$ technique was used to set a limit on the number of U boson candidates and thus exclude ε^2 larger than 8.6×10^{-7}–1.6×10^{-5} for 520 MeV $< m_U <$ 980 MeV [6]. The obtained exclusion is marked as KLOE$_{(2)}$ in Fig. 1.

A similar process, $e^+e^- \to U\gamma$, $U \to e^+e^-$ was studied at KLOE in order to explore the region of low U masses close to the di-electron mass threshold. In the analysis an explicit detection of both the electron and positron tracks and the photon was required. Dedicated analysis in order to distinguish electron-positron and muon couples was performed. Since the $e^+e^- \to e^+e^-\gamma$ process is dominant in the analysis acceptance, the total background was kept below 1 %. No peak in the e^+e^- invariant mass spectrum was observed and an upper limit on the kinetic mixing parameter was thus extracted. The obtained upper limit covers m_U between 2 MeV and 250 MeV and ranges from 10^{-7} to 10^{-5} [5] (see the KLOE$_{(3)}$ region in Fig. 1).

The last of the radiative processes investigated by KLOE was $e^+e^- \to U\gamma$ with $U \to \pi^+\pi^-$. The search was motivated by the lack of sensitivity of leptonic channels in the $\rho - \omega$ resonance mass region. The analysis was based on the one used for the measurements of the pion form factor [14] and a dedicated simulation was used based on the PHOKHARA generator [13] with the Gounaris-Sakurai pion form factor parametrization [15] to properly describe the $\rho - \omega$ region. No U boson signal was observed and an upper limit on the ε^2 parameter was set for m_U between 527 and 987 MeV. The excluded region indicated in Fig. 1 as KLOE$_{(4)}$ shows a more stringent limit than the existing ones in the $\rho - \omega$ region and above [7].

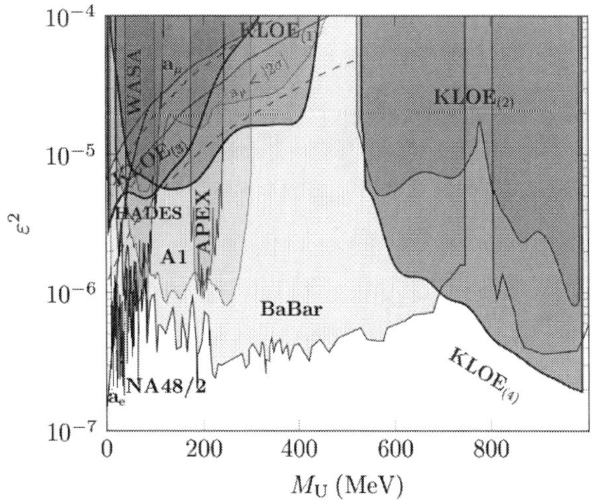

Figure 1. Exclusion limits (90 % CL) on the ε^2 parameter as a function of the U boson mass. Presented are KLOE limits obtained with $\phi \to \eta e^+e^-$ (KLOE$_{(1)}$), $e^+e^- \to \mu^+\mu^-\gamma$ (KLOE$_{(2)}$), $e^+e^- \to e^+e^-\gamma$ (KLOE$_{(3)}$), $e^+e^- \to \pi^+\pi^-\gamma$ (KLOE$_{(4)}$). For comparison the results of MAMI/A1 [16], APEX [17], WASA [18], HADES [19], NA48/2 [20] and BaBar [21] are given.

5 Search for dark Higgsstrahlung

Existence of the U boson, as a massive particle, would imply breaking of its associated U_D symmetry through a Higgs-like particle h' called the dark Higgs. KLOE has searched for the dark Higgsstrahlung process $e^+e^- \to Uh'$, $U \to \mu^+\mu^-$ in the "invisible dark Higgs" scenario where h' is lighter than U and escapes detection due to its long lifetime. The analysed data included a 1.65 fb^{-1} sample at the ϕ resonance peak and a 206 pb^{-1} sample collected off-peak (at 1000 MeV). After identification of two muon tracks and reconstruction of the missing mass the signal was searched in a 2-dimensional M_{miss} vs. $M_{\mu\mu}$ distribution. No resonance was observed and a bayesian upper limit on the number of candidate events was determined and translated to the $\alpha_D \times \varepsilon^2$ product where α_D is the coupling constant of the dark sector [10]. Resulting exclusion values depending on the U and h' masses are presented in Fig. 2.

6 Conclusions

The KLOE experiment has made a large contribution to the searches of the dark forces mediator, the U boson by exploring three different processes and a total of six final states. As no signal was observed by KLOE, upper limits on the kinetic mixing parameter ε^2 were set at the level between 10^{-7} and 10^{-5} (at 90 % CL) in the U boson mass range of 5–980 MeV. In the search for the dark Higgs with the $m_{h'} < m_U$ scenario, the $\alpha_D \times \varepsilon^2$ product was limited at a 90 % CL in the range of $2m_\mu < m_u < 1000$ MeV and $10 < m_{h'} < 500$ MeV.

The KLOE-2 experiment with new subdetectors is presently taking new data with the aim to collect a sample of about 5 fb^{-1}. As the results reported here are presently limited by statistical uncertainty, KLOE-2 is expected to twice improve their sensitivity.

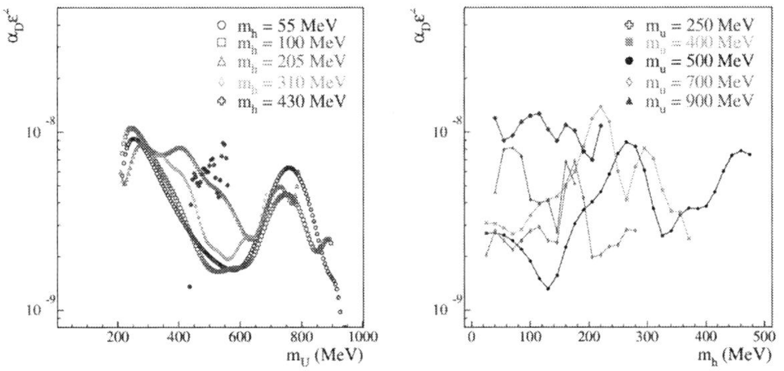

Figure 2. Upper limits (90 % CL) on $\alpha_D \times \varepsilon^2$ as a function of the U boson mass for various values of the h' mass (left) and as a function of the h' mass for various values of m_U (right) [10].

This work was supported in part by the EU Integrated Infrastructure Initiative Hadron Physics Project under contract number RII3-CT- 2004-506078; by the European Commission under the 7th Framework Programme through the 'Research Infrastructures' action of the 'Capacities' Programme, Call: FP7-INFRASTRUCTURES-2008-1, Grant Agreement No. 227431; by the Polish National Science Centre through the Grants No. 2011/03/N/ST2/02652, 2013/08/M/ST2/00323, 2013/11/B/ST2/04245, 2014/14/E/ST2/00262, 2014/12/S/ST2/00459.

References

[1] M. Aguilar *et al.* (AMS), Phys. Rev. Lett. **110**, 141102 (2013)
[2] O. Adriani *et al.* (PAMELA), Nature **458**, 607 (2009)
[3] P. Jean *et al.*, Astron. Astrophys. **407**, L55 (2003)
[4] R. Bernabei *et al.* (DAMA, LIBRA), Eur. Phys. J. C **67**, 39 (2010)
[5] A. Anastasi *et al.* (KLOE-2), Phys. Lett. B **750**, 633 (2015)
[6] D. Babusci *et al.* (KLOE-2), Phys. Lett. B **736**, 459 (2014)
[7] A. Anastasi *et al.* (KLOE-2), Phys. Lett. B **757**, 356 (2016)
[8] D. Babusci *et al.* (KLOE-2), Phys. Lett. B **720**, 111 (2013)
[9] F. Archilli *et al.* (KLOE-2), Phys. Lett. B **706**, 251 (2012)
[10] A. Anastasi *et al.* (KLOE-2), Phys. Lett. B **747**, 365 (2015)
[11] M. Adinolfi *et al.*, Nucl. Instrum. Meth. A **488**, 51 (2002)
[12] M. Adinolfi *et al.*, Nucl. Instrum. Meth. A **482**, 364 (2002)
[13] H. Czyz, A. Grzelinska, J.H. Kuhn, G. Rodrigo, Eur. Phys. J. C **39**, 411 (2005)
[14] D. Babusci *et al.* (KLOE), Phys. Lett. B **720**, 336 (2013)
[15] G.J. Gounaris, J.J. Sakurai, Phys. Rev. Lett. **21**, 244 (1968)
[16] H. Merkel *et al.*, Phys. Rev. Lett. **112**, 221802 (2014)
[17] S. Abrahamyan *et al.* (APEX), Phys. Rev. Lett. **107**, 191804 (2011)
[18] P. Adlarson *et al.* (WASA-at-COSY), Phys. Lett. B **726**, 187 (2013)
[19] G. Agakishiev *et al.* (HADES), Phys. Lett. B **731**, 265 (2014)
[20] J.R. Batley *et al.* (NA48/2), Phys. Lett. B **746**, 178 (2015)
[21] J.P. Lees *et al.* (BaBar), Phys. Rev. Lett. **113**, 201801 (2014)

EPJ Web of Conferences **130**, 07006 (2016) DOI: 10.1051/epjconf/201613007006
MESON 2016

Branching ratio estimates of $\tau^- \to \pi^-\eta^{(\prime)}\nu_\tau$ decays

Sergi Gonzàlez-Solís[1,*]

[1] *Institut de Física d'Altes Energies (IFAE), The Barcelona Institute of Science and Technology (BIST), Campus UAB, 08193 Bellaterra, Barcelona*

Abstract. We study the rare $\tau^- \to \pi^-\eta^{(\prime)}\nu_\tau$ decays. These processes occur via isospin violation and belong to the so-called second-class currents unseen in Nature so far. Our analysis is based on the framework of resonance chiral theory supplemented by dispersion relations. In this contribution we discuss the prospects for their discovery at forthcoming B-factories such Belle-II. While we find a total branching fraction for the $\pi^-\eta$ decay mode of $\sim 1.7 \times 10^{-5}$, well within the reach of Belle-II, the $\pi^-\eta'$ channel might be one or two orders of magnitude more suppressed.

1 Introduction

The τ is the only lepton heavy enough to decay into hadrons. Actually, its partial decay width involving hadrons in the final state is of $\sim 65\%$. At the exclusive level, hadronic τ decays represent a clean laboratory to access the non-perturbative regime of QCD i.e. they are useful to understand the hadronization of QCD currents, to study form factors and to extract resonance parameters. Rare decay channels of the τ are a very promising area to pursue at forthcoming B-factories since the interaction is suppressed and the sensitivity to new physics might eventually be enhanced. In the Standard Model (SM), the suppression of rare τ decays might be due to several reasons e.g. :

- Cabbibo suppression: strange hadronic final states are suppressed with respect to non-strange ones since the V_{us} element of the CKM matrix enters the description instead of V_{ud}. Examples of Cabbibo suppressed decays are $\tau^- \to K^-\eta\nu_\tau$ and $\tau^- \to K^-\eta'\nu_\tau$ that we have analyzed in Refs. [1, 2].
- Phase-space: decays involving kaons, η or η' mesons in the final state lead, usually, to phase-space suppression because of the large mass of these states.
- Second-class currents: decays occurring via G-parity violation i.e. the G-parity of the vector current is opposite to the G-parity of the hadronic system. In the isospin limit, G-parity is exact and these processes are forbidden in the SM. However, isospin is an $SU(2)$ approximate symmetry broken both by the up and down quark mass and electric charge differences leading the suppression.

In this contribution we address the study of the rare second-class currents $\tau^- \to \pi^-\eta^{(\prime)}\nu_\tau$ decays [3]. Our goal is two-fold: to describe the participating hadronic form factors, predict the decay spectra and estimate the corresponding branching ratios in order to stimulate the experimental collaborations to measure these decays for the first time.

*e-mail: sgonzalez@ifae.cat. This work has been supported by the FPI scholarship BES-2012-055371. The author thanks P. Roig for the reading of the manuscript.

2 Decay distribution and form factors

The expression for the invariant mass distribution has been derived in detail in Ref. [3] and reads

$$\frac{d\Gamma\left(\tau^- \to \pi^-\eta^{(\prime)}\nu_\tau\right)}{d\sqrt{s}} = \frac{G_F^2 M_\tau^3}{24\pi^3 s} S_{\rm EW} |V_{ud} F_+^{\pi^-\eta^{(\prime)}}(0)|^2 \left(1 - \frac{s}{M_\tau^2}\right)^2 \\ \times \left[\left(1 + \frac{2s}{M_\tau^2}\right) q^3_{\pi^-\eta^{(\prime)}}(s) |\widetilde{F}_+^{\pi^-\eta^{(\prime)}}(s)|^2 + \frac{3\Delta^2_{\pi^-\eta^{(\prime)}}}{4s} q_{\pi^-\eta^{(\prime)}}(s) |\widetilde{F}_0^{\pi^-\eta^{(\prime)}}(s)|^2\right], \tag{1}$$

where $q_{PQ}(s)$ is a kinematical factor and $\widetilde{F}_{+,0}^{\pi^-\eta^{(\prime)}}(s)$ are the vector (+) and scalar (0) form factors normalized to unity at the origin. Notice that $F_+^{\pi^-\eta^{(\prime)}}(0)$, an isospin-violating term (cf. Eq. (2)), factorizes in Eq. (1) explaining the smallness of these decays.

2.1 Form factors

- Vector form factor: Our initial approach to built the $\pi^-\eta^{(\prime)}$ vector form factor is through resonance chiral perturbation theory (RChT). It reads [3]

$$F_+^{\pi^-\eta^{(\prime)}}(s) = \varepsilon_{\pi\eta^{(\prime)}} \left(1 + \sum_V \frac{F_V G_V}{F^2} \frac{s}{M_V^2 - s}\right), \tag{2}$$

  where $\varepsilon_{\pi\eta} = 9.8(3) \cdot 10^{-3}$ and $\varepsilon_{\pi\eta'} = 2.5(1.5) \cdot 10^{-4}$ are isospin-violating prefactors we derived in Ref. [3], accounting for the π-η and π-η' mixing, respectively. In this framework, the term inside the parenthesis is the $\pi^-\pi^0$ vector form factor. Therefore, for our analysis we implement the pion form factor extracted from the experimental data on the well-known $\tau^- \to \pi^-\pi^0\nu_\tau$ decay mode [4].
- Scalar form factor: we have considered several parameterizations ordered according to their increasing fulfillment of analyticity and unitarity (See Ref. [3] for details and expressions).
  - Breit-Wigner: we start again with the RChT framework by considering first the exchange of the $a_0(980)$ resonance and then including the $a_0(1450)$ into the description. We have taken into account some (elastic and inelastic) final state hadronic interactions effects by resumming the imaginary part of the self-energy loop function into the propagator. However, this description does not incorporate the real part of the loop function and, therefore, it violates analyticity.
  - Elastic dispersion relation: to fulfill analyticity and elastic unitarity we rely on the Omnès integral. Unfortunately, we lack of any kind of experimental data either for the phase shift or for the decays spectra. Therefore, for our analysis we get a model for the phase from the elastic scattering amplitudes computed in Ref. [5] in RChT at one-loop unitarized through the N/D method.
  - Coupled channels: in Ref. [3] we have provided an alternative expression equivalent to the Omnès solution with one subtraction, given as a closed-form expression, for the single elastic case. We have also shown that one might accommodate inelasticities and obtain form factors in coupled-channels reasoning in the same way. In particular, for our analysis we have considered two-and three-coupled channels problem for the $\pi^-\eta$, K^-K^0 and $\pi^-\eta'$ scalar form factors.

3 Branching ratio predictions

In Fig. 1 we show the distribution for the $\tau^- \to \pi^-\eta\nu_\tau$ decay depending on the scalar form factor we have employed. The vector contribution dominates the distribution at low-energies with a clear

Table 1. Branching ratio predictions for $\tau^- \to \pi^-\eta^{(\prime)}\nu_\tau$ [3]. Three coupled channels for the scalar contribution.

Decay channel	Vector contribution	Scalar contribution	Branching ratio
$\tau^- \to \pi^-\eta\nu_\tau$	$0.26(2) \cdot 10^{-5}$	$1.41(9) \cdot 10^{-5}$	$1.67(9) \cdot 10^{-5}$
$\tau^- \to \pi^-\eta'\nu_\tau$	$[0.3, 5.7] \cdot 10^{-10}$	$[1 \cdot 10^{-7}, 1 \cdot 10^{-6}]$	$[1 \cdot 10^{-7}, 1 \cdot 10^{-6}]$

signal of the ρ while the scalar contribution might dominate at intermediate-and high-energies with a significant signal of the $a_0(980)$ or of the $a_0(1450)$ resonances. Our predictions are collected in Table 1 from where we observe that both the vector and scalar contributions are of the same order, though the scalar ones tend to dominate the decay, and that we respect the current experimental upper bounds of $9.9 \cdot 10^{-5}$ at 95% CL, $7.3 \cdot 10^{-5}$ at 90% CL and of $1.4 \cdot 10^{-4}$ at 95% CL reported by BaBar, Belle and CLEO collaborations, respectively [6–8]. Regarding the $\pi^-\eta'$ mode, the scalar form factor tends to dominate the decay instead. Apart from phase-space suppression, this channel is one or two orders of magnitude smaller than the $\pi^-\eta$ channel because of the smallness of the value of the π-η' mixing, $\varepsilon_{\pi\eta'}$.

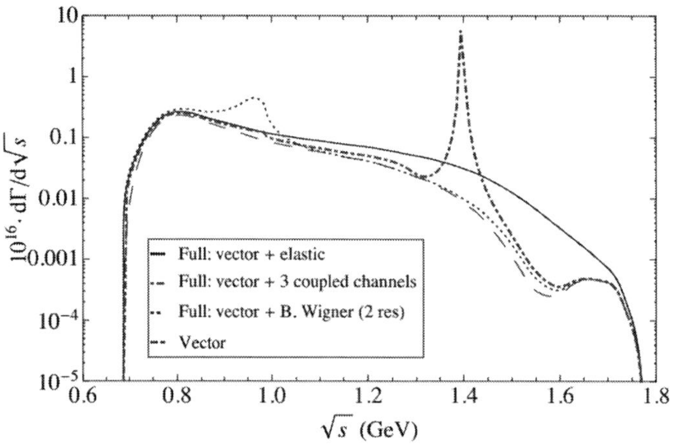

Figure 1. $\tau^- \to \pi^-\eta\nu_\tau$ decay distribution. The vector contribution (red-dashed curve) and the full distribution employing the scalar form factor in its elastic version (black solid curve), in the three coupled-channels analysis (green dot-dashed curve) and using a Breit-Wigner with two resonances (blue dotted curve) are displayed.

References

[1] R. Escribano, S. Gonzalez-Solis and P. Roig, JHEP **1310**, 039 (2013);
[2] R. Escribano, S. Gonzalez-Solis, M. Jamin and P. Roig, JHEP **1409**, 042 (2014)
[3] R. Escribano, S. Gonzalez-Solis and P. Roig, Phys. Rev. D **94**, no. 3, 034008 (2016)
[4] M. Fujikawa *et al.* [Belle Collaboration], Phys. Rev. D **78**, 072006 (2008)
[5] Z. H. Guo and J. A. Oller, Phys. Rev. D **84**, 034005 (2011)
[6] P. del Amo Sánchez *et al.* [BaBar Collaboration], Phys. Rev. D **83** (2011) 032002
[7] K. Hayasaka [Belle Collaboration], PoS EPS **-HEP2009**, 374 (2009)
[8] J. E. Bartelt *et al.* [CLEO Collaboration], Phys. Rev. Lett. **76** (1996) 4119

Experimental investigation of $dp \to ppn$ reaction at intermediate energies at Nuclotron

M. Janek[1,★], V. P. Ladygin[2], S. M. Piyadin[2], Yu. V. Gurchin[2], A. Yu. Isupov[2], J-T. Karachuk[3], A. N. Khrenov[2], A. K. Kurilkin[2], P. K. Kurilkin[2], A. N. Livanov[2], G. Martinska[4], S. G. Reznikov[2], G. Tarjanyiova[1], A. A. Terekhin[2], and I. E. Vnukov[5]
for the DSS Collaboration

[1] *Physics Department, University of Zilina, Zilina, Slovakia*
[2] *Joint Institute for Nuclear Research, Dubna, Russia*
[3] *Advanced Research Institute for Electrical Engineering, Bucharest, Romania*
[4] *University of P. J. Shafarik, Kosice, Slovakia*
[5] *Belgorod State University, Belgorod, Russia*

Abstract. There are still discrepancies between theory and experimental data in the polarisation observables of $dp \to ppn$ reaction in the low and intermediate energies, despite of significant process in the development of theoretical models which include three and more nucleon forces and relativistic effects. The data of $dp \to ppn$ reaction have been accumulated at 300, 400 and 500 MeV in the Nuclotron (Dubna, Russia) and partially processed for some kinematic configurations including few in which possible relativistic effects can appear. Kinematic simulation in the framework of ROOT and GEANT4 package have been performed before data processing. Part of the preliminary results are obtained in the form of energy deposit correlations of the two arms working in coincidence and few in the form of kinematic S curve.

1 Introduction

Relative smallness of the typical energies involved in the nuclear reactions in which protons and deuterons take part leads us firstly to non-relativistic Schrödinger equation that includes only the nucleon-nucleon (NN) potentials as a starting point. It is known that the data obtained from the NN scattering at low and intermediate energies (up to 350 MeV) can be described with high precision using the NN potentials only (e. g. CD-Bonn [1]). Modern NN potentials, however, underestimate the binding energies of the three-nucleon systems by about 0.5-1 MeV [2] and also do not reproduce the dp breakup and dp elastic scattering data.

The dp breakup reaction investigation at low and intermediate energies is very interesting. Quality of NN potentials in presence of third nucleon can be tested and the reaction has rich phase space. The effects from Coulomb interaction, relativistic effects and three nucleon forces ($3NF$) can be investigated in various regions of phase space. To get complementary information about the reaction mechanism and the structure of the objects involved in the dp breakup reaction the data accumulation

★e-mail: janek@fyzika.uniza.sk

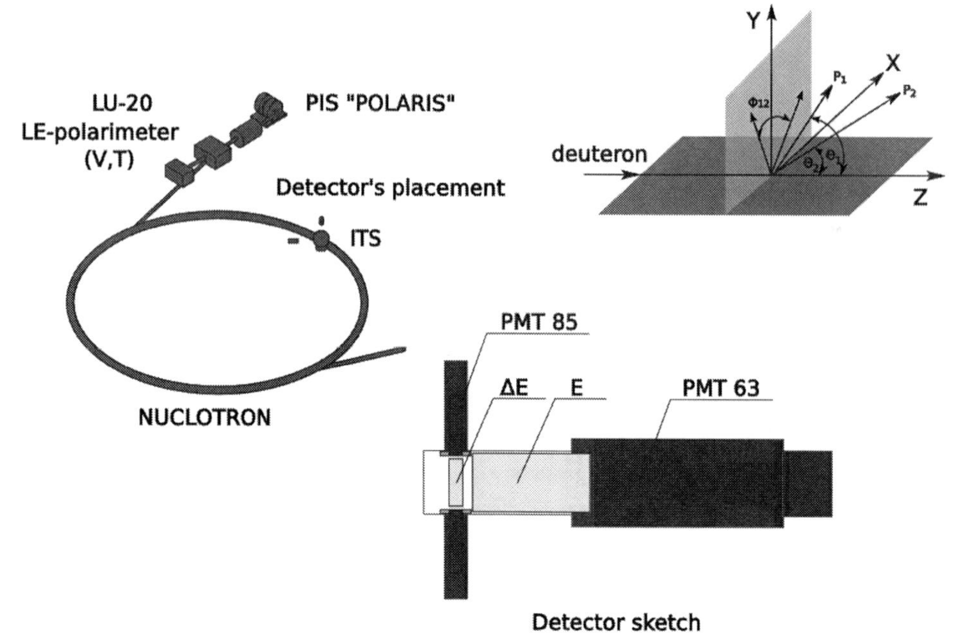

Figure 1. Nuclotron accelerator with sketched view of ΔE-E detector and angle description in case of dp reaction.

for the large region of the phase space is very desirable. It has been shown, that theoretical calculations based on Chiral Perturbation Theory, can satisfactorily describe data obtained for energies up to 100 MeV/nucleon. The worst situation is above 100 MeV/nucleon. Thus it is necessary to look for new types of these potentials or even for potentials involving more than three nucleons.

2 Experiment

The goal of the DSS experimental program is to obtain the information about two and three nucleon forces, including their spin dependent parts from dp-elastic scattering at the energies between 300 – 2000 MeV and dp-breakup with registration of two protons at deuteron energies of 300 – 500 MeV.

The dp breakup reaction is investigated at intermediate energies from 300 – 500 MeV of deuteron energy with eight $\Delta E - E$ scintillation detectors [3] (see Fig. 1) with PMT85 and PMT63 photomultiplier tubes. Special setup of the detectors has been chosen in the experiment performed at 2014 to follow inverse kinematics of the reaction in which relativistic effects can play a role [4]. Deuterons were accelerated up to required energy by Nuclotron (see Fig. 1). Polyethylene and carbon targets enclosed in a spherical hull of Internal Target Station (ITS) were used. Definition of mean angles of the detector placement with respect to the target is shown in Fig. 1. Solid angle of each of detector is 4.6°.

The kinematic variable S corresponds to the arc-length along the kinematic curve with zero point chosen at the minimal value of deposited energy in the second arm. Calculated kinematic S-curve with energy step of 5 MeV in case of dp breakup reaction for the detector arms placed at angles of $\Theta_1 = 35.0°$, $\Theta_2 = 43.0°$ and $\Phi_{12} = 180°$ at deuteron energy of 400 MeV is shown on the left panel in Fig. 2. The S is increasing in anti-clockwise direction. The number of breakup events in an interval $S - \Delta S/2$, and $S + \Delta S/2$ was obtained by projecting the events on a line perpendicular to the

DOI: 10.1051/epjconf/201613007007

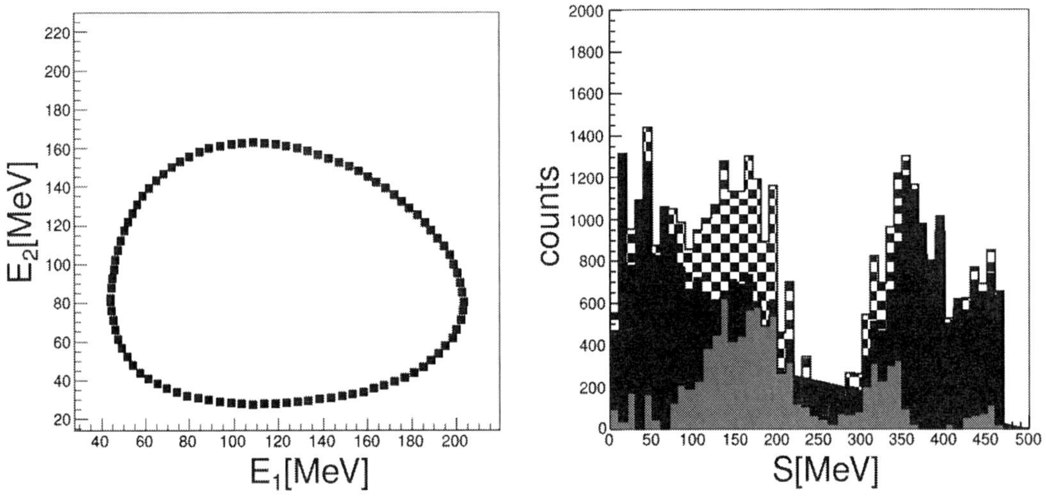

Figure 2. Left panel - calculated kinematic S-curve with energy step of 5 MeV in case of dp breakup reaction for the detector arms placed at angles of $\Theta_1 = 35.0°$, $\Theta_2 = 43.0°$ and $\Phi_{12} = 180°$ at deuteron energy of 400 MeV. Right panel - checked, black and grey histograms represent preliminary results of S distribution obtained on polyethylene, carbon and subtracted (dp) one (see text) for the same kinematic condition as for left the panel is used.

S-curve. Preliminary result for energy step $\Delta S = 5$ MeV and for the arm's polar angles of $\Theta_1 = 35°$, $\Theta_2 = 43°$ and azimuthal angle between them $\Phi_{12} = 180°$ for polyethylene (checked spectra), carbon (black spectra) and subtracted (grey spectra) at deuteron energy of 400 MeV is presented on the right panel in Fig. 2. Subtracted spectra is obtained by subtracting carbon content from polyethylene one. Carbon normalization was performed on the parts of polyethylene spectra where only carbon content is assumed.

3 Conclusion

Preliminary results of S distributions were obtained for some detector configuration at deuteron energy of 400 MeV. S distribution for particular detector configuration, $\Theta_1 = 35.0°$, $\Theta_2 = 43.0°$ and $\Phi_{12} = 180°$, is presented.

This work has been supported partly by the scientific cooperation program JINR-Slovak Republic in 2016 and by the Russian Foundation for Basic Research under grant RFBR 16-02-00203a.

References

[1] R. Machleidt *et al.*, Phys. Rev. C **53** 1483 (1996)
[2] A. Nogga *et al.*, Phys. Lett. B **409** 19 (1997)
[3] S. M. Piyadin *et al.*, Physics of Particles and Nuclei Letters **8** 107 (2011)
[4] H. Witala *et al.*, Few-Body Syst. **49** 61 (2011)

Three-point Green functions in the odd sector of QCD

T. Kadavý[1,*], K. Kampf[1], and J. Novotný[1]

[1] *Charles University, Faculty of Mathematics and Physics, Prague, Czech Republic*

Abstract. A review of familiar results of the three-point Green functions of currents in the odd-intrinsic parity sector of QCD is presented. Such Green functions include very well-known examples of VVP, VAS or AAP correlators. We also shortly present some of the new results for VVA and AAA Green functions with a discussion of their high-energy behaviour and its relation to the QCD condensates.

1 Introduction

The Green functions (also called as correlators) are defined as the vacuum expectation values of the time ordered products of the composite operators. In our case, such operators stand either for chiral vector and axial-vector currents $V^{\mu,a} = \bar{q}\gamma^\mu T^a q$, $A^{\mu,a} = \bar{q}\gamma^\mu\gamma_5 T^a q$, or scalar and pseudoscalar densities $S^a = \bar{q}T^a q$, $P^a = i\bar{q}\gamma_5 T^a q$, where q stands for the triplet of light quarks ($q = u, d, s$), T^a are the SU(3) generators ($T^a = \lambda^a/2$), and $a = 1\ldots 8$ are the group indices. There exist only five nontrivial three-point Green functions in the odd-intrinsic parity sector of QCD: VVP, VAS, AAP, VVA and AAA.

There are two regimes where the QCD dynamics of the current correlators is well understood. The first one is that of low external momenta where the dynamics is governed by the means of chiral perturbation theory (χPT) and resonance chiral theory (RχT). The second regime corresponds to the high energies where the asymptotic freedom allows us to use the perturbative approach in terms of the strong coupling α_s and where the asymptotics of the correlators for large euclidean momenta is given by the operator product expansion (OPE).

A consequence of the OPE is a presence of the vacuum avarages of composite gauge-invariant local operators, made of quark and gluon fields. Only the spin-zero operators, usually called the QCD condensates, contribute to the vacuum expectation value. Such nontrivial condensates with dimension $D \leq 6$ are shown here:

$$\begin{aligned} &O_3 = \langle 0|\bar{q}q|0\rangle\,, \qquad O_4 = \langle 0|G_{\mu\nu}G^{\mu\nu}|0\rangle\,, \qquad O_5 = \langle 0|\bar{q}\sigma_{\mu\nu}G^{\mu\nu}q|0\rangle\,, \\ &O_6^q = \langle 0|(\bar{q}\Gamma q)(\bar{q}\Gamma q)|0\rangle\,, \qquad O_6^G = \langle 0|G^a_{\mu\nu}G^{\nu,b}_{\sigma}G^{\sigma\mu,c}|0\rangle f_{abc}\,, \end{aligned} \tag{1}$$

where Γ stands for any combination of the 4×4 matrices $\{1, \gamma_5, \gamma^\mu, \gamma_5\gamma^\mu, \sigma^{\mu\nu}\}$ with 3×3 matrices $\{1, T^a\}$. In the following sections we will see that such QCD condensates play an important role in an evaluation of the OPE of the corresponding Green functions.

*e-mail: kadavy@ipnp.troja.mff.cuni.cz

2 Familiar results

Familiar Green functions that have been already studied extensively within the odd sector of QCD include VVP, VAS and AAP correlators. Since their tensor structure is similar, let us take the VVP correlator as an example and write down its structure in the form[1]

$$(\Pi_{VVP}(p,q;r))^{abc}_{\mu\nu} = \Pi_{VVP}(p^2,q^2;r^2)d^{abc}\varepsilon_{\mu\nu(p)(q)}\,. \tag{2}$$

The high-energy behaviour within the OPE framework can be written down as follows. If we consider all individual external momenta of all the three currents or densities to be large, it is easy to realize that the leading order contribution, where only the quark condensate contributes, corresponds to two contractions of the quark fields, contained in the respective currents and densities. Such contribution can be illustrated as

$$\langle 0| \longrightarrow\!\otimes\!\longrightarrow\!\otimes\!\longrightarrow\!\otimes\!\longrightarrow |0\rangle \neq 0\,, \tag{3}$$

where the crossed circle stands for an insertion either of the currents or the densities. The explicit calculation leads to the result [1]

$$\Pi_{VVP}((\lambda p)^2,(\lambda q)^2;(\lambda r)^2) = \frac{B_0F^2}{2\lambda^4}\frac{p^2+q^2+r^2}{p^2q^2r^2} + O\Big(\frac{1}{\lambda^6}\Big)\,. \tag{4}$$

Calculation of VVP (and also VAS) correlator in the odd sector up to $O(p^6)$ was provided in [2], together with a study of the OPE and other phenomenological applications.

3 OPE for VVA Green function with two large momenta

Before we advance, let us make a use of the known result of the OPE for the VVA correlator, where only two external momenta are considered to be large. The Ward identities restrict the general decomposition of the tensor part of the VVA into four terms,

$$(\Pi_{VVA}(p,q;r))^{abc}_{\mu\nu\rho} \equiv d^{abc}(w_L\varepsilon_{\mu\nu(p)(q)}r_\rho + w_T^{(1)}\Pi^{(1)}_{\mu\nu\rho} + w_T^{(2)}\Pi^{(2)}_{\mu\nu\rho} + w_T^{(3)}\Pi^{(3)}_{\mu\nu\rho})\,, \tag{5}$$

with the longitudinal part w_L fixed by the anomaly term and the transversal part defined in [3]. The formfactors w_L and $w_T^{(i)}$, $i = 1,2,3$, were obtained in the case of the odd-intrinsic parity sector up to $O(p^6)$ in [6]. Then, one can introduce the phenomenologically important formactor [3]

$$w_T(Q^2) = -16\pi^2[w_T^{(1)}(-Q^2,0,-Q^2) + w_T^{(3)}(-Q^2,0,-Q^2)]\,. \tag{6}$$

It turns out that the result for (6) up to $O(1/Q^8)$ can be also obtained from soft wall AdS/QCD and the OPE framework for two large external momenta, from which the outcome reads [4], [5]

$$w_T(Q^2) = \frac{N_c}{Q^2} + \frac{128\pi^3\alpha_s\chi\langle\bar{q}q\rangle^2}{9Q^6} + O\Big(\frac{1}{Q^8}\Big)\,. \tag{7}$$

Our result [6] for (6) can be expanded into a series in terms of Q^2 up to $O(1/Q^8)$ and compared with individual terms in (7). This procedure leads to several constraints for the coupling constants and also for the prediction for the deviation $\delta_{\rm BL}$, that propagates itself in the κ_3^{PV} constraint, if we take the Brodsky-Lepage behaviour of the transition form factor $\mathcal{F}_{\pi^0\gamma\gamma}$ into account [2], [7], [8]. Our calculations predict $\delta_{\rm BL} = -1.342$. However, such value is not in an agreement with the experimental data and one, therefore, has to rule out the OPE with two large momenta as a relevant OPE in the case of the VVA Green function (see [6] for a detailed discussion).

[1] In the following text we consider all momenta p,q,r as ingoing to the vertices, i.e. the law of momentum conservation is $p+q+r=0$. Also, a short-hand notation for the contractions of Levi-Civita tensor with the components of momenta is defined as $\varepsilon_{\mu\nu\alpha(p)} \equiv \varepsilon_{\mu\nu\alpha\beta}p^\beta$, $\varepsilon_{\mu\nu(p)(q)} \equiv \varepsilon_{\mu\nu\alpha\beta}p^\alpha q^\beta$.

4 OPE for VVA and AAA Green functions with three large momenta

Following the discussion above, one can try to calculate the OPE for VVA and also for AAA correlators and consider all three momenta to be large. The motivation is to find out if the results will become consistent with the experimental data. However, such an approach is not trivial in this case. None of these correlators have the leading order contribution to the OPE, i.e. the contributions of the topology (3) identically vanishes. Therefore, one is required to include the contributions to higher orders by means of diagrams with gluons coupled to the quark fields. Such diagrams naturally lead to contributions of higher QCD condensates. The scheme below shows the contributions of the quark condensate ($D = 3$), the gluon condensate ($D = 4$), the quark-gluon condensate ($D = 5$), the four-quark condensate and the three-gluon condensate ($D = 6$). This study is currently in progress [6].

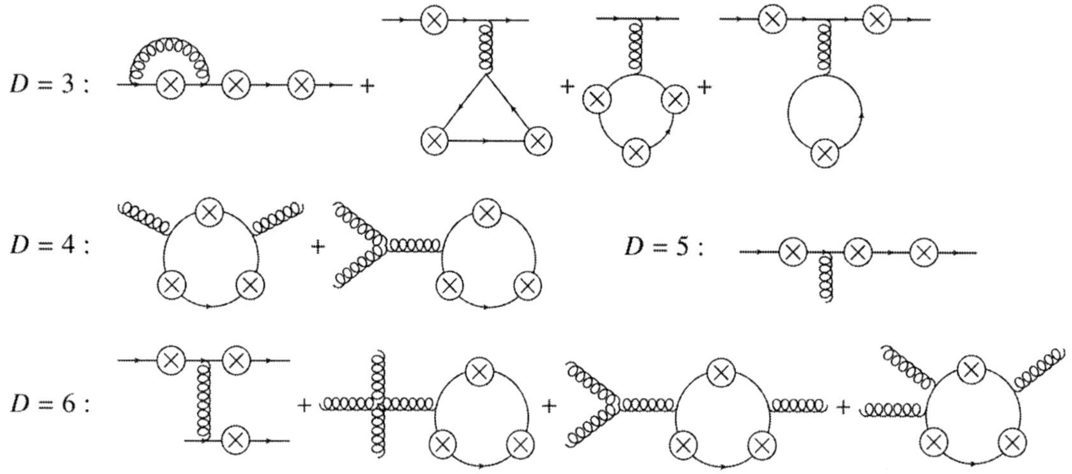

5 Conclusion

In this proceeding contribution, we summarized the known results for the high-energy behaviour of Green functions within the OPE framework, relevant in the odd-intrinsic parity sector. We also presented our approach in order to evaluate the OPE for the VVA and AAA correlators, where all momenta are large.

Acknowledgments. TK thanks the organizers for a pleasant conference. The support by the Charles University (project GA UK no. 700214) and by The Czech Science Foundation (project GA CR no. 15-18080S) is gratefully acknowledged.

References

[1] B. Moussallam, Phys. Rev. D **51**, 4939-4949 (1995)
[2] K. Kampf *et al.*, Phys. Rev. D **84**, 014036 (2011)
[3] M. Knecht *et al.*, JHEP **0403**, 035 (2004)
[4] P. Colangelo *et al.*, Phys. Rev. D **85**, 035013 (2012)
[5] J. J. Sanz-Cillero, PoS QNP **2012**, 118 (2012)
[6] T. Kadavý *et al.* (in preparation)
[7] G. P. Lepage *et al.*, Phys. Rev. D **22**, 2157 (1980)
[8] S. J. Brodsky *et al.*, Phys. Rev. D **24**, 1808 (1981)

Construction of the pion scalar form factor from few poles and zero

Robert Kamiński[1,*], Stanislav Dubnicka[2], Zuzana Dubnickova[3], and Andrej Liptaj[2]

[1] *Institute of Nuclear Physics PAS, Kraków, Poland*
[2] *Institute of Physics SAS, Bratislava, Slovak Republic*
[3] *Department of Theoretical Physics, Comenius University, Bratislava, Slovak Republic*

Abstract. Construction and analysis of the pion scalar-isoscalar form factor in the elastic region is presented. Precise S-wave $\pi\pi$ scattering phase shifts generated by dispersive analysis of experimental data with imposed crossing symmetry condition are used. Final result for values of the $f_0(500)$ meson mass and width, m_σ = (487 ± 31) MeV; Γ_σ = (542 ± 60) MeV is compatible with the results from dispersive analyses of the Bern and Madrid-Kraków groups to be considered now as the most reliable values of the $f_0(500)$ scalar meson parameters. Parameters of the $f_0(980)$, although lying almost on the $K\bar{K}$ threshold also agree with values predicted by these two groups.

1 Introduction

From the first identification of lightest hadronic resonance with vacuum quantum numbers 0^{++} i.e. $f_0(500)$ in 1974, a lot of work has been done and many papers to be concerned of this scalar meson are published up to now. However, only recently a clarification of this controversial situation with $f_0(500)$ has been achieved in the dispersive analyses done by the Bern [1, 2] and by the Madrid-Kraków group [3, 4], which are now considered to be the most reliable determinations of the $f_0(500)$ scalar meson parameters.

In the paper [5] another method for a determination of the $f_0(500)$ scalar meson parameters by means of the pion scalar-isoscalar form factor analysis in the elastic region has been proposed. The parameters were, however, fitted to the inaccurate, and in some region even contradicting experimental data on the S-wave isoscalar $\pi\pi$-scattering phase shift. So, the obtained mass and width of the $f_0(500)$ scalar meson can be considered only to demonstrate the existence of $f_0(500)$, however to be far away from the true values of the mass and width of this meson. The same concerns meson $f_0(980)$ whose parameters have been found to be very far from indicated by Particle Data Group [6].

Here, in the presented analysis, an ultimate analysis of the pion scalar form factor in the elastic region is proposed with true S-wave isoscalar $\pi\pi$ scattering phase shift data with theoretical errors, to be generated by the GKPY Roy-like equations derived and presented in [3].

Final result for values of the $f_0(500)$ meson mass and width, m_σ = (487±31) MeV; Γ_σ = (542±60) MeV is recovered to be consistent with Bern [2] and the Madrid-Kraków group [4], though determined in a completely different way.

*e-mail: robert.kaminski@ifj.edu.pl

2 Methodology

The pion scalar-isoscalar form factor (FF) $\Gamma_\pi(t)$ is defined by the parameterization of the matrix element of the scalar quark density

$$< \pi^i(p_2) \mid \widehat{m}(\bar{u}u + \bar{d}d) \mid \pi^j(p_1) >= \delta^{ij}\Gamma_\pi(t) \quad \text{where} \quad t = (p_2 - p_1)^2 \quad \text{and} \quad \widehat{m} = \frac{1}{2}(m_u + m_d). \quad (1)$$

It posses all known properties of the pion vector electromagnetic FF $F_\pi(t)$ like analyticity in t-plane (beside cuts on the positive real axis from two-pion threshold $t = 4m_\pi^2$ to $+\infty$), elastic unitarity, asymptotic behavior $\Gamma_\pi(t)_{|t|\to\infty} \sim \frac{1}{t}$, reality condition $\Gamma_\pi^*(t) = \Gamma_\pi(t^*)$ and normalization to the pion sigma term value $\Gamma(0) = (0.99 \pm 0.02)m_\pi^2$ to be predicted by the χPT [7].

The analyticity of the pion scalar FF in t-plane and asymptotic behavior (through the Cauchy formula) allow to derive dispersion relation with one subtraction

$$\Gamma_\pi(t) = P_n(t) exp\Big[\frac{t}{\pi}\int_{4m_\pi^2}^{\infty}\frac{\delta_0^0(t')}{t'(t'-t)}dt'\Big]. \quad (2)$$

In the elastic region ($\sqrt{t} < K\bar{K}$ threshold) one can express $\Gamma_\pi(t)$ in the form of a rational function

$$\Gamma_\pi(q) = \frac{\sum_{n=0}^{M} a_n q^n}{\prod_{i=1}^{N}(q - q_i)}, \quad (3)$$

which leads to the parameterization

$$\tan\delta_\Gamma(t) = \frac{A_1 q + A_3 q^3 + A_5 q^5 + A_7 q^7 + ...}{1 + A_2 q^2 + A_4 q^4 + A_6 q^6 + ...} \quad (4)$$

In the elastic region one may use $\delta_\Gamma \equiv \delta_0^0$ and then substitution of δ_0^0 from (4) into (2) leads to integrals which are suitable for an explicit calculation in the framework of the theory of residua (for details see [5]). Finally the explicit form of the pion scalar FF is:

$$\Gamma_\pi(t) = P_n(t)\frac{(q-q_1)}{(q+q_2)(q+q_3)(q+q_4)(q+q_5)} \times \frac{(i+q_2)(i+q_3)(i+q_4)(i+q_5)}{(i-q_1)} \quad (5)$$

where q_i are singularities of the FF to be placed in the upper or lower half q-plane and $P_n(t)$ is any polynomial normalized at $t = 0$ to one.

3 Results

Results of the fit to the output S-wave amplitude from [3] are presented on Fig. 1. Corresponding parameters A_i and positions of the poles and zeros are presented in Table 1. The parameter A_1 has been fixed to be equal to the $\pi\pi$ scalar-isoscalar scattering length $a_0^0 = 0.220$ found in [2] and [4]. As is seen from Table 1 only 5 parameters (in fact 4 free) were enough to describe the elastic region and in result only 2 complex poles and one zero in the complex momentum plane are enough to construct full mathematical structure of the FF in the elastic region. They are presented on Fig 2.

On Fig. 1 presented is also behavior of modulus of the pion scalar form factor in the region $-3\,\text{GeV}^2 < t < 3\,\text{GeV}^2$. Physically may be interpreted region only below 990 MeV ($t \approx 0.98\,\text{GeV}^2$)

Noteworthy is that value of the squared radius of the presented FF is $r^2 = 0.765^{+0.100}_{-0.087}\,\text{fm}^2$ what exceeds representative value $0.61 \pm 0.04\,\text{fm}^2$ presented in [8] and obtained on the basis of dispersion

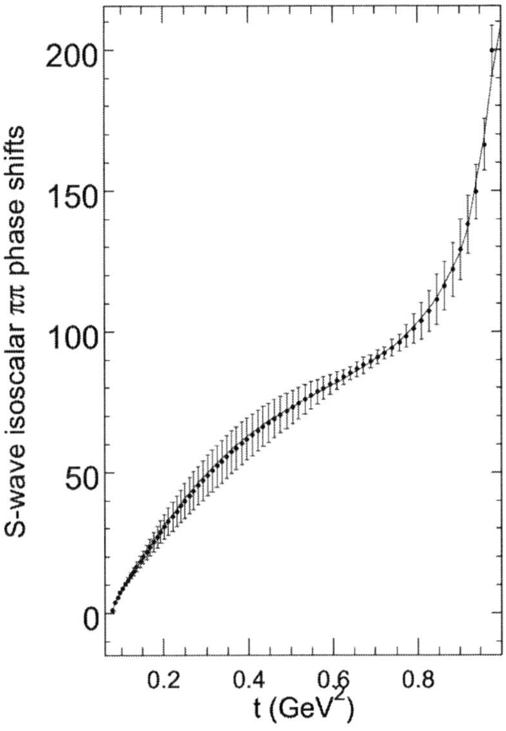

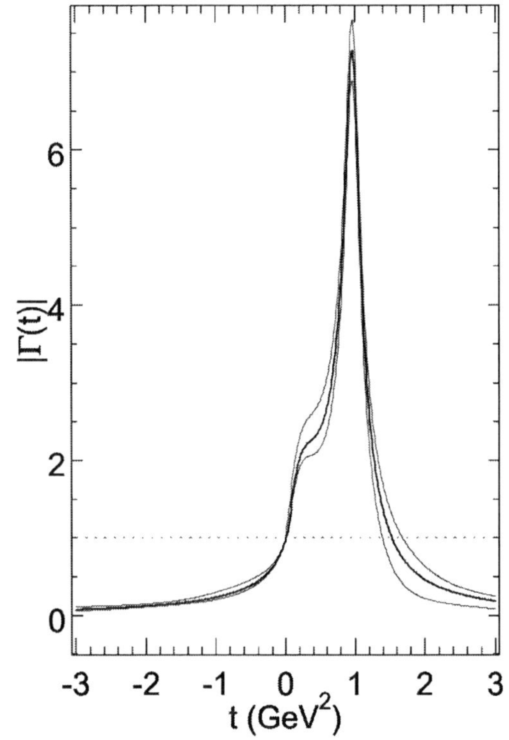

Figure 1. Left figure: the data on output $\delta_0^0(t)$ from [3] with theoretical errors to be generated by parameterization of the GKPY equations for the $S0$ wave in [3]. Solid line represents performed in the presented analysis fit. Right figure: behavior of the pion scalar form factor in the region $-3\,\text{GeV}^2 < t < 3\,\text{GeV}^2$. Lines represent central values and upper and lower limits. Physically may be interpreted region only below $t \approx 0.98\,\text{GeV}^2$.

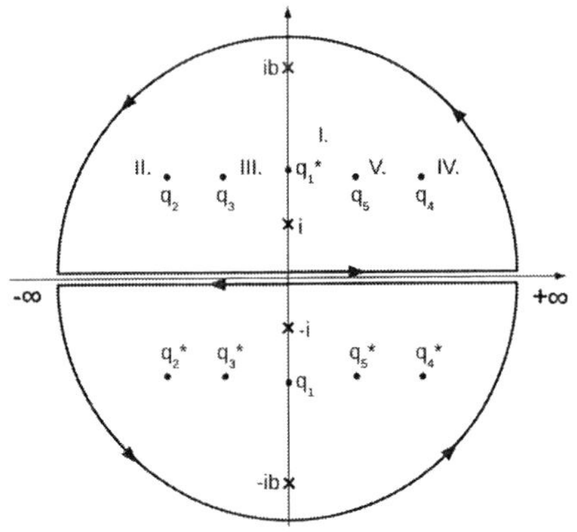

Figure 2. Schematic positions of poles (×) and branch points (•) of the $\Gamma_\pi(t)$ with contours of integrations in the upper and the lower half q-planes.

Table 1. Left table: results from fit to the output points phase shifts from [3]. Positions of poles are in MeV. Right table: positions of singularities of the pion-scalar FF in pion mass.

A_1	0.220 fixed
A_3	0.151 ± 0.020
A_5	-0.0144 ± 0.0016
A_2	-0.055 ± 0.038
A_4	-0.0089 ± 0.0048
$f_0(500)$	$451 \pm 14 - i260 \pm 33$
$f_0(980)$	$988 \pm 81 - i52 \pm 32$

	Req_i	Imq_i
q_1	0.00	-1.943 ± 0.20
q_2	3.397 ± 0.40	0.196 ± 0.04
q_3	-3.397 ± 0.40	0.196 ± 0.04
q_4	1.385 ± 0.10	1.085 ± 0.12
q_5	-1.385 ± 0.10	1.085 ± 0.12

theory. The value of the r^2 in presented here pion-scalar FF can be, however, significantly modified by for example adding two more parameters $A_{6,7}$ which lead to new pair of conjugated zeroes q_6 and q_7 or by simply multiplying FF by phase factor (nonzero only above the $K\bar{K}$ threshold) which leads to the phase of the $\Gamma_\pi(t)$ above the $K\bar{K}$ threshold very similar to that presented in [9] (see fig. 1 therein). Detailed analysis of these two methods will be presented in forthcoming publication.

In result first full and very simple construction of the pion scalar-isoscalar form factor has been presented. Only two poles (plus two conjugated ones) and one zero were needed to sufficiently precisely determine this form factor. Detailed analysis of the phase shifts used in performed fits (taken from dispersive analysis of the experimental data in the elastic $\pi\pi$ region) led to quite precise determination of positions of the $f_0(500)$ and $f_0(980)$ resonances presented in Table 1.

Interesting and noteworthy is correct energy dependence of the form factor just below the $K\bar{K}$ threshold where presented analysis ends. This greatly enhances the accuracy and reliability of the presented parameterization of the form factor.

Acknowledgements

The support of the Slovak Grant Agency for Sciences VEGA under Grant No. 2/0158/13 and of the Slovak Research and Development Agency under the contract No. APVV-0463-12 is acknowledged. This work has been also partially supported by the Polish Science Center (NCN) Grant No. Dec-2013/09/B/ST2/04382.

References

[1] G. Colangelo, J. Gasser, and H. Leutwyler, Nucl. Phys. B **603**, 125 (2001); B. Ananthanarayan, G. Colangelo, J. Gasser, and H. Leutwyler, Phys. Rep. **353**, 207 (2001)
[2] I. Caprini, G. Colangelo, H. Leutwyler, Phys. Rev. Lett. **96**, 132001 (2006)
[3] R. Garcia-Martin, R. Kamiński, J. R. Pela'ez, J. Ruiz de Elvira and F. J. Yndurain, Phys. Rev. D **83**, 074004 (2011)
[4] R. Garcia-Martin, R. Kamiński, J. R. Pela'ez and J. Ruiz de Elvira, Phys. Rev. Lett. **107**, 072001 (2011)
[5] S. Dubnicka, A. Z. Dubnickova, A. Liptaj, Phys. Rev. D **90**, 114003 (2014)
[6] K. A. Olive et al (Part. Data Group) Chin. Phys. C **38**, 090001 (2014)
[7] J. Gasser and U-G. Meissner, Nucl. Phys. B **357**, 90 (1991)
[8] J. F. Donoghue, J. Gasser, H. Leutwyler, Nucl. Phys. B **343**, 341 (1990)
[9] B. Ananthanarayan, I. Caprini, G. Colangelo, J. Gasser, and H. Leutwyler, Phys. Lett. B **602**, 218 (2004)

EPJ Web of Conferences **130**, 07010 (2016) DOI: 10.1051/epjconf/201613007010
MESON 2016

Experimental study of three-nucleon dynamics in the dp breakup collisions using the WASA detector

B. Kłos[1, a], I. Ciepał[2], G. Khatri[3], St. Kistryn[3], A. Kozela[2], A. Magiera[3], W. Parol[3], I. Skwira-Chalot[4], and E. Stephan[1]
for the WASA-at-COSY Collaboration

[1] *University of Silesia, PL-40007 Katowice, Poland*
[2] *Institute of Nuclear Physics PAN, PL-31342 Kraków, Poland*
[3] *Jagiellonian University, PL-30059 Kraków, Poland*
[4] *Faculty of Physics, University of Warsaw, Warsaw, Poland*

Abstract. An experiment to investigate the ^{1}H(d,pp)n breakup reaction using a deuteron beam of 340, 380 and 400 MeV and the WASA detector has been performed at the Cooler Synchrotron COSY-Jülich. The studied energy region, below but close to a pion production threshold, may provide information on various aspects of nuclear interactions, in particular on relativistic effects and their interplay with the three nucleon force (3NF). The almost 4π geometry of the WASA detector gives an unique possibility to study the different aspects of nucleon-nucleon dynamics in the three nucleon system. The main steps of the analysis, including energy calibration, PID and studied of efficiency, and their impact on final accuracy of the result, will be discussed.

1 Introduction

Few-nucleon system is an ideal laboratory to study nuclear forces. The system composed of three nucleons (3N) is the simplest nontrivial environment, in which various models of the nucleon-nucleon (NN) interaction can be tested. Three-nucleon system dynamics can be investigated quantitatively by comparing observables calculated with the use of Faddeev equations with precise measurements. The breakup observables can be calculated using modern realistic pairwise nucleon-nucleon NN interactions, combined with model of 3N forces [1]. Moreover, the two- and three-nucleon interactions can be modeled within the coupled-channel (CC) framework by an explicit treatment of the Δ-isobar [2]. Alternatively, the dynamics is generated by the chiral effective field theory (χEFT), so far at the next-to-next-to-leading order with all relevant NN and 3N contributions taken into account [3]. The new, improved version of ChPT is currently being developed [4]. The modern theoretical calculations include different pieces of nucleon-nucleon dynamics like the above mentioned three nucleon force but also the long-range Coulomb interaction or relativistic effects. In recent years the relativistic treatment of the breakup reaction in the 3N system has been developed using the NN potential [5] and this approach has also been extended for calculations including 3NF [6]. It was shown that in some particular regions of the breakup phase space, relativistic effects can increase or decrease the

[a] e-mail: barbara.klos@us.edu.pl

calculated breakup cross sections by up to 60%. At the same time the effects of 3NF may change certain observables by a similar factor. The relativistic effects and their interplay with 3NF become more important with increasing available energy in the three nucleon system. Therefore the energy range of interest is possibly close to pion production threshold (400 MeV for dp breakup reaction). We measured the differential cross section of the $^{1}H(d,pp)n$ breakup reaction at beam energies of 340, 380 and 400 MeV. The investigations will enable to study the evolution of the relativistic and 3NF effects in this energy, possibly setting constraints on the theoretical calculations.

2 Experiment

The experiment using the $^{1}H(\vec{d},pp)n$ breakup reaction at 340 MeV, 380 MeV and 400 MeV deuteron beam energy has been performed in January 2013 at the Cooler Synchrotron COSY-Jülich with the WASA-at-COSY detector [7], [8]. The WASA (Wide Angle Shower Apparatus) detector, covering almost full solid angle, gives a unique possibility of complex studies of the dp breakup reaction and the effects beyond non-relativistic NN-interactions, which appear in different phase-space regions. The WASA detector consists of four main components: Central Detector (CD), Forward Detector (FD), Pellet Target Device and the Scattering Chamber.

3 Data Analysis

Currently, the data analysis is focused on the proton-proton coincidences registered in the Forward Detector. It aims to determine the differential cross sections on dense angular grid of kinematical configurations defined by the emission angles of the two outgoing protons: two polar angles θ_1 and θ_2 (in t he range between 4° and 18°) and the relative azimuthal angle ϕ_{12}.

The first step of data analysis is the identification of events of interest, *i.e.* two protons from the breakup process and deuteron-proton pairs from elastic scattering channel in the Forward Detector. The particle identification is based on the ΔE-E technique .

The luminosity is determined on the basis of the number of the elastically scattered deuterons at a given θ angle and on the known cross section for elastic scattering at the studied energy [10]. After selection of deuterons registered in the Forward Detector and energy calibration, the correct energy vs angle relation for elastically scattered deuterons was obtained, see Fig. 1.

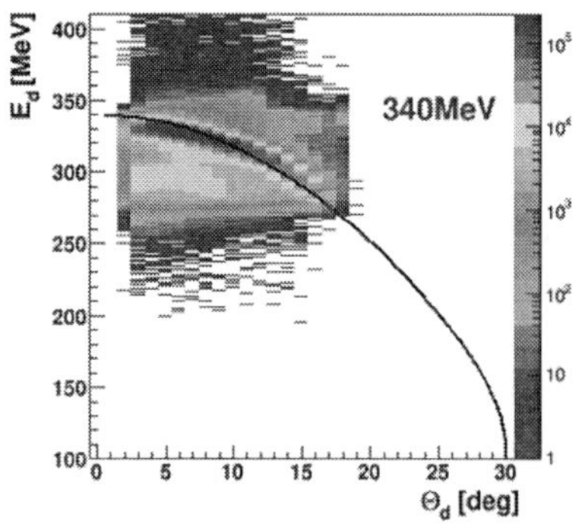

Figure 1. Selection of deuterons registered in FD. Solid curve corresponds to kinematics of elastically scattered deuterons.

After selection of the proton-proton coincidences and having performed the energy calibration, the differential cross section of the breakup reaction can be determined for a chosen kinematic configuration. The configuration is defined by the emission angles of the two outgoing protons: two polar

DOI: 10.1051/epjconf/201613007010

angles θ_1 and θ_2 and the relative azimuthal angle ϕ_{12}, with defined integration limits: $\Delta\theta_1$=$\Delta\theta_2$=1° and $\Delta\phi_{12}$=5°. The kinematical spectra E_1 vs E_2 are produced for each analyzed configuration.

Events belonging to the chosen configuration of interest are projected onto a kinematical curve in order to obtain the distribution as a function of the arc-length S measured along the kinematics. An example of the breakup event rate obtained for the chosen kinematical configurations is presented in Fig. 2. The normalization is preliminary: the analysis is ongoing to determine systematic uncertainties related to detection efficiency and to luminosity value obtained on the basis of existing data. It is clear that a very good control over data normalization is indispensable to conclude about relativistic effects.

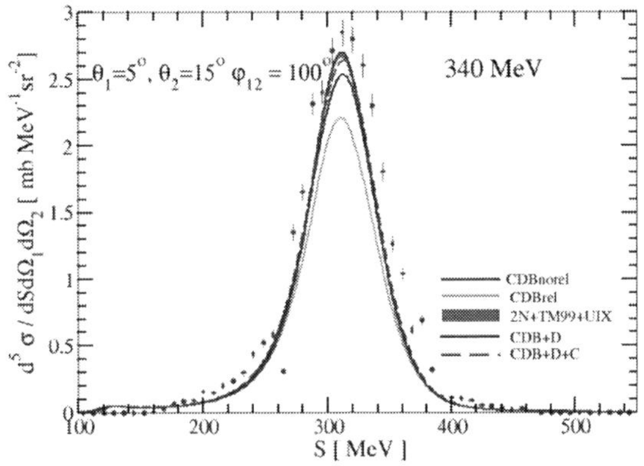

Figure 2. An example of the differential cross section of breakup reaction obtained for chosen kinematical configuration as a function of the S value. Magenta band represents calculations based on realistic potentials: 2N complemented with the TM99 3NF and the realistic AV18 potential combined with the Urbana IX. The calculations within the coupled-channel approach with the CD Bonn+ Δ potential without and with the Coulomb force included are represented by blue solid line and green dashed line, respectively. Violet and brown solid lines present theoretical calculations with CD Bonn potential in norelativistic and relativistic regime, respectively.

In this figure the theoretical calculations [1] [9] are also shown. The calculations in relativistic regime have recently been performed for the ^{1}H(*d,pp*)n breakup reaction at the beam energies of 340, 380 and 400 MeV [11], clearly indicating of relativistic description at these energies.

4 Summary and outlook

The analysis is continued with the aim to determine the differential cross sections for the the deuteron breakup process in the d+p system at energies of 340, 380 and 400 MeV, performed for a large set of kinematical configurations covering a significant part pf the reaction phase space. The data will be compared to the theoretical predictions for three nucleon systems with the aim to investigate relativistic effects. Currently, the analysis is focused on precise determination of efficiency of the detection system and its impact on final accuracy of the result, as well as on deterimination of systematic uncertainty of luminosity.

Acknowledgements

This work has been supported by FFE funds of the Forschunszentrum Jülich; by the European Commision under the 7th Framework Programme through the "Research Infrastructures" action of the Capacities Programme (Call: FP7-INFRASTRUCTURES-2008-1) and by Polish National Science Center from grant DEC-2012/05/B/ST2/02556.

References

[1] W. Glöckle , H. Witała, D. Hüber, H. Kamada, J. Golak, Phys. Rep. **274**, 107 (1996)
[2] A. Deltuva, K. Chmielewski and P.U. Sauer, Phys. Rev. C **67**, 034001 (2003);
A. Deltuva, R. Machleidt and P.U. Sauer, Phys. Rev. C **68**, 024005 (2003)
[3] E. Epelbaum, Rep. Prog. Nucl. Phys. **57**, 654 (2006)
[4] R. Machleidt and F. Sammarruca, Phys. Scripta **91**, 083007 (2016);
D. Djukanovic, E. Epelbaum, J. Gegelia, H. Krebs and U. G. Meißner, Eur. Phys. J. A **51**, 101 (2015);
E. Epelbaum, H. Krebs and U. G. Meißner, Phys. Rev. Lett. **115**, 122301 (2015);
S. Binder *et al.*, Phys. Rev. C **93**, 044002 (2016)
[5] H. Witała, R. Skibiński and J. Golak, The European Physical Journal A **30**, 369 (2006)
[6] H. Witała *et al.*, Phys. Rev. C **83**, 044001 (2011)
[7] Ch. Bargholtz *et al.*, Nucl. Instrum. Meth. A **594**, 339 (2008)
[8] H. H. Adam *et al.*, arXiv:nucl-ex/0411038 (2004)
[9] A. Deltuva *et al.*, Phys. Rev. C **72**, 054004 (2005)
[10] K. Ermisch *et al.*, Phys. Rev. C **68**, 051001(R) (2003)
[11] H. Witała, privite communication.

 DOI: 10.1051/epjconf/201613007011

Inclusive reconstruction of hadron resonances in elementary and heavy-ion collisions with HADES

Georgy Kornakov[1,*]
for the HADES Collaboration

[1] *Technische Universität Darmstadt, Germany*

Abstract. The unambiguous identification of hadron modifications in hot and dense QCD matter is one of the important goals in nuclear physics. In the regime of 1 - 2 GeV kinetic energy per nucleon, HADES has measured rare and penetrating probes in elementary and heavy-ion collisions. The main creation mechanism of mesons is the excitation and decay of baryonic resonances throughout the fireball evolution. The reconstruction of short-lived (≈ 1 fm/c) resonance states through their decay products is notoriously difficult. We have developed a new iterative algorithm, which builds the best hypothesis of signal and background by distortion of individual particle properties. This allows to extract signals with signal-to-background ratios of <1%.

1 Introduction

The mechanism of mass generation of hadrons stays an open question. Although the Higgs mechanism successfully explains mass generation of the three families of fermions, leptons and quarks, and the electroweak gauge bosons $W^{\pm}$ and Z, the way how matter, i.e. neutrons and protons, obtains its mass remains unclear. One of the scenario relates the QCD chiral condensate $\langle \bar{q}q \rangle$ to the mass. In vacuum, the chiral symmetry is spontaneously broken [1]. In AA collisions, depending on the energy and system size, chiral symmetry can be partially restored. These changes are reflected in the modification of the spectral functions of hadrons. It was shown that from relativistic (SIS18, SPS, AGS) to ultra-relativistic collisions (RHIC, LHC) baryons and anti-baryons play an important role in the modification of the spectral functions of the ρ, ω and ϕ mesons [2]. One can observe these modifications best in the di-lepton decay channel, although it might be that reconstruction of $\pi^+\pi^-$ decays could provide valuable information as well [3].

For this reason, the HADES (High Acceptance Di-Electron Spectrometer) [4] makes systematic measurements of elementary and heavy-ion collisions (pN, pA, π^-N, π^-A) at intermediate temperatures and high baryon chemical potential. HADES is located at the SIS-18 accelerator at GSI, Darmstadt, Germany. Accelerated ion beams, as well as protons and a secondary pion beam are available to be collided with a fixed target consisting of a liquid hydrogen, polyethylene, carbon, Au or W. The detector has low mass tracking detectors and a superconducting toroidal magnet, both divided in equal sextants, providing a momentum resolution better than 2 % for 1 GeV/c protons and pions. Hadron and lepton reconstruction and identification is based on the measurement of the specific energy loss

*e-mail: g.kornakov@gsi.de

 DOI: 10.1051/epjconf/201613007011

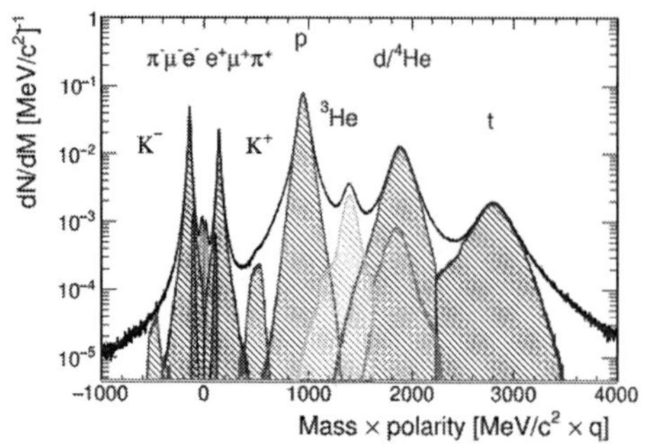

Figure 1. Reconstructed mass from momentum and velocity in HADES after particle identification in Au+Au collisions at 1.23 GeV/u.

and the time of flight technique. An example of the mass distribution from combined time of flight and momentum measurements is shown in Fig. 1.

2 Method, results and conclusions

The reconstruction of short-lived resonances by means of measurement of the four-momentum of their decay products into hadrons is obfuscated by random combinations, called combinatorial background. In order to overcome the limitations methods commonly used for that purpose, as the different event-mixing techniques [5], we have developed an iterative algorithm that prevents shape distortions and need of weakly-defined normalization factors [6]. The method uses random rotations of particles within the same event, allowing to reduce statistical fluctuations by increasing the number of rotations of one particle. The combinatorial background hypothesis is improved in every iteration step. Correlations appearing from a limited acceptance, occupancy and efficiency are taken into account as well as correlations due to the event reaction plane.

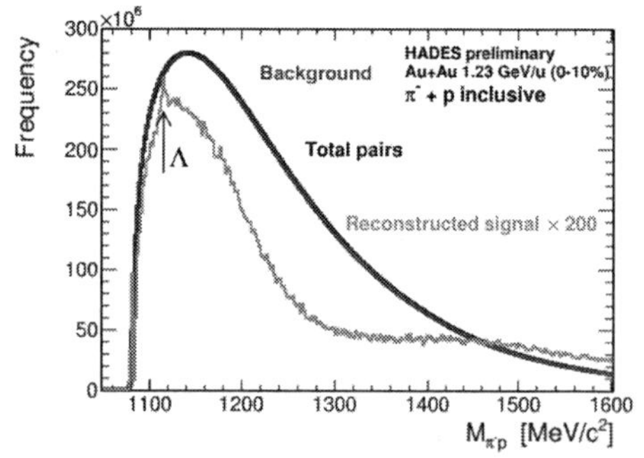

Figure 2. Reconstructed inclusive $\pi^- p$ signal measured by HADES in the 10 % most central Au+Au collisions at 1.23 GeV/u. The total measured signal is shown with a black line together with the obtained combinatorial background after applying the iterative technique. Their difference is represented in blue after scaling by a factor of 200 for visual comparison with the total measured distribution. The method performs efficiently in environments with signal-to-background rations smaller than 1 %.

An example of a reconstructed inclusive signal of $\pi^- p$ pairs from the 10 % most central Au+Au at 1.23 GeV/u collisions is shown in Fig. 2. The narrow Λ peak followed by a broad continuum that can be attributed to resonance decays, e.g. Δ, N^*, are identifiable. As the transformations used preserve energy and momentum, the iterative method works efficiently for elementary reactions as well. In this case, the algorithm behaves similarly to random phase-space generators [7], but includes the detector response obtained from real data, that is to say acceptance, efficiency and granularity. The technique was successfully tested with data from π^- induced reactions on a carbon target at an incident momentum of 1.7 GeV/c measured by HADES. The reconstructed signal in the invariant-mass - pair transverse momentum plane of $\pi^+\pi^-$ and $\pi^- p$ are shown in Fig. 3.

In conclusion, the developed method for calculating the combinatorial background is suitable for multi-differential analysis of heavy-ion collisions as well as elementary reactions. The method was tested with real data measured by HADES. The reconstruction of the inclusive production of resonances in the $\pi^{\pm}p$ and $\pi^{+}\pi^{-}$ channels in both Au+Au and elementary reactions will help to understand better the production mechanisms, medium and nuclear effects and also sets constraints on some hadron sources of the di-lepton cocktail.

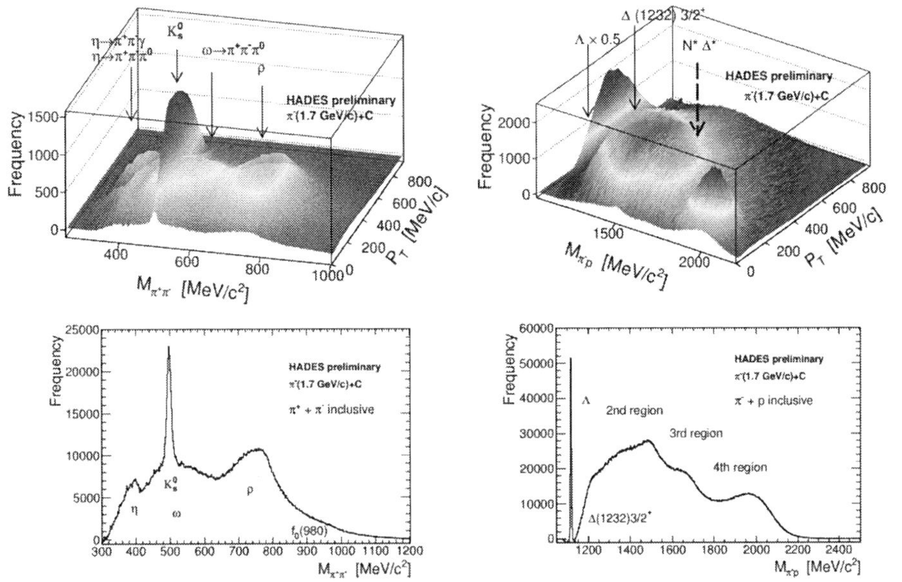

Figure 3. Inclusive $\pi^{+}\pi^{-}$ (left) and $\pi^{-}p$ (right) raw spectra measured by HADES in pion induced reactions on a carbon target at an incident momentum of 1.7 GeV/c. Upper panels: invariant mass - pair P_T distributions. Lower panels: one-dimensional projections of the invariant masses.

The HADES Collaboration gratefully acknowledges the support by grants PTDC/FIS/113339/2009 LIP Coimbra; NCN 2013/10/M/ST2/00042 SIP JUC Cracow; 2013 /10/M/ST2/00042 Helmholtz Alliance HA216/EMMI GSI Darmstadt; VH-NG-823, Helmholtz Alliance HA216/EMMI TU Darmstadt; 05P12CRGHE HZDR Dresden; Helmholtz Alliance HA216/EMMI, HIC for FAIR (LOEWE), GSI F&E Goethe-University Frankfurt; MLL, DFG ECLUST 153, VH-NG-330, BMBF 06MT7180, GSI TMLRG1316F, BMBF 05P15WOFCA, TU Munchen; BMBF:05P12RGGHM JLU Giessen; UCY/3411-2311 University Cyprus; CNRS/IN2P3 IPN Orsay; MSMT LG 12007, AS CR M100481202, GACR 13-06759S NPI AS CR, Rez and EU Contract No. HP3-283286.

References

[1] S. Weinberg, Phys. Rev. Lett. **18**, 507 (1967)
[2] R. Rapp, H. van Hees, Eur. Phys. J. A **52**, 257 (2016)
[3] R. Rapp, Nucl. Phys. A **725**, 254 (2003)
[4] G. Agakishiev *et al.*, [HADES Collaboration], Eur. Phys. J. A **41**, 243 (2009)
[5] D. L. Hôte, Nucl. Instr. Meth. Phys. Res. A **337**, 544 (1994)
[6] G. Kornakov, Manuscript in preparation
[7] F. James, *Monte Carlo Phase Space*, CERN 68-15, (1968)

Thermodynamic instabilities in hot and dense nuclear matter

A. Lavagno[1,2,*]

[1] *Department of Applied Science and Technology, Politecnico di Torino, Torino, Italy*
[2] *Istituto Nazionale di Fisica Nucleare (INFN), Sezione di Torino, Torino, Italy*

Abstract. We study the presence of thermodynamic instabilities in a hot and dense nuclear medium where a nuclear phase transition can take place. Similarly to the low density nuclear liquid-gas phase transition, we show that such a phase transition is characterized by pure hadronic matter with both mechanical instability (fluctuations on the baryon density) that by chemical-diffusive instability (fluctuations on the strangeness concentration). The analysis is performed by requiring the global conservation of baryon number and zero net strangeness in the framework of an effective relativistic mean field theory with the inclusion of the $\Delta(1232)$-isobars, hyperons and the lightest pseudoscalar and vector meson degrees of freedom. It turns out that in this situation hadronic phases with different values of strangeness content may coexist, altering significantly meson-antimeson ratios.

We investigate the nuclear medium in the context of relativistic mean field approach, where the nuclear force is mediated by the exchange of isoscalar-scalar (σ), isoscalar-vector (ω) and isovector-vector (ρ) mesons fields in the so-called TM1 parameter set [1, 2]. Hyperon degrees of freedom are included taking into account of the determination of the corresponding meson-hyperon coupling constants that have been fitted to hypernuclear properties [3, 4].

In regime of finite values of temperature and density, a state of high density resonance matter may be formed and the $\Delta(1232)$-isobar degrees of freedom are expected to play a central role [5–8]. In particular, the formation of resonances matter contributes essentially to baryon stopping, hadronic flow effects and enhanced strangeness. Following Ref.s [2, 7], we take into account of the Δ-isobar degree of freedoms.

We are dealing with the study of a multi-component system at finite temperature and density with two conserved charges: baryon (B) number and zero net strangeness (S) number ($r_S = \rho_S/\rho_B = 0$). For what concern the electric charge (Q), we work in symmetric nuclear matter with a fixed value of $Z/A = 0.5$ and we do not consider fluctuations in the electric charge fraction, due to the high temperature regime considered in the present investigation. Therefore, the electric charge results to be separately conserved in each phase during the phase transition and the chemical potential of particle of index i can be written as $\mu_i = b_i\mu_B + s_i\mu_S$, where b_i and s_i are, respectively, the baryon and the strangeness quantum numbers of i-th hadronic species.

By increasing the temperature and the baryon density during the high energy heavy ion collisions, a multi-particle system may take place and the formation of antiparticles become much more relevant. In analogy with the liquid-gas case [9], we are going to investigate the existence of a possible phase transition in the nuclear medium by studying the presence of instabilities (mechanical and/or chemical) in the nuclear equation of state.

*e-mail: andrea.lavagno@polito.it

For such a system, the Helmholtz free energy density F can be written as

$$F(T,\rho_B,\rho_S) = -P(T,\mu_B,\mu_S) + \mu_B\rho_B + \mu_S\rho_S\,, \tag{1}$$

with

$$\mu_B = \left(\frac{\partial F}{\partial \rho_B}\right)_{T,\rho_S}\,, \quad \mu_S = \left(\frac{\partial F}{\partial \rho_S}\right)_{T,\rho_B}\,. \tag{2}$$

In a system with N different particles, the particle chemical potentials are expressed as the linear combination of the two independent chemical potentials μ_B and μ_S and, as a consequence, $\sum_{i=1}^{N} \mu_i\rho_i = \mu_B\rho_B + \mu_S\rho_S$.

Assuming the presence of two phases (denoted as I and II, respectively), the system is stable against the separation in two phases if the free energy of a single phase is lower than the free energy in all two phases configuration. The phase coexistence is given by the Gibbs conditions

$$\mu_B^I = \mu_B^{II}\,, \qquad \mu_S^I = \mu_S^{II}\,, \tag{3}$$

$$P^I(T,\mu_B,\mu_S) = P^{II}(T,\mu_B,\mu_S)\,. \tag{4}$$

At a given baryon density ρ_B and at a given zero net strangeness density $r_S = \rho_S/\rho_B = 0$, the chemical potentials μ_B are μ_S are univocally determined by the following equations

$$\rho_B = (1-\chi)\rho_B^I(T,\mu_B,\mu_S) + \chi\rho_B^{II}(T,\mu_B,\mu_S)\,, \tag{5}$$

$$\rho_S = (1-\chi)\rho_S^I(T,\mu_B,\mu_S) + \chi\rho_S^{II}(T,\mu_B,\mu_S)\,, \tag{6}$$

where $\rho_B^{I(II)}$ and $\rho_S^{I(II)}$ are, respectively, the baryon and strangeness densities in the low density (I) and in the higher density (II) phase and χ is the volume fraction of the phase II in the mixed phase ($0 \le \chi \le 1$). An important feature of this conditions is that, unlike the case of a single conserved charge, baryon and strangeness densities can be different in the two phases, although the total ρ_B and ρ_S are fixed.

For a system in thermal equilibrium, the possible phase transition can be characterized by mechanical (fluctuations in the baryon density) and chemical instabilities (fluctuations in the strangeness number). The mechanical and chemical stability conditions ($r_S = \rho_S/\rho_B$) are satisfied if [7]:

$$\rho_B\left(\frac{\partial P}{\partial \rho_B}\right)_{T,\rho_C} > 0\,, \qquad \left(\frac{\partial \mu_S}{\partial r_S}\right)_{T,P} > 0\,. \tag{7}$$

In Fig. 1, we show the pressure as a function of baryon density at fixed temperature. The continuous (dashed) lines correspond to the solution obtained with (without) the Gibbs construction. In this case the conditions of mechanical and chemical stability are not satisfied and the system goes to a phase transition. Let us observe that, although the system has globally zero net strangeness ($r_S = 0$), the mixed phase is realized with two phase with different and finite values of strangeness and different baryon-antibaryon content (baryon and strangeness charges are globally conserved but are different in the two hadronic phases).

The above property can imply remarkably consequences in the meson-antimeson production and in the strangeness content at different temperatures and baryon densities. This matter of fact can be observed, for example, in Fig. 2, where the K^+/K^- ratios, as a function of the baryon density, are reported for different temperatures ($x_{\sigma\Delta} = g_{\sigma\Delta}/g_{\sigma N} = 1.2$ is the used σ meson-Δ ratio coupling constant and $x_{\omega\Delta} = g_{\omega\Delta}/g_{\omega N} = 1$). The black points in the isothermal curves stand for the beginning and the end of the phase transition. We observe that at lower ($T \le 100$ MeV) and higher ($T \ge 150$ MeV) temperatures, the phase transition does not occur, while, in the intermediate range of temperature

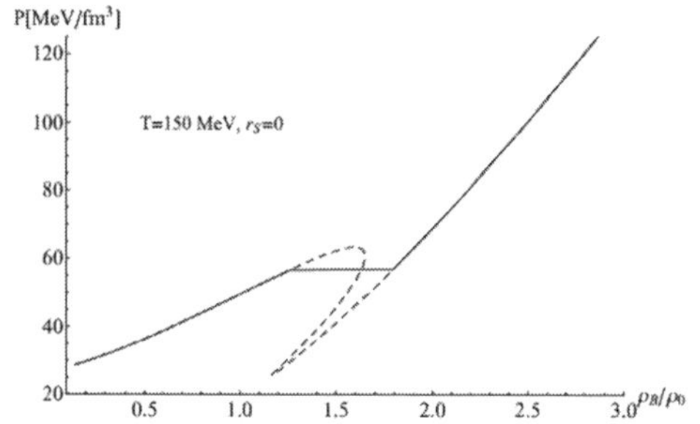

Figure 1. Pressure as a function of baryon density at fixed temperature. The continuous (dashed) lines correspond to the solution obtained with (without) the Gibbs construction.

(100 < T < 150 MeV), thermodynamic instabilities take place and this feature can imply an abrupt variation of the ratio during the phase transition region. Such behavior could be phenomenological relevant in order to identify the presence of such phase transition in the future compressed baryonic matter experiments [10].

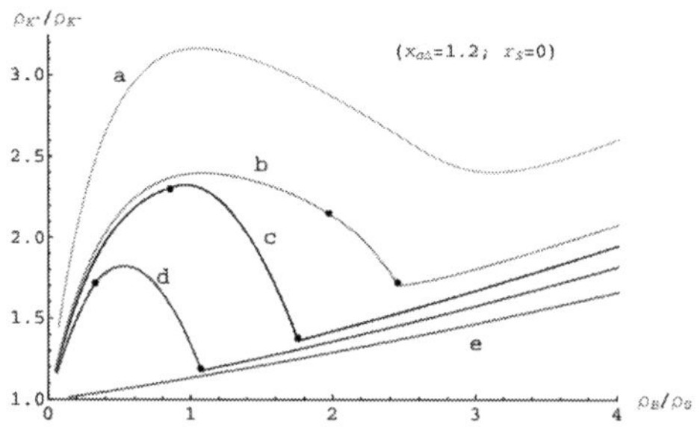

Figure 2. Kaon to antikaon ratios as a function of baryon density at different temperatures (labels a to e correspond to T=120, 127, 135, 140, 150 MeV, respectively). The black points stand for the beginning and the end of the phase transition.

References

[1] N.K. Glendenning and S.A. Moszkowski, Phys. Rev. Lett. **67** 2414 (1991)
[2] A. Lavagno, Phys. Rev. C **81** 044909 (2010)
[3] N.K. Glendenning and J. Schaffner-Bielich, Phys. Rev. C **60** 025803 (1999)
[4] A. Drago, A. Lavagno, G. Pagliara, Phys. Rev. D **89** 043014 (2014)
[5] M. Hofmann, R. Mattiello, H. Sorge, H. Stocker, and W. Greiner, Phys. Rev. C **51**, 2095 (1995)
[6] Z. Li, G. Mao, Y. Zhuo, and W. Greiner, Phys. Rev. C **56**, 1570 (1997)
[7] A. Lavagno, D. Pigato, Phys. Rev. C **86**, 024917 (2012)
[8] A. Drago, A. Lavagno, G. Pagliara, D. Pigato, Phys. Rev. C **90**, 065809 (2014)
[9] H. Müller, B.D. Serot, Phys. Rev. C **52**, 2072 (1995)
[10] P. Senger, Nucl. Phys. A **862-863**, 139 (2011)

Commissioning and initial experimental program of the BGO-OD experiment at ELSA

S. Alef[1], P. Bauer[1], D. Bayadilov[2], R. Beck[2], M. Becker[2], A. Bella[1], P. Bielefeldt[1], S. Böse[2], A. Braghieri[3], K. Brinkmann[4], P. Cole[1,5], R. Di Salvo[6], H. Dutz[1], D. Elsner[1], A. Fantini[6,7], O. Freyermuth[1], S. Friedrich[1], F. Frommberger[1], V. Ganenko[8], D. Geffers[1], G. Gervino[9,10], F. Ghio[11,12], S. Görtz[1], A. Gridnev[13], E. Gutz[4], D. Hammann[1], J. Hannappel[1], W. Hillert[1], A. Ignatov[14], R. Jahn[2], R. Joosten[2], T.C. Jude[1], F. Klein[1], J. Knaust[1], K. Kohl[1], K. Koop[2], B. Krusche[15], A. Lapik[14], P. Levi Sandri[16,★], I.V. Lopatin[13], G. Mandaglio[17,18], F. Messi[1], R. Messi[6,7], V. Metag[4], D. Moricciani[6], A. Mushkarenkov[14], M. Nanova[4], V. Nedorezov[14], D. Novinskiy[13], P. Pedroni[3], B. Reitz[1], M. Romaniuk[6], T. Rostomyan[1], N. Rudnev[14], C. Schaerf[6,7], G. Scheluchin[1], H. Schmieden[1], A. Stugelev[13], V. Sumachev[13], V.Tarakanov[13], V. Vegna[1], D. Walther[1], D. Watts[19], H. Zaunick[2 4], and T. Zimmermann[1]

[1]*Physikalishes Institut, Nußallee 12, 53115, Bonn, Germany*
[2]*Helmoltz-Institut für Strahlen- und Kernphysik, Nußallee 14-16, 53115 Bonn, Germany*
[3]*INFN sezione di Pavia, Via Agostino Bassi, 6 - 27100 Pavia, Italy*
[4]*Justus-Liebig-Universität, II. Physikalisches Institut, Heinrich-Buff-Ring 16, 35392 Gießen, Germany*
[5]*Idaho State University, 921 S. 8th Ave., Pocatello, Idaho 83209, USA*
[6]*INFN Roma Tor Vergata, Via della Ricerca Scientifica 1, 00133 Roma, Italy*
[7]*Universitá di Roma Tor Vergata, Via della Ricerca Scientifica 1, 00133 Roma, Italy*
[8]*Kharkov Institute of Physics and Technology, Akademicheskaya St. 1, Kharkov, 61108, Ukraine*
[9]*INFN sezione di Torino, Via P.Giuria 1, 10125 Torino, Italy*
[10]*Universitá di Torino, via P. Giuria 1, 10125 Torino, Italy*
[11]*INFN sezione di Roma, P.le Aldo Moro, 2, 00185 Roma, Italy*
[12]*Istituto Superiore di Sanitá, Viale Regina Elena 299, 00161 Roma, Italy*
[13]*Petersburg Nuclear Physics Institute, Gatchina, Leningrad District, 188300, Russia*
[14]*Russian Academy of Sciences Institute for Nuclear Research, prospekt 60-letiya Oktyabrya 7a, Moscow 117312, Russia*
[15]*Institut für Physik, Klingelbergstraße 82, 4056 Basel, Switzerland*
[16]*INFN Laboratori Nazionali di Frascati, Via E. Fermi 40, 00044 Frascati, Italy*
[17]*INFN sezione Catania, 95129 Catania, Italy*
[18]*Universitá degli Studi di Messina, Via Consolato del Mare 41, 98121 Messina, Italy*
[19]*The University of Edinburgh, James Clerk Maxwell Building, Mayfield Road, Edinburgh EH9 3JZ, UK*

Abstract. BGO-OD is a new meson photoproduction experiment at the ELSA facility of Bonn University. It aims at the investigation of non strange and strange baryon excitations, and is especially designed to be able to detect weekly bound meson-baryon type structures. The setup for the BGO-OD experiment is presented, the characteristics of the photon beam and the detector performances are shown and the initial experimental program is discussed.

★e-mail: Paolo.Levisandri@lnf.infn.it

DOI: 10.1051/epjconf/201613007013

1 Introduction

The BGO-OD Collaboration was formed at the end of 2010 with the aim of studying meson photoproduction off the nucleon (and eventually off light nuclei) at Bonn ELSA. The photon beam covers the region between 0.3 to 2.9 GeV, and substantial linear polarization can be obtained up to 1.8 GeV via coherent *Bremsstrahlung*. The detector covers $\sim$ 95% of 4π and is optimized for the measurement of final states with both charged and neutral particles. We present here the beam characteristics, the detector performances, as obtained in the commissioning runs of the experiment, and we discuss the main scientific goals and the initial experimental program.

2 The photon beam

The γ-ray beam for BGO-OD is obtained through *Bremsstrahlung* of the 3.2 GeV electrons extracted from the ELSA stretcher ring [1]. Amorphous copper radiators of different thicknesses are available and a diamond radiator can be inserted as well, allowing production of a linearly polarized photon beam. The energy of the photons is determined by the tagging technique: the electrons that have emitted *Bremsstrahlung* photons are momentum-analyzed by a tagger magnet and their position is measured by a tagger counter formed by 120 scintillators (160 ps time resolution, 0.5% - 2.0% energy resolution, and capable of handling up to 50 MHz event rate). The tagger is part of the event trigger. In order to have a more precise measurement of the photon energy (and thus of its polarization) under the coherent peak, the tagger is complemented with the *ARGUS* hodoscope formed by 480 scintillating fibers. The energy resolution of the *ARGUS* is $\sim$ 0.08%. In Fig. 1 we show the relative rate (diamond/copper) as measured by the tagger and by *ARGUS*.

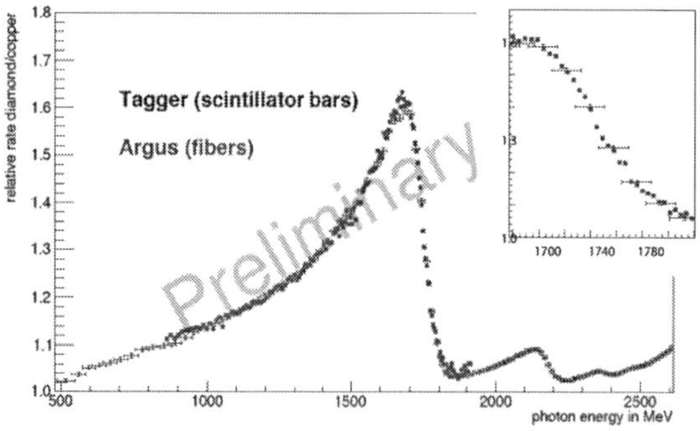

Figure 1. Relative count rate (diamond/copper) as measured by the tagger (scintillator bars, black points), and by the *ARGUS* hodoscope (scintillating fibers, red squares). In the insert, the detail of the coherent edge region.

3 The detector

The BGO-OD detector is formed by coupling the *Rugby Ball* calorimeter previously used at the GrAAL experiment with the *Open Dipole* magnetic spectrometer. The detector, shown in Fig. 2, can be divided into three parts according to the different polar angle coverage. The central part ($155° \leq \theta \leq 25°$) is covered by the *Rugby Ball* BGO calorimeter, formed by 480 crystals with PMT readout and equipped with sampling ADCs. The target (liquid H2 or D2 or solid) sits in the center of the calorimeter and is surrounded by ancillary detectors (cylindrical MWPC and scintillator *Barrel*) for particle tracking and identification. The polar angles between 8° and 25° are covered by a ring of scintillating counters (*SciRi*) and by MRPC chambers (*Daisy*) with excellent time resolution. Finally, the forward region is covered by the *Open Dipole* forward spectrometer. Its central part is a 0.45 T large aperture dipole magnet (provided by DESY on permanent loan basis). Tracking is performed by position detectors in front of the magnet (*MoMo* and *Scifi2*), both consisting of scintillating fibers, and 8 double layer Drift Chambers followed by 3 ToF scintillating walls behind it.

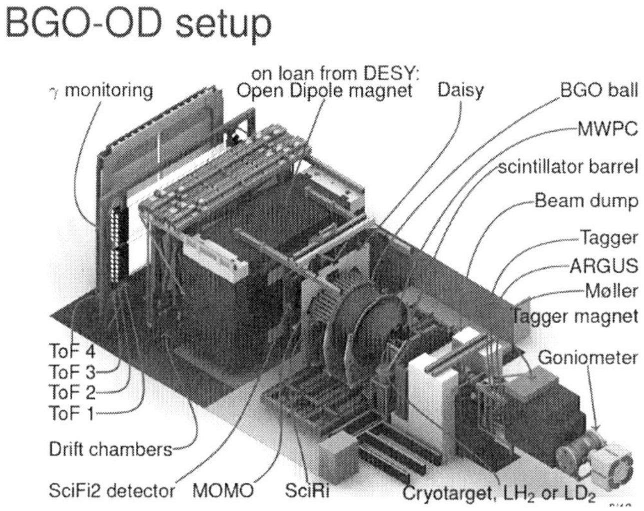

Figure 2. Schematic view of the BGO-OD experimental setup. Electron beam enters from the bottom-right corner.

The performance of the BGO *Rugby Ball* is summarized in Fig. 3 where we show the two-photon invariant mass spectrum from a proton target obtained with kinematical fit and confidence cut. The π^0, η, ω and η' peaks are all clearly visible. The energy calibration is obtained using a Na-22 source with peak at 1.275 MeV.

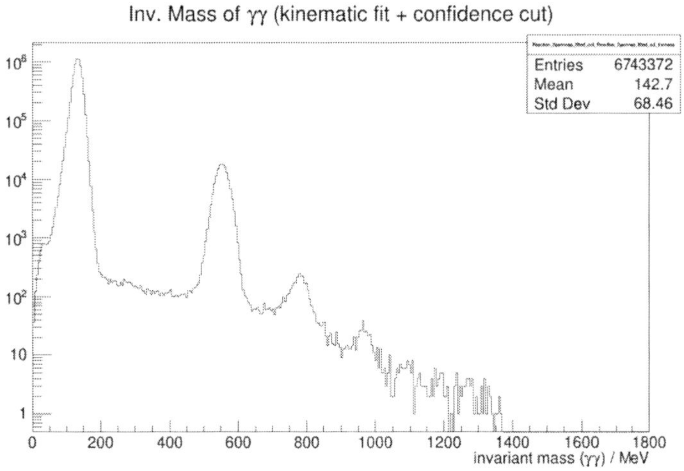

Figure 3. The two-photons invariant mass from a proton target measured by the *Rugby Ball* calorimeter.

The *Rugby Ball* can also measure protons[2] and neutrons[3], and by using sampling ADCs with ~2ns time resolution it can also measure K^+ following a method originally developed in Ref.[4]. Positive kaons are stopped within the crystals of the *Rugby Ball* up to a kinetic energy of approximately

400 MeV. The time resolution allows to distinguish the signal of the stopped K^+ from the signal of the subsequent weak decay $K^+ \rightarrow \mu^+\nu$. (see Fig. 4).

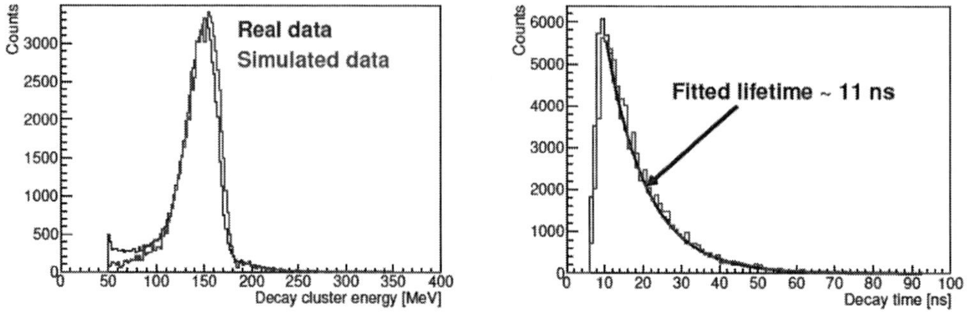

Figure 4. Left: energy of the delayed signal from the decay $K^+ \rightarrow \mu^+\nu$. The peak at 153 MeV corresponds to the μ^+ energy deposition from a K^+ decay at rest. Right: reconstructed K^+ lifetime.

In the forward region, particle identification is performed by looking at the bidimensional plot of particle velocity versus particle momentum (Fig. 5). By selecting protons or charged kaons in this plot we can obtain the two missing mass plots shown in Fig. 6.

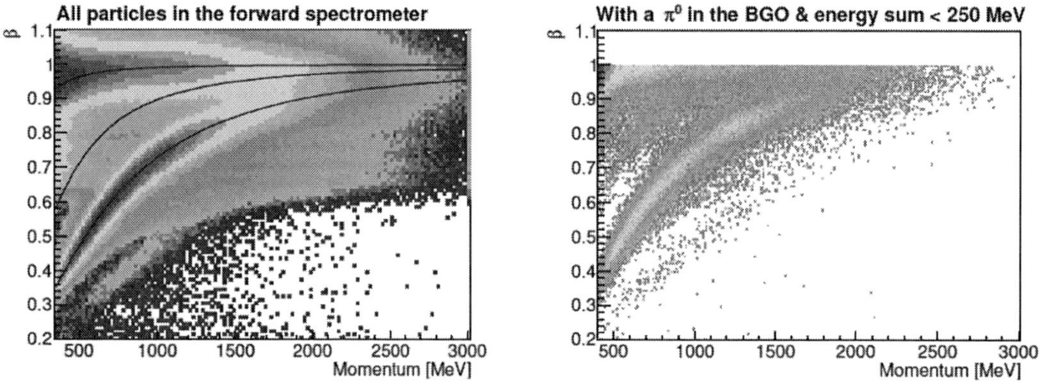

Figure 5. Left: β vs. momentum for all particles detected in the *Open Dipole* forward spectrometer. Right: same plot in coincidence with a π^0 in the calorimeter and with energy sum in the calorimeter less than 250 MeV. The K^+ line is accentuated and is clearly visible.

4 Experimental program

The setup is ideal for the investigation of reactions dominated by t-channel mechanisms due to the acceptance at forward angles with high momentum resolution. The unique combination of excellent calorimetry coupled with high resolution forward tracking is well suited to investigate reaction channels with open strangeness and complicated final states. Strangeness photoproduction off the proton (with final states $K^+\Lambda$, $K^+\Sigma^0$, $K^0\Sigma^+$...) will be accessed, and neutral particle identification in the central region helps to determine final states of mixed charge. The cusp-like structure observed by the

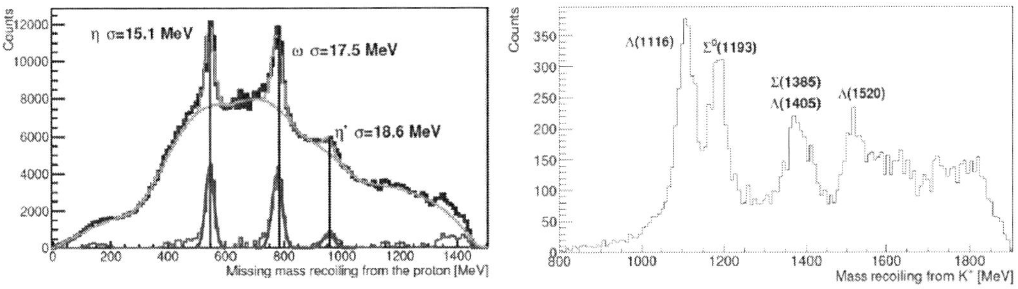

Figure 6. Missing mass spectrum in the *Open Dipole* forward spectrometer. Left: from the proton. η, ω and also η' peaks are clearly visible. Right: from the K^+. Λ and Σ^0 peaks are very clear and also other hyperons peaks are present.

CBELSA/TAPS collaboration[5] will be investigated by measuring the differential cross section in $\gamma p \rightarrow K^0\Sigma^+$ and the beam asymmetry over the K^* threshold region, achieving high statistics by including data of both neutral and charged decays of the K^0 and Σ^+. The acceptance of the *Open Dipole* forward spectrometer is also ideal to measure data at centre-of-mass angles $\theta_{cm} \leq 15°$ for K^+Y, where data are scarce and incompatible. A central goal is to disentangle the structure of hyperon excitations, e.g. the $\Lambda(1405)$ or hypothesised Σ^* states of $J^P = 1/2^-$ in the same mass range.
Pseudoscalar and vector meson photoproduction will be studied too, both off the proton and off the neutron, measuring cross sections and beam asymmetries. Thanks to the high energy resolution for the incoming photon energy provided by the *ARGUS* hodoscope, the beam asymmetry in η' photoproduction will be measured at threshold, allowing for a more detailed study of the strong energy dependence observed by the GrAAL collaboration[6], and at higher energy (up to 1800 MeV) where no experimental data are available yet. The high efficiency of the *Rugby Ball* for neutron detection will allow the measurements of η and ω off the quasi-free neutron with liquid D2 target. Finally, the Φ photoproducion will be investigated as well.

5 Conclusion

The BGO-OD experiment is commissioned; the performances of the beam and of the apparatus are as expected, production runs are underway and data analysis is ongoing for many reaction channels. The experimental program includes meson photoproduction from proton, deuteron and light nuclei.

References

[1] W. Hillert, Eur. Phys. J. A **28**, 139 (2006)
[2] A. Zucchiatti *et al.*, Nucl. Instr. and Meth. A **321**, 219 (1992)
[3] O. Bartalini *et al.*, Nucl. Instr. and Meth. A **562**, 85 (2006)
[4] T.C. Jude, D.I. Glazier and D.P. Watts, Phys. Lett. B **735**, 112 (2014)
[5] R. Ewald *et al.*, Phys. Lett. B **713**, 180 (2014)
[6] P. Levi Sandri *et al.*, Eur. Phys. J. A **51**, 77 (2015)

 DOI: 10.1051/epjconf/201613007014

New results of the antiproton-carbon annihilation cross section measurement at low energies

H. Aghai–Khozani[1], M. Corradini[2,3], R. Hayano[4], M. Hori[1,4], M. Leali[2,3], E. Lodi–Rizzini[2,3], V. Mascagna[2,3,★], Y. Murakami[4], M. Prest[5,6], L. Solazzi[7,3], E. Vallazza[8], L. Venturelli[2,3], and H. Yamada[4]

[1]*Max-Planck-Institut für Quantenoptik, D-85748 Garching, Germany*
[2]*Dipartimento di Ingegneria dell'Informazione, Università degli Studi di Brescia, I-25123 Brescia, Italy*
[3]*Istituto Nazionale di Fisica Nucleare - Pavia, I-27100 Pavia, Italy*
[4]*Department of Physics, University of Tokyo, Bunkyo-ku, Tokyo 113-0033, Japan*
[5]*Dipartimento di Scienza e Alta Tecnologia, Università degli Studi dell'Insubria, I-21100 Varese, Italy*
[6]*Istituto Nazionale di Fisica Nucleare - Milano Bicocca, I-20126 Milano, Italy*
[7]*Dipartimento di Ingegneria Meccanica e Industriale, Università degli Studi di Brescia, I-25123 Brescia, Italy*
[8]*Istituto Nazionale di Fisica Nucleare - Trieste, I-34127 Trieste, Italy*

Abstract. The antinucleon-nuclei annihilation cross section at very low energy has been measured at the LEAR facility (CERN) during the 80's and 90's and recently at the Antiproton Decelerator. The results have raised some interesting questions that still need answers. The ASACUSA collaboration performed an experiment to measure the σ_{ann} of 100 MeV/c antiprotons on carbon target and the preliminary results are presented here.

1 Introduction

The interest for the very low energy $\bar{p}$- and $\bar{n}$-nuclei interaction is due to their influence on both fundamental cosmology and nuclear physics. The study of the annihilation cross section on matter can give some contribution to the universe matter-antimatter asymmetry puzzle, as well as helping in the definition of strong interaction model parameters.

So far, the collected data for the antiproton and antineutron cross section show some discrepancies: at momenta of 300-400 MeV/c they both result [1] larger than predicted by the optical potential model which well reproduces the existing data at higher energies [2]. Moreover, the $\bar{n}$ annihilation cross section on several nuclei measured in the 70-400 MeV/c momentum range [1] seems to follow a 1/p^2 trend rather than a 1/p that is the expected one for a neutral projectile.

A comparison between the $\bar{p}$ and $\bar{n}$ data can represent a step forward in the clarification of the unsolved issues. Antiproton annihilation cross section data on proton are available for momenta below 80 MeV/c and on Sn for 100 MeV/c [3]. The ASACUSA Collaboration recently performed the measurement of the antiproton annihilation cross section in the ~100 MeV/c momentum region and the first preliminary results will be shown here.

★e-mail: valerio.mascagna@unibs.it

2 Experimental apparatus and technique

The experimental setup is presented in Fig. 1. A vacuum chamber composed by two cylindrical sections (total length 300 cm, diameters 120 cm and 60 cm) is put into the existing ASACUSA beam line which is equipped with two GEM profile beam monitors for the beam alignment and focusing. A system of 4 movable holders allows the positioning of 13 mm diameter circular frames (with or without targets) on the beam axis in different positions [4]. The targets consist in very thin self-supporting diamond-like carbon layers by Micromatter[1] with a thickness of 700 nm and 1000 nm.

Outside the chamber, in a position corresponding to the target, an arrangement of planar detectors is installed to detect the passage of the charged particles from $\bar{p}$ annihilations. They are made of plastic scintillator bars readout by multi-anode PMTs and a custom designed electronics system [6]. A Cherenkov counter is placed at the end of the beam line where all the $\bar{p}$s annihilate.

Bunches of $\sim 5\times 10^6$ antiprotons, 50 ns long (rise time $\sim$15 ns) are provided by the AD at a kinetic energy of 5.3 MeV. The $\bar{p}$s travel in the vacuum chamber at a speed of 3 cm/ns and almost all of them arrive at the end of the beam line after $\sim$120 ns, while a small fraction, of the order of 1-10 every 10^6 incoming, can undergo in-flight annihilation in the target (the physics process under study). The $\bar{p}$s scattered on the lateral walls can represent a background for the measurement. Actually their annihilation is delayed of more than 20 ns with respect to the early annihilations in the target since the chamber radius is 60 cm. Therefore, a fiducial time interval to count the annihilation events is selected: it corresponds to the first 20 ns region of the annihilations.

The data acquisition system performs a sampling at 300 MHz of the scintillator channel signal providing a hit time profile for every crossing bunch (see Fig. 2a). The time resolution at the ns level of the acquisition chain is enough to distinguish the hits coming from the target region and the ones from the chamber walls (time separation of $\geq$20 ns).

The calibration of the Cherenkov detector which monitors the beam intensity is performed with dedicated runs: a second circular empty frame is put 15 cm downstream of the target in order to intercept a known angular fraction of the scattered $\bar{p}$s that can be calculated using the Rutherford cross section and compared to the acquired data in this configuration.

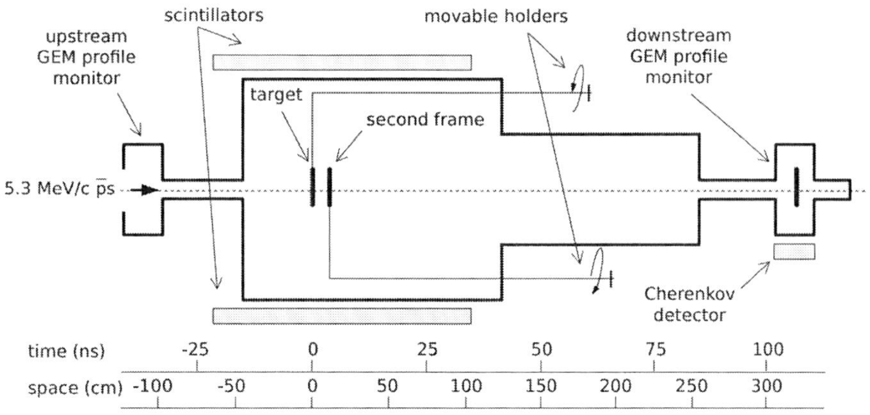

Figure 1: Scheme of the experimental setup. The time scale is based on $\bar{p}$ speed of 3 cm/ns.

[1] http://www.micromatter.com/dlc.php

3 Results

The plots in Fig. 2a show the time profiles, for different types of runs, detected while the $\bar{p}$ bunch crosses the target. The time separation between the annihilation in the target region (defined as the 520–540 ns window), from the lateral walls (after 20 ns or more) and from the end wall (after 100 ns) is possible. Our measurements lead to a preliminary evaluation of: $\sigma_{ann} = 2.32 \pm 0.35$ barn (statistical contribution to the error is 6%). Fig. 2b presents the obtained result superimposed to the existing $\bar{p}$ and $\bar{n}$ annihilation data on carbon [1, 7] and calculation from [2]. The total error can be reduced from ~15% to ~10% by implementing a better synchronization in the detector data.

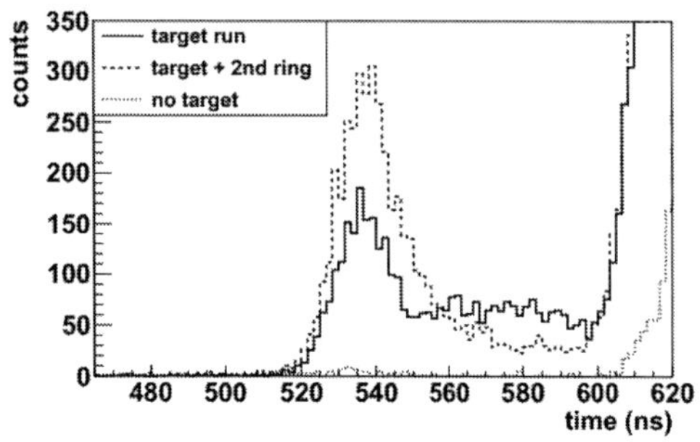

(a)

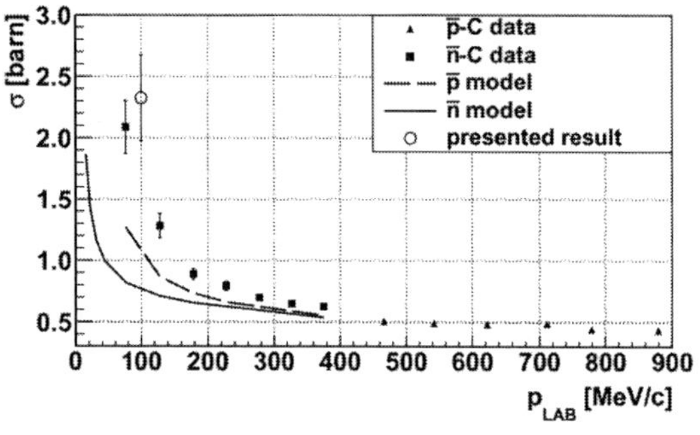

(b)

Figure 2: (a) Normalized hit profiles of one of the scintillator plane for different runs. (b) Result of the $\bar{p}$ annihilation cross section on carbon measurement superimposed to the existing data for p̄, n̄ and the corresponding theoretical predictions [2].

4 Conclusions

We reported the preliminary result for the $\bar{p}$-C annihilation cross section measurements at 5.3 MeV. The number is larger (1.5 times) than the $\bar{n}$-C one as expected by a Coulomb focusing mechanism; nevertheless it is twice the optical model prediction [2], a still unexplained issue (the same occurs in the $\bar{n}$ case). In the near future, when the ELENA facility will be operative at the AD, the same measurements will be possible at 100 keV as already demonstrated in [8].

References

[1] M. Astrua *et al.*, Nucl. Phys. A **697**, 209 (2002)
[2] E. Friedman *et al.*, Nucl. Phys. A **925**, 141 (2014)
[3] A. Bianconi *et al.*, Phys. Lett. B **704**, 461 (2011)
[4] M. Corradini *et al.*, Nucl. Instr. Meth. A **711**, 12 (2013)
[5] H. Aghai-Kohzani *et al.*, submitted to J. Phys. Soc. Jpn. Conf. Proceedings of LEAP 2016
[6] M. Corradini *et al.*, Hyp. Int. **233**, 53 (2015)
[7] K. Nakamura *et al.*, Phys. Rev. Lett. **52**, 731 (1984)
[8] H. Aghai-Kohzani *et al.*, Eur. Phys. J. Plus **127**, 55 (2012)

Studies of discrete symmetries in a purely leptonic system using the Jagiellonian Positron Emission Tomograph

P. Moskal[1,*], D. Alfs[1], T. Bednarski[1], P. Białas[1], C. Curceanu[2], E. Czerwiński[1], K. Dulski[1], A. Gajos[1], B. Głowacz[1], N. Gupta-Sharma[1], M. Gorgol[3], B. C. Hiesmayr[4], B. Jasińska[3], D. Kamińska[1], O. Khreptak[1], G. Korcyl[1], P. Kowalski[5], W. Krzemień[6], N. Krawczyk[1], E. Kubicz[1], M. Mohammed[1], Sz. Niedźwiecki[1], M. Pawlik-Niedźwiecka[1], L. Raczyński[5], Z. Rudy[1], M. Silarski[1], J. Smyrski[1], A. Wieczorek[1], W. Wiślicki[5], B. Zgardzińska[3], and M. Zieliński[1]

[1]*Faculty of Physics, Astronomy and Applied Computer Science, Jagiellonian University, Cracow, Poland*
[2]*INFN, Laboratori Nazionali di Frascati, 00044 Frascati, Italy*
[3]*Institute of Physics, Maria Curie-Skłodowska University, 20-031 Lublin, Poland*
[4]*Faculty of Physics, University of Vienna, 1090 Vienna, Austria*
[5]*Świerk Computing Centre, National Centre for Nuclear Research, 05-400 Otwock-Świerk, Poland*
[6]*High Energy Department, National Centre for Nuclear Research, 05-400 Otwock-Świerk, Poland*

Abstract. Discrete symmetries such as parity (P), charge-conjugation (C) and time reversal (T) are of fundamental importance in physics and cosmology. Breaking of charge conjugation symmetry (C) and its combination with parity (CP) constitute necessary conditions for the existence of the asymmetry between matter and antimatter in the observed Universe. The presently known sources of discrete symmetries violations can account for only a tiny fraction of the excess of matter over antimatter. So far CP and T symmetries violations were observed only for systems involving quarks and they were never reported for the purely leptonic objects. In this article we describe briefly an experimental proposal for the test of discrete symmetries in the decays of positronium atom which is made exclusively of leptons. The experiments are conducted by means of the Jagiellonian Positron Emission Tomograph (J-PET) which is constructed from strips of plastic scintillators enabling registration of photons from the positronium annihilation. J-PET tomograph together with the positronium target system enable to measure expectation values for the discrete symmetries odd operators constructed from (i) spin vector of the ortho-positronium atom, (ii) momentum vectors of photons originating from the decay of positronium, and (iii) linear polarization direction of annihilation photons. Linearly polarized positronium will be produced in the highly porous aerogel or polymer targets, exploiting longitudinally polarized positrons emitted by the sodium ^{22}Na isotope. Information about the polarization vector of ortho-positronium will be available on the event by event basis and will be reconstructed from the known position of the positron source and the reconstructed position of the ortho-positronium annihilation. In 2016 the first tests and calibration runs are planned, and the data collection with high statistics will commence in the year 2017.

*e-mail: p.moskal@uj.edu.pl

DOI: 10.1051/epjconf/201613007015

1 Introduction

The violation of the combined symmetries of charge conjugation and spatial parity (CP) was discovered 52 years ago in the decays of neutral K mesons [1]. CP symmetry is connected with the time reversal symmetry (T) via the CPT theorem [2, 3]. Time reversal concept is exceptionally attractive epistemologically [4] and its description is challenging [5], but nevertheless it took almost half a century till it was seen in a way independent of CP and CPT symmetries [6]. The T symmetry violation was observed in decays of the quantum entangled neutral B mesons [6, 7] using a method proposed in references [8, 9]. At present mesons, which are systems of quark and antiquark bound via strong interaction, remain the only objects in which violations of T and CP symmetries were observed. Therefore, it appears natural to ask a question about the degree of the violation of these symmetries in systems made of leptons. An ideal object for this kind of studies is the positronium atom: a system of lepton and antilepton bound via electromagnetic interaction. A test of the discrete symmetries in the decays of positronium into photons is limited by the photon-photon interaction which is expected to mimic the symmetry violation at the level of about 10^{-9} only [10, 11]. This is still about six orders of magnitude smaller than the present experimental limits equal to about 3×10^{-3} [12, 13].

In this article we briefly describe a new proposal for tests of the T, CP as well as CPT symmetries in the decays of the positronium. The experiments are conducted using the Jagiellonian Positron Emission Tomograph (J-PET). For the detailed description of the motivation to perform such experiments the interested reader is referred to the recent article [15].

2 Experimental method

J-PET tomograph consists of three cylindrical layers of plastic scintillator strips. The diameter of the inner layer is equal to 85 cm and the length of the scintillators amounts to 50 cm. Results of the first tests of single and double module prototypes are described in references [16–19]. For further details about the data acquisition system and analysis, simulation of the detector response, as well as hit-time and hit-position reconstruction methods the interested reader is referred to references [20–26]. Finally, the details about detector geometry along with its properties relevant for the registration of photons from the decays of ortho-positronium atoms are described in the recent article [27].

Ortho-positronium atoms will be produced by the positrons thermalised in the porous aerogels or polymers [28]. Positrons will be emitted by the ^{22}Na isotope positioned in the center of the cylinder covered with the porous material. Due to the parity violation by the weak interaction, positrons emitted in the decay of sodium: $^{22}Na \rightarrow ^{22}Ne^{*}\, e^{+}\, \nu_{e} \rightarrow ^{22}Ne\, \gamma\, e^{+}\, \nu_{e}$, are longitudinally polarized with the average polarization equal to about 67% [13, 14]. This value decreases only by about 8% during the thermalisation process [29, 30] and enables to produce ortho-positronium atoms polarized along the direction of the positron emission with the polarization degree of about 41% when using sodium ^{22}Na [13, 31]. The polarization axis varies from event to event, but it can be determined from the known points of the positron emission and positronium annihilation. The annihilation place will be reconstructed based on the trilateration method [32]. The positronium target will be inserted inside the J-PET detector. Photons from the ortho-positronium decay and from the deexcitation of ^{22}Ne are detected predominantly via the Compton interaction in the scintillator strips. A typical event useful for the studies of the discrete symmetries will include four or more signals: three signals from photons originating from the decay of ortho-positronium atom, one signal from the photon emitted by the excited ^{22}Ne isotope, and possibly signals from the secondary scattering in the detectors. The measurement of energy loss of photons, their time of interaction, as well as their relative angles will allow to disentangle signals from annihilation, deexcitation and secondary photons. The measurement of the time difference between the registration of annihilation and deexcitation photons will enable to disentangle the annihilation of ortho-positronium from the direct electron-positron annihilation and from the annihilation of para-positronium.

In the case of the Compton scattering there is no unique relation between the energy loss and the energy of the interacting photon. Nevertheless, in the case of the registration of three photons from the ortho-positronium decay, their energies may be reconstructed based on the measurement of relative angles between the photons and the application of the energy and momentum conservation [27]. The angular resolution of the J-PET is equal to about 0.4 degree. This results in the energy resolution of about 5 keV for the registration of photons originating from the decay of ortho-positronium [27]. Based on the tests of prototype detector with the 30 cm long scintillators [17, 18] and simulations [19], the expected hit-time resolution of the J-PET detector is in the range from about 100 ps(σ) to about 250 ps(σ) for deposited energy range from 340 keV to 50 keV, respectively. High time and angular resolution enables to suppress by nine orders of magnitude the background due to the electron-positron annihilation into two photons followed by the secondary scatterings in the scintillator strips [27]. The secondary photon scattering in the detector can be turned into advantage since it allows to determine a linear polarization ($\vec{\epsilon}$) of the primary photon at the moment of the scattering [15].

The above brief description shows that the J-PET tomograph, equipped with the dedicated positronium target, enables to measure photons momenta, their polarizations and the spin of ortho-positronium. This opens possibilities to test discrete symmetries in the decays of ortho-positronium via measurement of the expectation values of the various symmetry odd operators which may be constructed from the spin polarization vector of ortho-positronium ($\vec{S}$) as well as from momentum vectors ($\vec{k}$) and polarization directions ($\vec{\epsilon}$) of photons originating from the decay of ortho-positronium. Exemplary operators are shown in Tab. 1.

Table 1. Operators for the $o\text{-}Ps \to 3\gamma$ process and their properties with respect to the C, P, T, CP and CPT symmetries. New operators (including $\vec{\epsilon}$) not explored before are shown in the last three rows. The decay plane is oriented by the ordering of photons according to the descending momentum: $k_1 > k_2 > k_3$. The table is adapted from reference [15].

Operator	C	P	T	CP	CPT
$\vec{S} \cdot \vec{k_1}$	+	−	+	−	−
$\vec{S} \cdot (\vec{k_1} \times \vec{k_2})$	+	+	−	+	−
$(\vec{S} \cdot \vec{k_1})(\vec{S} \cdot (\vec{k_1} \times \vec{k_2}))$	+	−	−	−	+
$\vec{k_2} \times \vec{\epsilon_1}$	+	−	−	−	+
$\vec{S} \cdot \vec{\epsilon_1}$	+	+	−	+	−
$\vec{S} \cdot (\vec{k_2} \times \vec{\epsilon_1})$	+	−	+	−	−

3 Summary

Jagiellonian Positron Emission Tomograph is a multi-purpose detector which enables broad interdisciplinary investigations including, among others: (i) medical imaging [19], (ii) studies of discrete symmetries in the decays of positronium atoms [15, 27], (iii) quantum entanglement of high energy photons originating from the decay of ortho-positronium, as well as (iv) research in the field of life- and material-sciences [33–35]. In this article properties of the J-PET detector relevant for the studies of the discrete symmetries were briefly described.

References

[1] J. H. Christenson, J. W. Cronin, V. L. Fitch, R. Turlay, Phys. Rev. Lett. **13**, 138 (1964)
[2] J. S. Bell, Proc. R. Soc. London Ser. A **231**, 479 (1955)
[3] G. Lüders, Ann. Phys. **2**, 1 (1957); G. Lüders, Ann. Phys. (N.Y.) **281**, 1004 (2000)
[4] J. Gołosz, J. Gen. Philos. Sci. (2016)
[5] T. Durt, A. Di Domenico, B. Hiesmayr, arXiv:1512.0843 (2015)
[6] J. P. Lees *et al.*, Phys. Rev. Lett. **109**, 211801 (2012)
[7] J. P. Lees *et al.*, Phys. Rev. D **94**, 011101(R) (2016)
[8] J. Bernabeu *et al.*, JHEP **1208**, 64 (2012)
[9] J. Bernabeu, A. Di Domenico, P. Villanueva-Perez, Nucl. Phys. B **868**, 102 (2013)
[10] W. Bernreuther *et al.*, Z. Phys. C **41**, 143 (1988)
[11] B. K. Arbic *et al.*, Phys. Rev. A **37**, 3189 (1988)
[12] T. Yamazaki *et al.*, Phys. Rev. Lett. **104**, 083401 (2010)
[13] P. A. Vetter, S. J. Freedman, Phys. Rev. Lett. **91**, 263401 (2003)
[14] P. Coleman, Positron Beams and Their Applications, World Scientific (2000)
[15] P. Moskal *et al.*, Acta Phys. Polon. B **47**, 509 (2016)
[16] P. Moskal *et al.*, Bio-Algorithms and Med-Systems **7**, 73 (2011)
[17] P. Moskal *et al.*, Nucl. Instrum. and Meth. A **764**, 317 (2014)
[18] P. Moskal *et al.*, Nucl. Instrum. and Meth. A **775**, 54 (2015)
[19] P. Moskal *et al.*, Phys. Med. Biol. **61**, 2025 (2016)
[20] L. Raczyński *et al.*, Nucl. Instrum. and Meth. A **764**, 186 (2014)
[21] L. Raczyński *et al.*, Nucl. Instrum. and Meth. A **786**, 105 (2015)
[22] N. G. Sharma *et al.*, Nukleonika **60**, 765 (2015)
[23] P. Moskal *et al.*, Acta Phys. Polon. A **127**, 1495 (2015)
[24] P. Kowalski *et al.*, Acta Phys. Polon. B **47**, 549 (2016)
[25] G. Korcyl *et al.*, Acta Phys. Polon. B **47**, 491 (2016)
[26] W. Krzemień *et al.*, Acta Phys. Polon. B **47**, 561 (2016)
[27] D. Kamińska *et al.*, Eur. Phys. J. C **76**, 445 (2016)
[28] B. Jasińska *et al.*, Acta Phys. Polon. B **47**, 453 (2016)
[29] J. Van House, P. W. Zitzewitz, Phys. Rev. A **29**, 96 (1984)
[30] P. W. Zitzewitz *et al.*, Phys. Rev. Lett. **43**, 1281 (1979)
[31] J. Yang *et al.*, Jpn. J. Appl. Phys. **36**, 3764 (1997)
[32] A. Gajos *et al.*, Nucl. Instrum. and Meth. A **819**, 54 (2016)
[33] E. Kubicz *et al.*, Nukleonika **60**, 749 (2015)
[34] A. Wieczorek *et al.*, Acta Phys. Polon. A **127**, 1487 (2015)
[35] A. Wieczorek *et al.*, Nukleonika **60**, 777 (2015)

EPJ Web of Conferences **130**, 07016 (2016) DOI: 10.1051/epjconf/201613007016
MESON 2016

Partial wave analysis of $\pi\pi$ scattering below 2 GeV

V.Nazari[1,⋆], P. Bydžovský[2], and R. Kamiński[1]

[1] *Institute of Nuclear Physics, Polish Academy of Sciences, Kraków, Poland*
[2] *Nuclear Physics Institute, ASCR, Řež/Prague, Czech Republic*

Abstract. All resonances included in multi-channel D ($\pi\pi$, 4π, $K\bar{K}$ and $\eta\eta$) and F ($\pi\pi$, 4π, $\omega\pi$ and $K\bar{K}$) waves are analyzed. Results of the analysis are based on a set of well determined $\pi\pi$ scattering amplitudes for S - F waves in a unitary multi-channel approach. The amplitudes are refined and re-fitted to the dispersion relations up to 1.1 GeV, and to the experimental data in the effective two pion mass from the threshold to 2.7 GeV and 1.9 GeV for D and F waves, respectively.

Many existing parameterizations of the $\pi\pi$ scattering amplitudes still significantly do not satisfy the crossing symmetry condition or do not describe the $\pi\pi$ threshold region. Since they are used in some analyses, having a correct determination of $\pi\pi$ scattering amplitudes is essential to achieve a proper result.

In our previous work [1] we presented a general method of refining the $\pi\pi$ amplitudes by fitting them to the GKPY equations [2, 3] which are once-subtracted dispersion relations with imposed crossing symmetry condition. The method was demonstrated on refining of the unitary multi-channel S ($\pi\pi$, $K\bar{K}$ and $\eta\eta$) and P ($\pi\pi$, $\rho 2\pi$, and $\rho\sigma$) wave amplitudes from [4] mostly below 1.1 GeV with no change in their original mathematical structure. In Ref. [5] we included the multi-channel D ($\pi\pi$, effective $(2\pi)(2\pi)$, $K\bar{K}$ and $\eta\eta$) and F ($\pi\pi$, effective $(2\pi)(2\pi)$, $\omega\pi$ and $K\bar{K}$) wave amplitudes from [6]. The S-F amplitudes were refined and then re-fitted to the available experimental data and to the GKPY equations [2, 3]. In Ref. [5] we presented some part of the results of analysis for the D and F wave amplitudes especially, details on formalism of the analysis and precise determination of $\pi\pi$ scattering amplitudes for D and F waves. In this paper we discuss on the resonances included in the analysis and show their importance separately in the corresponding partial waves.

The method of analysis is generally the same as in Ref. [1]. To construct the $D0$ and $F1$ waves, we have used the same formalism as in Refs. [4, 6] and updated the list of contributing resonant states for the $D0$ wave according to the latest issue of PDG [7]. Then, a correct behavior of the phase shift near the threshold was controlled by the pseudo data (see Secs.II in [5]). The initial amplitudes for the $S0$ and $P1$ waves were from [1], both with added polynomial near the threshold (the "extended" amplitudes in [1]). Since there is no any resonance for $S2$ and $D2$ waves, the amplitudes for these isotensor waves were taken from [2] and kept unchanged during the analysis.

In the modification of the original amplitudes only those parameters that were expected to contribute significantly to the dispersive integrals, i.e. the low-energy resonances and the $\pi\pi$ background were allowed to change. The results provided precise determination of $D0$ and $F1$ partial wave amplitudes which satisfy the crossing symmetry condition and describe the experimental data very well.

⋆e-mail: vnazari@ifj.edu.pl

In the construction of the partial waves, apart from background part which is relatively small, the resonance contribution described by the multi-channel Breit-Wigner form determines the main structure of the amplitudes. Concerning the dominant, important and insignificant resonances of the *D*0 wave in the multi-channel analysis approach, various fits have been performed. Results show that not all first four lightest states, i.e. $f_2(1270)$, $f_2(1430)$, $f_2(1525)$, and $f_2(1640)$, are the most important states in the analysis, as it was expected at the beginning. For this, Table 1 demonstrates the significance of each resonance in the analysis separately. It shows the χ^2 for the fits when one particular resonance was omitted. The χ^2 was only composed of the χ^2_{Data} for *D*0 wave in the Eq. (16) in [8] and χ^2_{DR} for all partial waves in the Eq. (17) in [8]. Nevertheless, conditions are the same for all fits and they just differ by omitted resonances. Values of the χ^2 achieved from fits 9, 11 and 12 are almost equal to the value in the fit 1 where all resonance states are included. Accordingly, from the point of view of the χ^2 one can omit the $f_2(2010)$, $f_2(2300)$ and $f_2(2340)$ resonances and conclude that these resonances are insignificant in the description. Value of the χ^2 in the fits 3, 5, 7 and 8 show that the $f_2(1430)$, $f_2(1640)$, $f_2(1910)$ and $f_2(1950)$ don't play a very important role on the results but one should keep them. Additionally, the $f_2(1525)$, $f_2(1810)$, and $f_2(2150)$ resonances which strongly influence behavior of the amplitude and therefore value of the χ^2 are the most important resonances in the *D*0 wave and obviously the dominant resonance which dominate the behaviour of the amplitude is $f_2(1270)$ resonance.

From two resonance states $\rho_3(1690)$ and $\rho_3(1990)$ listed in PDG [7] for the isovector *F*1 wave only the $\rho_3(1690)$ resonance state has been considered in the analysis. For the $\rho_3(1990)$ resonance state, the partial widths are not known. In Refs. [5, 8] we have found that enough is only one resonance state $\rho_3(1690)$ which can satisfactorily describe the available experimental data in [9, 10] and give a reasonable description in the threshold region.

Table 1. Values of χ^2 after fitting when some specific resonance states are omitted in the *D*0 wave.

Fit No.		$\chi^2/n.d.f.$
1	All stated are included	326.21/330=0.9885
	Omitted Resonances	
2	$f_2(1270)$	5838.8/330=17.69
3	$f_2(1430)$	327.39/330=0.9921
4	$f_2(1525)$	336.97/330=1.0211
5	$f_2(1640)$	327.76/330=0.9932
6	$f_2(1810)$	348.07/330=1.0548
7	$f_2(1910)$	327.36/330=0.9920
8	$f_2(1950)$	327.26/330=0.9917
9	$f_2(2010)$	326.68/330=0.9899
10	$f_2(2150)$	368.81/330=1.1176
11	$f_2(2300)$	326.54/330=0.9895
12	$f_2(2340)$	326.64/330=0.9898

A figurative comparison of the phase shift of the resonances in $\pi\pi \to \pi\pi$ channel is presented in the Figs. 1-3 for *D*0 and *F*1 waves individually. The *D*0 wave consists of 11 resonances ($f_2(1270)$, $f_2(1430)$, $f_2(1525)$, $f_2(1640)$, $f_2(1810)$, $f_2(1910)$, $f_2(1950)$, $f_2(2010)$, $f_2(2150)$, $f_2(2300)$, $f_2(2340)$) from which the $f_2(1270)$ is the most dominant one. This can be seen in the Fig. 1 where the $f_2(1270)$ determines structure of the full phase shift. Effect of other phase shifts are comparably small and influence of the background part on the full phase shift notably appears above ~1600 MeV which describe the data very well. Removing the full phase shift of the *D*0 wave (Full phase), $f_2(1270)$, background (Bkg.) and experimental data in the Fig. 1 one can see the differences between other phase shifts in the Fig. 2. Influence of these phases on the full phase shift is less than 1 degree. However, one can conclude that effect of the $f_2(1640)$, $f_2(1910)$, $f_2(1950)$, $f_2(2010)$ and $f_2(2150)$ phase shifts is almost negligible. Note that the scarce data above ~1600 MeV indicate almost no resonant structure.

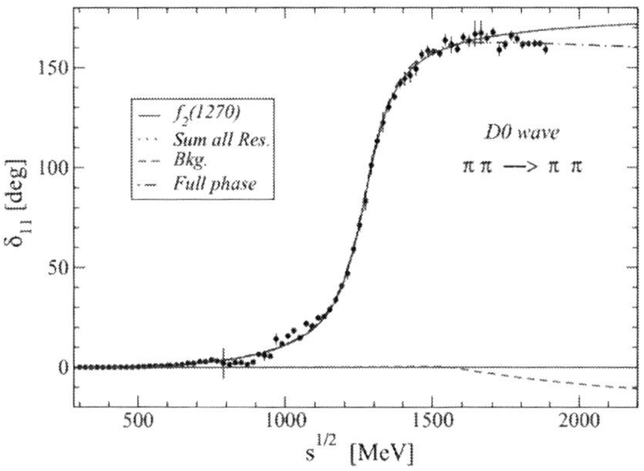

Figure 1. Illustrative comparison of the phase shifts of the resonances in the $D0$ wave. The $f_2(1270)$ resonance is overlapped by sum of all resonances (Sum of all Res.) when background part (Bkg.) is omitted. Except $f_2(1270)$ contributions of other resonances are relatively very small (less than 1 degree).

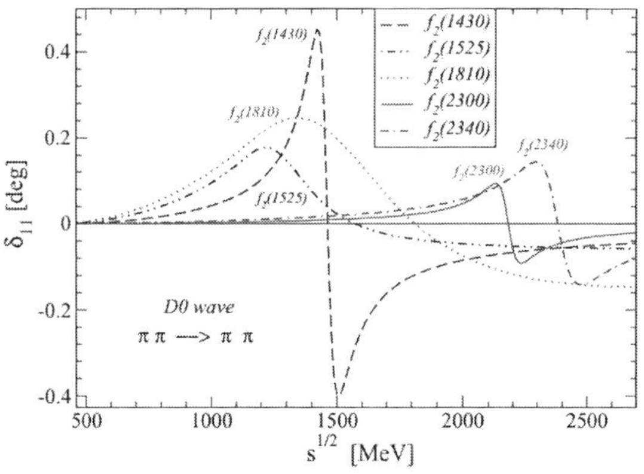

Figure 2. We show contributions to the elastic phase shift from other less important $D0$ wave resonances. Phase shifts of the $f_2(1640)$, $f_2(1910)$, $f_2(1950)$, $f_2(2010)$ and $f_2(2150)$ resonances are almost close to zero.

Finally for the $F1$ wave which has only one resonance ($\rho_3(1690)$), description of the phase shift is much simpler than $D0$ wave. Figure 3 shows the influence of the resonance and background parts on the full phase shift. Due to the scattered data below ~1700 MeV one can omit the background part, since only the $\rho_3(1690)$ phase shift is enough to describe the data. Effect of the background part is seen above ~1700 MeV to describe the data.

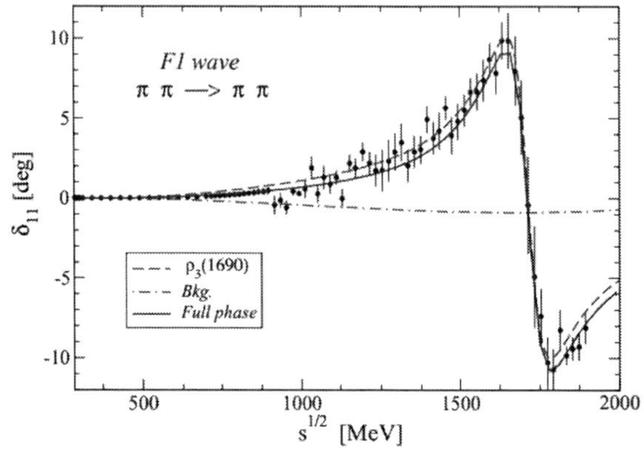

Figure 3. Illustrative comparison of the resonance and background (Bkg.) with the full phase shift of the $F1$ wave (Full phase) and experimental data [9, 10].

This work has been funded by the Polish National Science Centre (NCN) Grants No. DEC-2014/15/N/ST2/03504 and DEC-2013/09/B/ST2/04382 and partially supported by the Grant Agency of the Czech Republic under the grant No. P203/15/04301.

References

[1] P. Bydžovský, R. Kamiński, V. Nazari, Phys. Rev. D **90**, 116005 (2014)

[2] R. García-Martín and R. Kamiński and J. R. Peláez and J. Ruiz de Elvira and F. J. Yndurain, Phys. Rev. D **83**, 074004 (2011)

[3] R. Kamiński, Phys. Rev. D **83**, 076008 (2011)

[4] Yu. S. Surovtsev, P. Bydžovský, R. Kamiński, and M. Nagy, Phys. Rev. D **81**, 016001 (2010)

[5] V. Nazari, P. Bydžovský, R.Kamiński, Acta Phys. Pol. B Proc. Suppl. **9**, 609 (2016)

[6] Yu. S. Surovtsev et al., arXiv:1104.0538 [hep-ph]

[7] K. A. Olive et al (Part. Data Group) Chin. Phys. C **38**, 090001 (2014)

[8] V. Nazari, R.Kamiński, Acta Phys. Polon. B **46**, 7, 1355 (2015)

[9] B. Hyams et al., Nucl. Phys. B **64**, 134 (1973)

[10] S.J. Lindenbaum and R.S. Longacre, Phys. Lett. B **274**, 492 (1992); R.S. Longacre et al., Phys. Lett. B **177**, 223 (1986)

Dependence of mass in comparison to full width of mesons

Anatoliy Nedolya[1]*

[1]*Zaporizhzhya National University, 66 Zhukovsky str., 69600 Zaporizhzhya, Ukraine*

Abstract. The mass of mesons in comparison to their full width (lifetime) using the Review of Particle Physics (Particle Data Group) was phenomenologically investigated. The dependence showed that most of mesons were located inside the range, the upper edge of which varied periodically and looked like waves' superposition. Similar distribution was observed for baryons. Such uneven distribution of mesons and baryons explains discrete distribution of particles' mass and energy in the Universe. The results may be useful to clarify the range of mass for searching new particles.

1 Introduction

It is known that the lifetime of relativistic particles depends on their energy. For example, the lifetime of muons in cosmic rays increases by Lorentz factor $\gamma = (1-\beta^2)^{-1/2}$ times. Based on uncertainty principle, for the short-lived particles the lifetime is determined from the full width ($\Gamma = \hbar/\tau$) of energy distribution of the decaying particles. However, there is a problem of formal description of the unstable states using quantum mechanics and quantum field theory, because such description with the finite lifetime of unstable particle contradicts the notion of asymptotic state [1]. Currently the dependence of elementary particle's lifetime on its mass (rest energy) is not determined, although the particles' lifetime depends on the type of their fundamental interaction. Actually the particles that decay due to strong interaction normally do so in less than 10^{-20} s; the decay caused by electromagnetic interaction generally have lifetimes in the order of 10^{-16} s, the weak interaction decay takes longer than 10^{-10} s [2, 3].

2 Results

Description of unstable particles by methods of quantum field theory is associated with the value of Γ/M, which for stable and quasi-stable particles is very small, but it has essential value for most of hadrons. That is why in this study the full width (lifetime) of mesons in comparison to their masses was investigated using the last Particle Data Group [4].

Figure 1 shows that most of mesons were located inside range, upper edge of which varied periodically and the dependence was similar to a result of waves' superposition. The values of masses with a maximum of full widths relate to each other as 1, 2, 3, and possibly 6 with a period near to 2 GeV. Here were denoted: 1 - unflovered + strange maximum; 2 - $c\bar{c}$ maximum (charmonium); 3 -

*e-mail: avnedolya@hotmail.com

 DOI: 10.1051/epjconf/201613007017

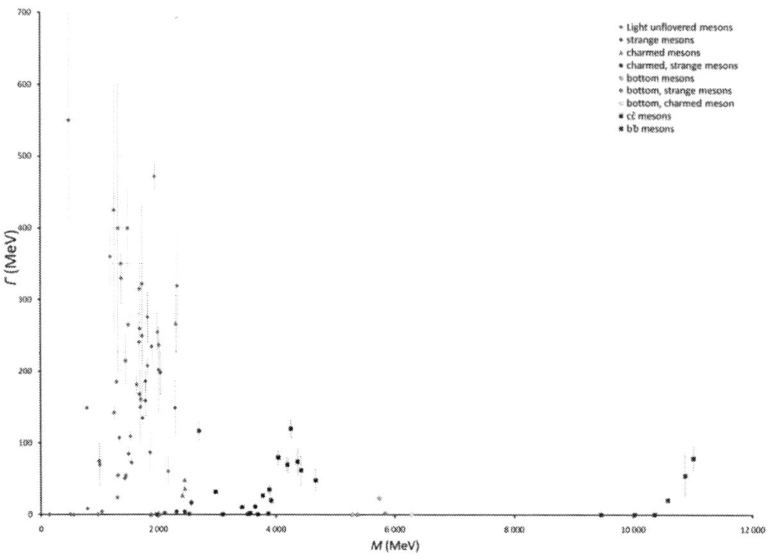

Figure 1: Masses of mesons in comparison to their full widths.

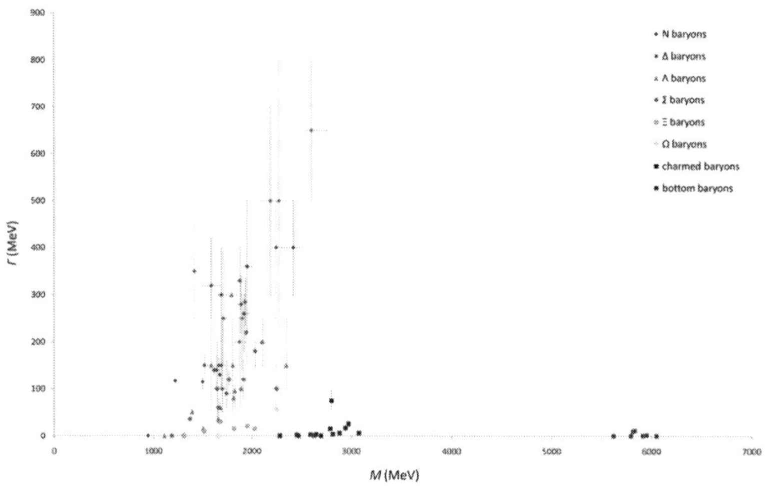

Figure 2: Masses of baryons in comparison to their full widths.

bottom maximum; 6 - $b\bar{b}$ possible maximum (bottonium). No mesons have been found in the mass range from 6300 to 9400 MeV. Only $f_0(500)$ unflavored meson was located outside that range and its error values were big for mass and full width. Their data should be revised. Similar distribution was observed for baryons (see Fig. 2).

Figure 2 includes the data of Breit-Wigner mass and Breit-Wigner full width for N and Δ baryons, but their description may be only approximate and may contain some contradiction since PDG lists in addition another value, which is called pole position [5]. But the overal $\Gamma(M)$ dependence for baryons is periodical and also resembles waves' superposition, where the mass and full width values

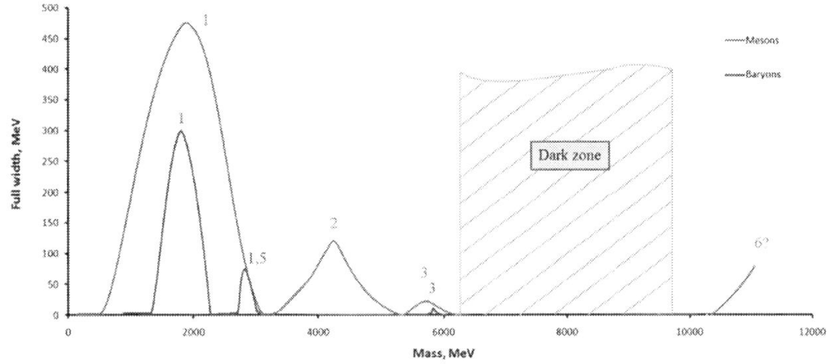

Figure 3: $\Gamma(M)$ schematic result of waves' superposition for hardrons.

are related to each other as 1, 1.5, 3 and where 1 is a baryons maximum; 1.5 - charmed baryons, 3 - bottom baryons maximum respectively.

Taking into account these data a tentative scheme of $\Gamma(M)$ dependence (see Fig. 3) was offered, which illustrated the general distribution of hadrons masses vs. their full widths. As a result the maximums are not always matched to flavor of the resonances and also may include the tetra- and pentaquark moleculas data [6, 7]. There is a wide range of unobserved values of mass (absorption band of spectrum), which is described as the Dark zone.

3 Summary

Thus it is possible that the hadrons are decaying with most probability at some values of mass, which are coinciding with the superposition's maximums of $\Gamma(M)$ dependence. It is a result of Higgs field interaction. It shows that most likely, mass and energy are distributed unevenly in the Universe. In addition, the $\Gamma(M)$ dependence can also explain different interaction radius of elementary particles which have a mass. Thus, conceptually the Higgs boson, which gives mass to elementary particles, may be more universal. The existence of a dark zone remains an open question and its can be regarded as absorption band of masses or range in which the particles may decay so quickly that modern equipment can not detected them. The results could be useful to clarify the mass range for searching new particles.

References

[1] V.I. Kuksa, Phys. Part. Nuclei **45 (3)**, 568 (2014)

[2] D. Griffiths, *Introduction to Elementary Particles* (Wiley-VCH GmbH & Co. KGaA, Weinheim, 2004) 72

[3] S.T. Thornton, A. Rex, *Modern Physics for Scientists and Engineers, Forth Edition* (Books/Cole, Boston, 2013) 530

[4] K.A. Olive *et al.* Chinese Physics C **38 (9)**, 090001 (2014)

[5] A.R. Bohm, Y. Sato, Phys. Rev. D **71**, (2005) 085018

[6] G. Cowan (2016) CERN see https://cds.cern.ch/record/2156233?ln=en

[7] H.X. Chen *et al.* Physics Report **639**, 1 (2016)

D-meson observables in heavy-ion collisions at LHC with EPOSHQ model

Vitalii Ozvenchuk[1,⋆], Joerg Aichelin[2], Pol-Bernard Gossiaux[2], Benjamin Guiot[3], Marlene Nahrgang[2], and Klaus Werner[2]

[1] *Institute of Nuclear Physics, Polish Academy of Science, Krakow, Poland*
[2] *Subatech, UMR 6457, IN2P3/CNRS, Universite de Nantes, Ecole des Mines de Nantes, Nantes, France*
[3] *Universidad Tecnica Federico Santa Maria, Valparaiso, Chile*

Abstract. We study the propagation of charm quarks in the quark-gluon plasma (QGP) created in ultrarelativistic heavy-ion collisions at LHC within EPOSHQ model. The interactions of heavy quarks with the light partons in ultrarelativistic heavy-ion collisions through the collisional and radiative processes lead to a large suppression of final D-meson spectra at high transverse momentum and a finite D-meson elliptic flow. Our results are in a good agreement with the available experimental data.

1 Introduction

At LHC the QGP is produced in ultrarelativistic heavy-ion collisions and the nuclear modification factor, R_{AA}, and the elliptic flow, $v2$, of D mesons are measured [1, 2]. The R_{AA} of intermediate- and high-p_T D mesons is significantly below unity. It indicates a large energy loss of high-p_T heavy quarks in the QGP. Many theoretical models describing the D-meson observables in heavy-ion collisions [3–6]. It remains, however, a challenge to describe R_{AA} and v_2 within the same model input.

In this contribution we present the nuclear modification factor and the elliptic flow of D mesons in heavy-ion collisions at LHC calculated using the different iterations of the EPOSHQ model, in particular, MC@sHQ+EPOS2, MC@sHQ+EPOS3 and EPOSHQ models.

2 The model

MC@sHQ+EPOS2 model couples a Monte Carlo propagation of heavy quarks [7], MC@sHQ, to the 3+1 dimensional ideal fluid dynamical evolution of the QGP from EPOS2 initial conditions [8], which combine perturbative QCD (pQCD) calculations of the hard scattering with the Gribov-Regge theory. In MC@sHQ+EPOS3 model we couple MC@sHQ with the recent upgrade EPOS3 that includes a viscous fluid dynamical evolution, based on [9].

We initialize the heavy quarks at the original nucleon-nucleon scattering points according to the p_T-distribution from pQCD results in fixed order plus next-to-leading logarithm (FONLL) [10]. In EPOSHQ model we implement the heavy quarks directly from EPOS3 initial conditions with initial

⋆e-mail: Vitalii.Ozvenchuk@ifj.edu.pl

state shadowing included. The heavy quarks in EPOS3 can be produced during the spacelike cascade, the born process and the partonic shower. Heavy quarks interact with plasma partons by either elastic or radiative collisions. The evolution of heavy quarks is described by the Boltzmann equation. The heavy quarks form hadrons via coalescence or fragmentation on the hypersurface of constant temperature $T = 155$ MeV.

3 The nuclear modification factor and elliptic flow

We perform 10000 heavy-quark events per EPOS2 ideal or EPOS3 viscous fluid dynamical event. In Figs. 1 and 2 we show our results for D-meson nuclear modification factor and elliptic flow in heavy-ion collisions at LHC energies within MC@sHQ+EPOS3 model (red lines) as well as the comparison to the MC@sHQ+EPOS2 results (blue lines) taken from [11]. In each model we perform calculations

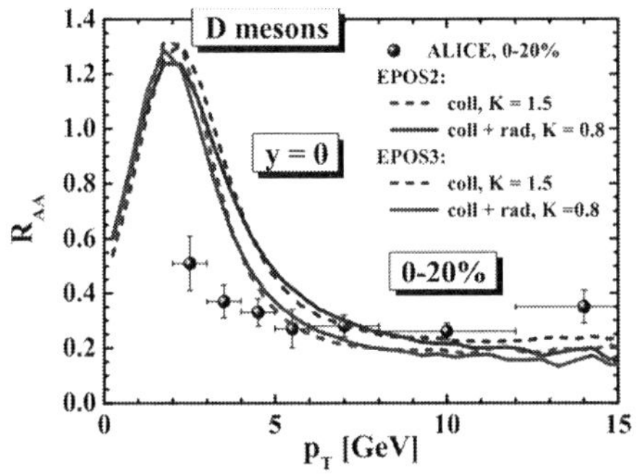

Figure 1. D-meson nuclear modification factor calculated in MC@sHQ+EPOS2 model (blue lines) in comparison to the MC@sHQ+EPOS3 results (red lines). Experimental data are taken from [1].

for two scenarios, with purely collisional energy loss (dashed lines) or with collisional+radiative energy loss (solid lines). Both models show a reasonable agreement with the experimental data for the R_{AA} of D mesons at $p_T > 5$ GeV. The larger suppression of D-meson yield is seen for the

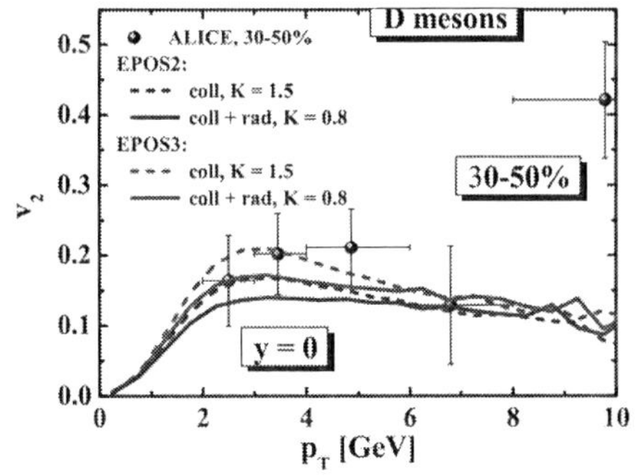

Figure 2. D-meson elliptic flow calculated in MC@sHQ+EPOS2 model (blue lines) in comparison to the MC@sHQ+EPOS3 results (red lines). Experimental data are taken from [2].

MC@sHQ+EPOS3 results at intermediate p_T. If the initial state shadowing is not included the results from both models are overestimated at low p_T. MC@sHQ+EPOS2 and MC@sHQ+EPOS3 results for the elliptic flow of D mesons are in a good agreement with the experimental data for the whole range of p_T. MC@sHQ+EPOS3 results show the enhancement of v_2 at intermediate p_T in comparison to MC@sHQ+EPOS2 results. The final elliptic flow of D mesons can be sensitive to the D-meson rescattering in the hadronic phase [12], we need to include the hadronic contribution for further study.

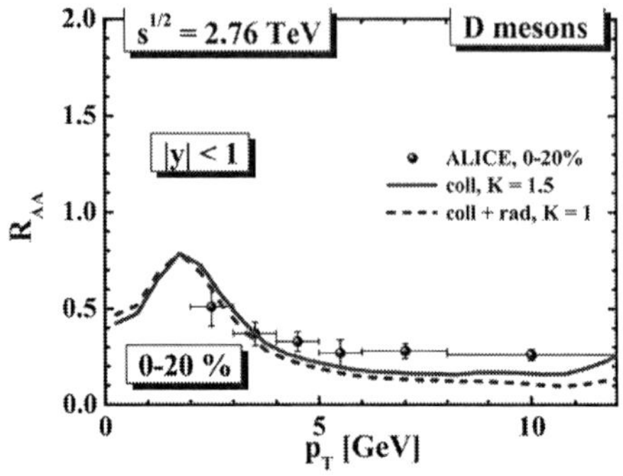

Figure 3. D-meson R_{AA} obtained within EPOSHQ model in comparison to the experimental data [1].

In Fig. 3 we present the D-meson nuclear modification factor calculated in EPOSHQ model for the central heavy-ion collisions at LHC. The calculations show a good agreement with the experimental data for whole range of p_T. For two scenarios (pure collisional and collisional+radiative) we obtained almost the same behavior. The presence of the initial state shadowing helped us to describe well the R_{AA} of D mesons at low p_T.

4 Conclusions

In this contribution we have shown the results for D-meson observables in ultrarelativistic heavy-ion collisions at LHC calculated within EPOSHQ model. EPOSHQ model is able to describe the nuclear modification factor and elliptic flow of D mesons in heavy-ion collisions at LHC energies with the same initial setup. For the central lead-lead collisions the importance of including initial state shadowing was presented.

We thank Jan Steinheimer, Iurii Karpenko and Andrzej Rybicki for fruitful discussions. This work was supported by the Project TOGETHER (Pays de la Loire) and by the National Science Center, Poland under Grant No. 2014/14/E/ST2/00018.

References

[1] B. Abelev *et al.*, JHEP **09**, 112 (2012)
[2] B. Abelev *et al.*, Phys. Rev. Lett. **111**, 102301 (2013)
[3] M. Nahrgang *et al.*, Nucl. Phys. A **931**, 575 (2014)
[4] T. Song *et al.*, Phys. ReV. C **93**, 034906 (2016)
[5] S. K. Das *et al.*, Phys. Lett. B **747**, 260 (2015)
[6] S. Cao *et al.*, Phys. Rev. C **88**, 044907 (2013)
[7] P.-B. Gossiaux *et al.*, Phys. Rev. C **78**, 014904 (2008)
[8] K. Werner *et al.*, Phys. Rev. C **82**, 044904 (2010)
[9] I. Karpenko *et al.*, Comput. Phys. Commun. **185**, 3016 (2014)
[10] M. Cacciari *et al.*, JHEP **10**, 137 (2012)

[11] M. Nahrgang M *et al.*, Phys. Rev. C **89**, 014905 (2014)
[12] V. Ozvenchuk *et al.*, Phys. Rev. C **90**, 054909 (2014)

EPJ Web of Conferences **130**, 07019 (2016) DOI: 10.1051/epjconf/201613007019
MESON 2016

Contribution of three nucleon force investigated in deuteron-proton breakup reaction

W. Parol[1,2,★], A. Kozela[1], I. Ciepał[1], K. Bodek[2], B. Jamroz[3], N. Kalantar-Nayestanaki[4], G. Khatri[2], St. Kistryn[2], B. Kłos[3], J. Kuboś[1], P. Kulessa[1], A. Magiera[2], I. Mazumdar[5], J.G. Messchendorp[4], D. Rozpędzik[2], A. Rusnok[3], I. Skwira-Chalot[6], E. Stephan[3], A. Wilczek[3], B. Włoch[1], A. Wrońska[2], and J. Zejma[2]

[1] *Institute of Nuclear Physics Polish Academy of Sciences, PL-31342 Krakow, Poland*
[2] *Instytute of Physics, Jagiellonian University, Krakow, Poland*
[3] *Institute of Physics, University of Silesia, Katowice, Poland*
[4] *KVI-CART, University of Groningen, Groningen, the Netherlands*
[5] *Tata Institute of Fundamental Research, Mumbai, India*
[6] *Faculty of Physics, University of Warsaw, Warszawa, Poland*

Abstract. The elastic scattering and deuteron breakup data were collected in the experiment performed at KVI (Groningen) with use of unpolarized deuteron beam with energy of 80 MeV per nucleon, impinging on hydrogen target. The procedure applied to determine total integrated luminosity is presented. The result will be used for normalization of the differential cross section for the deuteron-proton breakup reaction.

1 Introduction

3NF, Coulomb interactions [1] or relativistic component [2]. All effects reveal in different parts of the phase space with different magnitude what can be noticed in the observables.

Investigation of three-nucleon system dynamics provides basis for understanding interaction between nucleons, going also beyond simple pairwise nucleon-nucleon (NN) forces. Such additional dynamics is called three-nucleon force (3NF). It arises in the meson exchange picture as an intermediate excitation of a nucleon to a Δ-isobar or it appears in a certain order of expansion from Chiral Effective Field Theory [1]. Modern models of 3NF, like Tucson-Melbourn 99 [2] or Urbana IX [3], are combined with the adequate realistic NN potential. Alternatively, the Δ-isobar can be included explicitly in the coupled channel framework [4]. Precise measurement of observables like the cross-section in three-nucleon system is the way to validate these models.

2 Experimental set-up

The experiment was performed with the use of the deuteron beam provided by AGOR cyclotron at Kernfysisch Versneller Instituut in Groningen (the Netherlands). Unpolarized deuteron beam of

★ e-mail: wiktor.parol@ifj.edu.pl

160 MeV energy impinged on 3.3 mm thick liquid hydrogen target. Charged products of the reaction were detected by the BINA detection system (Big Instrument for Nuclear Analysis). BINA apparatus has been specially designed to investigate few-nucleon systems in the range of intermediate energies. It covers almost 4π geometry and is composed of two main parts: forward Wall (ϑ: 13°- 40°) and backward Ball (ϑ: 40°-165°). In the forward Wall particles are registered in three detectors. At first, they are detected in the multi-wire proportional chamber (for reconstruction of angles of the scattered charged particles), next – in one of 12 vertical thin plastic scintillators (stripes) and finally, in the hodoscope made of 10 horizontal thick plastic scintillator (slabs). The plastic stripes and slabs form 240 telescopes for the ΔE-E method of particle identification. Ball is made up of 149 triangular detector elements working in a phoswich mode. The Ball at the same time plays the role of the reaction chamber as well as the detector. Detailed description of the detector can be found in [5].

3 Data analysis

The analysis presented in this report focuses on the particles scattered forward from the reaction point and being detected in the WALL part of the detector. The details of the most basic steps of the data analysis including energy calibration, particle identification (PID) and detector efficiency calculation were described in previous publications [6–9]. To perform reliable PID the linearization method was applied. The ΔE–E spectra were transformed to the new energy dependent variable as described in [7]. Differential cross-section distributions for the deuteron breakup reaction has been calculated as:

$$\sigma\left(\vartheta_1, \vartheta_2, \varphi_{12}, S\right) = \frac{N_{pp}\left(\vartheta_1, \vartheta_2, \varphi_{12}, S\right)}{L \Delta\Omega_1 \Delta\Omega_2 \Delta S\, \epsilon^{\vartheta_1 \vartheta_2 \varphi_{12}}} \quad (1)$$

where: $N_{pp}\left(\vartheta_1, \vartheta_2, \varphi_{1,2}, S\right)$ is the number of proton–proton pairs, registered within a bin of 1° for polar angles $(\vartheta_1, \vartheta_2)$, of 10° for relative azimuthal angle (φ_{12}) and of 8 MeV for the S variable standing for length along the corresponding kinematical curve. L is the total integrated luminosity; $\epsilon^{\vartheta_1 \vartheta_2 \varphi_{12}}$ represents the detection efficiency for a proton pair emitted at these angles; $\Delta\Omega_{1,2}$ - the corresponding solid angles. The normalization factor L is obtained on the basis of elastic scattering data analyzed for each $\Delta\vartheta = 1°$ bin in the range (20°, 29°). The final luminosity is taken as an average value of:

$$L(\vartheta_p) = \frac{N^{el}(\vartheta_p)}{\sigma^{el}(\vartheta_p) \Delta\Omega(\vartheta_p) \epsilon^{\vartheta_p, \varphi_p}} \quad (2)$$

Here $\sigma^{el}(\vartheta_p)$ is cross-section for proton–deuteron elastic scattering, $\epsilon^{\vartheta_p, \varphi_p}$ represents the efficiency of registering of a proton and $\Delta\Omega(\vartheta_p)$ is the solid angle. $N^{el}(\vartheta_p)$ is the number of elastically scattered protons registered at the ϑ_p angle, after subtracting the background component (see Fig. 1) It is obtained by fitting the gaussian function and integrating within $\pm 3\sigma$. Since there is no experimental data for the cross-section for the elastic scattering at 80 MeV/A, $\sigma^{el}(\vartheta_k)$ was obtained from the interpolation between all the data measured at energies in wide range 30–190 MeV/A. For each energy, polynomial was fitted to angular distribution of the data in the center of mass frame (ϑ^{CM}). Then, for each value of ϑ^{CM} the polynomial was fitted to the energy distribution (see Fig. 2). Finally, the transformation of the interpolated values for 80 MeV/A to laboratory frame was done. The systematic uncertainty connected to particle identification procedure, mentioned earlier, is illustrated in Fig. 2. Since different particles tend to group around well defined particle kind specific values, one may use gaussian distribution to control the quality of particle type assignment. The systematical error was calculated as difference between values of luminosities obtained by accepting ranges of two and three σ around the proton peak in PID method. The error due to this cut was estimated to be about 7%.

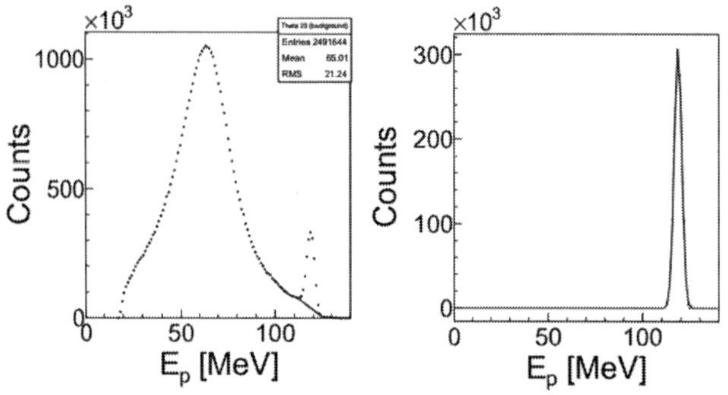

Figure 1. *Left panel:* Energy distribution of protons registered with the minimum-bias trigger in chosen $\vartheta_p = 23°$ bin. Straight red line represents the background model. *Right panel:* Energy distribution of protons from elastic scattering channel obtained after the background subtraction.

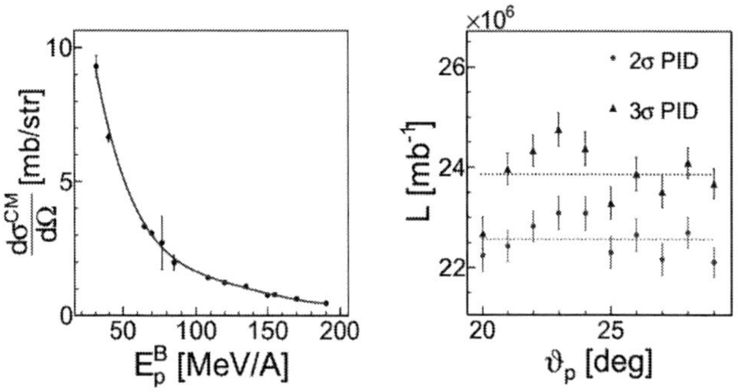

Figure 2. *Left panel:* Polynomial fit to the elastic scattering cross-section data in function of the beam energy example for $\vartheta^{CM} = 132°$. *Right panel:* Integrated luminosity obtained for each of the studied ϑ_p angle. Horizontal lines represent weighted averages of the data points.

We acknowledge support by the Małopolskie Centrum Przedsiębiorczości Project "Doctus - Małopolski fundusz stypendialny dla doktorantów". This work was supported by the European Commission within the Seventh Framework Programme through IA-ENSAR (contract no. RII3-CT-2010-262010).

References

[1] E. Epelbaum, Prog. Part. Nucl. Phys. **57**, 654 (2006)
[2] S.A. Coon and H.K. Han, Few-Body Syst. **30**, 131 (2001)
[3] B.S. Pudliner *et al.*, Phys. Rev. C **56**, 1720 (1997)
[4] A. Deltuva, Phys. Rev. C **80**, 064002 (2009)
[5] H. Mardanpour-Mollalar, Ph.D. Thesis, KVI, Groningen, the Netherlands (2008)
[6] W. Parol *et al.*, Acta Phys. Pol. B **45**, 527 (2014)
[7] W. Parol *et al.*, EPJ Web Conf. **81**, 06007 (2014)
[8] G. Khatri *et al.*, Few-Body Syst. **55**, 1035 (2014)
[9] G. Khatri *et al.*, EPJ Web Conf. **81**, 06006 (2014)

J-PET detector system for studies of the electron-positron annihilations

M. Pawlik-Niedźwiecka[1], O. Khreptak[1,*], A. Gajos[1], A. Wieczorek[1,2], D. Alfs[1], T. Bednarski[1], P. Białas[1], C. Curceanu[3], E. Czerwiński[1], K. Dulski[1], B. Głowacz[1], N. Gupta-Sharma[1], M. Gorgol[4], B. C. Hiesmayr[5], B. Jasińska[4], D. Kamińska[1], G. Korcyl[1], P. Kowalski[6], W. Krzmień[7], N. Krawczyk[1], E. Kubicz[1], M. Mohammed[1], Sz. Niedźwiecki[1], L. Raczyński[6], Z. Rudy[1], M. Silarski[1], W. Wiślicki[6], B. Zgardzińska[4], M. Zieliński[1], and P. Moskal[1]

[1] *Faculty of Physics, Astronomy and Applied Computer Science, Jagiellonian University, S. Łojasiewicza 11, 30-348 Cracow, Poland*
[2] *Institute of Metallurgy and Materials Science of Polish Academy of Sciences, 30-059 Cracow, Poland*
[3] *INFN, Laboratori Nazionali di Frascati, 00044 Frascati, Italy*
[4] *Department of Nuclear Methods, Institute of Physics, Maria Curie-Sklodowska University, 20-031 Lublin, Poland*
[5] *Faculty of Physics, University of Vienna, 1090 Vienna, Austria*
[6] *Świerk Computing Centre, National Centre for Nuclear Research, 05-400 Otwock-Świerk, Poland*
[7] *High Energy Department, National Centre for Nuclear Research, 05-400 Otwock-Świerk, Poland*

Abstract. Jagiellonian Positron Emission Tomograph (J-PET) has been recently constructed at the Jagiellonian University as a prototype of a cost-effective scanner for the metabolic imaging of the whole human body. J-PET detector is optimized for the measurement of momentum and polarization of photons from the electron-positron annihilations. It is built out of strips of plastic scintillators, forming three cylindrical layers. As detector of gamma quanta it will be used for studies of discrete symmetries and multiparticle entanglement of photons originating from the decays of ortho-positronium atoms.

1 Introduction

The Jagiellonian Positron Emission Tomograph (J-PET) has been constructed as a prototype of the novel cost-effective scanner for the metabolic imaging of the whole human body [1]. Using long strips of plastic scintillators as radiation detectors it has been optimized for the detection of photons from the electron-positron annihilation with high time- and high angular-resolution. J-PET as a multi-purpose detector provides new opportunities for studying decays of ortho-positronium atoms (*o*-Ps). Studies of the three-gamma *o*-Ps decays allow for searches of discrete symmetries violation and physics beyond the Standard Model (e.g. extra dimensions, dark matter and a new light vector gauge boson) [1, 2]. Moreover, determination of the angles between *o*-Ps spin and the scattering planes of photons will

*e-mail: alexander.khreptak@gmail.com

 DOI: 10.1051/epjconf/201613007020

allow for the multi-particle entanglement studies of high energy photons originating from the positronium decay. In this article we briefly describe the concept of the J-PET detector and its potential in the studies of the fundamental properties of leptonic systems. Detailed research program is described in Ref. [1].

2 Jagiellonian Positron Emission Tomograph (J-PET)

A novel concept of PET scanner, referred to as Jagiellonian Positron Emission Tomograph, was developed at the Jagiellonian University in Cracow. In the J-PET detector, axially arranged organic scintillators strips form a barrel. Light signals, generated by gamma quanta interaction with scintillator, are read out by pairs of photomultipliers connected to the opposite ends of each strip. Scheme of two-strips J-PET setup is presented in Fig. 1.

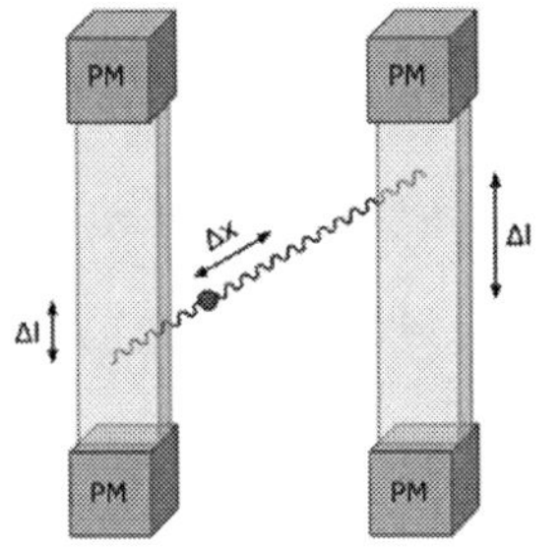

Figure 1. Scheme of the two-strips J-PET setup. Place of 511 keV γ quanta interaction with scintillator material (Δl) measured relative to the strip center is determined from the difference between light signals arrival times to both photomultipliers. Place of the two-photon annihilation event (Δx), along the line of response (LOR), is defined by time difference between two modules as described in detail in references [3–5].

First prototype of the multi-purpose J-PET scanner consists of three layers of EJ-230 scintillators strips read-out by Hamamatsu R9800 photomultipliers and forming a cylinder of diameter equal to 85 cm and its active inner part is equal to 50 cm, as one can see in right panel of Fig. 2. Scintillators are wrapped with Vikuiti foil and additionally covered with light-tight black foil. Electrical signals from each photomultiplier are probed by multi-threshold digital electronics with accuracy equal to 30 ps [4].

Figure 2. (Left) Visualization of the J-PET detector prototype. One can see three layers of scintillator strips connected with photomultipliers and arranged in a barrel. (Right) Photograph of the first prototype of the J-PET detector together with few members of the collaboration.

Data from electronics is collected by trigger-less acquisition system. Both the dedicated front-end electronics, as well as the acquisition system, were designed by the J-PET collaboration [6]. New methods for reconstruction of position and time of interaction of γ quanta based on compressive sensing theory and library of model signals has been also developed [7, 8]. The J-PET detector is optimized for the detection of gamma quanta originating from the electron–positron annihilation, but it can be also used for studies of ortho-positronium decays, as well as studies of discrete symmetries, which are briefly described in the next section.

3 Discrete symmetries in decays of positronium atoms

Positronium is leptonic object build out of electron and anti-electron (positron), so it is an eigenstate of charge (C) and spatial (P) parity, as well as their combination (CP) [9]. Therefore, it is an unique to test discrete symmetries with precision limited in principle only by the weak and photon-photon interactions [1]. The search for a possible violation of these symmetries may be done by measurement of the expectation values of various operators, odd with respect to the studied symmetry, constructed from momenta of photons and the spin of the ortho-positronium [1]. Polarized pozitronium is produced in the layer of porous material (eg. aerogel or polymer [10]) surrounding radioactive source providing longitudinally polarized positrons in a parity violating beta-plus decay (eg. sodium ^{22}Na). Positronium created this way decays into three photons which can interact via Compton effect with scintillator material. Position of the positronium annihilation can be determined based on time and positions of interaction of photons with scintillator using trilateration-based reconstruction [11]. Registration of de-excitation photon from the excited daughter nucleus created in the beta-plus decay enables also estimation of positronium lifetime.

4 Summary

J-PET being a novel solution in the Positron Emmision Tomography provides also new opportunities to study fundamental physics in the leptonic sector [1, 2, 11, 12] as well as in the life- and material-sciences [5, 13–15]. In particular it has a potential for improvement of present limits [16, 17] of the decays of positronium [1].

References

[1] P. Moskal *et al.*, Acta Phys. Polon. B **47**, 509 (2016)
[2] D. Kamińska *et al.*, Eur. Phys. J. C **76**, 445 (2016)
[3] P. Moskal *et al.*, Nucl. Instr. Meth. Phys. Res. A **775**, 54 (2015)
[4] P. Moskal *et al.*, Acta Phys. Pol. B **47**, 509 (2016)
[5] P. Moskal *et al.*, Phys. Med. Biol. **61**, 2025 (2016)
[6] G. Korcyl *et al.*, Acta Phys. Pol. B **47**, 491 (2016)
[7] L. Raczyński *et al.*, Nucl. Instrum. Meth. A **786**, 105 (2015)
[8] N. G. Sharma *et al.*, Nukleonika **60**, 765 (2015)
[9] M. Deutsch, Phys. Rev. **82**, 455 (1951)
[10] B. Jasińska *et al.*, Acta Phys. Pol. B **47**, 453 (2016)
[11] A. Gajos *et al.*, Nucl. Instrum. Meth. A **819**, 54 (2016)
[12] D. Kamińska *et al.*, Nukleonika **60**, 729 (2015)
[13] E. Kubicz *et al.*, Nukleonika **60**, 749 (2015)
[14] A. Wieczorek *et al.*, Acta Phys. Pol. A **127**, 1487 (2015)
[15] A. Wieczorek *et al.*, Nukleonika **60**, 777 (2015)
[16] T. Yamazaki *et al.*, Phys. Rev. Lett. **104**, 083401 (2010)
[17] P. A. Vetter, S. J. Freedman, Phys. Rev. Lett. **91**, 263401 (2003)

EPJ Web of Conferences **130**, 07021 (2016) DOI: 10.1051/epjconf/201613007021
MESON 2016

Studying ρ–N couplings with HADES in pion-induced reactions

Federico Scozzi[1,2,*] for the HADES Collaboration

[1] *TU Darmstadt, Darmstadt, Germany*
[2] *IPN Orsay, Orsay, France*

Abstract. The High-Acceptance Di-Electron Spectrometer (HADES) operates at the GSI Helmholtzzentrum für Schwerionenforschung in Darmstadt with pion, proton and heavy-ion beams provided by the synchrotron SIS18. In summer 2014, HADES took data using a pion beam on carbon and polyethylene targets. A large part of the data was taken at a pion beam momentum of 0.69 GeV/c in order to explore di-pion and di-electron production in the second resonance region and the sub-threshold coupling of the ρ to baryonic resonances. In this contribution lepton identification will be discussed as well the purity of reconstructed e^+e^-. Finally we will show the preliminary di-electon raw spectra.

1 Introduction

Previous measurements made by the HADES [1] showed an enhancement in di-lepton spectra below the vector meson pole both in proton-nucleus and in nucleus-nucleus collisions [2] in comparison with a cocktail of known hadronic sources. The main interpretation of this effect is based on the strong coupling of the ρ meson to the baryonic resonances (see for example [3]). Exploiting the possibility of GSI to provide a secondary pion beam, HADES has the purpose to study the coupling of the ρ with the baryonic resonances in the second resonance region by means of the reactions $\pi^-p \rightarrow \pi^+\pi^-n$ and $\pi^-p \rightarrow \pi^0\pi^-p$. In addition, the role of the ρ meson in electromagnetic transitions in the time-like region is investigated in the reaction $\pi^-p \rightarrow e^+e^-n$. The validity of the Vector Meson Dominance model will hence be clarified by the combined use of both hadronic and di-electron channels.

2 HADES detector

HADES is placed at GSI Helmholtzzentrum für Schwerionenforschung in Darmstadt. It is a fixed target high interaction rate experiment suitable to take high intensity beams (protons, ions and pions) and it is characterized by a large acceptance (almost full azimuthal angle and between $18-85°$ in polar angle). Particle identification is provided by time of flight measurements using TOF and RPC detectors and energy loss measurements in MDCs, that, together with a toroidal magnet, provide also track and momentum reconstruction. In addition, to separate leptons from hadrons, a hadron-blind Ring Imaging CHerenkov detector (RICH) and a pre-shower detector are used. The data analyzed in this work were taken in summer 2014 in π^--PE (polyethylene) and in $\pi^- - C$ collisions at a pion momentum around 0.69 GeV/c [4].

*e-mail: f.scozzi@gsi.de

DOI: 10.1051/epjconf/201613007021

3 Experimental conditions

SIS18 is able to provide a secondary π^- beam produced in the interaction between a primary beam of N_2 with an intensity of 8x10^{10} ions/spill (spill = 4s) and a thick beryllium target. The pions are then transported to HADES by means of a beam line composed of 7 quadrupole and 2 dipole magnets (Fig.1). Since the distribution of produced pions is broad, a system called CERBEROS, composed of two fast silicon detectors, was installed along the beam line to reconstruct the pion momentum with a precision of about 0.3%.

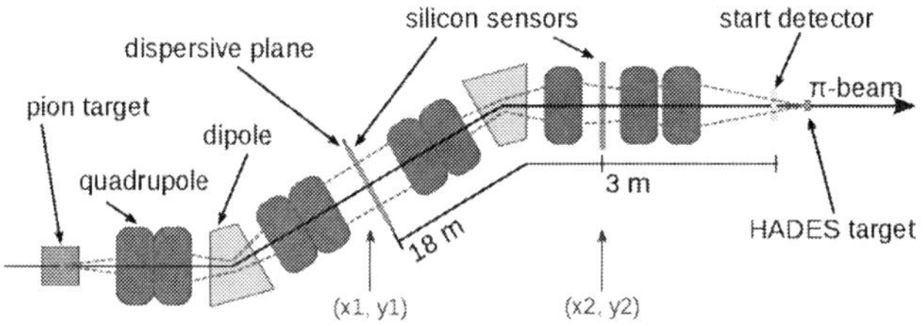

Figure 1. Schematic view of the pion beam line.

4 Lepton identification and purity

4.1 Lepton identification

The measurements of leptons is not an easy task: they are rare probes and they are produced in an environment dominated by pions. Therefore one would like to increase the efficiency of their detection, but at the same time it is necessary to be very cautious to avoid pion contamination. In order to increase the reconstruction efficiency of the di-electron channel an effort was made to analyse these data, characterized by low multiplicity, without information from RICH, relying only on β (=v/c) vs momentum and energy loss information for lepton identification. This was shown to be unsuccessful [5]. Therefore the RICH information was used trying two different approaches: the standard ring finder based on a pattern recognition algorithm to select the good rings [1] and a new approach called backtracking [6], that was developed in order to increase ring identification capabilities taking advantage of tracking information: when a track is found with velocity, momentum and energy loss that correspond to a lepton, it is propagated back to the RICH detector in order to fix the possible ring center on the pad plane. At this point it is therefore possible to be looser in the requirements for a ring. As written in [6], backtracking algorithm succeeded to have similar background rejection and achieves higher efficiency than the standard ring finder.

4.2 Purity determination

Detailed understanding of the purity of the lepton sample is fundamental; for this reason HADES has developed a method to establish purity, when the standard ring finder approach is adopted, using the so-called rotated RICH technique. The RICH detector is rotated (softwarewise) by 60° and then the tracks are matched with the rings. In this way any correlation between tracks and rings is lost and therefore one gets only random matches. Now it is possible to estimate the purity using the angular correlation between the tracks and the ring, i.e. $\Delta\phi\ sin\theta$ as shown in Fig.2. Now the data obtained after applying the rotated RICH technique (and all PID cuts) are considered as the background and the signal is obtained after subtraction of this background from the standard data sample (again after all PID cuts). The purity can be calculated as:

$$Purity = 1 - \frac{rot.\ RICH\ data}{standard\ data}$$

DOI: 10.1051/epjconf/201613007021

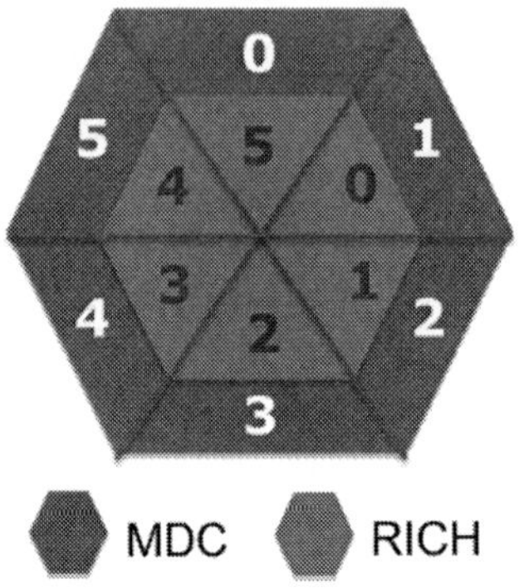

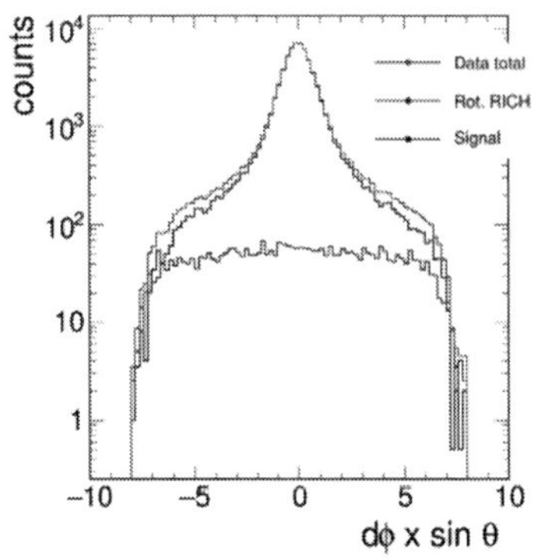

Figure 2. Left, schematic view of the rotation technique. Right, $\Delta\phi$ (difference between ϕ angle of the reconstructed track and ϕ angle of the reconstructed ring) $sin\theta$.

Table 1. Purity values.

p[MeV/c]	e^+	e^-
p < 100	99.7 %	99.1 %
100 < p < 200	99.4 %	98.7 %
200 < p < 300	97.7 %	93.7 %
300 < p < 400	97.3 %	89.0 %

This method has to be applied separetely for electrons and positrons in different momentum ranges, since the pion contamination depends on the momentum. In Table 1 the purity values are listed. The purity is very high for positrons in all momentum regions, while for electrons it decreases with increasing of momentum, but stays at a satisfactory level also for the highest momentum region. This trend is due to negative pions originating from elastic and quasi-elastic scattering.

5 Raw spectra

After identifying leptons, they are combined into pairs in order to get the invariant mass spectrum, however, in addition to the signal pairs, also uncorrelated pairs stemming from different production vertices (for example two different π^0 decays) are present in the event sample. Therefore all possible combinations have to be taken into account and then an estimation of the uncorrelated pairs, that form the combinatorial background (CB), is needed. CB is calculated by means of the same event like-sign technique: like-sign pairs (N_{++}, N_{--}) from the same events are formed and then the geometrical average is calculated: $CB = 2\sqrt{N_{++}N_{--}}$. In case one of the two numbers is 0, the arithmetic average is used. Finally to obtain the signal, formed by correlated pairs, the CB is subtracted from the unlike-sign pair sample. An additional cut is applied in order to reduce the CB originating from the conversion of photons from π^0 decay, requesting that the opening angle must be larger than 9°. In Fig. 3 the raw inclusive invariant mass is shown. The number of pairs with invariant mass larger than 140 MeV/c^2 is $\sim$ 1205 for standard ring finder and $\sim$ 3360 for backtracking. So the efficiency is increased by a factor 2.8 with the backtracking method and it will allow multi-differential analysis.

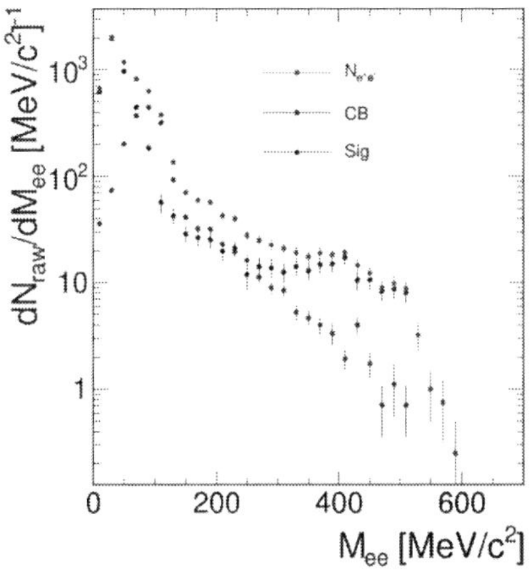

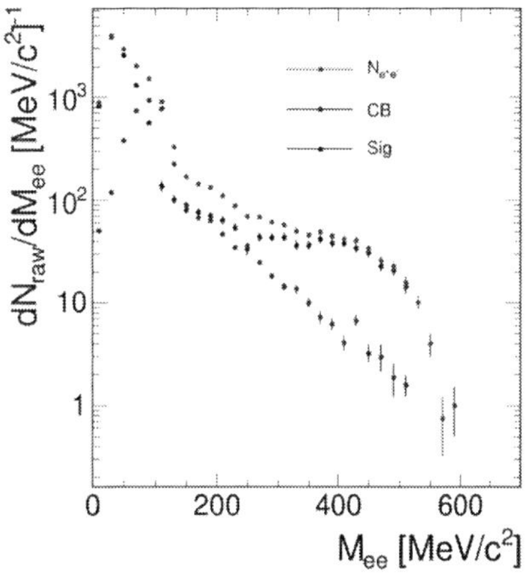

Figure 3. Raw inclusive invariant mass spectra using: left ring finder and right backtracking. Red solid circles show all e^+e^- combinations, blue solid circles show CB estimated using geometrical mean of same-event like-sign pairs, black solid circles show signal Sig = $N_{e^+e^-}$ - CB.

6 Conclusions

The HADES experiment measured for the first time di-electon production in pion-induced reactions in the second resonance region. Lepton identification was discussed and it was shown that the purity is high in all momenta for e^+ and e^-. The new backtracking algorithm increases in a significant way the efficiency of lepton identification in comparison with the standard ring finder. The collected statistics should permit to test different theoretical predictions and, in combination with the two pion production, get information on the role of rho meson in time-like electromagnetic processes.

The HADES collaboration gratefully acknowledges the support by the grants PTDC/FIS/113339/2009 LIP Coimbra, NCN grant 2013/10/M/ST2/00042 SIP JUC Cracow, Helmholtz Alliance HA216/EMMI GSI Darmstadt, VH-NG-823, Helmholtz Alliance HA216/EMMI TU Darmstadt, 283286, 05P12CRGHE HZDR Dresden, BMBF 06MT7180 TU München, Garching BMBF:05P12RGGHM JLU Giessen, Giessen UCY/3411-23100, University Cyprus CNRS/IN2P3, CNRS/IN2P3 (France), MSMT LG 12007, AS CR M100481202, GACR 13-06759S NPI AS CR, Rez, EU Contract No. HP3-283286.

References

[1] G. Agakichiev *et al.* (HADES collaboration), Eur. Phys. J. A **41**, 243 (2009)
[2] T. Galatyuk *et al.*, Int. J. Mod. Phys. Conf. Ser. **26**, 1460052 (2014)
[3] J. Weil, H. van Hees, U. Mosel, Eur. Phys. J. A **48**, 111 (2012)
[4] P. Salabura, J. Stroth, L. Fabbietti (HADES collaboration), Nucl. Phys. News **25**, 22 (2015)
[5] F. Scozzi et al. (HADES collaboration), GSI Scientific Report 2014, 10.15120/GR-2015-1
[6] P. Sellheim (HADES collaboration), J. Phys. Conf. Ser. **599**, 012027 (2015)

 DOI: 10.1051/epjconf/201613007022

Study of the lowest tensor and scalar resonances in the $\tau \to \pi\pi\pi\nu_\tau$ decay

Olga Shekhovtsova[1,2,*,**], Juan José Sanz-Cillero[3], and T. Przedziński[4]

[1] *Institute of Nuclear Physics PAN, Cracow, Poland*
[2] *NSC KIPT Institute for Theoretical Physics, Kharkov, Ukraine*
[3] *Departamento de Física Teórica and Instituto de Física Teórica, IFT-UAM/CSIC, Universidad Autónoma de Madrid, Cantoblanco, 28049 Madrid, Spain*
[4] *The Faculty of Physics, Astronomy and Applied Computer Science, Jagellonian University, Cracow, Poland*

Abstract. In this note we present a new parametrization of the hadronic current for the decay $\tau \to \pi\pi\pi\nu_\tau$ derived from the chiral lagrangian with explicit inclusion of resonances. We have included both scalar, vector and axial-vector resonances. For the first time, the lowest tensor resonance ($f_2(1270)$) is included as well. Both single and double-resonance contributions to the hadronic form factors are taken into account. To satisfy the correct high energy behaviour of the hadronic form factors, constraints on numerical values of the vertex constants are obtained.

1 Introduction

Hadronic decay modes of τ-lepton gives information about the hadronization mechanism and resonance dynamics in the energy region, where the pQCD methods are not applicable. In the last years substantial progress for the simulation of the process $\tau \to 3\pi\nu_\tau$ was achieved.The progress [1] was related to a new parametrization of the hadronic current based on the Resonance Chiral Lagrangian (RChL) and to the recent availability of the unfolded distributions from preliminary BaBar analysis [2] for all invariant hadronic masses for the three-prong mode. The lowest-energy scalar resonance was added phenomenologically and, as a result, the corresponding hadronic current does not reproduce the correct chiral low-energy behaviour and the $\pi^0\pi^0\pi^-$ and $\pi^-\pi^-\pi^+$ amplitudes do not reproduce the isospin relation [3]. Comparison with the data has demonstrated also a hint on the missing tensor resonance ($f_2(1270)$).

The goal of this note is to outline a consistent model to describe the tau-lepton decys into three pions based on RChL with scalar ($J^{PC} = 0^{++}$) and tensor resonances ($J^{PC} = 2^{++}$) and that fulfill the high-energy QCD and low-energy chiral limits for the hadronic form-factors. The detail description of the model and calculation of the hadronic form-factor will be presented [4].

*Speaker, e-mail: oshekhov@ifj.edu.pl
**IFJPAN-IV-2016-22

2 Three pion hadronic current. Axial-vector form-factors related with scalar and tensor resonances.

The most general Lorenz invariant current for $\tau^- \to \pi^{-(0)}(p_1)\pi^{-(0)}(p_2)\pi^{-(+)}(p_3)\nu_\tau$

$$H^{3\pi}_\mu(p_1,p_2,p_3) = iP_T(q)^\alpha_\mu\left((p_1^\mu - p_3)^\mu\mathcal{F}_2(s_1,s_2,q^2) + (p_2^\mu - p_3)^\mu\mathcal{F}_1(s_1,s_2,q^2)\right) + iq^\alpha\mathcal{F}_P(s_1,s_2,q^2)$$

with $s_i = (p_j - p_k)^2$, $q^2 = (p_1+p_2+p_3)^2$ and, due to the Boson symmetry, the hadronic form-factors are related: $\mathcal{F}_2(s_1,s_2,q^2) = \mathcal{F}_1(s_2,s_1,q^2)$. The longitudinal form-factor $\mathcal{F}_P$ is suppressed by m_π^2/q^2 compared to $\mathcal{F}_{1,2}$ and in this note we will neglect it. In Fig. 1, we show the three relevant diagrams

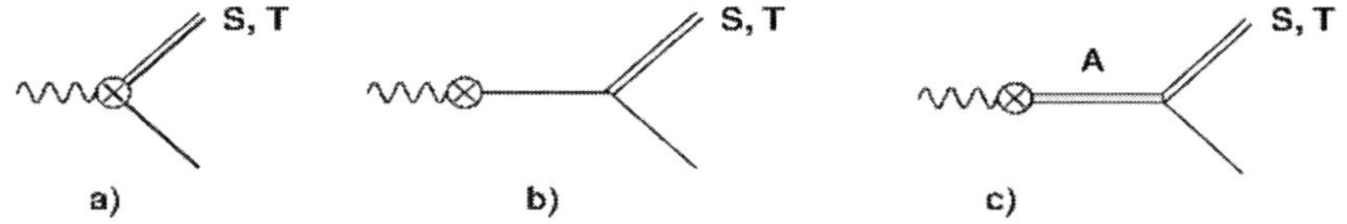

Figure 1. Relevant diagrams for the τ-lepton decay into $S\pi$ and $T\pi$. Single straight lines are for pions, the wave line is for an incoming W^-.

that must be taken into account: a) the direct production; b) the intermediate π^- production and c) the double resonance production through the intermediate a_1 axial-vector resonance. To calculate the corresponding diagrams we use the RChL approach [5] for the vector and axial-vector (A) resonances combined wothw the Lagrangian including interaction of a tensor (T) multiplet and pions [6]. Moreover we add the operators with two resonances:

- $AS\pi$ interaction $\Delta\mathcal{L}_{AS\pi} = \lambda_1^{AS}\langle\{\nabla_\mu S, A^{\mu\nu}\}u_\nu\rangle$
- $AT\pi$ interaction $\Delta\mathcal{L}_{AT\pi} = \lambda_1^{AT\pi}\langle\{A_{\alpha\beta}, \nabla^\alpha T^\mu_\mu\}u^\beta\rangle + \lambda_2^{AT\pi}\langle\{A_{\alpha\beta}, \nabla_\mu T^{\mu\alpha}\}u^\beta\rangle$,

where for the axial-vector field $A_{\alpha\beta}$ we apply the antisymmetric tensor representation [5], S is the scalar field, a tensor multiplet is $T^{\mu\nu} = f_2^{\mu\nu}/\sqrt{2} * diag(1,1,0)$ (we will assume the ideal mixing in the tensor nonet and that the $f_2(1270)$ resonance is pure $u\bar{u} + d\bar{d}$).

The $\pi^0\pi^0\pi^-$ and $\pi^-\pi^-\pi^+$ amplitudes obey the isospin relation [3] that leads to

$$\mathcal{F}_1^{--+}(s_1,s_2,q^2) = \mathcal{F}_1^{00-}(s_3,s_2,q^2) - \mathcal{F}_1^{00-}(s_3,s_1,q^2) - \mathcal{F}_1^{00-}(s_1,s_3,q^2)\,. \tag{1}$$

For the three-pion form-factor caused by the intermediate σ-resonance we have:

$$\mathcal{F}_1(s_1,s_2,q^2)^{00-} = \frac{2}{3}\mathcal{F}^a_{S\pi}(q^2;s_3)\mathcal{G}_{S\pi\pi}(s_3)\,, \tag{2}$$

where the $AS\pi$ form-factor and propagation of the σ-resonance and its decay into $\pi\pi$ are

$$\mathcal{F}^a_{S\pi}(q^2;k^2) = \frac{2c_d}{F_\pi} + \frac{\sqrt{2}F_A\lambda_1^{AS}}{F_\pi}\frac{q^2}{M_A^2 - q^2}\,,\quad \mathcal{G}_{S\pi\pi}(s_3) = \frac{\sqrt{2}c_d}{F_\pi^2}\frac{(s_3 - 2m_\pi^2)}{M_S^2 - s_3}$$

and $qp_j = (m_\pi^2 + q^2 - s_j)/2$. Requiring $\mathcal{F}^a_{S\pi}(q^2;k^2) \to 0$ for $q^2 \to \infty$ we got $F_A\lambda_1^{AS} = \sqrt{2}c_d$.

To include a σ–$f_0(980)$ splitting and non-zero width of the resonances we follow [7]

$$\frac{1}{M_S^2 - s} \longrightarrow \frac{\cos^2\phi_S}{M_\sigma^2 - s - f_\sigma(s) - iM_\sigma\Gamma_\sigma(s)} + \frac{\sin^2\phi_S}{M_{f_0}^2 - s - iM_{f_0}\Gamma_{f_0}}\,, \tag{3}$$

where ϕ_S is the scalar mixing angle. For the f_0 parameters we will use the numerical values M_{f_0} = 980 MeV, $\phi_S = -8°$ [7]. As a first approach we also consider the Breit-Wigner function for the σ-propagator in our numerical study.

Schematically the form-factor related with the intermediate tensor resonance state is written as

$$\mathcal{F}_1(s_1, s_2, q^2)^{00-} = \frac{H_1(q^2, s_1, s_2)}{(M_A^2 - q^2)(M_{f_2}^2 - s_3)} + \frac{H_2(q^2, s_1, s_2)}{(M_A^2 - q^2)} + \frac{H_3(q^2, s_1, s_2)}{(M_{f_2}^2 - s_3)} + H_4(q^2, s_1, s_2)\,, \quad (4)$$

where $H_i(q^2, s_1, s_2)$ are non-singular functions. We would like to stress that for $q^2 = M_A^2$ and $s_3 = M_{f_2}^2$ our expression (4) reproduces the corresponding contribution of Eq. (A.3) of [8] and that in [9]. However, for an arbitrary off-shell momentum of the intermediate tensor resonance we have a more general momentum structure of the hadronic current, which also ensures the right low energy behaviour and the transversality of the matrix element in the chiral limit. As a result it brings three additional functions $H_{2,3,4}(q^2, s_1, s_4)$ in (4) (see for discussion [4]).

To obtain the $\pi^-\pi^-\pi^+$ form-factors we apply the relation (1) for (2) and (4). Exact formulae are presented in [4].

The hadronic form-factors (2) and (4) have been implemented in the Monte Carlo Tauola [1]. To get the model parameters the one-dimentional spectra $d\Gamma/ds_1$, $d\Gamma/ds_3$ and $d\Gamma/dq^2$ with the hadronic form-factors (2) and (4) in addition to [10] have been fitted to the preliminary $\pi^-\pi^-\pi^+$ BaBar data [2]. The results are presented in Fig. 2 (as an example we present the result for the Breit-Wigner σ-meson propagator). For the first approach we have fixed the tensor resonance parameters to their PDG values. The difference between the data and the theoretical distributions is less than 5−7%, except for the low- and high-energy tails, where the statistics is low. The inclusion of the tensor resonance contribution in the fit and the study of the fit stability and systematic uncertainities are in progress.

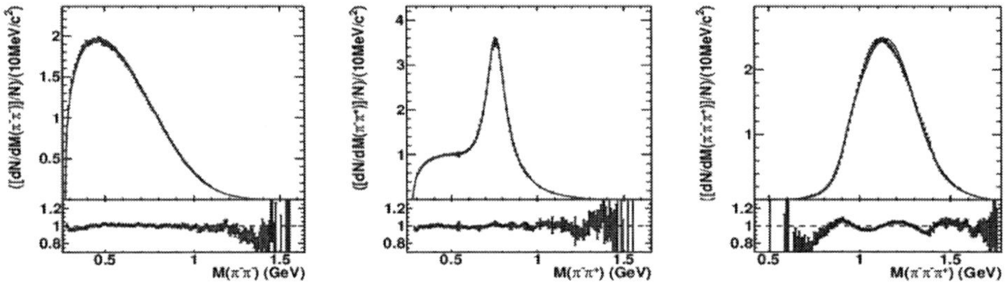

Figure 2. The $\tau^- \to \pi^-\pi^-\pi^+\nu_\tau$ decay invariant mass distribution. The preliminary BaBar data [2] are presented by points and the line corresponds to the model.

The work of J.J.S.C is partially supported by grant FPA2013-44773-P and the Centro de Excelencia Severo Ochoa Programme (Spanish Ministry MINECO) SEV-2012-0249, the research of O.Sh. was supported in part by funds of the Foundation of Polish Science grant POMOST/2013-7/12.

References

[1] I. M. Nugent, T. Przedzinski, P. Roig, O. Shekhovtsova and Z. Was, Phys. Rev. D **88**, 093012 (2013)

[2] I. M. Nugent [BaBar Collaboration], Nucl. Phys. Proc. Suppl. **253-255** , 38 (2014)

[3] L. Girlanda and J. Stern, Nucl. Phys. B **575**, 285 (2000)

[4] J.J. Sanz-Cillero, O. Shekhovtsova, in preparation

[5] G. Ecker, J. Gasser, A. Pich and E. de Rafael, Nucl. Phys. B **321**, 311 (1989)

[6] G. Ecker and C. Zauner, Eur. Phys. J. C **52**, 315 (2007)

[7] R. Escribano, P. Masjuan and J. J. Sanz-Cillero, JHEP **1105**, 094 (2011)

[8] D. M. Asner *et al.* [CLEO Collaboration], Phys. Rev. D **61**, 012002 (2000)

[9] G. L. Castro and J. H. Munoz, Phys. Rev. D **83**, 094016 (2011)

[10] O. Shekhovtsova, T. Przedzinski, P. Roig, Z. Was, Phys. Rev. D **86**, 113008 (2012)

Mass and width of the Delta resonance using complex-mass renormalization

Y. Ünal[1,2,★], T. Bauer[1], A. Küçükarslan[2], and S. Scherer[1]

[1]*PRISMA Cluster of Excellence, Institut für Kernphysik, Johannes Gutenberg-Universität Mainz, D-55099 Mainz, Germany*
[2]*Physics Department, Çanakkale Onsekiz Mart University, 17100 Çanakkale, Turkey*

Abstract. The pole mass and width of the Delta resonance are calculated in the relativistic chiral effective field theory approach. We choose a systematic power-counting procedure by applying the complex-mass scheme (CMS).

1 Introduction

At low energies, the chiral perturbation theory (ChPT) [1–3] provides a successful description of the strong interaction of pion and nucleon-based effective field theory (EFT) of quantum choromodynamics (QCD). Inclusion of heavy and especially resonant degrees of freedom in EFT becomes more complicated to set up a systematic counting scheme [4]. Power counting violating imaginary terms cannot be absorbed by the use of traditional renormalization schemes. Therefore, we can apply CMS to restore a systematic counting scheme [5].

In the current study, we determine the mass and width of the delta resonance in relativistic chiral EFT up to order $O(q^3)$. We write the most general effective Lagrangians as [6, 7]

$$\mathcal{L} = \mathcal{L}_0 + \mathcal{L}_{\pi\Delta} + \mathcal{L}_{\pi N\Delta} + \mathcal{L}_{\Delta}^{(2)}, \tag{1}$$

$$\begin{aligned}\mathcal{L}_0 =&\bar{\Psi}(i\not{D} - m_{N0})\Psi - \bar{\psi}_\mu P^{\frac{3}{2}}\big[(i\not{D} - m_{\Delta 0})g^{\mu\nu} \\ &- i(\gamma^\mu D^\nu + \gamma^\nu D^\mu) + i\gamma^\mu \not{D}\gamma^\nu + m_{\Delta 0}\gamma^\mu\gamma^\nu\big]P^{\frac{3}{2}}\psi_\nu,\end{aligned} \tag{2}$$

where N denotes nucleon isospin doublet with bare mass m_{N0}. ψ_ν are the vector-spinor isovector-isospinor Rarita Schwinger fields of the Δ resonance [8] with bare mass $m_{\Delta 0}$, $P^{\frac{3}{2}}$ is the isospin-3/2 projector [9].

★e-mail: yaseminunal@comu.edu.tr

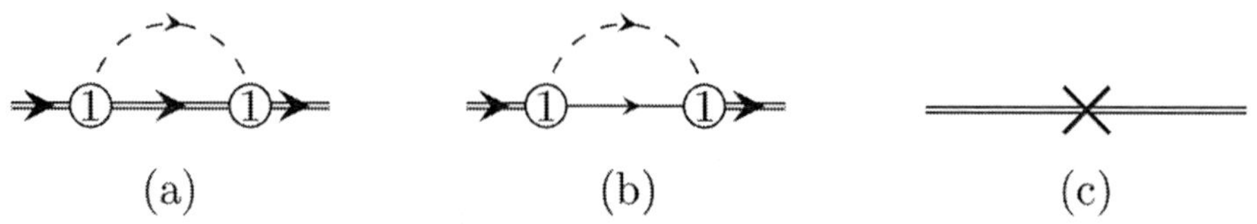

Figure 1. One-loop self energy diagrams of the delta. The dashed, straight, and double lines correspond to pion, nucleon and delta, respectively.

$$\begin{aligned}\mathcal{L}^{(1)}_{\pi\Delta} &= -\bar{\psi}^{\mu}\, P^{\frac{3}{2}}\, \Lambda^{A,n}_{\mu\nu}\, P^{\frac{3}{2}}\psi^{\nu},\\ &\equiv -\bar{\psi}^{\mu}\, P^{\frac{3}{2}}\Big\{(i\not{D} - m_{\Delta 0})g^{\mu\nu} + iA(\gamma^{\mu}D^{\nu} + \gamma^{\nu}D^{\mu})\\ &+ \frac{i}{n-2}[(n-1)A^2 + 2A + 1]\gamma_{\mu}\not{D}\gamma_{\nu}\\ &+ \frac{m_{\Delta 0}}{(n-2)^2}[n(n-1)A^2 + 4(n-1)A + n]\gamma_{\mu}\gamma_{\nu}\\ &+ \frac{g_{\Delta 1}}{2}\not{u}\gamma_5 g_{\mu\nu} + \frac{g_{\Delta 2}}{2}(\gamma_{\mu}u_{\nu} + u_{\mu}\gamma_{\nu})\gamma_5\\ &+ \frac{g_{\Delta 3}}{2}\gamma_{\mu}\not{u}\gamma_5\gamma_{\nu}\Big\}\, P^{\frac{3}{2}}\psi^{\nu},\end{aligned} \tag{3}$$

$$\mathcal{L}^{(1)}_{\pi N\Delta} = g_{\Delta}\bar{\psi}_{\mu,i}\, P^{\frac{3}{2}}_{ij}\, (g^{\mu\nu} + \tilde{z}\gamma_{\mu}\gamma\nu)u_{\nu,j}\, \Psi + \text{h.c}, \tag{4}$$

$$\mathcal{L}^{(2)}_{\Delta} = -c^{\Delta}_{1}\, \text{Tr}(\chi_{+})\, \bar{\psi}_{\mu,i}\, P^{\frac{3}{2}}_{ij}\, g^{\mu\nu}\, \psi_{\nu,j}. \tag{5}$$

We can write the bare parameters of the Lagrangians (3), (4) and (5) in renormalized parameters and counterterms as

$$m_{\Delta 0} = z_{\Delta} + \delta z_{\Delta}, \quad m_0 = m + \delta m, \quad c^{\Delta}_{1,0} = c^{\Delta}_{1} + \delta c^{\Delta}_{1}, \quad ...$$

where z_{Δ} is the complex pole of the delta propagator in the chiral limit, and m is the mass of the nucleon in the chiral limit. We use the renormalized mass parameters in the free propagators and improve the counterterms perturbatively.

The contributions to the delta self-energy up to order $O(q^3)$ from loop diagrams are shown in Fig. 1.

To determine the complex pole position z for the delta propagator we solve the equation (where s corresponds to square root of z in Ref. [9])

$$z - z_{\Delta} + 4c^{\Delta}_{1}M^2 - \tilde{\Sigma}_1(z^2) - z\Sigma_6(z^2) = 0, \tag{6}$$

the pole of delta resonance up to third order is given by

$$z = z_{\Delta} - 4c^{\Delta}_{1}M^2 + (\Sigma^{(a)} + \Sigma^{(b)})\big|_{\not{p}=z_{\Delta}} - (\Sigma^{(a)}_{\text{Sub}} + \Sigma^{(b)}_{\text{Sub}})\big|_{\not{p}=z_{\Delta}}. \tag{7}$$

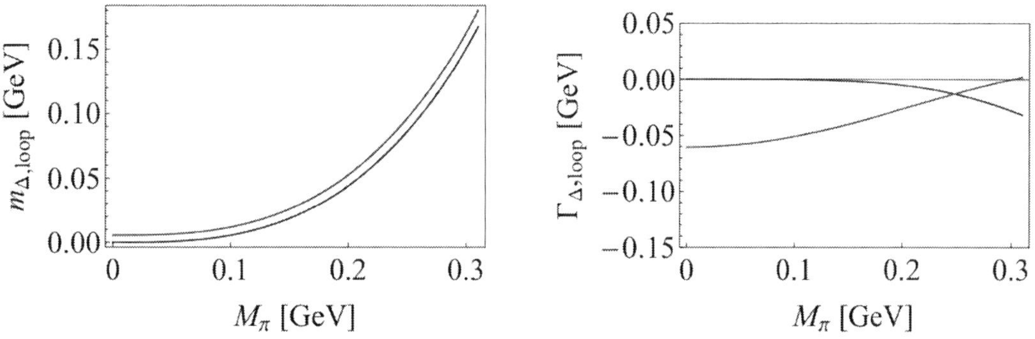

Figure 2. Contributions of the renormalized loop diagrams for real and imaginary part of the pole of delta resonance depending on the pion mass.

We use the numerical values of the parameters as $m_N = 0.94$ GeV, $M_\pi = 0.14$ GeV, $z_\Delta = (1.21 - \frac{i}{2}\ 0.10)$ GeV, $F_\pi = 0.09$ GeV, $g_A = 1.26$ GeV, $g_1 = \frac{9}{5}\ g_A$ GeV and $g_\Delta = 1.127$ GeV. Explicit contributions to the pole of delta resonance are

$$z = \left[\left(1.210 - \frac{i}{2}0.100\right) - 0.078\ c_1^\Delta + \left(0.015 - \frac{i}{2}\ 0.0024\right)\right]\mathrm{GeV}. \tag{8}$$

We compare the two power counting schemes the CMS and the small-scale expansion (SSE) [6] in Fig. 2. The contributions are represented depending on pion mass to the real and imaginary part of the pole. The red line stands for the results in SSE and the blue line is for the results in CMS. If one uses the SSE the chiral corrections to the real part of the pole are slightly larger in the entire $0 < \mathrm{M}_\pi < 0.3$ GeV range.

The authors would like to thank J. Gegelia and D. Djukanovic for valuable contributions to the work. The work of Y. Ü was supported by the Scientific and Technological Research Council of Turkey (TÜBİTAK).

References

[1] S. Weinberg, Physica A **96**, 327 (1979)
[2] J. Gasser and H. Leutwyler, Annals Phys. **158**, 142 (1984)
[3] J. Gasser and H. Leutwyler, Nucl. Phys. B **250**, 465 (1985)
[4] J. Gasser, M. E. Sainio and A. Svarc, Nucl. Phys. B **307**, 779 (1988)
[5] D. Djukanavic *et al.*, Phys. Lett. B **680**, 235 (2009)
[6] T. R. Hemmert, B. R. Holstein and J. Kambor, J. Phys. G **24**, 1831 (1998)
[7] L. M. Nath, B. Etemadi and J. D. Kimel, Phys. Rev. D **3**, 2153 (1971)
[8] W. Rarita and J. S. Schwinger, Phys. Rev. **60**, 61 (1941)
[9] C. Hacker *et al.*, Phys. Rev. C **72**, 055203 (2005)

AUTHOR INDEX

AUTHOR INDEX

AUTHOR INDEX

AUTHOR INDEX

AUTHOR INDEX